Springer-Verlag Berlin Heidelberg GmbH

A.P. S. Selvadurai

Partial Differential Equations in Mechanics 2

The Biharmonic Equation, Poisson's Equation

With 215 Figures

Springer

Professor Dr. A.P. S. Selvadurai
McGill University
Department of Civil Engineering and Applied Mechanics
Macdonald Engineering Building
817 Sherbrooke Street West
Montreal, QC   H3A 2K6
Canada
e-mail: apss@civil.lan.mcgill.ca
and
Humboldt-Forschungspreisträger
Institut A für Mechanik
Universität Stuttgart
Pfaffenwaldring 9
70550 Stuttgart, Germany

Library of Congress Cataloging-in-Publication Data. Selvadurai, A.P.S. Partial differential equations in mechanics/A.P.S. Selvadurai. p. cm. Includes bibliographical references and indexes. Contents: 1. Fundamentals, Laplace's equation, diffusion equation, wave equation – 2. The biharmonic equation, Poisson's equation.
1. Mechanics, Analytic. 2. Differential equations, Partial. I. Title. QA 805.S45 2000   531'.01'515353–dc21   00-044024

ISBN 978-3-642-08667-0   ISBN 978-3-662-09205-7 (eBook)
DOI 10.1007/978-3-662-09205-7

Typesetting: Data conversion by LE-TeX, Leipzig
Cover design: de'blik, Berlin

Printed on acid-free paper
SPIN: 10764517   62/3141/ba – 5 4 3 2 1 0

To the memory of my parents

K.S. Selvadurai and W.M.A. Selvadurai

and to

Sally, Emily, Paul, Mark

and

Elizabeth

Preface

> *"For he who knows not mathematics cannot know any other sciences;*
> *what is more, he cannot discover his own ignorance or find its proper*
> *remedies."* [Opus Majus]   Roger Bacon (1214-1294)

The material presented in these monographs is the outcome of the author's
long-standing interest in the analytical modelling of problems in mechanics
by appeal to the theory of partial differential equations. The impetus for writing these volumes was the opportunity to teach the subject matter to both
undergraduate and graduate students in engineering at several universities.
The approach is distinctly different to that which would adopted should
such a course be given to students in pure mathematics; in this sense, the
teaching of partial differential equations within an engineering curriculum
should be viewed in the broader perspective of *"The Modelling of Problems
in Engineering"*. An engineering student should be given the opportunity to
appreciate how the various combination of balance laws, conservation equations, kinematic constraints, constitutive responses, thermodynamic restrictions, etc., culminates in the development of a partial differential equation,
or sets of partial differential equations, with potential for applications to engineering problems. This ability to distill all the diverse information about
a physical or mechanical process into partial differential equations is a particular attraction of the subject area. A second aspect of the teaching of
partial differential equations to engineering students must cover topics that
should enable them to pose an engineering problem as a correct mathematical statement. This process can include the mathematical structuring of
the physical boundary conditions, regularity conditions, initial conditions
and uniqueness theorems to generate a well-posed problem in partial differential equations. Thirdly, the presentation should include an introduction
to the solution schemes that will highlight the basic structure of the solutions associated with the different classes of partial differential equations, by
appeal to suitable idealizations of engineering problems. These volumes are
also intended to illustrate the extensive range of applicability of the basic

linear partial differential equations in a *multidisciplinary sense*, with special emphasis on applications to problems in mechanics encountered in civil engineering, mechanical engineering, theoretical and applied mechanics, chemical engineering, geological engineering, earth sciences, etc., covering topics such as fluid flow, diffusion and mass transport in porous media, pressure transients and moisture diffusion in porous geomaterials, heat conduction in solids, waves in elastic solids, fluids and membranes, elasto-mechanics of solids and structural elements and mechanics of viscous fluids.

These companion volumes contain a total of nine chapters, which introduce the basic concepts of partial differential equations with the central theme of modelling and applications in mechanics. The division of the presentation into two companion volumes with chapter continuity achieves two purposes. The material contained in Volume I, chapters 1 to 7, can form the subject matter of a senior level undergraduate course or a graduate level course in applications of partial differential equations in mechanics for engineering students. The material contained in Volume II, chapters 8 and 9, is ideally suited for a graduate level course devoted to applications of partial differential equations in mechanics of solids. The introductory chapter gives a review of the mathematical preliminaries, including vector calculus, Fourier series and integral transforms. This chapter is not intended to provide an exhaustive coverage of these topics and the interested reader can further review this material by consulting the bibliography cited at the end of these volumes. The presentation in chapter 1 is kept to a reasonable length by introducing as brief a derivation of the salient results and procedures as possible. Chapter 2 introduces partial differential equations and definitions of order, linearity, homogeneity, and well posedness. Chapter 3 provides a brief account of first-order partial differential equations with applications that involve characteristic equations. Applications of integral transform techniques to the solution of elementary problems involving transport in porous media are also discussed. Chapter 4 deals with the classification of partial differential equations of the second-order with procedures for their reduction to canonical forms. The application of Laplace's equation to problems of steady state heat conduction, ideal fluid flow, flow in porous media and applications to the study of deflections of stretched membranes are detailed in chapter 5. Chapter 6 examines the diffusion equation in relation to transient heat conduction, mass transport in porous media and pressure transients in porous media. Chapter 7 deals with the wave equation and its application to wave propagation in infinite and finite strings, vibrations of membranes and elementary one-dimensional vibration problems in solid mechanics. This chapter also examines the application of the wave equation to the study of shallow water waves. Chapter 8 presents a very complete exposition of the application of the biharmonic equation to problems in mechanics. A perusal of many existing texts on partial differential equations reveals the conspicuous absence of any detailed treatment of the biharmonic equation. Yet, the biharmonic equation is one of the most important partial differential equations in applied mechanics, with applications in the theory of elasticity, mechanics of elastic plates and the theory of slow flows of viscous fluids, all subject areas of fundamental importance to the engineering sciences. Chapter 9 deals with Poisson's equation. Many expositions of Poisson's equation present themselves as appendages to Laplace's equation. The objective of this chapter is to demonstrate that Poisson's equation has significance in its own right and has extensive applications to engineering problems dealing with steady state heat conduction in heat generating media, groundwater flow with recharge or depletion, mechanics of stretched, loaded membranes and in the study of the theory of torsion of prismatic elastic bodies.

Each chapter contains a detailed discussion of the application of the relevant partial differential equation, its derivation in a generalized fashion and the formulation of consistent boundary and/or initial conditions required for their well posedness. The proof of the relevant uniqueness theorems, maximum principles and other topics of general interest to identifying the qualitative aspects of the specific partial differential equations are also discussed. Worked examples within the text and problem sets at the end of each chapter highlight engineering applications of the theories and relevant analytical developments. The volumes are reasonably self-contained, in the sense that all necessary material, including developments of the governing partial differential equations, proofs of general theorems and applications, are presented in their entirety without recourse to references within the text. There is, of course, a wealth of information available in other texts and treatises devoted to the subject of partial differential equations and the reader, and students in particular, should at least aware of these developments. To this end, these volumes contain an extensive bibliography divided into topics covering general engineering mathematics, ordinary differential equations, partial differential equations, Fourier series, integral transforms and special functions, boundary value problems and mathematical methods. The bibliography also contains titles of interest to the separate chapters, including first-order partial differential equations, Laplace's equation, the diffusion equation, the wave equation, the biharmonic equation and Poisson's equation. Although not required for the subject matter covered in these volumes, the bibliography also contains titles related to non-linear partial differential equations, numerical methods for the solution of partial differential equations, applications of computer based symbolic manipulation methods to mathematics and references to historical material in both mechanics and mathematics. Symbolic computer methods are gaining popularity in many engineering curricula; they should be considered as a useful complement to carrying out mathematical operations in a time effective manner, particularly as computer laboratory exercises of examples given in these volumes.

These monographs, by design, emphasize analytical procedures for the solution of the various partial differential equations. This should not be construed as an opportunity to de-emphasize numerical and computational techniques. On the contrary, any realistic engineering application of even the simplest of theories, such as those described by the linear partial differential equations treated in these monographs will invariably require access to sophisticated numerical schemes. These schemes can include finite difference, finite element and boundary integral equation techniques that are gaining considerable popularity in engineering curricula, both at the undergraduate and graduate levels. The teaching of these numerical techniques can only benefit by instilling in undergraduate engineering students the confidence in both the mathematical aspects of partial differential equations and their applications potential. Indeed, in advanced finite element formulations based on the Galerkin technique, the governing partial differential equations are a prerequisite; similarly, knowledge of the Green's function applicable to a particular partial differential equation is a requirement for the application of boundary integral equation techniques. The subject of partial differential equations has a long and rich tradition in mathematics and mechanics, and, as the bibliography demonstrates, has at its disposal an extensive collection of texts and treatises devoted to the subject. The bibliography is certainly not meant to be all encompassing and up to date. The texts cited are considered sufficient as supplementary reading material. The present volumes, however, differ from many traditional presentations in that the subject matter is introduced within the context of mathematical rigour, modelling in mechanics and the applications of elementary solutions to problems in engineering mechanics.

The need for engineers to understand fundamental concepts associated with partial differential equations and techniques available for their solution becomes apparent when mathematical modelling and analysis of engineering problems are set in their modern context. Engineers are constantly exposed to new engineering software tools, usually involving numerical methods that enable them to carry out engineering computations for practical problems with speed and accuracy. In a majority of these cases, the numerical schemes are designed to solve partial differential equations with complicated coupled phenomena, which usually have a non-linear character. The validity and success of the numerical procedures in such software tools cannot be confirmed in a universal sense. The opportunity, however, exists for engineers to conduct calibrations of such numerical schemes by recourse to solutions developed for classes of linearized problems, albeit simplified. The modelling of an engineering problem as a mathematical statement can be regarded as the modern equivalent of a design exercise. These companion volumes highlight the point that even elementary partial differential equations are a powerful tool for posing and solving practical problems in engineering. The mathematical rigour of the presentations is balanced by an extensive variety of applications in mechanics. The practical nature of the problems examined emphasizes mathematical modelling as an important aspect of the training of engineering students. In this sense, the subject of partial differential equations has much to offer to enhance the quality of the broader mathematical education of students in engineering and to effectively integrate mathematics into the mainstream of modelling of engineering problems.

Acknowledgements

It is a pleasure for me to record my appreciation to a number of individuals who have helped me either directly or indirectly, in achieving the task of initiating and completing these volumes. I am deeply grateful to Alexander von Humboldt Stiftung for the award of a Preistrager Fellowship, which enabled me to visit the University of Stuttgart during the completion of this work. I would like to thank Professor Lothar Gaul, Director of the Institut A für Mechanik, Universität Stuttgart, his colleagues and staff for their kind hospitality and for the productive intellectual environment that I enjoyed at the Institute. The support of Professor Pieter Vermeer of the Institut für Geotechnik, Universität Stuttgart is gratefully acknowledged. A significant part of the latter chapters and revisions to the manuscript were done during my stay at the University of Canterbury, Christchurch, New Zealand. I am grateful to the University for the award of an Erskine Fellowship and to Professor Rob Davis, of the Department of Civil Engineering, for making the visit possible. His critical comments on many of the chapters were invaluable. I record my thanks to Professor Marc Boulon and his colleagues at the Laboratoire 3S, Université Joseph Fourier, Grenoble, France, for the valuable interactions and for the hospitality during my brief sojourn in Grenoble.

The debt I owe to my *gurus* Professor A.J.M. Spencer FRS, Emeritus Professor of Theoretical Mechanics, University of Nottingham, and Professor R.E. Gibson, Emeritus Professor, University of London, is particularly substantial. Quite apart from their mentoring, support and encouragement, their contributions to continuum mechanics and theoretical geomechanics continue to be a source of inspiration to me personally, and to many other researchers who enjoy practicing the art of applying mathematics to the modelling and solution of problems in engineering. These volumes are in some small way a tribute to their substantial scholarly achievements. I would like to thank John Adjeleian, Professor Emeritus at Carleton University, Ottawa and Branko Ladanyi FRSC, Professor Emeritus at École Polytechnique,

Montreal, for the encouragement and support that I have received over the many years. It is also a pleasure to record the support of my friends in the geomechanics fraternity, who appreciate my continuing preoccupation with mathematical modelling, but most importantly, recognize the valuable role mathematics and modelling plays in the education of engineers. The opportunity to teach the material contained in the first seven chapters as parts of a required course to undergraduate engineering students and an elective course to graduate students in civil engineering at McGill and parts of chapter eight as a graduate course at Carleton is appreciated. The presentation of the material in these volumes has greatly benefitted from this teaching activity.

I would like to thank Dr. Dietrich Merkle, Engineering Editor, at Springer-Verlag, Heidelberg and his staff for their very positive support for the project and for the assistance they have provided in allowing me to present the manuscript in a form that would be of benefit to the readership. Although I had the pleasure of "writing" the manuscript the arduous task of preparing several versions of it into LaTeX forms was performed by Ms. Livia Nardini. Mr. Nick Vannelli converted my "sketches" to Corel Draw equivalents. I am also grateful to a number of technical assistants, notably, Marwan Al-Habbal and Todd Hirtle who were involved in preparing the final manuscripts to conform to Springer guidelines. As former students in the PDEs course, their familiarity with the course material was a great asset in this exercise. Every attempt has been made to check the accuracy of the material contained in these volumes. Some errors are bound to exist despite my best efforts. I would therefore be most grateful to the readers if they would be kind enough to bring these to my attention.

Last but not least, my appreciation goes to my wife Sally, not only for her editorial assistance, but also for her willingness to attend to the demands of a family when I was engrossed, absent or preoccupied with the task of completing these volumes against self-imposed deadlines. It is fair to say that all the family appreciates the completion of the project.

Montréal, April 2000   *A.P.S. Selvadurai*

Contents

8.   The biharmonic equation   1
  8.1   The concept of a continuum   3
  8.2   Displacements and strains in continua   5
        8.2.1   Physical interpretations of the strain matrix   12
        8.2.2   Physical interpretation of the rotation matrix   16
        8.2.3   Indicial notation representations
                of strain and rotation   18
        8.2.4   Transformation of the strain matrix   19
        8.2.5   Principal strains and strain invariants   24
        8.2.6   Compatibility of strains   26
  8.3   Stresses in a continuum   36
        8.3.1   The stress dyadic and the stress matrix   37
        8.3.2   Tractions on an arbitrary plane   39
        8.3.3   Equations of equilibrium   41
        8.3.4   Symmetry of the stress matrix   43
        8.3.5   Sign convention for stresses   45
        8.3.6   Transformation of the stress matrix   47
        8.3.7   Principal stresses and stress invariants   49
  8.4   Constitutive equations for linear elastic solids   57
        8.4.1   Generalized Hooke's Law   57
        8.4.2   The strain energy density   58
        8.4.3   Symmetry of the elasticity matrix   60
        8.4.4   Isotropic elastic material   62
        8.4.5   Thermodynamic constraints on the elastic constants .   66
        8.4.6   Boundary conditions   68
        8.4.7   Time derivatives   71
  8.5   Uniqueness theorem in the classical theory of elasticity   72
  8.6   Plane problems in classical elasticity   81
  8.7   The Airy stress function   92
  8.8   Methods of solution of the biharmonic equation   104

8.8.1   Series solution in Cartesian coordinates 105
8.8.2   Variables separable solution -
         rectangular Cartesian coordinates 118
8.8.3   Integral transform solution
         of the biharmonic equation governing plane problems 130
8.8.4   Line load problems for an infinite plane 138
8.8.5   Application of complex variable methods 145
8.9   Polar coordinate formulation
      of the plane problem in elasticity 151
      8.9.1   Definition of stresses in plane polar coordinates 152
      8.9.2   Definition of strains in plane polar coordinates 154
      8.9.3   Elastic stress-strain relations in polar coordinates ... 158
      8.9.4   Transformation from Cartesian equations 159
      8.9.5   The Airy stress function in plane polar coordinates . 161
      8.9.6   Boundary conditions in plane polar coordinates..... 162
      8.9.7   Development of the Airy stress function
               in polar coordinates 164
      8.9.8   An application of complex variable techniques 176
8.10 Biharmonic function formulation
      of three-dimensional problems in elasticity 181
      8.10.1   The strain potential 182
      8.10.2   The Galerkin vector 184
      8.10.3   General properties of biharmonic functions 186
      8.10.4   Love's strain function 193
      8.10.5   Axisymmetric problem formulation 197
      8.10.6   Polynomial solutions of the biharmonic equation;
               the interior solution 199
      8.10.7   Love's strain function approach: spherical polar
               coordinate formulation for unbounded domains 209
      8.10.8   Kelvin's problem 213
      8.10.9   A centre of dilatation 222
      8.10.10 Boussinesq's problem........................... 225
      8.10.11 The spherical cavity problem 233
      8.10.12 Application of integral transforms................ 238
      8.10.13 Mixed boundary value problems 258
8.11 Flexure of thin elastic plates 267
      8.11.1   Deformation of the plate region.................. 269
      8.11.2   Flexural stresses and stress resultants 279
      8.11.3   Plate stress-strain relations 282
      8.11.4   Equation of equilibrium for the plate 286
      8.11.5   Strain energy of a plate........................ 289

8.11.6   The principle of virtual work 291
8.11.7   Boundary conditions for plate problems 293
8.11.8   The classical theory of thin plates................. 297
8.11.9   Flexure of thin circular plates 305
8.11.10 Complex variable method for circular plates 325
8.11.11 Green's function for a thin plate 330
8.11.12 Flexure of rectangular plates 332
8.11.13 Certain general solutions of the biharmonic equation 339
8.11.14 Navier and Levy solutions for rectangular plates 345
8.11.15 Application of integral transform techniques........ 366
8.11.16 Complex variable methods for the solution
         of plate problems 374
8.11.17 Uniqueness of solution governing deflections
         of a plate 386
8.11.18 Uniqueness of solution for plates
         with clamped boundaries 390
8.12 Slow viscous flow 393
8.12.1   Kinematics of fluid flow 394
8.12.2   Equation of motion 397
8.12.3   Constitutive equations for a viscous fluid 397
8.12.4   Thermodynamic constraints
         on the viscosity coefficients 399
8.12.5   Navier-Stokes equation 403
8.12.6   Biharmonic formulations of problems
         in slow viscous flow 404
8.12.7   Planar problems in slow viscous flow 407
8.12.8   Axisymmetric problems in slow viscous flow 409
8.12.9   Boundary conditions 411
8.12.10 Stokes' paradox and related problems 419
8.12.11 Slow viscous flow past a sphere 424
8.12.12 Application of integral transform techniques........ 430
8.12.13 Diffusive motions in viscous fluids................ 440
8.13 The compatibility conditions 454
8.14 A proof of Stokes' paradox 459
8.15 A uniqueness theorem for viscous flows 462
8.16 PROBLEM SET 8 465

9. Poisson's equation 503
9.1 Flow through porous media with internal sources 504
9.2 Heat conduction in the presence of heat sources 507

9.3   Transverse deflections
of a stretched transversely loaded membrane 508

9.4   Boundary conditions 508

    9.4.1   Boundary conditions for flow in porous media 509

    9.4.2   Boundary conditions for heat conduction 510

    9.4.3   Boundary conditions for loaded membranes 511

9.5   Generalized results 513

9.6   Green's function for Poisson's equation 526

    9.6.1   Integral transform techniques 531

    9.6.2   Series solution techniques 535

    9.6.3   Method of eigenfunctions 540

    9.6.4   Symmetry of the Green's function 551

    9.6.5   The method of images 553

9.7   A monotonicity result for Poisson's equation 557

9.8   Viscous flow in conduits 559

9.9   Torsion of prismatic elastic solids 565

    9.9.1   General formulation of the torsion problem 568

    9.9.2   Torsion of prismatic elements
with multiply connected cross-sections 577

    9.9.3   Solutions for torsion problems
derived from equations of boundaries 581

    9.9.4   Variables separable solutions for the torsion problem 587

    9.9.5   A complex variable formulation
of the torsion problem 600

9.10 An alternative derivation of the elastic torsion problem 608

9.11 The gravitational potential 614

    9.11.1 Spatial distribution of matter 620

9.12 PROBLEM SET 9 634

Bibliography 649

Index 679

Chapter 8

The biharmonic equation

An important common theme in the developments presented in connection with Laplace's equation, the diffusion equation and the wave equation is that they are all of the second-order and represent the fundamental equations which govern elliptic, parabolic and hyperbolic partial differential equations, respectively. A further general observation in previous expositions is that as the phenomena that are being modelled becomes either more complex or encompasses more complicated fundamental processes, the partial differential equations which describe such phenomena are expected to acquire a higher order. This was evident in the description of advection-diffusion phenomena governing the transport of chemicals in porous media. In the presence of only advective phenomena the transport process can be described by a first-order partial differential equation; when diffusive processes are taken into consideration, the transport process can be described by a second-order partial differential equation. The biharmonic equation is one such partial differential equation which arises as a result of modelling more complex phenomena encountered in problems in science and engineering. The term biharmonic is indicative of the fact that the function describing the processes satisfies Laplace's equation twice explicitly. The exact first usage of the biharmonic equation is not entirely clear since every harmonic function which satisfies Laplace's equation is also a biharmonic function. Many of the applications of the biharmonic equation stem from the consideration of more complex mechanical and physical processes involving solids and fluids. One of the earliest applications of the biharmonic equation deals with the classical theory of flexure of elastic plates developed, among others, by J. Bernoulli (1667-1748), Euler (1707-1831), Lagrange (1736-1813), Germain (1776-1831), Poisson (1781-1840), Navier (1785-1836), Cauchy (1789-1857) and Lamé (1795-1870). Developments to the mathematical modelling of the theory of plates continued with contributions by Kirchhoff (1824-1887), Levy

(1838-1910), J.C. Maxwell (1831-1879) and Sir Horace Lamb (1849-1934). The mathematical theory of thin plates, generally attributed to Poisson and Kirchhoff, has been extensively applied to the stress analysis of structural plates composed of both metallic and non-metallic materials.

The reduction of the analysis of the two-dimensional problem in the classical theory of elasticity to the solution of the biharmonic equation is due to Airy (1801-1892), who used the calculations in the design of a structural support system for an astronomical telescope. The work of Green (1793-1841), Sir G.G. Stokes (1819-1903), Kelvin (1824-1907), J.C. Maxwell (1831-1879), Boussinesq (1842-1929), Hertz (1857-1894), Michell (1863-1940), Love (1863-1940), Morera (1856-1909), Beltrami (1835-1900) and Galerkin (1871-1945) in the areas relating to three-dimensional problems in the mathematical theory of elasticity also touch upon formulations which involve the biharmonic operator. These developments essentially lay the foundations to the study of the mathematical theory of elasticity which forms an important aspect of the mechanics of deformable media.

When considering the flow of ideal fluids, the mechanical behaviour of the fluid was considered to be incompressible (or volume preserving or divergence free) and inviscid (i.e. the fluid possessed no viscosity effects). This results in the classical Laplace's equation which governs the flow potential. The classical theory for the flow of an incompressible viscous fluid is governed by the Navier-Stokes equations. As indicated in an elementary example given in previously (section 6.1.6), the assumption of Newtonian viscous behaviour where the "viscous shear stresses" are linearly dependent on the "shear strains", is central to the development of the Navier-Stokes equations. The partial differential equations governing the Navier-Stokes equations are non-linear (the non-linearity arising from a convective term of the type $\mathbf{v}.\nabla\mathbf{v}$, where $\mathbf{v}$ is the velocity vector) and are extremely difficult to solve. Solutions can be obtained for limited time durations after which flows may become unstable leading to the developments of processes such as turbulence, and the Navier-Stokes equations then become inapplicable. The theory of slow viscous flows is a particular simplification of the Navier-Stokes equations which is applicable to steady laminar flows with low "Reynolds Number" (The Reynolds Number (Re) is a non-dimensional parameter which depends on the mass density of the fluid (ρ), its kinematic viscosity (μ) , a characteristic velocity (U) and a length parameter of the problem (L); $\mathrm{Re} = \rho U L / \mu$). As the Reynolds Number becomes small ($<< 1$) the slow viscous flow problem can be reduced to much simpler forms which correspond to the biharmonic equation for two-dimensional problems and Stokes' equation for problems with axial symmetry. Stokes' operator is closely related to the biharmonic operator. The biharmonic equation can therefore be successfully used to model slow viscous flow problems involving Newtonian viscous flows. These developments have been considered by a number of mathematical physicists including Navier (1785-1836), Sir G.G. Stokes (1819-1903), Sir Osborne Reynolds (1842-1912), Sir Horace Lamb (1849-1934) and Sir G.I. Taylor (1886-1975). The solutions developed for slow viscous flow problems have found extensive applications in many industrial problems including flow of molten metals, flow of particulate suspensions and in the modelling of bio-fluid dynamics.

In this Chapter, we shall introduce in a formal sense the classical definition of a continuum, and definitions of strain and stress applicable to deformable media. The constitutive equations governing linear relationships between stress and strain applicable to elastic media and those governing stress and strain rate in Newtonian viscous fluids are also introduced. These relationships along with the other relevant field equations are used to develop the biharmonic equations governing plane problems in elasticity theory, and slow viscous flow. The extensions of the approach to axisymmetric problems in elasticity theory is also briefly introduced. The Chapter also develops the biharmonic equation governing flexure of thin plates, described by the Germain-Poisson-Kirchhoff thin plate theory.

8.1 The concept of a continuum

The concept of a continuum was implicit in many of the developments associated with the derivation of the partial differential equations discussed in previous chapters. In the continuum representation of a physical system we can assume that although the material bodies through which fluids percolate, heat conduction takes place, chemicals diffuse, waves propagate, etc., possess an affine structure, the quantities of interest such as density, temperature, concentration, displacements, internal forces, etc., can be defined in relation to the abstract mathematical concept of a *point* within the medium and that the spatial and time derivatives of these are assumed to exist up to any required order. For example in the derivation of the partial differential equations, specification of boundary conditions, proving the uniqueness theorems, etc., it is implicitly assumed that such derivatives do in fact exist without ambiguity. If the derivatives gives rise to discontinuities in their definition, this will render much of what we have discussed and developed of limited value, particularly in terms of mathematical correctness. In reality, the continuum concept is not satisfied by any material body which has a discrete or particulate internal structure. This limitation of the continuum assumption become abundantly clear by evaluating a simple property such as the density of a material body which has a porous structure. If the concept of a density at a *point* (P) within the porous medium exists then we can compute the density by taking the limiting definition

$$\rho_P(\mathbf{x}, t) = \lim_{\Delta V \to 0} \left(\frac{\Delta M}{\Delta V} \right) \tag{8.1}$$

where ΔM is the mass of the medium contained in the region ΔV enclosing the point P.

As the material volume ΔV which is used to compute the density decreases, the estimate for the density at point P obtained through (8.1) will exhibit fluctuations of scale. It is evident that to derive the conventional measure of the density at a "point", the dimensions of the "point" have to be defined. If the density is assumed to be constant, the dimensions of the limiting volume V^* must be such that the affine structure of the material has no influence on definition of the density of the porous medium, i.e.

$$\rho_P(\mathbf{x}, t) = \lim_{\Delta V \to \Delta V^*} \left(\frac{\Delta M}{\Delta V} \right) \tag{8.2}$$

Similar comments apply to the definitions of other variables such as temperature, concentration, displacements, stresses, etc. To define these dependent variables in material bodies in a continuum sense, a limiting spatial region needs to be specified. The continuum formulations of physical problems in which the continuum definition of the dependent variables is restricted to a finite region rather than a mathematical notion of a "point" requires an inordinate amount of effort. Furthermore in many instances the scale of application of the continuum based solutions is much greater than the internal scale(s) associated with the material microstructure within ΔV^*. For this reason, the continuum concept advocated in the definitions such as (8.1) is retained with the proviso that the scale of applicability of the results derived from continuum solutions must reflect requirements such as those imposed by definitions of the type (8.2). Furthermore, the mathematical notion of a continuum with point-wise measures of dependent variables such as displacements, tractions, stresses, temperatures, body forces, etc., has become an integral part of modern continuum mechanics. These definitions in particular are essential to developments in classical solid and fluid mechanics which

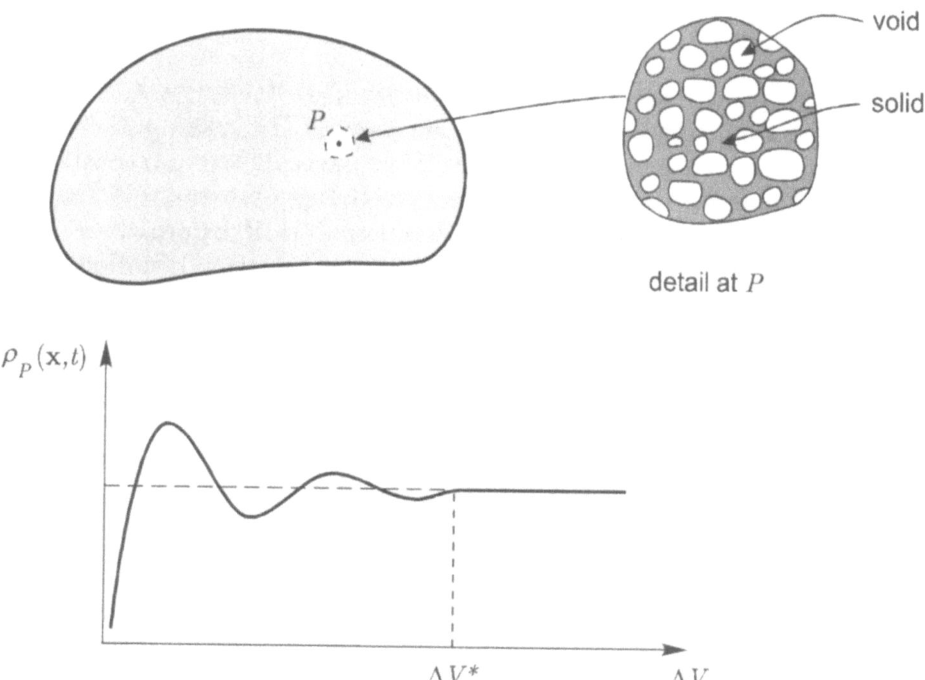

Figure 8.1: Influence of sample volume on the continuum measure of the density of a porous medium.

gives rise to partial differential equations in general and to the biharmonic partial differential equations discussed in this Chapter.

8.2 Displacements and strains in continua

Deformable bodies are characterized by their ability to change their size and shape when acted upon by forces. The types of forces that can act on deformable bodies can include external tractions, gravitational loads, thermal loads, pressures due to percolating fluids, etc. When acted upon by these forces, regions of a deformable body can experience both rigid body *motion*, which consists of rigid body translations and rotations and *strains* of the medium which consists of relative movement between points within the body. This latter category is of particular interest to the study of deformable continua since mathematical relationship for the *strains* within a continuum can be developed by considering the relative displacements between points within a continuum. When a continuum region experiences strain the internal constitution of the medium responds to the forces by allowing relative displacements between points within the continuum but maintaining the continuum intact.

The deformations which occur in a continuum, be it solid or a fluid, therefore consists of rigid body motions and strains. The magnitude of these deformations will depend on the ability of the internal material constitution of the continuum to resist such deformations. In the context of this Chapter, attention will be restricted to the consideration of small deformations where the term 'small' will be given a more precise mathematical definition in the ensuing discussion. In order to proceed with the study of the deformations of a continuum we consider a continuum region V_0, the points of which are referred to the rectangular Cartesian coordinate system shown in Figure 8.2.

A generic point of interest in V_0 is $P_0(\mathbf{x})$. After deformation, the new configuration of the body is denoted by V and the position of the generic point is denoted by $P(\mathbf{X})$. The displaced position of P_0 can be related to displacement vector $\mathbf{u}$ (Figure 8.2) by the relationship

$$\mathbf{X} = \mathbf{x} + \mathbf{u} \tag{8.3}$$

Since the deformations must maintain the continuum character of the medium, the Jacobian of the *deformation gradient matrix* $\mathbf{F}$, must be non-zero; i.e.

$$|J| = \det \mathbf{F} = \left|\frac{\partial \mathbf{X}}{\partial \mathbf{x}}\right| \neq 0 \tag{8.4}$$

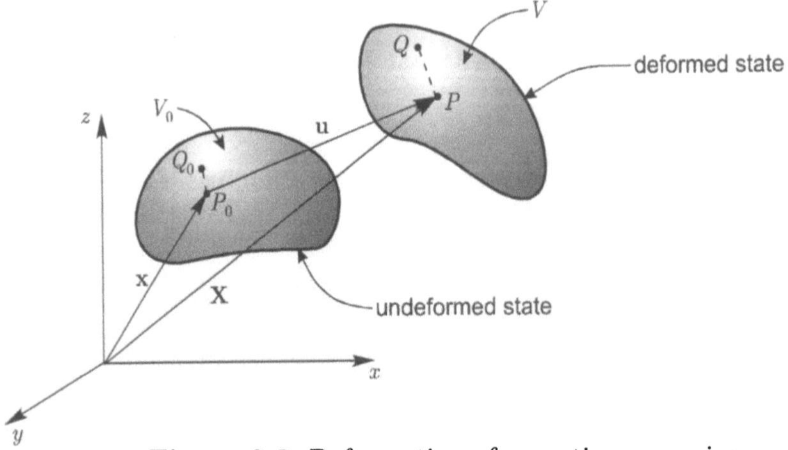

Figure 8.2: Deformation of a continuum region.

Using (8.3) in (8.4) and identifying the Cartesian components of the vectors $\mathbf{x}, \mathbf{X}$ and $\mathbf{u}$ by their components as

$$\mathbf{x} = [x \quad y \quad z]^T$$
$$\mathbf{X} = [X \quad Y \quad Z]^T \tag{8.5}$$
$$\mathbf{u} = [u \quad v \quad w]^T$$

where $[\quad]^T$ refers to the transpose, we can write

$$|\mathbf{F}| = \begin{vmatrix} (1 + \dfrac{\partial u}{\partial x}) & \dfrac{\partial u}{\partial y} & \dfrac{\partial u}{\partial z} \\[2ex] \dfrac{\partial v}{\partial x} & (1 + \dfrac{\partial v}{\partial y}) & \dfrac{\partial v}{\partial z} \\[2ex] \dfrac{\partial w}{\partial x} & \dfrac{\partial w}{\partial y} & (1 + \dfrac{\partial w}{\partial z}) \end{vmatrix} \tag{8.6}$$

If the points of the continuum are not displaced, the vector u is zero and in this case the deformation gradient is the unit matrix. If we restrict attention to small deformations of a continuum, the determinant of the deformation gradients matrix cannot be negative by undergoing a continuous deformation without passing through a zero value. Hence a necessary and sufficient condition for a deformation in a continuum region to be physically admissible, is that the determinant of $\mathbf{F}$ must be non-zero for any choice of $\mathbf{u}$. We can write the deformation gradients matrix as

$$\mathbf{F} = \mathbf{I} + \widehat{\mathbf{F}} \tag{8.7}$$

where $\mathbf{I}$ is the identity matrix (3×3) and

$$\widehat{\mathbf{F}} = \nabla \mathbf{u} = \begin{bmatrix} \dfrac{\partial u}{\partial x} & \dfrac{\partial u}{\partial y} & \dfrac{\partial u}{\partial z} \\[2ex] \dfrac{\partial v}{\partial x} & \dfrac{\partial v}{\partial y} & \dfrac{\partial v}{\partial z} \\[2ex] \dfrac{\partial w}{\partial x} & \dfrac{\partial w}{\partial y} & \dfrac{\partial w}{\partial z} \end{bmatrix} \tag{8.8}$$

and $\nabla \mathbf{u}$ is the *displacements gradients matrix* which is defined according to (8.8). The quantity $\widehat{\mathbf{F}}$ can also be represented as the sum of a symmetric matrix ϵ and an antisymmetric matrix $\boldsymbol{\omega}$; i.e.

$$\epsilon = \frac{1}{2}\left[\nabla \mathbf{u} + (\nabla \mathbf{u})^T\right] = \begin{bmatrix} \epsilon_{xx} & \epsilon_{xy} & \epsilon_{xz} \\ \epsilon_{yx} & \epsilon_{yy} & \epsilon_{yz} \\ \epsilon_{zx} & \epsilon_{zy} & \epsilon_{zz} \end{bmatrix} = \epsilon^T \tag{8.9}$$

and

$$\boldsymbol{\omega} = \frac{1}{2}\left[\nabla \mathbf{u} - (\nabla \mathbf{u})^T\right] = \begin{bmatrix} 0 & \omega_{xy} & \omega_{xz} \\ \omega_{yx} & 0 & \omega_{yz} \\ \omega_{zx} & \omega_{zy} & 0 \end{bmatrix} = -\boldsymbol{\omega}^T \tag{8.10}$$

where $(\ )^T$ denotes the transpose of $(\ )$.

In other mathematical representations of the state of deformation in a continuum the term $\nabla \mathbf{u}$, which is the *gradient of a vector*, is introduced as the "*dyadic*". A dyadic quantity is a *second-order tensor* quantity. Dyadic notations are compact *symbolic representations* of the various second-order quantities which are used rather infrequently in continuum mechanics due to conflicts with conventions associated with standard vector notation. (The physicist-mathematician Josiah Willard Gibbs (1839-1903) was an ardent proponent of the dyadic notation.) In the ensuing, where relevant, the dyadic approach will also be presented with the proviso that the analogous results can also be obtained in the more familiar indicial form. If we consider the gradient operator

$$\nabla = \mathbf{i}\frac{\partial}{\partial x} + \mathbf{j}\frac{\partial}{\partial y} + \mathbf{k}\frac{\partial}{\partial z} \tag{8.11}$$

and

$$\mathbf{u} = u\mathbf{i} + v\mathbf{j} + w\mathbf{k} \tag{8.12}$$

then the dyadic $\mathfrak{D} = \nabla \mathbf{u}$ is given by the sum of nine quantities; i.e.

$$\mathfrak{D} = \nabla \mathbf{u} = \mathbf{ii}\frac{\partial u}{\partial x} + \mathbf{ij}\frac{\partial v}{\partial x} + \mathbf{ik}\frac{\partial w}{\partial x}$$
$$+\mathbf{ji}\frac{\partial u}{\partial y} + \mathbf{jj}\frac{\partial v}{\partial y} + \mathbf{jk}\frac{\partial w}{\partial y}$$
$$+\mathbf{ki}\frac{\partial u}{\partial z} + \mathbf{kj}\frac{\partial v}{\partial z} + \mathbf{kk}\frac{\partial w}{\partial z} \tag{8.13}$$

It is noted that the order of the unit vectors in each term of (8.13) must be preserved such that

$$\mathbf{ij} \neq \mathbf{ji} \quad ; \quad \mathbf{ik} \neq \mathbf{ki} \quad ; \quad \mathbf{jk} \neq \mathbf{kj} \tag{8.14}$$

We can also define a *conjugate dyadic* $\mathfrak{D}_c = \mathbf{u}\nabla$ such that

$$\mathfrak{D}_c = \mathbf{u}\nabla = \mathbf{ii}\frac{\partial u}{\partial x} + \mathbf{ji}\frac{\partial v}{\partial x} + \mathbf{ki}\frac{\partial w}{\partial x}$$
$$+\mathbf{ij}\frac{\partial u}{\partial y} + \mathbf{jj}\frac{\partial v}{\partial y} + \mathbf{kj}\frac{\partial w}{\partial y}$$
$$+\mathbf{ik}\frac{\partial w}{\partial z} + \mathbf{jk}\frac{\partial v}{\partial z} + \mathbf{kk}\frac{\partial w}{\partial z} \tag{8.15}$$

If we define the *strain vectors*

$$\mathbf{e}_x = \epsilon_{xx}\mathbf{i} + \epsilon_{xy}\mathbf{j} + \epsilon_{xz}\mathbf{k}$$
$$\mathbf{e}_y = \epsilon_{yx}\mathbf{i} + \epsilon_{yy}\mathbf{j} + \epsilon_{yz}\mathbf{k} \tag{8.16}$$
$$\mathbf{e}_z = \epsilon_{zx}\mathbf{i} + \epsilon_{zy}\mathbf{j} + \epsilon_{zz}\mathbf{k}$$

then we can write

$$\mathbf{e}_x = \frac{\partial u}{\partial x}\mathbf{i} + \frac{1}{2}\left(\frac{\partial v}{\partial x} + \frac{\partial u}{\partial y}\right)\mathbf{j} + \frac{1}{2}\left(\frac{\partial w}{\partial x} + \frac{\partial u}{\partial z}\right)\mathbf{k} \tag{8.17}$$

or

$$\mathbf{e}_x = \frac{1}{2}\left(\nabla\mathbf{u} + \frac{\partial \mathbf{u}}{\partial x}\right) \tag{8.18}$$

Similarly we have

$$\mathbf{e}_y = \frac{1}{2}\left(\nabla\mathbf{u} + \frac{\partial\mathbf{u}}{\partial y}\right) \quad ; \quad \mathbf{e}_z = \frac{1}{2}\left(\nabla\mathbf{u} + \frac{\partial\mathbf{u}}{\partial z}\right) \tag{8.19}$$

We now introduce the *strain dyadic* $\mathfrak{E}$ where

$$\mathfrak{E} = \mathbf{i}\mathbf{e}_x + \mathbf{j}\mathbf{e}_y + \mathbf{k}\mathbf{e}_z \tag{8.20}$$

Then the equations (8.18) and (8.19) can be written as the *dyadic equation for the strain*

$$\mathfrak{E} = \frac{1}{2}\left(\nabla\mathbf{u} + \mathbf{u}\nabla\right) = \frac{1}{2}(\mathfrak{D} + \mathfrak{D}_c) \tag{8.21}$$

Similarly the *dyadic equation for the rotation* is given by

$$\mathfrak{R} = \frac{1}{2}\left(\nabla\mathbf{u} - \mathbf{u}\nabla\right) = \frac{1}{2}(\mathfrak{D} - \mathfrak{D}_c) \tag{8.22}$$

where

$$\nabla\mathbf{u} = \mathbf{i}\frac{\partial\mathbf{u}}{\partial x} + \mathbf{j}\frac{\partial\mathbf{u}}{\partial y} + \mathbf{k}\frac{\partial\mathbf{u}}{\partial z} \quad ; \quad \mathbf{u}\nabla = \frac{\partial\mathbf{u}}{\partial x}\mathbf{i} + \frac{\partial\mathbf{u}}{\partial y}\mathbf{j} + \frac{\partial\mathbf{u}}{\partial z}\mathbf{k} \tag{8.23}$$

So far we have discussed the concept of a deformation gradient and strain in an abstract sense without illustrating the notion of "*straining*" of the continuum region. The next step is to show that the measures (8.9) and (8.10) correspond to strains and rotations of infinitesimal elements within the continuum region. Consider the point Q_0 which is located at distance $(\mathbf{x} + d\mathbf{x})$ from P_0, in the undeformed configuration. Upon deformation, the point Q_0 moves to the location Q with coordinates $(\mathbf{X} + d\mathbf{X})$. We obtain from (8.3)

$$d\mathbf{X} = d\mathbf{x} + \frac{\partial\mathbf{u}}{\partial\mathbf{x}}d\mathbf{x} = \left(\mathbf{I} + \widehat{\mathbf{F}}\right)d\mathbf{x} \tag{8.24}$$

or

$$d\mathbf{X} = \mathbf{F}d\mathbf{x} \tag{8.25}$$

The notion of straining involves either a change in the length of elements or the change in shape of elements within the continuum region. We can first examine the length of the line element $P_0 Q_0$ in the undeformed state, which can be obtained in the form

$$(P_0 Q_0)^2 = (\delta L_0)^2 = (dx)^2 + (dy)^2 + (dz)^2 = (d\mathbf{x})^T (d\mathbf{x}) \tag{8.26}$$

Similarly, the length of the line element PQ in the deformed configuration is given by

$$(PQ)^2 = (\delta L)^2 = (dX)^2 + (dY)^2 + (dZ)^2 = (d\mathbf{X})^T (d\mathbf{X}) \tag{8.27}$$

Considering (8.25), we can write (8.27) as

$$(\delta L)^2 = (d\mathbf{x})^T \mathbf{F}^T \mathbf{F} (d\mathbf{x}) \tag{8.28}$$

Considering (8.7) we can write (8.28) as

$$(\delta L)^2 = (d\mathbf{x})^T \left[\mathbf{I} + \widehat{\mathbf{F}} + \widehat{\mathbf{F}}^T + \widehat{\mathbf{F}}\widehat{\mathbf{F}}^T \right] d\mathbf{x} \tag{8.29}$$

The measure of "straining" in the continuum can be obtained by taking the difference in the lengths of the line element in the *deformed* and *undeformed* configurations. We have

$$(\delta L)^2 - (\delta L_0)^2 = 2(d\mathbf{x})^T [\mathbf{D}] d\mathbf{x} \tag{8.30}$$

where

$$\mathbf{D} = \frac{1}{2} \left[\widehat{\mathbf{F}} + \widehat{\mathbf{F}}^T + \widehat{\mathbf{F}}\widehat{\mathbf{F}}^T \right] = \frac{1}{2} \left[\nabla \mathbf{u} + \mathbf{u} \nabla + (\mathbf{u}\nabla)(\nabla \mathbf{u}) \right] \tag{8.31}$$

is a measure of the "strain" in the continuum region. Since we are restricting attention to *small deformations* of the continuum region it can be assumed that

$$\nabla \mathbf{u} \ll 1 \quad \text{and} \quad (\mathbf{u}\nabla)(\nabla \mathbf{u}) \to 0 \tag{8.32}$$

and (8.31) gives

$$\mathbf{D} = \frac{1}{2}\left[\nabla\mathbf{u} + (\nabla\mathbf{u})^T\right] = \boldsymbol{\epsilon} \qquad (8.33)$$

and $\boldsymbol{\epsilon}$ is referred to as the strain matrix governing small strain behaviour of the continuum region. The components of $\boldsymbol{\epsilon}$, referred to the rectangular Cartesian coordinate system takes the form

$$\boldsymbol{\epsilon} = \begin{bmatrix} \epsilon_{xx} & \epsilon_{xy} & \epsilon_{xz} \\ \epsilon_{xy} & \epsilon_{yy} & \epsilon_{yz} \\ \epsilon_{xz} & \epsilon_{yz} & \epsilon_{zz} \end{bmatrix}$$

$$= \begin{bmatrix} \dfrac{\partial u}{\partial x} & \dfrac{1}{2}\left(\dfrac{\partial u}{\partial y} + \dfrac{\partial v}{\partial x}\right) & \dfrac{1}{2}\left(\dfrac{\partial u}{\partial z} + \dfrac{\partial w}{\partial x}\right) \\ \dfrac{1}{2}\left(\dfrac{\partial u}{\partial y} + \dfrac{\partial v}{\partial x}\right) & \dfrac{\partial v}{\partial y} & \dfrac{1}{2}\left(\dfrac{\partial v}{\partial z} + \dfrac{\partial w}{\partial y}\right) \\ \dfrac{1}{2}\left(\dfrac{\partial u}{\partial z} + \dfrac{\partial w}{\partial x}\right) & \dfrac{1}{2}\left(\dfrac{\partial v}{\partial z} + \dfrac{\partial w}{\partial y}\right) & \dfrac{\partial w}{\partial z} \end{bmatrix} \qquad (8.34)$$

and we note that

$$\boldsymbol{\epsilon} = \boldsymbol{\epsilon}^T \qquad (8.35)$$

8.2.1   Physical interpretations of the strain matrix

The physical interpretation of the elements of the *strain matrix* defined by (8.34) is important from the point of view of examining specific situations similar to those illustrated in Figures 8.3 and 8.4. The Figure 8.3 illustrates the idealized one-dimensional analogue of the state of deformation of a bar subjected to uniaxial stress.

Considering the elements $P_0 Q_0$ and PQ of the bars in the undeformed and deformed configurations respectively

$$P_0 Q_0 = dx \quad ; \quad PQ = dx + \frac{\partial u}{\partial x} dx \qquad (8.36)$$

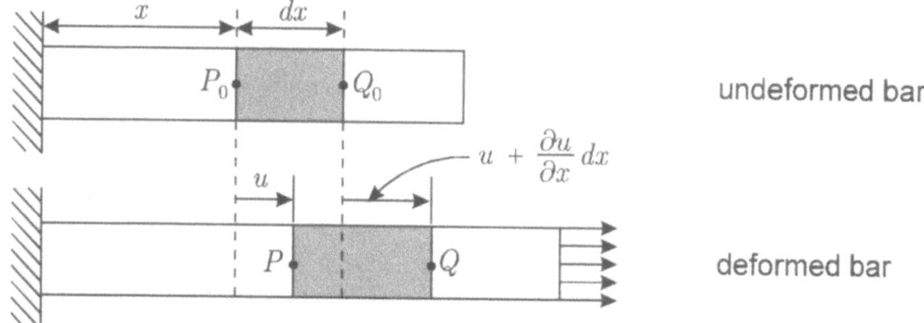

Figure 8.3: Uniaxial deformation of a bar element.

If we define the normal strain ϵ_{xx} as the unit change in length we have, for small deformations

$$\epsilon_{xx} = \frac{dx + \dfrac{\partial u}{\partial x}dx - dx}{dx} = \frac{\partial u}{\partial x} \tag{8.37}$$

Similar definitions can be derived for ϵ_{yy} and ϵ_{zz} the axial strains in the y- and z-directions respectively. We consider the deformations which can occur in a two-dimensional region (Figure 8.4), and restrict attention to rectangular elements $A_0B_0C_0D_0$ and $ABCD$ in the undeformed and deformed configurations respectively. The deformations are drawn to an exaggerated scale purely for purposes of clarity. We note that the region exhibits rigid body movements, changes in length of the sides and changes in shape as the sides rotate with respect to each other.

We can define the axial strain of the element A_0B_0 as

$$\epsilon_{xx} = \frac{AB - A_0B_0}{A_0B_0} = \frac{AB}{dx} - 1 \tag{8.38}$$

Similarly

$$\epsilon_{yy} = \frac{AD - A_0D_0}{A_0D_0} = \frac{AD}{dy} - 1 \tag{8.39}$$

Some sign conventions need to be defined; we shall consider length increases by *tensile strains* which are regarded as *positive*; conversely length decreases indicate *compressive* strains which are regarded as *negative*.

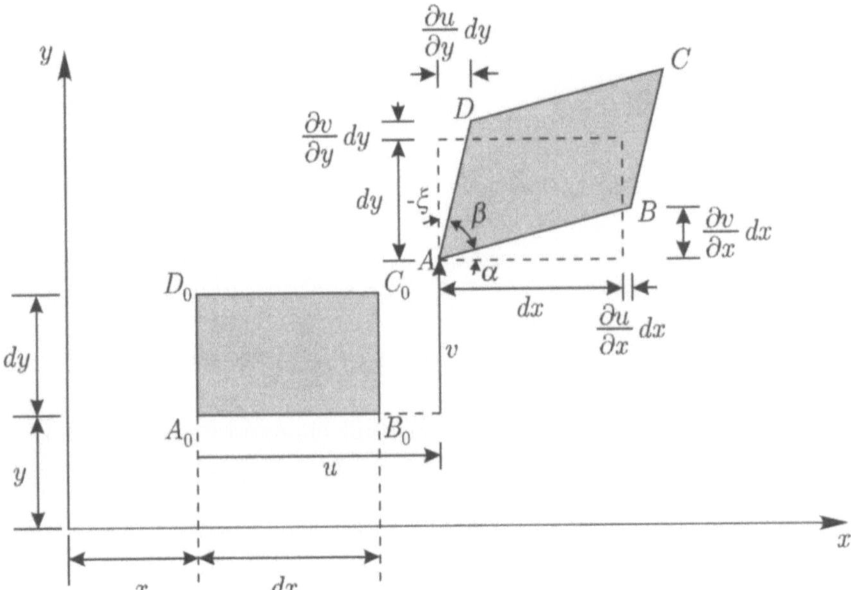

Figure 8.4: Two-dimensional deformations of a continuum element.

Shear strains are representative of changes in shape of regions during deformation. We can define the shear strain γ_{xy} as the change in the original right angle contained by line elements A_0B_0 and A_0D_0. As a sign convention, shear strain can be regarded as positive if the right angle between positive directions of the two axes decreases. The sign of the shear strain is therefore linked to the orientation of reference coordinate axes. The shear strain of the rectangular element during deformation is given by

$$\gamma_{xy} = \frac{\pi}{2} - \beta = \alpha - \xi \tag{8.40}$$

The sign of ξ is taken as negative since counter-clockwise angles of rotation are defined as positive. The shear strain defined by (8.40) is referred to as engineering strain which is related to the mathematical shear strain ϵ_{xy} by the relation

$$\epsilon_{xy} = \frac{1}{2}\gamma_{xy} = \frac{1}{2}(\alpha - \xi) \tag{8.41}$$

We can now proceed to make the appropriate reductions to (8.38), (8.39) and (8.41). From geometry we have

$$(AB)^2 = \left(dx + \frac{\partial u}{\partial x}dx\right)^2 + \left(\frac{\partial v}{\partial x}dx\right)^2 \tag{8.42}$$

From (8.38) we have

$$(AB)^2 = (1 + \epsilon_{xx})^2(dx)^2 \tag{8.43}$$

Equations (8.42) and (8.43) give

$$\epsilon_{xx}^2 + 2\epsilon_{xx} + 1 = 1 + 2\frac{\partial u}{\partial x} + \left(\frac{\partial u}{\partial x}\right)^2 + \left(\frac{\partial v}{\partial x}\right)^2 \tag{8.44}$$

Since we are restricting attention to only small deformations as expressed by (8.32), we can neglect the higher-order terms in (8.44). This gives

$$\epsilon_{xx} = \frac{\partial u}{\partial x} \tag{8.45}$$

similarly we can show that

$$\epsilon_{yy} = \frac{\partial v}{\partial y} \tag{8.46}$$

Considering the geometry of the deformed configuration we have

$$\tan \alpha = \frac{\left(\dfrac{\partial v}{\partial x}\right)dx}{\left[dx + \left(\dfrac{\partial u}{\partial x}\right)dx\right]} \quad ; \quad \tan(-\xi) = \frac{\left(\dfrac{\partial u}{\partial y}\right)dy}{\left[dy + \left(\dfrac{\partial v}{\partial y}\right)dy\right]} \tag{8.47}$$

Since we are dealing with small deformations, (8.47) can be written as

$$\tan \alpha \simeq \alpha = \frac{\partial v}{\partial x} \quad ; \quad \tan(-\xi) = -\xi = \frac{\partial u}{\partial y} \tag{8.48}$$

Hence the mathematical measure of the shear strain reduces to

$$\epsilon_{xy} = \frac{1}{2}\left(\frac{\partial u}{\partial y} + \frac{\partial v}{\partial x}\right) \tag{8.49}$$

Similar expressions can be derived for the other component of the strain matrix ϵ (i.e. $\epsilon_{zz}, \epsilon_{xz}$ and ϵ_{yz}) by considering appropriate elements referred to the $x - z$ and $y - z$ planes.

8.2.2   Physical interpretation of the rotation matrix

When the displacement components in a continuum are specified (e.g. $\mathbf{u} = \mathbf{u}(\mathbf{x})$), the strain and geometry of any element within the continuum region can be completely defined. Conversely, if the strain components within the continuum region are specified we cannot completely define the displacement field $\mathbf{u}$. The integration of the strain-displacement relationships of the type (8.33) or (8.34) will give rise to certain constants of integration which can be identified as rigid body translations and rotations. Consider an infinitesimal element in the continuum which is subjected to a two-dimensional state of deformation which includes strains and rigid body movements. We define

$$\omega_{xy} = \frac{1}{2}\left(\frac{\partial u}{\partial y} - \frac{\partial v}{\partial x}\right) = -\omega_{yx} \tag{8.50}$$

which represents the average of the angular displacements of dx and the angular displacement of dy and is referred to as the *rotation*. Referring to Figure 8.4, if the x-component of the displacement of A_0 is $u(x, y)$, then at C_0 we have the displacement $u + du$ such that

$$du = \frac{\partial u}{\partial x}dx + \frac{\partial u}{\partial y}dy \tag{8.51}$$

We can rewrite (8.51) as

$$du = \frac{\partial u}{\partial x}dx + \frac{1}{2}\left(\frac{\partial u}{\partial y} + \frac{\partial v}{\partial x}\right)dy + \frac{1}{2}\left(\frac{\partial u}{\partial y} - \frac{\partial v}{\partial x}\right)dy \tag{8.52}$$

or

$$du = \epsilon_{xx}dx + \epsilon_{xy}dy + \omega_{xy}dy \tag{8.53}$$

The first two terms of (8.53) represents the x-component of the displacement of C_0 relative to A_0, due to ϵ_{xx} and ϵ_{xy} due to the state of strain (see Figure 8.5).

The third term in (8.53) corresponds to a displacement derived from a rotation. If we combine all three effects, the element will acquire its final position. Similarly, we can write the y-component of the displacement of C_0 as

$$dv = \epsilon_{yy}dy + \epsilon_{xy}dx - \omega_{xy}dx \tag{8.54}$$

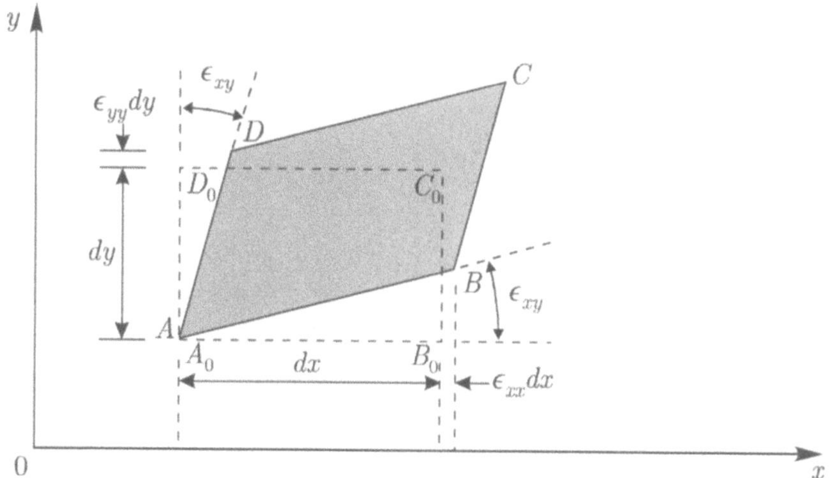

Figure 8.5: Pure two-dimensional deformation of a continuum element.

Since equations (8.53) and (8.54) are total differentials, they can be integrated (subject to the integrability conditions identified as the compatibility equations) to generate $u(x,y)$ and $v(x,y)$. It is sufficient to note that the final expressions will contain results of the form

$$u(x,y) = u^*(x,y) + u_0 + \Omega^0_{xy}y \tag{8.55}$$

$$v(x,y) = v^*(x,y) + v_0 - \Omega^0_{xy}x \tag{8.56}$$

where $u^*(x,y)$ and $v^*(x,y)$ are derived from the integration of the first two terms each of (8.53) and (8.54); u_0 and v_0 are rigid body translations of the region and Ω^0_{xy} refers to the rotation of the entire body. We can interpret the concept of a rotation in three dimensions. The graphical illustrations are cumbersome; the mathematical operations however can be extended to give

$$du = \epsilon_{xx}dx + \epsilon_{xy}dy + \epsilon_{xz}dz + \omega_{xy}dy - \omega_{xz}dz$$
$$dv = \epsilon_{yy}dy + \epsilon_{xy}dx + \epsilon_{yz}dz + \omega_{yz}dz - \omega_{xy}dx \qquad (8.57)$$
$$dw = \epsilon_{zz}dz + \epsilon_{xz}dx + \epsilon_{yz}dy + \omega_{xz}dx - \omega_{yz}dy$$

where

$$\omega_{xy} = \frac{1}{2}\left(\frac{\partial u}{\partial y} - \frac{\partial v}{\partial x}\right) = -\omega_{yx}$$
$$\omega_{yz} = \frac{1}{2}\left(\frac{\partial v}{\partial z} - \frac{\partial w}{\partial y}\right) = -\omega_{zy} \qquad (8.58)$$
$$\omega_{xz} = \frac{1}{2}\left(\frac{\partial u}{\partial z} - \frac{\partial w}{\partial x}\right) = -\omega_{zx}$$

and

$$\boldsymbol{\omega} = -\boldsymbol{\omega}^T \qquad (8.59)$$

8.2.3   Indicial notation representations of strain and rotation

As is evident from the preceding developments, it is also possible to present the definitions of strain by appeal to an indicial notation, where, the indices can range from 1 to 3 in the case of a generalized three-dimensional formulations and 1 to 2 in the case of two-dimensional formulations. We shall briefly summarize the definitions. The rectangular Cartesian coordinates are denoted by x_i where

$$x_1 = x \quad ; \quad x_2 = y \quad ; \quad x_3 = z \qquad (8.60)$$

and the displacement components in the x, y and z directions are denoted by u_i where

$$u_1 = u, \quad ; \quad u_2 = v \quad ; \quad u_3 = w \qquad (8.61)$$

The displacement gradients matrix is given by

$$F_{ij} = \frac{\partial u_i}{\partial x_j} = u_{i,j} \tag{8.62}$$

and the strain and rotation matrices are given by

$$\epsilon_{ij} = \frac{1}{2}(u_{i,j} + u_{j,i}) = \frac{1}{2}(F_{ij} + F_{ji}) \tag{8.63}$$

and

$$\omega_{ij} = \frac{1}{2}(u_{i,j} - u_{j,i}) = \frac{1}{2}(F_{ij} - F_{ji}) \tag{8.64}$$

The components of ϵ_{ij} and ω_{ij} can be written as

$$\epsilon_{ij} = \begin{bmatrix} \epsilon_{xx} & \epsilon_{xy} & \epsilon_{xz} \\ \epsilon_{yx} & \epsilon_{yy} & \epsilon_{yz} \\ \epsilon_{zx} & \epsilon_{zy} & \epsilon_{zz} \end{bmatrix} \tag{8.65}$$

and

$$\omega_{ij} = \begin{bmatrix} 0 & \omega_{xy} & \omega_{xz} \\ \omega_{yx} & 0 & \omega_{yz} \\ \omega_{zx} & \omega_{zy} & 0 \end{bmatrix} \tag{8.66}$$

with

$$\epsilon_{ij} = \epsilon_{ji} \qquad \text{and} \qquad \omega_{ij} = -\omega_{ji} \tag{8.67}$$

The definitions of the individual components of ϵ_{ij} and ω_{ij} referred to the rectangular Cartesian coordinate system are given by (8.34) and (8.58).

8.2.4 Transformation of the strain matrix

In many situations involving the analysis of stress and strain in deformable continua the directions of boundary constraint, directions of application of the external forces, etc., may not coincide with the chosen reference coordinate system. In these circumstances it is convenient to transform the strain matrix at the point of interest to conform to a new coordinate system in

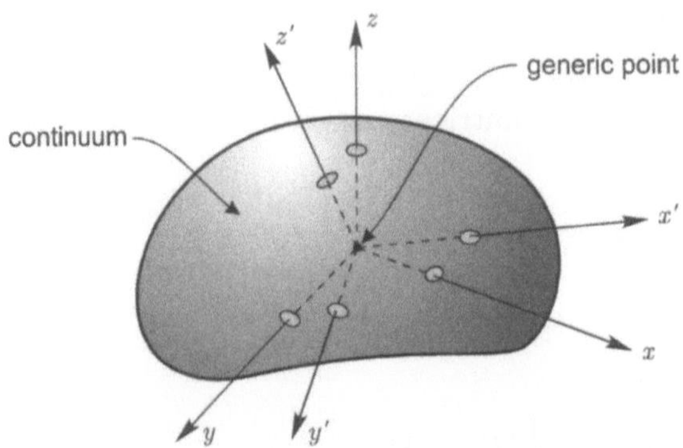

Figure 8.6: Rotation of coordinates for transformation of strain matrix.

which the relevant boundary constraints acquire a simplified form. We define the state of strain in the medium referred to the reference rectangular Cartesian coordinate system (x, y, z) by the strain matrix ϵ; i.e.

$$\epsilon = \begin{bmatrix} \epsilon_{xx} & \epsilon_{xy} & \epsilon_{xz} \\ \epsilon_{yx} & \epsilon_{yy} & \epsilon_{yz} \\ \epsilon_{zx} & \epsilon_{zy} & \epsilon_{zz} \end{bmatrix} = \epsilon^T \tag{8.68}$$

We now consider a new system of Cartesian coordinates (x', y', z') which is obtained by a generalized rotation of the (x, y, z) coordinate system (see Figure 8.6).

The strain matrix ϵ' referred to the new coordinate system is defined by

$$\epsilon' = \begin{bmatrix} \epsilon_{x'x'} & \epsilon_{x'y'} & \epsilon_{x'z'} \\ \epsilon_{y'x'} & \epsilon_{y'y'} & \epsilon_{y'z'} \\ \epsilon_{z'x'} & \epsilon_{z'y'} & \epsilon_{z'z'} \end{bmatrix} = (\epsilon')^T \tag{8.69}$$

The relationship between the elements of ϵ and the elements of ϵ' can be obtained by the matrix transformation operation in linear algebra, which gives

$$\epsilon' = \mathbf{H}^T \epsilon \mathbf{H} \tag{8.70}$$

where $\mathbf{H}$ is the directions cosines matrix between the two orthogonal coordinate systems which is defined by

$$\mathbf{H} = \begin{bmatrix} \cos(xox') & \cos(xoy') & \cos(xoz') \\ \cos(yox') & \cos(yoy') & \cos(yoz') \\ \cos(zox') & \cos(zoy') & \cos(zoz') \end{bmatrix} \tag{8.71}$$

and $\mathbf{H}^T$ denotes the transpose of $\mathbf{H}$.

The operation involving transformation of the strain matrix can also be demonstrated by considering the two-dimensional configuration of the diagonal line $A_0 C_0$ in the undeformed configuration and the corresponding line AC in the deformed configuration of the type of two-dimensional elemental area described in Figure 8.4. For convenience first we shall focus on the deformation of the element with reference to the rotation of the coordinate system.

Referring to Figure 8.7a, the dimensions of the elemental region are chosen such that

$$\frac{dy}{ds} = \sin \alpha \quad ; \quad \frac{dx}{ds} = \cos \alpha \tag{8.72}$$

Considering the deformed configuration of the element, if the displacements of the point A_0 in the directions x and y are u and v respectively, then the displacement of the point C_0 in the directions x and y are given by

$$\left(u + \frac{\partial u}{\partial x}dx + \frac{\partial u}{\partial y}dy \right) \text{ and } \left(v + \frac{\partial v}{\partial y}dy + \frac{\partial v}{\partial x}dx \right)$$

respectively.

We can now rigid bodily translate the deformed element such that A coincides with A_0, which illustrates the relative deformation of A with respect to C. Considering the geometry of the deformation we have

$$MC_0 = \frac{\partial u}{\partial x}dx + \frac{\partial u}{\partial y}dy \quad ; \quad MC = \frac{\partial v}{\partial y}dy + \frac{\partial v}{\partial y}dx \tag{8.73}$$

We can now project these displacements on AC to determine the change in length NC, i.e.

$$NC = \frac{(MC_0)\cos \alpha + (MC)\sin \alpha}{\cos \varphi} \tag{8.74}$$

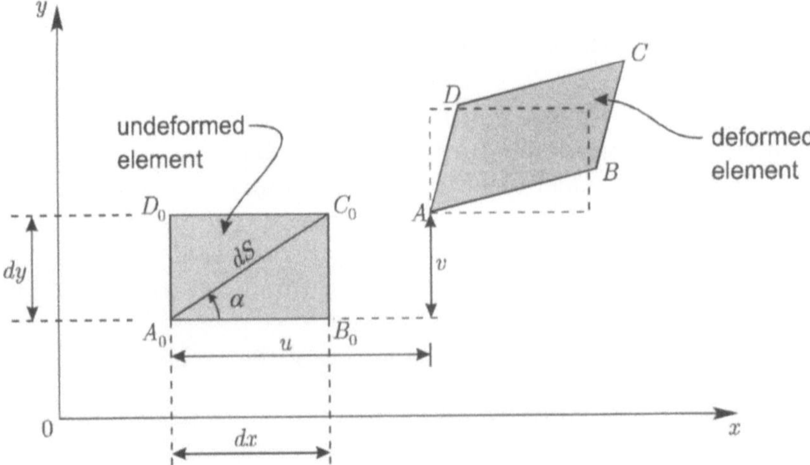

(a) geometry of deformation

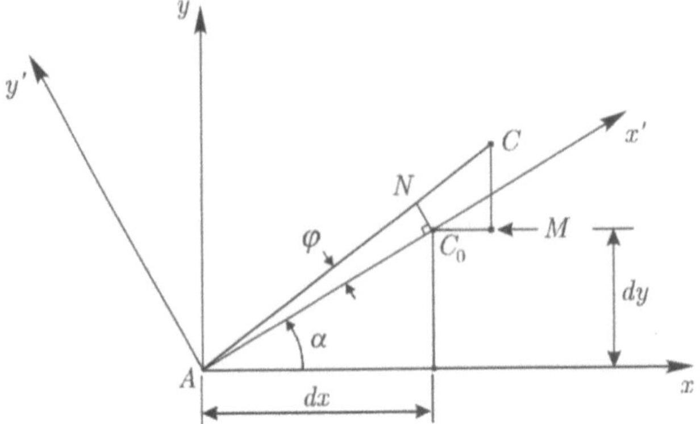

(b) deformation of diagonal element

Figure 8.7: Transformation of the strain matrix.

Since for small deformations φ can be assumed to be a small angle, $\cos\varphi \simeq 1$, and the small deformation measure of the normal strain in the x'-direction is defined as

$$\epsilon_{x'x'} = \frac{NC}{ds} \tag{8.75}$$

Substituting (8.73) in (8.74) and using the result in (8.75) we obtain

$$\epsilon_{x'x'} = \left(\frac{\partial u}{\partial x} \frac{dx}{ds} + \frac{\partial u}{\partial y} \frac{dy}{ds} \right) \cos \alpha + \left(\frac{\partial v}{\partial y} \frac{dy}{ds} + \frac{\partial v}{\partial x} \frac{dx}{ds} \right) \sin \alpha \qquad (8.76)$$

Using (8.72) we can write (8.76) as

$$\epsilon_{x'x'} = \left(\frac{\epsilon_{xx} + \epsilon_{yy}}{2} \right) + \left(\frac{\epsilon_{xx} - \epsilon_{yy}}{2} \right) \cos 2\alpha + \epsilon_{xy} \sin 2\alpha \qquad (8.77)$$

The normal strain $\epsilon_{y'y'}$ can be calculated either by selecting the diagonal direction in the configuration given in Figure 8.7a to coincide with one y'-direction or simply by substituting $(\alpha + \pi/2)$ for α in the result (8.77). We can show that

$$\epsilon_{y'y'} = \left(\frac{\epsilon_{xx} + \epsilon_{yy}}{2} \right) - \left(\frac{\epsilon_{xx} - \epsilon_{yy}}{2} \right) \cos 2\alpha - \epsilon_{xy} \sin 2\alpha \qquad (8.78)$$

The angular displacement φ is indicative of *a part* of engineering shear strain; to determine this we consider the geometry of the deformation (Figure 8.7b) which gives

$$\tan \varphi \simeq \varphi = \frac{NC_0}{ds} = \frac{(MC) \cos \alpha - (MC_0) \sin \alpha - (AC)\varphi}{ds} \qquad (8.79)$$

We note that

$$(AC)\varphi = (\epsilon_{x'x'} ds)\varphi \qquad (8.80)$$

which can be neglected since we are dealing with small deformation analysis, where products of the strains are assumed to be small. This reduces (8.79) to

$$\varphi = \left(\frac{\partial v}{\partial y} \frac{dy}{ds} + \frac{\partial v}{\partial x} \frac{dx}{ds} \right) \cos \alpha - \left(\frac{\partial u}{\partial x} \frac{dx}{ds} + \frac{\partial u}{\partial y} \frac{dy}{ds} \right) \sin \alpha \qquad (8.81)$$

or

$$\varphi = -(\epsilon_{xx} - \epsilon_{yy}) \sin \alpha \cos \alpha + \frac{\partial v}{\partial x} \cos^2 \alpha - \frac{\partial u}{\partial y} \sin^2 \alpha \qquad (8.82)$$

The analysis can now be extended to calculate the angular displacement of

y', which, by inspection will be equal to the value of φ evaluated by replacing α in (8.82) by $(\alpha + \pi/2)$. This gives

$$\varphi^* = -(\epsilon_{yy} - \epsilon_{xx})\sin\alpha\cos\alpha + \frac{\partial v}{\partial x}\sin^2\alpha - \frac{\partial u}{\partial y}\cos^2\alpha \tag{8.83}$$

The engineering shear strain referred to the rotated coordinate system is given by

$$\gamma_{x'y'} = 2\epsilon_{x'y'} = \varphi - \varphi^* \tag{8.84}$$

The negative sign of φ^* indicates, as in the case of (8.40) a clockwise change in the angle. From (8.82), (8.83) and (8.84) we obtain

$$\epsilon_{x'y'} = (\epsilon_{yy} - \epsilon_{xx})\sin\alpha\cos\alpha + \epsilon_{xy}\cos 2\alpha \tag{8.85}$$

The results for $\epsilon_{x'x'}$, $\epsilon_{y'y'}$ and $\epsilon_{x'y'}$ can also be obtained from the matrix transformation result (8.70) which, for the two-dimensional case reduces to

$$\begin{bmatrix} \epsilon_{x'x'} & \epsilon_{x'y'} \\ \epsilon_{x'y'} & \epsilon_{y'y'} \end{bmatrix} = \begin{bmatrix} \cos\alpha & -\sin\alpha \\ \sin\alpha & \cos\alpha \end{bmatrix} \begin{bmatrix} \epsilon_{xx} & \epsilon_{xy} \\ \epsilon_{xy} & \epsilon_{yy} \end{bmatrix} \begin{bmatrix} \cos\alpha & \sin\alpha \\ -\sin\alpha & \cos\alpha \end{bmatrix} \tag{8.86}$$

8.2.5 Principal strains and strain invariants

In the transformation of the strain matrix described in the previous section, it is possible to find an orientation of the axes x', y', z' such that in the transformed strain matrix ϵ', only the diagonal terms are present. This specific orientation of the rotated system of axes will give the *principal strains* of ϵ or the *eigenvalues* of the strain matrix and the particular orientation of the x', y', z' system will correspond to the *eigen-vectors*. The eigenvalues $\epsilon_n(n = 1, 2, 3)$ are given by the roots of the determinantal equation

$$|\epsilon - \epsilon\mathbf{I}| = 0 \tag{8.87}$$

or, alternatively, from the determinantal equation

$$|\epsilon_{ij} - \epsilon\delta_{ij}| = 0 \tag{8.88}$$

where δ_{ij} is the delta function due to Leopold Kronecker (1823-1891) [i.e. $\delta_{ij} = 1$ for $i = j$; $\delta_{ij} = 0$ for $i \neq j$;]. If we expand these equations, we obtain the Cayley-Hamilton equation [after A. Cayley (1821-1895) and Sir William Rowan Hamilton (1805-1865)]:

$$\epsilon^3 - \epsilon^2 J_1 + \epsilon J_2 - J_3 = 0 \tag{8.89}$$

where J_1, J_2 and J_3 are the *principal strain invariants* which are given by

$$\begin{aligned}
J_1 &= tr(\epsilon) \\
J_2 &= \frac{1}{2}\left[(tr\,\epsilon)^2 - tr\,\epsilon^2\right] \\
J_3 &= \det\,\epsilon
\end{aligned} \tag{8.90}$$

where 'tr' is the trace of appropriate matrix and 'det ' denotes the determinant. The results (8.90) can also be written in the indicial form

$$\begin{aligned}
J_1 &= \epsilon_{ii} \\
J_2 &= \frac{1}{2}\left[\epsilon_{ii}\epsilon_{jj} - \epsilon_{ij}\epsilon_{ij}\right] \\
J_3 &= |\epsilon_{ij}| = \frac{1}{6}\left[\epsilon_{ij}\epsilon_{jj}\epsilon_{kk} - 3\epsilon_{ij}\epsilon_{ji}\epsilon_{kk} + 2\epsilon_{ij}\epsilon_{jk}\epsilon_{ki}\right]
\end{aligned} \tag{8.91}$$

In (8.91), following the convention proposed by Albert Einstein (1874-1955), summation is carried out over repeated indices, for $i, j = (x, y, z)$. The term *invariant* implies that the magnitude of the quantities remain unchanged for all rotations of the coordinate system; e.g.

$$tr\,\epsilon = tr\,\epsilon'$$

$$(tr\,\epsilon)^2 - tr\,\epsilon^2 = (tr\,\epsilon')^2 - tr(\epsilon')^2 \tag{8.92}$$

$$\det\,\epsilon = \det\,\epsilon'$$

8.2.6 Compatibility of strains

The definitions of strain in a continuum region given by (8.34) can be regarded as a set of first-order linear partial differential equations which relate the displacements in the continuum to the resulting strains. If the displacements are specified within the continuum region, it is a relatively straightforward exercise to obtain the distribution of strains within the continuum. The converse problem of determining the displacements of the medium when strains are defined within, is not so routine. We encounter six linear first-order partial differential equations which need to be solved to obtain the three displacement components. At the outset it would appear that the system is overdetermined. Therefore to carry out the integration of the strain-displacement equations to obtain a unique solution for the displacements, certain conditions must be satisfied by the strain components. The issues concerning the integrability of the strain-displacement relationships for the evaluation of the displacement components was investigated by a number of eminent mathematicians and elasticians including Saint-Venant (1797-1886), Boussinesq (1842-1929), Beltrami (1835-1900) Michell (1863-1940), Cesaro (1859-1906) and Morera (1856-1909). The results of these and other investigations culminated in the development of the *compatibility equations* which must be satisfied by the components of ϵ if we are to obtain *single-valued* and *continuous* functions for displacements within the continuum. In this section we shall present derivations of the compatibility equations using both the *strain dyadic* and indicial conventions.

We recall that for the strain dyadic $\mathfrak{E}$ (see e.g. (8.20)) expressed in terms of the three vectors $\mathbf{e}_x, \mathbf{e}_y$ and $\mathbf{e}_z$,

$$div\ \mathfrak{E} = \nabla.\mathfrak{E} = \left(\mathbf{i}\frac{\partial}{\partial x} + \mathbf{j}\frac{\partial}{\partial y} + \mathbf{k}\frac{\partial}{\partial z}\right)\left(\mathbf{i}e_x + \mathbf{j}e_y + \mathbf{k}e_z\right) \tag{8.93}$$

or

$$\nabla.\mathfrak{E} = \frac{\partial \mathbf{e}_x}{\partial x} + \frac{\partial \mathbf{e}_y}{\partial y} + \frac{\partial \mathbf{e}_z}{\partial z} \tag{8.94}$$

Similarly we can define the *curl* of $\mathfrak{E}$ as

$$\text{curl } \mathfrak{E} = \nabla \times \mathfrak{E} = \begin{vmatrix} \mathbf{i} & \mathbf{j} & \mathbf{k} \\ \dfrac{\partial}{\partial x} & \dfrac{\partial}{\partial y} & \dfrac{\partial}{\partial z} \\ \mathbf{e}_x & \mathbf{e}_y & \mathbf{e}_z \end{vmatrix} \tag{8.95}$$

or

$$\nabla \times \mathfrak{E} = \left(\mathbf{i} \times \frac{\partial \mathfrak{E}}{\partial x} \right) + \left(\mathbf{j} \times \frac{\partial \mathfrak{E}}{\partial y} \right) + \left(\mathbf{k} \times \frac{\partial \mathfrak{E}}{\partial z} \right) \tag{8.96}$$

Similarly the *conjugate curl* is defined as

$$\mathfrak{E} \times \nabla = \left(\frac{\partial \mathfrak{E}}{\partial x} \times \mathbf{i} \right) + \left(\frac{\partial \mathfrak{E}}{\partial y} \times \mathbf{j} \right) + \left(\frac{\partial \mathfrak{E}}{\partial z} \times \mathbf{k} \right) \tag{8.97}$$

Using (8.96) and (8.97) it can be shown that

$$\nabla \times \nabla \mathfrak{E} + \mathfrak{E} \nabla \times \nabla = 0 \tag{8.98}$$

Since

$$\mathfrak{E} + \mathfrak{R} = \nabla \mathbf{u} \tag{8.99}$$

we can show that

$$\nabla \times \mathfrak{E} + \nabla \times \mathfrak{R} = \nabla \times \nabla \mathbf{u} = 0 \tag{8.100}$$

From (8.22)

$$\nabla \times \mathfrak{R} = -\frac{1}{2} \nabla \times (\mathbf{u} \nabla) = -\frac{1}{2} (\nabla \times \mathbf{u}) \nabla \tag{8.101}$$

Using this result and applying the conjugate curl to the result (8.100) we have

$$\nabla \times \mathfrak{E} \times \nabla = 0 \tag{8.102}$$

The equations (8.102) are the compatibility conditions that must be satisfied by $\mathfrak{E}$ in order that they can be integrated to generate single-valued and continuous functions for the displacements.

The compatibility equations can also be stated as constraints on the components of the strain matrix ϵ (see e.g. (8.34)) by deriving additional equations relating the strain components. For example, considering the components of strain $\epsilon_{xx}, \epsilon_{yy}$ and ϵ_{xy} we can write

$$\frac{\partial^2 \epsilon_{xx}}{\partial y^2} + \frac{\partial^2 \epsilon_{yy}}{\partial x^2} = \frac{\partial^2}{\partial x \partial y}\left(\frac{\partial u}{\partial y} + \frac{\partial v}{\partial x}\right) \tag{8.103}$$

Similarly, we have

$$2\frac{\partial^2 \epsilon_{xy}}{\partial x \partial y} = \frac{\partial^2}{\partial x \partial y}\left(\frac{\partial u}{\partial y} + \frac{\partial v}{\partial x}\right) \tag{8.104}$$

Hence the strain components must satisfy the condition

$$\frac{\partial^2 \epsilon_{xx}}{\partial y^2} + \frac{\partial^2 \epsilon_{yy}}{\partial x^2} = 2\frac{\partial^2 \epsilon_{xy}}{\partial x \partial y} \tag{8.105}$$

Similarly we can develop five other relationships between the individual components of ϵ. The resulting six compatibility equations which were originally derived by Saint Venant are as follows:

$$\begin{aligned}
\frac{\partial^2 \epsilon_{xx}}{\partial y^2} + \frac{\partial^2 \epsilon_{yy}}{\partial x^2} &= 2\frac{\partial^2 \epsilon_{xy}}{\partial x \partial y} \\[2ex]
\frac{\partial^2 \epsilon_{yy}}{\partial z^2} + \frac{\partial^2 \epsilon_{zz}}{\partial y^2} &= 2\frac{\partial^2 \epsilon_{yz}}{\partial y \partial z} \\[2ex]
\frac{\partial^2 \epsilon_{zz}}{\partial x^2} + \frac{\partial^2 \epsilon_{xx}}{\partial z^2} &= 2\frac{\partial^2 \epsilon_{xz}}{\partial x \partial z} \\[2ex]
\frac{\partial^2 \epsilon_{xx}}{\partial y \partial z} &= \frac{\partial}{\partial x}\left\{-\frac{\partial \epsilon_{yz}}{\partial x} + \frac{\partial \epsilon_{xz}}{\partial y} + \frac{\partial \epsilon_{xy}}{\partial z}\right\} \\[2ex]
\frac{\partial^2 \epsilon_{yy}}{\partial z \partial x} &= \frac{\partial}{\partial y}\left\{\frac{\partial \epsilon_{yz}}{\partial x} - \frac{\partial \epsilon_{xz}}{\partial y} + \frac{\partial \epsilon_{xy}}{\partial z}\right\} \\[2ex]
\frac{\partial^2 \epsilon_{zz}}{\partial x \partial y} &= \frac{\partial}{\partial z}\left\{\frac{\partial \epsilon_{yz}}{\partial x} + \frac{\partial \epsilon_{xz}}{\partial y} - \frac{\partial \epsilon_{xy}}{\partial z}\right\}
\end{aligned} \tag{8.106}$$

All of the above compatibility conditions can also be written in the compact indicial notation as

$$\frac{\partial^2 \epsilon_{i\ell}}{\partial x_j \partial x_k} + \frac{\partial^2 \epsilon_{jk}}{\partial x_i \partial x_\ell} = \frac{\partial^2 \epsilon_{j\ell}}{\partial x_i \partial x_k} + \frac{\partial^2 \epsilon_{ik}}{\partial x_j \partial x_\ell} \tag{8.107}$$

with $i, j, k.\ell$ taking the values x, y, z and summation is carried out over the repeated indices ($i = x, y, z, \quad j = x, y, z$, etc.). The *six* compatibility equations (8.106) which represent independent, linear, second-order partial differential equations can be reduced to *three*, fourth-order partial differential equations (e.g. The first equations of (8.106)) can be differentiated twice with respect to z; the second, twice with respect to x and the third, twice with respect to y. It can be observed that these results are respectively equal to those obtained by differentiating the fourth equation with respect to y and z; the fifth with respect to z and x and the sixth with respect to x and y. The advantage of using the set of six compatibility equations (8.106) is that the solution of the *second-order* partial differential equations will yield a smaller number of arbitrary functions than the solution of the *fourth-order* partial differential equations. Up to now we have established additional conditions which must be satisfied by the strains in a continuum region in order that we can obtain a unique displacement field. The proof that the compatibility conditions are both necessary and sufficient for such purposes is presented in Section 8.13.

Example 8.1

The displacements of a continuum region referred to a rectangular Cartesian system of coordinates (x, y, z) are given by

$$u = C(x^2 + y^2) \quad ; \quad v = 2Cxz \quad ; \quad w = 2Cyz \tag{8.108}$$

where $C \ll 1$ is a constant.

Determine:

(i)   the strain matrix $\boldsymbol{\epsilon}$ and the rotation matrix $\boldsymbol{\omega}$,

(ii) the strain matrix $\boldsymbol{\epsilon}'$ where the x', y', z' axes are obtained by rotating the x, y, z system $45°$ about the z-axis in the counter clockwise direction, and

(iii) the principal components of ϵ at the location $(1, 1, 1)$.

Solution

(i) From (8.8), the displacement gradients matrix $\widehat{\mathbf{F}}$ is given by

$$\widehat{\mathbf{F}} = \nabla\mathbf{u} = \begin{bmatrix} 2Cx & 2Cy & 0 \\ 2Cz & 0 & 2Cx \\ 0 & 2Cz & 2Cy \end{bmatrix} \tag{8.109}$$

From (8.9) and (8.10) the strain and rotation matrices are given by

$$\epsilon = \frac{1}{2}\left[\nabla\mathbf{u} + (\nabla\mathbf{u})^T\right] = C \begin{bmatrix} 2x & (y+z) & 0 \\ (y+z) & 0 & (x+z) \\ 0 & (x+z) & 2y \end{bmatrix} \tag{8.110}$$

and

$$\omega = \frac{1}{2}\left[\nabla\mathbf{u} - (\nabla\mathbf{u})^T\right] = C \begin{bmatrix} 0 & (y-z) & 0 \\ (z-y) & 0 & (x-z) \\ 0 & (z-x) & 2y \end{bmatrix} \tag{8.111}$$

respectively. It will also be noted that since the components of ϵ are linear functions of the spatial variables, the compatibility equations (8.106) are satisfied, trivially.

(ii) The transformation matrix $\mathbf{H}$ defined by (8.71) is obtained by considering the rotation of the (x, y, z) system to obtain the (x', y', z') system (Figure 8.8).

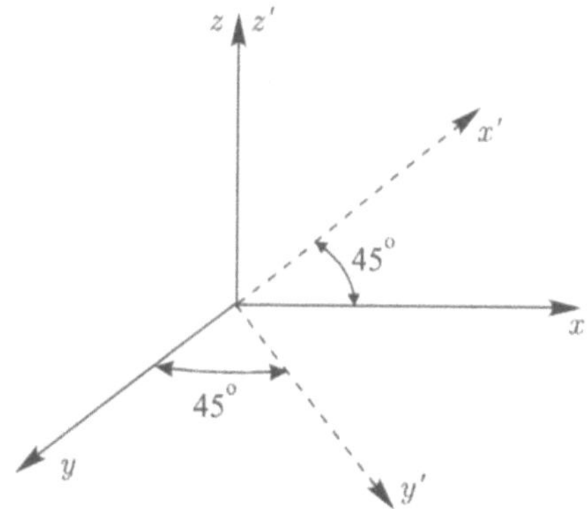

Figure 8.8: Rotation of coordinates.

$$
\mathbf{H} = \begin{bmatrix} \cos 45^\circ & \cos 45^\circ & \cos 90^\circ \\ \cos(135^\circ) & \cos 45^\circ & \cos 90^\circ \\ \cos 90^\circ & \cos 0^\circ & \cos 0^\circ \end{bmatrix} = \frac{1}{\sqrt{2}} \begin{bmatrix} 1 & 1 & 0 \\ -1 & 1 & 0 \\ 0 & 0 & \sqrt{2} \end{bmatrix} \tag{8.112}
$$

The strain matrix $\boldsymbol{\epsilon}'$ referred to the (x', y', z') is obtained from the transformation operation (8.70), i.e.

$$
\boldsymbol{\epsilon}' = \frac{C}{2} \begin{bmatrix} 1 & -1 & 0 \\ 1 & 1 & 0 \\ 0 & 0 & \sqrt{2} \end{bmatrix} \begin{bmatrix} 2x & (y+z) & 0 \\ (y+z) & 0 & (x+z) \\ 0 & (x+z) & 2y \end{bmatrix} \begin{bmatrix} 1 & 1 & 0 \\ -1 & 1 & 0 \\ 0 & 0 & \sqrt{2} \end{bmatrix} \tag{8.113}
$$

or

$$
\boldsymbol{\epsilon}' = \begin{bmatrix} 2(x-y-z) & 2x & -\sqrt{2}(x+z) \\ 2x & 2(x+y+z) & \sqrt{2}(x+z) \\ -\sqrt{2}(x+z) & \sqrt{2}(x+z) & 4y \end{bmatrix} \tag{8.114}
$$

(iii) The strain matrix evaluated at $(1,1,1)$ gives

$$\epsilon = 2C \begin{bmatrix} 1 & 1 & 0 \\ 1 & 0 & 1 \\ 0 & 1 & 1 \end{bmatrix} \tag{8.115}$$

The principal components of ϵ are obtained from the determinantal equation (8.87); i.e.

$$\begin{vmatrix} (2C - \epsilon) & 2C & 0 \\ 2C & -\epsilon & 2C \\ 0 & 2C & 2C - \epsilon \end{vmatrix} = 0 \tag{8.116}$$

which gives

$$(2C - \epsilon)(\epsilon + 2C)(\epsilon - 4C) = 0 \tag{8.117}$$

Hence the principal strains at $(1,1,1)$ are

$$\epsilon_1 = 4C \quad ; \quad \epsilon_2 = 2C \quad ; \quad \epsilon_3 = -2C \tag{8.118}$$

Considering (8.92) it can be verified that (8.118) and (8.115) satisfies the invariance requirements.

Example 8.2

The strains in a continuum region referred to a rectangular Cartesian coordinate system are given by

$$\epsilon_{xx} = -Kxy \quad ; \quad \epsilon_{yy} = \nu Kxy \quad ; \quad \epsilon_{xy} = -K\frac{(1+\nu)}{2}(a^2 - x^2) \tag{8.119}$$

where a, ν and K are constants, $(K \ll 1)$. If the strains $\epsilon_{xz}, \epsilon_{yz}$ and the displacement component w are identically zero, determine the displacement components u and v subject the conditions:

$$u(0,0,0) = v(0,0,0) = \omega_{xy}(0,0,0) = 0 \tag{8.120}$$

Solution

Since the strains $\epsilon_{xz}, \epsilon_{yz}$ and the displacement w are identically zero, the state of strain is two-dimensional and restricted to the $x - y$ plane. From (8.34) and (8.119) we have

$$
\begin{aligned}
\epsilon_{xx} &= \frac{\partial u}{\partial x} = -Kxy \\
\epsilon_{yy} &= \frac{\partial v}{\partial y} = \nu Kxy \\
\epsilon_{xy} &= \frac{1}{2}\left(\frac{\partial u}{\partial y} + \frac{\partial v}{\partial x}\right) = \frac{-K(1+\nu)}{2}(a^2 - x^2)
\end{aligned}
\tag{8.121}
$$

Substituting (8.121) into the compatibility equations (8.106) it is evident that they are trivially satisfied, and the state of deformation maintains the medium as a continuum. Integrating the first two equations of (8.121) we have

$$u(x,y) = -K\frac{x^2 y}{2} + f(y) \tag{8.122}$$

$$v(x,y) = \nu K\frac{xy^2}{2} + g(x) \tag{8.123}$$

where $f(y)$ and $g(x)$ are arbitrary functions of integration. From (8.122) and (8.123)

$$\epsilon_{xy} = \frac{1}{2}\left(\frac{\partial u}{\partial y} + \frac{\partial v}{\partial x}\right) = \frac{1}{2}\left[-\frac{Kx^2}{2} + \frac{\nu Ky^2}{2} + \frac{df}{dy} + \frac{dg}{dx}\right] \tag{8.124}$$

If the compatibility conditions are satisfied, then (8.124) must be identical to the third equation of (8.121), i.e.

$$-\frac{K(1+\nu)}{2}(a^2 - x^2) = \frac{1}{2}\left[-\frac{Kx^2}{2} + \frac{\nu K y^2}{2} + \frac{df}{dy} + \frac{dg}{dx}\right] \qquad (8.125)$$

We can write (8.125) as

$$\frac{df}{dy} + \frac{\nu K y^2}{2} = -\frac{dg}{dx} - K(1+\nu)a^2 + K\left(\frac{3}{2}+\nu\right)x^2 = c_1 \qquad (8.126)$$

where c_1 is an arbitrary constant. From (8.126) we obtain the two ordinary differential equations

$$\frac{df}{dy} = -\frac{\nu K y^2}{2} + c_1 \qquad (8.127)$$

and

$$\frac{dg}{dx} = K\left(\frac{3}{2}+\nu\right)x^2 - K(1+\nu)a^2 - c_1 \qquad (8.128)$$

Integrating (8.127) and (8.128), we have

$$f(y) = -\nu K\frac{y^3}{6} + c_1 y + c_2$$

$$\qquad (8.129)$$

$$g(x) = K\left(\frac{3}{2}+\nu\right)\frac{x^3}{3} - K(1+\nu)a^2 x - c_1 x + c_3$$

where c_2 and c_3 are constants of integration. Hence, from (8.122), (8.123) and (8.129) we have

$$u(x,y) = -\frac{Kx^2y}{2} - \frac{\nu Ky^3}{6} + c_1 y + c_2$$

$$v(x,y) = \nu K\frac{xy^2}{2} + K\left(\frac{3}{2}+\nu\right)\frac{x^3}{3} - K\left(1+\nu\right)a^2x - c_1 x + c_3$$

$$(8.130)$$

The displacement field (8.130) is indeterminate to within the constants c_1, c_2 and c_3. We can make use of the boundary conditions (8.120) to determine these. From the conditions

$$u(0,0,0) = v(0,0,0) = 0 \tag{8.131}$$

we have

$$c_2 = c_3 = 0 \tag{8.132}$$

The third condition gives

$$\Omega_{xy} = \frac{1}{2}\left(\frac{\partial u}{\partial y} - \frac{\partial v}{\partial x}\right)_{x=0,y=0} = c_1 + K\left(1+\nu\right)a^2 = 0$$

or

$$c_1 = -K\left(1+\nu\right)a^2 \tag{8.133}$$

The final expressions for the displacement components u and v are

$$u(x,y) = -K \left[\frac{x^2 y}{2} + \frac{\nu y^3}{6} + (1+\nu)a^2 y \right] \tag{8.134}$$

$$v(x,y) = K \left[\frac{\nu x y^2}{2} + \left(\frac{3}{2} + \nu\right)\frac{x^3}{3} \right] \tag{8.135}$$

It can be verified that (8.134) and (8.135) yields the strain components (8.119).

8.3   Stresses in a continuum

The application of external forces to a deformable continuum region will result in the development of stresses within it. The manner in which such internal stresses will vary within the medium will depend on the heterogeneity of the material constitution of the deformable medium. The *measure* used to describe the internal state of stress at a *point* within the medium should, however, be *independent* of the internal constitution of the medium. This notion of the measure of stress within a continuum region, which was first postulated by Augustin Louis Cauchy (1789-1857), is considered to be one of the key developments in classical continuum mechanics. Cauchy's assertion was that the internal *tractions* which develop within the deformable medium are similar in character to tractions which can be applied externally to create the internal state of stress.

Consider the body V which is subjected to external forces as shown in Figure 8.9. The internal surface S^* divides the region V into two separate regions V_1 and V_2. If we consider the interaction of region V_1 with region V_2, we can denote the force vector acting on the elemental area dA located at a point on S^* of V_1 by $\Delta \mathbf{F}$. There will be an equal an opposite elemental force which acts in the corresponding elemental area in V_2. The stress vector $\mathbf{T}$ at the point P is defined by the limit

$$\mathbf{T} = \operatorname*{Lim}_{\Delta A \to 0} \left(\frac{\Delta \mathbf{F}}{\Delta A} \right) \tag{8.136}$$

In (8.136), ΔA is the current area of the element under consideration. When considering small deformations of the continuum, the current area $\Delta A = \Delta A_0$, where ΔA_0 is the area of the element prior to the application of the traction vector $\Delta \mathbf{F}$. Here again, the notion of a continuum is important to ensuring that the limit is attainable without ambiguity. If we denote the

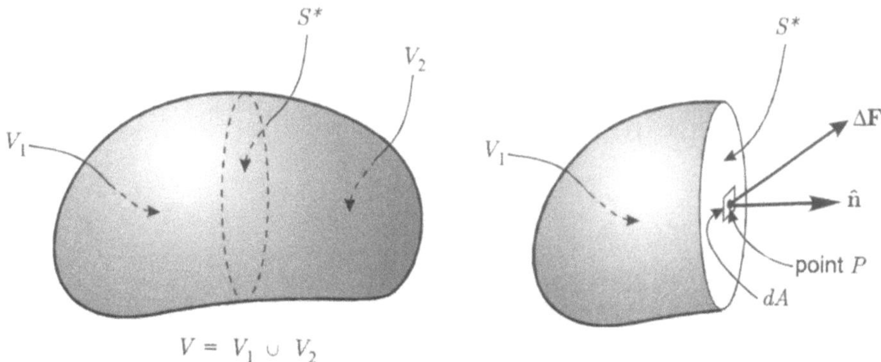

Figure 8.9: Internal tractions within a continuum.

outward unit normal to the surface at P by $\widehat{\mathbf{n}}$, then Cauchy showed that we can determine $\mathbf{T}$ from the result

$$\mathbf{T} = \boldsymbol{\sigma}^T \widehat{\mathbf{n}} \tag{8.137}$$

where $\boldsymbol{\sigma}$ is referred to as the *stress matrix* and

$$\widehat{\mathbf{n}} = \begin{bmatrix} n_x & n_y & n_z \end{bmatrix}^T \tag{8.138}$$

8.3.1 The stress dyadic and the stress matrix

The generalized results of Cauchy can be further utilized to develop the *stress dyadic* and the stress matrix in a manner similar to that adopted in the previous section dealing with strains. Consider, for example, the particular case when the orientation of the internal surface S^* of the body V is such that the unit normal to the elemental area coincides with the x-direction of a system of rectangular Cartesian coordinates x, y, z (Figure 8.10).

We shall denote the *stress vector* acting on the elemental area by $\mathbf{t}_x$. The components of $\mathbf{t}_x$ in the x, y and z directions are σ_{xx}, σ_{xy} and σ_{xz} respectively, such that

$$\mathbf{t}_x - \sigma_{xx}\mathbf{i} + \sigma_{xy}\mathbf{j} + \sigma_{xz}\mathbf{k} \tag{8.139}$$

In a similar way we can select internal surfaces within the medium such that the normal to elemental surfaces located on them are oriented in the directions of the Cartesian axes y and z. The equivalent expressions for the *stress vectors* on these elements are given, respectively, by

$$\mathbf{t}_y = \sigma_{yx}\mathbf{i} + \sigma_{yy}\mathbf{j} + \sigma_{yz}\mathbf{k} \tag{8.140}$$

and

$$\mathbf{t}_z = \sigma_{zx}\mathbf{i} + \sigma_{zy}\mathbf{j} + \sigma_{zz}\mathbf{k} \tag{8.141}$$

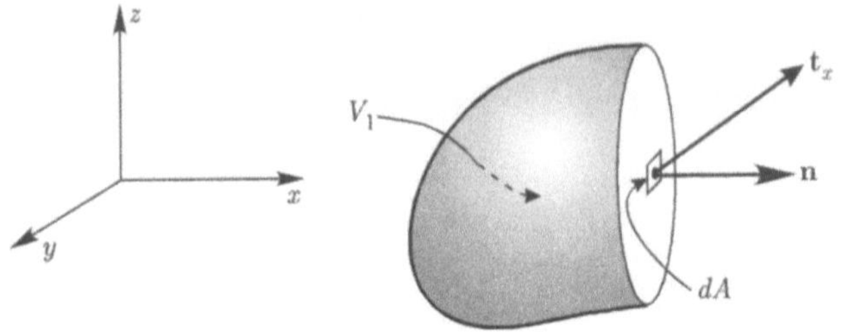

Figure 8.10: Stress vector on elemental area with specific orientation.

The stress dyadic $\mathfrak{S}$ can be expressed in terms of the three vectors (8.139) to (8.141) such that

$$\mathfrak{S} = \mathbf{i}\mathbf{t}_x + \mathbf{j}\mathbf{t}_y + \mathbf{k}\mathbf{t}_z \tag{8.142}$$

which can be also written in the form

$$\begin{aligned}
\mathfrak{S} = \quad & \mathbf{ii}\sigma_{xx} + \mathbf{ij}\sigma_{xy} + \mathbf{ik}\sigma_{xz} \\
& +\mathbf{ji}\sigma_{yx} + \mathbf{jj}\sigma_{yy} + \mathbf{jk}\sigma_{yz} \\
& +\mathbf{ki}\sigma_{zx} + \mathbf{kj}\sigma_{zy} + \mathbf{kk}\sigma_{zz}
\end{aligned} \tag{8.143}$$

We can also structure the *stress matrix* or the *stress tensor* referred to the system of rectangular Cartesian coordinates as

$$\boldsymbol{\sigma} = \begin{bmatrix} \sigma_{xx} & \sigma_{xy} & \sigma_{xz} \\ \sigma_{yx} & \sigma_{yy} & \sigma_{yz} \\ \sigma_{zx} & \sigma_{zy} & \sigma_{zz} \end{bmatrix} \tag{8.144}$$

In indicial notation, the stress matrix is given by

$$\boldsymbol{\sigma} = \sigma_{ij} \qquad i = x, y, z \quad ; \quad j = x, y, z \qquad (8.145)$$

8.3.2 Tractions on an arbitrary plane

The definitions of the traction vectors $\mathbf{t}_i(i = x, y, z)$ can also be used to evaluate the stress resultants on any arbitrary plane and to show that the relationship between the state of stress $\boldsymbol{\sigma}$ at a point and the tractions on an arbitrary plane through this point is independent of any body forces or forces due to dynamic or other effects. The relationship also becomes useful when examining boundary conditions which relate externally applied tractions to the internal stress state within the body.

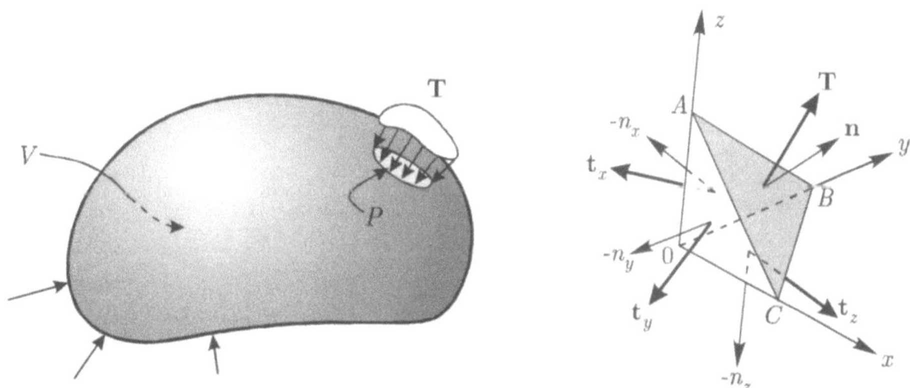

Figure 8.11: Traction vectors on an arbitrary plane.

Consider a point P on the surface of a body which is subjected to external tractions $\mathbf{T}$. The outward unit normal to the Euclidean plane at P is given by

$$\mathbf{n} = n_x\mathbf{i} + n_y\mathbf{j} + n_z\mathbf{k} \qquad (8.146)$$

The Figure 8-11 shows tetrahedral element where the plane ABC is the Euclidean plane (named after the illustrious mathematician Euclid (b.c.330 - b.c.260)) at P where the tractions $\mathbf{T}$ are applied. The planes OAB, OAC and OBC are located within the body and the traction vectors acting on those planes are indicated in the Figure. As a generalization of the analysis, we assume that "body force" vectors act on the element; these body forces could be due to gravity, effects of fluids percolating through the medium, etc. If the body force has intensities f_x^b, f_y^b and f_z^b (measured as force per unit volume) then the vector of forces due to "body force effects" $\mathbf{F}^b$ is given by

$$\mathbf{F}^b = \mathbf{f}^b \Delta V \tag{8.147}$$

where ΔV is the volume of the tetrahedral element. Similarly, if the accelerations in the three coordinate directions are $\partial^2 \mathbf{u}/\partial t^2$, then the dynamic forces on the element are

$$\mathbf{F}^d = (\rho \Delta V) \frac{\partial^2 \mathbf{u}}{\partial t^2} \tag{8.148}$$

where ρ is the mass density of the medium. Considering the tetrahedral element, if the area of ABC is ΔS, the areas of AOB, AOC and BOC are given, respectively, by

$$\Delta S_x = n_x(\Delta S) \quad ; \quad \Delta S_y = n_y(\Delta S) \quad ; \quad \Delta S_z = n_z(\Delta S) \tag{8.149}$$

The *force* vectors acting on the surfaces of the tetrahedral element, ABC, AOB, AOC and BOC are given, respectively, by

$$\mathbf{F} = \mathbf{T} \Delta S$$
$$\tag{8.150}$$
$$\mathbf{F}_x = -n_x(\Delta S)\mathbf{t}_x \quad ; \quad \mathbf{F}_y = -n_y(\Delta S)\mathbf{t}_y \quad ; \quad \mathbf{F}_z = -n_z(\Delta S)\mathbf{t}_z$$

(note that the outward unit normal to the surfaces AOB, AOC and BOC are oriented along the negative directions of the coordinate axes). Applying Newton's second law to the tetrahedral element, we have

$$\mathbf{F} + \mathbf{F}_x + \mathbf{F}_y + \mathbf{F}_z + \mathbf{f}^b \Delta V = (\rho \Delta V) \frac{\partial^2 \mathbf{u}}{\partial t^2} \tag{8.151}$$

Substituting (8.148) and (8.150) in (8.151) we have

$$\mathbf{T} = n_x \mathbf{t}_x + n_y \mathbf{t}_y + n_z \mathbf{t}_z + \frac{\Delta V}{\Delta S} \left(\rho \frac{\partial^2 \mathbf{u}}{\partial t^2} - \mathbf{f}^b \right) \tag{8.152}$$

We now fix the point P but reduce the dimensions of the tetrahedral element to zero. Since ΔV has the dimensions of $(\text{length})^3$ and ΔS has dimensions of $(\text{length})^2$, as we reduce the dimensions of the tetrahedral element to zero, $(\Delta V/\Delta S) \rightarrow 0$. As a result we have

$$\mathbf{T} = n_x \mathbf{t}_x + n_y \mathbf{t}_y + n_z \mathbf{t}_z \tag{8.153}$$

or

$$T_x = n_x \sigma_{xx} + n_y \sigma_{yx} + n_z \sigma_{zx}$$

$$T_y = n_x \sigma_{xy} + n_y \sigma_{yy} + n_z \sigma_{zy} \tag{8.154}$$

$$T_z = n_x \sigma_{xz} + n_y \sigma_{yz} + n_z \sigma_{zz}$$

In vector form we can write these as

$$\mathbf{T} = \sigma^T \mathbf{n} \tag{8.155}$$

or in indicial notation, (8.155) can be written as

$$T_i = \sigma_{ji} n_j \tag{8.156}$$

8.3.3 Equations of equilibrium

We now develop the equations of equilibrium which are applicable to the entire region V. It is assumed that the region V with surface S is subjected to external tractions $\mathbf{T}$ which results in the state of stress σ, and the acceleration $\partial^2 \mathbf{u}/\partial t^2$. In addition, body forces $\mathbf{f}^b$ act within the medium. We also assume that under equilibrium, the couple of the resultant force about the origin of a fixed frame of reference is zero. Considering the dynamic equilibrium of the entire body V we have

$$\iint_S \mathbf{T} \, dS + \iiint_V \mathbf{f}^b dV = \iiint_V \rho \frac{\partial^2 \mathbf{u}}{\partial t^2} dV \tag{8.157}$$

where ρ is the mass density for the continuum. Considering (8.155) we can write, with the aid of the divergence theorem

$$\int\int_S \mathbf{T}\, dS = \int\int_S \boldsymbol{\sigma}^T \mathbf{n}\, dS = \int\int\int_V (\nabla.\mathfrak{S})dV \tag{8.158}$$

where

$$\nabla.\mathfrak{S} = \left(\mathbf{i}\frac{\partial}{\partial x} + \mathbf{j}\frac{\partial}{\partial y} + \mathbf{k}\frac{\partial}{\partial z}\right)\cdot\left(\mathbf{i t}_x + \mathbf{j t}_y + \mathbf{k t}_z\right)$$

$$= \frac{\partial \mathbf{t}_x}{\partial x} + \frac{\partial \mathbf{t}_y}{\partial y} + \frac{\partial \mathbf{t}_z}{\partial z} \tag{8.159}$$

and $\mathbf{t}_i(i = x, y, z)$ are defined by (8.139) to (8.141).

From (8.157) and (8.158) we obtain

$$\int\int\int_V \left[\nabla.\mathfrak{S} + \mathbf{f}^b - \rho\frac{\partial^2 \mathbf{u}}{\partial t^2}\right] dV = 0 \tag{8.160}$$

Since the control volume V is arbitrary, by virtue of the Dubois-Reymond lemma, the equation of dynamic equilibrium for the continuum region reduces to

$$\nabla.\mathfrak{S} + \mathbf{f}^b = \rho\frac{\partial^2 \mathbf{u}}{\partial t^2} \tag{8.161}$$

Using the result (8.159) and the expressions for $\mathbf{t}_x, \mathbf{t}_y$ and $\mathbf{t}_z$ we can write the components of the vector equation of equilibrium (8.161) as

$$\frac{\partial \sigma_{xx}}{\partial x} + \frac{\partial \sigma_{yx}}{\partial y} + \frac{\partial \sigma_{zx}}{\partial z} + f_x^b = \rho\frac{\partial^2 u}{\partial t^2}$$

$$\frac{\partial \sigma_{xy}}{\partial x} + \frac{\partial \sigma_{yy}}{\partial y} + \frac{\partial \sigma_{zy}}{\partial z} + f_y^b = \rho\frac{\partial^2 v}{\partial t^2} \tag{8.162}$$

$$\frac{\partial \sigma_{xz}}{\partial x} + \frac{\partial \sigma_{yz}}{\partial y} + \frac{\partial \sigma_{zz}}{\partial z} + f_z^b = \rho\frac{\partial^2 w}{\partial t^2}$$

and indicial notation, we have

$$\sigma_{ji,j} + f_i^b = \rho \frac{\partial^2 u_i}{\partial t^2} \tag{8.163}$$

where u_i are the components of the displacement vector in the rectangular Cartesian coordinate directions.

8.3.4 Symmetry of the stress matrix

In developing the measure of stress in a deformable continuum region in its various forms, we have not applied any constraints the individual components of $\boldsymbol{\sigma}$ must satisfy to preserve overall equilibrium of the body. The single constraint on the overall equilibrium of the body pertained to the vanishing moment of the resultant forces about a fixed origin. We can now examine the implications of this assumption by considering moment equilibrium of a body V which is subjected to external tractions $\mathbf{T}$ on the surface S, body forces $\mathbf{f}^b$ within V and dynamic forces $\rho \partial^2 \mathbf{u}/\partial t^2$ within V. The moment equilibrium about the origin 0 can be written as

$$\int\int\int_V \left[\mathbf{r} \times \left(\mathbf{f}^b - \rho \frac{\partial^2 \mathbf{u}}{\partial t^2} \right) \right] dV + \int\int_S \mathbf{r} \times \mathbf{T}\, dS = 0 \tag{8.164}$$

where $\mathbf{r}$ is the position vector of dV or dS with respect to the arbitrary but *fixed* origin. We can rewrite the integrand of the surface integral in (8.164) as

$$\mathbf{r} \times \mathbf{T} = \mathbf{R}^* \mathbf{T} \tag{8.165}$$

where $\mathbf{R}^*$ is a skew-matrix defined by

$$\mathbf{R}^* = \begin{bmatrix} 0 & -z & y \\ x & 0 & -x \\ -y & z & 0 \end{bmatrix} = -\left(\mathbf{R}^*\right)^T \tag{8.166}$$

The surface integral in (8.164) can now be written as

$$\int\int_S \mathbf{r} \times \mathbf{T}\, dS = \int\int_S \mathbf{R}^* (\boldsymbol{\sigma}^T \mathbf{n}) dS \tag{8.167}$$

Using the divergence theorem, we can rewrite the integral on the right hand side of (8.167) as

$$\int\int_S \mathbf{R}^*(\boldsymbol{\sigma}^T\mathbf{n})dS = \int\int\int_V \nabla.(\mathbf{R}^*\mathfrak{S})dV \tag{8.168}$$

We can show that the integrand of the volume integral of (8.168) can be represented in the form

$$\nabla.(\mathbf{R}^*\mathfrak{S}) = \mathbf{R}^*\nabla.\mathfrak{S} + \langle\mathfrak{B}\rangle = \mathbf{r}\times(\nabla.\mathfrak{S}) + \langle\mathfrak{B}\rangle \tag{8.169}$$

where $\langle\mathfrak{B}\rangle$ is the rotation vector of the stress dyadic which is defined by

$$\langle\mathfrak{B}\rangle = \mathbf{i}(\sigma_{zy} - \sigma_{yz}) + \mathbf{j}(\sigma_{zx} - \sigma_{xz}) + \mathbf{k}(\sigma_{yx} - \sigma_{xy}) \tag{8.170}$$

Substituting (8.169) into (8.168) and using the result in (8.167) and (8.164) we obtain

$$\int\int\int_V \left[\mathbf{r}\times\left\{\nabla.\mathfrak{S} + \mathbf{f}^b - \rho\frac{\partial^2\mathbf{u}}{\partial t^2}\right\}\right]dV + \int\int\int_V \langle\mathfrak{B}\rangle\,dV = 0 \tag{8.171}$$

By virtue of the equation of equilibrium (8.161) the first integral of (8.171) is identically zero; and since the control volume considered is arbitrary the second integral of (8.171), by virtue of the Dubois-Reymond lemma, gives

$$\langle\mathfrak{B}\rangle = 0 \tag{8.172}$$

Considering (8.170), (8.172) implies that

$$\sigma_{xy} = \sigma_{yx} \quad ; \quad \sigma_{xz} = \sigma_{zx} \quad ; \quad \sigma_{yz} = \sigma_{zy} \tag{8.173}$$

or the stress matrix is symmetric; i.e.

$$\mathfrak{S} = \mathfrak{S}^C \quad ; \quad \boldsymbol{\sigma} = \boldsymbol{\sigma}^T \tag{8.174}$$

where $\mathfrak{S}^C$ is the conjugate of the stress dyadic. Also in indicial notation

$$\sigma_{ij} = \sigma_{ji} \tag{8.175}$$

The resulting simplifications can now be implemented in the expressions (8.144), (8.156) and (8.161).

8.3.5 Sign convention for stresses

The problems in elasticity examined in this Chapter will require the solution of the biharmonic equation. To solve specific problems involving the biharmonic equation it is necessary to prescribe appropriate boundary conditions. In terms of the partial differential equations discussed in previous Chapters such boundary conditions can be invariably posed as either Dirichlet, Neumann or of the mixed type directly in terms of the dependent variable. With problems in stress analysis, however, the boundary conditions should be related to physical constraints indicative of traction and displacement constraints. The specification of displacement boundary conditions is straight forward since these can indicate surfaces on which the displacements are prescribed. The specification of the traction boundary conditions, however, requires the development of a sign convention which will take into consideration the application of boundary conditions in a consistent manner.

In the case of normal stresses, the sign convention can be developed by considering either the *tensile* or *compressive* nature of the stress state. In elasticity problems that are discussed in this Chapter we shall assume that the *tensile stresses are positive*. A useful interpretation of this sign convention can be illustrated by appeal to the two-dimensional state of stress in isolated elements shown in Figure 8.12.

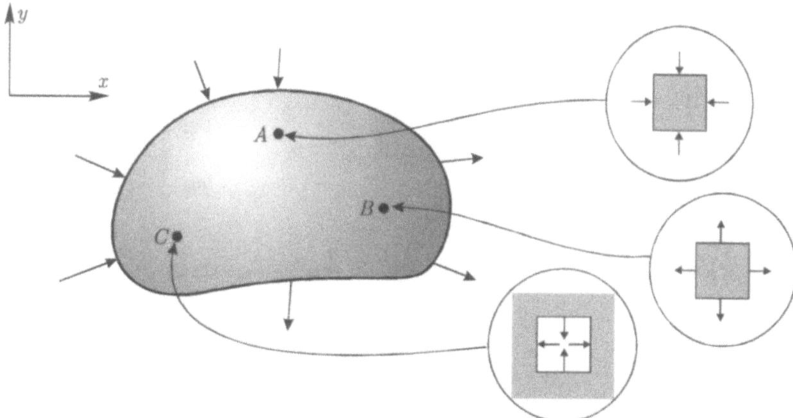

Figure 8.12: Sign convention for normal stresses.

The normal stresses acting on a surface can be considered positive when the stresses are directed outward from the surface and negative if they are directed inward towards the surface. Referring to elements at A, B and C shown in Figure 8.12, the state of stress A is compressive in all directions; the state of stress at B is tensile in all directions; the state of stress at C is compressive in the x-direction and tensile in the y–direction.

In contrast to the sign convention for normal stresses, the sign convention for shear stresses cannot be easily specified in relation to mechanical actions imposed on surfaces on which the shear stresses act. When prescribing a sign convention for the shear stresses, it is necessary to adopt a sign convention which utilizes both the orientation of the plane on which the shear stress acts together with the direction of action of the shear stress. A *positive shear stress* is defined as a shear stress which acts along a positive (or negative) coordinate axis and on a surface, the unit normal to which is also directed along a positive (or negative)coordinate axis. Referring to Figure 8.13, the shear stresses indicated on the element at A positive. Conversely, if *either* the direction of the shear stress, *or* the unit normal to the surface are aligned in an opposite sense, the shear stresses are considered to be negative. For example, the shear stresses acting on the surfaces of the exterior region indicated at element B are negative.

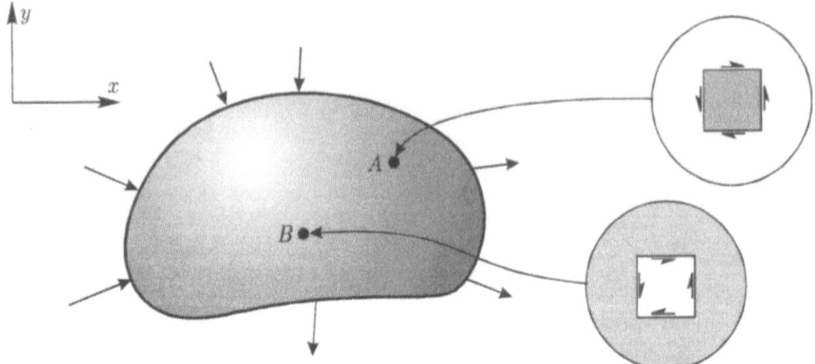

Figure 8.13: Sign convention for shear stresses.

It should also be noted that since the stress matrix is symmetric, the shear stresses are conjugate, i.e. the direction of action of a shear stress on any one surface fixes the directions of the shear stresses on conjugate surfaces. We can now consider situations where a combination of normal and shear tractions are applied to a surface of a continuum region. Referring to Figure 8.14a, the applied tractions $\mathbf{T}$ will induce a positive normal stress and a negative shear stress. In Figure 8.14b, the tractions $\mathbf{T}$ will induce a negative normal stress and a positive shear stress. In Figure 8.14c, the tractions **T** will induce both positive normal stresses and positive shear stresses.

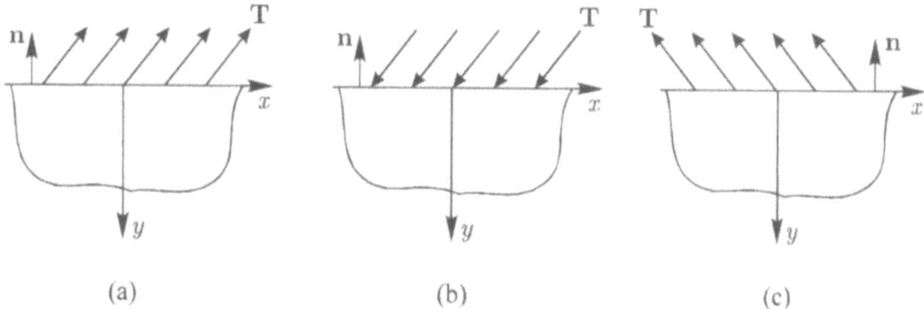

Figure 8.14: Tractions on a boundary.

8.3.6 Transformation of the stress matrix

The transformation of the stress matrix involves the representation of the stress state at a point P which is referred to the rectangular Cartesian coordinates (x, y, z) by an equivalent stress matrix but referred to a *rotated* system (x', y', z'). The basic procedure involved is similar to that used for the calculation of tractions on an arbitrary oriented plane through P.

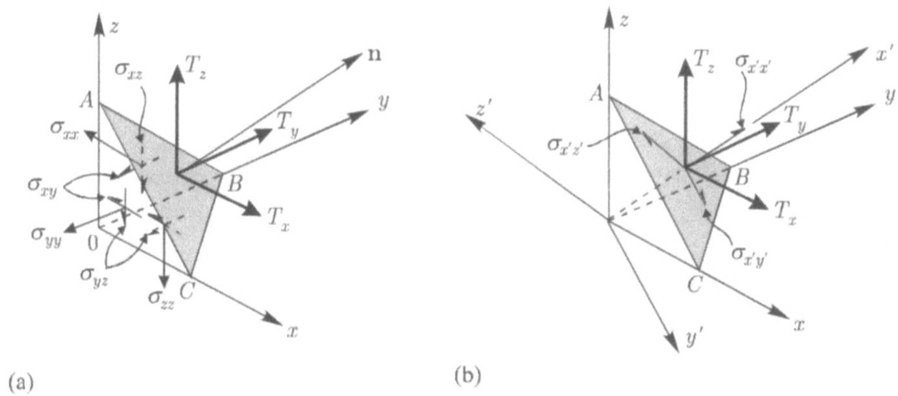

Figure 8.15: Transformations of the stress matrix.

Referring to Figure 8.15a and the result (8.154), the components of the resultant traction on the oblique plane are given by

$$T_x = \sigma_{xx} n_x + \sigma_{xy} n_y + \sigma_{xz} n_z$$

$$T_y = \sigma_{xy} n_x + \sigma_{yy} n_y + \sigma_{yz} n_z \tag{8.176}$$

$$T_z = \sigma_{xz} n_x + \sigma_{yz} n_y + \sigma_{zz} n_z$$

where n_x, n_y, n_z are the direction cosines of the unit normal to the oblique plane. We can now identify the direction of the unit normal to the oblique plane as the x' direction in the rotated coordinate system (Figure 8.15b). The projection of the tractions $\mathbf{T}$ in the x' direction will give the normal stress $\sigma_{x'x'}$; i.e.

$$\sigma_{x'x'} = T_x n_x + T_y n_y + T_z n_z \tag{8.177}$$

which is equivalent to

$$\sigma_{x'x'} = \sigma_{xx} n_x^2 + \sigma_{yy} n_y^2 + \sigma_{zz} n_z^2$$
$$+ 2\sigma_{xy} n_x n_y + 2\sigma_{yz} n_y n_z + 2\sigma_{xz} n_x n_z \tag{8.178}$$

We can also project the components of the traction on the oblique plane in the direction of y' and obtain

$$\sigma_{x'y'} = T_x m_x + T_y m_y + T_z m_z \tag{8.179}$$

where (m_x, m_y, m_z) are the direction cosines of the y' direction. The result (8.179) is equivalent to

$$\sigma_{x'y'} = \sigma_{xx} m_x n_x + \sigma_{yy} m_y n_y + \sigma_{zz} m_z n_z + \sigma_{xy}(m_x n_y + m_y n_x)$$
$$+ \sigma_{yz}(m_y n_z + m_z n_y) + \sigma_{xz}(m_x n_z + m_z n_x) \tag{8.180}$$

Similarly, by projecting the tractions $\mathbf{T}$ in the z' direction with direction cosines (p_x, p_y, p_z) we obtain

$$\sigma_{x'z'} = \sigma_{xx} p_x n_x + \sigma_{yy} p_y n_y + \sigma_{zz} p_z n_z + \sigma_{xy}(p_x n_y + p_y n_x)$$
$$+ \sigma_{yz}(p_y n_z + p_z n_y) + \sigma_{xz}(p_x n_z + p_z n_x) \tag{8.181}$$

This procedure can be repeated by selecting orientations of the rotated axes such that the normal to the oblique plane coincides respectively with the y' and z' axes, to determine $\sigma_{y'y'}$ and $\sigma_{z'z'}$. By virtue of the symmetry of the stress matrix, all components of $\boldsymbol{\sigma}'$ are now defined. The transformation procedure is equivalent to the operation outlined in Section 8.2.4 in connection with the transformation of the strain matrix. We can therefore state that the stress matrix $\boldsymbol{\sigma}'$ referred to the (x', y', z') coordinate system is given by the transformation operation.

$$\boldsymbol{\sigma}' = \mathbf{H}^T \boldsymbol{\sigma} \mathbf{H} \tag{8.182}$$

where the transformation matrix $\mathbf{H}$ is defined by (8.71).

8.3.7 Principal stresses and stress invariants

The orientation of the coordinate system (x', y', z') defined in section (8.3.6) could be chosen in such a way that the shear stresses $\sigma_{x'y'}$, $\sigma_{y'z'}$ and $\sigma_{x'z'}$ could be made to vanish for a specific orientation $\mathbf{x}'^{(P)}$. This specific orientation gives the principal stresses of the stress matrix $\boldsymbol{\sigma}$ or its eigenvalues. The values of the principal components of $\boldsymbol{\sigma}$ can be obtained from the determinantal equation

$$|\boldsymbol{\sigma} - \mathbf{I}\sigma| = 0 \tag{8.183}$$

which gives the Cayley-Hamilton equation

$$\sigma^3 - I_1\sigma^2 + I_2\sigma - I_3 = 0 \tag{8.184}$$

where $I_n(n = 1, 2, 3)$ are the principal invariants of $\boldsymbol{\sigma}$, which are given by

$$I_1 = tr\,\boldsymbol{\sigma}$$
$$I_2 = \frac{1}{2}\left[(tr\,\boldsymbol{\sigma})^2 - tr\,\boldsymbol{\sigma}^2\right] \tag{8.185}$$
$$I_3 = \det \boldsymbol{\sigma}$$

In indicial notation, (8.185) can be written as

$$I_1 = \sigma_{ii}$$

$$I_2 = \frac{1}{2}\left[\sigma_{ii}\sigma_{jj} - \sigma_{ij}\sigma_{ij}\right] \tag{8.186}$$

$$I_3 = |\sigma_{ij}| = \frac{1}{6}\left[\sigma_{ii}\sigma_{jj}\sigma_{kk} - 3\sigma_{ij}\sigma_{ji}\sigma_{kk} + 2\sigma_{ij}\sigma_{jk}\sigma_{ki}\right]$$

with summation over (x, y, z) being carried out over repeated indices.

Example 8.3

The two-dimensional state of stress in a semi-infinite continuum region $(-\infty < x < \infty \; ; \; 0 \le y < \infty)$ in static equilibrium is such that $\boldsymbol{\sigma} = \boldsymbol{\sigma}(x, y)$ and the non-zero components are given by

$$[\sigma_{xx} \; ; \; \sigma_{yy} \; ; \; \sigma_{xy}] = -\frac{2P}{\pi(x^2 + y^2)^2}\left[x^2 y \; ; \; y^3 \; ; \; xy^2\right] \tag{8.187}$$

where P is a constant. Interpret this solution in relation to the half-plane region.

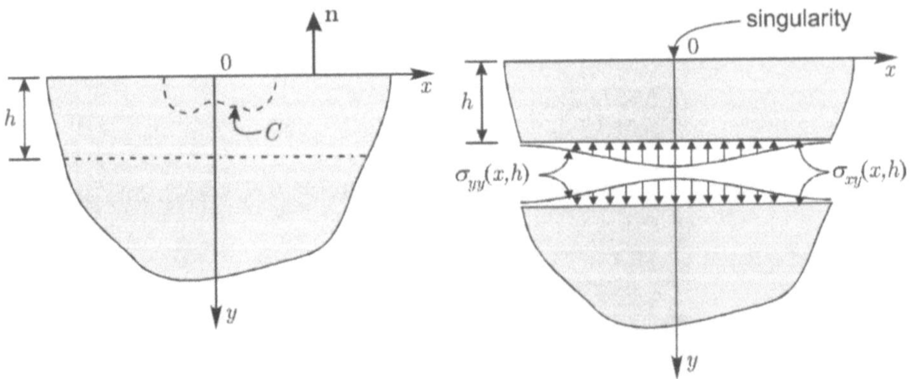

Figure 8.16: Stress state in a half-plane region.

Solution

For the stress components to constitute an acceptable state of stress, they must satisfy the equations of static equilibrium; i.e. from (8.161) we have

$$\nabla.\mathfrak{S} + \mathbf{f}^b = \mathbf{0} \tag{8.188}$$

In two-dimensions, this gives

$$\frac{\partial \sigma_{xx}}{\partial x} + \frac{\partial \sigma_{xy}}{\partial y} + f_x^b = 0 \quad ; \quad \frac{\partial \sigma_{xy}}{\partial x} + \frac{\partial \sigma_{yy}}{\partial y} + f_y^b = 0 \tag{8.189}$$

Substituting of the expressions for the stress components (8.187) in (8.189) will indicate that the equations of static equilibrium will be satisfied in a continuum region in which the body forces $\mathbf{f}^b$ are identically zero. Having established the conditions necessary for static equilibrium, we need to examine the boundary stresses that are produced by the stress state (8.187) in relation to the half-plane region. The only boundary of the region is $y = 0$ and the remaining "boundary conditions" constitutes a regularity condition as $(x^2 + y^2)^{\frac{1}{2}} \rightarrow \infty$. On the boundary of the half-plane region $y = 0$ (Figure 8.16), we have $\mathbf{n} = (0\mathbf{i} + 1\mathbf{j})$. The traction boundary conditions (8.155) on $y = 0$ give

$$T_x(x, 0) = -\sigma_{xy}(x, 0) = 0$$

$$\tag{8.190}$$

$$T_y(x, 0) = -\sigma_{yy}(x, 0) = 0$$

for all values of x, except at the origin. If we evaluate $\boldsymbol{\sigma}$ at $x = 0$ we note that

$$\sigma_{yy}(0, y) \rightarrow 0 \left(\frac{1}{y}\right) \tag{8.191}$$

and as $y \rightarrow 0$, the stress σ_{yy} becomes singular. Also as $(x^2 + y^2)^{\frac{1}{2}} \rightarrow \infty$, $\boldsymbol{\sigma} \rightarrow 0$. Hence, it is reasonable to assume that at all locations on the boundary, excluding the origin, the traction boundary conditions are satisfied.

The nature of the singularity which contributes to the stress distribution can be evaluated in a number of ways. One possibility is to consider a contour C

(Figure 8.16) and to evaluate the traction resultants on this surface. (This will be discussed in a subsequent example). An alternative is to consider any arbitrary location $y = h$ (which excludes the origin) and to evaluate the traction resultants on this surface. The "strength" of the singularity can be obtained by evaluating these resultants as $h \to 0$. Considering the stress distribution (8.187) we have at $y = h$

$$\sigma_{yy}(x, h) = -\frac{2P}{\pi}\frac{h^3}{(x^2 + h^2)^2}$$

$$\sigma_{xy}(x, h) = -\frac{2P}{\pi}\frac{xh^2}{(x^2 + h^2)^2}$$

(8.192)

The normal (compressive) traction on the surface $y = h$ is

$$T_y = -\frac{2P}{\pi}\frac{h^3}{(x^2 + h^2)^2}$$

(8.193)

The resultant of all normal tractions acting on $y = h$ over $-\infty < x < \infty$ is given by

$$R_y^* = \int_{-\infty}^{\infty} -\frac{2P}{\pi}\frac{h^3 dx}{(x^2 + h^2)^2} = -P$$

(8.194)

Hence, irrespective of the location of the plane $y = h$, the resultant of the normal tractions is a constant. At any distance $y = h$, the normal stress distribution is, however, non zero. Hence on the plane $y = 0$, the normal tractions are zero everywhere except at the origin, where it corresponds to a singularity of magnitude $-P$. (The negative sign implies that the normal stresses induced by P are compressive). It can also be shown that the resultant of shear tractions on the plane $y = h$, given by

$$R_x^* = \int_{-\infty}^{\infty} -\frac{2P}{\pi}\frac{xh^2 dx}{(x^2 + h^2)^2} \equiv 0$$

(8.195)

Example 8.4

Consider the two-dimensional state of stress given by (8.187) and obtain the components of

$$\boldsymbol{\sigma}^C = \begin{bmatrix} \sigma_{rr} & \sigma_{r\theta} \\ \sigma_{r\theta} & \sigma_{\theta\theta} \end{bmatrix} \tag{8.196}$$

which represents the stress state referred to the plane polar coordinate system (r,θ).

Solution

Since $x = r\cos\theta$; $y = r\sin\theta$; $(x^2 + y^2) = r^2$, we can rewrite the state of stress (8.187) as

$$[\sigma_{xx} \; ; \; \sigma_{yy} \; ; \; \sigma_{xy}] = -\frac{2P\sin\theta}{r}\left[\cos^2\theta \; ; \; \sin^2\theta \; ; \; \sin\theta\cos\theta\right] \tag{8.197}$$

The transformation between the coordinate systems (x,y) and (r,θ) is achieved through the transformation matrix

$$[\mathbf{H}] = \begin{bmatrix} \cos\theta & \sin\theta \\ -\sin\theta & \cos\theta \end{bmatrix} \tag{8.198}$$

From the transformation rule (8.182) we have

$$\begin{bmatrix} \sigma_{rr} & \sigma_{r\theta} \\ \sigma_{r\theta} & \sigma_{\theta\theta} \end{bmatrix} = \begin{bmatrix} \cos\theta & \sin\theta \\ -\sin\theta & \cos\theta \end{bmatrix}\left(-\frac{2P\sin\theta}{r}\right)$$
$$\cdot \begin{bmatrix} \cos^2\theta & \sin\theta\cos\theta \\ \sin\theta\cos\theta & \sin^2\theta \end{bmatrix}\begin{bmatrix} \cos\theta & -\sin\theta \\ \sin\theta & \cos\theta \end{bmatrix} \tag{8.199}$$

Evaluating (8.199) we have

$$[\sigma_{rr} \ ; \ \sigma_{\theta\theta} \ ; \ \sigma_{r\theta}] = -\frac{2P\sin\theta}{r}\,[1 \ ; \ 0 \ ; \ 0] \qquad (8.200)$$

which represents a purely radial state of stress, which exhibits singular behaviour as $r \to 0$ and the radial stress vanishes, as $r \to \infty$. The magnitude of the 'force' responsible for the singular state of stress can be calculated by evaluating the resultant of tractions, acting in the x and y-directions, on any cylindrical contour $r = a$ (Figure 8.17).

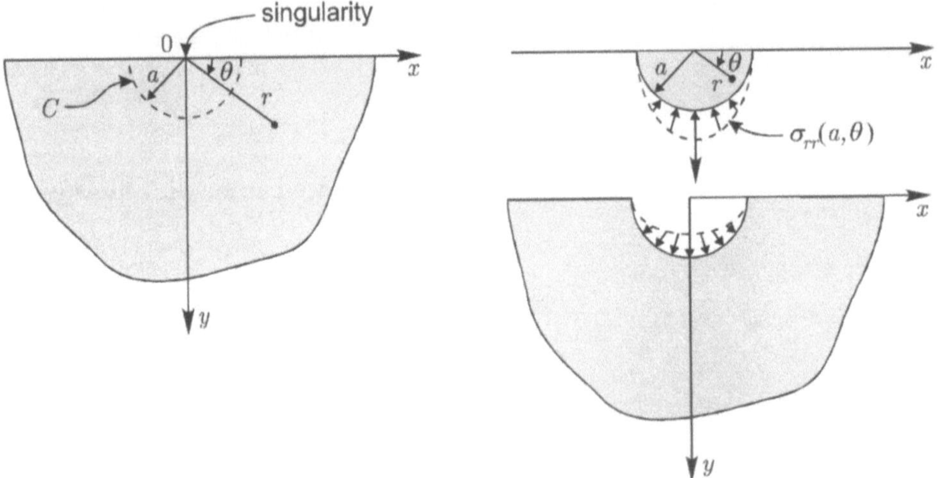

Figure 8.17: Traction resultants on a cylindrical surface C.

The resultant of tractions on the surface C acting of the y-direction is

$$R_y^* = \int_0^\pi \left[-\frac{2P\sin\theta}{\pi r}\right]_{r=a} \sin\theta\, ad\theta = -P \qquad (8.201)$$

Similarly, it can be shown that the resultant of tractions acting in the x-direction

$$R_x^* = \int_0^\pi \left[-\frac{2P\sin\theta}{\pi r}\right]_{r=a} \cos\theta\, ad\theta = 0 \qquad (8.202)$$

Example 8.5

A long bar with an elliptical cross-section S_0 is subjected to a uniform axial stress σ_0. Obtain the tractions on a surface (S_1) obtained by rotating the elliptical cross-section about the x-axis such that the resulting surface is circular (Figure 8.18). Evaluate the resultant of the tractions on S_1 along the axis of the bar and show that the total axial force on the bar is $\pi\sigma_0 ab$.

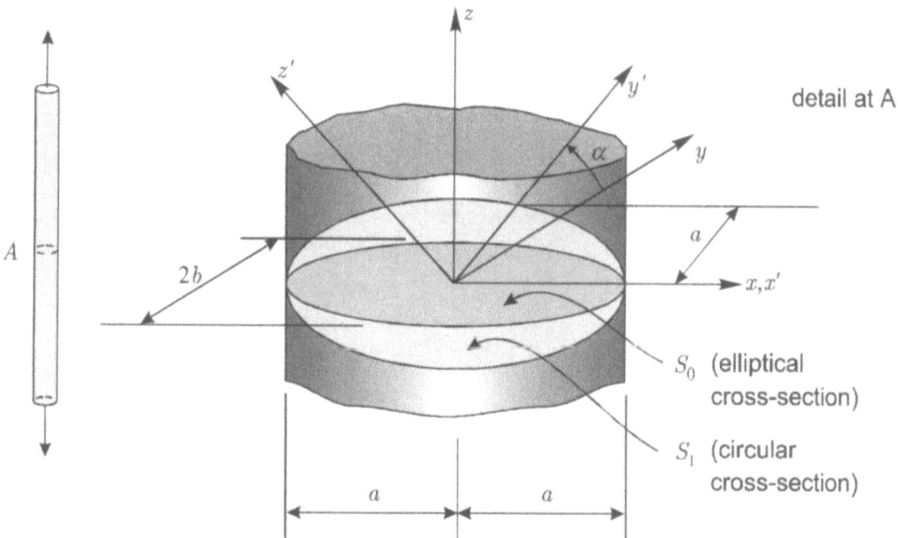

Figure 8.18: Axial loading of an elliptical bar.

Solution

The state of stress in the bar referred to the x, y, z coordinate system is given by

$$\boldsymbol{\sigma} = \begin{bmatrix} 0 & 0 & 0 \\ 0 & 0 & 0 \\ 0 & 0 & \sigma_0 \end{bmatrix} \tag{8.203}$$

The transformation matrix associated with the rotation of the plane S_0 is given by

$$\mathbf{H} = \begin{bmatrix} 1 & 0 & 0 \\ 0 & \cos\alpha & \sin\alpha \\ 0 & -\sin\alpha & \cos\alpha \end{bmatrix} \tag{8.204}$$

From the geometry of the rotation, if S_1 is a circle

$$\cos\alpha = \frac{b}{a} \quad ; \quad \sin\alpha = \frac{\sqrt{a^2 - b^2}}{a} \tag{8.205}$$

The state of stress in elliptical bar referred to the rotation system of coordinates x', y', z' is given by

$$\boldsymbol{\sigma}' = \begin{bmatrix} 1 & 0 & 0 \\ 0 & \cos\alpha & \sin\alpha \\ 0 & -\sin\alpha & \cos\alpha \end{bmatrix} \begin{bmatrix} 0 & 0 & 0 \\ 0 & 0 & 0 \\ 0 & 0 & \sigma_0 \end{bmatrix} \begin{bmatrix} 1 & 0 & 0 \\ 0 & \cos\alpha & -\sin\alpha \\ 0 & \sin\alpha & \cos\alpha \end{bmatrix} \tag{8.206}$$

Evaluating (8.206) we have

$$\boldsymbol{\sigma}' = \begin{bmatrix} 0 & 0 & 0 \\ 0 & \sigma_0 \sin^2\alpha & \sigma_0 \sin\alpha\cos\alpha \\ 0 & \sigma_0 \sin\alpha\cos\alpha & \sigma_0 \cos^2\alpha \end{bmatrix} \tag{8.207}$$

It may be noted that (8.207) satisfies the symmetry and invariance requirements. To evaluate the total axial force on S_1 in the z-direction, we need to compute the resultant of tractions on the $x' - y'$ plane. The tractions acting on S_1 are in general due to $\sigma_{z'z'}, \sigma_{x'z'}$ and $\sigma_{y'z'}$. Since $\sigma_{x'z'} = 0$, the resultant of tractions is given by

$$R_z^* = \int\!\!\int_{S_1} \{\sigma_{z'z'}\cos\alpha + \sigma_{y'z'}\sin\alpha\}\, dx'dy' \tag{8.208}$$

Substituting the expressions for $\sigma_{z'z'}$ and $\sigma_{y'z'}$ in (8.208) we have

$$R_z^* = \sigma_0\cos\alpha \int\!\!\int_{S_1} r'dr'd\theta = \sigma_0\cos\alpha \int_0^{2\pi}\!\!\int_0^a r'dr'd\theta \tag{8.209}$$

which reduces to

$$R_z = \pi ab \tag{8.210}$$

8.4 Constitutive equations for linear elastic solids

In order to complete the formulation of equations governing the mechanics of a continuum region, it is necessary to specify relationships between the stresses and strains for the material, or its *constitutive equations*. In this Chapter attention will be primarily focussed on linear elastic constitutive relationships which characterize the small deformation behaviour of most engineering materials. The observations concerning linear elastic behaviour of materials date back to the classical studies by many scientists and engineers including Hooke (1635-1703), Mariotte (1620-1684), Young (1773-1829) and Navier (1785-1836). Constitutive behaviour of elastic materials, in general, are characterized by their ability to store work done during a deformation, in the form of potential energy. This "strain energy" is assumed to be totally recoverable when the body is unloaded and restored to its original state without any permanent deformation. In real materials, however, a part of the work can be transformed into heat and dissipated by the development of some irreversible deformations. In the development of the classical theory of elasticity, the energy lost is assumed to be negligible in comparison to the recoverable energy. The development of constitutive equations for most elastic materials will have to involve experimentation combined with constraints that are imposed on such equations by virtue of material symmetry and thermodynamical arguments.

The biharmonic partial differential equation which is discussed in this Chapter in connection with the mechanics of an elastic continuum is a consequence of the assumption of linear elastic constitutive behaviour of the continuum region.

8.4.1 Generalized Hooke's Law

The most general form of the linear elastic constitutive relationship assumes that each component in the strain matrix ϵ can be expressed as a linear combination of the components of the stress matrix σ, and vice versa. The resulting linear relationship can be written in the form

$$\overset{*}{\sigma} = \mathbf{C}\,\overset{*}{\epsilon} \tag{8.211}$$

where

$$\overset{*}{\boldsymbol{\sigma}} = \begin{bmatrix} \sigma_{xx} & \sigma_{yy} & \sigma_{zz} & \sigma_{xy} & \sigma_{yz} & \sigma_{zx} \end{bmatrix}^T \tag{8.212}$$

$$\overset{*}{\boldsymbol{\epsilon}} = \begin{bmatrix} \epsilon_{xx} & \epsilon_{yy} & \epsilon_{zz} & \epsilon_{xy} & \epsilon_{yz} & \epsilon_{zx} \end{bmatrix}^T \tag{8.213}$$

and the matrix of elastic constants $\mathbf{C}$ is given by

$$\mathbf{C} = \begin{bmatrix} C_{11} & C_{12} & C_{13} & C_{14} & C_{15} & C_{16} \\ C_{21} & C_{22} & C_{23} & C_{24} & C_{25} & C_{26} \\ C_{31} & C_{32} & C_{33} & C_{34} & C_{35} & C_{36} \\ C_{41} & C_{42} & C_{43} & C_{44} & C_{45} & C_{46} \\ C_{51} & C_{52} & C_{53} & C_{54} & C_{55} & C_{56} \\ C_{61} & C_{62} & C_{63} & C_{64} & C_{65} & C_{66} \end{bmatrix} \tag{8.214}$$

where the suffixes are taken from 1 to 6 for ease of presentation. In general, 36 elastic constants are necessary to characterize a linear elastic material. The linear elastic constitutive behaviour defined by (8.211), invokes no assumptions concerning relationships that can exist between the components of $\mathbf{C}$ as a result of thermodynamical or symmetry requirements. We can show that certain constraints can be imposed on $\mathbf{C}$ by considering the "*elastic energy*" that is stored in the body during deformation.

8.4.2   The strain energy density

Consider a body V with surface area S which is in static equilibrium under the action of tractions $\mathbf{T}$ which act on S and body forces $\mathbf{f}^b$ which act in V. The resulting state of stress in the body is given by $\boldsymbol{\sigma}$ and the incremental displacements in the neighbourhood of a generic point P are denoted by $d\mathbf{u}$. We shall assume that "strain energy" U is equal to the work done (W) by the applied tractions $\mathbf{T}$ and the body forces $\mathbf{f}^b$ as they transform the body from an *undeformed* to a *deformed* configuration. We also assume that in the computation of W, the displacements change from a zero value to its final value. The total work W is given by

$$W = \int\int\int_V dV \int_0^{FV} \mathbf{f}^b.d\mathbf{u} + \int\int_S dS \int_0^{FV} \mathbf{T}.d\mathbf{u} \tag{8.215}$$

where the upper limit 'FV' refers to the *final value*. Using (8.158), we can transform the surface integral in (8.215) to a volume integral, which reduces (8.215) to the following:

$$W = \int \int \int_V dV \int_0^{FV} \left[\mathbf{f}^b.d\mathbf{u} + \nabla.(\mathfrak{S}.d\mathbf{u}) \right] \tag{8.216}$$

We note that

$$\nabla.(\mathfrak{S}.d\mathbf{u}) = (\nabla.\mathfrak{S}).d\mathbf{u} + \mathfrak{S} : \nabla d\mathbf{u} = (\nabla.\mathfrak{S}).d\mathbf{u} + \mathfrak{S} : d(\nabla\mathbf{u}) \tag{8.217}$$

where $(:)$ denotes the double scalar product

$$\mathfrak{S} : d(\nabla\mathbf{u}) = \mathfrak{S} : d(\mathfrak{E} + \mathfrak{R}) = tr[\boldsymbol{\sigma}.d(\boldsymbol{\epsilon} + \boldsymbol{\omega})] \tag{8.218}$$

where $\boldsymbol{\epsilon}$ and $\boldsymbol{\omega}$ are respectively the strain and rotation matrices defined by (8.9) and (8.10). Since $\boldsymbol{\sigma} = \boldsymbol{\sigma}^T$ and $\boldsymbol{\omega} = -\boldsymbol{\omega}^T$,

$$tr(\boldsymbol{\sigma}.d(\boldsymbol{\omega})) = 0 \tag{8.219}$$

and (8.216) gives

$$W = \int \int \int_V dV \int_0^{FV} \left\{ \left[\nabla.\mathfrak{S} + \mathbf{f}^b \right] d\mathbf{u} + tr(\boldsymbol{\sigma}.d\boldsymbol{\epsilon}) \right\} \tag{8.220}$$

By virtue of the assumed static equilibrium (see e.g. (8.161))

$$\nabla.\mathfrak{S} + \mathbf{f}^b = 0 \tag{8.221}$$

and since $\boldsymbol{\sigma}$ and $\boldsymbol{\epsilon}$ are linearly related

$$\int_0^{FV} \text{trace } (\boldsymbol{\sigma}.d\boldsymbol{\epsilon}) = \frac{1}{2} tr(\boldsymbol{\sigma}.\boldsymbol{\epsilon}) \tag{8.222}$$

which reduces (8.220) to

$$W = \frac{1}{2} \int \int \int_V tr(\boldsymbol{\sigma}.\boldsymbol{\epsilon}) dV \tag{8.223}$$

If we define the strain energy in the body V by

$$\mathcal{U} = \int \int \int_V U \, dV \tag{8.224}$$

where U is the strain energy density or strain energy per unit volume of the medium, then we have

$$U = \frac{1}{2} tr(\boldsymbol{\sigma}.\boldsymbol{\epsilon}) \tag{8.225}$$

In indicial notation (8.225) can be written in the form

$$U = \frac{1}{2}\sigma_{ij}\epsilon_{ij} \tag{8.226}$$

where $i, j = x, y, z$, and summation is carried out over the repeated indices.

8.4.3 Symmetry of the elasticity matrix

We shall now examine the constraints that will be imposed on the elasticity matrix $\mathbf{C}$ defined by (8.214) by virtue of the strain energy stored in a linearly elastic material. We can write (8.226) as

$$dU = \overset{*}{\boldsymbol{\sigma}}^T d \overset{*}{\boldsymbol{\epsilon}} \tag{8.227}$$

where $\overset{*}{\boldsymbol{\sigma}}$ and $d \overset{*}{\boldsymbol{\epsilon}}$ are respectively the components of $\boldsymbol{\sigma}$ and $d\boldsymbol{\epsilon}$ arranged in vector form (see e.g. (8.212) and (8.213)). If $U = U(\overset{*}{\boldsymbol{\epsilon}})$ then we can expand U as a Taylor series (named after Brooke Taylor (1685-1731)) about $\overset{*}{\boldsymbol{\epsilon}}= 0$ such that

$$dU = U_0 + \frac{\partial U}{\partial \overset{*}{\boldsymbol{\epsilon}}} \cdot d \overset{*}{\boldsymbol{\epsilon}} +.... \tag{8.228}$$

Comparing (8.227) and (8.228) we have

$$\overset{*}{\boldsymbol{\sigma}} = \left(\frac{\partial U}{\partial \overset{*}{\boldsymbol{\epsilon}}}\right)^T \tag{8.229}$$

In indicial notation, (8.229) can be written as

$$\sigma_{ij} = \frac{\partial U}{\partial \epsilon_{ij}} \qquad ; \qquad i,j = x, y, z \tag{8.230}$$

Expanding (8.230) we have

$$\sigma_{xx} = \frac{\partial U}{\partial \epsilon_{xx}} = c_{11}\epsilon_{xx} + c_{12}\epsilon_{yy} + c_{13}\epsilon_{zz} + c_{14}\epsilon_{xy} + \dots + c_{16}\epsilon_{xz}$$

$$\sigma_{yy} = \frac{\partial U}{\partial \epsilon_{yy}} = c_{21}\epsilon_{xx} + c_{22}\epsilon_{yy} + c_{23}\epsilon_{zz} + c_{24}\epsilon_{xy} + \dots + c_{26}\epsilon_{xz}$$

$$\vdots \tag{8.231}$$

$$\sigma_{xz} = \frac{\partial U}{\partial \epsilon_{xz}} = c_{61}\epsilon_{xx} + c_{62}\epsilon_{yy} + c_{63}\epsilon_{zz} + c_{64}\epsilon_{xy} + \dots + c_{66}\epsilon_{xz}$$

Since U is a continuously differentiable function of its arguments, we can differentiate (8.231) to obtain

$$\frac{\partial^2 U}{\partial \epsilon_{yy}\partial \epsilon_{xx}} = c_{12} \qquad ; \qquad \frac{\partial^2 U}{\partial \epsilon_{xx}\partial \epsilon_{yy}} = c_{21}$$

$$\frac{\partial^2 U}{\partial \epsilon_{zz}\partial \epsilon_{xx}} = c_{13} \qquad ; \qquad \frac{\partial^2 U}{\partial \epsilon_{xx}\partial \epsilon_{zz}} = c_{31}$$

$$\vdots \tag{8.232}$$

$$\frac{\partial^2 U}{\partial \epsilon_{xz}\partial \epsilon_{yy}} = c_{26} \qquad ; \qquad \frac{\partial^2 U}{\partial \epsilon_{yy}\partial \epsilon_{xz}} = c_{62}$$

...etc. Since the differentiations in (8.232) commute we have

$$\mathbf{C} = \mathbf{C}^T \tag{8.233}$$

or

$$c_{ij} = c_{ji} \tag{8.234}$$

and the elasticity matrix is *symmetric*. This reduces the number of *independent elastic constants* required to characterize a linear elastic material from 36 to 21. Such a material is classified as an *anisotropic* elastic material. Both Enrico Betti (1823-1892) and J.C. Maxwell (1831-1879) have given proofs of the applicability of recipocity relationships of the type (8.233), respectively, to linear elastic continua and elastic structural systems.

8.4.4   Isotropic elastic material

In this Chapter, we shall focus attention on the biharmonic equation which results from the consideration of linearly elastic materials which exhibit both *material isotropy* and *material homogeneity*. In the case of an isotropic linearly elastic material, the form of the elasticity matrix should be independent of the orientation of the reference coordinate system. For isotropy, the constitutive matrix $\mathbf{C}$ reduces to

$$\mathbf{C} = \begin{bmatrix} (\lambda + 2\mu) & \lambda & \lambda & 0 & 0 & 0 \\ \lambda & (\lambda + 2\mu) & \lambda & 0 & 0 & 0 \\ \lambda & \lambda & (\lambda + 2\mu) & 0 & 0 & 0 \\ 0 & 0 & 0 & 2\mu & 0 & 0 \\ 0 & 0 & 0 & 0 & 2\mu & 0 \\ 0 & 0 & 0 & 0 & 0 & 2\mu \end{bmatrix} \tag{8.235}$$

where λ and μ are referred to as Lamé's constants (named after Gabriel Lamé (1795-1870)). For homogeneous linearly elastic materials λ and μ are constants at all points within the medium. The specific constitutive equations involving components of $\boldsymbol{\sigma}$ and $\boldsymbol{\epsilon}$ can be written in the form

$$\boldsymbol{\sigma} = \lambda \operatorname{div}(\mathbf{u})\mathbf{I} + 2\mu\boldsymbol{\epsilon} \tag{8.236}$$

where

$$\operatorname{div} \mathbf{u} = \nabla.\mathbf{u} = \frac{\partial u}{\partial x} + \frac{\partial v}{\partial y} + \frac{\partial w}{\partial z} = \operatorname{trace} \boldsymbol{\epsilon} \tag{8.237}$$

The result (8.236) can also be represented in the *dyadic* form

$$\mathfrak{S} = \lambda\, \mathfrak{I}(\nabla.\mathbf{u}) + 2\mu\mathfrak{E} \tag{8.238}$$

where $\mathfrak{S}$ is the *stress dyadic* defined by (8.143), $\mathfrak{E}$ is the *strain dyadic* defined by (8.21) and $\mathfrak{I}$ is the *unit dyadic* defined by

$$\mathfrak{I} = \mathbf{ii} + \mathbf{jj} + \mathbf{kk} \tag{8.239}$$

The result (8.236) can also be written in the indicial form

$$\sigma_{ij} = \lambda e \delta_{ij} + 2\mu\epsilon_{ij} \tag{8.240}$$

where

$$e = tr\ \epsilon_{ij} \tag{8.241}$$

Since we assume that no energy is lost during deformation, a complement to the strain energy function namely the complementary energy function exists. As a consequence, the inversion of the stress-strain relationship always gives a unique result of the form

$$\epsilon = \frac{\sigma}{2\mu} - \frac{\lambda}{2\mu(3\lambda + 2\mu)}\mathbf{I}\ tr\ \sigma \tag{8.242}$$

The dyadic form of (8.242) is

$$\mathfrak{E} = \frac{\mathfrak{S}}{2\mu} - \frac{\lambda}{2\mu(3\lambda + 2\mu)}\mathfrak{I}(\mathfrak{S}) \tag{8.243}$$

and the indicial form of (8.242) is

$$\epsilon_{ij} = \frac{\sigma_{ij}}{2\mu} - \frac{\lambda}{2\mu(3\lambda + 2\mu)}\delta_{ij}\, p \tag{8.244}$$

where

$$p = tr\ \sigma_{ij} \tag{8.245}$$

We can identify the relationships between the various elastic constants by examining the behaviour of the isotropic linear elastic material under elementary states of stress. Green (1793-1841) was the first to prove that an isotropic linear elastic material can be characterized by two independent elastic constants. Prior to Green's studies, the work of Navier (1785-1836) required only a single elastic constant to describe the mechanical behaviour of an isotropic linearly elastic material. Furthermore Navier's result was accepted by many mechanicians including Cauchy (1789-1857), Poisson (1741-1840), Lamé (1795-1870) and Clapeyron (1799-1864).

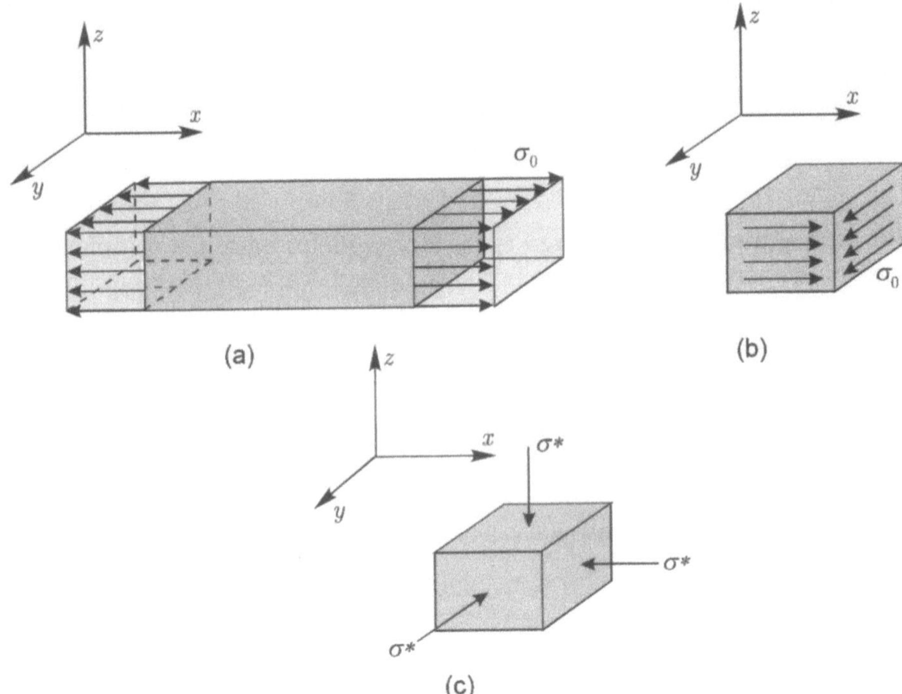

Figure 8.19: Homogeneous states of stress.

Referring to Figure 8.19a we consider the problem where an element of an isotropic elastic material is subjected to a uniaxial stress σ_0 such that

$$\sigma = \begin{bmatrix} \sigma_0 & 0 & 0 \\ 0 & 0 & 0 \\ 0 & 0 & 0 \end{bmatrix} \tag{8.246}$$

Using (8.246) in (8.242) we obtain

$$\epsilon = \frac{\sigma_0}{2\mu(3\lambda + 2\mu)} \begin{bmatrix} 2(\lambda + \mu) & 0 & 0 \\ 0 & -\lambda & 0 \\ 0 & 0 & -\lambda \end{bmatrix} \tag{8.247}$$

In the description of the uniaxial stress state it is also convenient to introduce two elastic constants, namely Young's modulus (E) and Poisson's ratio (ν) such that

$$\epsilon = \frac{\sigma_0}{E} \begin{bmatrix} 1 & 0 & 0 \\ 0 & -\nu & 0 \\ 0 & 0 & -\nu \end{bmatrix} \tag{8.248}$$

From (8.247) and (8.248) we have

$$E = \frac{\mu(3\lambda + 2\mu)}{(\lambda + \mu)} \quad ; \quad \nu = \frac{\lambda}{2(\lambda + \mu)} \tag{8.249}$$

Considering the state of pure shear (of magnitude σ_0) in x-y plane (Figure 8.19(b)) we obtain from (8.242)

$$2\epsilon_{xy} = \frac{\sigma_0}{\mu} = \frac{\sigma_0}{G} \tag{8.250}$$

and $G(=\mu)$ is the proportionality constant between the shear stress and the corresponding shear strain $(2\epsilon_{xy})$ or the "*shear modulus*" of the material. (In certain texts, the shear modulus is denoted G). We now examine the problem where, an element of the isotropic elastic material is subjected to an isotropic state of compression (Figure 8.19c)

$$\boldsymbol{\sigma} = -\sigma^* \, \mathbf{I} \tag{8.251}$$

Considering (8.242) we have

$$\epsilon_{kk} = tr\,\epsilon = -\frac{3\sigma^*}{(3\lambda + 2\mu)} \tag{8.252}$$

We can define a "Bulk Modulus", K which relates the isotropic stress σ^* to the corresponding *volumetric strain* ϵ_{kk} such that

$$\sigma^* = -K\epsilon_{kk} \tag{8.253}$$

or

$$K = \left(\lambda + \frac{2}{3}\mu\right) \tag{8.254}$$

From (8.249) and (8.254) we can show that

$$\lambda = \frac{\nu E}{(1+\nu)(1-2\nu)} = \frac{2\nu\mu}{(1-2\nu)} = \frac{2\nu G}{(1-2\nu)} \tag{8.255}$$

$$K = \frac{E}{3(1-2\nu)} \tag{8.256}$$

$$G = \mu = \frac{E}{2(1+\nu)} \tag{8.257}$$

We note that as $\nu \to \frac{1}{2}$; $K \to \infty$, which is indicative of an *"incompressible"* elastic material.

8.4.5   Thermodynamic constraints on the elastic constants

In the previous section we have introduced the model of a linear elasticity by examining the behaviour of a material largely from a mathematical perspective. Except for the existence of a stored energy function during deformations of the body, no other constraints were imposed on the parameters, defined by the elements of $\mathbf{C}$ (equation (8.214)). It is natural to enquire whether the elements of $\mathbf{C}$ are bounded in anyway or whether they can take any arbitrary values. It will be shown that the elements of $\mathbf{C}$ have to satisfy certain constraints of a thermodynamic nature if the strain energy stored in an elastic material during a deformation is constrained to be *positive definite*. We shall restrict attention to the case of an isotropic elastic medium the constitutive equation for which is given by (8.236).

We introduce the *deviatoric* components of the stress matrix ($\boldsymbol{\sigma}^D$) and the *deviatoric* component of the strain matrix ($\boldsymbol{\epsilon}^D$) which care defined through

$$\boldsymbol{\sigma} = \boldsymbol{\sigma}^D + \frac{1}{2}tr(\boldsymbol{\sigma})\mathbf{I} \tag{8.258}$$

and

$$\boldsymbol{\epsilon} = \boldsymbol{\epsilon}^D + \frac{1}{3}tr(\boldsymbol{\epsilon})\mathbf{I} \tag{8.259}$$

respectively. The expression (8.225) for the strain energy function can now be written as

$$U = \frac{1}{2}tr\left\{\left[\boldsymbol{\sigma}^D + \frac{1}{3}tr(\boldsymbol{\sigma})\mathbf{I}\right]\left[\boldsymbol{\epsilon}^D + \frac{1}{3}tr(\boldsymbol{\epsilon})\mathbf{I}\right]\right\} \tag{8.260}$$

From (8.258) and (8.259) we note that

$$tr(\boldsymbol{\sigma}^D) = 0 \quad ; \quad tr(\boldsymbol{\epsilon}^D) = 0 \tag{8.261}$$

and that $tr(\boldsymbol{\sigma}^D\boldsymbol{\epsilon}^D)$ *need not be zero*. We can take the trace of both sides of the constitutive equation (8.236) which gives

$$tr\,\boldsymbol{\sigma} = (3\lambda + 2\mu)\,tr(\boldsymbol{\epsilon}) \tag{8.262}$$

The constitutive equation (8.236) can also be written as

$$\boldsymbol{\sigma}^D + \frac{1}{3}tr(\boldsymbol{\sigma})\mathbf{I} = \lambda\left\{tr(\boldsymbol{\epsilon})\right\} + 2\mu\left\{\boldsymbol{\epsilon}^D + \frac{1}{3}tr(\boldsymbol{\epsilon})\mathbf{I}\right\} \tag{8.263}$$

Using (8.262) and (8.263) we obtain

$$\boldsymbol{\sigma}^D = 2\mu\boldsymbol{\epsilon}^D \tag{8.264}$$

Expanding (8.260) and making us of (8.261) we have

$$U = \frac{1}{2}\left[tr(\boldsymbol{\sigma}^D\boldsymbol{\epsilon}^D) + \frac{1}{3}tr(\boldsymbol{\sigma})tr(\boldsymbol{\epsilon})\right] \tag{8.265}$$

Substituting (8.262) and (8.264) in (8.265) gives

$$U = \mu\, tr\left(\left[\epsilon^D\right]^2\right) + \frac{1}{6}(3\lambda + 2\mu)tr(\epsilon^2) \tag{8.266}$$

From the form of (8.266) we note that both $tr(\left[\epsilon^D\right]^2)$ and $tr(\epsilon)^2$ will be positive and will be zero only if $\epsilon = 0$. If the stored energy is to be recoverable then U must be positive definite. Since ϵ^D and $tr\,\epsilon$ are independent, for U to be positive definite we require

$$\mu > 0 \quad ; \quad (3\lambda + 2\mu) > 0 \tag{8.267}$$

Considering the results (8.255) to (8.257) it can be shown that the conditions indicated by (8.267) are equivalent to

$$K > 0 \quad ; \quad \mu > 0 \quad ; \quad E > 0 \tag{8.268}$$

and

$$-1 < \nu < \frac{1}{2} \tag{8.269}$$

From the point of view of applications to elastic materials which will not experience *lateral expansion during axial extension*, the constraint (8.269) can be revised to the form

$$0 < \nu < \frac{1}{2} \tag{8.270}$$

8.4.6 Boundary conditions

In the solution of the partial differential equations governing problems in linear elasticity theory, it is necessary to prescribe appropriate boundary conditions. In many instances, these boundary conditions relate to displacements and/or tractions which are prescribed on various surfaces, either internal or external,of an elastic body. Consider an elastic body V with surface S, which is subjected to prescribed displacements over the region S_u and prescribed tractions over the region S_T, and both S_u and S_T are subsets of S.

The boundary conditions related to the surface S can be stated in terms of the boundary conditions on S_u and S_T .

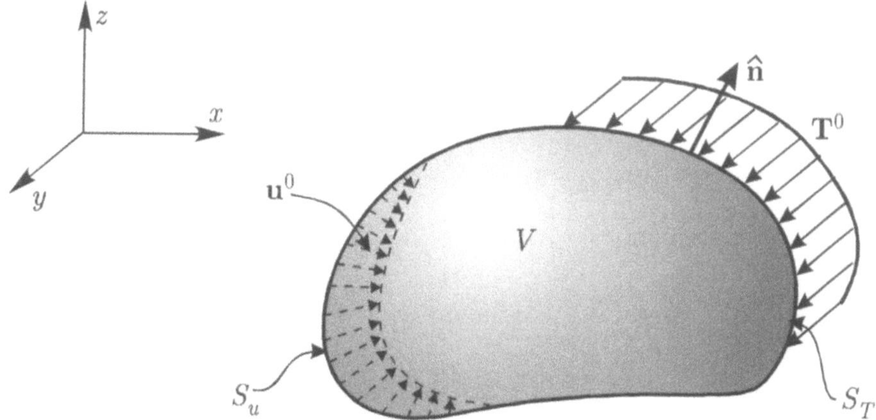

Figure 8.20: Boundary Conditions.

i. The displacement boundary conditions are

$$\mathbf{u}(\mathbf{x}) = \mathbf{u}^0(\mathbf{x}) \qquad ; \qquad \mathbf{x} \in S_u \tag{8.271}$$

ii. The traction boundary conditions are

$$\boldsymbol{\sigma}^T \widehat{\mathbf{n}} = \mathbf{T}^0(\mathbf{x}) \qquad ; \qquad \mathbf{x} \in S_T \tag{8.272}$$

other "mixed" boundary conditions can also be prescribed; for example if a region $S_M \in S$ is "elastically supported", such that the tractions at a point in S_M are proportional to the displacements at that point, then the appropriate form of the boundary condition has a "coupled" form.

iii. The boundary conditions for elastic support are

$$\boldsymbol{\sigma}^T \widehat{\mathbf{n}} = \mathbf{K}\mathbf{u} \qquad ; \qquad \mathbf{x} \in S_M \tag{8.273}$$

where $\mathbf{K}$ is referred to as a "stiffness matrix", which depends on the stiffness characteristics of the elastic support. Usually $\mathbf{K}$ is a diagonal matrix.

As an example, consider the two-dimensional problem of an elastic body with a rectangular shape (Figure 8.21) which is

(i) fixed to an immovable support at $x = 0$,

(ii) subjected to tractions p_0 on the surface $y = H$

(iii) traction free on the surface $x = L$, and

(iv) elastically supported, continuously at the plane $y = 0$, along the entire length L.

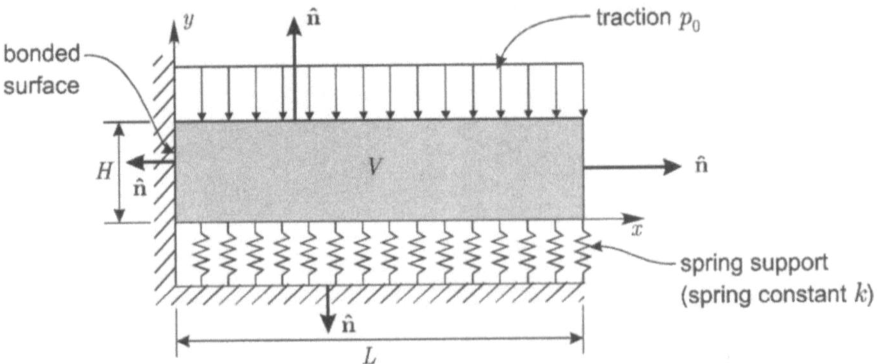

Figure 8.21: Elastically supported region.

The two-dimensional state of deformation (i.e. the displacements, strains and stresses are independent of the coordinate z) is characterized by displacements u and v in the x and y directions respectively, and the non-zero components of the stress matrix are σ_{xx}, σ_{yy} and σ_{xy}.

(i) The displacement boundary conditions are

$$u(0, y) = 0 \quad ; \quad 0 \leq y \leq H$$

$$(8.274)$$

$$v(0, y) = 0 \quad ; \quad 0 \leq y \leq H$$

(ii) The traction boundary conditions are

$$\sigma_{yy}(x, H) = -p_0 \quad ; \quad 0 < x < L$$

$$\sigma_{xx}(L, y) = 0 \quad ; \quad 0 < y < H \tag{8.275}$$

$$\sigma_{xy}(L, y) = 0 \quad ; \quad 0 < y < H$$

(iii) The "mixed" boundary condition is

$$\sigma_{yy}(x, 0) = kv(x, 0) \quad ; \quad 0 < x < L \tag{8.276}$$

where k is the stiffness of the elastic support.

8.4.7 Time derivatives

In describing the time derivatives of the dependent variables we have not made a distinction as to whether these derivatives are either total or partial derivatives. It should be noted that since we are interested in the motion of a point within the body V, we should use the total derivative, which is the derivative with respect to time computed for the given point, whereas the partial derivative is computed at a fixed point of space. In the case of small deformations of the elastic medium, however, a point is assumed to be close to its mean position and we can neglect the small difference between the two derivatives.

The *material time derivative* (D/Dt) is related to the *local time derivative* $(\partial/\partial t)$ by the relationship

$$\frac{D}{Dt} = \frac{\partial}{\partial t} + \mathbf{v}.\nabla \tag{8.277}$$

where $\mathbf{v}$ is the velocity vector of a particular particle which passes through $\mathbf{x}$ at time t. Taking the *material time derivative* of $\mathbf{u}$ we have

$$\frac{D\mathbf{u}}{Dt} = \frac{\partial \mathbf{u}}{\partial t} + \mathbf{v}.\nabla \mathbf{u} \tag{8.278}$$

For infinitesimal deformations we have

$$\mathbf{v}.\nabla\mathbf{u} \simeq \mathbf{0} \qquad (8.279)$$

and (8.278) reduces to

$$\frac{D\mathbf{u}}{Dt} \simeq \frac{\partial\mathbf{u}}{\partial t} \qquad (8.280)$$

and gives

$$\frac{D\mathbf{v}}{Dt} \simeq \frac{\partial^2\mathbf{u}}{\partial t^2} \qquad (8.281)$$

The results (8.280) and (8.281) are a consequence of the small deformation theory and additional terms need to be incorporated if large deformations are to be considered.

8.5 Uniqueness theorem in the classical theory of elasticity

In this section, we shall present the uniqueness theorem governing problems in the classical small deformation theory of elasticity. A discussion of aspects of uniqueness of the solution at this stage has some advantages since the development of allied theorems applicable to both plane and three-dimensional problems in elasticity theory makes use of general uniqueness theorems applicable to both elastostatic and elastodynamic problems in classical elasticity. A proof of the uniqueness of solution of the elastostatic problem was first presented by Kirchhoff (1824-1887). The concepts were later extended by Neumann (1832-1925) and others include the elastodynamic problem. The approaches adopted in the proof of uniqueness range from the consideration of *"thermodynamic arguments"* to approaches based on *"properties of analytic functions"*. In this section we shall present expositions which incorporate the thermodynamical arguments. We shall restrict attention to the proof of uniqueness of the initial boundary value problem in classical elastodynamics. The proof of uniqueness of the boundary value problem in classical elastostatics is left as an exercise.

THEOREM 8.1

Consider the isotropic homogeneous elastic region V with surface S which undergoes time-dependent deformations. The surface of the elastic medium is subjected to surface tractions over the sub-region S_T and displacements are prescribed over the sub-region S_u.

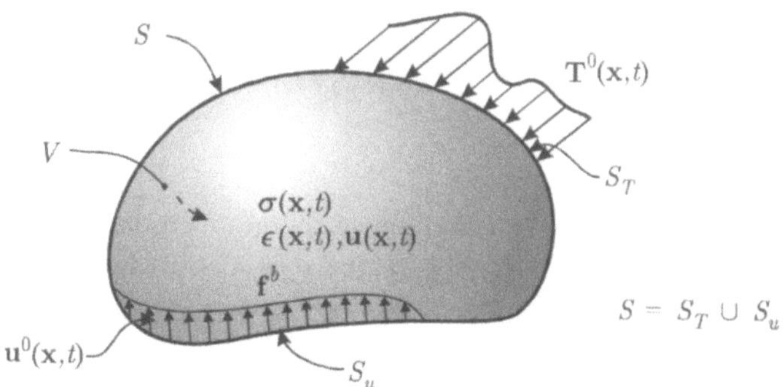

Figure 8.22: Generalized loading of an elastic body.

The elastic region is also subjected to a body force field $\mathbf{f}^b$. The displacement, strain and stress fields in the region V are denoted by $\mathbf{u}(\mathbf{x}, t)$, $\epsilon(\mathbf{x}, t)$ and $\sigma(\mathbf{x}, t)$ respectively. These satisfy the equations of motion

$$\nabla.\mathfrak{S} + \mathbf{f}^b = \rho\frac{\partial^2 \mathbf{u}}{\partial t^2} \quad ; \quad \mathbf{x} \in V \quad ; \quad 0 < t < \infty \tag{8.282}$$

the stress-strain relations

$$\sigma = 2\mu\,\epsilon + \lambda(tr\,\epsilon)\mathbf{I} \quad ; \quad \mathbf{x} \in V \quad ; \quad 0 < t < \infty \tag{8.283}$$

and boundary conditions

$$\mathbf{T} = \sigma\,\mathbf{n} = \mathbf{T}^0(\mathbf{x}, t) \quad ; \quad \mathbf{x} \in S_T \quad ; \quad 0 < t < \infty \tag{8.284}$$

$$\mathbf{u} = \mathbf{u}^0(\mathbf{x}, t) \quad ; \quad \mathbf{x} \in S_u \quad ; \quad 0 < t < \infty \tag{8.285}$$

Since we are considering the dynamic equilibrium of the elastic medium we shall also impose the initial conditions on the displacements and velocities in the medium as follows:

$$\mathbf{u}(\mathbf{x}, 0) = \xi_0(\mathbf{x}) \quad ; \quad \mathbf{x} \in V \tag{8.286}$$

$$\frac{\partial \mathbf{u}}{\partial t}(\mathbf{x}, 0) = \xi_1(\mathbf{x}) \quad ; \quad \mathbf{x} \in V \tag{8.287}$$

where either $\xi_0(\mathbf{x})$ or $\xi_1(\mathbf{x})$ is a non-zero function for well-posedness of the initial boundary value problem. We shall also implicitly assume that the strains in the continuum region satisfy the compatibility equations

$$\nabla \times \mathfrak{E} \times \nabla = 0 \tag{8.288}$$

where $\mathfrak{E}$ is the strain dyadic. The uniqueness theorem in classical elasticity assures that the solutions to $\mathbf{u}(\mathbf{x}, t)$, $\boldsymbol{\epsilon}(\mathbf{x}, t)$ and $\boldsymbol{\sigma}(\mathbf{x}, t)$ obtained by satisfying the equations governing the well posed initial boundary value problem are unique.

PROOF

Let us assume that the initial boundary value problem defined by (8.282) to (8.288) admits two sets of solutions given by the displacement, strain and stress fields $\mathbf{u}^{(1)}(\mathbf{x}, t)$, $\boldsymbol{\epsilon}^{(1)}(\mathbf{x}, t)$, $\boldsymbol{\sigma}^{(1)}(\mathbf{x}, t)$ and $\mathbf{u}^{(2)}(\mathbf{x}, t)$, $\boldsymbol{\epsilon}^{(2)}(\mathbf{x}, t)$, $\boldsymbol{\sigma}^{(2)}(\mathbf{x}, t)$. Then the difference of these two sets of solutions i.e.

$$\overset{*}{\mathbf{u}}(\mathbf{x}, t) = \mathbf{u}^{(1)}(\mathbf{x}, t) - \mathbf{u}^{(2)}(\mathbf{x}, t)$$

$$\overset{*}{\boldsymbol{\epsilon}}(\mathbf{x}, t) = \boldsymbol{\epsilon}^{(1)}(\mathbf{x}, t) - \boldsymbol{\epsilon}^{(2)}(\mathbf{x}, t) \tag{8.289}$$

$$\overset{*}{\boldsymbol{\sigma}}(\mathbf{x}, t) = \boldsymbol{\sigma}^{(1)}(\mathbf{x}, t) - \boldsymbol{\sigma}^{(2)}(\mathbf{x}, t)$$

satisfies the governing equations

$$\nabla \cdot \overset{*}{\mathfrak{S}} = \rho \frac{\partial^2 \overset{*}{\mathbf{u}}}{\partial t^2} \quad ; \quad \mathbf{x} \in V \quad ; \quad 0 < t < \infty \tag{8.290}$$

$$\overset{*}{\boldsymbol{\sigma}} = 2\mu\, \overset{*}{\boldsymbol{\epsilon}} + \lambda\, tr(\overset{*}{\boldsymbol{\epsilon}})\mathbf{I} \quad ; \quad \mathbf{x} \in V \quad ; \quad 0 < t < \infty \tag{8.291}$$

the boundary conditions

$$\overset{*}{\mathbf{T}} = \overset{*}{\boldsymbol{\sigma}}\,\mathbf{n} = \mathbf{0} \quad ; \quad \mathbf{x} \in S_T \quad ; \quad 0 < t < \infty \tag{8.292}$$

$$\overset{*}{\mathbf{u}} = \mathbf{0} \quad ; \quad \mathbf{x} \in S_u \quad ; \quad 0 < t < \infty \tag{8.293}$$

and the initial conditions

$$\overset{*}{\mathbf{u}}(\mathbf{x}, 0) = \mathbf{0} \quad ; \quad \mathbf{x} \in V \tag{8.294}$$

$$\left[\frac{\partial \overset{*}{\mathbf{u}}}{\partial t} \right]_{t=0} = \mathbf{0} \quad ; \quad \mathbf{x} \in V \tag{8.295}$$

Considering the *theorem of power and energy* we note that the rate at which work is done by surface forces and body forces is equal to the rate of change of total energy of the system. Considering the initial boundary value problem posed by (8.289) to (8.295) we note that either $\overset{*}{\mathbf{T}}$ is zero on S_T, or $\overset{*}{\mathbf{u}}$ is zero on S_u or either $\overset{*}{\mathbf{T}}$ or $\overset{*}{\mathbf{u}}$ is zero on the remainder of the surface S. Therefore form the above theorem we obtain (since $\overset{*}{\mathbf{f}}^b \equiv \mathbf{0}$)

$$\int_S \overset{*}{\mathbf{T}}\,\frac{\partial \overset{*}{\mathbf{u}}}{\partial t}\,dS = 0 \tag{8.296}$$

The condition (8.296) implies that the total energy of the system $\mathcal{U}(t)$ is constant, i.e.

$$\mathcal{U}(t) = \mathcal{U}(0) \quad ; \quad 0 < t < t_0 \tag{8.297}$$

where t_0 is a finite time. Integrating (8.296) with respect to time over a finite interval we can write

$$\int_0^t \left[\int \overset{*}{\mathbf{T}}\,\frac{\partial \overset{*}{\mathbf{u}}}{\partial t}\,dS \right] dt = \mathcal{U}(t) - \mathcal{U}(0) \tag{8.298}$$

The left hand side of the expression is the total work done over a finite time interval $(0, t)$. From the conditions (8.294) and (8.295), the body is initially unstrained and in a state of rest i.e.

$$\mathcal{U}(0) = 0 \tag{8.299}$$

From (8.297) and (8.299) we have

$$\mathcal{U}(t) = 0 \tag{8.300}$$

We will now show that the total energy consists of kinetic energy and strain energy components. Replacing $\overset{*}{\mathbf{T}}$ in (8.296) by $\overset{*}{\boldsymbol{\sigma}}\,\mathbf{n}$ and applying the divergence theorem to resulting equation we have

$$\int\int_S \overset{*}{\mathbf{T}} \cdot \frac{\partial \overset{*}{\mathbf{u}}}{\partial t} dS + \int\int\int_V \nabla \cdot \left\{ \overset{*}{\mathbb{G}} \cdot \frac{\partial \overset{*}{\mathbf{u}}}{\partial t} \right\} dV = 0 \tag{8.301}$$

Using the result (8.217) we can express (8.301) in the form

$$\int\int\int_V \left\{ \nabla \cdot \overset{*}{\mathbb{G}} \right\} \cdot \frac{\partial \overset{*}{\mathbf{u}}}{\partial t} dV + \int\int\int_V tr\left\{ \overset{*}{\boldsymbol{\sigma}} \cdot \nabla \left(\frac{\partial \overset{*}{\mathbf{u}}}{\partial t} \right) \right\} dV = 0 \tag{8.302}$$

Using the equation of equilibrium (8.290), and noting that

$$\nabla \left(\frac{\partial \overset{*}{\mathbf{u}}}{\partial t} \right) = \frac{d}{dt} \left(\overset{*}{\boldsymbol{\epsilon}} + \overset{*}{\boldsymbol{\omega}} \right) \tag{8.303}$$

the result (8.302) can be expressed in the form

$$\frac{d}{dt} \left\{ \int\int\int_V \left[\frac{1}{2}\rho \left(\frac{\partial \overset{*}{\mathbf{u}}}{\partial t} \right)^2 + tr\left(\overset{*}{\boldsymbol{\sigma}} \cdot \overset{*}{\boldsymbol{\epsilon}} \right) \right] dV \right\} = 0 \tag{8.304}$$

It was shown in section (8.4.5) that the second term in the integrand of (8.304) is positive definite provided $\mu > 0$ and $(3\lambda + 2\mu) > 0$. Also, the first term in the integrand of (8.304) is positive definite provided $\rho > 0$. (It must be pointed out that there are constraints based on the consideration of the static problem). Hence, by virtue of the Dubois-Reymond lemma, the integrand of (8.304) will vanish if and only if

$$\overset{*}{\boldsymbol{\epsilon}} = 0 \quad ; \quad \frac{\partial \overset{*}{\mathbf{u}}}{\partial t} = 0 \qquad \text{for} \quad \mathbf{x} \in V \quad ; \quad t > 0 \tag{8.305}$$

From the second equation of (8.289) we have

$$\boldsymbol{\epsilon}^{(1)}(\mathbf{x}, t) \equiv \boldsymbol{\epsilon}^{(2)}(\mathbf{x}, t) \tag{8.306}$$

Also, if $\boldsymbol{\epsilon}(\mathbf{x}, t)$ satisfies the compatibility conditions, the strains can be integrated to obtain unique values of $\mathbf{u}(\mathbf{x}, t)$. Hence (8.306) implies that

$$\mathbf{u}^{(1)}(\mathbf{x}, t) \equiv \mathbf{u}^{(2)}(\mathbf{x}, t) \tag{8.307}$$

and the displacement field is indeterminate only to within a rigid body motion.

From the constitutive equations governing linear elastic materials, (8.306) also implies that

$$\boldsymbol{\sigma}^{(1)}(\mathbf{x}, t) \equiv \boldsymbol{\sigma}^{(2)}(\mathbf{x}, t) \tag{8.308}$$

and the uniqueness theorem is proved.

$$\odot \ \odot \ \odot$$

It should be noted that, in the above development of the proof of uniqueness of the dynamical problem in classical elasticity we have implicitly assumed that the displacement field $\mathbf{u}(\mathbf{x}, t)$ is continuous with derivatives in both $\mathbf{x}$ and t at least to within the second order. This will clearly exclude situations where the influence of dynamic effects will result in the development of "shock type" discontinuities in the displacement fields. The uniqueness arguments can be extended to cover these types of situations. Such considerations are however beyond the scope of this introductory exposition of the classical theory. The uniqueness theorem presented is expected to cover a wider class of problems involving both static and dynamic effects in homogeneous isotropic elastic solids which experience small deformations.´

In the development of the proof of Theorem 8.1, however, we have implicitly assumed that the necessary and sufficient conditions for the positive definiteness of the strain energy density correspond to the thermodynamic constraints

$$\mu > 0 \quad ; \quad -1 < \nu < \frac{1}{2} \tag{8.309}$$

derived via the consideration of the statical formulation. Although the lower limit of -1 for Poisson's ratio is unlikely to be satisfied by actual elastic materials, the question arises as to whether the constraints (8.309) are necessary and sufficient for uniqueness for problems in elastodynamics. It can be shown that in instances where displacement boundary conditions are prescribed (i.e. the first boundary value problem), the constraints can be relaxed without loss of uniqueness. The relaxed constraints on the elastic constants applicable to the dynamic problem are

$$\mu > 0 \quad ; \quad -\infty < \nu < \frac{1}{2} \quad , \quad 1 < \nu < \infty \tag{8.310}$$

These are derived from the wave speeds corresponding to equivoluminal (P-waves) and irrotational (S-waves) waves encountered in elastodynamic problems. These waves are characterized by the wave speeds

$$c_P^2 = \frac{\mu}{\rho} \geq 0 \quad ; \quad c_S^2 = \frac{2(1-\nu)\mu}{(1-2\nu)\rho} \geq 0 \tag{8.311}$$

The constraints (8.310) ensure that the wave speeds are real.

The objective of the ensuing development is to show that for the first boundary value problem, the constraints (8.310) are sufficient to ensure uniqueness of the elastodynamic problem. Considering the strain-displacement equations (8.21) and the strain relations (8.238), we can express the equations of dynamic equilibrium (8.161) in terms of the displacements $\mathbf{u}(\mathbf{x}, t)$ as follows

$$\mu\nabla^2\mathbf{u} + (\lambda + \mu)\nabla\nabla.\mathbf{u} + \mathbf{f}^b = \rho\frac{\partial^2\mathbf{u}}{\partial t^2} \tag{8.312}$$

and note that

$$(\lambda + \mu) = \frac{\mu}{(1 - 2\nu)} \tag{8.313}$$

THEOREM 8.2

Consider a bounded isotropic, homogeneous elastic region V with boundary S and assume that the elastic constants satisfy the constraints

$$\mu > 0 \quad ; \quad -\infty < \nu < \frac{1}{2} \quad , \quad 1 < \nu < \infty \tag{8.314}$$

Then there exists a unique solution for the displacements $\mathbf{u}(\mathbf{x}, t)$ which are continuously differentiable in V and S ($0 < t < \infty$), which satisfies the equations of dynamic equilibrium (8.312), the displacement boundary conditions

$$\mathbf{u}(\mathbf{x}, t) = \mathbf{u}^0(\mathbf{x}, t) \quad ; \quad \mathbf{x} \in S \quad ; \quad 0 < t < \infty \tag{8.315}$$

and the initial conditions

$$\mathbf{u}(\mathbf{x}, 0) = \xi_0(\mathbf{x}) \quad ; \quad \mathbf{x} \in V \tag{8.316}$$

$$\left[\frac{\partial \mathbf{u}}{\partial t} \right]_{t=0} = \xi_1(\mathbf{x}) \quad ; \quad \mathbf{x} \in V \tag{8.317}$$

where $\xi_0(\mathbf{x})$ and $\xi_1(\mathbf{x})$ are prescribed vector-valued initial displacements and velocities.

PROOF

Again, let us assume the existence of two sets of solutions
$\{\mathbf{u}^{(1)}(\mathbf{x}, t), \boldsymbol{\epsilon}^{(1)}(\mathbf{x}, t), \boldsymbol{\sigma}^{(1)}(\mathbf{x}, t)\}$ and $\{\mathbf{u}^{(2)}(\mathbf{x}, t), \boldsymbol{\epsilon}^{(2)}(\mathbf{x}, t), \boldsymbol{\sigma}^{(2)}(\mathbf{x}, t)\}$
such that the differences between the two sets of solutions
$\{\overset{*}{\mathbf{u}}(\mathbf{x}, t), \overset{*}{\boldsymbol{\epsilon}}(\mathbf{x}, t), \overset{*}{\boldsymbol{\sigma}}(\mathbf{x}, t)\}$ satisfies

$$\mu \nabla^2 \overset{*}{\mathbf{u}} + \frac{\mu}{(1 - 2\nu)} \nabla \nabla \cdot \overset{*}{\mathbf{u}} = \rho \frac{\partial^2 \overset{*}{\mathbf{u}}}{\partial t^2} \quad ; \quad \mathbf{x} \in V \quad ; \quad 0 < t < \infty \tag{8.318}$$

together with

$$\overset{*}{\mathbf{u}}\,(\mathbf{x},0) = 0 \quad ; \quad \mathbf{x} \in V \tag{8.319}$$

$$\left[\frac{\partial \overset{*}{\mathbf{u}}}{\partial t}\right]_{t=0} = 0 \quad ; \quad \mathbf{x} \in V \tag{8.320}$$

and

$$\overset{*}{\mathbf{u}}\,(\mathbf{x},t) = 0 \quad ; \quad \mathbf{x} \in S \quad ; \quad 0 < t < \infty \tag{8.321}$$

Considering the *kinetic energy* of the body we have

$$\mathcal{K}(t) = \frac{\rho}{2} \int \int \int_V \left[\frac{\partial \overset{*}{\mathbf{u}}}{\partial t}\right]^2 dV \tag{8.322}$$

and define the strain energy in the body through the result (8.225). This result can be expressed in the form

$$\mathcal{E}(t) = \frac{\mu}{2} \int \int \int_V \left[\frac{2(1-\nu)}{(1-2\nu)}\left(\nabla.\overset{*}{\mathbf{u}}\right)^2 + (\nabla \times \overset{*}{\mathbf{u}})^2\right] dV \tag{8.323}$$

Using the displacement equation of equilibrium (8.318) and the divergence theorem, the rate of change of total energy in the body and can be expressed in the form

$$\frac{d\mathcal{K}(t)}{dt} + \frac{d\mathcal{E}(t)}{dt} = \mu \int \int_S \left[\frac{2(1-\nu)}{(1-2\nu)}\nabla.\overset{*}{\mathbf{u}} + \frac{\partial \overset{*}{\mathbf{u}}}{\partial t} \times (\nabla \times \overset{*}{\mathbf{u}})\right] \mathbf{n}\,dS \tag{8.324}$$

where $\mathbf{n}$ is the outward unit normal to S.

From the condition (8.321) we note that the right hand side of (8.324) is zero, which is equivalent to

$$\frac{d\mathcal{K}(t)}{dt} + \frac{d\mathcal{E}(t)}{dt} = \text{const.} \quad ; \quad 0 < t < \infty \tag{8.325}$$

From (8.319), (8.320), (8.322) and (8.323) we note that

$$\mathcal{K}(0) + \mathcal{E}(0) = 0 \tag{8.326}$$

Hence, for continuity of the total energy

$$\mathcal{K}(t) + \mathcal{E}(t) = 0 \tag{8.327}$$

If

$$\frac{2(1-\nu)}{(1-2\nu)} \geq 0 \tag{8.328}$$

then both $\mathcal{K}(t)$ and $\mathcal{E}(t)$ are either non-negative or non-positive such that (8.327) gives

$$\mathcal{K}(t) = 0 \quad ; \quad \mathcal{E}(t) = 0 \tag{8.329}$$

If $\mathcal{K}(t) = 0$, then the result (8.322) gives

$$\frac{\partial \overset{*}{\mathbf{u}}}{\partial t} = 0 \quad \text{in} \quad \mathbf{x} \in V \quad ; \quad 0 \leq t < \infty \tag{8.330}$$

From (8.330) and the initial conditions (8.319) and (8.320) we note that

$$\overset{*}{\mathbf{u}}(\mathbf{x}, t) = \overset{*}{\mathbf{u}}(\mathbf{x}, 0) = 0 \quad ; \quad \mathbf{x} \in V \quad ; \quad 0 \leq t < \infty \tag{8.331}$$

The result (8.331) implies that

$$\mathbf{u}^{(1)}(\mathbf{x}, t) = \mathbf{u}^{(2)}(\mathbf{x}, t) \tag{8.332}$$

and the theorem is proved.

8.6 Plane problems in classical elasticity

Many problems of engineering interest which can be modelled by appeal to the classical theory of elasticity have, in general, three-dimensional configurations. The development of solutions to such general three-dimensional problems require the use of more advanced mathematical techniques and computational procedures which are beyond the scope of this Chapter. The modelling of problems in elasticity as *plane problems* or *two-dimensional problems* is a simplification that is extensively utilized to reduce the complexity of analysis associated with three-dimensional problems in classical elasticity. It must be emphasized, at the outset, that plane problems are mathematical idealizations and not a physical reality. There are, however, configurations of three-dimensional problems which lend themselves to approximation by the plane problem.

In this section we shall present a relatively general exposition of the *plane problem* in classical elasticity related to either a *state of plane strain* or a *state of generalized plane stress* in an isotropic, homogeneous elastic material.

We consider a plane region $\mathbb{R}$ with generators which are parallel to the z-axis and the plane boundaries of the region correspond to $z = \pm a$ (Figure 8.23). We consider the generalized surface S of the plane region and assume that displacements $\mathbf{u}^0(\mathbf{x})$ are prescribed on a subregion S_u and tractions $\mathbf{T}^0(\mathbf{x})$ are prescribed on a subregion S_T. The plane problem in elasticity involves the determination of the displacements, strains and stresses in the elastic medium (with elastic constants λ and μ and body forces $\mathbf{f}^b(x, y)$) satisfying the boundary conditions

$$\mathbf{u}(x, y) = \mathbf{u}^0(x, y) \quad ; \quad (x, y) \in S_u \tag{8.333}$$

$$\boldsymbol{\sigma}(x, y)\mathbf{n}(x, y) = \mathbf{T}^0(x, y) \quad ; \quad (x, y) \in S_T \tag{8.334}$$

$$T_z(x, y) = 0 \quad\quad ; \quad z = \pm a \tag{8.335}$$

Since the surface displacements, surface tractions and body forces are *independent* of z and parallel to the x, y plane, we can conclude that $\mathbf{u}(x, y)$ in the region $\mathbb{R}$ will also be independent of z. Considering this aspect of the plane problem we introduce the *state of plane strain*. We define an elastic stress state with displacements $\mathbf{u}$, strains $\boldsymbol{\epsilon}$ and stresses $\boldsymbol{\sigma}$ as a state of plane strain if

$$u = u(x, y) \quad ; \quad v = v(x, y) \quad ; \quad w = 0 \tag{8.336}$$

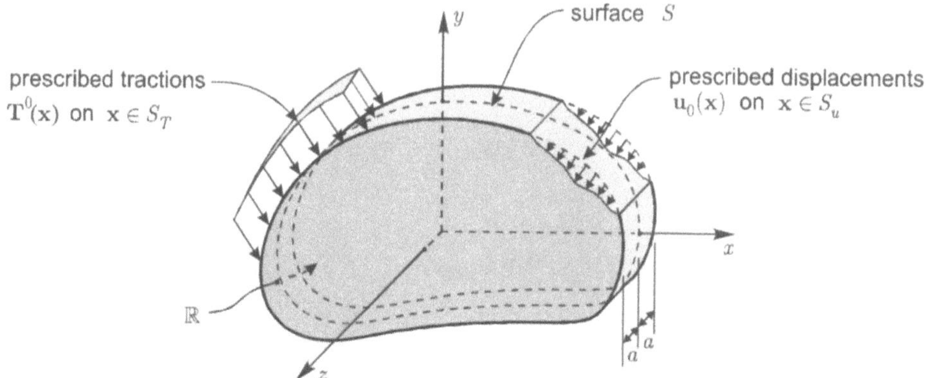

Figure 8.23: Plane region for modelling the two-dimensional problem.

which implies that

$$\boldsymbol{\epsilon} = \boldsymbol{\epsilon}(x, y) \quad ; \quad \boldsymbol{\sigma} = \boldsymbol{\sigma}(x, y) \tag{8.337}$$

and

$$\epsilon_{xz} = \epsilon_{yz} = \epsilon_{zz} = 0 \tag{8.338}$$

The stress-strain relationships are given by

$$\boldsymbol{\sigma} = 2\mu\,\boldsymbol{\epsilon} + \lambda\,tr(\boldsymbol{\epsilon})\mathbf{I} \tag{8.339}$$

with

$$\sigma_{xz} = \sigma_{yz} = 0 \quad ; \quad \sigma_{zz} = \lambda\,tr\,\boldsymbol{\epsilon} = \lambda(\epsilon_{xx} + \epsilon_{yy}) \tag{8.340}$$

Considering the expressions for σ_{xx} and σ_{yy} derived from (8.339) and using these in the result for σ_{zz} given by (8.340) we obtain

$$\sigma_{zz} = \frac{\lambda}{2(\lambda + \mu)}\,(\sigma_{xx} + \sigma_{yy}) = \nu(\sigma_{xx} + \sigma_{yy}) \tag{8.341}$$

The complete set of field equations for the plane strain problem is summarized in the following: the equations of equilibrium are

$$\frac{\partial \sigma_{xx}}{\partial x} + \frac{\partial \sigma_{xy}}{\partial y} + f_x^b = 0 \quad ; \quad \frac{\partial \sigma_{xy}}{\partial x} + \frac{\partial \sigma_{yy}}{\partial y} + f_y^b = 0 \tag{8.342}$$

The stress-strain relationships are

$$\sigma_{xx} = 2\mu\, \epsilon_{xx} + \lambda(\epsilon_{xx} + \epsilon_{yy})$$
$$\sigma_{yy} = 2\mu\, \epsilon_{yy} + \lambda(\epsilon_{xx} + \epsilon_{yy}) \tag{8.343}$$
$$\sigma_{xy} = 2\mu\, \epsilon_{xy}$$

supplemented by

$$\sigma_{xz} = 0 \quad ; \quad \sigma_{yz} = 0 \quad ; \quad w = 0 \quad ; \quad \sigma_{zz} = \nu(\sigma_{xx} + \sigma_{yy}) \tag{8.344}$$

The boundary conditions are

$$\mathbf{u}(x,y) = \mathbf{u}_0(x,y) \quad ; \quad (x,y) \in S_u \tag{8.345}$$

$$\boldsymbol{\sigma}(x,y)\mathbf{n}(x,y) = \mathbf{T}^0(x,y) \quad ; \quad (x,y) \in S_T \tag{8.346}$$

with additional constraints

$$\sigma_{xz} = \sigma_{yz} = \sigma_{zz} = 0 \quad \text{on} \quad z = \pm a \tag{8.347}$$

The system of equations defined by (8.342), (8.343) and subject to boundary conditions (8.345) and (8.346) constitutes a two-dimensional boundary value problem which has a *unique* solution which is referred to as the "*plane strain solution*" associated with the plane problem. From (8.344) and (8.347), it is evident that the solution to the *plane strain problem* also constitutes a solution to the *plane problem* when either

$$\nu = 0 \quad \text{or} \quad (\sigma_{xx} + \sigma_{yy}) = 0 \tag{8.348}$$

The plane strain solution, however, is an exact solution to the problem if the condition (8.344) is replaced by the mixed-mixed boundary condition

$$\sigma_{xz} = \sigma_{yz} = w = 0 \quad ; \quad z = \pm a \tag{8.349}$$

The *plane strain solution* to the plane problem can be used to model material regions in which the dimension in one direction is substantially larger than in the other two directions and the displacements in that direction are constrained to be zero. Examples of these could include tunnel shaped cavities in elastic bodies, water retaining structures such as concrete gravity dams, surface loading of semi-infinite media by strip loads and other contact mechanics problems as indicated in Figure 8.24.

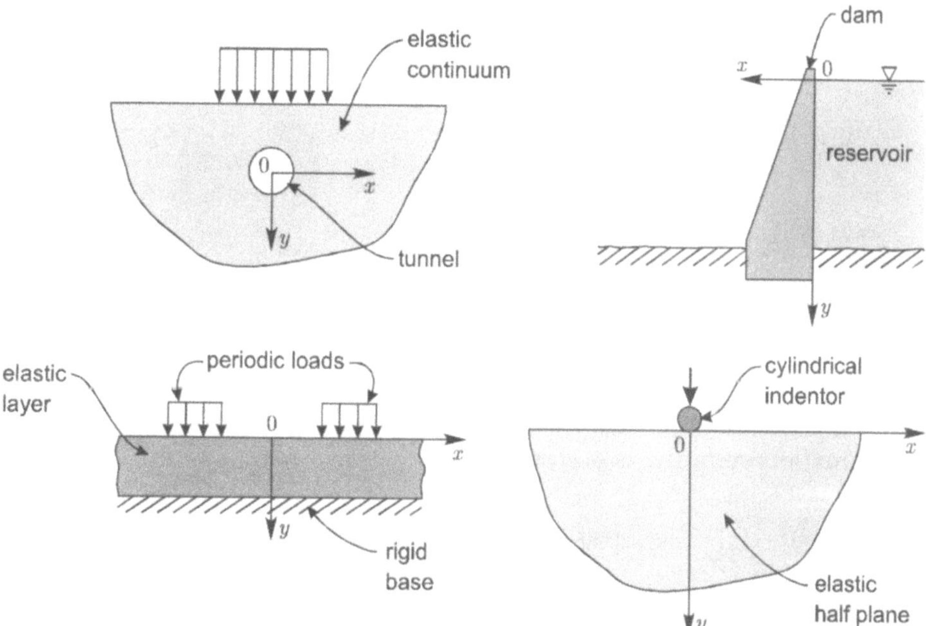

Figure 8.24: Plane strain problems in elasticity.

We now consider *"generalized plane stress solutions"* of the plane problem which characterizes the thickness averages of $\mathbf{u}, \boldsymbol{\epsilon}$, and $\boldsymbol{\sigma}$ under the approximative assumption

$$\sigma_{zz} \equiv 0 \tag{8.350}$$

From the symmetry associated with the plane problem we note that

$$u(x, y, z) = u(x, y, -z) \quad ; \quad v(x, y, z) = v(x, y, -z)$$

$$w(x, y, z) = -w(x, y, -z) \tag{8.351}$$

If this symmetry condition is not satisfied, a second solution

$$\widehat{u}(x, y, z) = \widehat{u}(x, y, -z) \quad ; \quad \widehat{v}(x, y, z) = \widehat{v}(x, y, -z)$$

$$\widehat{w}(x, y, z) = -w(x, y, -z) \tag{8.352}$$

would violate the uniqueness theorem for elastostatics, which can be proved by adopting the procedures outlined in Section 8.5. For the *generalized plane stress problem*, we consider thickness averages of $\mathbf{u}$, $\boldsymbol{\epsilon}$ and $\boldsymbol{\sigma}$ such that

$$\overline{\mathbf{u}}(x, y) = \frac{1}{2a} \int_{-a}^{a} \mathbf{u}(x, y) dz \tag{8.353}$$

$$\overline{\boldsymbol{\epsilon}}(x, y) = \frac{1}{2a} \int_{-a}^{a} \boldsymbol{\epsilon}(x, y) dz \tag{8.354}$$

$$\overline{\boldsymbol{\sigma}}(x, y) = \frac{1}{2a} \int_{-a}^{a} \boldsymbol{\sigma}(x, y) dz \tag{8.355}$$

Since the thickness average of an odd function is zero and since the derivative of a function even in z is odd we have

$$\overline{w} = 0 \quad ; \quad \overline{\epsilon}_{xz} = 0 \quad ; \quad \overline{\epsilon}_{yz} = 0 \tag{8.356}$$

and from the stress-strain relations (8.339) we have

$$\overline{\sigma}_{xz} = 0 \quad ; \quad \overline{\sigma}_{yz} = 0 \tag{8.357}$$

Considering the non-zero averaged components of $\overline{\boldsymbol{\epsilon}}$, and the non-zero averaged stresses $\overline{\boldsymbol{\sigma}}$, the non-zero expressions for the *averaged stress-strain relations* give

$$\overline{\sigma}_{xx} = 2\mu\overline{\epsilon}_{xx} + \overline{\lambda}(\overline{\epsilon}_{xx} + \overline{\epsilon}_{yy})$$

$$\overline{\sigma}_{yy} = 2\mu\overline{\epsilon}_{yy} + \overline{\lambda}(\overline{\epsilon}_{xx} + \overline{\epsilon}_{yy}) \tag{8.358}$$

$$\overline{\sigma}_{xy} = 2\mu\overline{\epsilon}_{xy}$$

and since $\overline{\sigma}_{zz} \equiv 0$ we note that

$$\overline{\lambda} = \frac{2\mu\lambda}{(\lambda + 2\mu)} \tag{8.359}$$

Inverting the averaged form of the stress-strain relationships (8.358) we have

$$2\mu\,\overline{\epsilon}_{xx} = \overline{\sigma}_{xx} - \overline{\nu}(\overline{\sigma}_{xx} + \overline{\sigma}_{yy})$$
$$2\mu\,\overline{\epsilon}_{yy} = \overline{\sigma}_{yy} - \overline{\nu}(\overline{\sigma}_{xx} + \overline{\sigma}_{yy}) \tag{8.360}$$
$$2\mu\,\overline{\epsilon}_{xy} = \overline{\sigma}_{xy}$$

where

$$\overline{\nu} = \frac{\nu}{(1 + \nu)} \tag{8.361}$$

If we consider the thermodynamic constraint on ν that is given by (8.269) we note that $\overline{\nu}$ must lie between the limits

$$-\infty < \overline{\nu} < \frac{1}{3} \tag{8.362}$$

The more realistic value of the lower limit of ν (i.e. $\nu = 0$), however, gives

$$0 < \overline{\nu} < \frac{1}{3} \tag{8.363}$$

The *generalized plane stress* solution to the plane problem can be used to examine the state of stress in bodies such as prismatic plates which are subjected to *in-plane loadings*. The state of stress in the vicinity of defects such as cracks and cavities in plates can be examined provided the dimensions of the defect are large in comparison with the thickness of the plate.

Furthermore, the plates should be loaded in such a way that the symmetries imposed by (8.351) are satisfied and local buckling of the plate does not occur in regions subjected to compressional states of stress. Examples of configurations to which the state of *generalized plane stress* can be applied, are shown in Figure 8.25.

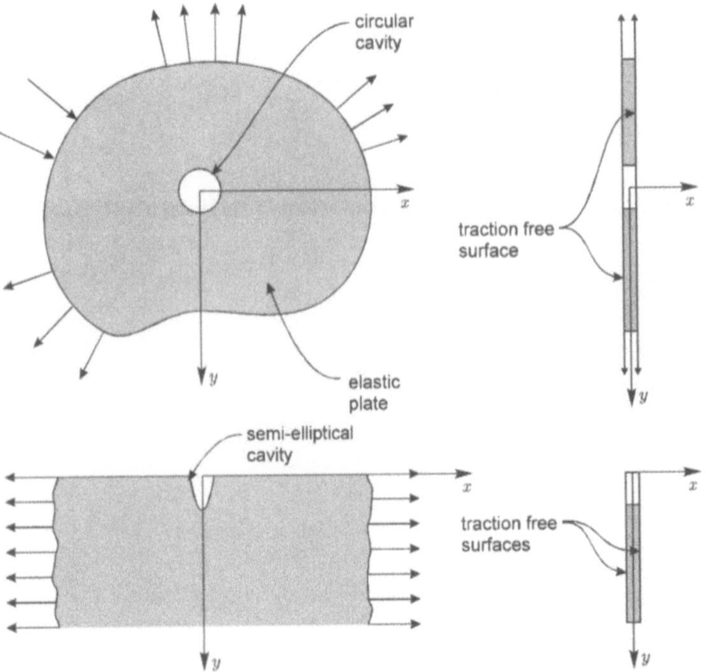

Figure 8.25: Generalized plane stress problems in elasticity.

To specify a state of plane strain or generalized plane stress it is only necessary to specify the plane components of $\mathbf{u}$ and either $\boldsymbol{\sigma}$ or $\overline{\boldsymbol{\sigma}}$ defined by (8.339), together with either (8.343) and (8.344) or (8.356) and (8.357).The developments presented previously can be summarized in the following way. Let the region $\mathbb{R}$ have an intersection S with the $x - y$ plane. Let S_u and S_T be regular sub-segments of S the boundary of $\mathbb{R}$ and denote by $\mathbf{n}$ the outward unit normal to S. We assume that the elastic region is homogeneous and isotropic with $\mu \neq 0$; $\nu \neq \frac{1}{2}$, 1. To specify a state of plane strain or generalized plane stress it is only necessary to specify the plane components of $\mathbf{u} = (u, v)$ and stresses $\boldsymbol{\sigma} = (\sigma_{xx}, \sigma_{yy}, \sigma_{xy})$. We can define a plane elastic state if the displacement and stress fields are smooth and continuous in the region $\mathbb{R}$ and satisfy the field equations (8.342) and (8.343). If we note that every plane elastic state in conjunction with (8.344) leads to a three-dimensional state of plane strain then, the following points could be noted.

(i) Uniqueness theorem for the plane problem

THEOREM 8.3

If $(\mathbf{u}^{(1)}, \boldsymbol{\sigma}^{(1)})$ and $(\mathbf{u}^{(2)}, \boldsymbol{\sigma}^{(2)})$ are plane elastic states which utilize the same elastic constants μ and ν being positive definite and utilize the same boundary conditions

$$\mathbf{u}^{(1)} = \mathbf{u}^{(2)} = \mathbf{u}^0(\mathbf{x}) \quad ; \quad \mathbf{x} \in S_u \tag{8.364}$$

$$\boldsymbol{\sigma}^{(1)}\mathbf{n} = \boldsymbol{\sigma}^{(2)}\mathbf{n} = \mathbf{T}^0(\mathbf{x}) \quad ; \quad \mathbf{x} \in S_T \tag{8.365}$$

then

$$\boldsymbol{\sigma}^{(1)} \equiv \boldsymbol{\sigma}^{(2)} \quad ; \quad \mathbf{x} \in \mathbb{R} \tag{8.366}$$

PROOF

The proof of the uniqueness theorem for the plane problem is simply a special case of the more general theorem proved in Section 8.5.

⊙ ⊙ ⊙

(ii) Compatibility relationship for stresses

THEOREM 8.4

Consider a state of *plane strain* in an elastic medium with elastic constants μ and ν and body forces $\mathbf{f}^b$ which is characterized by displacements $\mathbf{u}$ and stresses $\boldsymbol{\sigma}$. Then the stresses should satisfy the *compatibility relation for stresses*

$$\nabla \cdot \left\{ \nabla (tr\,\widehat{\boldsymbol{\sigma}}) + \frac{1}{(1-\nu)}\mathbf{f}^b \right\} = 0 \tag{8.367}$$

where

$$tr\,\widehat{\boldsymbol{\sigma}} = (\sigma_{xx} + \sigma_{yy}) \tag{8.368}$$

PROOF

We shall assume that $\mathbf{u}$ and $\boldsymbol{\sigma}$ are analytic in $\mathbb{R}$. Accordingly, $\mathbf{u}$ and $\boldsymbol{\sigma}$ satisfy the compatibility condition

$$\frac{\partial^2 \epsilon_{xx}}{\partial y^2} + \frac{\partial^2 \epsilon_{yy}}{\partial x^2} = 2\frac{\partial^2 \epsilon_{xy}}{\partial x \partial y} \tag{8.369}$$

and the equilibrium conditions

$$\frac{\partial \sigma_{xx}}{\partial x} + \frac{\partial \sigma_{xy}}{\partial y} + f_x = 0 \quad ; \quad \frac{\partial \sigma_{xy}}{\partial x} + \frac{\partial \sigma_{yy}}{\partial y} + f_y = 0 \tag{8.370}$$

We can invert the stress-strain relationship (8.343) to obtain

$$2\mu\,\epsilon_{xx} = (1 - \nu)\sigma_{xx} - \nu\,\sigma_{yy}$$

$$2\mu\,\epsilon_{yy} = (1 - \nu)\sigma_{yy} - \nu\,\sigma_{xx} \tag{8.371}$$

$$2\mu\,\epsilon_{xy} = \sigma_{xy}$$

From (8.369) and (8.371) we obtain

$$\frac{\partial^2 \sigma_{xx}}{\partial y^2} + \frac{\partial^2 \sigma_{yy}}{\partial x^2} - \nu\nabla^2(\sigma_{xx} + \sigma_{yy}) = 2\frac{\partial^2 \sigma_{xy}}{\partial x \partial y} \tag{8.372}$$

From the equilibrium equations (8.370) we have

$$\frac{\partial^2 \sigma_{xx}}{\partial x^2} = -\frac{\partial^2 \sigma_{xy}}{\partial x \partial y} - \frac{\partial f_x^b}{\partial x} \quad ; \quad \frac{\partial^2 \sigma_{yy}}{\partial y^2} = -\frac{\partial^2 \sigma_{xy}}{\partial x \partial y} - \frac{\partial f_y^b}{\partial y} \tag{8.373}$$

or

$$\frac{\partial^2 \sigma_{xx}}{\partial x^2} + \frac{\partial^2 \sigma_{yy}}{\partial y^2} = -2\frac{\partial^2 \sigma_{xy}}{\partial x \partial y} - \left(\frac{\partial f_x^b}{\partial x} + \frac{\partial f_y^b}{\partial y}\right) \tag{8.374}$$

Combining (8.372) and (8.374) we obtain

$$\nabla^2(\sigma_{xx} + \sigma_{yy}) = -\frac{1}{(1-\nu)}\left(\frac{\partial f_x^b}{\partial x} + \frac{\partial f_y^b}{\partial y}\right) \tag{8.375}$$

which is identical to (8.367). The converse of the theorem implies that if $\boldsymbol{\sigma}$ is continuously differentiable in $\mathbb{R}$ such that (8.375) and the equilibrium equations (8.370) are satisfied then there exists a strain field $\boldsymbol{\epsilon}$ which satisfies the compatibility equation (8.369). If the compatibility equations are satisfied then there exists a unique displacement field which is continuous in $\mathbb{R}$ (indeterminate to within a rigid body displacement).

$$\odot \ \odot \ \odot$$

(iii) Equivalence of plane elastic stress states- Levy's Theorem

THEOREM 8.5

Assume that the plane region $\mathbb{R}$ is simply connected, and let $\mathbf{u}$ and $\boldsymbol{\sigma}$ correspond to a plane elastic state with the positive definite elastic constants μ and ν. Similarly let $\mathbf{u}'$ and $\boldsymbol{\sigma}'$ be an elastic state corresponding to positive definite elastic constants μ' and ν'. The stress states satisfy the traction boundary conditions

$$\boldsymbol{\sigma}\mathbf{n} = \boldsymbol{\sigma}'\mathbf{n} = \mathbf{T}^0(\mathbf{x}) \ \ ; \ \ \mathbf{x} \in S \tag{8.376}$$

Then

$$\boldsymbol{\sigma} = \boldsymbol{\sigma}' \ \ ; \ \ \mathbf{x} \in \mathbb{R} \tag{8.377}$$

PROOF

Since $\boldsymbol{\sigma}'$ satisfies the compatibility relationship (8.375) (with $\mathbf{f}^b = 0$) and the equations of equilibrium (8.370), then there exists a displacement field $\widehat{\mathbf{u}}$ such that $(\widehat{\mathbf{u}}, \boldsymbol{\sigma}')$ is a plane elastic state corresponding to positive definite elastic constants μ and ν. Therefore $(\mathbf{u}, \boldsymbol{\sigma})$ and $(\widehat{\mathbf{u}}, \boldsymbol{\sigma}')$ correspond to the same elastic constants and the same surface forces. We can conclude from the uniqueness theorem, with $S_u = 0$ and $S_T = S$, that

$$\boldsymbol{\sigma} = \boldsymbol{\sigma}' \quad ; \quad \mathbf{x} \in \mathbb{R} \tag{8.378}$$

Hence in plane problems in classical elasticity theory, the stress state characterized by $\{\sigma_{xx}, \sigma_{yy}, \sigma_{xy}\}$ is independent of the elastic constants of the material.

8.7 The Airy stress function

In the previous section we have presented the general formulation and theorems describing the plane problem in the classical theory of elasticity. The plane problem is characterized by the equations of equilibrium (8.342), the strain-displacement relationships of the type (8.63) and the appropriate form of the stress-strain relationship (e.g. (8.343)). Together, these equations provide *eight linear partial differential equations* of the *first-order* for the *eight* unknowns $\mathbf{u}, \boldsymbol{\epsilon}$ and $\boldsymbol{\sigma}$. These equations can be further reduced by adopting two approaches. In the first, the strain-displacement equations are replaced by a single compatibility equation which then reduces the system of equations to *six linear partial differential equations* of the *second-order*, for the six unknowns $\boldsymbol{\epsilon}$ and $\boldsymbol{\sigma}$. The procedure can be continued to replace the compatibility equation in terms of the stresses and this will yield *three linear partial differential equations* up to *second-order* for the components of $\boldsymbol{\sigma}$. For example, in plane strain we have

$$\nabla . \widehat{\mathfrak{S}}^P + \mathbf{f}^b = \mathbf{0} \tag{8.379}$$

and

$$\nabla . \left\{ \nabla \left(tr \, \widehat{\boldsymbol{\sigma}}^P \right) + \frac{1}{(1-\nu)} \mathbf{f}^b \right\} = \mathbf{0} \tag{8.380}$$

where

$$\nabla . \widehat{\mathfrak{S}}^P = \left(\mathbf{i} \frac{\partial}{\partial x} + \mathbf{j} \frac{\partial}{\partial y} \right) \left(\mathbf{i} \, \mathbf{t}_x + \mathbf{j} \, \mathbf{t}_y \right) \tag{8.381}$$

$$tr \, \widehat{\boldsymbol{\sigma}}^P = (\sigma_{xx} + \sigma_{yy}) \tag{8.382}$$

All the governing equations have now been reduced to their simplest forms, in terms of only stresses.

An alternative approach is to express the stress-strain relations in terms of the displacements which, when combined with the equations of equilibrium, gives rise to *five linear partial differential equations* of the *first-order* for the unknown stresses $\boldsymbol{\sigma}$ with components $\{\sigma_{xx}, \sigma_{yy}, \sigma_{xy}\}$ and the unknown displacements $\mathbf{u}$ with components (u, v). These equations can be further reduced to *two second-order linear partial differential equations* for the displacement components; i.e.

$$\mu \nabla^2 \mathbf{u} + (\lambda + \mu)\nabla\nabla.\mathbf{u} + \mathbf{f}^b = \mathbf{0} \tag{8.383}$$

where ∇^2 is Laplace's operator.

All the governing equations have now been reduced to their simplest forms in terms of displacements only. To proceed to solve either the governing equations in terms of stresses (8.379) and (8.380) or the governing equations in terms of displacements (8.383) it becomes necessary to identify representations where either the stresses can be expressed in terms of certain unknown function(s) or the displacements can be expressed in terms of unknown function(s). In this section we shall focus on the procedure for the representation of stresses in terms of an unknown function, which was introduced by Sir G.B.Airy (1801-1892). The stress function introduced by Airy, is assumed to be a scalar function with continuous derivatives to the third-order within the region $\mathbb{R}$. We further assume that the discussion is restricted to a conservative body force field for which the body forces are derived from a potential function V. It is assumed that the stresses can be expressed in terms of the Airy stress function $\varphi(x, y)$ and the potential V in the forms

$$\sigma_{xx} = \frac{\partial^2 \varphi}{\partial y^2} - V \quad ; \quad \sigma_{yy} = \frac{\partial^2 \varphi}{\partial x^2} - V \quad ; \quad \sigma_{xy} = -\frac{\partial^2 \varphi}{\partial x \partial y} \tag{8.384}$$

and

$$f_x^b = \frac{\partial V}{\partial x} \quad ; \quad f_y^b = \frac{\partial V}{\partial y} \tag{8.385}$$

Substituting these representations in the equations of equilibrium (8.342)(or (8.379)), it is evident that they are identically satisfied for all choices of φ and V. The stress compatibility equation (8.380) gives

$$\nabla^4 \varphi + \frac{(1-2\nu)}{(1-\nu)}\nabla^2 V = 0 \tag{8.386}$$

In the instance when either the body forces $\mathbf{f}^b$ are constant or if V is a harmonic function or if the plane strain problem deals with an incompressible elastic material (8.386) reduces to

$$\nabla^4 \varphi = 0 \tag{8.387}$$

which is the biharmonic equation for Airy's stress function. In this instance the traction boundary value problem reduces to that of finding a biharmonic function $\varphi(x,y)$ such that the traction boundary condition is satisfied: i.e.

$$\widehat{\sigma}^P \mathbf{n} = \mathbf{T}^0(\mathbf{x}) \quad ; \quad \mathbf{x} \in S_T \tag{8.388}$$

where $\mathbf{T}^0(\mathbf{x})$ are prescribed boundary tractions.

The solution of the biharmonic equation can be achieved in a variety of ways. All harmonic functions, naturally, are acceptable solutions. The biharmonic equation is, however, expected to have a wider class of solutions. The *complex potentials approach* is an alternative representation of the solution of the plane problem in classical elasticity. The theorem associated with the representation is due to Goursat (1858-1936) and the applications of the complex potentials approach to plane problems in elasticity is attributed to Kolosov (1867-1936) and Muskhelishvili (1891-1976).

THEOREM 8.6

Let φ be a biharmonic function in $\mathbb{R}$. Then there exists analytic complex functions ψ and χ in $\mathbb{R}$ such that

$$\varphi(z) = \mathrm{Re}\,\{\,\bar{z}\,\psi\,(z) + \chi(z)\} \tag{8.389}$$

PROOF

The problem deals with the classical biharmonic equation (8.387) which is applicable to the Airy stress function formulation in the absence of body forces. The complex notation is

$$z = (x + iy) \quad ; \quad \overline{z} = (x - iy) \tag{8.390}$$

and

$$\frac{\partial}{\partial x} = \frac{\partial}{\partial z} + \frac{\partial}{\partial \overline{z}} \quad ; \quad \frac{\partial}{\partial y} = i\left(\frac{\partial}{\partial z} - \frac{\partial}{\partial \overline{z}}\right) \tag{8.391}$$

As a matter of convention we shall retain this definition of the complex variables z and $\overline{z}$, with the understanding that for two-dimensional problems, the state of deformation and stress is independent of the coordinate variable z.

Let

$$P = \nabla^2 \varphi = (\sigma_{xx} + \sigma_{yy}) \tag{8.392}$$

We note from (8.380), that P is harmonic and, as such, will have a conjugate harmonic function Q. The function $(P+iQ)$ is therefore an analytic function of z such that

$$f(z) = (P + iQ) \tag{8.393}$$

The integral of $f(z)$ with respect to z is therefore another analytic function which we can denote by, say, $4\psi(z)$. If p and q are the real and imaginary parts of $\psi(z)$, then

$$\psi(z) = p + iq = \frac{1}{4}\int f(z)dz \tag{8.394}$$

or

$$\frac{d\psi}{dz} = \frac{1}{4}f(z) \tag{8.395}$$

If we represent the function ψ and f in terms of the respective complex and real parts we can write

$$\frac{\partial \psi}{\partial x} = \frac{\partial p}{\partial x} + i\frac{\partial q}{\partial x} = \frac{1}{4}f(z) = \frac{1}{4}(P + iQ) \tag{8.396}$$

This is equivalent to

$$\frac{\partial p}{\partial x} = \frac{P}{4} \tag{8.397}$$

Since p and q must satisfy the *Cauchy-Riemann* equations (named after Augustin Louis Cauchy (1789-1857) and G.F.B. Riemann (1826-1866))

$$\frac{\partial p}{\partial x} = \frac{\partial q}{\partial y} \quad ; \quad \frac{\partial p}{\partial y} = -\frac{\partial q}{\partial x} \tag{8.398}$$

we have, from (8.397) and (8.398),

$$\frac{\partial q}{\partial y} = \frac{P}{4} \tag{8.399}$$

Since $P(= \nabla^2\varphi)$, is harmonic, from (8.397) and (8.399)

$$\nabla^2(\varphi - xp - yq) = \nabla^2\varphi - 2\frac{\partial p}{\partial x} - 2\frac{\partial q}{\partial y} = 0 \tag{8.400}$$

Hence, in general

$$\varphi = xp + yq + \varphi_0 \tag{8.401}$$

where p and q are conjugate harmonic functions and φ_0 is a harmonic function. It may be observed that the stress function φ can be constructed by selecting either of the functions p or q and a harmonic function φ^*. Therefore, instead of (8.401), we can write

$$\nabla^2(\varphi - 2xp) = \nabla^2\varphi - 4\frac{\partial p}{\partial x} = 0$$

$$\nabla^2(\varphi + 2yp) = \nabla^2\varphi + 4\frac{\partial q}{\partial y} = 0 \tag{8.402}$$

which is equivalent to the functions

$$\varphi = 2xp + \varphi_p \quad ; \quad \varphi = -2yq + \varphi_q \tag{8.403}$$

where p, q, φ_p and φ_q are suitably chosen harmonic functions. Considering (8.401) we can introduce the function p_1 and its conjugate harmonic function q_1 such that

$$\chi(z) = p_1 + iq_1 \tag{8.404}$$

We can verify, from (8.394) and (8.403), that the *real part* of $(\overline{z}\psi(z) + \chi(z))$ is identical to the right hand side of (8.401). Hence the stress function $\varphi(z)$ can be represented in the form

$$\varphi(z) = \mathrm{Re}\left[\overline{z}\psi(z) + \chi(z)\right] \tag{8.405}$$

where Re is the real part of the complex function and $\psi(z)$ and $\chi(z)$ are suitably chosen analytic functions.

◉ ◉ ◉

The analysis of the problem becomes complete when the displacements u and v corresponding to the state of plane deformation of the region $\mathbb{R}$ are determined. Since the problem is formulated in terms of the Airy stress function, naturally the solution of the biharmonic equation gives the stress field in $\mathbb{R}$, which can be used to determine displacements, through the integration of the stress-strain relations. The equations of compatibility (see e.g. (8.106)) permits us to perform the integrations and to derive unique values for the displacement field (at least in a *simply connected domain*), which is indeterminate to within an arbitrary rigid displacement. The following representation theorem deals with the determination of the displacement through a knowledge of the Airy stress function $\varphi(x, y)$.

THEOREM 8.7

Consider the solution to a plane strain problem in classical elasticity in a simply connected domain $\mathbb{R}$, characterized by the displacement field $\mathbf{u} = (u, v)$ and the stress field $\boldsymbol{\sigma}^P = (\sigma_{xx}, \sigma_{yy}, \sigma_{xy})$ with elastic constants μ and ν (with $\mu > 0$) and with zero body forces. Let $\varphi(z)$ be an Airy stress function from with the stresses $\boldsymbol{\sigma}^P$ are determined and let the analytic complex function ψ in (8.405) be represented in the form

$$4\psi(z) = p(x, y) + iq(x, y) \tag{8.406}$$

where p and q are real. Then the displacement components in a simply connected domain are given by

$$2\mu\, u(x, y) = \left[-\frac{\partial \varphi}{\partial x} + (1 - \nu)p \right] + u_0$$

$$\tag{8.407}$$

$$2\mu\, v(x, y) = \left[-\frac{\partial \varphi}{\partial y} + (1 - \nu)q \right] + v_0$$

where u_0 and v_0 are components of a plane rigid body displacement of $\mathbb{R}$.

PROOF

Considering the stress-strain relationships (8.343) and the representations of the stress components of $\boldsymbol{\sigma}^P$ in terms of the stress function $\varphi(x, y)$ (see e.g. (8.384) with $V = 0$) we have

$$2\mu\, \frac{\partial u}{\partial x} = -\frac{\partial^2 \varphi}{\partial x^2} + (1 - \nu)\nabla^2 \varphi \tag{8.408}$$

$$2\mu\, \frac{\partial v}{\partial y} = -\frac{\partial^2 \varphi}{\partial y^2} + (1 - \nu)\nabla^2 \varphi \tag{8.409}$$

$$\mu\left(\frac{\partial u}{\partial y} + \frac{\partial v}{\partial x} \right) = -\frac{\partial^2 \varphi}{\partial x \partial y} \tag{8.410}$$

Considering the representation (8.405) and (8.406) we have

$$\varphi(x, y) = \frac{1}{4}[xp + yq] + \mathrm{Re}\ \chi(z) \tag{8.411}$$

Hence

$$\nabla^2 \varphi = \frac{1}{4}\left[\frac{\partial p}{\partial x} + \frac{\partial q}{\partial y}\right] \tag{8.412}$$

since p and q are conjugate harmonic functions, they satisfy the Cauchy-Riemann equations (8.398) and

$$\nabla^2 \varphi = \frac{\partial p}{\partial x} = \frac{\partial q}{\partial y} \tag{8.413}$$

The equations (8.408) and (8.409) can be re-written in the form

$$2\mu\,\frac{\partial u}{\partial x} = -\frac{\partial^2 \varphi}{\partial x^2} + (1 - \nu)\frac{\partial p}{\partial x} \tag{8.414}$$

$$2\mu\,\frac{\partial v}{\partial y} = -\frac{\partial^2 \varphi}{\partial y^2} + (1 - \nu)\frac{\partial q}{\partial y} \tag{8.415}$$

Integrating (8.414) and (8.415) we have

$$2\mu\,u(x, y) = -\frac{\partial \varphi}{\partial x} + (1 - \nu)p + 2\mu\,u_0(y) \tag{8.416}$$

$$2\mu\,v(x, y) = -\frac{\partial \varphi}{\partial y} + (1 - \nu)q + 2\mu\,v_0(x) \tag{8.417}$$

Substituting (8.416) and (8.417) into (8.410) we obtain

$$\frac{du_0}{dy} + \frac{dv_0}{dx} = 0 \tag{8.418}$$

which implies that

$$u_0(y) = -v_0(x) = \text{const.} \tag{8.419}$$

Hence equations (8.407) are the representations for the displacement components in terms of φ and ψ.

$$\odot \;\; \odot \;\; \odot$$

The foregoing proof explicitly assumes that the region $\mathbb{R}$ is simply-connected, which assures a single-valued displacement field $\mathbf{u}$ in $\mathbb{R}$. If the region is multiply-connected, additional conditions need to be imposed on $\varphi(x, y)$ to ensure that the displacement field in the multiply-connected domain $\mathbb{R}_M$ is single-valued. These relationships were developed by Michell (1863-1940). In the ensuing we shall outline briefly, the conditions that should be satisfied by the Airy stress function in order to ensure single-valuedness in the displacement field.

THEOREM 8.8

Consider a plane multiply connected region $\mathbb{R}$ such that it is internally bounded by a set of closed curves C_m, (see Figure 8.26) and we denote the "internal boundary" of $\mathbb{R}$ by $\partial \mathbb{R}_m$ where

$$\partial \mathbb{R}_M = C_1 \cup C_2 \cup \,.....\, \cup C_M \tag{8.420}$$

Let φ be the Airy stress function applicable to the region $\mathbb{R}_m$ and we assume that φ has continuous derivatives at least up to the third order.

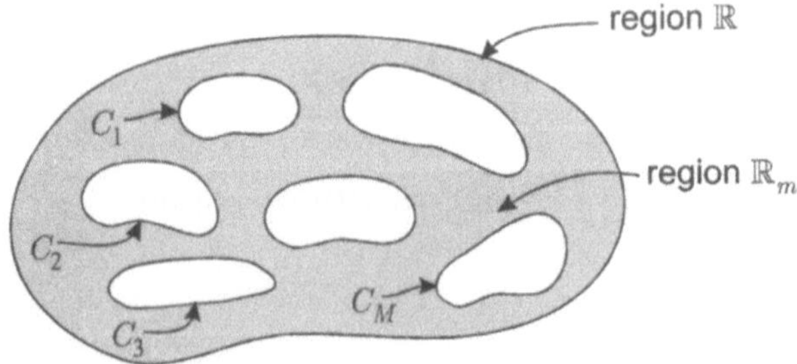

Figure 8.26: Multiply connected elastic region.

Let

$$2\mu\, u = -\frac{\partial\varphi}{\partial x} + (1-\nu)p + u_0$$

$$\text{(8.421)}$$

$$2\mu\, v = -\frac{\partial\varphi}{\partial y} + (1-\nu)q + v_0$$

where, in Theorem 8.7 we have shown that u_0 and v_0 correspond to a plane rigid displacement. Then, the necessary and sufficient conditions for the single-valuedness in the displacement field in $\mathbb{R}$ is that φ satisfy the $3M$-3 conditions

$$\int_{C_m} \frac{\partial}{\partial n}(\nabla^2\varphi)d\ell = 0 \tag{8.422}$$

$$\int_{C_m}\left[x\frac{\partial}{\partial n}(\nabla^2\varphi) + y\frac{\partial}{\partial \ell}(\nabla^2\varphi)\right]d\ell = -\frac{\xi_1^{(m)}}{(1-\nu)} \tag{8.423}$$

$$\int_{C_m}\left[x\frac{\partial}{\partial \ell}(\nabla^2\varphi) - y\frac{\partial}{\partial n}(\nabla^2\varphi)\right]d\ell = -\frac{\xi_2^{(m)}}{(1-\nu)} \tag{8.424}$$

for $m = 2, 3, ...M$, where ℓ is measured along C_m and n is measured along the outward unit normal to C_m, and

$$\xi_1^{(m)} = \int_{C_m}(\sigma_{xx}n_x + \sigma_{xy}n_y)d\ell \tag{8.425}$$

$$\xi_2^{(m)} = \int_{C_m}(\sigma_{xy}n_x + \sigma_{yy}n_y)d\ell \tag{8.426}$$

PROOF

We note that the displacement field $\mathbf{u}$ will be single-valued if and only if

$$\int_{C_m}\left(\frac{du}{d\ell}\right)d\ell = 0 \quad ; \quad \int_{C_m}\left(\frac{dv}{d\ell}\right)d\ell = 0 \quad ; \quad m = 2, 3, ...M \tag{8.427}$$

From the definitions of u and v given (8.421) the condition (8.427) is equivalent to

$$\int_{C_m} \left(\frac{dp}{d\ell} \right) d\ell = \frac{1}{(1-\nu)} \int_{C_m} \frac{d}{d\ell} \left(\frac{\partial \varphi}{\partial x} \right) d\ell \quad ; \quad m = 2, 3, ... M \quad (8.428)$$

$$\int_{C_m} \left(\frac{dq}{d\ell} \right) d\ell = \frac{1}{(1-\nu)} \int_{C_m} \frac{d}{d\ell} \left(\frac{\partial \varphi}{\partial x} \right) d\ell \quad ; \quad m = 2, 3, ... M \quad (8.429)$$

If $\varphi, (\partial \varphi / \partial x)$ and $(\partial \varphi / \partial y)$ do not necessarily vanish at some point on C_m, then (8.428) and (8.429) can be written as

$$\int_{C_m} \left(\frac{dp}{d\ell} \right) d\ell = - \frac{\xi_2^{(m)}}{(1-\nu)} \quad ; \quad \int_{C_m} \left(\frac{dq}{d\ell} \right) d\ell = \frac{\xi_1^{(m)}}{(1-\nu)} \qquad (8.430)$$

where $m = 2, 3, ... M$. Also,

$$\frac{dp}{d\ell} = \frac{\partial p}{\partial x} \frac{dx}{d\ell} + \frac{\partial p}{\partial y} \frac{dy}{d\ell} \quad ; \quad \frac{dq}{d\ell} = \frac{\partial q}{\partial x} \frac{dx}{d\ell} + \frac{\partial q}{\partial y} \frac{dy}{d\ell} \qquad (8.431)$$

on C_m where

$$\mathbf{x}(\ell) = (x(\ell), y(\ell)) \quad ; \quad 0 \le \ell \le \ell_0 \qquad (8.432)$$

is the parametric representation for C_m.

Let P and Q be the real and imaginary parts of $f(z)$ defined by Theorem 8.6, i.e.

$$f(z) = P(x, y) + i\, Q(x, y) \qquad (8.433)$$

Then, from (8.396) to (8.399)

$$\frac{\partial p}{\partial x} = \frac{P}{4} \quad ; \quad \frac{\partial q}{\partial x} = \frac{Q}{4}$$

$$\qquad (8.434)$$

$$\frac{\partial p}{\partial y} = - \frac{Q}{4} \quad ; \quad \frac{\partial q}{\partial y} = \frac{P}{4}$$

Using (8.431) and (8.434), (8.430) can be written as

$$\int_{C_m} \left[Q\frac{dx}{d\ell} + P\frac{dy}{d\ell} \right] d\ell = \frac{4\xi_1^{(m)}}{(1-\nu)} \tag{8.435}$$

$$\int_{C_m} \left[P\frac{dx}{d\ell} - Q\frac{dy}{d\ell} \right] d\ell = -\frac{4\xi_2^{(m)}}{(1-\nu)} \tag{8.436}$$

Integrating the left hand side of (8.435) and (8.436) by parts we have

$$[Qx + Py]_{\ell=0}^{\ell=\ell_0} - \int_{C_m} \left[x\frac{dQ}{d\ell} + y\frac{dP}{d\ell} \right] d\ell = \frac{4\xi_1^{(m)}}{(1-\nu)}; \tag{8.437}$$

$$[Px - Qy]_{x=0}^{x=\ell} - \int_{C_m} \left[x\frac{dP}{d\ell} - y\frac{dQ}{d\ell} \right] d\ell = -\frac{4\xi_2^{(m)}}{(1-\nu)}. \tag{8.438}$$

From the definition of the Airy stress function, $P(=\nabla^2\varphi)$ is single-valued. Hence for (8.437) and (8.438) to be valid it is sufficient that

$$\int_{C_m} \frac{dQ}{d\ell} d\ell = 0 \tag{8.439}$$

$$\int_{C_m} \left[x\frac{dQ}{d\ell} + y\frac{dP}{d\ell} \right] d\ell = -\frac{4\xi_1^{(m)}}{(1-\nu)} \tag{8.440}$$

$$\int_{C_m} \left[x\frac{dQ}{d\ell} - y\frac{dP}{d\ell} \right] d\ell = \frac{4\xi_2^{(m)}}{(1-\nu)} \tag{8.441}$$

Equation (8.439) indicates that Q is single-valued and (8.440) and (8.441) are equivalent to (8.437) and (8.438) when both P and Q are single-valued.

From the definitions of displacements given by (8.407) we can show that

$$Q = -\frac{4\mu}{(1-\nu)} \left[\frac{\partial u}{\partial y} - \frac{\partial v}{\partial x} - \omega_0 \right] \tag{8.442}$$

where

$$\omega_0 = \frac{\partial u_0}{\partial y} - \frac{\partial v_0}{\partial x} \tag{8.443}$$

is a plane rigid displacement. In order to prove that (8.439) to (8.441) are also *necessary* for (8.437) and (8.438) to be valid we only need to prove that Q is single-valued whenever $\mathbf{u}$ is single-valued. We have

$$\frac{dQ}{d\ell} = \frac{\partial Q}{\partial x}\frac{dx}{d\ell} + \frac{\partial Q}{\partial y}\frac{dy}{d\ell} \tag{8.444}$$

Using the Cauchy-Riemann equations and noting that

$$n_x = \frac{dy}{d\ell} \quad ; \quad n_y = -\frac{dx}{d\ell} \tag{8.445}$$

(8.444) gives

$$\frac{dQ}{d\ell} = \frac{\partial P}{\partial x}n_x + \frac{\partial P}{\partial y}n_y = \frac{\partial}{\partial n}(P) \tag{8.446}$$

Since $P = \nabla^2 \varphi$, the conditions (8.439) to (8.441) reduce to the required conditions for the single-valuedness of the displacements.

8.8   Methods of solution of the biharmonic equation

In the ensuing we shall outline briefly the use of biharmonic equation for the stress analysis of problems in plane elasticity. The range of examples associated with problems in plane elasticity is quite extensive and the techniques for their solution will invariably depend on the specific problem definition and the region of interest. Attention is therefore focussed first on the study of plane problems referred to rectangular Cartesian and plane polar coordinate systems. Even with the restriction to these two coordinate systems, a variety of approaches can be adopted for the solution of the plane problem in elasticity theory. These include, power series solutions, variables separation techniques with a restricted form, integral transform techniques and complex variable techniques. At the outset it cannot be recommended that any one particular approach will be the most efficient technique for obtaining solutions to plane problems in elasticity theory; each problem should be examined by taking into consideration the domain of applicability of the solution, the symmetries associated with the problem, the boundary conditions that are applicable to the problem and to the results of importance. In the developments presented here, we shall focus attention primarily on problems where the body force term $\mathbf{f}^b$ is assumed to be zero. The analyses can be extended to cover the influence of the body force terms. Also, since the governing partial differential equation is linear, the effects of the body force terms can, in most cases, be regarded as an additive solution.

8.8.1 Series solution in Cartesian coordinates

The separation of variables method was a very effective method for obtaining generalized solutions to the second-order partial differential equations discussed in previous Chapters. In the case of the biharmonic equation referred to the rectangular Cartesian coordinate system, however, it can be easily verified that a solution scheme based on a generalized variables separation scheme is inadmissible. The solutions obtained for the Laplace's equation via a variables separation scheme (see e.g. Sections 5.4.2 and 5.4.3) are, of course, also solutions of the biharmonic equation. The simplest form of the solution of the biharmonic equation utilizes a power series representation of the form

$$\varphi(x,y) = \sum_{m=0}^{\alpha} \sum_{n=0}^{\beta} C_{mn} x^m y^n \tag{8.447}$$

Admittedly, the values of α and β have a finite range in order to satisfy the governing partial differential equation (8.387). For example, considering (8.447) and (8.387) it is clear that the complete polynomial

$$\begin{aligned}
\varphi(x,y) = {}& C_{00} + C_{10}x + C_{20}x^2 + C_{30}x^3 + C_{40}x^4 \\
& + C_{01}y + C_{11}xy + C_{21}x^2y + C_{31}x^3y + C_{41}x^4y \\
& + C_{02}y^2 + C_{12}xy^2 + C_{22}x^2y^2 + C_{32}x^3y^2 + C_{42}x^4y^2 \\
& + C_{03}y^3 + C_{13}xy^3 + C_{23}x^2y^3 + C_{33}x^3y^3 + C_{43}x^4y^3 \\
& + C_{04}y^4 + C_{14}xy^4 + C_{24}x^2y^4 + C_{34}x^3y^4 + C_{44}x^4y^4
\end{aligned} \tag{8.448}$$

will be an Airy stress function provided

$$\begin{aligned}
C_{14} = C_{41} &= C_{23} = C_{32} = C_{24} \\
&= C_{42} = C_{43} = C_{34} = C_{33} = C_{44} = 0
\end{aligned} \tag{8.449}$$

and

$$3(C_{04} + C_{40}) + C_{22} = 0 \tag{8.450}$$

Similarly, the polynomial

$$\varphi(x,y) = C_{14}xy^4 + C_{41}x^4y + C_{23}x^2y^3 + C_{32}x^3y^2$$
$$+C_{50}x^5 + C_{05}y^5 \tag{8.451}$$

in an Airy stress function provided

$$C_{14} + C_{23} + 5C_{05} = 0 \quad ; \quad C_{41} + C_{32} + 5C_{50} = 0 \tag{8.452}$$

Since the stresses are functions of the second derivative of $\varphi(x,y)$, the constant and linear terms in (8.448) will have no influence on the stresses and the remaining terms can be used to determine the two-dimensional stress field. It is clear that the stresses are derived from the reduced form of (8.448), i.e.

$$\sigma_{xx} = 2C_{02} + 6C_{03}y + 6C_{04}y^2 + 6C_{13}xy - 6(C_{40} + C_{04})x^2$$
$$\sigma_{yy} = 2C_{20} + 6C_{30}x + 6C_{40}x^2 + 6C_{31}xy - 6(C_{40} + C_{04})y^2 \tag{8.453}$$
$$\sigma_{xy} = -C_{11} - \{2(C_{12} + C_{21}) - 12(C_{04} + C_{40})\}\,xy$$
$$-3C_{13}y^2 - 3C_{31}x^2$$

satisfy the equations of equilibrium for all choices of the constants C_{mn}. The applicability of the solution developed will largely be governed by the specific stress analysis problem under consideration. If the stress field (8.453) is to remain finite, these solutions are clearly inapplicable to regions which are either *semi-infinite* or *infinite* in extent. Also since the stress field developed is more applicable to domains with plane boundaries, it is clear that the solutions (8.453) are applicable, in general, to regions with rectangular and square regions and other regular regions with plane boundaries. The use of the solution (8.453) will be demonstrated by appeal to specific examples involving such plate shaped regions. The use of solutions of the type (8.448) and (8.451) will be demonstrated by appeal to specific examples involving plate shaped regions which are subjected to boundary tractions.

Example 8.6

A plane elastic region in the form of square plate is subjected to the boundary stress field as shown in Figure 8.27. Assuming that a state of generalized plane stress exists in the plate, determine the stress field in the plate region

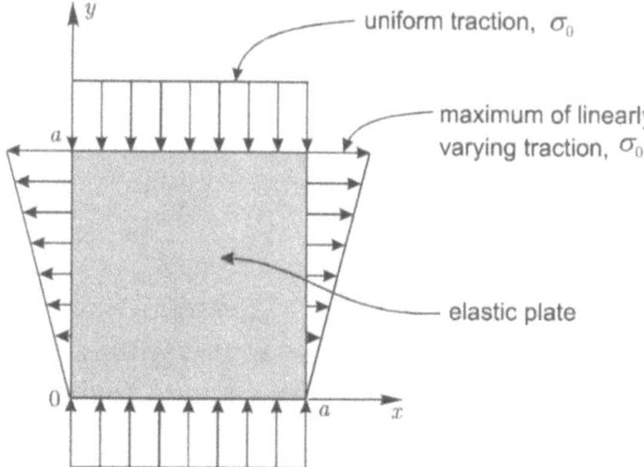

Figure 8.27: Boundary tractions acting on an elastic plate region.

Solution

Considering the boundary conditions on the edges of the plate we have

$$\sigma_{xx}(0,y) = \frac{\sigma_0\, y}{a} \quad ; \quad \sigma_{xx}(a,y) = \frac{\sigma_0\, y}{a}$$

$$\sigma_{yy}(x,0) = -\sigma_0 \quad ; \quad \sigma_{yy}(x,a) = -\sigma_0 \tag{8.454}$$

$$\sigma_{xy}(0,y) = \sigma_{xy}(a,y) = \sigma_{xy}(x,0) = \sigma_{xy}(x,a) = 0$$

Noting the form of the boundary tractions we can select a stress function of the type

$$\varphi(x,y) = C_{20}x^2 + C_{30}y^3 \tag{8.455}$$

We can show that all the boundary conditions are identically satisfied if we set

$$C_{20} = -\frac{\sigma_0}{2} \quad ; \quad C_{30} = \frac{\sigma_0}{6a} \tag{8.456}$$

The corresponding stress field in the square plate is given by

$$\sigma_{xx} = \frac{\sigma_0\, y}{a} \quad ; \quad \sigma_{yy} = -\sigma_0 \quad ; \quad \sigma_{xy} = 0 \tag{8.457}$$

where σ_0 has units of stress.

From the uniqueness theorem associated with problems in classical elasticity theory (Theorem 8.2), the stress state (8.457) constitutes the exact solution to the problem.

Example 8.7

An isotropic elastic region shown in Figure 8.28 exhibits a state of plane strain such that $w = 0$. It is subjected to boundary tractions which vary linearly along the edges $x = 0$ and $x = L$. Determine the displacements of the region at the location $x = L$, $y = 0$ assuming that the displacements and rotation of the region are zero at $x = 0$, $y = 0$.

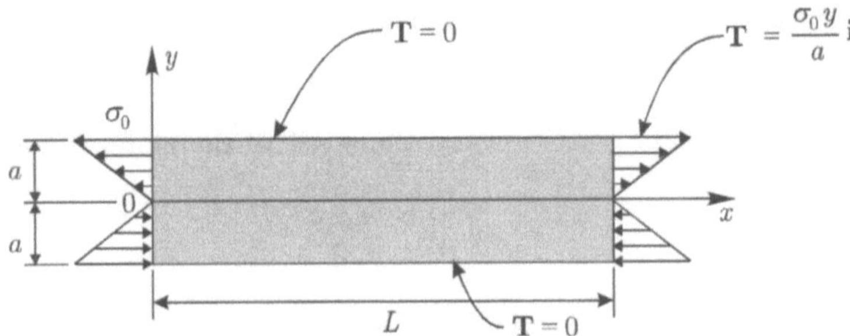

Figure 8.28: Boundary tractions on a rectangular plane which induces flexure.

Solution

From the results presented in Example 8.6, it is evident that the boundary conditions

$$\sigma_{xx}(0, y) = \sigma_{xx}(L, y) = \frac{\sigma_0\, y}{a}$$

$$\sigma_{xy}(0, y) = \sigma_{xy}(L, y) = 0 \tag{8.458}$$

$$\sigma_{yy}(x, a) = \sigma_{yy}(x, -a) = \sigma_{xy}(x, a) = \sigma_{xy}(x, -a) = 0$$

can be satisfied by selecting the stress function

$$\varphi(x, y) = \frac{\sigma_0\, y^3}{6a} \tag{8.459}$$

which gives

$$\sigma_{xx} = \frac{\sigma_0 y}{a} \quad ; \quad \sigma_{yy} = 0 \quad ; \quad \sigma_{xy} = 0 \tag{8.460}$$

where σ_0 has units of stress. Considering the stress-strain relationships for plane strain deformations (see e.g. (8.371)) of the elastic medium we have

$$\epsilon_{xx} = \frac{\partial u}{\partial x} = \frac{(1 - \nu^2)}{E}\sigma_{xx} = \frac{(1 - \nu^2)}{E}\frac{\sigma_0\, y}{a} \tag{8.461}$$

$$\epsilon_{yy} = \frac{\partial v}{\partial y} = -\frac{\nu(1 + \nu)}{E}\sigma_{xx} = -\frac{\nu(1 + \nu)}{E}\frac{\sigma_0\, y}{a} \tag{8.462}$$

Integrating (8.461) and (8.462) we obtain

$$u(x, y) = \frac{(1 - \nu^2)}{E}\sigma_0\, xy + f(y) \tag{8.463}$$

$$v(x, y) = \frac{\nu(1 + \nu)}{E}\sigma_0\frac{y^2}{2} + g(x) \tag{8.464}$$

where $f(y)$ and $g(x)$ are arbitrary functions of the respective arguments. Considering the shear stress-shear strain relationship of (8.371) and (8.463) and (8.464) we have

$$\frac{\sigma_{xy}}{G} = 2\epsilon_{xy} = \frac{\partial u}{\partial y} + \frac{\partial v}{\partial x} = \frac{\sigma_0(1 - \nu^2)}{Ea}x + \frac{df}{dy} + \frac{dg}{dx} = 0 \tag{8.465}$$

We can write (8.465) as

$$\frac{dg}{dx} + \frac{\sigma_0(1 - \nu^2)}{Ea}x = -\frac{df}{dy} = c_1 = \text{const.} \tag{8.466}$$

Integrating (8.466) we obtain

$$g(x) = -\frac{\sigma_0(1 - \nu^2)x^2}{2Ea} + c_1 x + c_2$$

$$\tag{8.467}$$

$$f(y) = -c_1 y + c_3$$

where c_1, c_2 and c_3 are arbitrary constants. To determine these constants we make use of the conditions imposed on the displacements and the rotation at the origin, i.e.

$$u(0,0) = 0 \quad ; \quad v(0,0) = 0 \quad ; \quad \omega_{xy}(0,0) = 0 \tag{8.468}$$

which give

$$c_1 = c_2 = c_3 = 0 \tag{8.469}$$

The final expressions for the displacements are given by

$$u(x,y) = \frac{\sigma_0(1-\nu^2)xy}{Ea}$$

$$\tag{8.470}$$

$$v(x,y) = -\frac{\sigma_0\nu(1+\nu)y^2}{2Ea} - \frac{\sigma_0(1-\nu^2)x^2}{2Ea}$$

We can express the tractions acting at the ends of the rectangular region in terms of the moment per unit length in the z-direction by

$$M_0 = \int_{-a}^{a} y\,\sigma_{xx}(0,y)\,dy = \frac{2\sigma_0\,a^3}{3} \tag{8.471}$$

Also, the flexural moment of the inertia of the section about which M_0 acts is defined by

$$I = \frac{2a^3}{3} \tag{8.472}$$

The deflection of the rectangular region in the y-direction at $(L,0)$ can be expressed in the form

$$v(L,0) = -\frac{M_0(1-\nu^2)L^2}{2EI} \tag{8.473}$$

which agrees with that which can be obtained by considering the deflections of a thin prismatic beam where the flexural deflections are described by the *"Bernoulli-Euler"* beam theory; i.e.

$$\frac{EI}{(1-\nu^2)}\frac{d^2v}{dx^2} = M(x) = -M_0 \tag{8.474}$$

Integrating (8.474) and making use of the boundary conditions

$$v(0) = \left[\frac{dv}{dx}\right]_{x=0} = 0 \tag{8.475}$$

we obtain

$$v(x) = -\frac{M_0(1-\nu^2)x^2}{2EI} \tag{8.476}$$

It should be noted that the *exact closed form solution* for the bending of a rectangular region is obtained only in the very special case when the boundary shear tractions are zero and the boundary normal tractions vary in a linear fashion as specified by (8.458). For any other distribution of bending stresses, the flexure of the rectangular region can induce normal stresses σ_{yy} and shear stresses σ_{xy} which can vary within the beam region.

Example 8.8

The state of plane stress in a thin plate occupying the region $x \in (-a, a)$, $y \in (0, a)$ is given by

$$\sigma_{xx} = \sigma_0 \frac{x^3 y}{a^4} - 2C_1 \frac{xy}{a^2} + C_2 \frac{y}{a}$$

$$\sigma_{yy} = \sigma_0 \frac{xy^3}{a^4} - 2\sigma_0 \frac{x^3 y}{a^4} \tag{8.477}$$

$$\sigma_{xy} = -\frac{3}{2}\sigma_0 \frac{x^2 y^2}{a^4} + C_1 \frac{y^2}{a^2} + \frac{\sigma_0}{2}\frac{x^4}{a^4} + C_3$$

where σ_0 has units of stress and C_1, C_2 and C_3 are arbitrary constants. Determine the Airy stress function corresponding to the above stress state and the specific form of the stresses which gives zero shear tractions on $x = \pm a$ and zero normal traction on $x = -a$.

Solution

From the relationship between the Airy stress function and the stresses we have

$$\frac{\partial^2 \varphi}{\partial x^2} = \sigma_0 \frac{xy^3}{a^4} - 2\sigma_0 \frac{x^3 y}{a^4} \tag{8.478}$$

$$\frac{\partial^2 \varphi}{\partial y^2} = \sigma_0 \frac{x^3 y}{a^4} - 2C_1 \frac{xy}{a^2} + C_2 \frac{y}{a} \tag{8.479}$$

$$\frac{\partial^2 \varphi}{\partial x \partial y} = \frac{3}{2}\sigma_0 \frac{x^2 y^2}{a^4} - C_1 \frac{y^2}{a^2} - \frac{\sigma_0}{2}\frac{x^4}{a^4} - C_3 \tag{8.480}$$

Integrating (8.478) to (8.480) we obtain

$$\varphi(x, y) = \frac{\sigma_0 x^3 y^3}{6a^4} - \frac{\sigma_0 x^5 y}{10a^4} + x\widehat{F}(y) + F(y) \tag{8.481}$$

$$\varphi(x, y) = \frac{\sigma_0 x^3 y^3}{6a^4} - \frac{C_1 x y^3}{3a^2} + \frac{C_2 y^3}{6a} + y\widehat{G}(x) + G(x) \tag{8.482}$$

$$\varphi(x, y) = \frac{\sigma_0 x^3 y^3}{6a^4} - \frac{\sigma_0 x^5 y}{10a^4} - \frac{C_1 x y^3}{3a^2} - C_3 xy + L(x) + M(y) \tag{8.483}$$

where $\widehat{F}(y)$, $F(y)$, $\widehat{G}(x)$, $G(x)$, etc. are arbitrary functions . From an inspection of the results (8.481) to (8.483) it is evident that the appropriate form of the Airy stress function is

$$\varphi(x, y) = \frac{\sigma_0 x^3 y^3}{6a^4} - \frac{\sigma_0 x^5 y}{10a^4} - \frac{C_1 x y^3}{3a^2} + \frac{C_2 y^3}{6a} - C_3 xy \tag{8.484}$$

where the arbitrary functions are set equal to zero. Considering the zero shear traction boundary conditions on $x = \pm a$ we have

$$\sigma_{xy}(\pm a, y) = \left(-\frac{3}{2}\sigma_0 + C_1\right)\frac{y^2}{a^2} + \left(\frac{\sigma_0}{2} + C_3\right) = 0 \tag{8.485}$$

Similarly, the zero normal traction boundary condition on $x = a$ gives

$$\sigma_{xx}(-a, y) = (-\sigma_0 + 2C_1 + C_2)\frac{y}{a} = 0 \tag{8.486}$$

From (8.485) and (8.486) we have

$$C_1 = \frac{3}{2}\sigma_0 \quad ; \quad C_2 = -2\sigma_0 \quad ; \quad C_3 = -\frac{1}{2}\sigma_0 \tag{8.487}$$

and the final form of the stress components are

$$\sigma_{xx} = \sigma_0 \left[\frac{x^3 y}{a^4} - 3\frac{xy}{a^2} - 2\frac{y}{a} \right]$$

$$\sigma_{yy} = \sigma_0 \left[\frac{xy^3}{a^4} - 2\frac{x^3 y}{a^4} \right] \tag{8.488}$$

$$\sigma_{xy} = \sigma_0 \left[-\frac{3}{2}\frac{x^2 y^2}{a^4} + \frac{1}{2}\frac{x^4}{a^4} + \frac{3}{2}\frac{y^2}{a^2} - \frac{1}{2} \right]$$

It can be verified that (8.488) satisfies the stress compatibility equation

$$\nabla^2(\sigma_{xx} + \sigma_{yy}) = 0 \tag{8.489}$$

$$\bullet \; \bullet \; \bullet$$

It may be noted that many early solutions to elasticity problems were found this way; i.e. the class of solutions were matched with the appropriate problem by essentially following the procedure outlined in this example.

Example 8.9

An elastic *cantilever* of infinite extent in the z-direction is "supported" at the end $x = L$ such that the displacements $u(L, 0)$ and $v(L, 0)$ are zero at the supported end and the slope of the central plane of the cantilever $(y = 0)$ at the location $x = L$ is specified by the condition

$$\frac{\partial v}{\partial x} = 0 \qquad \text{at} \qquad x = L \quad , \quad y = 0 \tag{8.490}$$

The end $x = 0$ is subjected to shear tractions which are equivalent to a resultant shear force P/unit length in the z-direction (Figure 8.29). Considering the Airy stress function

$$\varphi(x,y) = C_{11}xy + C_{13}xy^3 \tag{8.491}$$

determine the displacement field in the region corresponding to the zero displacement and zero slope boundary condition specified by (8.490).

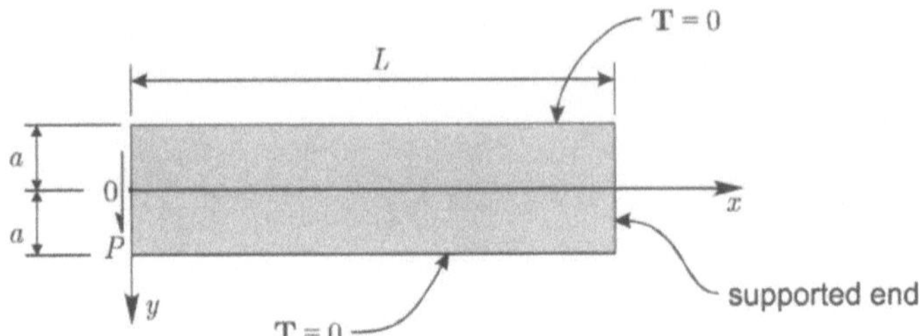

Figure 8.29: Edge loading of a cantilever region.

Solution

It should be noted that the cantilever region is "supported" along the edge $x = L$ to preserve overall equilibrium (both force and moment equilibrium) of the rectangular region. The nature of the support is unspecified; the rigid body displacements and rotations are utilized to satisfy the specified boundary conditions at the "supported" end $x = L$.

The traction boundary conditions applicable to the cantilever are as follows:

$$\sigma_{xx}(0,y) = 0 \tag{8.492}$$

$$\sigma_{yy}(x,a) = \sigma_{yy}(x,-a) = 0 \tag{8.493}$$

$$\sigma_{xy}(x,a) = \sigma_{xy}(x,-a) = 0 \tag{8.494}$$

Furthermore, the resultant of shear tractions on $x = 0$ must be equivalent to the shear force P, i.e.

$$\int_{-a}^{a} \sigma_{xy}(0,y)dy = -P \tag{8.495}$$

The negative sign of P indicates that applied shear stresses which contribute to P are negative (see e.g. the sign convention in Section 8.3.5). The stresses derived from the Airy stress function are

$$\sigma_{xx} = 6C_{13}xy \quad ; \quad \sigma_{yy} = 0$$
$$\sigma_{xy} = -(C_{11} + 3C_{13}y^2) \tag{8.496}$$

The plane stress field (8.496) exactly satisfies boundary conditions (8.492) and (8.493); the boundary conditions (8.494) yield

$$C_{11} + 3C_{13}a^2 = 0 \quad \text{or} \quad C_{13} = -\frac{C_{11}}{3a^2} \tag{8.497}$$

The constant C_{11} can be determined by considering the stress resultant boundary condition (8.495); i.e.

$$-P = \int_{-a}^{a} -C_{11}\left(1 - \frac{y^2}{a^2}\right) dy = -\frac{4}{3}C_{11}a \tag{8.498}$$

The stress state (8.496) can now be written as

$$\sigma_{xx} = -\frac{Pxy}{I} \quad ; \quad \sigma_{yy} = 0 \quad ; \quad \sigma_{xy} = -\frac{P}{2I}(a^2 - y^2) \tag{8.499}$$

where

$$I = \frac{2}{3}a^3 \tag{8.500}$$

is the moment of inertia, of a unit length of the cantilever in the z-direction, about the axis of flexure (i.e. the z-axis). Considering the stress state (8.499) and the stress-strain relations for plane strain deformations we have

$$\epsilon_{xx} = \frac{\partial u}{\partial x} = -\frac{P(1 - \nu^2)}{EI}xy \tag{8.501}$$

$$\epsilon_{yy} = \frac{\partial v}{\partial y} = \frac{P\nu(1 + \nu)}{EI}xy \tag{8.502}$$

Integrating (8.501) and (8.502) we have

$$u(x,y) = -\frac{P(1-\nu^2)}{2EI}x^2y + f(y) \tag{8.503}$$

$$v(x,y) = \frac{P\nu(1+\nu)}{2EI}xy^2 + g(x) \tag{8.504}$$

Considering (8.503), (8.504) and the shear strain-displacement relation we have

$$\frac{\partial u}{\partial y} + \frac{\partial v}{\partial x} = -\frac{P(1-\nu^2)x^2}{2EI} + \frac{P\nu(1+\nu)y^2}{2EI} + \frac{df}{dy} + \frac{dg}{dx}$$

$$= -\frac{P(1+\nu)}{EI}(a^2 - y^2) = \frac{2\sigma_{xy}(1+\nu)}{E} \tag{8.505}$$

Rearranging (8.505) we can write

$$G(x) + F(y) = C = -\frac{P(1+\nu)a^2}{EI} \tag{8.506}$$

where

$$G(x) = \frac{dg}{dx} - \frac{P(1-\nu^2)x^2}{2EI} = c_1 \tag{8.507}$$

$$F(y) = \frac{df}{dy} - \frac{P(1+\nu)(2-\nu)}{2EI}y^2 = c_2 \tag{8.508}$$

where

$$c_1 + c_2 = C \tag{8.509}$$

Integrating (8.507) and (8.508) we have

$$g(x) = \frac{P(1-\nu^2)x^3}{6EI} + c_1x + c_3 \tag{8.510}$$

$$f(y) = \frac{P(1+\nu)(2-\nu)y^3}{6EI} + c_2y + c_4 \tag{8.511}$$

where c_3 and c_4 are arbitrary constants. The complete expressions for the displacements can be written as

$$u(x,y) = -\frac{P(1-\nu^2)x^2y}{2EI} + \frac{P(1+\nu)(2-\nu)y^3}{6EI} + c_2y + c_4 \qquad (8.512)$$

$$v(x,y) = \frac{P\nu(1+\nu)xy^2}{2EI} + \frac{P(1-\nu^2)x^3}{6EI} + c_1x + c_3 \qquad (8.513)$$

To determine the constants c_1, c_2, c_3 and c_4 we make use of (8.509) and the displacement and rotation boundary conditions at $x = L, y = 0$.

$$u(L,0) = c_4 = 0$$

$$v(L,0) = \frac{P(1-\nu^2)L^3}{6EI} + c_1L + c_3 = 0 \qquad (8.514)$$

$$\left[\frac{\partial v}{\partial x}\right]_{\substack{x=L \\ y=0}} = \frac{P(1-\nu^2)L^2}{2EI} + c_1 = 0$$

These equations yield

$$c_1 = -\frac{P(1-\nu^2)L^2}{2EI} \qquad ; \qquad c_2 = -\frac{P(1+\nu)a^2}{EI} + \frac{P(1-\nu^2)L^2}{2EI}$$
$$\qquad (8.515)$$
$$c_3 = \frac{P(1-\nu^2)L^3}{3EI} \qquad ; \qquad c_4 = 0$$

The final expressions for the displacement components are given by

$$u(x,y) = \frac{P(1+\nu)}{EI}\left[-\frac{(1-\nu)x^2y}{2} + \frac{(2-\nu)y^3}{6}\right.$$
$$\left. -\left\{a^2 + (1-\nu)\frac{L^2}{2}\right\}y\right] \qquad (8.516)$$

$$v(x,y) = \frac{P(1+\nu)}{EI}\left[\frac{\nu x^2y}{2} + (1-\nu)\left\{\frac{x^3}{6} - \frac{xL^2}{2} + \frac{L^3}{3}\right\}\right] \qquad (8.517)$$

The displacement in the y-direction at the loaded end at the neutral axis is given by

$$v(0,0) = \frac{P(1-\nu^2)L^3}{3EI} \tag{8.518}$$

This is in agreement with the equivalent result obtained by considering the flexural deflections of a "*Bernoulli-Euler*" thin prismatic plate which is fixed against displacement and rotation at the location $x = L$ and subjected to a shear force P at the location $x = 0$. The appropriate equation governing the deflection $v(x)$ is given by

$$\frac{EI}{(1-\nu^2)}\frac{d^2v}{dx^2} = M(x) = Px \tag{8.519}$$

Integrating (8.519) we have

$$\frac{EI}{(1-\nu^2)}v(x) = \frac{Px^3}{6} + d_1 x + d_2 \tag{8.520}$$

where d_1 and d_2 are constants. The boundary conditions

$$v(L) = \left[\frac{dv}{dx}\right]_{x=L} = 0 \tag{8.521}$$

gives

$$d_1 = -\frac{PL^2}{2} \quad ; \quad d_2 = \frac{PL^3}{3} \tag{8.522}$$

The deflection of the prismatic plate is given by

$$v(x) = \frac{P(1-\nu^2)}{EI}\left[\frac{x^3}{6} - \frac{xL^2}{2} + \frac{L^3}{3}\right] \tag{8.523}$$

which corresponds exactly to the second term of (8.517) (i.e. $v(x,0)$).

8.8.2 Variables separable solution - rectangular Cartesian coordinates

As remarked previously, the biharmonic equation (8.387) can be solved by a repeated application of variables separation procedures. The details of such a technique will be discussed in section 8.11.11. For the present purposes it is sufficient to note that solutions can, however, be derived for specified forms of $\varphi(x, y)$ which lend themselves to a direct solution via a variables separation technique. For example, if we assume that $\varphi(x, y)$ admits a solution of the form

$$\varphi(x, y) = \sin\left(\frac{m\pi x}{L}\right) F(y) \tag{8.524}$$

where m is an integer, L is a constant and $F(y)$ is only a function of y. The biharmonic equation now yields the following:

$$\left\{ \frac{d^4 F}{dy^4} - 2\alpha_m^2 \frac{d^2 F}{dy^2} + \alpha_m^4 F \right\} \sin(\alpha_m x) = 0 \tag{8.525}$$

where $\alpha_m = m\pi/L$. The solution for the ordinary differential equation for $F(y)$ can be obtained by a variety of techniques including the D-operator method; the resulting solution can be written in the form

$$F(y) = \left[A_m e^{\alpha_m y} + B_m e^{-\alpha_m y} + C_m y e^{\alpha_m y} + D_m y e^{-\alpha_m y} \right] \tag{8.526}$$

where A_m, B_m,etc., are arbitrary constants. The Airy stress function can be written as

$$\varphi(x, y) = \left[A_m e^{\alpha_m y} + B_m e^{-\alpha_m y} + C_m y e^{\alpha_m y} \right.$$
$$\left. + D_m y e^{-\alpha_m y} \right] \sin(\alpha_m x) \tag{8.527}$$

or, alternatively, the exponential terms can be replaced by hyperbolic trigonometric functions which gives

$$\varphi(x, y) = \left[A_m^* \cosh(\alpha_m y) + B_m^* \sinh(\alpha_m y) + C_m^* y \cosh(\alpha_m y) \right.$$
$$\left. + D_m^* y \sinh(\alpha_m y) \right] \sin(\alpha_m x) \tag{8.528}$$

where A_m^*, B_m^*, etc., are constants.

The stresses derived from (8.527) take the forms

$$\sigma_{xx} = \left[\alpha_m^2 A_m e^{\alpha_m y} + \alpha_m^2 B_m e^{-\alpha_m y} + \alpha_m(2 + \alpha_m y)C_m e^{\alpha_m y}\right.$$
$$\left. - \alpha_m(2 - \alpha_m y)D_m e^{-\alpha_m y}\right]\sin(\alpha_m x) \tag{8.529}$$

$$\sigma_{yy} = -\alpha_m^2\left[A_m e^{\alpha_m y} + B_m e^{-\alpha_m y} + C_m y e^{\alpha_m y}\right.$$
$$\left. + D_m y e^{-\alpha_m y}\right]\sin(\alpha_m x) \tag{8.530}$$

$$\sigma_{xy} = -\alpha_m\left[\alpha_m A_m e^{\alpha_m y} - \alpha_m B_m e^{-\alpha_m y} + (1 + \alpha_m y)C_m e^{-\alpha_m y}\right.$$
$$\left. + (1 - \alpha_m y)D_m e^{-\alpha_m y}\right]\cos(\alpha_m x) \tag{8.531}$$

Similarly, it can be shown that

$$\varphi(x, y) = \left[\widetilde{A}_m e^{\alpha_m y} + \widetilde{B}_m e^{-\alpha_m y} + \widetilde{C}_m y e^{\alpha_m y}\right.$$
$$\left. + \widetilde{D}_m y e^{-\alpha_m y}\right]\cos(\alpha_m x) \tag{8.532}$$

is also an Airy stress function which yields the stress components

$$\sigma_{xx} = \left[\alpha_m^2 \widetilde{A}_m e^{\alpha_m y} + \alpha_m^2 \widetilde{B}_m e^{-\alpha_m y} + \alpha_m(2 + \alpha_m y)\widetilde{C}_m e^{\alpha_m y}\right.$$
$$\left. - \alpha_m(2 - \alpha_m y)\widetilde{D}_m e^{-\alpha_m y}\right]\cos(\alpha_m x) \tag{8.533}$$

$$\sigma_{yy} = -\alpha_m^2\left[\widetilde{A}_m e^{\alpha_m y} + \widetilde{B}_m e^{-\alpha_m y} + \widetilde{C}_m y e^{\alpha_m y}\right.$$
$$\left. + \widetilde{D}_m y e^{-\alpha_m y}\right]\cos(\alpha_m x) \tag{8.534}$$

$$\sigma_{xy} = \alpha_m\left[\alpha_m \widetilde{A}_m e^{\alpha_m y} - \alpha_m \widetilde{B}_m e^{-\alpha_m y} + (1 + \alpha_m y)\widetilde{C}_m y e^{\alpha_m y}\right.$$
$$\left. + (1 - \alpha_m y)\widetilde{D}_m e^{-\alpha_m y}\right]\sin(\alpha_m x) \tag{8.535}$$

where $\widetilde{A}_m, \widetilde{B}_m$, etc., are arbitrary constants.

As is evident, the two-dimensional states of stress defined by (8.529) to (8.531) and (8.533) to (8.535) are valid for regions of finite extent where the stresses are expected to be bounded, except at points where the stress state can experience singular behaviour (e.g. localized tractions, discontinuities in the displacement boundary conditions, etc. For semi-infinite media which occupy $x \in (-\infty, \infty)$; $y \in (0, \infty)$, the boundedness of the stresses at infinity is assured if we set $A_m, C_m, \widetilde{A}_m$ and $\widetilde{C}_m$ equal to zero in the previous solutions. Alternatively, for a semi-infinite region where $y \in (-\infty, 0)$, the constants $B_m, D_m, \widetilde{B}_m$ and $\widetilde{D}_m$ can be set to zero to recover bounded stress fields as $y \to -\infty$. It should also be noted that in the development of these variables separable solutions, we have assumed a specific form where the dependence in x has a trigonometric form which automatically gives rise to biharmonic operator is symmetric in x and y, the procedure can be reversed to generate Airy stress functions of the form

$$\varphi(x, y) = \left[A_n e^{\alpha_n x} + B_n e^{-\alpha_n x} + C_n x e^{\alpha_n x} \right.$$
$$\left. + D_n x e^{-\alpha_n x} \right] \sin(\alpha_n y) \tag{8.536}$$

and

$$\varphi(x, y) = \left[\widetilde{A}_n e^{\alpha_n x} + \widetilde{B}_n e^{-\alpha_n x} + \widetilde{C}_n x e^{\alpha_n x} \right.$$
$$\left. + \widetilde{D}_n x e^{-\alpha_n x} \right] \cos(\alpha_n y) \tag{8.537}$$

The corresponding expressions for the stress components can be obtained by simply interchanging the variables x and y in the expressions (8.529) to (8.531) and (8.533) to (8.535) respectively. The application of the separation of variable technique involving harmonic functions of a single variable can be best illustrated through certain elementary problems involving domains of a semi-infinite $(x \in (-\infty, \infty); y \in (0, \infty))$ and strip $(x \in (-\infty, \infty); y \in (-a, a))$ type.

Example 8.10

The Figure 8.30 shows a semi-infinite elastic region where the surface is subjected to tractions

$$\mathbf{T}(x, 0) = -\sigma_0 \cos(\lambda x)\mathbf{i} \tag{8.538}$$

where λ is a constant. Determine the displacements of the semi-infinite region assuming plane strain behaviour of the region and vanishing displacements as $y \to \infty$ and bounded displacements as $x \to \infty$.

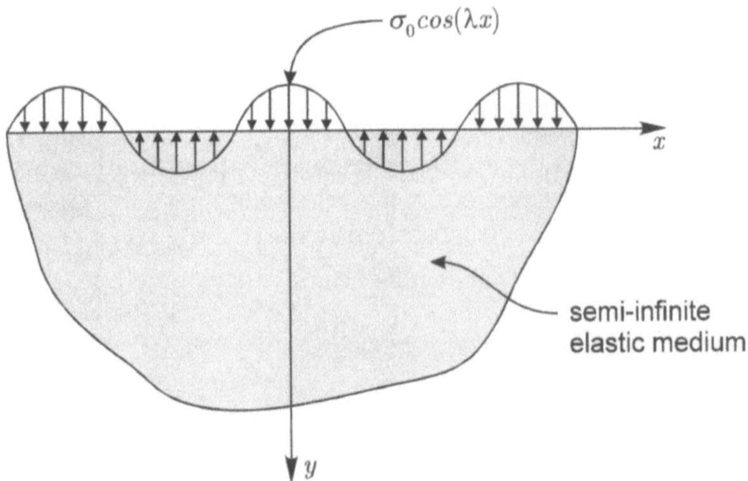

Figure 8.30: Sinusoidal surface loading of an elastic half-plane.

Solution

Since the region is semi-infinite and considering the form of the traction boundary conditi ons, the appropriate form of the Airy stress function is

$$\varphi(x,y) = \left[\widetilde{B} + \widetilde{D}y\right] e^{-\lambda y} \cos(\lambda x) \tag{8.539}$$

where $\widetilde{B}$ and $\widetilde{D}$ are to be determined by satisfying the traction boundary conditions (8.538) which are equivalent to

$$\sigma_{yy}(x,0) = -\sigma_0 \cos(\lambda x) \tag{8.540}$$

$$\sigma_{xy}(x,0) = 0 \tag{8.541}$$

The stresses derived from (8.539) are

$$\sigma_{xx} = \left[\lambda^2\widetilde{B} + \lambda(-2 + \lambda y)\widetilde{D}\right] e^{-\lambda y} \cos(\lambda x)$$

$$\sigma_{yy} = \left[-\lambda^2\widetilde{B} - \lambda^2 y\widetilde{D}\right] e^{-\lambda y} \cos(\lambda x) \tag{8.542}$$

$$\sigma_{xy} = \left[-\lambda^2\widetilde{B} + \lambda(1 - \lambda y)\widetilde{D}\right] e^{-\lambda y} \sin(\lambda x)$$

Considering (8.542), the boundary conditions (8.540) and (8.541) give

$$\widetilde{B} = \frac{\sigma_0}{\lambda^2} \quad ; \quad \widetilde{D} = \frac{\sigma_0}{\lambda} \tag{8.543}$$

The stress components (8.542) now reduce to

$$\sigma_{xx} = \sigma_0 \left[-1 + \lambda y \right] e^{-\lambda y} \cos(\lambda x)$$

$$\sigma_{yy} = \sigma_0 \left[-1 - \lambda y \right] e^{-\lambda y} \cos(\lambda x) \tag{8.544}$$

$$\sigma_{xy} = \sigma_0 \left[-\lambda y \right] e^{-\lambda y} \sin(\lambda x)$$

Considering the stress field (8.544) and the stress-strain relations for plane strain (see e.g. (8.371)) we have

$$\epsilon_{xx} = \frac{\partial u}{\partial x} = \frac{\sigma_0(1+\nu)}{E} \left[-(1-2\nu) + \lambda y \right] e^{-\lambda y} \cos(\lambda x) \tag{8.545}$$

Integrating (8.545) we have

$$u(x,y) = \frac{\sigma_0(1+\nu)}{\lambda E} \left[-(1-2\nu) + \lambda y \right] e^{-\lambda y} \sin(\lambda x) + f(y) \tag{8.546}$$

where $f(y)$ is an arbitrary function of y. Similarly integrating the expression for ϵ_{yy} we obtain

$$v(x,y) = \frac{\sigma_0(1+\nu)}{\lambda E} \left[2(1-\nu) + \lambda y \right] e^{-\lambda y} \cos(\lambda x) + g(x) \tag{8.547}$$

where $g(x)$ is an arbitrary function of x.

Considering the shear stress vs. shear strain relationship of (8.371) together with (8.546) and (8.547) gives

$$\left(\frac{\partial u}{\partial y} + \frac{\partial v}{\partial x} \right) = \frac{\sigma_0(1+\nu)}{E} \left[-2\lambda y e^{-\lambda y} \sin(\lambda x) \right] + \frac{df}{dy} + \frac{dg}{dx} \tag{8.548}$$

Comparing this result with the equivalent expression for $2\epsilon_{xy}$ derived from the last equation of (8.544) we note that

$$\frac{df}{dy} + \frac{dg}{dx} = 0 \tag{8.549}$$

Integrating (8.549) we have

$$f(y) = C_1 y + C_2 \quad ; \quad g(x) = -C_1 x + C_3 \tag{8.550}$$

where C_1, C_2 and C_3 are constants. For the displacements to be bounded as $x, y \to \infty$ we require $C_1 = 0$. Also for the displacements to vanish as $y \to \infty$ we require $C_3 = 0$. Finally, from the symmetric nature of the applied stress, $u(0, y) = 0$, and one concludes that $C_2 = 0$. The displacements of the semi-infinite region are given by

$$u(x, y) = \frac{\sigma_0(1 + \nu)}{\lambda E}\left[-(1 - 2\nu) + \lambda y\right] e^{-\lambda y} \sin(\lambda x)$$

$$\tag{8.551}$$

$$v(x, y) = \frac{\sigma_0(1 + \nu)}{\lambda E}\left[2(1 - \nu) + \lambda y\right] e^{-\lambda y} \cos(\lambda x)$$

The general result developed for the stresses within the half-plane region due to harmonic dependency in the stress function in one of the variables, can also be used to generate stress distributions induced by a periodic sequence of tractions distributed on the surface of the half-plane.

Example 8.11

The surface of a half-plane region is subjected to a train of normal stresses which are applied over a finite interval $2c$ (Figure 8.31). Derive an expression for the stress distribution within the half-plane region.

Solution

The period traction applied with periodicity $2a$ gives rise to the following traction boundary conditions

$$\sigma_{yy}(x, 0) = \begin{cases} -\sigma_0 \; ; \; -c < x < c \\ \\ 0 \; ; \; |x| > c \end{cases} \tag{8.552}$$

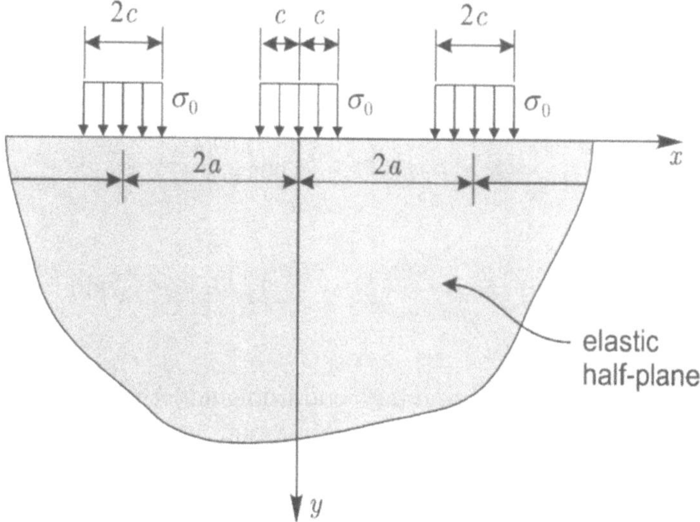

Figure 8.31: Periodic loading of the surface of an elastic half-plane.

and

$$\sigma_{xy}(x,0) = 0 \tag{8.553}$$

The discontinuous periodic function for $\sigma_{yy}(x,0)$ shown in Figure 8.31 can be expressed in the form of a Fourier series as follows (see e.g. (1.197)):

$$\sigma_{yy}(x,0) = -\frac{\sigma_0 c}{a}\left[1 + \frac{a}{\pi c}\sum_{n=1,2,}^{\infty}\frac{1}{n}\sin\left(\frac{n\pi c}{a}\right)\cos\left(\frac{n\pi x}{a}\right)\right] \tag{8.554}$$

Considering the Airy stress function applicable to the half-space region, and possessing the correct symmetry (e.g. (8.539)), we can generalize the result to form a Fourier series representation

$$\varphi(x,y) = \sum_{n=1,2,3}\left[\widetilde{B}_n + \widetilde{D}_n y\right]e^{-\frac{n\pi y}{a}}\cos\left(\frac{n\pi x}{a}\right) \tag{8.555}$$

where $\lambda_n = n\pi/a$.

The corresponding stress components are given by

$$\sigma_{xx} = \sum_{n=1,2,}^{\infty} \left[\left(\frac{n\pi}{a}\right)^2 \tilde{B}_n + \left(\frac{n\pi}{a}\right)\left(-2 + \frac{n\pi y}{a}\right)\tilde{D}_n \right] e^{-\frac{n\pi y}{a}} \cos\left(\frac{n\pi x}{a}\right)$$

$$\sigma_{yy} = \sum_{n=1,2,}^{\infty} \left(\frac{n\pi}{a}\right)^2 \left[-\tilde{B}_n - y\tilde{D}_n \right] e^{-\frac{n\pi y}{a}} \cos\left(\frac{n\pi x}{a}\right) \tag{8.556}$$

$$\sigma_{xy} = \sum_{n=1,2,}^{\infty} \left[-\left(\frac{n\pi}{a}\right)^2 \tilde{B}_n + \frac{n\pi}{a}\left(1 - \frac{n\pi y}{a}\right)\tilde{D}_n \right] e^{-\frac{n\pi y}{a}} \sin\left(\frac{n\pi x}{a}\right)$$

Considering the form of the boundary condition and the stresses derived from (8.555) we note that a further stress function

$$\varphi(x, y) = C_{20} x^2 \tag{8.557}$$

must be added to satisfy the constant stress part of the boundary condition (8.554). From (8.556) and the shear stress boundary condition (8.553) we have

$$\tilde{D}_n = \frac{n\pi}{a} \tilde{B}_n \tag{8.558}$$

Similarly, considering (8.556) and the traction boundary condition (8.554) we can show that

$$\sigma_{yy}(x, 0) = 2C_{20} - \sum_{n=1,2,}^{\infty} \left(\frac{n\pi}{a}\right)^2 \tilde{B}_n \cos\left(\frac{n\pi x}{a}\right)$$

$$= -\frac{\sigma_0 c}{a}\left[1 + \frac{a}{\pi c} \sum_{n=1,2,}^{\infty} \frac{1}{n} \sin\left(\frac{n\pi c}{a}\right) \cos\left(\frac{n\pi x}{a}\right) \right] \tag{8.559}$$

Equation (8.559) gives

$$C_{20} = -\frac{\sigma_0 c}{2a} \quad ; \quad \tilde{B}_n = \frac{\sigma_0}{a}\left(\frac{a}{n\pi}\right)^3 \sin\left(\frac{n\pi c}{a}\right) \tag{8.560}$$

The stress components in the half-plane can be evaluated in the form

$$\sigma_{xx} = \sigma_0 \sum_{n=1,2,}^{\infty} \left[-\frac{1}{n\pi} + \frac{y}{a} \right] e^{-\frac{n\pi y}{a}} \sin\left(\frac{n\pi c}{a}\right) \cos\left(\frac{n\pi x}{a}\right)$$

$$\sigma_{yy} = -\frac{\sigma_0 c}{2a} - \sigma_0 \sum_{n=1,2,}^{\infty} \left[\frac{1}{n\pi} + \frac{y}{a} \right] e^{-\frac{n\pi y}{a}} \sin\left(\frac{n\pi c}{a}\right) \cos\left(\frac{n\pi x}{a}\right) \quad (8.561)$$

$$\sigma_{xy} = -\sigma_0 \sum_{n=1,2,}^{\infty} \frac{y}{a} e^{-\frac{n\pi y}{a}} \sin\left(\frac{n\pi c}{a}\right) \cos\left(\frac{n\pi x}{a}\right)$$

It can be verified that the stress components (8.561) identically satisfy the equations of equilibrium (8.370) with $\mathbf{f} = 0$, and the resulting stress compatibility condition (8.375).

$$\bullet \ \bullet \ \bullet$$

The general form of the solutions for the Airy stress function presented previously can be applied to both semi-infinite and finite plane regions. In the following example we apply the general solution (e.g. of the form (8.528)) to examine the state of stress in a region of finite extent.

Example 8.12

Consider the problem of an elastic plane of finite dimensions $x \in (0, \ell)$ and $y \in (-a, a)$ (Figure 8.32) which is subjected only to the normal tractions

$$\sigma_{yy}(x, \pm a) = \sigma_0 \sin\left(\frac{2\pi x}{\ell}\right) \quad (8.562)$$

Determine the state of stress in the plane rectangular region.

Solution

The boundary conditions governing the problem include (8.562) and the additional conditions

$$\sigma_{xy}(x, a) = 0 \quad ; \quad \sigma_{xy}(x, -a) = 0 \quad (8.563)$$

$$\sigma_{xx}(0, y) = 0 \quad ; \quad \sigma_{xx}(\ell, y) = 0 \quad (8.564)$$

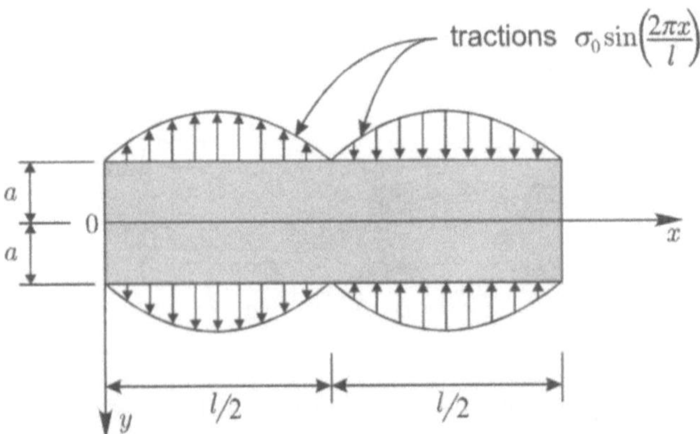

Figure 8.32: Harmonic loading of a plane rectangular region.

We have specified three sets of the traction boundary conditions on the edges of the rectangular region. The fourth boundary condition relates to the shear stresses acting on the planes $x = 0$ and $x = \ell$. If we observe the tractions that are applied on the surfaces $y = \pm a$, we note that the overall equilibrium of the region in the y-direction is preserved since the applied tractions are self equilibrating. Similarly, since no tractions are applied on the boundaries, in the x-direction, (i.e. boundary conditions (8.563) and (8.564)), the overall equilibrium of the body in the x-direction is also preserved. This leaves the tractions on the boundary of the region to satisfy the moment equilibrium condition. We note that since the actions of the applied tractions on $y = \pm a$ are equal and opposite, these tractions give a zero moment resultant at any point within the region. Hence the resultant of shear tractions on $x = 0$ and $x = \ell$ must be identically equal to zero; otherwise the moment of equilibrium condition would be violated. Hence the stress state derived must also satisfy the conditions

$$\int_{-a}^{a} \sigma_{xy}(0,y)dy = 0 \quad ; \quad \int_{-a}^{a} \sigma_{xy}(\ell,y)dy = 0 \tag{8.565}$$

i.e. the shear stresses on the planes $x = 0$ and $x = \ell$ themselves need not be zero but the resultants of the shear stresses must vanish to preserve the moment equilibrium condition. Considering the Airy stress function (8.527) the stresses can be obtained in the forms

$$\sigma_{xx} = \Big[\lambda^2 A e^{\lambda y} + \lambda^2 B e^{-\lambda y} + \lambda(2 + \lambda y)C e^{\lambda y}$$
$$-\lambda(2 - \lambda y)e^{-\lambda y}\Big]\sin(\lambda x) \tag{8.566}$$

$$\sigma_{yy} = -\lambda^2 \Big[A e^{\lambda y} + B e^{-\lambda y} + Cy e^{\lambda y} + Dy e^{-\lambda y}\Big]\sin(\lambda x) \tag{8.567}$$

$$\sigma_{xy} = -\lambda \Big[\lambda A e^{\lambda y} - \lambda B e^{-\lambda y} + (1 + \lambda y)C e^{\lambda y}$$
$$+(1 - \lambda y)D e^{-\lambda y}\Big]\cos(\lambda x) \tag{8.568}$$

where $\lambda = 2\pi/\ell$. We note that the boundary conditions (8.564) are identically satisfied by the stress component (8.566). The boundary conditions (8.563) and the expression (8.568) for σ_{xy}, give

$$\lambda A e^{\lambda a} - \lambda B e^{-\lambda a} + (1 + \lambda a)C e^{\lambda a} + (1 - \lambda a)D e^{-\lambda a} = 0$$
$$\tag{8.569}$$
$$\lambda A e^{-\lambda a} + \lambda B e^{\lambda a} + (1 - \lambda a)C e^{-\lambda a} + (1 + \lambda a)D e^{\lambda a} = 0$$

Similarly, the boundary conditions (8.562) and the expression (8.567) for σ_{yy}, give

$$A e^{\lambda a} + B e^{-\lambda a} + Ca e^{\lambda a} + Da e^{-\lambda a} = -\frac{\sigma_0}{\lambda^2}$$
$$\tag{8.570}$$
$$A e^{-\lambda a} + B e^{\lambda a} - Ca e^{-\lambda a} - Da e^{\lambda a} = -\frac{\sigma_0}{\lambda^2}$$

The sets of equations (8.569) and (8.570) can be solved for the unknown constants A, B, C and D. The results for the shear stress (8.568) can be evaluated in the form

$$\sigma_{xy} = 2\sigma_0 \cos(\lambda x) \tag{8.571}$$
$$\cdot \left[\frac{\lambda a(e^{\lambda a} + e^{-\lambda a})(e^{\lambda y} - e^{-\lambda y}) - \lambda y(e^{\lambda a} - e^{-\lambda a})(e^{\lambda y} + e^{-\lambda y})}{(e^{2\lambda a} - e^{-2\lambda a} - 4\lambda a)}\right]$$

Considering this result for σ_{xy} it can be shown that the conditions (8.565) are exactly satisfied.

8.8.3 Integral transform solution of the biharmonic equation governing plane problems

We have shown that a variables separable solution of the biharmonic equation governing two-dimensional plane problems can be obtained when the dependency in one of the coordinate variables is harmonic. This would suggest that Fourier transform techniques can be used to develop alternative representations of the solution. We define the *exponential Fourier transform* of the Airy stress function $\varphi(x, y)$ with respect to the variable x as follows:

$$\mathcal{F}\{\varphi(x, y); \xi\} = \widetilde{\varphi}(\xi, y) = \frac{1}{\sqrt{2\pi}} \int_{-\infty}^{\infty} \varphi(x, y) e^{i\xi x} dx \tag{8.572}$$

The corresponding inverse transform is given by

$$\mathcal{F}^{-1}\{\widetilde{\varphi}(\xi, y); x\} = \varphi(x, y) = \frac{1}{\sqrt{2\pi}} \int_{-\infty}^{\infty} \widetilde{\varphi}(\xi, y) e^{-i\xi x} d\xi \tag{8.573}$$

Considering the results given in Sections 1.12.2 and 5.5, and the homogeneous form of the biharmonic equation (e.g. (8.387)), we have

$$\mathcal{F}\left\{\left\{\frac{\partial^4 \varphi}{\partial y^4} + 2\frac{\partial^4 \varphi}{\partial x^2 \partial y^2} + \frac{\partial^4 \varphi}{\partial x^4}\right\}; \xi\right\}$$

$$= \frac{d^4 \widetilde{\varphi}(\xi, y)}{dy^4} - 2\xi^2 \frac{d^2 \widetilde{\varphi}(\xi, y)}{dy^2} + \xi^4 \widetilde{\varphi}(\xi, y)$$

$$= 0 \tag{8.574}$$

The solutions of $\widetilde{\varphi}(\xi, y)$ can be written as

$$\widetilde{\varphi}(\xi, y) = (A + By)e^{-|\xi|y} + (C + Dy)e^{+|\xi|y} \tag{8.575}$$

where A, B, C and D are generally functions of ξ and can be determined by applying the boundary conditions and regularity conditions applicable to a particular problem. From the Fourier inversion theorem, we can write

$$\varphi(x, y) = \frac{1}{\sqrt{2\pi}} \int_{-\infty}^{\infty} \left[(A + By)e^{-|\xi|y} + (C + Dy)e^{+|\xi|y}\right] e^{-i\xi x} d\xi \tag{8.576}$$

and the stress components can be obtained in the form

$$\sigma_{xx} = \frac{\partial^2 \varphi}{\partial y^2} = \frac{1}{\sqrt{2\pi}} \int_{-\infty}^{\infty} \frac{d^2 \widetilde{\varphi}}{dy^2} e^{-i\xi x} d\xi$$

$$\sigma_{yy} = \frac{\partial^2 \varphi}{\partial x^2} = -\frac{1}{\sqrt{2\pi}} \int_{-\infty}^{\infty} \xi^2 e^{-i\xi x} d\xi \qquad (8.577)$$

$$\sigma_{xy} = -\frac{\partial^2 \varphi}{\partial x \partial y} = \frac{1}{\sqrt{2\pi}} \int_{-\infty}^{\infty} i\xi \frac{d\widetilde{\varphi}}{dy} e^{-i\xi x} d\xi$$

Considering, for example, the plane strain deformations of the elastic medium, we have

$$\epsilon_{xx} = \frac{\partial u}{\partial x} = \frac{(1-\nu^2)}{E}\sigma_{xx} - \frac{\nu(1+\nu)}{E}\sigma_{yy} \qquad (8.578)$$

Applying the exponential Fourier transform to (8.578) we have

$$\mathcal{F}\left\{\frac{\partial u}{\partial x}; \xi\right\} = \frac{(1-\nu^2)}{E}\mathcal{F}\left\{\sigma_{xx}(x,y); \xi\right\}$$

$$-\frac{\nu(1+\nu)}{E}\mathcal{F}\left\{\sigma_{yy}(x,y); \xi\right\} \qquad (8.579)$$

If $u(x,y) \to 0$ as $|x| \to \infty$, then we can show that

$$\mathcal{F}\left\{\frac{\partial u}{\partial x}; \xi\right\} = -\frac{1}{\sqrt{2\pi}} \int_{-\infty}^{\infty} i\xi u(x,y)e^{i\xi x} dx = -i\xi \widetilde{u}(\xi, y) \qquad (8.580)$$

Obtaining the Fourier transforms of σ_{xx} and σ_{yy} from (8.577) and substituting them in (8.579) we can write

$$-i\xi \widetilde{u}(\xi, y) = \frac{(1-\nu^2)}{E} \frac{d^2 \widetilde{\varphi}}{dy^2} + \frac{\nu(1+\nu)}{E}\xi^2 \widetilde{\varphi} \qquad (8.581)$$

Inverting (8.581) we obtain

$$u(x,y) = \frac{(1+\nu)}{\sqrt{2\pi}E} \int_{-\infty}^{\infty} \left[(1-\nu)\frac{d^2\widetilde{\varphi}}{dy^2} + \nu\xi^2\widetilde{\varphi}\right] \frac{ie^{-i\xi x}}{\xi} d\xi \qquad (8.582)$$

Considering the shear stress-shear strain relationship we can show that

$$\frac{\partial v}{\partial x} = \frac{2(1+\nu)}{E}\sigma_{xy} - \frac{\partial u}{\partial y} \tag{8.583}$$

Again, assuming that $v(x,y) \to 0$ as $|x| \to \infty$, we have

$$-i\xi\widetilde{v}(\xi,y) = \frac{2(1+\nu)}{E}i\xi\frac{d\widetilde{\varphi}}{dy}$$
$$-\frac{i}{\xi}\left[\frac{(1-\nu^2)}{E}\frac{d^3\widetilde{\varphi}}{dy^3} + \frac{\nu(1+\nu)}{E}\xi^2\frac{d\widetilde{\varphi}}{dy}\right] \tag{8.584}$$

Applying the Fourier inversion theorem to (8.584) we obtain

$$v(x,y) = \frac{(1+\nu)}{\sqrt{2\pi}E}\int_{-\infty}^{\infty}\left[(1-\nu)\frac{d^3\widetilde{\varphi}}{dy^3} - (2-\nu)\xi^2\frac{d\widetilde{\varphi}}{dy}\right]\frac{e^{-i\xi x}}{\xi^2}d\xi \tag{8.585}$$

The developments presented here, in terms of the exponential Fourier transform can be used to recover specialized cases involving the Fourier sine transform and the Fourier cosine transform. These can be used to examine problems which display either asymmetry or symmetry, respectively, about the y-axis.

Example 8.13

The surface of a semi-infinite elastic plane is subjected to a normal traction $p(x)$ (Figure 8.33). Use the Fourier transform technique to determine formal integral expressions for the stresses and displacements of the half-plane.

Solution

Since the region is semi-infinite, traction boundary conditions can be prescribed only along the surface $y = 0$. The traction boundary conditions are

$$\sigma_{yy}(x,0) = -p(x) \quad ; \quad -\infty < x < \infty \tag{8.586}$$

$$\sigma_{xy}(x,0) = 0 \quad ; \quad -\infty < x < \infty \tag{8.587}$$

In addition to these boundary conditions, we shall impose the regularity condition that the stresses vanish as $|x| \to \infty$ or as $y \to \infty$ (The condition that the stresses vanish as $|x| \to \infty$ obviously constrains the form of $p(x)$). This does not imply that the resultant(s) of these stresses need vanish as e.g. $y \to \infty$. On any surface $y =$constant, the stress field should provide a non-zero total force which should balance the total force resultant of $p(x)$, to preserve equilibrium. Since the stresses decay to zero as $|x| \to \infty$ or $y \to \infty$, we can only be certain that the appropriate spatial derivatives of the displacements, which contribute to the strains also reduce to zero as $|x| \to \infty$ and/or $y \to \infty$.

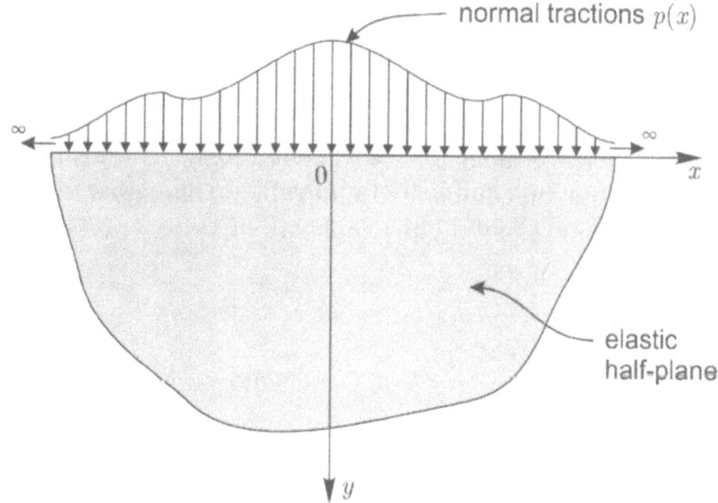

Figure 8.33: Surface loading of a half-plane.

To satisfy the requirement for vanishing stresses as $y \to \infty$, we require $C = D = 0$ in the result (8.575) for $\widetilde{\varphi}(\xi, y)$. Therefore the appropriate form for $\widetilde{\varphi}(\xi, y)$ is

$$\widetilde{\varphi}(\xi, y) = [A(\xi) + B(\xi)y]\, e^{-|\xi|y} \tag{8.588}$$

Considering the expression for the shear stress, in (8.577), we have

$$\sigma_{xy}(x, y) = \frac{1}{\sqrt{2\pi}} \int_{-\infty}^{\infty} [-|\xi|\, A(\xi) + B(\xi)$$
$$- |\xi|\, yB(\xi)]\, i\xi e^{-|\xi|y - i\xi x} d\xi \tag{8.589}$$

To satisfy the boundary condition (8.587) we require

$$|\xi|\,A(\xi) = B(\xi) \tag{8.590}$$

The stresses can now be written as

$$\sigma_{xx} = -\frac{1}{\sqrt{2\pi}}\int_{-\infty}^{\infty} \xi^2 A(\xi)\left[1 - |\xi|\,y\right]e^{-|\xi|y-i\xi x}d\xi$$

$$\sigma_{yy} = -\frac{1}{\sqrt{2\pi}}\int_{-\infty}^{\infty} \xi^2 A(\xi)\left[1 + |\xi|\,y\right]e^{-|\xi|y-i\xi x}d\xi \tag{8.591}$$

$$\sigma_{xy} = -\frac{i}{\sqrt{2\pi}}\int_{-\infty}^{\infty} \xi^3 A(\xi)ye^{-|\xi|y-i\xi x}d\xi$$

The solution for the stress field is now indeterminate to within a single function $A(\xi)$. This function can be determined by making use of the remaining boundary condition (8.586). The exponential Fourier transform of the applied stress is defined by

$$\widetilde{p}(\xi) = -\frac{1}{\sqrt{2\pi}}\int_{-\infty}^{\infty} p(x)e^{i\xi x}dx \tag{8.592}$$

Considering the value of $\sigma_{yy}(x,0)$ from (8.591) we have

$$\sigma_{yy}(x,0) = -\frac{1}{\sqrt{2\pi}}\int_{-\infty}^{\infty} \xi^2 A(\xi)e^{-i\xi x}d\xi \tag{8.593}$$

or

$$\widetilde{\sigma}_{yy}(\xi,0) = -\xi^2 A(\xi) = -\widetilde{p}(\xi) \tag{8.594}$$

The stress field (8.591) can now be written as

$$\sigma_{xx}(x,y) = -\frac{1}{\sqrt{2\pi}} \int_{-\infty}^{\infty} \widetilde{p}(\xi)\left[1 - |\xi|\,y\right] e^{-|\xi|y - i\xi x} d\xi$$

$$\sigma_{yy}(x,y) = -\frac{1}{\sqrt{2\pi}} \int_{-\infty}^{\infty} \widetilde{p}(\xi)\left[1 + |\xi|\,y\right] e^{-|\xi|y - i\xi x} d\xi \qquad (8.595)$$

$$\sigma_{xy}(x,y) = -\frac{1}{\sqrt{2\pi}} \int_{-\infty}^{\infty} iy\xi\widetilde{p}(\xi) e^{-|\xi|y - i\xi x} d\xi$$

Also if we assume that the expressions for the displacements given by (8.582) and (8.585) hold, then,

$$u(x,y) = \frac{(1+\nu)}{\sqrt{2\pi}E} \int_{-\infty}^{\infty} \widetilde{p}(\xi)\left[-(1-2\nu) + |\xi|\,y\right] \frac{ie^{-|\xi|y - i\xi x}}{\xi} d\xi \qquad (8.596)$$

and

$$v(x,y) = \frac{(1+\nu)}{\sqrt{2\pi}E} \int_{-\infty}^{\infty} \widetilde{p}(\xi)\left[2(1-\nu)\,|\xi| + \xi^2 y\right] \frac{e^{-|\xi|y - i\xi x}}{\xi^2} d\xi \qquad (8.597)$$

These results can be used to generate the stress and displacements in semi-infinite regions which can be subjected to generalized normal surface tractions. These solutions can also be used to evaluate separately cases where the surface normal loads exhibit either symmetry or asymmetry about the y-axis.

Example 8.14

Use the Fourier cosine transform approach to develop the stress distribution in a half-plane region, $x \in (-\infty, \infty)$, $y \in (0, \infty)$ where the surface $y = 0$ is subjected to a uniform compressive normal traction over the region $x(-a, a)$ (Figure 8.34).

Solution

Although the results developed in the previous example can be used to develop the appropriate results, we shall formulate the problem by considering

at the outset a development in terms of Fourier cosine transforms. We note
that the boundary conditions applicable to the problem are

$$\sigma_{yy}(x,0) = \begin{cases} -\sigma_0 \; ; \; -a < x < a \\[2mm] 0 \; ; \; |x| > a \end{cases} \tag{8.598}$$

$$\sigma_{xy}(x,0) = 0 \; ; \; -\infty < x < \infty \tag{8.599}$$

In addition, the stresses should satisfy the regularity condition, in that they
should vanish at infinity but contributes to a resultant force in the y-direction
of magnitude $2\sigma_0 a$, to preserve overall equilibrium of the region.

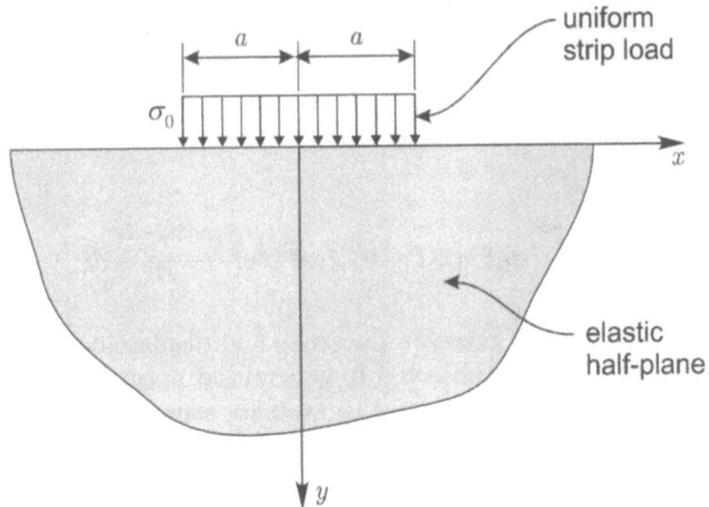

Figure 8.34: Loading of the surface of a half-plane by a uniform surface
load of finite width.

The Fourier cosine transform of the Airy stress function is defined by

$$\widetilde{\varphi}_c(\xi,y) = \mathcal{F}_c\left\{\varphi(x,y);\xi\right\} = \sqrt{\frac{2}{\pi}}\int_0^\infty \varphi(x,y)\cos(\xi x)dx \tag{8.600}$$

and the corresponding inversion theorem is

$$\varphi(x,y) = \mathcal{F}_c^{-1}\left\{\widetilde{\varphi}(\xi,y);x\right\} = \sqrt{\frac{2}{\pi}}\int_0^\infty \widetilde{\varphi}(\xi,y)\cos(\xi x)d\xi \tag{8.601}$$

Applying the Fourier cosine transform to the biharmonic equation, it can be shown that the general solution can be obtained in the form

$$\varphi(x,y) = \sqrt{\frac{2}{\pi}} \int_0^\infty \Big[\{A(\xi) + B(\xi)y\}e^{-\xi y}$$

$$+ \{C(\xi) + D(\xi)y\}e^{\xi y}\Big] \cos(\xi x)d\xi \tag{8.602}$$

where A, B, C and D are either constants or functions of ξ. It may be noted that since the transform parameter is defined over the interval $(0, \infty)$, the representation in terms of $|\xi|$ is not required. Considering the half-plane problem where $y \in (0, \infty)$ it is clear that for the stresses to vanish as $y \to \infty$, we require, as before, $C = D = 0$.

The boundary condition (8.599) governing the shear stresses on $y = 0$ gives

$$\sigma_{xy}(x,0) = \left[\sqrt{\frac{2}{\pi}} \int_0^\infty \xi\left[\xi A(\xi) - B(\xi)(1 - \xi y)\right] e^{-\xi y} \cos(\xi x)d\xi\right]_{y=0}$$

$$= 0 \tag{8.603}$$

which can be satisfied by setting

$$\xi A(\xi) = B(\xi) \tag{8.604}$$

The expressions for the stress components can now be written as

$$\sigma_{xx}(x,y) = \sqrt{\frac{2}{\pi}} \int_0^\infty \xi^2 A(\xi) \left[-1 + \xi y\right] e^{-\xi y} \cos(\xi x)d\xi$$

$$\sigma_{yy}(x,y) = \sqrt{\frac{2}{\pi}} \int_0^\infty \xi^2 A(\xi) \left[-1 - \xi y\right] e^{-\xi y} \cos(\xi x)d\xi \tag{8.605}$$

$$\sigma_{xy}(x,y) = \sqrt{\frac{2}{\pi}} \int_0^\infty \xi^2 A(\xi) \left[-\xi y\right] e^{-\xi y} \sin(\xi x)d\xi$$

The unknown function $A(\xi)$ can be determined by making use of the traction boundary condition (8.598). Considering the normal tractions on $y = 0$ to be a general symmetric function $p(x)$, the Fourier cosine transform can be written as

$$\widetilde{p}(\xi) = \sqrt{\frac{2}{\pi}} \int_0^\infty p(x)\cos(\xi x)\,dx \tag{8.606}$$

with its inverse

$$p(x) = \sigma_{yy}(x,0) = \sqrt{\frac{2}{\pi}} \int_0^\infty \widetilde{p}(\xi)\cos(\xi x)\,d\xi \tag{8.607}$$

Comparing (8.607) with the expression for $\sigma_{yy}(x,0)$ given by the second equation of (8.605) we obtain

$$A(\xi) = -\frac{p(\xi)}{\xi^2} \tag{8.608}$$

Using the expression for $\sigma_{yy}(x,0)$ defined by (8.598) in (8.606) we have

$$\widetilde{p}(\xi) = \sqrt{\frac{2}{\pi}} \int_0^\infty -\sigma_0\cos(\xi x)\,dx = -\sigma_0\sqrt{\frac{2}{\pi}}\frac{\sin(\xi a)}{\xi} \tag{8.609}$$

The expressions for the stresses, (8.605), can now be written as

$$\sigma_{xx} = \frac{2\sigma_0}{\pi} \int_0^\infty [-1+\xi y]e^{-\xi y}\frac{\sin(\xi a)}{\xi}\cos(\xi x)\,d\xi$$

$$\sigma_{yy} = \frac{2\sigma_0}{\pi} \int_0^\infty [-1-\xi y]e^{-\xi y}\frac{\sin(\xi a)}{\xi}\cos(\xi x)\,d\xi \tag{8.610}$$

$$\sigma_{xy} = \frac{2\sigma_0}{\pi} \int_0^\infty [-\xi y]e^{-\xi y}\frac{\sin(\xi a)}{\xi}\sin(\xi x)\,d\xi$$

8.8.4   Line load problems for an infinite plane

We consider the problem of an elastic infinite plane ($x \in (-\infty,\infty); y \in (-\infty,\infty)$) which is subjected to a concentrated force which acts at the origin of coordinates. In general, any such concentrated force which acts in an arbitrary orientation can be resolved into components which act in the coordinate directions. (Referring to Figure 8.35, it should be noted that the line loads P_x and P_y are expressed in force/unit length; the length being measured in the z-direction.)

The solutions to these interior loading problems represent the *influence functions* or *Green's functions* for the infinite plane which can be used to generate, by integration, solutions to both uniform and non-uniform force distributions within the infinite medium. In the interpretation of the solutions to these problems as Green's functions it should be understood that the solutions are not derived from an analysis of the inhomogeneous form of a biharmonic equation where the right hand side corresponds to a product of delta functions :(see e.g. (5.231) and (7.129)). This is due to the fact that both displacements and stress components involve derivatives of the Airy stress function and the direct solution of an inhomogeneous form of the bi-

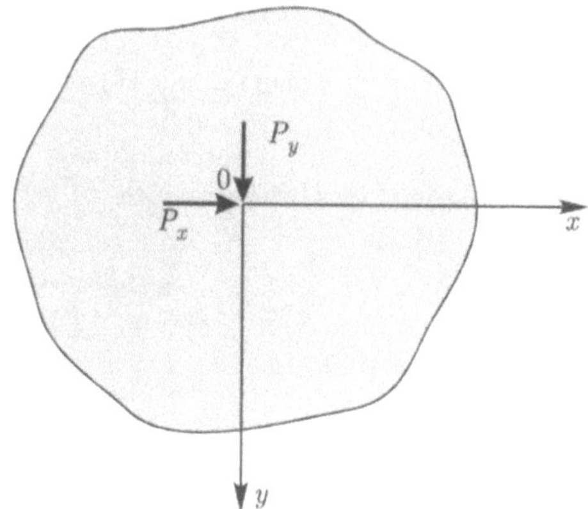

Figure 8.35: Concentrated line loads acting at the interior of an elastic plane of infinite extent.

harmonic equation does not lend itself to any physically meaningful interpretation of either the displacements or the stresses at the point of application of the delta functions. The type of problems which involve interior loading of a half-plane region can be examined by adopting a variety of solution schemes. In this exposition, however, we use the Fourier transform approach which is applied to the basic equations of equilibrium and compatibility. In considering the equations of equilibrium, we retain the contribution from the body forces and represent the localized forces as concentrated body forces of known magnitude. Since the elastic plane is of infinite extent no boundary conditions can be specified; we can, however, impose the regularity condition that $\sigma' \to 0$ as $|x|$ or $|y| \to \infty$. Referring to Chapter 1, Section 1.12.2h, we introduce the two-dimensional or double Fourier transform $\widetilde{F}(\xi, \eta)$ of a function $F(x, y)$ such that the function and its inverse are obtained from the relationships:

$$\widetilde{F}(\xi,\eta) = \frac{1}{2\pi} \int_{-\infty}^{\infty} \int_{-\infty}^{\infty} F(x,y) e^{i(\xi x + \eta y)} \, dx\, dy \tag{8.611}$$

$$F(x,y) = \frac{1}{2\pi} \int_{-\infty}^{\infty} \int_{-\infty}^{\infty} \widetilde{F}(\xi,\eta) e^{-i(\xi x + \eta y)} \, d\xi\, d\eta \tag{8.612}$$

We can now define the double Fourier transform of the stresses $\boldsymbol{\sigma}$ and the body forces $\mathbf{f}^b$ as

$$\left[\widetilde{\boldsymbol{\sigma}}(\xi,\eta); \widetilde{\mathbf{f}}^b(\xi,\eta)\right] = \frac{1}{2\pi} \int_{-\infty}^{\infty} \int_{-\infty}^{\infty} \left[\boldsymbol{\sigma}(x,y); \mathbf{f}^b(x,y)\right]$$
$$\cdot e^{i(\xi x + \eta y)} \, dx\, dy \tag{8.613}$$

Applying the double Fourier transform to the equations of equilibrium (i.e. by multiplying both sides of (8.370) by $e^{i(\xi x + \eta y)}$ and integrating over the entire plane) we obtain

$$\xi \widetilde{\sigma}_{xx} + \eta \widetilde{\sigma}_{xy} + i \widetilde{f}_x^b = 0 \tag{8.614}$$

$$\xi \widetilde{\sigma}_{xx} + \eta \widetilde{\sigma}_{yy} + i \widetilde{f}_y^b = 0 \tag{8.615}$$

The equations of compatibility (8.369) and the stress-strain relations for plane strain deformations (8.371), similarly give

$$\widetilde{\sigma}_{xx}\left[(1-\nu)\eta^2 - \nu\xi^2\right] + \widetilde{\sigma}_{yy}\left[(1-\nu)\xi^2 - \nu\eta^2\right] = 2\xi\eta\widetilde{\sigma}_{xy} \tag{8.616}$$

If we assume that $\widetilde{f}_x$ and $\widetilde{f}_y$ are known, then we can solve the algebraic equations (8.614) to (8.616) for the components of $\widetilde{\boldsymbol{\sigma}}$, i.e. we can write these as

$$\widetilde{\sigma}_{xx} + \widetilde{\sigma}_{yy} = -\frac{i\left[\xi\widetilde{f}_x^b + \eta\widetilde{f}_y^b\right]}{(1-\nu)(\xi^2 + \eta^2)} \tag{8.617}$$

$$\tilde{\sigma}_{xx} - \tilde{\sigma}_{yy} = -\frac{(1-2\nu)i\left(\xi^2 - \eta^2\right)\left[\xi\tilde{f}_x^b + \eta\tilde{f}_y^b\right]}{(1-\nu)(\xi^2 + \eta^2)^2}$$
$$-\frac{4i\xi\eta\left[\eta\tilde{f}_x^b - \xi\tilde{f}_y^b\right]}{(\xi^2 + \eta^2)^2} \tag{8.618}$$

$$\tilde{\sigma}_{xy} = -\frac{i\left(\xi^2 - \eta^2\right)\left[\xi\tilde{f}_x^b - \eta\tilde{f}_y^b\right]}{(\xi^2 + \eta^2)^2}$$
$$-\frac{(1-2\nu)i\xi\eta\left[\eta\tilde{f}_x^b + \xi\tilde{f}_y^b\right]}{(1-\nu)(\xi^2 + \eta^2)^2} \tag{8.619}$$

We can apply the inversion theorem to (8.617) which gives

$$\sigma_{xx} + \sigma_{yy} = -\frac{1}{2\pi(1-\nu)}\int_{-\infty}^{\infty}\int_{-\infty}^{\infty}\frac{i\left[\xi\tilde{f}_x^b + \eta\tilde{f}_y^b\right]}{(\xi^2 + \eta^2)}e^{-i(\xi x + \eta y)}d\xi\, d\eta \tag{8.620}$$

Since

$$\int_{-\infty}^{\infty}\int_{-\infty}^{\infty}\frac{i\xi e^{-i(\xi x + \eta y)}}{(\xi^2 + \eta^2)}d\xi\, d\eta = 4\int_{0}^{\infty}\xi\sin(\xi x)d\xi\int_{0}^{\infty}\frac{\cos(\eta y)d\eta}{(\eta^2 + \xi^2)} \tag{8.621}$$

we can apply the Faltung theorem to express a part of the integral of (8.620) as

$$\int_{-\infty}^{\infty}\int_{-\infty}^{\infty}\frac{i\xi\tilde{f}_x^b(\xi,\eta)e^{-i(\xi x + \eta y)}}{(\xi^2 + \eta^2)}d\xi\, d\eta$$
$$= \int_{-\infty}^{\infty}\int_{-\infty}^{\infty}\frac{(x - x^*)f_x^b(x^*,y^*)dx^*dy^*}{[(x - x^*)^2 + (y - y^*)^2]} \tag{8.622}$$

Similarly we can show that

$$\int_{-\infty}^{\infty}\int_{-\infty}^{\infty}\frac{i\eta\tilde{f}_y^b(\xi,\eta)e^{-i(\xi x + \eta y)}}{(\xi^2 + \eta^2)}d\xi\, d\eta$$
$$-\int_{-\infty}^{\infty}\int_{-\infty}^{\infty}\frac{(y - y^*)f_y^b(x^*,y^*)dx^*dy^*}{[(x - x^*)^2 + (y - y^*)^2]} \tag{8.623}$$

Hence the result for (8.620) can be written as

$$\sigma_{xx} + \sigma_{yy} = -\frac{1}{2\pi(1-\nu)} \int_{-\infty}^{\infty}$$
$$\left\{ \int_{-\infty}^{\infty} \frac{\left[(x-x^*)f_x^b(x^*,y^*) + (y-y^*)f_y^b(x^*,y^*)\right] dx^*}{[(x-x^*)^2 + (y-y^*)^2]} \right\} dy^* \tag{8.624}$$

In a similar manner, we can invert (8.618) to give

$$\sigma_{xx} - \sigma_{yy} = -\frac{(1-2\nu)}{\pi(1-\nu)} \int_{-\infty}^{\infty}$$
$$\left\{ \int_{-\infty}^{\infty} \frac{(x-x^*)(y-y^*)\left[(y-y^*)f_x^b(x^*,y^*) - (x-x^*)f_y^b(x^*,y^*)\right] dx^*}{[(x-x^*)^2 + (y-y^*)^2]^2} \right\} dy^*$$
$$-\frac{1}{\pi} \int_{-\infty}^{\infty}$$
$$\left\{ \int_{-\infty}^{\infty} \frac{[(x-x^*)^2 - (y-y^*)^2]\left[(x-x^*)f_x^b(x^*,y^*) + (y-y^*)f_y^b(x^*,y^*)\right] dx^*}{[(x-x^*)^2 + (y-y^*)^2]^2} \right\} dy^* \tag{8.625}$$

and the inversion of (8.619) gives

$$\sigma_{xy} = -\frac{(1-2\nu)}{4\pi(1-\nu)} \int_{-\infty}^{\infty}$$
$$\left\{ \int_{-\infty}^{\infty} \frac{[(y-x^*)^2 - (x-y^*)^2]\left[(y-y^*)f_x^b(x^*,y^*) - (x-x^*)f_y^b(x^*,y^*)\right] dx^*}{[(x-x^*)^2 + (y-y^*)^2]^2} \right\} dy^*$$
$$-\frac{1}{\pi} \int_{-\infty}^{\infty} \int_{-\infty}^{\infty} \frac{(x-x^*)(y-y^*)\left[(x-x^*)f_x^b(x^*,y^*) + (y-y^*)f_y^b(x^*,y^*)\right] dx^* dy^*}{[(x-x^*)^2 + (y-y^*)^2]^2} \tag{8.626}$$

We can now consider the special case when the body force distributions $\mathbf{f}(x^*, y^*)$ correspond to the concentrated line forces which are acting at the origin (Figure 8.35). In the special case when the force P_x is acting at the origin and directed along the positive x-direction,

$$f_x^b(x^*, y^*) = C\delta(x^*)\delta(y^*) \quad ; \quad f_y^b(x^*, y^*) = 0 \tag{8.627}$$

where $\delta(x), \delta(y)$ are the delta functions and C is a constant, which can be evaluated by considering the condition

$$\int_{-\infty}^{\infty}\int_{-\infty}^{\infty} f_x^b(x^*, y^*)dx^*dy^* = P_x \tag{8.628}$$

or $C = P_x$. The expressions (8.624) to (8.626) along with (8.627) can now be used to evaluate the stress components $\boldsymbol{\sigma}$. Performing the reductions we obtain

$$\sigma_{xx} = -\frac{xP_x}{4\pi(1-\nu)r^2}\left[(1-2\nu) + \frac{2x^2}{r^2}\right]$$

$$\sigma_{yy} = -\frac{xP_x}{4\pi(1-\nu)r^2}\left[(1+2\nu) - \frac{2x^2}{r^2}\right] \tag{8.629}$$

$$\sigma_{xy} = -\frac{yP_x}{4\pi(1-\nu)r^2}\left[(1-2\nu) + \frac{2x^2}{r^2}\right]$$

where $r^2 = (x^2 + y^2)$. Similarly by considering the separate action of the concentrated force P_y which acts in the positive y-direction, it can be shown that the stress distribution in the elastic half plane is given by

$$\sigma_{xx} = -\frac{yP_y}{4\pi(1-\nu)r^2}\left[(1+2\nu) - \frac{2y^2}{r^2}\right]$$

$$\sigma_{yy} = -\frac{yP_y}{4\pi(1-\nu)r^2}\left[(1-2\nu) + \frac{2y^2}{r^2}\right] \tag{8.630}$$

$$\sigma_{xy} = -\frac{xP_y}{4\pi(1-\nu)r^2}\left[(1-2\nu) + \frac{2y^2}{r^2}\right]$$

Since the interpretation of the *influence function* or *Green's function* is in relation to a physical problem where either traction (or displacements) are prescribed at some location, we can develop other forms of such functions which are applicable to specific domains. For example, we can develop a fundamental solution to the problem of a *half-space region* which is subjected to a concentrated force which acts normal to the plane boundary (Figure 8.36).

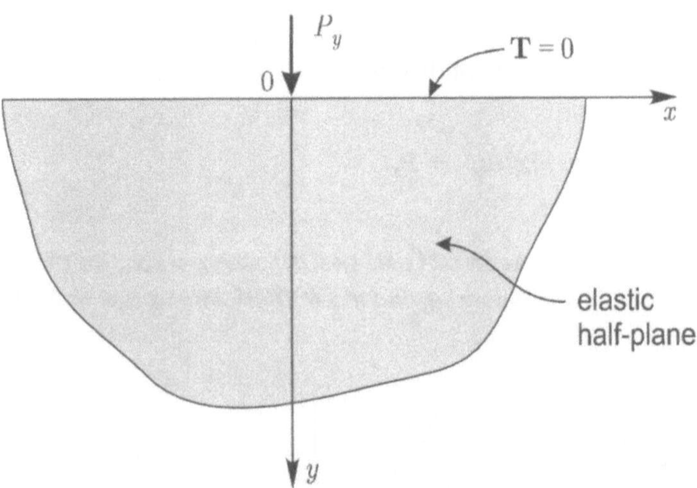

Figure 8.36: Concentrated normal load acting at the surface of a
half-plane.

The solution to this problem can be easily recovered from the results given
in Example 8.14 for the problem of a half-plane region the surface of which
is subjected to a uniform stress of intensity σ_0 and width $2a$. We consider
the limit of the Fourier transform of the loading given by (8.610) as $a \to 0$
such that $\sigma_0 = P_y/2a$, i.e. for the concentrated force P_y

$$\widetilde{p}(\xi) = \operatorname*{Lim}_{a \to 0} \frac{P_y}{2a} \sqrt{\frac{2}{\pi}} \frac{\sin(\xi a)}{\xi} = \frac{P_y}{\sqrt{2\pi}} \tag{8.631}$$

The integrals (8.611) for the stresses can now be reduced to the following
forms:

$$\sigma_{xx} = \frac{P_y}{\pi} \int_0^\infty [-1 + \xi y] e^{-\xi y} \cos(\xi x) d\xi = -\frac{2P_y\, x^2 y}{\pi(x^2 + y^2)^2}$$

$$\sigma_{yy} = \frac{P_y}{\pi} \int_0^\infty [-1 - \xi y] e^{-\xi y} \cos(\xi x) d\xi = -\frac{2P_y\, y^3}{\pi(x^2 + y^2)^2} \tag{8.632}$$

$$\sigma_{xy} = \frac{P_y}{\pi} \int_0^\infty [-\xi y] e^{-\xi y} \sin(\xi x) d\xi = -\frac{2P_y\, xy^2}{\pi(x^2 + y^2)^2}$$

Using a similar general approach, results can also be obtained for the problem
of a concentrated force of magnitude P_x which is directed along the x-axis
and localized at the origin of coordinates.

8.8.5 Application of complex variable methods

Complex variable methods outlined briefly in Section 8.7, together with the Theorems 8.6 to 8.8, presents a very effective method for obtaining solutions to plane problems in elasticity. The results presented in Section 8.7 can be summarized as follows; the Airy stress function $\varphi(z)$ can be presented in terms of two analytic function $\psi(z)$ and $\chi(z)$ such that

$$\sigma_{xx} + \sigma_{yy} = 2\left[\psi'(z) + \overline{\psi'(z)}\right]$$

$$\sigma_{yy} - \sigma_{xx} + 2i\sigma_{xy} = 2\left[\bar{z}\psi''(z) + \chi''(z)\right] \tag{8.633}$$

$$2\mu(u + iv) = \kappa\psi(z) - z\overline{\psi'(z)} - \overline{\chi'(z)}$$

where

$$\kappa = (3 - 4\nu) \qquad \text{for plane strain, and} \tag{8.634}$$

$$\kappa = \frac{(3 - \nu)}{(1 + \nu)} \qquad \text{for plane stress} \tag{8.635}$$

Here again, since the two analytic functions are quite general, we can let $\psi(z)$ and $\chi(z)$ be polynomials or power series in z or $(z)^{-1}$. The choice of the particular expansion will depend on the domain under consideration. For example, if the region includes the origin, only positive powers of z can be used if the functions are to remain bounded at the origin. Similarly, if the domain excludes the origin but extends to infinity, only negative powers can be used to ensure boundedness of the functions. It is instructive to examine certain special cases involving elementary forms of $\psi(z)$ and $\chi(z)$.

Example 8.15

Examine the state of stress and displacement field associated with the analytic functions

$$\psi(z) = c_1 z \quad ; \quad \chi(z) = c_2 z \tag{8.636}$$

where c_1 and c_2 are constants which can be complex-valued.

Solution

From (8.633) we have

$$\sigma_{xx} + \sigma_{yy} = 2\left[\psi'(z) + \overline{\psi'(z)}\right] = 2(c_1 + \bar{c}_1) \tag{8.637}$$

$$\sigma_{yy} - \sigma_{xx} + 2i\sigma_{xy} = 2\left[z\psi''(z) + \chi'(z)\right] = 2c_2 \tag{8.638}$$

since c_1 occurs in the form $(c_1 + \bar{c}_1)$, we can take the sum to be real, giving

$$\sigma_{xx} + \sigma_{yy} = 4c_1 \tag{8.639}$$

The results (8.638) and (8.639) indicate that the stress state is homogeneous and if the principal stresses σ_1 and σ_2 are such that σ_1 is inclined at β to the x-axis then

$$\sigma_{y'y'} - \sigma_{x'x'} + 2i\sigma_{x'y'} = \left[\sigma_{yy} - \sigma_{xx} + 2i\sigma_{xy}\right]e^{2i\beta} \tag{8.640}$$

Identifying $\sigma_{y'y'}$ and $\sigma_{x'x'}$ with σ_2 and σ_1 we have from (8.638) and (8.640),

$$c_1 = \frac{1}{4}(\sigma_1 + \sigma_2) \quad ; \quad c_2 = \frac{1}{2}(\sigma_2 - \sigma_1)e^{-2i\beta} \tag{8.641}$$

A number of special cases involving homogeneous stress states can be recovered from the above generalized result:

(i) For a hydrostatic state of stress of magnitude σ_0,

$$c_1 = \frac{\sigma_0}{2} \quad ; \quad c_2 = 0 \tag{8.642}$$

(ii) For a state of uniaxial stress σ_1 acting along the x-direction (i.e. $\beta = 0$)

$$c_1 = \frac{\sigma_1}{4} \quad ; \quad c_2 = -\frac{\sigma_1}{2} \tag{8.643}$$

(iii) For pure shear of magnitude τ_0

$$\sigma_2 = -\sigma_1 = \tau_0 \quad ; \quad \beta = \frac{\pi}{4} \quad ; \quad c_2 = i\tau_0 \tag{8.644}$$

From (8.633), the displacements are given by

$$2\mu(u + iv) = (\kappa - 1)c_1 z - \bar{c}_2 \bar{z}$$
(8.645)

Noting that c_1 is real, we have

$$2\mu u = (\kappa - 1)c_1 x - \mathrm{Re}(\bar{c}_2 \bar{z})$$

$$2\mu v = (\kappa - 1)c_1 y - \mathrm{Im}(\bar{c}_2 \bar{z})$$
(8.646)

In the special case when the axes are principal axes, we have $\beta = 0$ and (8.646) reduce to

$$8\mu\, u = \left[(\kappa + 1)\sigma_1 + (\kappa - 3)\sigma_2\right] x$$

$$8\mu\, v = \left[(\kappa - 3)\sigma_1 + (\kappa + 1)\sigma_2\right] y$$
(8.647)

Example 8.16

The surface of a half-plane region $(-\infty < y < \infty \;\; ; \; x \geq 0)$ is subjected to normal and tangential stresses σ_0 and τ_0 respectively (Figure 8.37). Use the complex potentials

$$\psi(z) = A\,\ln(z) \quad ; \quad \chi(z) = B\,\ln(z)$$
(8.648)

where A and B can be complex-valued constants, to determine the solution to the problem where surface of the half-plane is subjected to a concentrated line load with components P_x and P_y.

Solution

Considering the representation of the stresses σ_{xx}, σ_{yy} and σ_{xy} in terms of the Airy stress function $\varphi(x, y)$ we can write

$$\sigma_{yy} - \sigma_{xx} + 2i\sigma_{xy} = \frac{\partial^2 \varphi}{\partial x^2} - \frac{\partial^2 \varphi}{\partial y^2} - 2i\frac{\partial^2 \varphi}{\partial x \partial y}$$

$$(8.649)$$

$$\sigma_{xx} + \sigma_{yy} = \frac{\partial^2 \varphi}{\partial x^2} + \frac{\partial^2 \varphi}{\partial y^2}$$

Considering the tractions that are applied on the surface $x = 0$, we can evaluate the total load per unit length perpendicular to the $x - y$ plane, between the locations P and P_2 from the following:

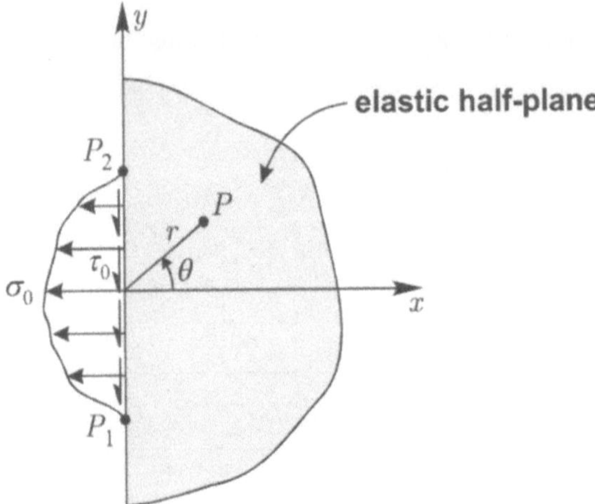

Figure 8.37: Surface loading of a half-plane region.

$$\int_{P_1}^{P_2} [\sigma_0 + i\tau_0]\,dy = \int_{P_1}^{P_2} [\sigma_{xx} - i\sigma_{xy}]\,dy \tag{8.650}$$

We can use the expressions (8.649) to develop an expression for the integrand on the right hand side of (8.650), in terms of the Airy stress function $\varphi(x, y)$, i.e.

$$\int_{P_1}^{P_2} [\sigma_0 + i\tau_0]\,dy = \int_{P_1}^{P_2} \left[\frac{\partial^2 \varphi}{\partial y^2} - i\frac{\partial^2 \varphi}{\partial x \partial y}\right]\,dy$$

$$= -i\left[\frac{\partial \varphi}{\partial x} + i\frac{\partial \varphi}{\partial y}\right]_{P_1}^{P_2} \tag{8.651}$$

Considering the representation of $\varphi(z)$ in terms of the complex potentials $\psi(z)$ and $\chi(z)$ we have

$$\varphi(z) = \mathrm{Re}[\bar{z}\psi(z) + \chi(z)] \tag{8.652}$$

or, alternatively

$$\varphi(z) = \frac{1}{2}\left[\bar{z}\psi(z) + z\overline{\psi(z)} + \chi(z) + \overline{\chi(z)}\right] \tag{8.653}$$

Performing the differentiation of $\varphi(z)$ given by (8.653) with respect to x and y, we obtain, respectively,

$$\frac{\partial\varphi}{\partial x} = \frac{1}{2}\left[\psi(z) + \bar{z}\psi'(z) + \overline{\psi(z)} + z\overline{\psi'(z)} + \chi'(z) + \overline{\chi'(z)}\right]$$
$$\tag{8.654}$$
$$\frac{\partial\varphi}{\partial y} = \frac{1}{2}\left[-i\psi(z) + i\bar{z}\psi'(z) + i\overline{\psi(z)} - iz\overline{\psi'(z)} + i\chi'(z) - i\overline{\chi'(z)}\right]$$

Substituting the expressions for $(\partial\varphi/\partial x)$ and $(\partial\varphi/\partial y)$ in (8.651) we obtain

$$\int_{P_1}^{P_2} [\sigma_0 + i\tau_0]\, dy = -i\left[\psi(z) + z\overline{\psi'(z)} + \overline{\chi(z)}\right]_{P_1}^{P_2} \tag{8.655}$$

Using the complex potentials given by (8.648) in (8.655) and identifying the locations P_1 and P_2 by the distances $r_1 e^{-i\pi/2}$ and $r_2 e^{i\pi/2}$, respectively, we obtain

$$\int_{P_1}^{P_2} [\sigma_0 + i\tau_0]\, dy = -i\left[A\ln\left(\frac{r_2}{r_1}\right) + i\pi A + \overline{B}\ln\left(\frac{r_2}{r_1}\right) - i\pi\overline{B}\right] \tag{8.656}$$

If the total force computed over the interval $P_1 P_2$ is to correspond to a force of finite magnitude, with components P_x and P_y, the terms containing $\ln(r_2/r_1)$ should vanish for all choices, i.e.

$$A + \overline{B} = 0 \quad ; \quad \pi(A - \overline{B}) = P_x + iP_y \tag{8.657}$$

Hence the complex potentials can be expressed in the forms

$$\psi(z) = \frac{(P_x + P_y)}{2\pi} \ln(z) \quad ; \quad \chi(z) = -\frac{(P_x - iP_y)}{2\pi} \ln(z) \tag{8.658}$$

In the particular instance when the surface of the half-plane region is subjected to a concentrated normal force P_x at the origin we have $P_y = 0$. To aid the evaluation of the expressions (8.649) we note that

$$\frac{\partial^2 \varphi}{\partial x^2} = \frac{1}{2} \left[2\psi'(z) + \bar{z}\psi''(z) + 2\overline{\psi'(z)} + z\overline{\psi''(z)} \right.$$

$$\left. + \chi''(z) + \overline{\chi''(z)} \right]$$

$$\frac{\partial^2 \varphi}{\partial y^2} = \frac{1}{2} \left[2\psi'(z) - \bar{z}\psi''(z) + 2\overline{\psi'(z)} - z\overline{\psi''(z)} \right. \tag{8.659}$$

$$\left. - \chi''(z) - \overline{\chi''(z)} \right]$$

$$\frac{\partial^2 \varphi}{\partial x \partial y} = \frac{1}{2} \left[i\bar{z}\psi''(z) - iz\overline{\psi''(z)} + i\chi''(z) - i\overline{\chi''(z)} \right]$$

Evaluating (8.649) we obtain

$$\sigma_{yy} - \sigma_{xx} + 2i\sigma_{xy} = \frac{P_x}{\pi r^2} \left\{ -\frac{(x - iy)^3}{r^2} - (x - iy) \right\}$$

$$\tag{8.660}$$

$$\sigma_{xx} + \sigma_{yy} = \frac{P_x}{\pi r^2} \left\{ 2x \right\}$$

where $r^2 = x^2 + y^2$. From these equations we obtain

$$[\sigma_{xx} \; ; \; \sigma_{yy} \; ; \; \sigma_{xy}] = \frac{2P_x}{\pi(x^2 + y^2)^2} \left[x^3 \; ; \; xy^2 \; ; \; x^2y \right] \tag{8.661}$$

The displacement components can be evaluated by using the complex potentials given by(8.658) and the third equation of (8.633). We can show that

$$2\mu(u + iv) = \frac{P_x}{2\pi} \left\{ (\kappa + 1)\ln r - \cos 2\theta + i[(\kappa - 1)\theta - \sin 2\theta] \right\} \tag{8.662}$$

These results are indeterminate to within an arbitrary rigid body displacement. It can be shown that

$$u(0,y) = \frac{(\kappa+1)P_x}{4\pi\mu}\ln(y) + \text{const.} \tag{8.663}$$

8.9 Polar coordinate formulation of the plane problem in elasticity

As remarked previously, it is advantageous to formulate the plane problem in elasticity in terms of orthogonal curvilinear coordinates especially when examining problems where the boundary conditions are prescribed on surfaces which correspond to constant values of the coordinate variables. Examples of curvilinear plane coordinate systems, include, plane polar coordinates, elliptical coordinates and bi-polar coordinates. These coordinate systems can be conveniently adopted to examine problems involving, respectively, circular boundaries, elliptical boundaries and combinations of plane and circular boundaries. Examples of typical applications are shown in Figure 8.38.

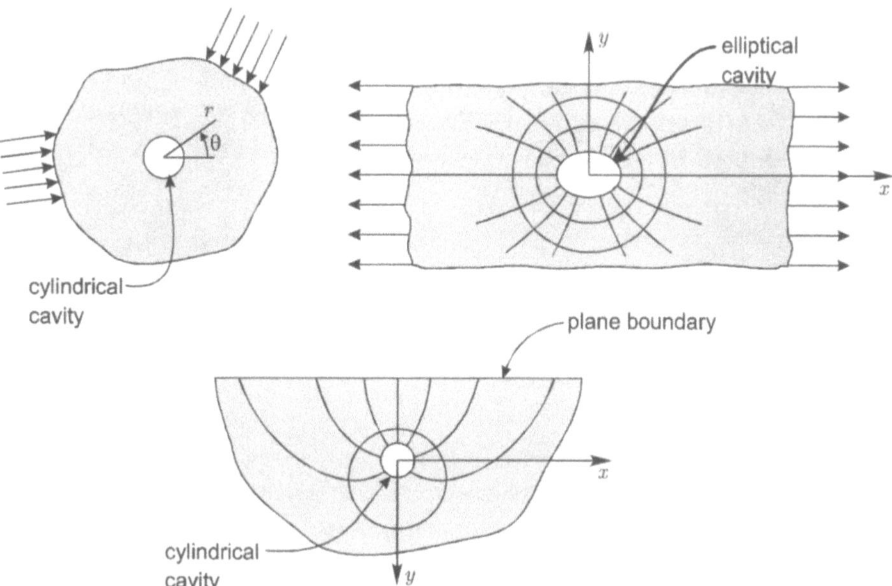

Figure 8.38: Application of curvilinear coordinate systems.

In this section we formulate the plane problem in elasticity by appeal to plane polar coordinates. There are several approaches to obtaining the governing equations applicable to a system of plane polar coordinates. One such method involves the transformation of the equations obtained previously with reference to the Cartesian coordinate system, by using the transformation rules applicable to orthogonal coordinate systems. This procedure is invariably lengthy but can be performed with speed and accuracy using currently available symbolic mathematical manipulation schemes such as MAPLE®, MATHEMATICA® and others. These transformation procedures themselves are insufficient to generate a physical understanding of the dependent variables such as stresses and strains referred to plane polar coordinates. For this reason, the alternative approach is to derive the relevant equation by considering descriptions of stresses and strains which are referred to the plane polar coordinate system. In the ensuing subsections, we shall present the physical interpretations of the stresses and strains referred to the plane polar coordinate system and present a brief outline of the transformation procedures use to generate certain specific results.

8.9.1   Definition of stresses in plane polar coordinates

The state of stress at a point within a continuum region which is referred to a plane polar coordinate system (r, θ) can be defined by considering an infinitesimal element bounded by surfaces where r and θ are constant. Figure 8.39 shows such an element and indicates the stresses acting on the respective surfaces. The stresses are again interpreted in limiting sense similar to the definition adopted in connection with the rectangular Cartesian coordinate system (8.136).

The state of stress in a two-dimensional plane associated with a state of plane strain is given by

$$\boldsymbol{\sigma}_P = \begin{bmatrix} \sigma_{rr} & \sigma_{r\theta} & 0 \\ \sigma_{\theta r} & \sigma_{\theta\theta} & 0 \\ 0 & 0 & \sigma_{zz} \end{bmatrix} \tag{8.664}$$

where the suffix 'P' designates reference to the system of plane polar coordinates (r, θ). In the absence of body couples or couples distributed within the region,

$$\boldsymbol{\sigma}_P = \boldsymbol{\sigma}_P^T \tag{8.665}$$

and the stress matrix is symmetric. For a state of generalized plane stress $\sigma_{zz} = 0$. The sign convention adopted for the definition of the state of stress (8.634) is identical to that described in Section 8.3.4, and all the stress

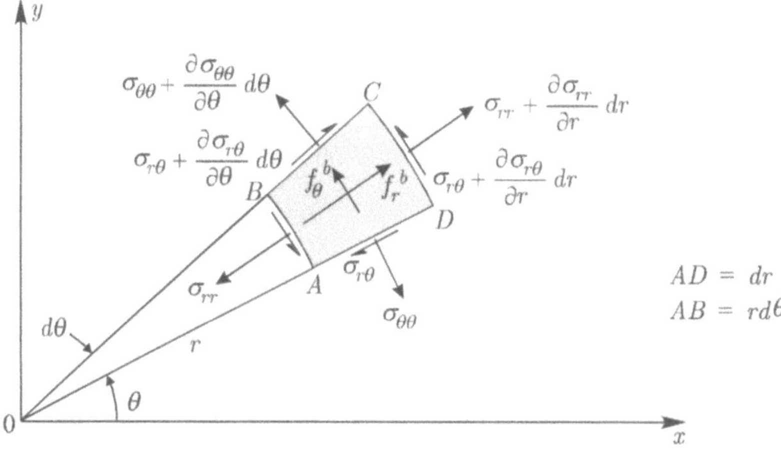

Figure 8.39: Stresses acting on an infinitesimal element; plane polar coordinates.

components shown in Figure 8.39 are positive. The resultants of the body force terms act in the radial (r) and tangential (θ) directions and are centered at the centroid of the element $ABCD$. The equations of equilibrium in the radial and tangential directions can be obtained by considering the equilibrium of *forces* in the respective directions. The summation of forces acting on an element of unit thickness in the z-direction and acting parallel to the radial direction through the center of the element gives

$$\left(\sigma_{rr} + \frac{\partial \sigma_{rr}}{\partial r} \right)(r + dr)\, d\theta - \sigma_{rr}(rd\theta) + f_r^b \, rdrd\theta$$
$$- \left(\sigma_{\theta\theta} + \frac{\partial \sigma_{\theta\theta}}{\partial \theta} d\theta \right) dr \sin\left(\frac{d\theta}{2} \right)$$
$$+ \left(\sigma_{r\theta} + \frac{\partial \sigma_{r\theta}}{\partial \theta} d\theta - \sigma_{r\theta} \right) dr \cos\left(\frac{d\theta}{2} \right) = 0 \tag{8.666}$$

If we introduce the approximations

$$\sin\left(\frac{d\theta}{2}\right) \simeq \frac{d\theta}{2} \quad ; \quad \cos\left(\frac{d\theta}{2}\right) \simeq 1 \tag{8.667}$$

and divide (8.666) throughout by $rdrd\theta$ we have

$$\frac{\sigma_{rr}}{r} + \frac{\partial\sigma_{rr}}{\partial r}\left(1 + \frac{dr}{r}\right) - \frac{\sigma_{\theta\theta}}{r} - \frac{\partial\sigma_{\theta\theta}}{\partial\theta}\frac{d\theta}{r} + \frac{1}{r}\frac{\partial\sigma_{r\theta}}{\partial\theta} + f_r^b = 0 \tag{8.668}$$

In the limit as dr and $d\theta$ reduce to zero (8.668) reduces to

$$\frac{\partial\sigma_{rr}}{\partial r} + \frac{1}{r}\frac{\partial\sigma_{r\theta}}{\partial\theta} + \frac{\sigma_{rr} - \sigma_{\theta\theta}}{r} + f_r^b = 0 \tag{8.669}$$

Similarly, by considering the equilibrium of *forces* in the θ-direction we obtain the second equation

$$\frac{\partial\sigma_{r\theta}}{\partial r} + \frac{1}{r}\frac{\partial\sigma_{\theta\theta}}{\partial\theta} + \frac{2\sigma_{r\theta}}{r} + f_\theta^b = 0 \tag{8.670}$$

It may be noted that the derivation of (8.668) is not strictly rigorous since we have summed all the forces as if they were aligned along the radial direction through the center of the element. This is a variance with the recognition of the absence of parallelism between the traction component acting on the sides corresponding to θ and $\theta + d\theta$. Nonetheless, the procedure yield the same result as a rigorous formulation based on derivatives related to orthogonal curvilinear coordinates. Also as with the previous definitions of the equations of equilibrium, the derivatives are with respect to the spatial coordinates since equilibrium occurs in the deformed configuration. In the classical theory of elasticity the differences between spatial and material coordinates are ignored and both coordinates are denoted by (r, θ). Finally, we note that for the dynamic case the right hand sides of (8.669) and (8.670) will be non-zero and will contain the momentum terms $\rho(\partial^2 u_r/\partial t^2)$ and $\rho(\partial^2 u_\theta/\partial t^2)$ which are the contributions from the acceleration terms.

8.9.2   Definition of strains in plane polar coordinates

For the definition of the strains components referred to the plane polar coordinate system, we make use of the displacement vector

$$\mathbf{u} = u_r\mathbf{i}_r + u_\theta\mathbf{i}_\theta + u_z\mathbf{i}_z \tag{8.671}$$

referred to the plane polar coordinate system, and its derivatives. For plane problems,

$$u_r = u_r(r, \theta) \quad ; \quad u_\theta = u_\theta(r, \theta) \tag{8.672}$$

and u_z will be constant or zero depending upon whether generalized plane stress or plane strain problems are being investigated. We consider an infinitesimal element of the continuum which has the configuration $ABCD$ in the undeformed state and moves to the position $A'B'C'D'$ during a deformation (Figure 8.40). During the deformation process, the plane element undergoes a change in shape as well as a change in area.

The components of the strains in the medium when referred to the plane polar coordinate system can be evaluated by considering the changes in lengths of the line elements and changes in shape of the element as shown in Figure 8.40. Attention is restricted to small deformations. The strain in the radial direction

$$\epsilon_{rr} = \frac{A'D' - AD}{AD} \tag{8.673}$$

Similarly, the strain in the circumferential direction

$$\epsilon_{\theta\theta} = \frac{A'B' - AB}{AB} \tag{8.674}$$

and the shear strain in the $r - \theta$ plane is

$$\epsilon_{r\theta} = \frac{1}{2}(\theta_1 + \theta_2) = \frac{1}{2}(\theta_4 - \theta_3 + \theta_2) \tag{8.675}$$

From geometry we have

$$A'M = dr + \frac{\partial u_r}{\partial r}dr \quad ; \quad B'L = \frac{\partial u_r}{\partial \theta}d\theta$$

$$D'M = \frac{\partial u_\theta}{\partial r}dr \tag{8.676}$$

$$A'L = PN + LN - A'P = (r + u_r)d\theta + u_\theta + \frac{\partial u_\theta}{\partial \theta}d\theta - u_\theta$$

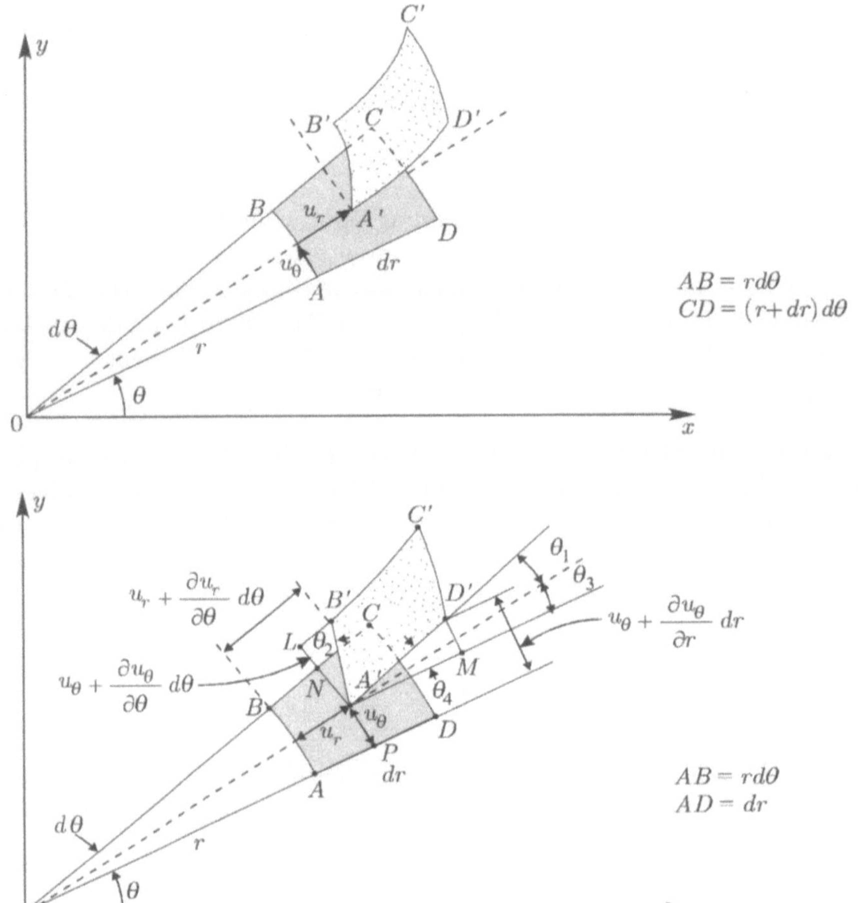

Figure 8.40: Deformations in plane polar coordinates.

Considering the right triangle $A'MD'$ we have

$$(A'D')^2 = \left(\frac{\partial u_\theta}{\partial r} dr\right)^2 + (dr)^2 \left(1 + \frac{\partial u_r}{\partial r}\right)^2 \tag{8.677}$$

and from (8.673),

$$\epsilon_{rr} = \frac{\left[(dr)^2 \left(1 + \frac{\partial u_r}{\partial r}\right)^2 + \left(\frac{\partial u_\theta}{\partial r} dr\right)^2\right]^{\frac{1}{2}} - dr}{dr} \tag{8.678}$$

Neglecting higher-order terms, (8.678) reduces to

$$\epsilon_{rr} = \frac{\partial u_r}{\partial r} \tag{8.679}$$

From the expression (8.674) for $\epsilon_{\theta\theta}$ we have

$$(A'B')^2 = (1 + \epsilon_{\theta\theta})^2 (AB)^2 = (1 + \epsilon_{\theta\theta})^2 (rd\theta)^2 \tag{8.680}$$

From the right triangle $A'LB'$ we have

$$(A'B')^2 = (B'L)^2 + (LA')^2 \tag{8.681}$$

Using the expressions (8.676), (8.681) can be reduced to

$$(A'B')^2 = \left(r^2 + 2ru_r + 2r\frac{\partial u_\theta}{\partial \theta} \right) (d\theta)^2 \tag{8.682}$$

Equating (8.680) and (8.682) we obtain

$$\epsilon_{\theta\theta} = \frac{1}{r}\frac{\partial u_\theta}{\partial \theta} + \frac{u_r}{r} \tag{8.683}$$

To calculate the shear strain we assume small angle changes such that

$$\theta_2 \simeq \tan\theta_2 = \frac{B'L}{A'L} = \frac{(\partial u_r/\partial\theta)\,d\theta}{(r + u_r)d\theta + (\partial u_\theta/\partial\theta)d\theta} \tag{8.684}$$

Neglecting $(u_r d\theta)$ and $(\partial u_\theta/\partial\theta)d\theta$ in the denominator of (8.684) we obtain

$$\theta_2 = \frac{1}{r}\frac{\partial u_\theta}{\partial r} \tag{8.685}$$

Similarly it can be shown that

$$\theta_4 = \frac{\partial u_\theta}{\partial r} \quad ; \quad \theta_3 = \frac{u_\theta}{r} \tag{8.686}$$

Using (8.685) and (8.686) in (8.675) we have

$$\epsilon_{r\theta} = \frac{1}{2}\left(\frac{\partial u_\theta}{\partial r} - \frac{u_\theta}{r} + \frac{1}{r}\frac{\partial u_r}{\partial \theta}\right) \tag{8.687}$$

For plane strain deformation, the strain matrix related to the plane polar coordinate system is given by

$$\boldsymbol{\epsilon}_P = \begin{bmatrix} \left(\dfrac{\partial u_r}{\partial r}\right) & \dfrac{1}{2}\left(\dfrac{\partial u_\theta}{\partial r} - \dfrac{u_\theta}{r} + \dfrac{1}{r}\dfrac{\partial u_r}{\partial \theta}\right) \\ \dfrac{1}{2}\left(\dfrac{\partial u_\theta}{\partial r} - \dfrac{u_\theta}{r} + \dfrac{1}{r}\dfrac{\partial u_r}{\partial \theta}\right) & \left(\dfrac{1}{r}\dfrac{\partial u_\theta}{\partial \theta} + \dfrac{u_r}{r}\right) \end{bmatrix} \tag{8.688}$$

and $\boldsymbol{\epsilon}_P = \boldsymbol{\epsilon}_P^T$.

Eliminating u_r and u_θ from the components of $\boldsymbol{\epsilon}_P$, we obtain the compatibility equation

$$\frac{\partial^2 \epsilon_{\theta\theta}}{\partial r^2} + \frac{1}{r^2}\frac{\partial^2 \epsilon_{rr}}{\partial \theta^2} + \frac{2}{r}\frac{\partial \epsilon_{\theta\theta}}{\partial r} - \frac{1}{r}\frac{\partial \epsilon_{rr}}{\partial r} = \frac{2}{r}\frac{\partial^2 \epsilon_{r\theta}}{\partial r\partial \theta} + \frac{2}{r^2}\frac{\partial \epsilon_{r\theta}}{\partial \theta} \tag{8.689}$$

8.9.3 Elastic stress-strain relations in polar coordinates

When considering *orthogonal curvilinear coordinates*, the physical components of measures such as stresses and strains at a point are simply Cartesian components in a *local rectangular Cartesian coordinate* system at that point with its axes tangent to the coordinate curves. Any algebraic relationships between physical components, such as stresses and strains, are therefore identical in form to the relationships applicable to the Cartesian coordinate formulation. For example, considering the result (8.343) for the elastic stress-strain relationship for plane strain deformations of the medium which are expressed in relation to a rectangular Cartesian coordinate system, we can write down the equivalent expressions in plane polar coordinates as

$$\sigma_{rr} = \lambda(\epsilon_{rr} + \epsilon_{\theta\theta}) + 2\mu\epsilon_{rr}$$

$$\sigma_{\theta\theta} = \lambda(\epsilon_{rr} + \epsilon_{\theta\theta}) + 2\mu\epsilon_{\theta\theta} \tag{8.690}$$

$$\sigma_{r\theta} = 2\mu\epsilon_{r\theta}$$

$$\sigma_{zz} = \nu(\sigma_{rr} + \sigma_{\theta\theta})$$

where x is replaced by r and y is replaced by θ. Extra terms will appear only when an operation such as differentiation is performed, as is the case with the equations of equilibrium and in the definition of strains in terms of the displacement components. Similarly, the strain energy density function for plane strain behaviour which can be expressed as

$$U = \frac{(1-\nu^2)}{2E}(\sigma_{xx}^2 + \sigma_{yy}^2) - \frac{\nu(1+\nu)}{E}\sigma_{xx}\sigma_{yy} + \frac{1}{2\mu}\sigma_{xy}^2 \qquad (8.691)$$

with reference to the rectangular Cartesian coordinate system, can be used to express the equivalent result in plane polar coordinates; i.e.

$$U = \frac{(1-\nu^2)}{2E}(\sigma_{rr}^2 + \sigma_{\theta\theta}^2) - \frac{\nu(1+\nu)}{E}\sigma_{rr}\sigma_{\theta\theta} + \frac{1}{2\mu}\sigma_{r\theta}^2 \qquad (8.692)$$

8.9.4 Transformation from Cartesian equations

The representations derived previously for the plane polar coordinate system can also be obtained by considering a transformation of the results from rectangular Cartesian coordinates to plane polar coordinates. In the ensuing, we shall present as an illustration, the transformation of the equations of equilibrium from rectangular Cartesian coordinates to plane polar coordinates. Considering the plane rectangular Cartesian coordinates (x, y) and the plane polar coordinates (r, θ), we have the relationships

$$x = r\cos\theta \qquad ; \qquad y = r\sin\theta \qquad (8.693)$$

and

$$r^2 = (x^2 + y^2) \qquad ; \qquad \theta = \tan^{-1}\left(\frac{y}{x}\right) \qquad (8.694)$$

Also

$$\frac{\partial r}{\partial x} = \frac{x}{r} = \cos\theta \qquad ; \qquad \frac{\partial r}{\partial y} = \frac{y}{r} = \sin\theta$$

$$\qquad (8.695)$$

$$\frac{\partial \theta}{\partial x} = -\frac{y}{r^2} = -\frac{\sin\theta}{r} \qquad ; \qquad \frac{\partial \theta}{\partial y} = \frac{x}{r^2} = \frac{\cos\theta}{r}$$

The derivatives with respect to x and y in the Cartesian coordinate system can be transformed with respect to r and θ by using chain rule differentiation, i.e.

$$\frac{\partial}{\partial x} = \frac{\partial}{\partial r}\frac{\partial r}{\partial x} + \frac{\partial}{\partial \theta}\frac{\partial \theta}{\partial x} = \cos\theta\frac{\partial}{\partial r} - \frac{\sin\theta}{r}\frac{\partial}{\partial \theta}$$

$$\frac{\partial}{\partial y} = \frac{\partial}{\partial r}\frac{\partial r}{\partial y} + \frac{\partial}{\partial \theta}\frac{\partial \theta}{\partial y} = \sin\theta\frac{\partial}{\partial r} + \frac{\cos\theta}{r}\frac{\partial}{\partial \theta}$$

(8.696)

We can now express the components of the stress matrix $\boldsymbol{\sigma}$ referred to the rectangular Cartesian coordinate system in terms of the components of the stress matrix $\boldsymbol{\sigma}_P$ by using the transformation rule (see, e.g. (8.182));

$$\boldsymbol{\sigma} = \mathbf{A}^T\boldsymbol{\sigma}_P\mathbf{A} \tag{8.697}$$

where

$$\mathbf{A} = \begin{bmatrix} \cos\theta & \sin\theta \\ -\sin\theta & \cos\theta \end{bmatrix} \tag{8.698}$$

Evaluating (8.697) we obtain

$$\sigma_{xx} = \sigma_{rr}\cos^2\theta + \sigma_{\theta\theta}\sin^2\theta - 2\sin\theta\cos\theta\,\sigma_{r\theta}$$

$$\sigma_{yy} = \sigma_{rr}\sin^2\theta + \sigma_{\theta\theta}\cos^2\theta + 2\sin\theta\cos\theta\,\sigma_{r\theta} \tag{8.699}$$

$$\sigma_{xy} = (\sigma_{rr} - \sigma_{\theta\theta})\sin\theta\cos\theta + (\cos^2\theta - \sin^2\theta)\sigma_{r\theta}$$

Consider the equations of equilibrium (8.370) expressed in rectangular Cartesian coordinates and purely for purposes of the illustration of the transformation procedure we shall assume that the body forces are zero. Transforming the derivatives $\partial/\partial x$ and $\partial/\partial y$ according to (8.696) and transforming the stresses σ_{xx} and σ_{xy} according to (8.699) we obtain from the first equation of equilibrium

$$\left[\frac{\partial\sigma_{rr}}{\partial r} + \frac{1}{r}\frac{\partial\sigma_{r\theta}}{\partial \theta} + \frac{(\sigma_{rr} - \sigma_{\theta\theta})}{r}\right]\cos\theta$$

$$- \left[\frac{1}{r}\frac{\partial\sigma_{\theta\theta}}{\partial \theta} + \frac{\partial\sigma_{r\theta}}{\partial r} + \frac{2\sigma_{r\theta}}{r}\right]\sin\theta = 0 \tag{8.700}$$

Since this equation is valid for all values of θ we require

$$\frac{\partial \sigma_{rr}}{\partial r} + \frac{1}{r}\frac{\partial \sigma_{r\theta}}{\partial \theta} + \frac{(\sigma_{rr} - \sigma_{\theta\theta})}{r} = 0 \qquad (8.701)$$

$$\frac{\partial \sigma_{r\theta}}{\partial r} + \frac{1}{r}\frac{\partial \sigma_{\theta\theta}}{\partial \theta} + \frac{2\sigma_{r\theta}}{r} = 0 \qquad (8.702)$$

Also, since the choice of the x-direction, which corresponds to the direction in which equilibrium of forces is considered is arbitrary, (8.701) and (8.702) must be valid for all θ. These results are identical to the equations of equilibrium in the r and θ-directions, (8.669) and (8.670) respectively obtained by considering the equilibrium of forces acting on an infinitesimal element (with $\mathbf{f}^b = \mathbf{0}$).

8.9.5 The Airy stress function in plane polar coordinates

The transformation procedure described in Section 8.9.4. can also be used to express the stress components $\boldsymbol{\sigma}_P$ in terms of the Airy stress function $\varphi(r,\theta)$ applicable to the plane polar coordinate formulation. In the absence of body forces, these representations take the forms

$$\sigma_{rr} = \frac{1}{r^2}\frac{\partial^2 \varphi}{\partial \theta^2} + \frac{1}{r}\frac{\partial \varphi}{\partial r}$$

$$\sigma_{\theta\theta} = \frac{\partial^2 \varphi}{\partial r^2} \qquad (8.703)$$

$$\sigma_{r\theta} = -\frac{1}{r}\frac{\partial^2 \varphi}{\partial r \partial \theta} + \frac{1}{r^2}\frac{\partial \varphi}{\partial \theta} = -\frac{\partial}{\partial r}\left(\frac{1}{r}\frac{\partial \varphi}{\partial \theta}\right)$$

and

$$\sigma_{zz} = \nu(\sigma_{rr} + \sigma_{\theta\theta}) \qquad (8.704)$$

It can be verified that the representations (8.703) identically satisfy the equations of equilibrium (8.669) and (8.670) in plane polar coordinates. Using the stress-strain relations (8.690), the strain components can be expressed in terms of the derivatives of the Airy stress function $\varphi(r,\theta)$. Substituting the resulting expressions in the equation of compatibility applicable to plane deformations (8.689), yields the biharmonic equation for the stress function $\varphi(r,\theta)$; i.e.

$$\nabla^4\varphi(r,\theta) = \nabla^2\nabla^2\varphi(r,\theta) = 0 \tag{8.705}$$

where

$$\nabla^2 = \frac{\partial^2}{\partial r^2} + \frac{1}{r}\frac{\partial}{\partial r} + \frac{1}{r^2}\frac{\partial^2}{\partial\theta^2} \tag{8.706}$$

is Laplace's operator in plane polar coordinates.

8.9.6 Boundary conditions in plane polar coordinates

The solution of the biharmonic equation (8.705) is usually subject to displacement and traction boundary conditions which act on specified surfaces with either r=const., or $\theta=$ const. The displacement boundary conditions can be expressed in the general forms

$$\mathbf{u}(r_0,\theta) = \widehat{u}_r(r_0,\theta)\mathbf{i}_r + \widehat{u}_\theta(r_0,\theta)\mathbf{i}_\theta = \widehat{\mathbf{U}}(\theta) \tag{8.707}$$

or

$$\mathbf{u}(r,\theta_0) = u_r^*(r,\theta_0)\mathbf{i}_r + u_\theta^*(r,\theta_0)\mathbf{i}_\theta = \mathbf{U}^*(r) \tag{8.708}$$

where $\widehat{\mathbf{U}}(\theta)$ and $\mathbf{U}^*(r)$ are specified vector valued functions of the arguments.

The traction boundary conditions are specified in relation to a known surface $(F^*(r,\theta))$ such that

$$\boldsymbol{\sigma}_P\mathbf{n} = \mathbf{T}_P^* \tag{8.709}$$

where $\mathbf{T}_P^*$ are prescribed tractions and $\mathbf{n}$ is the outward unit normal to $F^*(r,\theta)$ on which $\mathbf{T}_P^*$ are prescribed. These equations can be written in the expanded form

$$T_r = \sigma_{rr}n_r + \sigma_{r\theta}n_\theta \tag{8.710}$$

$$T_\theta = \sigma_{r\theta}n_r + \sigma_{\theta\theta}n_\theta \tag{8.711}$$

For example, considering the circular boundary $r = a$ shown in Figure 8.41, we have

$$F^*(r, \theta) = r - a = 0 \tag{8.712}$$

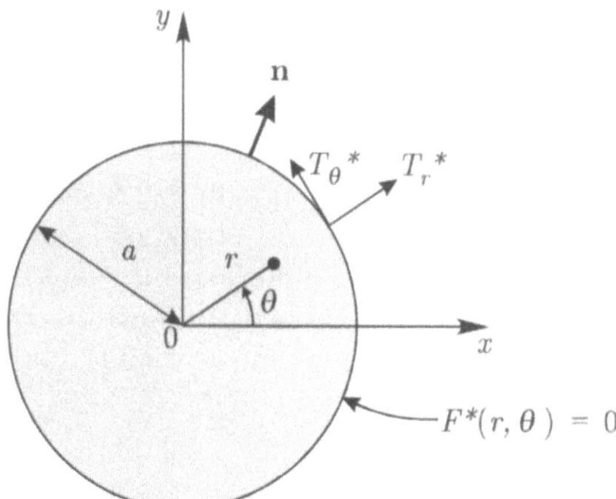

Figure 8.41: Traction vectors on circular boundary.

The direction cosines of the unit outward normal to the boundary $F^*(r, \theta)$ are

$$n_r = \frac{\partial F^*/\partial r}{\left[\left(\dfrac{\partial F^*}{\partial r}\right)^2 + \left(\dfrac{1}{r}\dfrac{\partial F^*}{\partial \theta}\right)^2\right]^{\frac{1}{2}}}$$

$$n_\theta = \frac{\partial F^*/r\partial \theta}{\left[\left(\dfrac{\partial F^*}{\partial r}\right)^2 + \left(\dfrac{1}{r}\dfrac{\partial F^*}{\partial \theta}\right)^2\right]^{\frac{1}{2}}} \tag{8.713}$$

which reduce to

$$n_r = 1 \quad ; \quad n_\theta = 0 \tag{8.714}$$

and the traction boundary conditions (8.710) and (8.711) reduce to

$$T_r^*(a,\theta) = \sigma_{rr}(a,\theta) \tag{8.715}$$

$$T_\theta^*(a,\theta) = \sigma_{r\theta}(a,\theta) \tag{8.716}$$

8.9.7 Development of the Airy stress function in polar coordinates

The general solution of the biharmonic equation in polar coordinates applicable to plane elastic regions with an annular shape or a solid circle was found by Michell (1863 -1940) who also established the completeness of the solution through rigorous mathematical proofs. To illustrate the procedure for obtaining a part of this solution, we assume a variables separable form of $\varphi(r,\theta)$ which admits a representation

$$\varphi(r,\theta) = F(r)G(\theta) \tag{8.717}$$

where

$$G(\theta) = \begin{cases} \sin n\theta \\ \cos n\theta \end{cases} \tag{8.718}$$

The form of (8.718) indicates that $G(\theta)$ may be taken as either $\sin(n\theta)$ or $\cos(n\theta)$. Applying the Laplace operator to (8.717) we have

$$\nabla^2\varphi(r,\theta) = \left(\frac{d^2F}{dr^2} + \frac{1}{r}\frac{dF}{dr} - \frac{n^2F}{r^2}\right)\begin{Bmatrix} \sin(n\theta) \\ \cos(n\theta) \end{Bmatrix} \tag{8.719}$$

and the biharmonic equation (8.705) gives

$$\nabla^4\varphi(r,\theta) = \left(\frac{d^2}{dr^2} + \frac{1}{r}\frac{d}{dr} - \frac{n^2}{r^2}\right)\left(\frac{d^2F}{dr^2} + \frac{1}{r}\frac{dF}{dr} - \frac{n^2F}{r^2}\right)\begin{Bmatrix} \sin(n\theta) \\ \cos(n\theta) \end{Bmatrix}$$
$$= 0 \tag{8.720}$$

Since θ is arbitrary, we require

$$\frac{d^4 F}{dr^4} + \frac{2}{r}\frac{d^3 F}{dr^3} - \left(\frac{1+2n^2}{r^2}\right)\frac{d^2 F}{dr^2} + \left(\frac{1+2n^2}{r^3}\right)\frac{dF}{dr}$$
$$+ \frac{n^2(n^2-4)}{r^4}F = 0 \tag{8.721}$$

The equation (8.721) is an ordinary differential equation of the Euler-type. We can change the independent variable to t such that

$$t = \ln r \tag{8.722}$$

and reduce the equation (8.721) to an ordinary differential equation with constant coefficients; i.e.

$$\frac{d^4 F}{dt^4} - 4\frac{d^3 F}{dt^3} + 2(2-n^2)\frac{d^2 F}{dt^2} + (3+4n^2)\frac{dF}{dt}$$
$$+ n^2(n^2-4)F = 0 \tag{8.723}$$

Further, if we assume that (8.723) has solutions of the form

$$F = Ce^{\zeta t} \tag{8.724}$$

the resulting polynomial for the determination of the admissible values of ζ is

$$\zeta^4 - 4\zeta^3 + 2(2-n^2)\zeta^2 + (3+4n^2)\zeta + n^2(n^2-4) = 0 \tag{8.725}$$

The equation (8.725) has roots

$$\zeta_1 = n \quad ; \quad \zeta_2 = -n \quad ; \quad \zeta_3 = (2+n) \quad ; \quad \zeta_4 = (2-n) \tag{8.726}$$

which corresponds to the four required solutions. It can be shown that if $n \geq 2$, the four solutions represented by (8.726) will be the general solution of the equation (8.725). If $n = 0$ or $n = 1$, double roots exist and the general solution will have either a second or third form, i.e.

$$\text{for} \quad n \geq 2; \quad F(r) = a_n r^n + b_n r^{-n} + c_n r^{2+n} + d_n r^{2-n} \tag{8.727}$$

$$\text{for} \quad n = 0; \quad F(r) = a_0 + b_0 \ln r + c_0 r^2 + d_0 r^2 \ln r \tag{8.728}$$

$$\text{for} \quad n = 1; \quad F(r) = a_1 r + \frac{b_1}{r} + c_1 r^3 + d_1 r \ln r \tag{8.729}$$

where $a_0, b_0, \ldots$etc. are constants.

We can construct the appropriate forms for $\varphi(r, \theta)$ with dependency in θ for $n = 1$ and $n \geq 2$, with the appropriate change in the constants $a_n, b_n (n = 1; n \geq 2)$ to reflect the form of independent solutions valid for $\sin(n\theta)$ and $\cos(n\theta)$.

Michell has also shown that

$$\varphi = A_0' \theta + D_0^* r^2 \theta + \frac{A_1}{2} r\theta \sin\theta - \frac{C_1}{2} r\theta \cos\theta \tag{8.730}$$

also constitutes a solution which has applications to regions in the form of annular sectors, and the combination of the solutions arising from the variables separation technique discussed previously and (8.730) constitutes the complete general solution of the Airy stress function $\varphi(r, \theta)$. The complete solution can be written as

$$\begin{aligned}
\varphi(r, \theta) =\ & \left[A_0 \ln r + B_0 r^2 + C_0 r^2 \ln r + D_0 \right] + D_0^* r^2 \theta + A_0' \theta \\
& + \frac{A_1}{2} r\theta \sin\theta + \left[A^* r + B_1 r^3 + \frac{A_1'}{r} + B_1' r \ln r \right] \cos\theta \\
& - \frac{C_1}{2} r\theta \cos\theta + \left[B^* r + D_1 r^3 + \frac{C_1'}{r} + D_1' r \ln r \right] \sin\theta \\
& + \sum_{n=2}^{\infty} \left[A_n r^n + B_n r^{2+n} + A_n' r^{-n} + B_n' r^{2-n} \right] \cos(n\theta) \\
& + \sum_{n=2}^{\infty} \left[C_n r^n + D_n r^{2+n} + C_n' r^{-n} + D_n' r^{2-n} \right] \sin(n\theta) \tag{8.731}
\end{aligned}$$

where $A_0, B_0, A_1, A_1', \ldots$etc. are all constants. Although (8.731) is the complete solution of the biharmonic equation referred to the plane polar coordinate system, obviously, not all solutions are required for the analysis of a specific boundary value problem. Additional constraints such as

(i) symmetry of the problem

(ii) boundedness of the stress field as $r \to \infty$

(iii) single-valuedness of the stress and displacement fields for regions where $\theta = 0, 2\pi$

(iv) boundedness of the stress and displacement fields as $r \to 0$

can be used to select the appropriate form of the stress function. In the ensuing, Airy stress functions selected from (8.731) will be used to develop solutions to certain problems of engineering interest.

Example 8.17

A long cylindrical elastic plug of radius $a(1+\delta)$ where $\delta \ll 1$, is fitted within a rigid cylindrical cavity of radius a. Assuming plane strain deformations of the fitted elastic plug, determined the state of stress within the elastic *"inclusion"*.

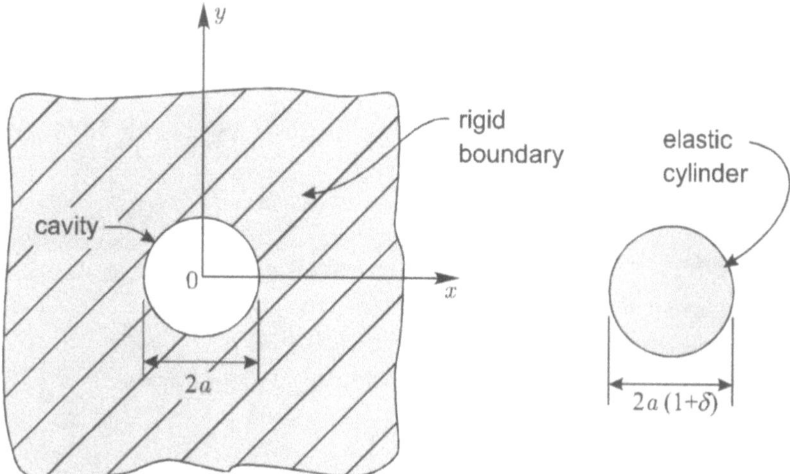

Figure 8.42: An elastic inclusion problem.

Solution

From the geometry of the problem it is evident that the state of stress in the fitted elastic "inclusion" exhibits radial symmetry, i.e. $\sigma_P = \sigma_P(r)$. Also a state of radial symmetry implies that $\sigma_{r\theta}$. The Airy stress function required to generate a radially symmetric stress field is given by

$$\varphi(r) = A_0 \ln r + B_0 r^2 + C_0 r^2 \ln r \tag{8.732}$$

Using (8.732), the stress components σ_{rr} and $\sigma_{\theta\theta}$ can be evaluated from (8.703) as follows: i.e.,

$$\sigma_{rr} = \frac{A_0}{r^2} + 2B_0 + C_0(1 + 2\ln r) \tag{8.733}$$

$$\sigma_{\theta\theta} = -\frac{A_0}{r^2} + 2B_0 + C_0(3 + 2\ln r)$$

From the physical considerations of the problem, the stresses should not be singular within the inclusion region. To satisfy this condition we require

$$A_0 = C_0 = 0 \tag{8.734}$$

Therefore the stress state within the solid cylinder is homogeneous and independent of the coordinates. The stress-strain relations for plane strain conditions give

$$2\mu\epsilon_{\theta\theta} = 2\mu\frac{u_r}{r} = (1 - \nu)\sigma_{\theta\theta} - \nu\sigma_{rr} = 2B_0(1 - 2\nu) \tag{8.735}$$

Hence the radial displacement within the inclusion is

$$u_r(r) = \frac{B_0 r(1 - 2\nu)}{\mu} \tag{8.736}$$

To fit the elastic plug within the cavity, the applied stress must be such that the radial displacement at $r = a(1 + \delta)$ must be δa, i.e.

$$\delta a = \frac{Ba(1 + \delta)(1 - 2\nu)}{\mu} \quad \text{or} \quad B = \frac{\delta\mu}{(1 + \delta)(1 - 2\nu)} \simeq \frac{\delta\mu}{1 - 2\nu} \tag{8.737}$$

Therefore the stress state within the fitted inclusion is

$$\sigma_{rr} = \sigma_{\theta\theta} = \frac{2\delta\mu}{(1 - 2\nu)} \tag{8.738}$$

It may be noted that the homogeneous stress state becomes infinite as $\nu \to 1/2$. This implies that *if* the elastic material is *incompressible*, it cannot be fitted into a cavity of smaller radius, particularly for plane strain deformations which prevents displacements of the elastic material in the $z-$ or axial direction.

● ● ●

For the preceding problem, as with all problems involving radial symmetry, the dependence in the θ-variable is absent, as such, the formulations involve total derivatives of the radial coordinate and can be solved by considering at the outset the associated ordinary differential equation governing the biharmonic equation. In the ensuing we shall examine certain problems where the Airy stress function is dependent on both independent variables r and θ.

Example 8.18

A cylindrical channel located at the surface of an elastic geological medium conveys an ideal fluid of density ρ_0 (Figure 8.43). Assuming plane strain behaviour of the geological medium, determine the state of stress and the associated displacement field.

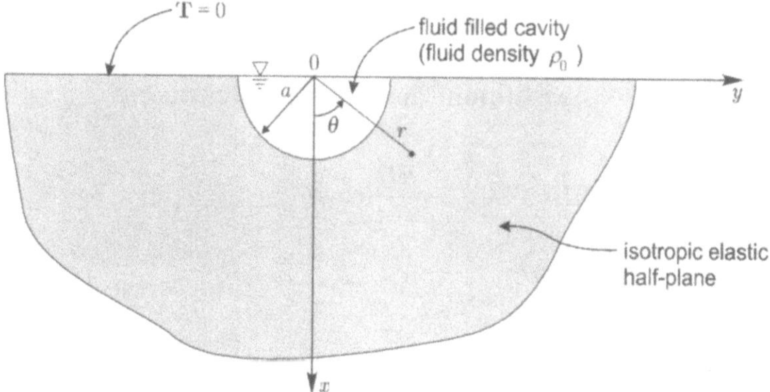

Figure 8.43: Fluid filled cavity at the surface of an elastic half-plane.

Solution

We consider plane strain conditions and assume that the dimensions of the geological medium are large in comparison with the radius of the fluid filled channel. Also since the ideal fluid has no viscosity, the fluid pressure at any depth from the free surface is proportional to x. Hence on the boundary of the channel the traction boundary condition are

$$\sigma_{rr}(a,\theta) = -\rho_0 g a \cos\theta$$

$$\sigma_{r\theta}(a,\theta) = 0$$

(8.739)

On the boundary of the half-plane exterior to the channel we require

$$\sigma_{\theta\theta}\left(r, \pm\frac{\pi}{2}\right) = 0$$

$$\sigma_{r\theta}\left(r, \pm\frac{\pi}{2}\right) = 0$$

(8.740)

In addition, since the geological medium is 'infinite' in extent, the stresses in the region should reduce to zero as $r \to \infty$.

Considering the form of the boundary conditions and the expressions for the stresses in terms of $\varphi(r,\theta)$ given by (8.703), we select $\varphi(r,\theta)$ to be of the form (see e.g. (8.731)),

$$\varphi(r,\theta) = \frac{A_1}{2} r\theta \sin\theta + \left[A^* r + B_1 r^3 + \frac{A'}{r} + B_1 r \ln r\right] \cos\theta \qquad (8.741)$$

From (8.741) and (8.703) we obtain the stress components as

$$\sigma_{rr} = \frac{A_1 \cos\theta}{r} + \left[2Br - \frac{2A'}{r^3} + \frac{B'}{r}\right] \cos\theta$$

$$\sigma_{\theta\theta} = \left[6Br + \frac{2A'}{r^3} + \frac{B'}{r}\right] \cos\theta \qquad (8.742)$$

$$\sigma_{r\theta} = \left[2Br - \frac{2A'}{r^3} + \frac{B'}{r}\right] \sin\theta$$

If we consider the regularity of the stress field as $r \to \infty$, we require $B \equiv 0$. Considering the shear stress boundary condition given by (8.739) we note that (with $B = 0$)

$$-\frac{2A'}{a^3} + \frac{B'}{a} = 0 \tag{8.743}$$

If this condition is satisfied by A' and B' at $r = a$, this will also render zero the radial stresses derived from solutions containing A' and B'. Hence the solutions corresponding to B, A' and B' are irrelevant to the problem, and the stress field reduces to

$$\sigma_{rr} = A_1 \frac{\cos\theta}{r} \quad ; \quad \sigma_{\theta\theta} = \sigma_{r\theta} = 0 \tag{8.744}$$

The arbitrary constant A_1 can be evaluated by considering the first boundary condition of (8.739); this gives

$$\sigma_{rr} = -\frac{\rho_0 g a^2}{r}\cos\theta \quad ; \quad \sigma_{\theta\theta} = \sigma_{r\theta} = 0 \tag{8.745}$$

and the boundary conditions (8.740) are trivially satisfied.

Considering the stress-strain relations for plane strain (see e.g. (8.371)) we can write with the help of (8.688)

$$\epsilon_{rr} = \frac{\partial u_r}{\partial r} = \frac{1}{E}\left[(1 - \nu^2)\sigma_{rr} - \nu(1 + \nu)\sigma_{\theta\theta}\right] \tag{8.746}$$

$$\epsilon_{\theta\theta} = \frac{1}{r}\frac{\partial u_\theta}{\partial \theta} + \frac{u_r}{r} = \frac{1}{E}\left[(1 - \nu^2)\sigma_{\theta\theta} - \nu(1 + \nu)\sigma_{rr}\right] \tag{8.747}$$

$$\epsilon_{r\theta} = \frac{1}{2}\left(\frac{\partial u_\theta}{\partial r} - \frac{u_\theta}{r} + \frac{1}{r}\frac{\partial u_r}{\partial \theta}\right) = \frac{(1 + \nu)\sigma_{r\theta}}{E} \tag{8.748}$$

Substituting (8.745) into (8.746) and integrating the result we obtain

$$u_r(r, \theta) = -\frac{(1 - \nu^2)}{E}\rho_0 g a^2 \ln r \,\cos\theta + f(\theta) \tag{8.749}$$

where $f(\theta)$ is an arbitrary function. Similarly from (8.747) we obtain

$$u_\theta(r,\theta) = \frac{(1+\nu)}{E}\rho_0 g a^2 \sin\theta \left[\nu + (1-\nu)\ln r\right]$$

$$- \int f(\theta)d\theta + g(r) \tag{8.750}$$

where $g(r)$ is an arbitrary function. Considering the shear stress - shear strain relation (8.748) and using the above expressions for u_r and u_θ we obtain the following ordinary differential equations for $f(\theta)$ and $g(r)$.

$$\frac{df(\theta)}{d\theta} + \int f(\theta)d\theta + \frac{(1+\nu)(1-2\nu)}{E}\rho_0 g a^2 \sin\theta = D$$

$$\tag{8.751}$$

$$-r\frac{dg(r)}{dr} + g(r) = D$$

where D is an arbitrary constant. Integrating these we obtain

$$f(\theta) = A\sin\theta + B\cos\theta - \frac{(1+\nu)(1-2\nu)\rho_0 g a^2}{2E}\theta\sin\theta$$

$$\tag{8.752}$$

$$g(r) = Cr + D$$

where A, B and C are arbitrary constants. We can now write the expressions for u_r and u_θ as

$$u_r(r,\theta) = -\frac{(1-\nu^2)\rho_0 g a^2 \ln r}{E}\cos\theta$$

$$- \frac{(1+\nu)(1-2\nu)\rho_0 g a^2}{2E}\theta\sin\theta + A\sin\theta + B\cos\theta \tag{8.753}$$

$$u_\theta(r,\theta) = \frac{\rho_0 g a^2(1+\nu)}{E}\sin\theta + \frac{(1-\nu^2)\rho_0 g a^2}{E}\ln r\ \sin\theta$$

$$- \frac{(1+\nu)(1-2\nu)\rho_0 g a^2}{2E}\theta\cos\theta + A\cos\theta$$

$$- B\sin\theta + Cr + D \tag{8.754}$$

The arbitrary constants A, B, C and D can be determined by considering the conditions imposed by symmetry of the deformation. For example on the plane $\theta = 0$, u_θ must be zero for all choices of r. This gives

$$A + D + Cr = 0 \tag{8.755}$$

To satisfy (8.755) we require $A+D = 0\,;\, C = 0$, since D is a rigid body translation we can, without loss of generality we can, without loss of generality, set it to zero. Similarly if we consider the displacements

$$u_r = B\cos\theta \quad;\quad u_\theta = -B\sin\theta \tag{8.756}$$

the resultant of these displacement components is equivalent to a rigid body displacement of the entire region in the x-direction and consequently can be set to zero.

Example 8.19

The determination of the state of stress in a uniformly stressed plate containing a circular hole in a classical problem in theory of elasticity and was examined independently by G. Kirsch in 1898 and Sir Charles Inglis (1875-1952). Consider the problem of a plate infinite extent, containing a circular hole of radius a, which is subjected to a state of stress which corresponds to a uniform stress σ_0 (Figure 8.44). Determine the state of stress in the plate.

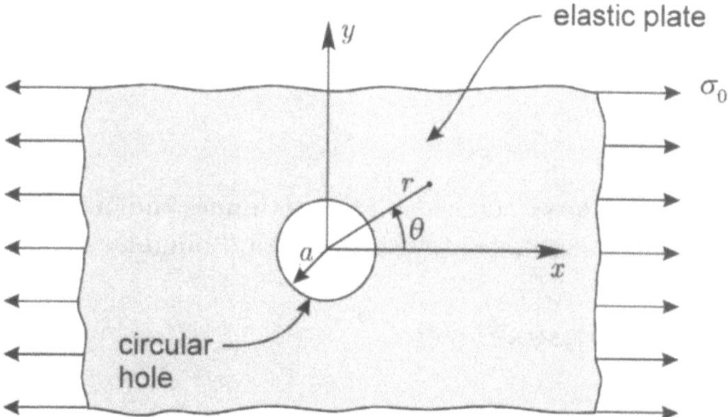

Figure 8.44: Uniformly stressed plate containing a circular hole.

Solution

Since the uniformly stressed plate is of infinite extent, we can choose the reference coordinate system such that the uniform stress field acts in the x-direction. Considering the state of stress in an infinite 'intact' plate we have

$$\sigma_{xx} = \sigma_0 \quad ; \quad \text{all other } \boldsymbol{\sigma} = 0 \tag{8.757}$$

Considering (8.384) (with $V = 0$), we note that the stress function $\varphi(x, y)$ required to generate the stress field (8.757) is

$$\varphi(x, y) = \frac{\sigma_0}{2} y^2 \tag{8.758}$$

The stress function (8.758) can be expressed in polar coordinates in the form

$$\varphi(r, \theta) = \frac{\sigma_0}{2} r^2 \sin^2 \theta = \frac{\sigma_0 r^2}{4}(1 - \cos 2\theta) \tag{8.759}$$

Using (8.703) and (8.759) we obtain

$$\begin{aligned}
\sigma_{rr} &= \frac{\sigma_0}{2}(1 + \cos 2\theta) \\
\sigma_{\theta\theta} &= \frac{\sigma_0}{2}(1 - \cos 2\theta) \\
\sigma_{r\theta} &= -\frac{\sigma_0}{2} \sin 2\theta
\end{aligned} \tag{8.760}$$

Considering the uniformly stretched plate containing the circular hole, the solution to the problem should satisfy the traction boundary conditions

$$\sigma_{rr}(a, \theta) = 0 \quad ; \quad \sigma_{r\theta}(a, \theta) = 0 \tag{8.761}$$

at the boundary of the circular hole and reduce to the uniaxial state of stress (8.760) as $|x| \to \infty$. Considering the form of the stress field (8.760) and the relationship between the Airy stress function $\varphi(r, \theta)$ and the stress components (see e.g. (8.703)), we seek a solution of $\varphi(r, \theta)$ of the form

$$\varphi(r, \theta) = F_1(r) + F_2(r) \cos 2\theta \tag{8.762}$$

We can substitute (8.762) in the biharmonic equation (8.705) and solve the resulting fourth-order ordinary differential equations for $F_1(r)$ and $F_2(r)$, the final results for $\varphi(r,\theta)$ can be written as (see also (8.731))

$$\varphi(r,\theta) = \left[A_0 \ln r + B_0 r^2 + C_0 r^2 \ln r + D_0\right]$$
$$+ \left[A_2 r^2 + B_2 r^4 + \frac{A_2'}{r^2} + B_2'\right] \cos 2\theta \qquad (8.763)$$

The corresponding stress components are

$$\sigma_{rr} = C_0\left\{1 + 2\ln r\right\} + 2B_0 + \frac{A_0}{r^2}$$
$$- \left\{2A_2 + 6\frac{A_2'}{r^4} + 4\frac{B_2'}{r^2}\right\}\cos 2\theta$$
$$\sigma_{\theta\theta} = C_0\left\{3 + 2\ln r\right\} + 2B_0 - \frac{A_0}{r^2} \qquad (8.764)$$
$$+ \left\{2A_2 + 12B_2 r^2 + 6\frac{A_2'}{r^4}\right\}\cos 2\theta$$
$$\sigma_{r\theta} = \left\{2A_2 + 6B_2 r^2 - \frac{6}{r^4}A_2 - \frac{2}{r^2}B_2'\right\}\sin 2\theta$$

where $A_0, B_0, C_0, ...$etc., are *seven* constants that need to be determined. Since the stress field derived form (8.763) should be bounded as $r \to \infty$, we require

$$C_0 = B_2 \equiv 0 \qquad (8.765)$$

the remaining *five* constants are determined by satisfying the traction boundary conditions (8.761) at the cavity boundary and the following traction boundary conditions as $r \to \infty$

$$\sigma_{rr}(\infty,\theta) \to \frac{\sigma_0}{2}(1 + \cos 2\theta) \quad ; \quad \sigma_{r\theta}(\infty,\theta) \to -\frac{\sigma_0}{2}\sin 2\theta \qquad (8.766)$$

These traction boundary conditions give

$$2B_0 + \frac{A_0}{2} = 0 \quad ; \quad 2A_2 + 6\frac{A_2'}{a^4} + 4\frac{B_2'}{a^2} = 0$$

$$2A_2 - 6\frac{A_2'}{a^4} - 2\frac{B_2'}{a^2} = 0 \tag{8.767}$$

$$-2B_0 = \frac{\sigma_0}{2} \quad ; \quad -2A_2 = \frac{\sigma_0}{2}$$

Solving the equations we obtain

$$B_0 = \frac{\sigma_0}{4} \quad ; \quad A_0 = \frac{a^2\sigma_0}{2} \quad ; \quad A_2 = -\frac{\sigma_0}{4}$$

$$\tag{8.768}$$

$$A_2' = -\frac{a^4\sigma_0}{4} \quad ; \quad B_2' = \frac{a^2\sigma_0}{2}$$

and the stress components are given by

$$\sigma_{rr}(r,\theta) = \frac{\sigma_0}{2}\left(1 - \frac{a^2}{r^2}\right) + \frac{\sigma_0}{2}\left(1 - 4\frac{a^2}{r^2} + 3\frac{a^4}{r^4}\right)\cos 2\theta$$

$$\sigma_{\theta\theta}(r,\theta) = \frac{\sigma_0}{2}\left(1 + \frac{a^2}{r^2}\right) - \frac{\sigma_0}{2}\left(1 + 3\frac{a^4}{r^4}\right)\cos 2\theta \tag{8.769}$$

$$\sigma_{r\theta}(r,\theta) = -\frac{\sigma_0}{2}\left(1 + 2\frac{a^2}{r^2} - 3\frac{a^4}{r^4}\right)\sin 2\theta$$

The displacement components u_r and u_θ can be obtained by considering the above stress field, the stress-strain relations (8.690) and the strain-displacement relations (8.688). This is left as an exercise for the reader.

8.9.8   An application of complex variable techniques

The complex variable technique discussed previously in connection with the solution of plane problems in elasticity, referred to the rectangular Cartesian system of coordinates can also be adopted to examine problems which are formulated in reference to a system of plane polar coordinates. The right hand side of (8.633) can be written as

$$\sigma_{rr} + \sigma_{\theta\theta} = \sigma_{xx} + \sigma_{yy} \tag{8.770}$$

$$\sigma_{\theta\theta} - \sigma_{rr} + 2i\sigma_{r\theta} = \left(\sigma_{yy} - \sigma_{xx} + 2i\sigma_{xy}\right)e^{2i\theta} \tag{8.771}$$

$$u_r + iu_\theta = \left(u_x + iu_y\right)e^{-i\theta} \tag{8.772}$$

We assume that $\psi(z)$ and $\chi(z)$ are functions of the variable z such that $z = re^{i\theta}$ (also $\bar{z} = -e^{-i\theta}$).

Considering (8.633), the expressions for (8.770) to (8.772) can be written in terms of $\psi(z)$ and $\chi(z)$ as follows:

$$\sigma_{rr} + \sigma_{\theta\theta} = 2\left[\psi'(z) + \overline{\psi'(z)}\right] = 4\,\mathrm{Re}\left[\psi'(z)\right] \tag{8.773}$$

$$\sigma_{\theta\theta} - \sigma_{rr} + 2i\sigma_{r\theta} = 2e^{2i\theta}\left[\bar{z}\psi''(z) + \chi'(z)\right] \tag{8.774}$$

$$2\mu(u_r + iu_\theta) = e^{-i\theta}\left[\kappa\psi(z) - z\overline{\psi'(z)} - \overline{\chi(z)}\right] \tag{8.775}$$

and κ is defined either by (8.634) or (8.635) depending upon whether the problem exhibits a state of plane strain or generalized plane stress, respectively. Therefore, once the functions $\psi(z)$ and $\chi(z)$ are determined by satisfying the appropriate boundary conditions of a problem (see e.g. section 8.8.5.) the stresses and displacements in the elastic medium referred to the plane polar coordinate system can be evaluated using (8.773) to (8.775).

Example 8.20

Examine the state of stress and displacements corresponding to the analytic functions

$$\psi(z) = 0 \quad ; \quad \chi(z) = \frac{C}{z} \tag{8.776}$$

where C is a real constant. Evaluate the solution for the special case where the elastic plane occupies the region $a \le r \le \infty$; $a \le \theta \le 2\pi$ where a is the radius of the cavity and the boundary of the cavity is subjected to traction boundary conditions

$$\sigma_{rr}(a,\theta) = -\sigma_c \quad ; \quad \sigma_{r\theta}(a,\theta) = 0 \tag{8.777}$$

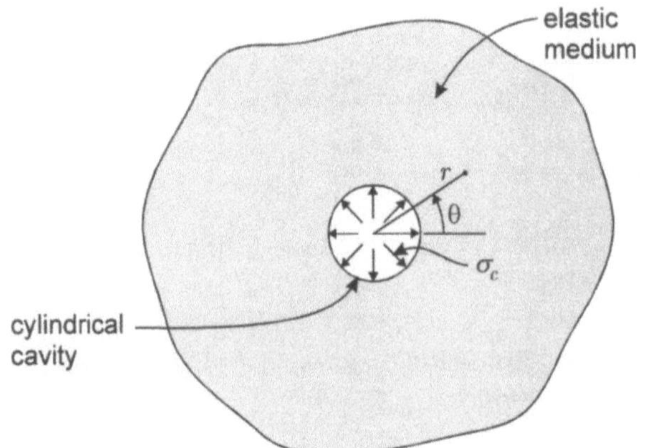

Figure 8.45: Cylindrical cavity in an elastic body subjected to internal pressure σ_c.

Solution

From (8.776), (8.773) and (8.774) we obtain

$$\sigma_{rr} + \sigma_{\theta\theta} = 0$$
$$\sigma_{\theta\theta} - \sigma_{rr} + 2i\sigma_{r\theta} = 2e^{2i\theta}\left[\chi'(z)\right] = -\frac{2C}{r^2} \tag{8.778}$$

These equations give

$$\sigma_{rr} = \frac{C}{r^2} \quad ; \quad \sigma_{\theta\theta} = -\frac{C}{r^2} \quad ; \quad \sigma_{r\theta} = 0 \tag{8.779}$$

From the boundary condition (8.777) we have $C = -\sigma_c a^2$, which gives

$$\sigma_{rr}(r) = -\sigma_c\frac{a^2}{r^2} \quad ; \quad \sigma_{\theta\theta}(r) = \sigma_c\frac{a^2}{r^2} \quad ; \quad \sigma_{r\theta} = 0 \tag{8.780}$$

Considering (8.775) and the analytic functions (8.776) we obtain

$$2\mu(u_r + iu_\theta) = e^{-i\theta}\left[-\overline{\chi(z)}\right] = e^{-i\theta}\left[\frac{\sigma_c a^2}{re^{-i\theta}}\right] \tag{8.781}$$

which gives, $u_\theta = 0$ and

$$u_r(r) = \frac{\sigma_c a^2}{2\mu r} \tag{8.782}$$

In this problem, it may be noted that $\sigma_{zz}(= \nu(\sigma_{rr} + \sigma_{\theta\theta})$; for the plane strain problem) is identically zero and the radially symmetric stress field (8.780) and the displacement field (8.782) are valid for both plane strain and generalized plane stress conditions.

Example 8.21

Consider the problem of a cylindrical cavity in a geological medium of infinite extent which is subjected to a uniaxial compressive stress field at infinity (Figure 8.44). Show that the solution to this problem can be obtained from the complex potentials

$$\psi(z) = -\frac{\sigma_0}{4}\left[z + \frac{C_1}{z}\right] \quad ; \quad \chi(z) = \frac{\sigma_0}{2}\left[z + \frac{C_2}{z} + \frac{C_3}{z^3}\right] \tag{8.783}$$

where σ_0 is the far field stress and C_1, C_2 and C_3 are undetermined real constants associated with a state of stress which vanishes as $z \to \infty$.

Solution

Considering the complex potentials $\psi(z)$ and $\chi(z)$ given by (8.783), it is evident that the first term in each expression corresponds to the state of uniaxial compression in the x-direction (see e.g. Example 8.15). The derivatives of the complex potentials relevant to the analysis are

$$\psi'(z) = -\frac{\sigma_0}{4}\left[1 - \frac{C_1}{z^2}\right] \quad ; \quad \psi''(z) = -\frac{\sigma_0}{4}\left[2\frac{C_1}{z^3}\right] \tag{8.784}$$

$$\chi'(z) = \frac{\sigma_0}{2}\left[1 - \frac{C_2}{z^2} - 3\frac{C_3}{z^4}\right] \tag{8.785}$$

Using these results in (8.773) we obtain

$$\sigma_{rr} + \sigma_{\theta\theta} = -\frac{\sigma_0}{2}\left[2 - C_1\frac{e^{-2i\theta}}{r^2} - C_1\frac{e^{2i\theta}}{r^2}\right] \tag{8.786}$$

Using the substitution $e^{\pm 2i\theta} = (\cos 2\theta \pm i \sin 2\theta)$, we obtain

$$\sigma_{rr} + \sigma_{\theta\theta} = -\sigma_0 \left[1 - \frac{C_1}{r^2} \cos 2\theta \right] \tag{8.787}$$

Similarly, using (8.784) and (8.785) in (8.774) we have

$$\sigma_{\theta\theta} - \sigma_{rr} + 2i\sigma_{r\theta} = -\sigma_0 \left[\frac{C_2}{r^2} - e^{2i\theta} - \left(\frac{C_1}{r^2} + \frac{3C_3}{r^4} \right) e^{-2i\theta} \right] \tag{8.788}$$

This gives

$$\sigma_{\theta\theta} - \sigma_{rr} = -\sigma_0 \left[\frac{C_2}{r^2} - \left(1 - \frac{C_1}{r^2} - \frac{3C_3}{r^4} \right) \cos 2\theta \right] \tag{8.789}$$

$$\sigma_{r\theta} = \frac{\sigma_0}{2} \left[1 + \frac{C_1}{r^2} + \frac{3C_3}{r^4} \right] \sin 2\theta \tag{8.790}$$

Using (8.786) and (8.789) we obtain

$$\sigma_{rr}(r,\theta) = -\frac{\sigma_0}{2} \left\{ 1 - \frac{C_2}{r^2} + \left(1 - \frac{2C_1}{r^2} - \frac{3C_3}{r^4} \right) \cos 2\theta \right\} \tag{8.791}$$

We can now make use of (8.790) and (8.791), and the boundary conditions

$$\sigma_{rr}(a,\theta) = 0 \quad ; \quad \sigma_{r\theta}(a,\theta) = 0 \tag{8.792}$$

applicable to the traction free cavity, to determine the constants C_1, C_2 and C_3. To satisfy (8.792) we require

$$1 - \frac{C_2}{a^2} = 0 \quad ; \quad 1 - \frac{2C_1}{a^2} - \frac{3C_3}{a^4} = 0 \quad ; \quad 1 + \frac{C_1}{a^2} + \frac{3C_3}{a^4} = 0 \tag{8.793}$$

The stress components can now be evaluated in the form

$$\sigma_{rr}(r,\theta) = -\frac{\sigma_0}{2}\left(1 - \frac{a^2}{r^2}\right) - \frac{\sigma_0}{2}\left(1 - \frac{4a^2}{r^2} + \frac{3a^4}{r^4}\right)\cos 2\theta$$

$$\sigma_{\theta\theta}(r,\theta) = -\frac{\sigma_0}{2}\left(1 + \frac{a^2}{r^2}\right) + \frac{\sigma_0}{2}\left(1 + \frac{3a^4}{r^4}\right)\cos 2\theta \qquad (8.794)$$

$$\sigma_{r\theta}(r,\theta) = \frac{\sigma_0}{2}\left(1 + \frac{2a^2}{r^2} - \frac{3a^4}{r^4}\right)\sin 2\theta$$

which, (except for a change in sign reflecting the compressive nature of σ_0) is identical to the result (8.769) derived using the directly obtained solution of $\varphi(r,\theta)$.

We can evaluate the displacement components u_r and u_θ by making use of (8.775) and the appropriate expressions in (8.783) to (8.785). We have

$$2\mu(u_r + iu_\theta) = e^{-i\theta}\left[\kappa\left(-\frac{\sigma_0}{4}\right)\left(z + \frac{C_1}{z}\right) - z\left(-\frac{\sigma_0}{4}\right)\left(1 - \frac{C_1}{\bar{z}^2}\right)\right.$$

$$\left. -\left(\frac{\sigma_0}{2}\right)\left(\bar{z} + \frac{C_2}{\bar{z}} + \frac{C_3}{\bar{z}^3}\right)\right] \qquad (8.795)$$

Substituting the values for C_1, C_2 and C_3 determined from (8.793) and performing the simplifications we obtain

$$u_r(r,\theta) = -\frac{a\sigma_0}{8\mu}\left[(\kappa - 1 + 2\cos 2\theta)\frac{r}{a}\right.$$

$$\left. + \frac{2a}{r}\left\{1 + \left(\kappa + 1 - \frac{a^2}{r^2}\right)\cos 2\theta\right\}\right] \qquad (8.796)$$

$$u_\theta(r,\theta) = -\frac{a\sigma_0}{8\mu}\left[-\frac{2r}{a} + \frac{2a}{r}\left\{1 - \kappa - \frac{a^2}{r^2}\right\}\right]\sin 2\theta$$

8.10 Biharmonic function formulation of three-dimensional problems in elasticity

The biharmonic function approach has also been successfully applied to the formulation of three-dimensional problems in elasticity. There is however a distinction between the basic approach that we have adopted in the formulation of the biharmonic equation for two-dimensional problems in terms of the Airy stress function and the approach adopted for the formulation of three-dimensional problems. In the Airy stress function approach we introduce at the outset a representation of the stresses in terms of a single function such that the equations of equilibrium are identically satisfied. The single equation of compatibility applicable to two-dimensional problems is then utilized to arrive at the biharmonic equation. With three-dimensional problems, we introduce representations of the displacement components in terms of a set of arbitrary functions and arrive at the partial differential equation(s) governing these functions by considering the equations of equilibrium. We note here that since the *displacement functions* are introduced explicitly to formulate the governing equations, the compatibility conditions are not necessary. The six strain-displacement relationships, the six stress-strain relationships and the three equations of equilibrium give the fifteen equations necessary to determine the three displacement components, the six strain components and the six stress components, provided the displacement fields derived from the *displacement functions* possess continuous derivatives up to the second-order. The use of a displacement function approach appears to be a more convenient approach than formulations which are based on stress function techniques. In situations where mixed boundary conditions are specified, there are certain advantages to be gained by adopting a formulation where the displacement components are obtained directly.

8.10.1   The strain potential

The strain potential approach proposed by Lamé (1795-1870) is of general interest here even though the formulation yields only a *potential equation* for the governing function. In the strain potential approach it is assumed that the displacement vector $\mathbf{u}$ can be represented in the form

$$\mathbf{u} = \frac{1}{2\mu} \nabla \varphi_0 \tag{8.797}$$

where φ_0 is scalar function of $\mathbf{x}$. Since φ_0 defines the displacements directly, and hence the strains, it is referred to as a *strain function* or a *strain potential*. We shall restrict attention to the situation where the body forces in the medium are assumed to be zero. The equations of equilibrium (8.383) in terms of $\mathbf{u}$ reduce to

$$\nabla^2\mathbf{u} + \frac{1}{(1-2\nu)}\nabla(\nabla.\mathbf{u}) = 0 \tag{8.798}$$

Substituting the expression for $\mathbf{u}$ given by (8.797) in (8.798) and noting that, since the operators commute

$$_, \quad \nabla(\nabla.\mathbf{u}) = \nabla\nabla^2\varphi_0 = \nabla^2\nabla\varphi_0 \tag{8.799}$$

we have

$$\frac{(2-2\nu)}{(1-2\nu)}\nabla(\nabla^2\varphi_0) = 0 \tag{8.800}$$

Equation (8.800) implies that

$$\nabla^2\varphi_0 = C_0 \tag{8.801}$$

where C_0 is a constant. The solution of (8.801) consists of a particular integral (φ_0^P) and a complementary function (φ_0^c). The particular integral can be identified as a homogeneous state of stress corresponding to

$$\nabla.\mathbf{u} = tr\,\epsilon = \text{const.} \tag{8.802}$$

In terms of the complementary function φ_0^c, the expressions for the stresses take the form

$$\boldsymbol{\sigma} = \frac{1}{2}\left[\nabla(\nabla\varphi_0^c) + \{\nabla(\nabla\varphi_0^c)\}^T\right] = sym(\nabla[\nabla\varphi_0^c]) \tag{8.803}$$

where sym ( ) indicates the symmetric part, and in indicial notation

$$\sigma_{ij} = (\varphi_0^c)_{,ij} \tag{8.804}$$

The representation (8.797), however, lacks generality; in the sense that the rotation matrix vanishes identically for all choices of φ_0; i.e.

$$\omega - \frac{1}{4\mu}\left[\nabla(\nabla\varphi_0) - \{\nabla(\nabla\varphi_0)\}^T\right] = \mathbf{0} \tag{8.805}$$

which implies that only irrotational deformation fields are accounted for by the representation (8.797).

8.10.2   The Galerkin vector

In the attempt to introduce more generality in the description of every possible deformation in terms of the chosen functions, Galerkin (1871-1945) assumed that the displacement vector $\mathbf{u}$ can be expressed in terms of a *vector-valued function* $\mathbf{F}$, in the form

$$2\mu\mathbf{u} = c\nabla^2\mathbf{F} - \nabla(\nabla.\mathbf{F}) \tag{8.806}$$

where, for the moment, c is assumed to be an arbitrary constant. The motivation for this representation could be traced to the decomposition theorem of Helmholz (1821-1894) which states that the displacement vector can be represented as the sum of a solenoidal component $\mathbf{u}_S$ (i.e. divergence free) and an irrotational component $\mathbf{u}_I$ (i.e. the curl is zero) such that

$$\mathbf{u} = \mathbf{u}_S + \mathbf{u}_I \tag{8.807}$$

Furthermore, we can assume that $\mathbf{u}_I$ could be derived from a scalar potential function φ and the solenoidal part $\mathbf{u}_S$ can be related to the curl of another vector-valued function $a\varPsi$ which itself is solenoidal, i.e.

$$\mathbf{u} = \nabla \times a\varPsi + \nabla\varphi \tag{8.808}$$

where a is a constant and

$$\nabla^2\varphi = 0 \quad ; \quad \nabla.\varPsi = 0 \tag{8.809}$$

If $\nabla.\varPsi = 0$, then $\varPsi$ could itself be derived from another vector $\mathbf{H}$ such that

$$\varPsi = -\nabla \times \mathbf{H} \tag{8.810}$$

The expression (8.808) for the displacement vector can now be written as

$$\mathbf{u} = \nabla\varphi + \nabla \times a(\nabla \times \mathbf{H}) = \nabla\varphi + a\nabla^2\mathbf{H} - a\nabla(\nabla.\mathbf{H}) \tag{8.811}$$

A common choice is to set $\nabla.\mathbf{H} = 0$; we can, however, set $\nabla.\mathbf{H} = \varphi$. Without loss of generality we can reduce (8.811) to

$$\mathbf{u} = a\nabla^2\mathbf{H} - b\nabla(\nabla.\mathbf{H}) \tag{8.812}$$

where $b = (1 - a)$. This result is representative of the form chosen by Galerkin as indicated in (8.806). Substituting the representation (8.806) in the displacement equations of equilibrium (8.798) we obtain

$$\nabla^2\left\{c\nabla^2\mathbf{F} - \nabla\nabla.\mathbf{F}\right\} + \frac{1}{(1-2\nu)}\nabla\nabla.\left\{c\nabla^2\mathbf{F} - \nabla\nabla.\mathbf{F}\right\} = 0 \tag{8.813}$$

By considering component expansion we can show that

$$\nabla^2[\nabla(\nabla.\mathbf{F})] \equiv \nabla[\nabla.\nabla^2\mathbf{F}] \equiv \nabla\{\nabla.[\nabla(\nabla.\mathbf{F})]\} \tag{8.814}$$

Using this result in (8.813) we obtain

$$c\nabla^2\nabla^2\mathbf{F} + \left\{\frac{c - 2(1 - \nu)}{(1 - 2\nu)}\right\}\nabla^2[\nabla(\nabla.\mathbf{F})] = 0 \tag{8.815}$$

since $0 < \nu < \frac{1}{2}$, we can select the arbitrary constant c as

$$c = 2(1 - \nu) \tag{8.816}$$

which reduces (8.815) to

$$\nabla^4\mathbf{F} = 0 \tag{8.817}$$

Hence, in the absence of body forces, the Galerkin vector function $\mathbf{F}$ satisfies the biharmonic equation and the components F_x, F_y and F_z are each biharmonic. It should be noted that if other coordinate systems are used, the unit vectors themselves are functions of the coordinates and the components of the Galerkin vector are *not* biharmonic. For example, in the cylindrical polar coordinate system (r, θ, z), we assume $\mathbf{F}$ to be of the form

$$\mathbf{F} = F_r\mathbf{i}_r + F_\theta\mathbf{i}_\theta + F_z\mathbf{i}_z \tag{8.818}$$

and we can show that F_r and F_θ satisfy the equations

$$\widetilde{\nabla}^4 F_r = \left(\widetilde{\nabla}^2 - \frac{1}{r^2}\right)^2 F_r - \frac{4}{r^4}\frac{\partial^2 F_r}{\partial\theta^2} - \frac{2}{r^2}\left[\frac{\partial}{\partial\theta}\left(\widetilde{\nabla}^2 - \frac{1}{r^2}\right)F_\theta\right]$$
$$-2\left(\widetilde{\nabla}^2 - \frac{1}{r^2}\right)\frac{1}{r^2}\frac{\partial F_\theta}{\partial\theta} \tag{8.819}$$

$$\widetilde{\nabla}^4 F_\theta = \left(\widetilde{\nabla}^2 - \frac{1}{r^2}\right)^2 F_\theta - \frac{4}{r^4}\frac{\partial^2 F_r}{\partial\theta^2} + \frac{2}{r^2}\frac{\partial}{\partial\theta}\left[\left(\widetilde{\nabla}^2 - \frac{1}{r^2}\right)F_r\right]$$
$$+2\left(\widetilde{\nabla}^2 - \frac{1}{r^2}\right)\frac{1}{r^2}\frac{\partial F_r}{\partial\theta} \tag{8.820}$$

where

$$\widetilde{\nabla}^2 = \frac{\partial^2}{\partial r^2} + \frac{1}{r}\frac{\partial}{\partial r} + \frac{1}{r^2}\frac{\partial^2}{\partial\theta^2} + \frac{\partial^2}{\partial z^2} \tag{8.821}$$

is Laplace's operator referred to the generalized system of cylindrical polar coordinates. Only F_z satisfies the biharmonic equation

$$\widetilde{\nabla}^2\widetilde{\nabla}^2 F_z = 0 \tag{8.822}$$

the completeness of the representation (8.806) has been investigated quite extensively and details of the proofs are given many standard texts, treatises and review articles devoted to the classical theory of elasticity. A detailed exposition of three-dimensional problems in classical elasticity which makes use of the biharmonic vector function $\mathbf{F}$, admittedly referred to a Cartesian coordinate system, is beyond the scope of this Chapter. An interested reader could consult the bibliography given at the end of the book. We shall, however, outline the biharmonic function approach to the solution of axisymmetric problems which are referred to the cylindrical and spherical polar coordinate systems.

8.10.3   General properties of biharmonic functions

Prior to the solution of certain axisymmetric problems in classical elasticity which utilizes the biharmonic formulation it is useful to establish certain general relationships between *harmonic functions* and *biharmonic functions*.

THEOREM 8.9

If $\varphi_n(\mathbf{x})$ $(n = 0, 1, ..., 4)$ are harmonic functions then $\left(\varphi_0 + x\varphi_1 + y\varphi_2 + z\varphi_3 + R^2\varphi_4\right)$ is a biharmonic function, where $R^2 = \left(|\mathbf{x}|\right)^2$.

PROOF

Consider the gradient of the product of two functions φ and χ (see e.g. Chapter 1 (1.27)) we have

$$\nabla\left(\varphi\chi\right) = \left(\nabla\varphi\right)\chi + \varphi(\nabla\chi) \tag{8.823}$$

By taking the divergence of both sides of (8.823) we have

$$\nabla^2(\varphi\chi) = \nabla.\nabla(\varphi\chi) = \left(\nabla^2\varphi\right)\chi + 2\left(\nabla\varphi\right)\cdot\left(\nabla\chi\right) + \varphi(\nabla^2\chi) \tag{8.824}$$

Setting $\chi = x$, we note from (8.824), that

$$\nabla^2(x\varphi) = x\nabla^2\varphi + 2\frac{\partial\varphi}{\partial x} \tag{8.825}$$

If φ is a harmonic function, then by applying Laplace's operator to both sides of (8.825) we have

$$\nabla^2\nabla^2(x\varphi) = 2\nabla^2\left(\frac{\partial\varphi}{\partial x}\right) = 2\frac{\partial}{\partial x}\left(\nabla^2\varphi\right) = 0 \tag{8.826}$$

Hence, if φ is a harmonic, then $x\varphi$, $y\varphi$ and $z\varphi$ are all biharmonic.

Also consider the function $\varphi\mathbf{R}$ where φ is harmonic and $\mathbf{R} = x\mathbf{i} + y\mathbf{j} + z\mathbf{k}$, is the position vector. We have

$$\nabla\left[\left(\mathbf{R}.\nabla\right)\varphi\right] = \Im.\nabla\varphi + \left(\mathbf{R}.\nabla\right)\nabla\varphi \tag{8.827}$$

where $\Im$ is the unit dyadic (see e.g. (8.239)). Taking the divergence of both sides of (8.827) we have

$$\nabla^2\left[\left(\mathbf{R}.\nabla\right)\varphi\right] = \nabla.\nabla\left[\left(\mathbf{R}.\nabla\right)\varphi\right] = \nabla^2\varphi + \left(\Im.\nabla\right).\nabla\varphi + \left(\mathbf{R}.\nabla\right)\nabla^2\varphi$$

$$= 2\nabla^2\varphi + (\mathbf{R}.\nabla)\nabla^2\varphi \tag{8.828}$$

Hence if φ is harmonic, then $(\mathbf{R}.\nabla)\varphi$ is also harmonic. The function $R^2\varphi$ is therefore biharmonic, e.g.

$$\frac{\partial}{\partial x}(R^2\varphi) = 2x\varphi + R^2\frac{\partial\varphi}{\partial x}$$

$$\tag{8.829}$$

$$\frac{\partial^2}{\partial x^2}(R^2\varphi) = 2\varphi + 4x\frac{\partial\varphi}{\partial x} + R^2\frac{\partial^2\varphi}{\partial x^2}$$

which can be generalized to give

$$\nabla^2(R^2\varphi) = 6\varphi + 4\left(x\frac{\partial\varphi}{\partial x} + y\frac{\partial\varphi}{\partial y} + z\frac{\partial\varphi}{\partial z}\right) + R^2\nabla^2\varphi \tag{8.830}$$

Hence

$$\nabla^2\nabla^2(R^2\varphi) = 20\nabla^2\varphi + 8\left(x\frac{\partial}{\partial x} + y\frac{\partial}{\partial y} + z\frac{\partial}{\partial z}\right)\nabla^2\varphi$$

$$+ R^2\nabla^2(\nabla^2\varphi) \tag{8.831}$$

Therefore if φ is harmonic then $R^2\varphi$ is biharmonic. We can generalize the result by combining (8.826) and (8.831). If φ_0, φ_1, φ_2, φ_3 and φ_4 are harmonic functions, then

$$F = \varphi_0 + x\varphi_1 + y\varphi_2 + z\varphi_3 + R^2\varphi_4 \tag{8.832}$$

is biharmonic.

◉ ◉ ◉

Not all of the terms of (8.832) are independent. Three of the terms φ_i ($i = 1, 2, 3, 4$) can be chosen arbitrarily. This relationship between harmonic functions and biharmonic functions becomes useful in constructing solution to many problems in elasticity of engineering interest. In particular, a wide class of harmonic functions can be used to construct appropriate biharmonic functions. Particularly with axisymmetric problems, involving localized loadings applied at the interior of elastic media of infinite extent (Kelvin's problem) and loadings applied at the surface of half-space regions (Boussinesq's problem), the appropriate biharmonic function can be constructed by appeal to the general relationship between harmonic and biharmonic functions. Similarly, solution for the harmonic equation derived via integral transform techniques can also be used to obtain appropriate biharmonic solutions particularly for problems with axial symmetry.

Two further results involving the relationship between *harmonic* and *biharmonic* functions can be deduced by considering the converse of the result proved in Theorem 8.9.

THEOREM 8.10

If φ is biharmonic in the region V and if every line parallel to the x-axis intersects S the boundary of V at least at two points, then there exist two harmonic functions φ_1 and φ_2 in V such that φ admits the representation

$$\varphi(\mathbf{x}) = x\varphi_1(\mathbf{x}) + \varphi_2(\mathbf{x}) \tag{8.833}$$

PROOF

If we assume that the theorem is correct, then there exists a function such that

$$\nabla^2 \varphi_1 = 0 \tag{8.834}$$

and

$$\nabla^2(x\varphi_1 - \varphi) = \nabla^2 \varphi_2 = 0 \tag{8.835}$$

Since φ_1 is harmonic, we have from (8.825)

$$\nabla^2 \varphi = \nabla^2(x\varphi_1) = 2\frac{\partial \varphi_1}{\partial x} \tag{8.836}$$

The particular integral of (8.836) can be written as

$$\overline{\varphi}_1(\mathbf{x}) = \frac{1}{2} \int_{x_0}^{x} \nabla_{\xi}^2 \varphi(\xi, y, z) d\xi \tag{8.837}$$

where ∇_{ξ}^2 is Laplace's operator in rectangular Cartesian coordinates with the variable x replaced by the integration variable ξ, and x_0 is an arbitrary point in V. The solution $\overline{\varphi}_1(x)$ is not necessarily harmonic, but since φ is biharmonic, we have

$$\frac{\partial}{\partial x}(\nabla^2 \overline{\varphi}_1) = \nabla^2 \left(\frac{\partial \varphi_1}{\partial x}\right) = \frac{1}{2} \nabla^4 \varphi = 0 \tag{8.838}$$

The result (8.838) implies that $\nabla^2 \overline{\varphi}_1$ is function of the variables y and z, which we shall denote by $f(y, z)$. We now consider a function $\varphi_1^*(y, z)$ such that

$$\overset{*}{\nabla}^2 \varphi_1^*(y, z) = -f(y, z) \tag{8.839}$$

and $\overset{*}{\nabla}^2$ is Laplace's operator referred to the two-dimensional system of coordinates (y, z). The equation (8.839) is the two-dimensional equivalent of the inhomogeneous form of either Laplace's equation or Poisson's equation, the solution of which can be derived by making use of the Green's function for Laplace's equation in two dimensions (see e.g. Chapters 5 and 9). For two dimensions the Green's function takes the form

$$G(y, z; \xi, \eta) = \frac{1}{2\pi} \ln \left[(y - \xi)^2 + (z - \eta)^2\right]^{\frac{1}{2}} \tag{8.840}$$

and the solution of (8.839) can be written as

$$\varphi_1^*(y, z) = -\frac{1}{2\pi} \int \int_{\mathbb{R}} \ln \left[(y - \xi)^2 + (z - \eta)^2\right]^{\frac{1}{2}} f(\xi, \eta) d\xi d\eta \tag{8.841}$$

and $\mathbb{R}$ is the region corresponding to the ranges of y and z. The function $\varphi_1 = \overline{\varphi}_1 + \varphi_1^*$ therefore satisfies the conditions (8.834) and (8.836) and the theorem is proved.

◎ ◎ ◎

The Theorem 8.10 can be generalized to include the case where the intersection of the boundary of V with a radius vector occurs at least at one point.

THEOREM 8.11

Consider a region V with boundary S, such that the origin of coordinates is located within V and each radius vector intersects the boundary S at least at one point. Let $\varphi(\mathbf{x})$ be a biharmonic function in V; then there exists two functions $\varphi_1(\mathbf{x})$ and $\varphi_2(\mathbf{x})$ which are harmonic in V such that

$$\varphi(\mathbf{x}) = (R^2 - R_0^2)\varphi_1(\mathbf{x}) + \varphi_2(\mathbf{x}) \tag{8.842}$$

where $R^2 = (x^2 + y^2 + z^2)$ and R_0 is a constant.

PROOF

We proceed as in Theorem 8.10 by assuming that the stated theorem is satisfied, then there exists a function $\varphi(\mathbf{x})$ which satisfies (8.842). Since $\varphi_1(\mathbf{x})$ and $\varphi_2(\mathbf{x})$ are harmonic we have

$$\nabla^2 \varphi_1 = 0 \tag{8.843}$$

and

$$\nabla^2 \left[\varphi - (R^2 - R_0^2)\varphi_1 \right] = 0 \tag{8.844}$$

Using the developments associated with the proof of Theorem 8.9, (8.844) can be reduced to

$$\nabla^2 \varphi = 6\varphi_1 + 4R\frac{\partial \varphi_1}{\partial R} \tag{8.845}$$

An integral of (8.845) can be written as

$$\varphi_1 = \frac{1}{R^{\frac{3}{2}}} \int_0^R \frac{\sqrt{\zeta}}{4}(\nabla^2 \varphi)d\zeta \tag{8.846}$$

To show that this integral satisfies (8.843), we utilize a representation of Laplace's operator with reference to a spherical polar coordinate system (R, α, Θ) such that

$$r = R\sin\Theta \quad ; \quad z = R\cos\Theta \tag{8.847}$$

and

$$\widehat{\nabla}^2_S = \frac{\partial^2}{\partial R^2} + \frac{2}{R}\frac{\partial}{\partial R} + \frac{\cot\Theta}{R^2}\frac{\partial}{\partial\Theta} + \frac{1}{R^2}\frac{\partial^2}{\partial\Theta^2} \tag{8.848}$$

Operating on both sides of (8.846) with $\widehat{\nabla}^2_S$ we have

$$\widehat{\nabla}^2_S\varphi_1 = \widehat{\nabla}^2_S\left\{\frac{1}{R^{\frac{3}{2}}}\int_0^R \frac{\sqrt{\zeta}}{4}\widehat{\nabla}^2_\zeta\varphi\,d\zeta\right\} \tag{8.849}$$

where $\widehat{\nabla}^2_\zeta$ is $\widehat{\nabla}^2_S$ where R is replaced by ζ. The operator $\widehat{\nabla}^2_S$ on the right hand side of (8.849) can be taken under the integral sign, and since $\widehat{\nabla}^2_S\widehat{\nabla}^2_S\varphi = 0$ we have

$$\left(\cot\Theta\frac{\partial}{\partial\Theta} + \frac{\partial^2}{\partial\Theta^2}\right)\widehat{\nabla}^2_S\varphi = -\frac{\partial}{\partial R}\left(R^2\frac{\partial}{\partial R}\right)\widehat{\nabla}^2_S\varphi \tag{8.850}$$

and (8.849) can be written as

$$\begin{aligned}
\widehat{\nabla}^2_S\varphi_1 = {}& \frac{1}{R^2}\frac{\partial}{\partial R}\left(R^2\frac{\partial}{\partial R}\right)\frac{1}{R^{\frac{3}{2}}}\int_0^R \frac{\sqrt{\zeta}}{4}\widehat{\nabla}^2_\zeta\varphi\,d\zeta \\
& -\frac{1}{R^{\frac{7}{2}}}\int_0^R \frac{\sqrt{\zeta}}{4}\frac{\partial}{\partial\zeta}\left(\zeta^2\frac{\partial}{\partial\zeta}\right)\widehat{\nabla}^2_\zeta\varphi\,d\zeta
\end{aligned} \tag{8.851}$$

Performing the differentiation of the first term and integrating the second term by parts twice, we can show that

$$\widehat{\nabla}^2_S \varphi_1 = \frac{1}{R^2} \left\{ \frac{3}{16R^{\frac{3}{2}}} \int_0^R \sqrt{\zeta}\,\widehat{\nabla}^2_\zeta \varphi\,d\zeta - \frac{3}{8}\widehat{\nabla}^2_S \varphi + \frac{1}{4}\widehat{\nabla}^2_S \varphi \right.$$
$$\left. + \frac{R}{4}\frac{\partial}{\partial R}\left(\widehat{\nabla}^2_S \varphi\right) \right\} - \frac{1}{4R}\frac{\partial}{\partial R}\left(\widehat{\nabla}^2_S \varphi\right) + \frac{1}{8}\widehat{\nabla}^2_S \varphi$$
$$- \frac{3}{16R^{\frac{7}{2}}} \int_0^R \sqrt{\zeta}\,\widehat{\nabla}^2_\zeta \varphi\,d\zeta = 0 \tag{8.852}$$

Hence φ_1 given by (8.846) satisfies all the requirements and the theorem is proved.

8.10.4 Love's strain function

In this section we shall outline a *strain function* approach proposed by Love (1863-1940) for the solution of axisymmetric problems in elasticity, which essentially results in a biharmonic equation for a scalar strain potential Φ, and precedes that proposed by Galerkin. In essence, the Galerkin vector approach can be regarded as a generalization of Love's approach for axisymmetric problems, referred to a cylindrical polar coordinate system (r, θ, z). When the Galerkin vector $\mathbf{F}$ is set equal to

$$\mathbf{F} = \Phi \mathbf{i}_z \tag{8.853}$$

we obtain, in the absence of body forces,

$$\widetilde{\nabla}^4 \Phi = 0 \tag{8.854}$$

If we assume that, in general, $\Phi = \Phi(r, \theta, z)$, then the components of the displacement vector

$$\mathbf{u} = u_r \mathbf{i}_r + u_\theta \mathbf{i}_\theta + u_z \mathbf{i}_z \tag{8.855}$$

referred to the cylindrical polar coordinate system can be expressed in terms of Φ, in the form

$$2\mu u_r = -\frac{\partial^2 \Phi}{\partial r \partial z} \quad ; \quad 2\mu u_\theta = -\frac{1}{r}\frac{\partial^2 \Phi}{\partial \theta \partial z}$$

$$\tag{8.856}$$

$$2\mu u_z = 2(1-\nu)\widetilde{\nabla}^2 \Phi - \frac{\partial^2 \Phi}{\partial z^2}$$

and $\widetilde{\nabla}^2$ is defined in (8.821).

The stress matrix $\boldsymbol{\sigma}$ referred to the cylindrical polar coordinate system has the following components:

$$\boldsymbol{\sigma} = \begin{bmatrix} \sigma_{rr} & \sigma_{r\theta} & \sigma_{rz} \\ \sigma_{r\theta} & \sigma_{\theta\theta} & \sigma_{\theta z} \\ \sigma_{rz} & \sigma_{\theta z} & \sigma_{zz} \end{bmatrix} = \boldsymbol{\sigma}^T \tag{8.857}$$

The components of $\boldsymbol{\sigma}$ can be expressed in terms of Φ as

$$\sigma_{rr} = \frac{\partial}{\partial z}\left[\nu\widetilde{\nabla}^2\Phi - \frac{\partial^2\Phi}{\partial r^2} \right]$$

$$\sigma_{\theta\theta} = \frac{\partial}{\partial z}\left[\nu\widetilde{\nabla}^2\Phi - \frac{1}{r}\frac{\partial\Phi}{\partial r} - \frac{1}{r^2}\frac{\partial^2\Phi}{\partial\theta^2} \right]$$

$$\sigma_{zz} = \frac{\partial}{\partial z}\left[(2-\nu)\widetilde{\nabla}^2\Phi - \frac{\partial^2\Phi}{\partial z^2} \right]$$

$$\sigma_{r\theta} = -\frac{\partial^3}{\partial r\partial\theta\partial z}\left[\frac{\Phi}{r} \right] \tag{8.858}$$

$$\sigma_{\theta z} = \frac{1}{r}\frac{\partial}{\partial\theta}\left[(1-\nu)\widetilde{\nabla}^2\Phi - \frac{\partial^2\Phi}{\partial z^2} \right]$$

$$\sigma_{rz} = \frac{\partial}{\partial r}\left[(1-\nu)\widetilde{\nabla}^2\Phi - \frac{\partial^2\Phi}{\partial z^2} \right]$$

In the remainder of the presentation we shall restrict attention to the application of Love's approach to axisymmetric problems in elasticity referred to the cylindrical polar coordinate system r, θ, z. In axisymmetric problem, the displacements, strains and stresses in the medium are independent of θ and due to symmetry $u_\theta \equiv 0$. The displacement vector is

$$\mathbf{u} = u_r(r, z)\mathbf{i}_r + u_z(r, z)\mathbf{i}_z \tag{8.859}$$

The components of the strain matrix $\boldsymbol{\epsilon}_c$ referred to the cylindrical polar coordinate system can be obtained by transforming (8.68) to the cylindrical polar coordinate system. We can show that

$$\boldsymbol{\epsilon}_c = \begin{bmatrix} \epsilon_{rr} & 0 & \epsilon_{rz} \\ 0 & \epsilon_{\theta\theta} & 0 \\ \epsilon_{rz} & 0 & \epsilon_{zz} \end{bmatrix} = \begin{bmatrix} \dfrac{\partial u_r}{\partial r} & 0 & \dfrac{1}{2}\left(\dfrac{\partial u_r}{\partial z} + \dfrac{\partial u_z}{\partial r} \right) \\ 0 & \dfrac{u_r}{r} & 0 \\ \dfrac{1}{2}\left(\dfrac{\partial u_r}{\partial z} + \dfrac{\partial u_z}{\partial r} \right) & 0 & \dfrac{\partial u_z}{\partial z} \end{bmatrix} \tag{8.860}$$

The state of stress in an elastic medium which exhibits axial symmetry is defined by

$$\boldsymbol{\sigma} = \begin{bmatrix} \sigma_{rr} & 0 & \sigma_{rz} \\ 0 & \sigma_{\theta\theta} & 0 \\ \sigma_{rz} & 0 & \sigma_{zz} \end{bmatrix} \tag{8.861}$$

and the stresses acting on an infinitesimal element are shown in Figure 8.46. The equations of equilibrium in the elastic body referred to the state of axial symmetry can be obtained by examining force equilibrium of the tractions acting in the radial and axial directions (Figure 8.46). (Note that due to axial symmetry, $\sigma_{\theta\theta}$ does not exhibit a variation in the θ-direction. Considering the equilibrium of forces in the r and z directions (and neglecting body forces) we have

$$\left(\sigma_{rr} + \frac{\partial \sigma_{rr}}{\partial r} dr \right) (r + dr)\, d\theta dz - \sigma_{rr} r\, d\theta dz - \sigma_{\theta\theta} d\theta dr dz$$

$$+ \left(\sigma_{rz} + \frac{\partial \sigma_{rz}}{\partial z} dz \right) r\, dr d\theta - \sigma_{rz} r\, dr d\theta = 0 \tag{8.862}$$

and

$$\left(\sigma_{zz} + \frac{\partial \sigma_{zz}}{\partial z} dz \right) r\, dr d\theta - \sigma_{zz} r\, dr d\theta$$

$$+ \left(\sigma_{rz} + \frac{\partial \sigma_{rz}}{\partial r} dr \right) (r + dr)\, d\theta dz + \sigma_{rz} r\, d\theta dz = 0 \tag{8.863}$$

respectively. These reduce to

$$\frac{\partial \sigma_{rr}}{\partial r} + \frac{\partial \sigma_{rz}}{\partial z} + \frac{\sigma_{rr} - \sigma_{\theta\theta}}{r} = 0 \tag{8.864}$$

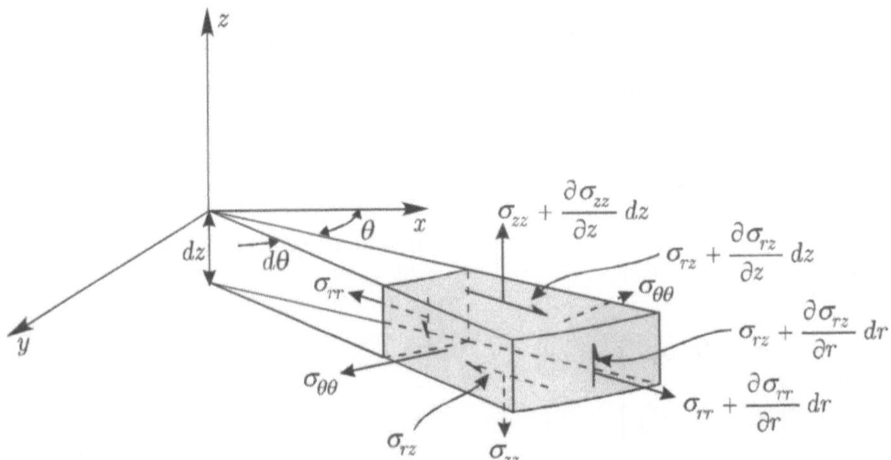

Figure 8.46: Definition of stresses in an elastic medium, exhibiting axial symmetry.

and

$$\frac{\partial \sigma_{zz}}{\partial z} + \frac{\partial \sigma_{rz}}{\partial r} + \frac{\sigma_{rz}}{r} = 0 \tag{8.865}$$

respectively.

The stress-strain relationships governing an axisymmetric state of stress take the forms

$$\epsilon_{rr} = \frac{1}{E}\left[\sigma_{rr} - \nu(\sigma_{\theta\theta} + \sigma_{zz})\right] \quad ; \quad \epsilon_{\theta\theta} = \frac{1}{E}\left[\sigma_{\theta\theta} - \nu(\sigma_{rr} + \sigma_{zz})\right] \tag{8.866}$$

$$\epsilon_{zz} = \frac{1}{E}\left[\sigma_{zz} - \nu(\sigma_{rr} + \sigma_{\theta\theta})\right] \quad ; \quad \epsilon_{rz} = \frac{\sigma_{rz}}{2\mu}$$

For axial symmetry, Love's strain function $\Phi = \Phi(r, z)$, is governed by the biharmonic equation

$$\widehat{\nabla}^2\widehat{\nabla}^2\Phi(r, z) = 0 \tag{8.867}$$

where

$$\widehat{\nabla}^2 = \frac{\partial^2}{\partial r^2} + \frac{1}{r}\frac{\partial}{\partial r} + \frac{\partial^2}{\partial z^2} \tag{8.868}$$

is the axisymmetric form of Laplace's operator. The representations of u_r, u_z, σ_{rr}, etc. in terms of $\Phi(r,z)$ can be obtained by setting $\partial/\partial\theta = 0$ and $\widetilde{\nabla}^2 \to \widehat{\nabla}^2$ in equations (8.856) and (8.858). This gives

$$2\mu u_r = -\frac{\partial^2\Phi}{\partial r\partial z}; \quad 2\mu u_z = 2(1-\nu)\widehat{\nabla}^2\Phi - \frac{\partial^2\Phi}{\partial z^2} \tag{8.869}$$

and

$$\sigma_{rr} = \frac{\partial}{\partial z}\left[\nu\widehat{\nabla}^2\Phi - \frac{\partial^2\Phi}{\partial r^2}\right]$$

$$\sigma_{\theta\theta} = \frac{\partial}{\partial z}\left[\nu\widehat{\nabla}^2\Phi - \frac{1}{r}\frac{\partial\Phi}{\partial r}\right]$$

$$\sigma_{zz} = \frac{\partial}{\partial z}\left[(2-\nu)\widehat{\nabla}^2\Phi - \frac{\partial^2\Phi}{\partial z^2}\right] \tag{8.870}$$

$$\sigma_{rz} = \frac{\partial}{\partial r}\left[(1-\nu)\widehat{\nabla}^2\Phi - \frac{\partial^2\Phi}{\partial z^2}\right]$$

respectively.

8.10.5 Axisymmetric problem formulation

In the absence of body forces, the analysis of the axisymmetric problem in classical elastostatics, referred to the cylindrical polar coordinate system (r,θ,z), is reduced to the solution of the biharmonic equation

$$\widehat{\nabla}^2\widehat{\nabla}^2\Phi(r,z) = 0 \quad ; \quad (r,z) \in V \tag{8.871}$$

where $\Phi(r,z)$ is Love's strain function, $\widehat{\nabla}^2$ is the axisymmetric form of Laplace's operator referred to (r,θ,z) and V is the domain of interest.

The solution of (8.871) is subject to displacement and traction boundary conditions specified on subsets of S the boundary of V. The displacement boundary conditions can be specified as

$$\mathbf{u}(r, z) = \mathbf{u}^0(r, z) \quad ; \quad (r, z) \in S_D \tag{8.872}$$

where $\mathbf{u}^0$ are prescribed displacements on the boundary S_D. The traction boundary conditions can be prescribed as

$$\mathbf{T}(r, z) = \boldsymbol{\sigma}\mathbf{n} = \mathbf{T}^0(r, z) \quad ; \quad (r, z) \in S_T \tag{8.873}$$

where $\mathbf{T}^0$ are prescribed tractions of the surface S_T and $\mathbf{n}$ is the outward unit normal to S_T.

In certain instances, particularly involving contact problems in the theory of elasticity, specific components of the traction vector and the displacement vector can be specified on the same surface. For example, the axisymmetric indentation of the surface of an elastic half-space by a smooth rigid indentor with a profile $w(r)$ (Figure 8.47) gives rise to the following mixed boundary conditions in the contact region.

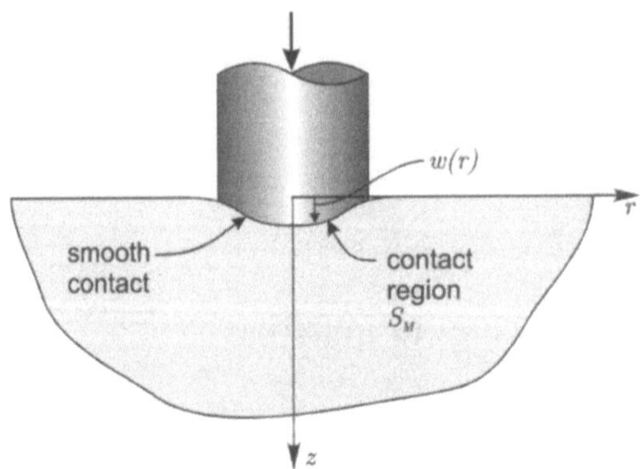

Figure 8.47: Indentation of elastic half-space by a smooth indentor with a profile $w(r)$.

The displacement boundary condition in the region S_M is

$$u_z(r, 0) = w(r) \quad ; \quad r \in S_M \tag{8.874}$$

The traction boundary condition in the region S_M is

$$\sigma_{rz}(r,0) = 0 \quad ; \quad r \in S_M \tag{8.875}$$

8.10.6 Polynomial solutions of the biharmonic equation; the interior solution

The development of polynomial solutions is facilitated if, at the outset, the axisymmetric for of Laplace's operator given by (8.848) is represented by its equivalent in the axisymmetric spherical polar coordinate system (R, ϑ, Θ) where $r = R \sin \Theta$ and $z = R \cos \Theta$. (This notation is chosen to avoid possible confusion with φ, which is also used in this section to designate a harmonic function). It can be shown that (e.g. Section 8.9.4) in the spherical polar coordinate system, the axisymmetric form of Laplace's operator is given by

$$\widehat{\nabla}_S^2 = \frac{1}{R^2} \frac{\partial}{\partial R} \left(R^2 \frac{\partial}{\partial R} \right) + \frac{1}{R^2 \sin \Theta} \frac{\partial}{\partial \Theta} \left(\sin \Theta \frac{\partial}{\partial \Theta} \right) \tag{8.876}$$

We seek solutions of

$$\widehat{\nabla}_S^2 \varphi(R, \Theta) = 0 \tag{8.877}$$

of the form

$$\varphi_n(R, \Theta) = R^n G_n(\Theta) \tag{8.878}$$

From the form of (8.878), it is clear that the solutions developed will be valid only for finite R which includes $R = 0$. Hence the solution developed is usually referred to as the interior solution. Substituting (8.878) in (8.777) we obtain

$$\frac{1}{\sin \Theta} \frac{d}{d\Theta} \left(\sin \Theta \frac{dG_n}{d\Theta} \right) + n(n+1)G_n = 0 \tag{8.879}$$

Introducing a change of variable such that

$$\xi = \cos \Theta \tag{8.880}$$

the ordinary differential equation (8.879) can be transformed into *Legendre's equation* (named after A.M. Legendre (1752-1833))

$$(1 - \xi^2)\frac{d^2 G_n}{d\xi^2} - 2\xi\frac{dG_n}{d\xi} + n(n+1)G_n = 0 \qquad (8.881)$$

The solution of (8.881) can be approached by representing G_n as a polynomial in ξ and using (8.881) to obtain a set of recurrence relations. The solution procedure can be found in a number of standard texts on special functions, including the general references cited at the end of the text. The two basic classes of solutions of (8.881) are by $P_n(\xi)$ the Legendre function of the first kind and $Q_n(\xi)$ the Legendre function of the second kind. For integer values of $n = 0, 1, 2, 3, ...$, the expressions for $P_n(\xi)$ take the form

$$P_0(\xi) = 1 \quad ; \quad P_1(\xi) = \xi \quad ; \quad P_2(\xi) = \frac{1}{2}\left[3\xi^2 - 1\right]$$

$$P_3(\xi) = \frac{1}{2}\left[5\xi^2 - 3\xi\right]$$

$$P_4(\xi) = \frac{1}{8}\left[35\xi^2 - 30\xi^2 + 3\right] \qquad (8.882)$$

$$P_5(\xi) = \frac{1}{8}\left[63\xi^5 - 70\xi^3 + 15\xi\right]$$

$$..... = ...$$

etc., and the general recurrence relationship is

$$(n + 1)P_{n+1}(\xi) - (2n + 1)\xi P_n(\xi) + nP_{n-1}(\xi) = 0 \qquad (8.883)$$

Similarly, for integer values of $n = 0, 1, 2, 3, ...$, the expressions for $Q_n(\xi)$ take the forms

$$Q_0(\xi) = \frac{1}{2}\ln(q) \quad ; \quad Q_1(\xi) = \frac{\xi}{2}\ln(q) - 1$$

$$Q_2(\xi) = \frac{1}{4}\left[3\xi^2 - 1\right]\ln(q) - \frac{3\xi}{2}$$

$$Q_3(\xi) = \frac{1}{4}\left[5\xi^2 - 3\xi\right]\ln(q) - \frac{5\xi^2}{2} + \frac{2}{3}$$

$$Q_4(\xi) = \frac{1}{16}\left[35\xi^4 - 30\xi^2 + 3\right]\ln(q) - \frac{35\xi^2}{8} + \frac{55\xi}{24} \tag{8.884}$$

$$\ldots = \ldots$$

etc., where

$$q = \left(\frac{1+\xi}{1-\xi}\right) \tag{8.885}$$

and the general recurrence relationship is

$$(n+1)Q_{n+1}(\xi) - (2n+1)\xi Q_n(\xi) + nQ_{n-1}(\xi) = 0 \tag{8.886}$$

We can convert the solutions in terms of Legendre functions to the (r, z) coordinate system by using the substitutions

$$\xi = \cos\Theta \quad ; \quad R\xi = z \quad ; \quad R^2 = (r^2 + z^2) \tag{8.887}$$

and the Legendre polynomial solutions for $\varphi_n(r, z)$ corresponding to $P_n(\xi)$ take the forms

$$\varphi_0(r, z) = C_0$$

$$\varphi_1(r, z) = C_1 z$$

$$\varphi_2(r, z) = C_2 \left[z^2 - \frac{1}{3}(r^2 + z^2) \right] \tag{8.888}$$

$$\varphi_3(r, z) = C_3 \left[z^3 - \frac{3}{5}z(r^2 + z^2) \right]$$

$$\varphi_4(r, z) = C_4 \left[z^4 - \frac{6}{7}z^2(r^2 + z^2) + \frac{3}{35}(r^2 + z^2)^2 \right]$$

$$\varphi_5(r, z) = C_5 \left[z^5 - \frac{10}{9}z^3(r^2 + z^2) + \frac{5z}{21}(r^2 + z^2)^2 \right]$$

...etc., where C_n are constants. These represent solutions of Laplace's equation for $\varphi_n(r, z)$. Following the proofs presented in Section 8.10.2, we note that $\Phi_n = R^2 \varphi_n(r, z)$ are biharmonic. The solutions of the biharmonic equation constructed this way take the forms

$$\Phi_2(r, z) = B_2(r^2 + z^2)$$

$$\Phi_3(r, z) = B_3 z(r^2 + z^2)$$

$$\Phi_4(r, z) = B_4(2z^2 - r^2)(r^2 + z^2) \tag{8.889}$$

$$\Phi_5(r, z) = B_5(2z^3 - 3r^2 z)(r^2 + z^2)$$

etc., where $B_2, B_3, \ldots$ are constants and the subscript of the constants in both (8.888) and (8.889) are indicative of the degree of the polynomial.

Similar procedures can be used to construct the sets of harmonic and biharmonic functions corresponding to $Q_n(\xi)$. These solutions will be discussed in relation to specific problems. It may, however, be noted that the functions $Q_n(\xi)$ contain the term $\ln\left[(1 + \xi)/(1 - \xi)\right]$ which can be written in the form

$$\ln\left(\frac{1+\xi}{1-\xi}\right) = \ln\left[\frac{\sqrt{r^2 + z^2} + z}{\sqrt{r^2 + z^2} - z}\right] \tag{8.890}$$

which is singular as $r \to 0$. (i.e. on the z-axis). Therefore the solutions of the biharmonic equation which are obtained from $Q_n(\xi)$ are admissible when such solutions exhibit the singular behaviour prescribed through traction and displacement boundary conditions of the problem (e.g. localized axial forces acting in the z-direction and located on the z-axis).

In the ensuing we shall apply these elementary polynomial solutions of the type (8.888) and (8.889) to examine elastic states of stress in finite regions.

Example 8.22

A circular plate shaped region on an elastic medium is subjected to the radial traction distribution shown in Figure 8.48. Determine the state of stress in the plate and the resulting displacement field assuming that at the origin of coordinates the axial displacement is zero.

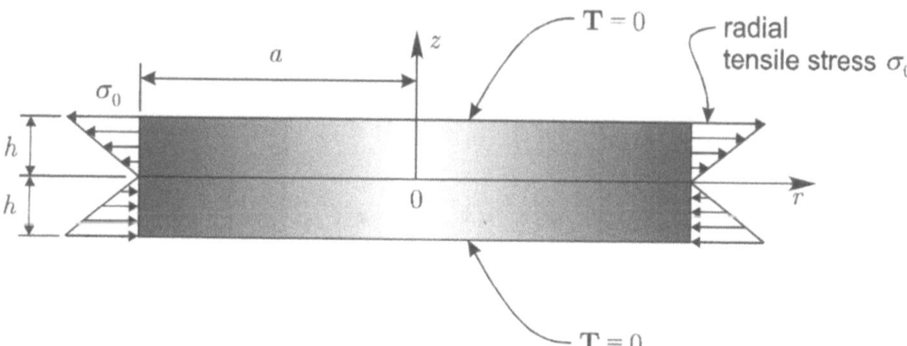

Figure 8.48: Pure bending of a circular plate.

Solution

This problem is the axisymmetric analogue of the Example 8.7 which examined the pure bending of a rectangular beam section. The traction boundary conditions governing the problem are as follows; on the plane surfaces we require

$$\sigma_{zz}(r, h) = \sigma_{zz}(r, -h) = 0$$

$$\sigma_{rz}(r, h) = \sigma_{rz}(r, -h) = 0$$

$$(8.891)$$

and on the cylindrical surface we require

$$\sigma_{rr}(a, z) = \sigma_0 \frac{z}{h}$$

$$\sigma_{rz}(a, z) = 0$$
(8.892)

We note that the radial stress is linear in z at the boundary $r = a$, and from (8.870), in order to achieve such a variation, a polynomial of the 4th degree is required. We can select solutions of the 4th degree polynomials which are harmonic as well as biharmonic and defined by (8.888) and (8.889) respectively. A trial solution can be of the form

$$\Phi(r, z) = C_4^*(8z^4 - 24r^2z^2 + 3r^4) + B_4^*(2z^4 + r^2z^2 - r^4)$$
(8.893)

where C_4^* and B_4^* are arbitrary constants. The term *trial solution* is used primarily because of the existence of two such solutions which can be combined in a linear fashion. The guesswork , of course, has been eliminated by identifying a priori the required degree of the polynomial solutions. Substituting this form of Love's strain function in (8.870) we obtain

$$\sigma_{rr}(r, z) = [96C_4^* + 4B_4^*(14\nu - 1)]\, z = \sigma_{\theta\theta}(r, z)$$

$$\sigma_{zz}(r, z) = [-192C_4^* + 4B_4^*(16 - 14\nu)]\, z$$
(8.894)

$$\sigma_{rz}(rz) = [96C_4^* - 2B_4^*(16 - 14\nu)]\, r$$

If we apply the boundary conditions (8.891), both these conditions can be simultaneously satisfied if we set

$$C_4^* = \frac{B_4^*}{24}(8 - 7\nu)$$
(8.895)

The stress field now reduces to

$$\sigma_{rr}(r, z) = \sigma_{\theta\theta}(r, z) = 28(1 + \nu)B_4^*z$$
(8.896)

$$\sigma_{rz} = \sigma_{zz} = 0$$

Considering (8.896) and the first boundary condition of (8.892) we have

$$B_4^* = \frac{\sigma_0}{28h(1+\nu)} \tag{8.897}$$

and

$$\sigma = \begin{bmatrix} \sigma_0 \dfrac{z}{h} & 0 & 0 \\[2ex] 0 & \sigma_0 \dfrac{z}{h} & 0 \\[2ex] 0 & 0 & 0 \end{bmatrix} \tag{8.898}$$

The displacement components associated with this pure bending mode can be obtained from the result (8.869); i.e.

$$u_r(r,z) = \frac{\sigma_0(1-\nu)}{2\mu h(1+\nu)}(rz)$$

$$\tag{8.899}$$

$$u_z(r,z) = \frac{\sigma_0}{2\mu h(1+\nu)}\left[-\nu z^2 - (1-\nu)\frac{r^2}{2}\right]$$

It may be noted that the solution to the problem of axisymmetric pure bending of a circular plate shaped region is an *exact* mathematical solution, within the context of the classical theory of elasticity. Similar remarks are applicable to the problem of pure bending of a rectangular plane region discussed in Problem 8.7. Also, in the case of plane strain deformations of the rectangular region subjected to pure bending, (8.470) give $tr\,\epsilon \equiv 0$, throughout the rectangular region irrespective of Poisson's ratio of the material. In the circular plate, however,

$$tr\,\epsilon = \frac{\sigma_0(1-2\nu)}{\mu(1+\nu)}\frac{z}{h} \tag{8.900}$$

and *isochoric* or *volume preserving* deformations at each point of the circular plate region will occur only when $\nu = \frac{1}{2}$. Volume is however preserved in a global sense during pure bending of the circular plate. (i.e. $\int\int\int_V (tr\,\epsilon)dV \equiv 0$).

$$\bullet\ \bullet\ \bullet$$

The axisymmetric problem concerning the pure bending of a circular plate shaped region by prescribed radial traction over the cylindrical surface is a very specific example where all the traction boundary conditions are *exactly* satisfied. In other applications involving polynomial solutions the procedure involved is somewhat inductive, in the sense that, various harmonic and biharmonic functions are combined in an effort to satisfy as many boundary conditions exactly and to ensure that those boundary conditions that cannot be satisfied exactly are satisfied in an integral sense. The major justification for this approximation is that the effect of any discrepancy in satisfying the boundary condition in an average or integral sense is limited to a local region. This is derived from a celebrated principle due to Saint-Venant (1797-1886) which can be invoked, at least for homogeneous and isotropic elastic solids, to render the mathematical analysis possible. Further discussion on this aspect is given in an Appendix at the end of Chapter 8. The primary objective of a polynomial based solution is, in general, to satisfy the boundary conditions exactly on surfaces where traction boundary conditions are prescribed (usually regions with comparatively larger dimensions) and to satisfy the boundary conditions on the remaining surfaces in an average or integral sense.

Example 8.23

The Figure 8.49 shows a circular plate shaped elastic solid which is subjected to uniform normal surface tractions on the boundary $z = h$. Develop a solution to the problem by assuming that the traction boundary conditions are satisfied exactly on the plane surfaces $z = \pm h$ and the traction boundary conditions on the cylindrical surface are satisfied in an integral sense.

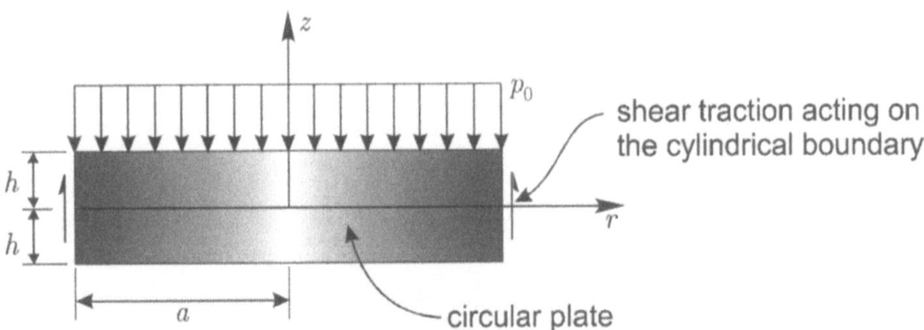

Figure 8.49: Uniformly loaded circular plate.

Solution

The traction boundary conditions on $z = \pm h$, that need to be *exactly satisfied* are as follows:

$$\sigma_{zz}(r, h) = -p_0 \quad ; \quad \sigma_{zz}(r, -h) = 0$$

$$\sigma_{rz}(r, \pm h) = 0 \tag{8.901}$$

The boundary conditions that need to be satisfied in an *integral* sense are as follows:

$$\int_{-h}^{h} \sigma_{rr}(a, z)dz = 0 \quad ; \quad \int_{-h}^{h} z\sigma_{rr}(a, z)dz = 0$$

$$\int_{-h}^{h} \sigma_{rz}(a, z)dz = \frac{p_0 a}{2} \tag{8.902}$$

The last equation of (8.902) is a condition for the overall equilibrium of the plate region.

Considering the form of the expression for σ_{zz} in (8.870) and the nature of the first two boundary conditions of (8.901), we require an *even-valued* function of Φ up to the sixth degree in z. We can select both harmonic and biharmonic functions to construct a Love strain potential as follows:

$$\Phi(r, z) = \frac{C_6^*}{3} \left\{ 16z^6 - 120z^4r^2 + 90z^2r^4 - 5r^6 \right\}$$
$$+ B_6^* \left\{ 8z^6 - 16z^4r^2 - 21z^2r^4 + 3r^6 \right\} \tag{8.903}$$

The stress components corresponding to (8.903) can be determined by making use of (8.870); i.e.

$$\sigma_{rr} = C_6^* \left\{ 320z^3 - 720r^2z \right\}$$
$$+ B_6^* \left\{ 64(2 + 11\nu)z^3 + (504 - 1056\nu)r^2z \right\}$$
$$\sigma_{zz} = C_6^* \left\{ -640z^3 + 960r^2z \right\}$$
$$+ B_6^* \left\{ [-960 + 704(2 - \nu)]z^3 + [384 - 1056(2 - \nu)]r^2z \right\} \tag{8.904}$$
$$\sigma_{rz} = C_6^* \left\{ 960rz^2 - 240r^3 \right\}$$
$$+ B_6^* \left\{ [-672 + 1056\nu]rz^2 + (432 - 264\nu)r^3 \right\}$$

We can add a state of stress defined by

$$\sigma_{rr} = \sigma_{\theta\theta} = 96C_4^* z \quad ; \quad \sigma_{zz} = -192C_4^* z \quad ; \quad \sigma_{rz} = 96C_4^* r \qquad (8.905)$$

which is essentially derived from (8.894) by setting $B_4^* = 0$. A further stress state which corresponds to

$$\begin{aligned}
\sigma_{rr} = \sigma_{\theta\theta} &= 6C_3^* + (10\nu - 2)B_3^* \\
\sigma_{zz} &= -12C_3^* + (14 - 10\nu)B_3^* \\
\sigma_{rz} &= 0
\end{aligned} \qquad (8.906)$$

and adjusted to give a uniform stress state

$$\sigma_{zz} = \widetilde{C}_3 \qquad (8.907)$$

can also be added. The combination of stress states defined by (8.904), (8.905) and (8.907) gives a stress state which is indeterminate to within the constants $\widetilde{C}_3, C_4^*, C_6^*$ and B_6^*. These constants can be determined by satisfying the boundary conditions (8.901).

It can be shown that the resulting expression for σ_{rr} is odd in z and as such will satisfy the integral boundary condition pertaining to zero resultant radial traction on $r = a$, but will yield a non-zero resultant moment boundary condition of (8.902) we can superimpose the solution corresponding to a constant moment developed in Example 8.22 but with the magnitude suitably adjusted to give a zero moment condition for the final stress state. The final expressions for the stresses take the forms

$$\sigma_{rr} = \frac{p_0 z}{h^3} \left[\frac{(2+\nu)}{40}(3h^2 - 5z^2) - \frac{3(3+\nu)}{32}(a^2 - r^2) \right]$$

$$\sigma_{\theta\theta} = \frac{p_0 z}{h^3} \left[\frac{(2+\nu)}{40}(3h^2 - 5z^2) - \frac{3}{32}\left\{(3+\nu)a^2 - (1+3\nu)r^2\right\} \right]$$

$$\qquad (8.908)$$

$$\sigma_{zz} = -\frac{p_0}{4h^3}(2h - z)(h + z)^2$$

$$\sigma_{rz} = \frac{3p_0 r}{8h^3}(h^2 - z^2)$$

$$\bullet \; \bullet \; \bullet$$

From the examples presented in this section it is evident that the polynomial solutions of Laplace's equation for the interior problem and the associated biharmonic solutions are valid for problems where R is finite and are clearly inapplicable to unbounded domains. The solutions however are bounded as $R \to 0$.

8.10.7 Love's strain function approach: spherical polar coordinate formulation for unbounded domains

We shall develop here, the general results required for the application of Love's strain potential to certain classical problems in the theory of elasticity where the domain of interest is of infinite extent. While integral transform techniques can be applied to the solution of some of these problems, we shall attempt to develop the solutions by seeking an explicit solution of the relevant biharmonic function. In particular, four problems of practical interest will be discussed. The first relates to the action of a concentrated force at the interior of an infinite space and in order to preserve axial symmetry the force is directed along the z-axis. This problem was first solved by Lord Kelvin (1824-1907) and is considered to be one of the key *fundamental solutions* in the classical theory of elasticity. The second problem extends the results developed by Kelvin to include the case where dipoles of self equilibrating collinear forces are acting along the three coordinate axes. This problem was first investigated by Love and gives rise to a solution referred to as either a center of compression (or dilation) depending on the sense of action of the dipoles. The third problem investigates Boussinesq's classical solution to the problem of loading of an elastic half-space region by a concentrated normal force which acts at the boundary. Finally we shall examine the axisymmetric problem of the uniform axial loading of an infinite region which is bounded internally by a spherical cavity. This problem was investigated by Sir R.V. Southwell (1888-1970) (with H.R. Gough) and by J.N. Goodier (1905-1969) in two celebrated papers dealing with the spherical defect and spherical inclusion problems.

In all these problems the medium is of infinite extent and the solutions can be developed by utilizing combinations of harmonic and biharmonic functions, both of which are acceptable forms for Love's strain function. They are selected to satisfy, where relevant, the singular behaviour of the stress field at the origin and regularity conditions at infinity.

The analyses of these problems is facilitated by adopting at the outset the formulation of the governing equations with reference to a spherical polar coordinate system (R, ϑ, Θ). This enables the effective utilization of spherical harmonic solutions based on Legendre polynomials and their adaptation to the formulation of biharmonic functions. Owing to the assumed axial symmetry, the displacements and stresses are independent of the azimuthal coordinate ϑ. The displacement vector referred to the spherical polar coordinate system takes the form

$$\mathbf{u} = u_R(R, \Theta)\mathbf{i}_R + u_\Theta(R, \Theta)\mathbf{i}_\Theta \tag{8.909}$$

The components of the strain tensor referred to the spherical polar coordinate system can be obtained by a coordinate transformation procedure similar to that indicated in previous sections (e.g. Section 8.9.4). The strain tensor takes the form

$$\epsilon = \begin{bmatrix} \epsilon_{RR} & 0 & \epsilon_{R\Theta} \\ 0 & \epsilon_{\vartheta\vartheta} & 0 \\ \epsilon_{\Theta R} & 0 & \epsilon_{\Theta\Theta} \end{bmatrix} = \epsilon^T \tag{8.910}$$

where

$$\epsilon_{RR} = \frac{\partial u_R}{\partial R} \quad ; \quad \epsilon_{\vartheta\vartheta} = \frac{u_R}{R} + \frac{u_\Theta}{R}\cot\Theta$$

$$\epsilon_{\Theta\Theta} = \frac{u_R}{R} + \frac{1}{R}\frac{\partial u_\Theta}{\partial\Theta} \quad ; \quad \epsilon_{R\Theta} = \frac{1}{2}\left(\frac{\partial u_\Theta}{\partial R} - \frac{u_\Theta}{R} + \frac{1}{R}\frac{\partial u_R}{\partial\Theta}\right) \tag{8.911}$$

For axial symmetry, the stress tensor referred to the spherical polar coordinate system takes the form

$$\sigma = \begin{bmatrix} \sigma_{RR} & 0 & \sigma_{R\Theta} \\ 0 & \sigma_{\vartheta\vartheta} & 0 \\ \sigma_{\Theta R} & 0 & \sigma_{\Theta\Theta} \end{bmatrix} = \sigma^T \tag{8.912}$$

and, in the absence of the body forces, the non-trivial equations of equilibrium are

$$\frac{\partial \sigma_{RR}}{\partial R} + \frac{1}{R}\frac{\partial \sigma_{R\Theta}}{\partial \Theta} + \frac{1}{R}(2\sigma_{RR} - \sigma_{\vartheta\vartheta} - \sigma_{\Theta\Theta} + \sigma_{R\Theta}\cot\Theta) = 0$$

$$\tag{8.913}$$

$$\frac{\partial \sigma_{R\Theta}}{\partial R} + \frac{1}{R}\frac{\partial \sigma_{\Theta\Theta}}{\partial \Theta} + \frac{1}{R}(3\sigma_{R\Theta} + (\sigma_{\Theta\Theta} - \sigma_{\vartheta\vartheta})\cot\Theta) = 0$$

Since the spherical polar coordinate system is orthogonal, the stress-strain relations for isotropic elastic materials simply follow from (8.866); i.e.

$$\epsilon_{RR} = \frac{1}{E}\left[\sigma_{RR} - \nu(\sigma_{\Theta\Theta} + \sigma_{\vartheta\vartheta})\right] \quad ; \quad \epsilon_{\vartheta\vartheta} = \frac{1}{E}\left[\sigma_{\vartheta\vartheta} - \nu(\sigma_{RR} + \sigma_{\Theta\Theta})\right]$$

$$\tag{8.914}$$

$$\epsilon_{\Theta\Theta} = \frac{1}{E}\left[\sigma_{\Theta\Theta} - \nu(\sigma_{RR} + \sigma_{\vartheta\vartheta})\right] \quad ; \quad \epsilon_{R\Theta} = \frac{\sigma_{R\Theta}}{2\mu}$$

The solution of the axisymmetric problem referred to the spherical coordinate system can be formulated either via a Lamé strain potential function formulation discussed in section 8.10.1 or via the Love strain function approach discussed in Section 8.10.3. We shall present the relevant results in the spherical polar coordinate system applicable to axisymmetric problems using both formulations.

In the Lamé strain potential approach, the displacement field is defined in terms of the gradient of a potential function $\varphi(R, \Theta)$. For axial symmetry, the result (8.797) gives

$$2\mu u_R = \frac{\partial \varphi}{\partial R} \quad ; \quad 2\mu u_\Theta = \frac{1}{R}\frac{\partial \varphi}{\partial \Theta} \tag{8.915}$$

and since φ is harmonic

$$2\mu\widehat{\nabla}.\mathbf{u} = 2\mu\widehat{\nabla}.\widehat{\nabla}\varphi = 0 \tag{8.916}$$

The corresponding stress components can be obtained by making use of the expressions for strains in terms of the displacements (8.911) and the stress strain relations (8.914). We have

$$\sigma_{RR} = \frac{\partial^2 \varphi}{\partial R^2} \quad ; \quad \sigma_{\vartheta\vartheta} = \frac{1}{R}\frac{\partial \varphi}{\partial R} + \frac{\cot\Theta}{R^2}\frac{\partial \varphi}{\partial\Theta}$$

$$\sigma_{\Theta\Theta} = \frac{1}{R}\frac{\partial \varphi}{\partial R} + \frac{1}{R^2}\frac{\partial^2 \varphi}{\partial\Theta^2} \quad ; \quad \sigma_{R\Theta} = \frac{\partial^2}{\partial R\partial\Theta}\left(\frac{\varphi}{R}\right) \tag{8.917}$$

and we note that

$$tr\,\boldsymbol{\sigma} = \widehat{\nabla}_S^2 \varphi = 0 \tag{8.918}$$

We now consider the solution of the axisymmetric problem which is based on Love's strain function, $\Phi(R,\Theta)$. As indicated in Section 8.10.2, the formulation of the problem in terms of $\Phi(R,\Theta)$ results in a biharmonic equation only when the Galerkin vector has a single component of the form (8.853). Therefore the results for $\mathbf{u}$ and $\boldsymbol{\sigma}$ given in relation to the cylindrical polar coordinate system, (8.869) and (8.870) respectively, can be transformed to the spherical polar coordinate system. The transformed equivalents of (8.869) takes the form

$$2\mu u_R = \cos\Theta\left[2(1-\nu)\widehat{\nabla}_S^2 - \frac{\partial^2}{\partial R^2}\right]\Phi + \frac{\sin\Theta}{R}\frac{\partial}{\partial\Theta}\left(\frac{\partial}{\partial R} - \frac{1}{R}\right)\Phi$$

$$2\mu u_\Theta = \sin\Theta\left[-2(1-\nu)\widehat{\nabla}_S^2 + \frac{1}{R}\frac{\partial}{\partial R} + \frac{1}{R^2}\frac{\partial^2}{\partial\Theta^2}\right]\Phi \tag{8.919}$$

$$+\frac{\cos\Theta}{R}\frac{\partial}{\partial\Theta}\left(-\frac{\partial}{\partial R} + \frac{1}{R}\right)\Phi$$

Also

$$\frac{2\nu\mu}{(1-2\nu)}\nabla.\mathbf{u} = \left(\nu\cos\Theta\frac{\partial}{\partial R} - \frac{\nu\sin\Theta}{R}\frac{\partial}{\partial\Theta}\right)\widehat{\nabla}_S^2\Phi \tag{8.920}$$

The stress components of (8.912) can be expressed in the form

$$\sigma_{RR} = \cos\Theta\,\frac{\partial}{\partial R}\left[(2-\nu)\widehat{\nabla}_S^2 - \frac{\partial^2}{\partial R^2}\right]\Phi$$
$$+\frac{\sin\Theta}{R}\frac{\partial}{\partial\Theta}\left[-\nu\widehat{\nabla}_S^2 + \frac{\partial^2}{\partial R^2} - \frac{2}{R}\frac{\partial}{\partial R} + \frac{2}{R^2}\right]\Phi$$

$$\sigma_{\vartheta\vartheta} = \left(\cos\Theta\,\frac{\partial}{\partial R} - \frac{\sin\Theta}{R}\frac{\partial}{\partial\Theta}\right)$$
$$\cdot\left[-(1-\nu)\widehat{\nabla}_S^2 + \frac{\partial^2}{\partial R^2} + \frac{1}{R}\frac{\partial}{\partial R} + \frac{1}{R^2}\frac{\partial^2}{\partial\Theta^2}\right]\Phi$$

$$\sigma_{\Theta\Theta} = \cos\Theta\,\frac{\partial}{\partial R}\left[\nu\widehat{\nabla}_S^2 - \frac{1}{R}\frac{\partial}{\partial R} - \frac{1}{R^2}\frac{\partial^2}{\partial\Theta^2}\right]\Phi \qquad (8.921)$$
$$+\frac{\sin\Theta}{R}\frac{\partial}{\partial\Theta}\left[-(2-\nu)\widehat{\nabla}_S^2 + \frac{3}{R}\frac{\partial}{\partial R} - \frac{2}{R^2} + \frac{1}{R^2}\frac{\partial^2}{\partial\Theta^2}\right]\Phi$$

$$\sigma_{R\Theta} = \frac{\cos\Theta}{R}\frac{\partial}{\partial\Theta}\left[(1-\nu)\widehat{\nabla}_S^2 - \frac{\partial^2}{\partial R^2} + \frac{2}{R}\frac{\partial}{\partial R} - \frac{2}{R^2}\right]\Phi$$
$$+\sin\Theta\,\frac{\partial}{\partial R}\left[-(1-\nu)\widehat{\nabla}_S^2 + \frac{1}{R}\frac{\partial}{\partial R} + \frac{1}{R^2}\frac{\partial^2}{\partial\Theta^2}\right]\Phi$$

8.10.8 Kelvin's problem

We shall define an isotropic elastic infinite space as a domain which is un-bounded in all directions. In the spherical polar coordinate system $R \in (0,\infty)$, $\vartheta \in (0,2\pi)$ and $\Theta \in (0,\pi)$. The elastic medium is subjected to a concentrated force of magnitude P which acts at a point in the domain (Figure 8.50). Kelvin's problem deals with the determination of the displacement and stress fields in the elastic medium resulting from the action of P.

Since the medium is of infinite extent we can satisfy the requirements for axial symmetry by selecting the cylindrical polar coordinate system such that the z-axis coincides with the line of action of P. The point of action of the concentrated force is taken as the origin of coordinates. There are several approaches to the solution of Kelvin's problem; here we shall concentrate on the application of Love's strain function formulation referred to the spherical polar coordinate system (R,ϑ,Θ). We seek a solution of

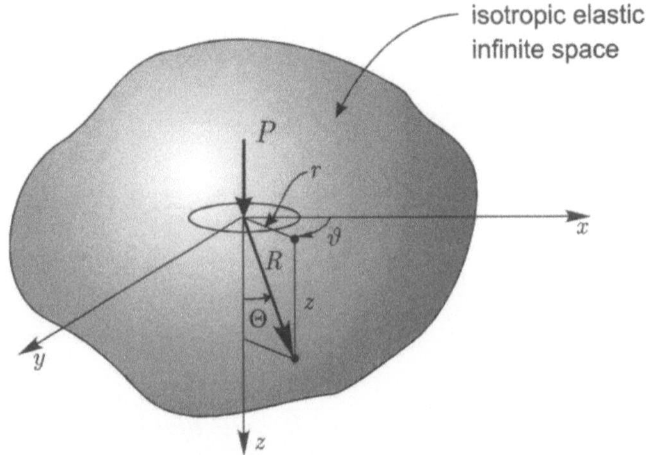

Figure 8.50: Kelvin's problem.

$$\widehat{\nabla}_S^2 \widehat{\nabla}_S^2 \Phi(R, \Theta) = 0 \tag{8.922}$$

such that the displacement and stress fields will vanish as $R \to \infty$ and the resultant of tractions in the z-direction on any closed surface S containing the origin is equivalent to a force P. The state of stress can exhibit singular behaviour at the origin.

Although the selection of an appropriate form of Love's strain function may seem difficult at first glance, we can guide this procedure by examining the physical attributes of Kelvin's problem and certain '*dimensional*' considerations. Since the concentrated force acts at a '*point*' within an infinite domain, there is no *length scale* associated with the problem. Yet, the solution of the problem through the use of Love's strain function approach should give the correct dimensions for stresses in the medium. Considering the form of (8.921) for the stresses expressed in terms of $\Phi(R, \Theta)$ it becomes clear that Love's biharmonic strain function should be of such an order in R that, when differentiated thrice should yield functions of order $1/R^2$.

Considering the results for

$$\widehat{\nabla}_S^2 \varphi(R, \Theta) = 0 \tag{8.923}$$

we can show that $\varphi(R) = C/R$, where C is a constant, is a harmonic function. Hence from Theorem 8.9, we note that

$$\Phi(R) = R^2\varphi = CR \tag{8.924}$$

is a biharmonic function which can serve as the appropriate Love's strain function.

We can also deduce the solution by considering '*exterior solutions*' of (8.923) which must display singular behaviour in the stress field derived from them and the regularity conditions in the stress field as $R \to \infty$. The relevant solutions can be obtained by examining the form of the solution assumed in (8.878). If we assume that n in (8.878) is specified and examine the solution when we replace n by $-(n+1)$; i.e.

$$\varphi_{-n-1}(R,\Theta) = \frac{G_{-n-1}(\Theta)}{R^{n+1}} \tag{8.925}$$

The coefficient of the G_n term in (8.879), which is obtained by differentiating (8.925) twice with respect to R becomes $-(n+1)(-n)$; i.e. it has the same value for $G_{-n-1}(\Theta)$ as it does for $G_n(\Theta)$. Therefore we can assume that

$$\varphi_{-n-1}(R,\Theta) = \frac{G_n(\Theta)}{R^{n+1}} \tag{8.926}$$

Also from the properties of Legendre functions we note that

$$P_{-(n+1)}(\xi) = P_n(\xi) \tag{8.927}$$

Using (8.927) to define $G_n(\Theta)$ we obtain the harmonic functions for exterior domains as

$$\varphi_1(R,\Theta) = \frac{A_1}{R} \quad ; \quad \varphi_2(R,\Theta) = A_2\frac{\cos\Theta}{R^2}$$

$$\varphi_3(R,\Theta) = A_3\left[\frac{3\cos^2\Theta}{R^3} - \frac{1}{R^3}\right] \tag{8.928}$$

$$\varphi_4(R,\Theta) = A_4\left[\frac{5\cos^3\Theta}{R^4} - 3\frac{\cos\Theta}{R^4}\right]$$

$$\dots = \dots$$

etc., where A_n are arbitrary constants. These solutions are of course biharmonic in a trivial sense. We can now make use of Theorem 8.9 to construct from (8.928) the corresponding biharmonic solutions. The relevant biharmonic solutions are

$$\Phi_1(R,\Theta) = B_1 R$$

$$\Phi_2(R,\Theta) = B_2 \cos\Theta$$

$$\Phi_3(R,\Theta) = B_3 \left[\frac{3\cos^2\Theta}{R} - \frac{1}{R}\right]$$

$$\Phi_3(R,\Theta) = B_4 \left[\frac{5\cos^3\Theta}{R^2} - 3\frac{\cos\Theta}{R^2}\right]$$

(8.929)

where B_n are arbitrary constants. Any of these harmonic or biharmonic functions or their linear combinations can be taken as Love's strain functions which satisfy (8.922). Although they give rise to singular stress states at the origin of coordinates and satisfy the regularity conditions, $\boldsymbol{\sigma} \to 0$ as $R \to \infty$; for any solution to be meaningful, it has to be interpreted in relation to the resultants of tractions or concentrated force systems which give rise to the singular behaviour.

As is evident, the Love strain function chosen at the outset by consideration of the physical attributes of Kelvin's problem is the first in the sequence of biharmonic functions (8.929), which are applicable to a region of infinite extent. Using the strain function (8.924) and the expressions for the stresses in terms of $\Phi(R,\Theta)$ given by (8.921) we can show that

$$\sigma_{RR} = -\frac{2C(2-\nu)}{R^2}\cos\Theta$$

$$\sigma_{\Theta\Theta} = \sigma_{\vartheta\vartheta} = \frac{C(1-2\nu)}{R^2}\cos\Theta$$

$$\sigma_{R\Theta} = \frac{C(1-2\nu)}{R^2}\sin\Theta$$

(8.930)

which is indeterminate to within the arbitrary constant C. To determine this constant we need to consider the equilibrium between the applied force P and the stresses generated in the medium. We can develop the equation of equilibrium by considering tractions in the z-direction on *any closed surface* which contains the origin, the point of application of P. The most convenient surface is a spherical surface centered about the origin (Figure 8.51).

Consider the elemental surface of width $ad\Theta$ and circumferential length $2\pi a \sin \Theta$. The component traction acting along the z-direction can be obtained by considering the stress components σ_{RR} and $\sigma_{R\Theta}$.

The axial component of tractions on any part of the surface is given by

$$F_v = (\sigma_{RR} \cos \Theta - \sigma_{R\Theta} \sin \Theta) \tag{8.931}$$

Considering the equilibrium between the applied force P and the resultants of tractions on the spherical surface we have

$$P + \int_0^{2\pi} \int_0^{\pi} [\sigma_{RR} \cos \Theta - \sigma_{R\Theta} \sin \Theta]_{R=a} \, a^2 \sin \Theta d\vartheta d\Theta = 0 \tag{8.932}$$

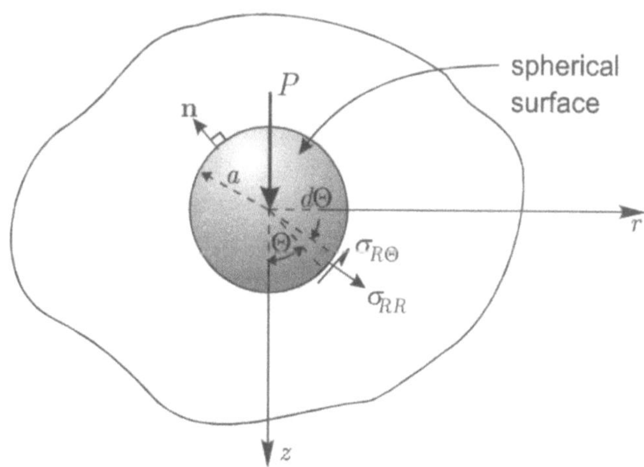

Figure 8.51: Tractions closed spherical surface.

Substituting for σ_{RR} and $\sigma_{R\Theta}$ from (8.930), the above gives

$$P + 2\pi C \int_0^{\pi} \left[-2(2-\nu)\cos^2\Theta - (1-2\nu)\sin^2\Theta\right] \sin \Theta d\Theta = 0 \tag{8.933}$$

The above equation gives

$$C = \frac{P}{8\pi(1-\nu)} \tag{8.934}$$

The final expressions for the displacements and stresses take the forms

$$2\mu u_R = \frac{P}{2\pi}\frac{\cos\Theta}{R} \quad ; \quad 2\mu u_\Theta = -\frac{P(3-4\nu)}{8\pi(1-\nu)}\frac{\sin\Theta}{R} \tag{8.935}$$

and

$$\sigma_{RR} = -\frac{(2-\nu)P}{4\pi(1-\nu)}\frac{\cos\Theta}{R^2}$$

$$\sigma_{\Theta\Theta} = \sigma_{\vartheta\vartheta} = \frac{(1-2\nu)P}{8\pi(1-\nu)}\frac{\cos\Theta}{R^2} \tag{8.936}$$

$$\sigma_{R\Theta} = \frac{(1-2\nu)P}{8\pi(1-\nu)}\frac{\sin\Theta}{R^2}$$

respectively.

The solution of Kelvin's problem can be obtained in a relatively simple form when the problem is formulated with reference to the spherical polar coordinate system where axial symmetry is enforced. The nature of the spherical polar coordinate system and the specific orientation of the concentrated force, however, makes the solution restrictive in terms of its applicability to generate displacement and stress distributions due to tractions which are applied over finite regions. A way of overcoming this deficiency is to reformulate the problem with reference to harmonic displacement functions similar to those proposed by P.F. Papkovich (1887-1946) and H. Neuber (1906-1989). The method is described in a number of standard advanced texts on elasticity and the result for the displacement vector $\mathbf{u}$ in the cylindrical polar coordinate system takes the form

$$\mathbf{u} = \frac{P}{16\pi\mu(1-\nu)}\left[\frac{rz}{R^2}\mathbf{i}_r + \left\{\frac{3-4\nu}{R} + \frac{z^2}{R^3}\right\}\mathbf{i}_z\right] \tag{8.937}$$

where P acts in the z-direction.

An alternative is to transform the results, such as (8.935) and (8.937) for the displacements, to obtain their equivalent in the rectangular Cartesian coordinate system. We shall cite here the relevant results for the displacement components; i.e.

$$u_x = \frac{P}{16\pi\mu(1-\nu)}\frac{xz}{R^3} \quad ; \quad u_y = \frac{P}{16\pi\mu(1-\nu)}\frac{yz}{R^3}$$

$$u_z = \frac{P}{16\pi\mu(1-\nu)}\left[\frac{(3-4\nu)}{R} + \frac{z^2}{R^3}\right] \tag{8.938}$$

where $R = (x^2 + y^2 + x^2)^{\frac{1}{2}}$ and the concentrated for P acts at the origin and is directed along the z-axis. If the point of application of P is at (x^*, y^*, z^*), the expressions corresponding to (8.938) can be obtained by a 'shift' of the origin.

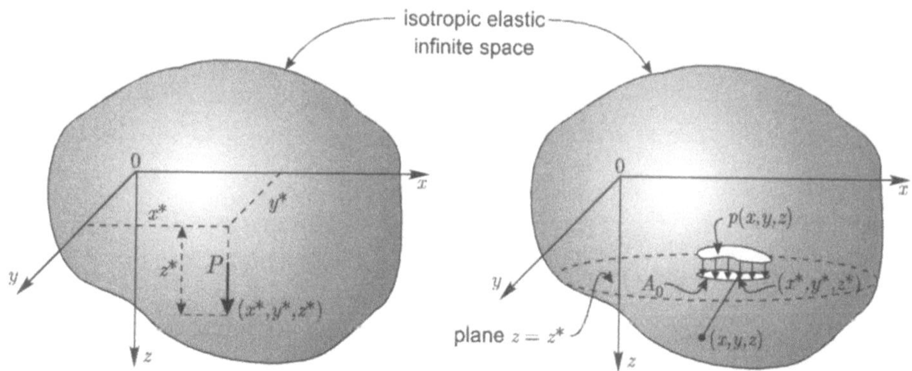

Figure 8.52: Distributed interior loading of an infinite space.

Referring to Figure 8.52, if the point of application of P is at (x^*, y^*, z^*), the expression for u_x, for example, takes the form

$$u_x(x,y,z) = \frac{P}{16\pi\mu(1-\nu)}\frac{(x-x^*)(z-z^*)}{[(x-x^*)^2 + (y-y^*)^2 + (z-z^*)^2]^{\frac{3}{2}}} \tag{8.939}$$

and similar expressions follow for the u_y and u_z. Results such as (8.939) can be used to develop general expressions for situations where a distribution of tractions is applied over a finite region. Referring to Figure 8.52, when the tractions are distributed over a region A_0 located on the plan $z = z^*$, the expression for the displacement u_x at any point within the medium can be obtained from the integral

$$u_x(x, y, z) = \frac{1}{16\pi\mu(1-\nu)}$$
$$\cdot \int\int_{A_0} \frac{p(x^*, y^*, z^*)(x - x^*)(z - z^*)dx^*dy^*}{[(x - x^*)^2 + (y - y^*)^2 + (z - z^*)^2]^{\frac{3}{2}}} \qquad (8.940)$$

Expressions similar to (8.940) can be developed to describe the variations of displacement and stress fields in the infinite elastic medium due to application of tractions which can be distributed over line elements, areas and volumes within the elastic solid. For loaded regions with an irregular shape the integrations can be performed by a quadrature technique.

A distinct advantage in representing the solution to Kelvin's problem with reference to a Cartesian coordinate system is that the solution can be readily adapted to cover situations where the concentrated force is directed along other coordinate directions, simply by re-ordering (with some care) the coordinate system. Consider, for example, the problem where a Kelvin force P^* is directed along the x-axis (Figure 8.53).

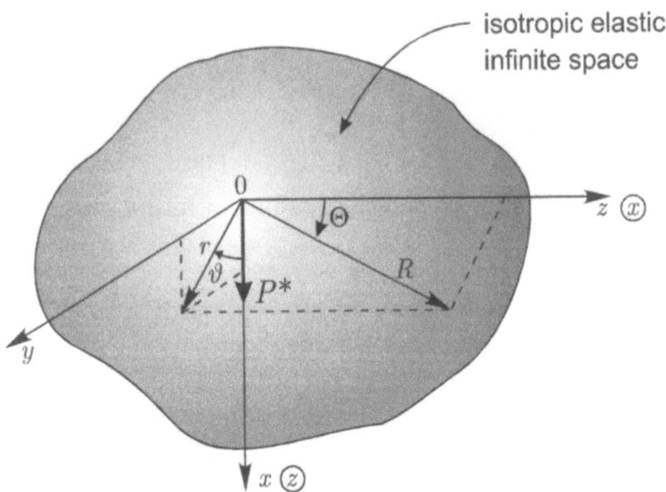

Figure 8.53: Kelvin force P^* directed along th z-axis.

The circled coordinate designations in Figure 8.53 are applicable to the case where the Kelvin force is directed along the z-axis.

From geometry, we have

$$r = R \sin \Theta \quad ; \quad z = R \cos \Theta$$

$$x = r \cos \vartheta = R \sin \Theta \cos \vartheta \tag{8.941}$$

$$y = r \sin \vartheta = R \sin \Theta \sin \vartheta$$

Considering the displacement component u_r in the cylindrical polar coordinate system we have

$$u_r = u_x \cos \vartheta + u_y \sin \vartheta$$

$$u_\vartheta = u_y \cos \vartheta - u_x \sin \vartheta \tag{8.942}$$

and u_z is unchanged between the cylindrical polar coordinate and rectangular Cartesian coordinate systems. We can now express u_R, u_ϑ and u_Θ referred to the spherical polar coordinate system in terms of u_x, u_y and u_z i.e.

$$u_R = u_x \sin \Theta \cos \vartheta + u_y \sin \Theta \sin \vartheta + u_z \cos \Theta$$

$$u_\vartheta = -u_x \sin \vartheta + u_y \cos \vartheta \tag{8.943}$$

$$u_\Theta = u_x \cos \Theta \cos \vartheta + u_y \cos \Theta \sin \vartheta - u_z \sin \Theta$$

Now considering the solution to Kelvin's problem derived previously where P acts in the z-direction we can develop the analogous solution for the case where P^* acts in the x-direction simply by interchanging x and z; e.g. the displacement components for P^* take the forms

$$u_x = \frac{P^*}{16\pi\mu(1-\nu)}\left[\frac{(3-4\nu)}{R} + \frac{x^2}{R^3}\right]$$

$$u_y = \frac{P^*}{16\pi\mu(1-\nu)}\left[\frac{xy}{R^3}\right] \quad ; \quad u_z = \frac{P^*}{16\pi\mu(1-\nu)}\left[\frac{xz}{R^3}\right] \tag{8.944}$$

Substituting these expressions in (8.943) we can obtain the relevant expressions for $\mathbf{u}$ in the spherical polar coordinate system; i.e.

$$\mathbf{u} = \frac{P^*(3-4\nu)}{16\pi\mu(1-\nu)R}$$

$$\cdot \left\{\left[\frac{4(1-\nu)}{(3-4\nu)}\sin\Theta\cos\vartheta\right]\mathbf{i}_R - \sin\vartheta\mathbf{i}_\vartheta + \cos\Theta\cos\vartheta\mathbf{i}_\Theta\right\} \tag{8.945}$$

The expressions for the stress components can be evaluated in the form

$$\sigma_{RR} = -\frac{(2-\nu)P^*}{4\pi(1-\nu)}\frac{\sin\Theta\cos\vartheta}{R^2}$$

$$\sigma_{\vartheta\vartheta} = \sigma_{\Theta\Theta} = \frac{(1-2\nu)P^*}{8\pi(1-\nu)}\frac{\sin\Theta\cos\vartheta}{R^2}$$

$$\sigma_{\Theta\vartheta} = 0 \quad ; \quad \sigma_{\vartheta R} = \frac{(1-2\nu)P^*}{8\pi(1-\nu)}\frac{\sin\vartheta}{R^2}$$

$$\sigma_{R\Theta} = -\frac{(1-2\nu)P^*}{8\pi(1-\nu)}\frac{\cos\Theta\cos\vartheta}{R^2}$$

$$(8.946)$$

The procedure can be extended to include the case where the Kelvin force is directed along the y-axis. The combination of the three problems can be used to obtain the solution to the problem where the Kelvin force is acting at an arbitrary orientation. The alternative approach to this latter problem is to consider transformation of coordinates where one orthogonal coordinate system is chosen in such a way that an axis (z'-axis) coincides with the line of action of the concentrated force.

8.10.9   A centre of dilatation

The concept of a centre of dilatation (or a centre of compression) can be visualized as a state of deformation and stress, in an isotropic elastic medium of infinite extent, which is purely radially symmetric. From a physical point of view we can visualize such a state of radial symmetry as arising from the pressurization of a spherical cavity which is located within an infinite space. A further example can include the deformation of a spherical inclusion due to thermal mismatch with the surrounding elastic medium. For spherical symmetry, the governing strain potential $\varphi(R)$ is

$$\frac{1}{R^2}\frac{d}{dR}\left(R^2\frac{d\varphi}{dR}\right) = 0 \qquad (8.947)$$

and the non-zero displacement and stress components expressed in terms of $\varphi(R)$ are as follows:

$$2\mu u_R = \frac{d\varphi}{dR} \quad ;$$

$$\sigma_{RR} = \frac{d^2\varphi}{dR^2} \quad ; \quad \sigma_{\vartheta\vartheta} = \sigma_{\Theta\Theta} = \frac{1}{R}\frac{d\varphi}{dR}$$

$$(8.948)$$

The Love strain function solutions which are interpreted through the Galerkin vector $\mathbf{F} = \mathbf{i}_z\Phi(r,z)$ of course do not apply for the radially symmetric case, as is evident from the results for the displacement and stress components given by (8.919) and (8.921) respectively.

The general solution of (8.947) is

$$\varphi(R) = \frac{A}{R} + B \tag{8.949}$$

where A and B are arbitrary constants. The displacement and stresses take the forms

$$2\mu u_R = -\frac{A}{R^2}$$

$$\sigma_{RR} = 2\frac{A}{R^2} \quad ; \quad \sigma_{\vartheta\vartheta} = \sigma_{\Theta\Theta} = -\frac{A}{R^3}$$

$$(8.950)$$

which are clearly applicable to exterior domains, and the singular behaviour of the radially symmetric stress field as $R \to 0$ is circumvented by bounding the infinite space internally by a spherical cavity of finite radius.

The centre of dilatation is one of the special cases of 'nuclei of strain' which Love attributes to J. Dougall (1867-1960). Love's analysis of the centre of dilatation commences with Kelvin's problem which was discussed in Section 8.10.8. By considering a 'doublet' of Kelvin forces (i.e. forces of equal magnitude P are located a distance 'd' apart (Figure 8.54).

The problem is no longer spherically or radially symmetric and the state of stress needs to be referred to the cylindrical polar coordinate system, (r,ϑ,z). If d is infinitesimal, the stress components referred to the (r,ϑ,z) system take the forms

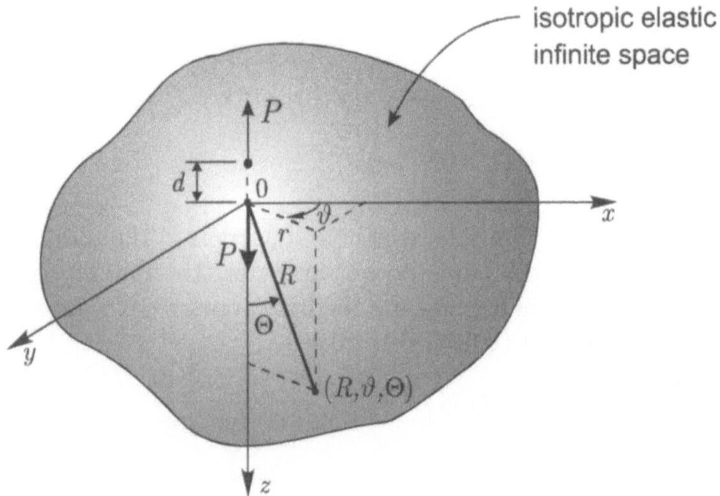

Figure 8.54: Doublet of Kelvin forces.

$$\sigma_{rr} = -\frac{Pd}{8\pi(1-\nu)}\frac{\partial}{\partial z}\left[(1-2\nu)\frac{z}{R^3} - \frac{3r^2 z}{R^5}\right]$$

$$\sigma_{\vartheta\vartheta} = -\frac{Pd}{8\pi(1-\nu)}\frac{\partial}{\partial z}\left[(1-2\nu)\frac{z}{R^3}\right]$$

(8.951)

etc. These stresses can be transformed to the spherical polar coordinate system to give

$$\sigma_{RR} = -\frac{Pd(1+\nu)}{4\pi(1-\nu)R^3}\left[-\sin^2\Theta + \frac{2(2-\nu)}{(1+\nu)}\cos^2\Theta\right]$$

$$\sigma_{R\Theta} = -\frac{Pd(1+\nu)}{4\pi(1-\nu)R^3}\left[\sin\Theta\cos\Theta\right]$$

(8.952)

etc. If two identical doublets of forces are applied in the x and y directions the resulting combined stress state is equivalent to

$$\sigma_{RR} = -\frac{Pd(1-2\nu)}{2\pi(1-\nu)R^3} \quad ; \quad \sigma_{\vartheta\vartheta} = \sigma_{\Theta\Theta} = \frac{Pd(1-2\nu)}{4\pi(1-\nu)R^3} \qquad (8.953)$$

and all other $\boldsymbol{\sigma} = 0$. The state of stress due to the mutually orthogonal system of doublets of equal strength is therefore spherically symmetric.

The solution for the centre of dilatation can be integrated over regions of the half-space (e.g. along lines, surfaces and volumes) to generate results that can be successfully employed to satisfy boundary conditions applicable to other problems associated with both half-space and infinite space problems.

8.10.10 Boussinesq's problem

Boussinesq (1842-1929) examined the problem concerning the determination of the displacements and stresses within a semi-infinite elastic region, the plane surface of which is subjected to a concentrated normal force of finite magnitude P_B. A semi-infinite elastic medium or an elastic half-space is defined as a region which is bounded by a plane surface where the coordinate axis orthogonal to the plane extends from 0 to ∞, the medium extends to infinity in the orthogonal directions.

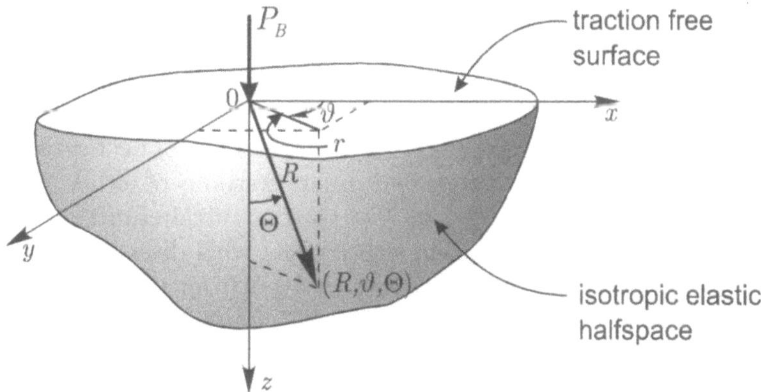

Figure 8.55: Boussinesq's problem for a half-space.

The solution of Boussinesq's problem as a well posed boundary value problem can be approached in a variety of ways. These include potential function techniques, integral transform methods and methods which utilize combinations of biharmonic functions and harmonic functions. In this section we shall focus primarily on an approach which makes use of solution derived for Kelvin's problem. When the force P_B is normal to the plane $z = 0$ with the point of action of the force as the origin of coordinates, the problem is symmetric about the z-axis (Figure 8.55). We can therefore formulate the problem either in relation to the cylindrical polar coordinate system (r, ϑ, z) or in relation to the spherical polar coordinate system (R, ϑ, Θ). The boundary value problem involves obtaining a solution of the biharmonic

$$\widehat{\nabla}_S^2 \widehat{\nabla}_S^2 \Phi(R, \Theta) = 0 \tag{8.954}$$

such that $\mathbf{u}$ and $\boldsymbol{\sigma}$ derived from Φ vanish as $R \to \infty$, and the tractions on the plane boundary with outward unit normal $\mathbf{n}$ satisfies the traction free boundary condition

$$\boldsymbol{\sigma}\mathbf{n} = \mathbf{0} \tag{8.955}$$

everywhere on the plane $z = 0$ except at the origin where the stress field can be singular. Finally the resultant of vertical tractions acting on a surface that encompasses the origin and the plane boundary (which can also include surfaces such as $r =$ constant; $z \in (0, \infty)$ or $z =$ constant; $r \in (0, \infty)$) is equivalent to a force of magnitude P_B.

Prior to the analysis of the problem it is useful to note that, since the force P_B is applied at a point along the surface of the half-space and here again, since the elastic medium is semi-infinite, there is no natural length scale associated with Boussinesq's problem. As remarked in connection with the analysis of Kelvin's problem, the relevant differentiations of either the Lamé potential $\varphi(R, \Theta)$ twice with respect to R or the Love strain function $\Phi(R, \Theta)$ thrice with respect to R should yield expressions with dimensions of $1/R^2$ in order for the dimensions of the stresses derived from these functions be correct.

We have already utilized Love's strain function $\varphi(R, \Theta) = CR$, in connection with the analysis of Kelvin's problem as the combination of a matching pair of half-spaces each of which is subjected to a concentrated force of magnitude $(P/2)$ in the z-direction (Figure 8.56). If we consider the traction boundary conditions on the plane $\Theta = \pi/2$ (or $z = 0$) of the half-space region I, we can express the tractions acting on this plane in terms of the general solution (8.930) as follows:

$$\sigma_{\Theta\Theta}^{(1)}(R, \pi/2) = 0 \quad ; \quad \sigma_{R\Theta}^{(1)}(R, \pi/2) = \frac{C(1 - 2\nu)}{R^2} \tag{8.956}$$

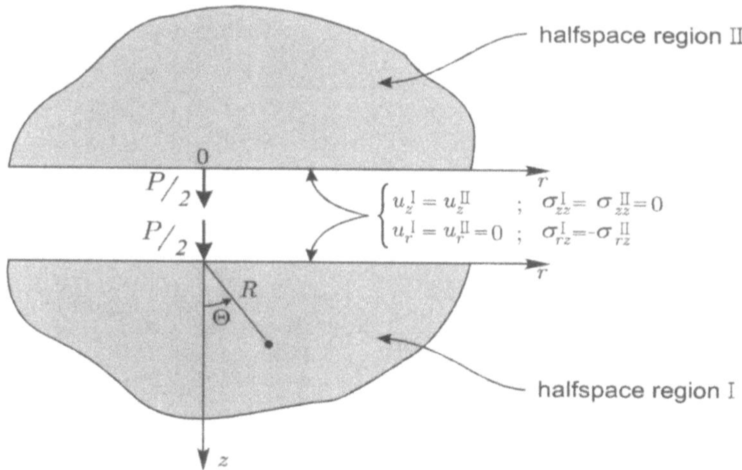

Figure 8.56: Half-space equivalents of Kelvin's problem.

and we shall maintain the solution in terms of the arbitrary constant C and $^{(1)}$ denotes the first part of the solution (and not the region I).

Also, we note that the displacement and stress components derived for Kelvin's problem also satisfy the regularity conditions as $R \to \infty$. Returning to Boussinesq's problem, the traction boundary conditions (8.955) are equivalent to

$$\sigma_{\Theta\Theta}(R, \pi/2) = 0 \quad ; \quad \sigma_{R\Theta}(R, \pi/2) = 0 \tag{8.957}$$

The solution to Kelvin's problem therefore satisfies the first boundary condition of (8.957), exhibits singular behaviour in the stresses as $R \to 0$ and regularity conditions as $R \to \infty$. In order to satisfy the second boundary condition of (8.957) we need an exterior solution applicable to $\varphi(R, \Theta)$ since we have exhausted the relevant exterior solution to $\Phi(R, \Theta)$ in developing Kelvin's solution. The form of $\varphi(R, \Theta)$ must be such that, when differentiated *twice* with respect to R, the resulting expression should be proportional to $1/R^2$. The function should be of the general form

$$\varphi(R, \Theta) = A \ln\left[R f(\Theta)\right] \tag{8.958}$$

where A is a constant and $f(\Theta)$ is an arbitrary function of Θ. (Note that $\ln R$ by itself is not harmonic in the (R, ϑ, Θ) space). Substituting (8.958) in

Laplace's equation for $\varphi(R, \Theta)$ we obtain the following ordinary differential equation for $f(\Theta)$;

$$\frac{d}{d\Theta}\left\{\frac{\sin\Theta}{f}\frac{df}{d\Theta}\right\} + \sin\Theta = 0 \tag{8.959}$$

If the differentiation of the first term of (8.959) is performed, the resulting second-order ordinary differential equation has an awkward non-linear form. Such a procedure is unnecessary, since (8.959) can be integrated directly. Integrating once we obtain

$$\frac{\sin\Theta}{f}\frac{df}{d\Theta} = \int_0^\Theta (-\sin\zeta)d\zeta = (\cos\Theta - 1) \tag{8.960}$$

Separating variables and integrating (8.960) we obtain

$$\ln f(\Theta) = \int_0^\Theta \left(\frac{\cos\zeta - 1}{\sin\zeta}\right) d\zeta \tag{8.961}$$

Note that

$$\int \frac{d\Theta}{\sin\Theta} = -\frac{1}{2}\ln\left\{\frac{1+\cos\Theta}{1-\cos\Theta}\right\} = -Q_0(\cos\Theta) \tag{8.962}$$

where Q_0 is the zeroth-order Legendre function of the second-kind which is admissible for the *half-space region* since the singularity associated with $\Theta = 0$ will cancel out. Solving (8.961) for $f(\Theta)$ we obtain

$$f(\Theta) = (1 + \cos\Theta) \tag{8.963}$$

The Lamé strain potential therefore can be written as

$$\varphi(R, \Theta) = A\ln[R(1 + \cos\Theta)] \tag{8.964}$$

The stress components $\sigma_{\Theta\Theta}^{(2)}$ and $\sigma_{R\Theta}^{(2)}$ derived from (8.964) take the forms

$$\sigma_{\Theta\Theta}^{(2)} = \frac{A\cos\Theta}{R^2(1 + \cos\Theta)} \quad ; \quad \sigma_{R\Theta}^{(2)} = \frac{A\sin\Theta}{R^2(1 + \cos\Theta)} \tag{8.965}$$

where the super-script $^{(2)}$ denotes the second set of stresses (and displacements) derived from (8.964). Evaluating (8.965) at $\Theta = \pi/2$ we have

$$\sigma^{(2)}_{\Theta\Theta}(R, \pi/2) = 0 \quad ; \quad \sigma^{(2)}_{R\Theta}(R, \pi/2) = \frac{A}{R^2} \tag{8.966}$$

We can now make use of the result for shear tractions $\sigma^{(1)}_{R\Theta}(R, \pi/2)$ given by (8.956) and $\sigma^{(2)}_{R\Theta}(R, \pi/2)$ given by (8.966) to obtain a relationship between A and C which will satisfy the zero shear traction boundary condition associated with Boussinesq's problem, i.e.

$$\frac{C(1 - 2\nu)}{R^2} + \frac{A}{R^2} = 0 \tag{8.967}$$

which gives

$$A = -C(1 - 2\nu) \tag{8.968}$$

The solution is still indeterminate to within an arbitrary constant. We can make use of the "equilibrium condition" pertaining to the resultant of vertical tractions acting on a closed surface (closure can include the plane boundary $\Theta = \pi/2$) enclosing the point of application of P_B. The most convenient surface is the hemispherical surface centred at the origin. Prior to evaluating the resultant of vertical tractions on an arbitrary hemispherical surface $R = a$, we note that

$$\sigma_{RR} = \sigma^{(1)}_{RR} + \sigma^{(2)}_{RR} = \frac{C}{R^2}\left[(1 - 2\nu) - 2(2 - \nu)\cos\Theta\right]$$

$$\tag{8.969}$$

$$\sigma_{R\Theta} = \sigma^{(1)}_{R\Theta} + \sigma^{(2)}_{R\Theta} = \frac{C(1 - 2\nu)}{R^2}\frac{\sin\Theta\cos\Theta}{(1 + \cos\Theta)}$$

To determine the relationship between the Boussinesq force P_B and the undetermined constant C, we follow a procedure similar to that used in connection with the development of Kelvin's solution.

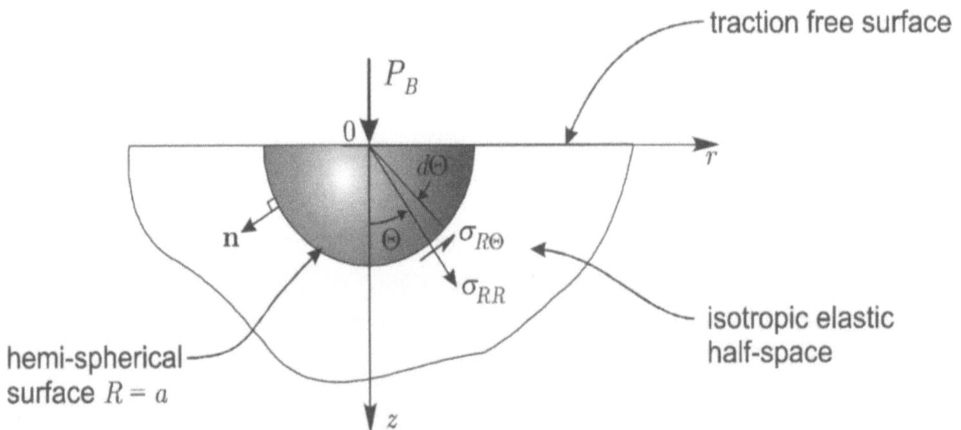

Figure 8.57: Tractions on a hemispherical surface.

Considering the equilibrium between the Boussinesq force P_B and the resultant of tractions on the hemi-spherical surface $R = a$ we have

$$P_B + \int_0^{2\pi} \int_0^{\pi/2} [\sigma_{RR} \cos \Theta - \sigma_{R\Theta} \sin \Theta]_{R=a}\, a^2 \sin \Theta d\Theta d\vartheta = 0 \quad (8.970)$$

Substituting for σ_{RR} and $\sigma_{R\Theta}$ from (9.969) and evaluating the integrals, we obtain

$$C = \frac{P_B}{2\pi} \qquad (8.971)$$

This formally completes the solution of Boussinesq's problem for a half-space region. The final expressions for $\mathbf{u}$ and $\boldsymbol{\sigma}$ can be obtained by combining the two solutions $^{(1)}$ and $^{(2)}$. The displacement components are

$$2\mu u_R = \frac{P_B}{2\pi R}[4(1 - \nu)\cos \Theta - (1 - 2\nu)]$$

$$2\mu u_\Theta = \frac{P_B \sin \Theta}{2\pi R}\left[-(3 - 4\nu) + \frac{(1 - 2\nu)}{(1 + \cos \Theta)}\right] \qquad (8.972)$$

and the stress components are

$$\sigma_{RR} = \frac{P_B}{2\pi R^2} \left[1 - 2\nu - 2(2 - \nu) \cos \Theta \right]$$

$$\sigma_{\Theta\Theta} = \frac{P_B(1 - 2\nu) \cos^2 \Theta}{2\pi R^2 (1 + \cos \Theta)}$$

$$\sigma_{\vartheta\vartheta} = \frac{P_B(1 - 2\nu)}{2\pi R^2} \left(\frac{\cos \Theta - \sin^2 \Theta}{1 + \cos \Theta} \right) \tag{8.973}$$

$$\sigma_{R\Theta} = \frac{P_B(1 - 2\nu)}{2\pi R^2} \frac{\sin \Theta \cos \Theta}{(1 + \cos \Theta)}$$

These results can be compared with the expressions for displacements and stresses which are expressed in relation to the cylindrical polar coordinate system (r, ϑ, z). The displacement components are

$$2\mu u_r = \frac{P_B}{2\pi R} \left[rz - \frac{(1 - 2\nu)r}{(R + z)} \right] \quad ; \quad u_\vartheta = 0$$

$$2\mu u_z = \frac{P_B}{2\pi R} \left[2(1 - \nu) + \frac{z^2}{R^2} \right] \tag{8.974}$$

and the stress components are given by

$$\sigma_{RR} = \frac{P_B}{2\pi R^2} \left[-\frac{3r^2 z}{R^3} + \frac{(1 - 2\nu)R}{(R + z)} \right]$$

$$\sigma_{\vartheta\vartheta} = \frac{(1 - 2\nu)P_B}{2\pi R^2} \left[\frac{z}{R} - \frac{R}{(R + z)} \right] \tag{8.975}$$

$$\sigma_{zz} = -\frac{3P_B z^3}{2\pi R^5} \quad ; \quad \sigma_{rz} = -\frac{3P_B r z^2}{2\pi R^5}$$

The relationship between these two sets of results can be verified by appeal to a transformation of coordinates. Although the results obtained for Boussinesq's problem with reference to the spherical polar coordinate system are in a convenient form, they are somewhat restrictive in their applicability for the analysis of distributed loads which are applied over the surface of a half-space. In essence, Boussinesq's solution is the fundamental solution, or Green's function, for the traction boundary value problem associated with an elastic half-space region. The fundamental solution can thus be integrated over the loaded region to recover distributions of displacement and stresses within the half-space region. (see e.g. the result (8.940) for the evaluation of displacements in an elastic infinite space due to a distributed loading).

In the special case when the elastic material is *incompressible*, $\nu = 1/2$, and the results for displacements and stresses in the half-space due to the Boussinesq force P_B, (8.972) and (8.973) respectively, reduce to

$$2\mu u_R = \frac{P_B \cos\Theta}{\pi R} \quad ; \quad 2\mu u_\Theta = -\frac{P_B \sin\Theta}{2\pi R} \tag{8.976}$$

and

$$\sigma_{RR} = -\frac{3 P_B \cos\Theta}{2\pi R^2} \quad ; \quad \sigma_{\vartheta\vartheta} = \sigma_{\Theta\Theta} = \sigma_{R\Theta} = 0 \tag{8.977}$$

respectively. Also in the limit of material *incompressibility* the expressions (8.935) and (8.936) for the displacements and stresses associated with Kelvin's problem reduce to

$$2\mu u_R = \frac{(P/2)\cos\Theta}{\pi R} \quad ; \quad 2\mu u_\Theta = \frac{(P/2)\sin\Theta}{2\pi R} \tag{8.978}$$

and

$$\sigma_{RR} = -\frac{3(P/2)\cos\Theta}{2\pi R^2} \quad ; \quad \sigma_{\vartheta\vartheta} = \sigma_{\Theta\Theta} = \sigma_{R\Theta} = 0 \tag{8.979}$$

respectively. If we identify P_B with $(P/2)$ them the two solutions are equivalent. Also the state of stress within the half-space and infinite space regions can be characterized as a purely "radial" stress field in relation to the spherical polar coordinate system.

Boussinesq's problem represents one of most fundamental results relevant to the application of the theory of elasticity to the calculation of displacements and stresses in earth masses which are subjected to surface loads. A generalization of both Kelvin's and Boussinesq's solutions was presented by R.D. Mindlin (1906-1987) who examined the problem of the action of a concentrated force at the interior of a half-space region. From Mindlin's exact closed form solution, the solutions of Kelvin and Boussinesq are recovered as special cases.

8.10.11   The spherical cavity problem

We now consider the problem of an isotropic elastic medium of infinite extent which is bounded internally by a spherical cavity of finite radius a (Figure 8.58). The elastic solid is subjected to a state of stress which corresponds to a uniaxial tensile stress field of magnitude σ_0 at distances remote from the spherical cavity. The surface of the cavity is free of traction. The objective of the study is to determine the perturbation in the uniform stress field which occurs in the vicinity of the spherical cavity. Since the traction free conditions are prescribed on the surface of the cavity, it is convenient to use a spherical polar coordinate formulation to examine the problem. The boundary value problem involves the use of solutions of the biharmonic equation for Love's strain function $\Phi(R,\Theta)$ which satisfies

$$\widehat{\nabla}_S^2 \widehat{\nabla}_S^2 \Phi(R,\Theta) = 0 \quad ; \quad R \in (a,\infty)$$

$$\Theta \in (0,\pi) \quad ; \quad \vartheta \in (0,2\pi)$$

(8.980)

and for Lamé's strain potential $\varphi(R,\Theta)$ which satisfies

$$\widehat{\nabla}_S^2 \varphi(R,\Theta) = 0 \quad ; \quad R \in (a,\infty)$$

$$\Theta \in (0,\pi) \quad ; \quad \vartheta \in (0,2\pi)$$

(8.981)

such that the final solution satisfies the zero traction boundary condition

$$\boldsymbol{\sigma}\mathbf{n} = \mathbf{0} \quad ; \quad R = a \quad ; \quad \Theta \in (0,\pi) \quad ; \quad \vartheta \in (0,2\pi) \tag{8.982}$$

where $\mathbf{n}$ is the outward unit normal to the surface of the spherical cavity. In addition, the state of stress should correspond to the uniform stress state as $R \to \infty$.

Considering the uniaxial state of stress shown in Figure 8.58, the far field stress referred to the cylindrical polar coordinate system (r,ϑ,z) is given by

$$\boldsymbol{\sigma} = \begin{bmatrix} 0 & 0 & 0 \\ 0 & 0 & 0 \\ 0 & 0 & \sigma_0 \end{bmatrix} \tag{8.983}$$

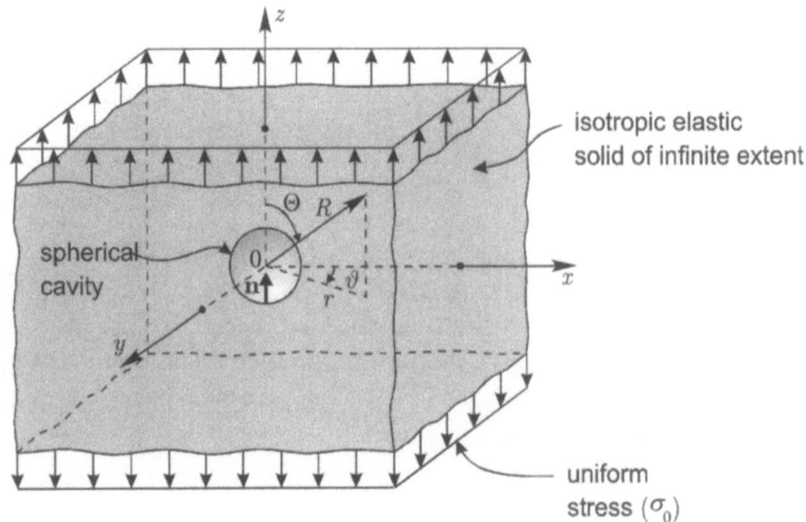

Figure 8.58: Spherical cavity in an isotropic elastic solid subjected to
uniaxial tension at infinity.

The method of solution of the spherical cavity problem is formally similar to
the procedure adopted in connection with the analysis of the circular cavity
problem for a thin plate of infinite extent subjected to a uniaxial stress field
at infinity (see e.g. Example 8.19). We first consider the state of stress in
the intact solid and express the uniaxial stress field (8.983) in terms of its
components referred to the spherical coordinate system. The transformation
matrix for the ordered system (R, ϑ, Θ) is given by

$$[\mathbf{H}] = \begin{bmatrix} \sin\Theta & 0 & \cos\Theta \\ 0 & 1 & 0 \\ \cos\Theta & 0 & -\sin\Theta \end{bmatrix} \tag{8.984}$$

The state of stress referred to the spherical polar coordinate system (Figure
8.59) is given by

$$\boldsymbol{\sigma}' = [\mathbf{H}]^T \boldsymbol{\sigma} [\mathbf{H}] = \begin{bmatrix} \dfrac{\sigma_0}{2}(1 + \cos 2\Theta) & 0 & -\dfrac{\sigma_0}{2}\sin 2\Theta \\ 0 & 0 & 0 \\ -\dfrac{\sigma_0}{2}\sin 2\Theta & 0 & \dfrac{\sigma_0}{2}(1 - \cos 2\Theta) \end{bmatrix} \tag{8.985}$$

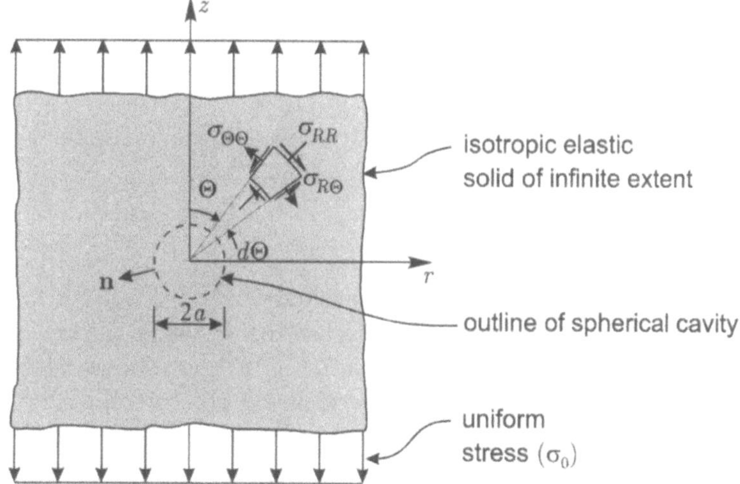

Figure 8.59: State of stress in an infinite space subjected to a uniaxial
state of stress σ_0.

If we consider a surface $R = a$ in the intact solid under a uniaxial state of
stress σ_0, the tractions acting on the spherical surface with unit normal $\mathbf{n}$
as indicated in Figure 8.59 are given by

$$\mathbf{T} = \boldsymbol{\sigma}'\mathbf{n} \tag{8.986}$$

For the spherical surface $R = a$

$$\mathbf{n} = \mathbf{i}_R \tag{8.987}$$

and the tractions on $R = a$ are

$$T_R(a, \Theta) = \sigma_{RR}(a, \Theta) = \frac{\sigma_0}{2}(1 + \cos 2\Theta)$$

$$\tag{8.988}$$

$$T_\Theta(a, \Theta) = \sigma_{R\Theta}(a, \Theta) = -\frac{\sigma_0}{2}\sin 2\Theta$$

The analysis of the spherical cavity problem involves obtaining "*exterior*"
solution in the appropriate spherical harmonics which can eliminate the
tractions (8.988) due to the uniaxial state of stress to render the surface of

the spherical cavity traction free. The "*exterior*" solutions imply the stresses and displacements due to the "*corrective*" solutions will vanish as $R \to \infty$, and the only admissible far field stress (and displacement) state corresponds to uniaxial state of stress due to σ_0. We need to select three independent solutions of (8.980) and (8.981) which will yield stresses of the form

$$\sigma_{RR} = f_1(R) + f_2(R) \cos 2\Theta \quad ; \quad \sigma_{R\Theta} = f_3(R) \sin 2\Theta \tag{8.989}$$

where $f_i(R) \quad (i = 1, 2, 3)$ are arbitrary functions of R which should vanish as $R \to \infty$. Considering the Legendre function solutions for $\varphi_n(R, \Theta)$ given by (8.928) and the solutions for $\Phi(R, \Theta)$ given by (8.929) and the expressions for σ in terms of these functions ((8.917) and (8.921) respectively) it is evident that the required functions are

$$\varphi(R, \Theta) = \frac{A_0}{R} + \frac{A_3}{R^3} \left(\frac{3}{2} \cos 2\Theta + \frac{1}{2} \right) \tag{8.990}$$

and

$$\Phi(R, \Theta) = B_2 \cos \Theta \tag{8.991}$$

The stress components σ_{RR} and $\sigma_{R\Theta}$ derived from these solutions using expression (8.917) and (8.921) are given by

$$\sigma_{RR}(R, \Theta) = 2\frac{A_0}{R^3} + \frac{A_3}{R^5} \{18 \cos 2\Theta + 6\}$$
$$+ \frac{B_2}{R^3} \{(5 - \nu) \cos 2\Theta + (3 - 3\nu)\} \tag{8.992}$$
$$\sigma_{R\Theta}(R, \Theta) = \frac{A_3}{R^5} \{12 \sin 2\Theta\} + \frac{B_2}{R^3} (1 + \nu) \{\sin 2\Theta\}$$

The constants A_0, A_3 and B_2 are to be determined by satisfying the conditions that the combination of the stresses σ_{RR} and $\sigma_{R\Theta}$ due to the uniaxial state of stress and due to the corrective solution results in zero tractions on $R = a$, i.e.

$$2\frac{A_0}{a^3} + \frac{\sigma_0}{2}(1 + \cos 2\Theta) + \frac{A_3}{a^5} \{18 \cos 2\Theta + 6\}$$
$$+ \frac{B_2}{a^3} \{(5 - \nu) \cos 2\Theta + (3 - 3\nu)\} = 0 \tag{8.993}$$

$$\frac{A_3}{a^5}\left\{12\sin 2\Theta\right\} + \frac{B_2}{a^3}(1+\nu)\left\{\sin 2\Theta\right\} - \frac{\sigma_0}{2}\sin 2\Theta = 0 \qquad (8.994)$$

For equations (8.993) and (8.994) to be satisfied for *any* choice of Θ, we require

$$2\frac{A_0}{a^3} + 6\frac{A_3}{a^5} + (3-3\nu)\frac{B_2}{a^3} = -\frac{\sigma_0}{2}$$

$$18\frac{A_3}{a^5} + (5-\nu)\frac{B_2}{a^3} = -\frac{\sigma_0}{2} \qquad (8.995)$$

$$12\frac{A_3}{a^5} + (1+\nu)\frac{B_2}{a^3} = \frac{\sigma_0}{2}$$

Solving these equations we obtain

$$A_0 = \frac{\sigma_0 a^3(1-5\nu)}{2(7-5\nu)} \quad ; \quad A_3 = \frac{\sigma_0 a^5}{2(7-5\nu)} \quad ; \quad B_2 = \frac{-5\sigma_0 a^3}{2(7-5\nu)} \qquad (8.996)$$

The final expressions for the stress components in the elastic medium can be evaluated by using the explicit expressions for $\varphi(R,\Theta)$ and $\Phi(R,\Theta)$ in (8.917) and (8.921) and combining the results with the uniaxial state of stress expressed in the (R,ϑ,Θ) system; we have

$$\sigma_{RR} = \overline{\sigma}_0 \left[\cos 2\Theta \left\{(7-5\nu) - 5(5-\nu)\frac{a^3}{R^3} + 18\frac{a^5}{R^5}\right\}\right.$$
$$\left. + \left\{(7-5\nu) - (13-5\nu)\frac{a^3}{R^3} + 6\frac{a^5}{R^5}\right\}\right]$$

$$\sigma_{\vartheta\vartheta} = 3\overline{\sigma}_0 \left[\cos 2\Theta \left\{\frac{5}{2}(1-2\nu)\frac{a^3}{R^3} - \frac{5}{2}\frac{a^5}{R^5}\right\} + \left\{\frac{1}{2}\frac{a^3}{R^3} - \frac{3}{2}\frac{a^5}{R^5}\right\}\right]$$

$$\qquad (8.997)$$

$$\sigma_{\Theta\Theta} = \overline{\sigma}_0 \left[\cos 2\Theta \left\{-(7-5\nu) + \left(\frac{5}{2}-5\nu\right)\frac{a^3}{R^3} - \frac{21}{2}\frac{a^5}{R^5}\right\}\right.$$
$$\left. + \left\{(7-5\nu) + \left(\frac{13}{2}-10\nu\right)\frac{a^3}{R^3} - \frac{3}{2}\frac{a^5}{R^5}\right\}\right]$$

$$\sigma_{R\Theta} = \overline{\sigma}_0 \left[-(7-5\nu) - 5(1+\nu)\frac{a^3}{R^3} + 12\frac{a^5}{R^5}\right]\sin 2\Theta$$

where

$$\bar{\sigma}_0 = \frac{\sigma_0}{2(7 - 5\nu)} \tag{8.998}$$

It can be verified that the stress state (8.997) exactly satisfies the traction free boundary conditions on $R = a$ and reduces to the uniaxial stress state as $R \to \infty$. A result of some importance to engineering applications relates to the variation of $\sigma_{\Theta\Theta}$ at the boundary of the spherical cavity. At the location $(a, \pi/2)$, we have

$$\frac{\sigma_{\Theta\Theta}(a, \pi/2)}{\sigma_0} = \frac{3(9 - 5\nu)}{2(7 - 5\nu)} \tag{8.999}$$

This stress corresponds to the stress in the z-direction at the boundary of the cavity. Since $0 < \nu \leq 1/2$,

$$\frac{27}{14} < \frac{\sigma_{\Theta\Theta}(a, \pi/2)}{\sigma_0} \leq \frac{13}{6} \tag{8.1000}$$

Therefore, irrespective of the value of Poisson's ratio, the applied stress σ_0 is magnified at the boundary of the spherical cavity. The failure limit of the material can thus be reached at stress levels significantly below that required for failure in far field regions. Such defects can affect the state and fatigue failure behaviour of brittle elastic materials.

8.10.12   Application of integral transforms

Integral transform techniques offer an efficient procedure for the solution of the biharmonic equation, particularly when the type of axisymmetric problems examined can be conveniently described in relation to a cylindrical polar coordinate system. Although the application of integral transform techniques extends to a wider class of three-dimensional problems in elasticity, we shall restrict attention to the special case where the problems are symmetric with respect to the z-axis and formulated in relation to Love's strain function $\Phi(r, z)$. In the absence of body forces $\Phi(r, z)$ is governed by

$$\widehat{\nabla}^2 \widehat{\nabla}^2 \Phi(r, z) = 0 \tag{8.1001}$$

where $\widehat{\nabla}^2$ is the axisymmetric form of Laplace's operator referred to the (r, θ, z) system. The displacement and stress components referred to the (r, θ, z) coordinate system, and expressed in terms $\Phi(r, z)$, are given by (8.869) and (8.870) respectively. Integral transforms offer more straightforward techniques for obtaining solutions to infinite media, semi-infinite media, solid regions of finite thickness and infinite lateral extent, infinite and semi-infinite cylindrical regions, etc. The formal nature of integral transform procedures makes for their convenient application to a variety of problems of engineering interest.

We first consider the class of axisymmetric problems where r ranges from 0 to ∞. Such problems include infinite space regions, half-space regions and elastic layer regions. From the results given in Section 1.12.3, in connection with Hankel transforms, we define the Hankel transform of Love's strain function, of order n, as

$$\overline{\Phi}(\xi, z) = \mathcal{H}_n \left\{ \Phi(r, z); \xi \right\} = \int_0^\infty r\Phi(r, z) J_n(\xi r) dr \tag{8.1002}$$

where $J_n(\xi r)$ is the first order Bessel function of order n. Prior to application the Hankel integral transform it is useful to obtain the definite integral

$$I = \int_0^\infty r\widehat{\nabla}^2 \Phi(r, z) J_0(\xi r) dr \tag{8.1003}$$

Substituting the expression for the Laplacian, (8.1003) can be written as

$$I = \int_0^\infty r \left\{ \frac{\partial^2 \Phi}{\partial r^2} + \frac{1}{r}\frac{\partial \Phi}{\partial r} \right\} J_0(\xi r) dr + \frac{\partial^2}{\partial z^2} \int_0^\infty r\Phi J_0(\xi r) dr \tag{8.1004}$$

The first integral in (8.1004) can be written as

$$\int_0^\infty \frac{\partial}{\partial r}\left(r\frac{\partial \Phi}{\partial r} \right) J_0(\xi r) dr = \left[r\frac{\partial \Phi}{\partial r} J_0(\xi r) \right]_0^\infty$$
$$-\xi \int_0^\infty r\frac{\partial \Phi}{\partial r} J_0'(\xi r) dr \tag{8.1005}$$

We assume that the first term on the right hand side of (8.1005) vanishes at both limits. This is a requirement that needs to be satisfied by the symmetry and regularity conditions associated with the axisymmetric problem in elasticity theory. The second integral of (8.1005) gives

$$-\xi \int_0^\infty r \frac{\partial \Phi}{\partial r} J_0'(\xi r) dr = -\xi \left[r\Phi J_0'(\xi r) \right]_0^\infty$$

$$+\xi \int_0^\infty \Phi \left\{ J_0'(\xi r) + \xi r J_0''(\xi r) \right\} d\xi \qquad (8.1006)$$

Again we assume that the first term on the right hand side vanishes at both limits. Also noting that $J_0(\xi r)$ satisfies the equation

$$J_0''(\xi r) + \frac{1}{\xi r} J_0'(\xi r) + J_0(\xi r) = 0 \qquad (8.1007)$$

we obtain from (8.1003) to (8.1006)

$$\int_0^\infty r \widehat{\nabla}^2 \Phi(r, z) J_0(\xi r) dr = \left(\frac{d^2}{dz^2} - \xi^2 \right) \overline{\Phi}(\xi, z) \qquad (8.1008)$$

Repeating the operations we can show that

$$\int_0^\infty r \widehat{\nabla}^2 \widehat{\nabla}^2 \Phi(r, z) J_0(\xi r) dr = \left(\frac{d^2}{dz^2} - \xi^2 \right)^2 \overline{\Phi}(\xi, z) \qquad (8.1009)$$

The application of the zeroth-order Hankel transform reduces the biharmonic equation to the following:

$$\left(\frac{d^2}{dz^2} - \xi^2 \right)^2 \overline{\Phi}(\xi, z) = 0 \qquad (8.1010)$$

The important observation is that the application of the Hankel transform efficiently reduces the fourth-order partial differential equation (8.1001) to a fourth-order ordinary differential equation for the zeroth-order Hankel transform of Love's strain function.

The general solution of (8.1010) is given by

$$\overline{\Phi}(\xi, z) = (A + Bz)e^{-\xi z} + (C + Dz)e^{\xi z} \qquad (8.1011)$$

where A, B, C and D are either arbitrary functions of ξ or arbitrary constants. These can be determined by making use of either displacement or traction boundary conditions applicable to a particular boundary value problem. The evaluation of the arbitrary functions is facilitated by the use of the expressions for the components of displacements and stresses which are expressed in terms of $\overline{\Phi}(\xi, z)$. Considering the expression for $u_r(r, z)$ given by (8.869), we multiply both sides of the equation by $rJ_0'(\xi r)$ and integrate over r from 0 to ∞; we have

$$2\mu \int_0^\infty r u_r(r, z) J_1(\xi r) dr = \xi \frac{d\overline{\Phi}}{dz} \tag{8.1012}$$

Using (1.231) we invert (8.1012) to obtain

$$2\mu u_r(r, z) = \int_0^\infty \xi^2 \frac{d\overline{\Phi}}{dz} J_1(\xi r) d\xi \tag{8.1013}$$

Considering the expression for $u_z(r, z)$ given in (8.869) we multiply both sides of the equation by $rJ_0(\xi r)$ and integrate over r from 0 to ∞; this gives

$$2\mu \int_0^\infty r u_z(r, z) J_0(\xi r) dr = (1 - 2\nu) \frac{d^2\overline{\Phi}}{dz^2} - 2(1 - \nu)\xi^2 \overline{\Phi} \tag{8.1014}$$

Inverting (8.1014) we have

$$2\mu u_z(r, z) = \int_0^\infty \xi \left[(1 - 2\nu) \frac{d^2\overline{\Phi}}{dz^2} - 2(1 - \nu)\xi^2 \overline{\Phi} \right] J_0(\xi r) d\xi \tag{8.1015}$$

Using similar procedures, we can develop expressions for the stress components. Two stress components that are encountered in specifying boundary conditions are σ_{zz} and σ_{rz}. We shall present here the integral expressions for these two stresses in terms of $\overline{\Phi}(\xi, z)$.

Considering the third equation of (8.870) we can write

$$\int_0^\infty r\sigma_{zz}(r, z) J_0(\xi r) dr = \int_0^\infty r \left[(2 - \nu) \frac{\partial}{\partial z} \left(\widehat{\nabla}^2 \Phi \right) - \frac{\partial^3 \Phi}{\partial z^3} \right] J_0(\xi r) dr$$

$$= (2 - \nu) \left\{ \frac{d^3\overline{\Phi}}{dz^3} - \xi^2 \frac{d\overline{\Phi}}{dz} \right\} - \frac{d^3\overline{\Phi}}{dz^3} \tag{8.1016}$$

Inverting (8.1016) we obtain

$$\sigma_{zz}(r,z) = \int_0^\infty \xi \left[(1-\nu)\frac{d^3\overline{\Phi}}{dz^3} - (2-\nu)\xi^2\frac{d\overline{\Phi}}{dz} \right] J_0(\xi r)d\xi \qquad (8.1017)$$

In the development of the equivalent expression for σ_{rz} we adopt the following procedure; considering the last equation of (8.870) we denote

$$(1-\nu)\widehat{\nabla}^2\Phi - \frac{\partial^2\Phi}{\partial z^2} = \Omega(r,z) \qquad (8.1018)$$

We can multiply both sides of the expression for σ_{rz} by $rJ_1(\xi r)$ and perform the integration with respect to r between limits 0 to ∞; we have

$$\int_0^\infty r\sigma_{zz}(r,z)J_1(\xi r)dr = \int_0^\infty r\frac{\partial\Omega}{\partial r}J_1(\xi r)dr \qquad (8.1019)$$

Evaluating the integral on the right hand side of (8.1019) we have

$$\int_0^\infty r\frac{\partial\Omega}{\partial r}J_1(\xi r)dr = [r\Omega J_1(\xi r)]_0^\infty$$

$$+ \int_0^\infty \Omega \left[J_0'(\xi r) + \xi r J_0''(\xi r) \right] dr \qquad (8.1020)$$

Again, assuming that the first term on the right hand side of (8.1020) vanishes at both limits, we have

$$\int_0^\infty r\frac{\partial\Omega}{\partial r}J_1(\xi r) = -\xi \int_0^\infty r\Omega(r,z)J_0(\xi r)dr \qquad (8.1021)$$

and (8.1019) can be written as

$$\int_0^\infty r\sigma_{rz}(r,z)J_1(\xi r)dr$$

$$= -\xi \int_0^\infty r\left[(1-\nu)\widehat{\nabla}^2\Phi - \frac{\partial^2\Phi}{\partial z^2} \right] J_0(\xi r)dr \qquad (8.1022)$$

Using (8.1008), (8.1022) can be reduced to

$$\int_0^\infty r\sigma_{rz}(r,z)J_1(\xi r)dr = \xi\left[\nu\frac{d^2\overline{\Phi}}{dz^2} + \xi^2(1-\nu)\overline{\Phi}\right] \qquad (8.1023)$$

Inverting (8.1023) we have

$$\sigma_{rz}(r,z) = \int_0^\infty \xi^2\left[\nu\frac{d^2\overline{\Phi}}{dz^2} + \xi^2(1-\nu)\overline{\Phi}\right]J_1(\xi r)d\xi \qquad (8.1024)$$

In the ensuing, we shall use the integral relationships developed for u_r, u_z, σ_{zz} and σ_{rz} to determine the solutions to typical traction and displacement boundary value problems associated with half-space and infinite space regions.

Example 8.24

An isotropic elastic half-space occupies the region $r \in (0,\infty)$ and $z \in (0,\infty)$. The surface $z = 0$ is reinforced with an inextensible membrane which prevents radial displacements in the plane of the membrane. The surface of the half-space is subjected to a concentrated force P_0 which acts in the z-direction at the origin (Figure 8.60). Use a Hankel transform approach to develop an expression for the surface displacement of the half-space region.

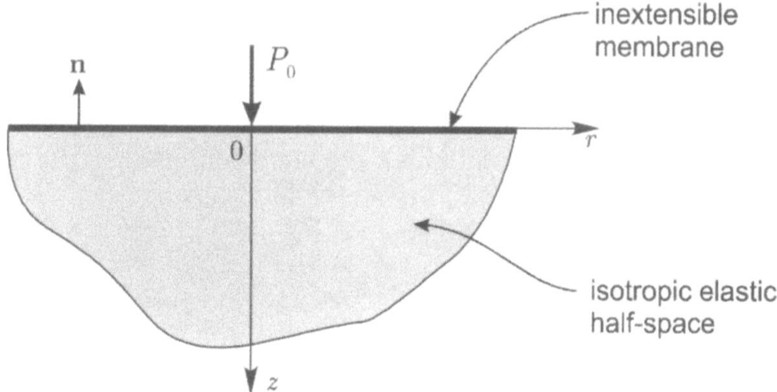

Figure 8.60: Boussinesq problem for a surface reinforced half-space.

Solution

For the purpose of developing a generalized solution, we shall assume that the surface of the reinforced half-space is subjected to an arbitrary axisymmetric load $p(r)$ over $r \in (0, \infty)$. The problem can be reduced to the determination of a solution of the biharmonic equation for Love's stress function which should satisfy the boundary conditions

$$\sigma_{zz}(r, 0) = -p(r) \qquad ; \qquad 0 < r < \infty \tag{8.1025}$$

$$u_r(r, 0) = 0 \qquad ; \qquad 0 < r < \infty \tag{8.1026}$$

In addition, we assume that the normal tractions $p(r)$ are such that the displacements and stresses derived from the Love strain function vanish as $r, z \to \infty$. Considering the general solution of the Hankel transform of the Love strain function, (8.1005), the regularity conditions are satisfied by the function

$$\overline{\Phi}(\xi, z) = [A(\xi) + B(\xi)z]\, e^{-\xi z} \tag{8.1027}$$

The Hankel integral relationship for the radial displacement can be written as

$$2\mu u_r(r, z) = \int_0^\infty \xi^2 \left[-\xi A(\xi) + B(\xi)(1 - \xi z)\right] e^{-\xi z} J_1(\xi r) d\xi \tag{8.1028}$$

For $u_r(r, z)$ to satisfy the boundary condition (8.1026) for all r, we require

$$B(\xi) = \xi A(\xi) \tag{8.1029}$$

Using this result, the integral relation for the normal stress $\sigma_{zz}(r, 0)$ can be evaluated (using (8.1017) and (8.1027)); in the form

$$\sigma_{zz}(r, 0) = \int_0^\infty 2(1 - \nu) A(\xi) \xi^4 J_0(\xi r) d\xi \tag{8.1030}$$

To determine $A(\xi)$ we make use of the boundary condition (8.1025) and we assume that $p(r)$ can be expressed in the form

$$p(r) = \int_0^\infty \xi \overline{p}(\xi) J_0(\xi r) d\xi \tag{8.1031}$$

where $\overline{p}(\xi)$ is the zeroth-order Hankel transform of $p(r)$. To satisfy the boundary condition (8.1025) we require

$$A(\xi) = - \frac{\overline{p}(\xi)}{2\xi^3(1-\nu)} \tag{8.1032}$$

This formally completes the solution to the problem. An explicit integral expression for $u_z(r, z)$ can be obtained by using the results given in these preceding equations in (8.1015). This gives

$$2\mu u_z(r, z) = \int_0^\infty \frac{\overline{p}(\xi)}{2(1-\nu)} \left[(3 - 4\nu) + \xi z\right] e^{-\xi z} J_0(\xi r) d\xi \tag{8.1033}$$

The surface displacement in the z-direction is given by

$$2\mu u_z(r, 0) = \frac{(3 - 4\nu)}{2(1-\nu)} \int_0^\infty \overline{p}(\xi) J_0(\xi r) d\xi \tag{8.1034}$$

This expression is valid for all choices of $p(r)$ provided the conditions invoked in the development of Hankel transforms (see e.g. (8.1005), (8.1006) etc.) are satisfied. An alternative constraint can be the requirement that the total force exerted on $z = 0$ by $p(r)$ is finite.

In the instance when

$$p(r) = \begin{cases} p_0 & 0 < r < a \\ \\ 0 & a < r < \infty \end{cases} \tag{8.1035}$$

where p_0 is a constant,

$$\overline{p}(\xi) = \int_0^\infty p_0 r J_0(\xi r) dr \tag{8.1036}$$

since

$$\int_0^a r J_0(\xi r)\,dr = \frac{a}{\xi} J_1(\xi a) \tag{8.1037}$$

we can write (8.1036) as

$$\bar{p}(\xi) = \frac{P_0}{\pi a \xi} J_1(\xi a) \tag{8.1038}$$

where P_0 is the total load acting within the circular region of radius a and stress intensity p_0. The particular case of a concentrated force can be recovered from (8.1038) by considering the limit

$$[\bar{p}(\xi)]_{\text{conc. force}} = \frac{P_0}{\pi} \operatorname*{Lim}_{a \to 0} \left(\frac{J_1(\xi a)}{a} \right) = \frac{P_0}{2\pi} \tag{8.1039}$$

The surface displacement of the surface reinforced half-space subjected to a concentrated force can be obtained by substituting (8.1039) in (8.1034); i.e.

$$2\mu u_z^c(r,0) = \frac{(3-4\nu)}{2(1-\nu)} \int_0^\infty \frac{P_0}{2\pi} J_0(\xi r)\,d\xi \tag{8.1040}$$

where $(\ )^c$ denotes the result for the concentrated force. Since

$$\int_0^\infty J_0(\xi r)\,d\xi = \frac{1}{r} \tag{8.1041}$$

we have

$$2\mu u_z^c(r,0) = \frac{(3-4\nu)P_0}{4\pi(1-\nu)r} \tag{8.1042}$$

It can be easily verified that this result for the *"surface reinforced"* half-space is in fact the solution that corresponds to Kelvin's result (see e.g. (8.935)) with $\Theta = \pi/2$; $R = r$ and note that u_Θ is positive in the direction of increase in $\Theta \in (0, \pi/2)$.

Example 8.25

An elastic layer occupies the region $r \in (0, \infty)$, $z \in (-H, H)$. The layer is subjected to compressive normal surface tractions $p_1(r)$ and $p_2(r)$ on the surfaces $z = H$ and $z = -H$ respectively (Figure 8.61). If no other tractions act on either the surfaces $z = \pm H$ or within the layer region, develop an integral expression for the variation of normal stress σ_{zz} at the mid-plane of the layer

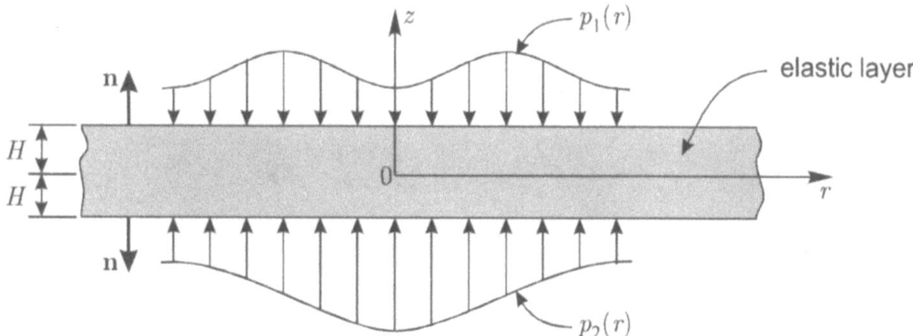

Figure 8.61: Axisymmetric loading of an elastic layer.

Solution

Since no other tractions act either on the surfaces of the layer or within it, the tractions $p_1(r)$ and $p_2(r)$ should preserve overall equilibrium of the layer. Hence p_1 and p_2 are related by the equilibrium constraint

$$\int_0^\infty r\left\{p_1(r) - p_2(r)\right\} dr = 0 \tag{8.1043}$$

The boundary value problem involves the solution of the biharmonic equation for Love's strain function $\Phi(r, z)$ such that the following traction boundary conditions are satisfied:

$$\sigma_{zz}(r, H) = -p_1(r) \qquad ; \qquad 0 < r < \infty \tag{8.1044}$$

$$\sigma_{zz}(r, -H) = -p_2(r) \qquad ; \qquad 0 < r < \infty \tag{8.1045}$$

$$\sigma_{rz}(r, H) = \sigma_{rz}(r, -H) = 0 \qquad ; \qquad 0 < r < \infty \tag{8.1046}$$

since $z \in (-H, H)$, the Hankel transform solution for $\Phi(r, z)$ can include both positive and negative exponential functions. In terms of the complete expression for $\overline{\Phi}(\xi, z)$ given by (8.1011), the stress components σ_{zz} and σ_{rz} take the forms

$$\sigma_{zz}(r, z) = \int_0^\infty \xi \left[\xi^3 e^{-\xi z} A + \left\{ (1 - 2\nu)\xi^2 + \xi^3 z \right\} e^{-\xi z} B - \xi^3 e^{\xi z} C \right.$$
$$\left. + \left\{ (1 - 2\nu) \xi^2 - \xi^3 z \right\} e^{\xi z} D \right] J_0(\xi r) d\xi \tag{8.1047}$$

and

$$\sigma_{rz}(r, z) = \int_0^\infty \xi \left[\xi^2 e^{-\xi z} A + \left\{ -2\nu\xi - \xi^2 z \right\} e^{-\xi z} B \right.$$
$$\left. + \xi^2 e^{\xi z} C + \left\{ 2\nu\xi + \xi^2 z \right\} e^{\xi z} D \right] J_1(\xi r) d\xi \tag{8.1048}$$

respectively. Considering the boundary conditions (8.1044) and (8.1045) we can write

$$p_1(r) = \int_0^\infty \xi \overline{p}_1(\xi) J_0(\xi r) d\xi \tag{8.1049}$$

$$p_2(r) = \int_0^\infty \xi \overline{p}_2(\xi) J_0(\xi r) d\xi \tag{8.1050}$$

To satisfy the boundary conditions (8.1044) to (8.1046) we require

$$A\xi^3 e^{-\xi H} + B\{(1-2\nu)\xi^2 + \xi^3 H\}e^{-\xi H} - C\xi^3 e^{\xi H}$$
$$+D\{(1-2\nu)\xi^2 - \xi^3 H\}e^{\xi H} = -\bar{p}_1(\xi)$$

$$A\xi^3 e^{\xi H} + B\{(1-2\nu)\xi^2 - \xi^3 H\}e^{\xi H} - C\xi^3 e^{-\xi H}$$
$$+D\{(1-2\nu)\xi^2 + \xi^3 H\}e^{-\xi H} = -\bar{p}_2(\xi)$$

$$\text{(8.1051)}$$

$$A\xi^2 e^{-\xi H} + B\{-2\nu\xi - \xi^2 H\}e^{-\xi H} + C\xi^2 e^{\xi H}$$
$$+D\{2\nu\xi + \xi^2 H\}e^{\xi H} = 0$$

$$A\xi^2 e^{\xi H} + B\{-2\nu\xi + \xi^2 H\}e^{\xi H} + C\xi^2 e^{-\xi H}$$
$$+D\{2\nu\xi - \xi^2 H\}e^{-\xi H} = 0$$

These equations can be solved to determine the expressions for the arbitrary functions A, B, etc. The inversion can be performed by using symbolic computation software such as MAPLE® and MATHEMATICA®. Formal integral relationships can thus be obtained for the distribution of displacements and stresses within the elastic layer. In the special case when the elastic layer is compressed to equal distributions of tractions with

$$p_1(r) = p_2(r) = p(r) \tag{8.1052}$$

The distribution of axial stresses σ_{zz} reduces to

$$\sigma_{zz}(r,z) = -2\int_0^\infty \xi F(\xi)\bar{p}(\xi)J_0(\xi r)d\xi \tag{8.1053}$$

where $\bar{p}(\xi)$ is the zeroth-order Hankel transform $p(r)$ and

$$F(\xi) = \frac{1}{2\xi H + \sinh(2\xi H)}\left[\xi H\cosh(\xi z)\cosh(\xi H)\right.$$
$$\left. -\xi z\sinh(\xi z)\sinh(\xi H) + \sinh(\xi H)\cosh(\xi z)\right] \tag{8.1054}$$

On the mid-plane of the elastic layer we have

$$\sigma_{zz}(r,0) = \frac{-2}{H^2}\int_0^\infty \eta\bar{p}\left(\frac{\eta}{H}\right)\left\{\frac{\eta\cosh(\eta) + \sinh(\eta)}{2\eta + \sinh(\eta)}\right\}$$
$$\cdot J_0\left(\eta\frac{r}{H}\right)d\eta \tag{8.1055}$$

Integrals of the type (8.1055) can be best evaluated by adopting a numerical integration scheme where the quadrature scheme is sufficiently refined to take into consideration the oscillatory nature of the Bessel function.

$$\bullet\ \bullet\ \bullet$$

We next consider the class of problems which examine the stress analysis of cylindrical regions governed by the biharmonic equation for a Love strain function $\Phi_c(r, z)$. Typical problems can include situations where cylindrical bars and cylindrical cavities are subjected to tractions on cylindrical surfaces, maintaining the axially symmetric nature of the resulting problem. Some examples of such problems are illustrated in Figure 8.62.

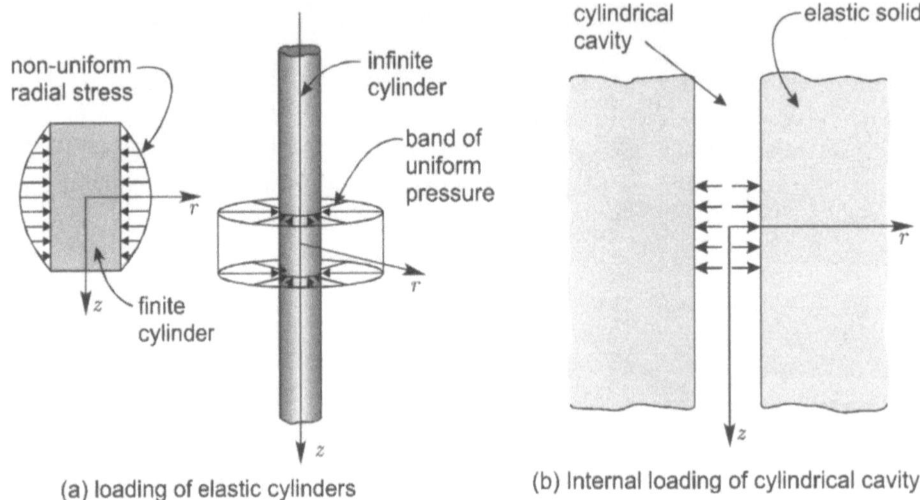

Figure 8.62: Axisymmetric deformation of elastic solids with cylindrical boundaries.

It is evident that the variations in the application tractions occurs in the z-direction. Therefore in the development of integral transform solutions to this class of traction boundary value problems, we seek solutions of

$$\widehat{\nabla}^2\widehat{\nabla}^2\Phi^*(r, z) = 0 \tag{8.1056}$$

of the form

$$\Phi^*(r, z) = F_\alpha(r)\sin(\xi z) \quad \text{or} \quad \Phi^*(r, z) = F_\beta(r)\cos(\xi z) \tag{8.1057}$$

where $F_n(r)$ $(n = \alpha, \beta)$ are arbitrary functions and ξ is an arbitrary parameter. We can develop more generalized solutions of (8.1056) with the appropriate distributions in z, involving both cosine and sine terms, by using an exponential Fourier transform approach where

$$\overline{\Phi}^*(r, \xi) = \int_{-\infty}^{\infty} e^{i\xi z} \Phi^*(r, z)\, dz \tag{8.1058}$$

and ξ has a positive imaginary part.

Consider the integral

$$L = \int_{-\infty}^{\infty} e^{i\xi z} \widehat{\nabla}^2 \Phi^*(r, z)\, dz \tag{8.1059}$$

Substituting the expression for the Laplacian, (8.1059) can be written as

$$L = \left\{ \frac{d^2}{dr^2} + \frac{1}{r}\frac{d}{dr} \right\} \int_{-\infty}^{\infty} e^{i\xi z} \Phi^*(r, z)\, dz + \int_{-\infty}^{\infty} e^{i\xi z} \frac{\partial^2 \Phi^*}{\partial z^2}\, dz \tag{8.1060}$$

Integrating the second integral of (8.1060) by parts we have

$$\int_{-\infty}^{\infty} e^{i\xi z} \frac{\partial^2 \Phi^*}{\partial z^2}\, dz = \left[e^{i\xi z} \frac{\partial \Phi^*}{\partial z} \right]_{-\infty}^{\infty} - i\xi \int_{-\infty}^{\infty} e^{i\xi z} \frac{\partial \Phi^*}{\partial z}\, dz \tag{8.1061}$$

$$= \left[e^{i\xi z} \frac{\partial \Phi^*}{\partial z} - i\xi e^{i\xi z} \Phi^* \right]_{-\infty}^{\infty} + (i\xi)^2 \int_{-\infty}^{\infty} e^{i\xi z} \Phi^*\, dz$$

We assume that the first term of the right hand side of the equation (8.1061) vanishes at both limits. Again, this is a requirement that needs to be satisfied by the regularity conditions imposed on Φ^* and $(\partial \Phi^*/\partial z)$ as $|z| \to \infty$. Therefore, we can write

$$\int_{-\infty}^{\infty} e^{i\xi z} \frac{\partial^2 \Phi^*}{\partial z^2}\, dz = (i\xi)^2\, \overline{\Phi}^*(r, \xi) \tag{8.1062}$$

or, in general

$$\int_{-\infty}^{\infty} e^{i\xi z} \frac{\partial^n \Phi^*}{\partial z^n}\, dz = (i\xi)^n\, \overline{\Phi}^*(r, \xi) \tag{8.1063}$$

Hence

$$\int_{-\infty}^{\infty} e^{i\xi z}\widehat{\nabla}^2 \Phi^*(r,z)dz = \overline{\nabla}^2\overline{\Phi}^*(r,\xi) \tag{8.1064}$$

where the operator $\overline{\nabla}^2$ is defined by

$$\overline{\nabla}^2 = \frac{d^2}{dr^2} + \frac{1}{r}\frac{d}{dr} - \xi^2 \tag{8.1065}$$

Application of the exponential Fourier transform to (8.1056) gives the following ordinary differential equation for $\overline{\Phi}^*(r,\xi)$;

$$\overline{\nabla}^2\overline{\nabla}^2\overline{\Phi}^*(r,\xi) = 0 \tag{8.1066}$$

The general solution of (8.1066) can be written as

$$\overline{\Phi}^*(r,\xi) = AK_0(\xi r) + BrK_1(\xi r) + CI_0(\xi r) + DrI_1(kr) \tag{8.1067}$$

where A, B, C, D are either constants or functions of ξ, K_n and I_n are modified Bessel functions of the first and second-kind respectively of order n. Inverting the Fourier transform we can write

$$\Phi^*(r,z) = \frac{1}{2\pi}\int_{i\gamma-\infty}^{i\gamma+\infty} \overline{\Phi}^*(r,\xi)e^{-i\xi z}d\xi \tag{8.1068}$$

where $\gamma > 0$.

It can also be shown that

$$\Phi^*(r,z) = \frac{1}{\sqrt{2\pi}}\int_{-\infty}^{\infty} \sin(\xi z)\overline{\Phi}^*(r,\xi)d\xi \tag{8.1069}$$

and

$$\Phi^*(r,z) = \frac{1}{\sqrt{2\pi}}\int_{-\infty}^{\infty} \cos(\xi z)\overline{\Phi}^*(r,\xi)d\xi \tag{8.1070}$$

are also solutions of (8.1056). The choice of the functions I_0, I_1, K_0 and K_1 will depend on the domain of the particular problem under consideration. For example, the modified Bessel functions I_0 and I_1 are zero at $r = 0$ but approach infinity as $r \to \infty$. As such, a solution containing these functions is expected to be valid in regions where r is finite. Similarly, the functions K_0 and K_1 approach infinity as $r \to 0$, but decay to zero as $r \to \infty$. These functions are therefore suitable for obtaining solutions to exterior regions where the origin is excluded. In the ensuing, we shall consider some typical problems associated with circular bars of infinite length and cylindrical cavity regions in elastic solids, which can be examined by appeal to the general solution (8.1068) presented earlier.

For purposes of future reference, we shall record here the expressions for the transforms of the displacements and stresses in terms of the integral transform solution involving $\overline{\Phi}^*(r, \xi)$. The expressions for the displacements are

$$\overline{u}_r(r, \xi) = \frac{i\xi}{2\mu} \frac{d\overline{\Phi}^*}{dr}$$

$$\overline{u}_z(r, \xi) = \frac{1}{2\mu} \left\{ (1 - 2\nu)\overline{\nabla}^2 + \frac{d^2}{dr^2} + \frac{1}{r}\frac{d}{dr} \right\} \overline{\Phi}^*$$

$$(8.1071)$$

and the expressions for the stresses are

$$\overline{\sigma}_{rr}(r, \xi) = -i\xi \left\{ \nu\overline{\nabla}^2 - \frac{d^2}{dr^2} \right\} \overline{\Phi}^*$$

$$\overline{\sigma}_{\theta\theta}(r, \xi) = -i\xi \left\{ \nu\overline{\nabla}^2 - \frac{1}{r}\frac{d}{dr} \right\} \overline{\Phi}^*$$

$$\overline{\sigma}_{zz}(r, \xi) = -i\xi \left\{ (2 - \nu)\overline{\nabla}^2 + \xi^2 \right\} \overline{\Phi}^*$$

$$\overline{\sigma}_{rz}(r, \xi) = \frac{d}{dr} \left\{ (1 - \nu)\overline{\nabla}^2 + \xi^2 \right\} \overline{\Phi}^*$$

$$(8.1072)$$

Example 8.26

A cylindrical bar of finite radius and infinite length (i.e. $r \in (0, a)$; $z \in (-\infty, \infty)$) is subjected to a non-uniform radial stress $F_0^*(z)$ over the cylindrical surface $r = a$; $z \in (0, \infty)$. Use the Love strain function approach to determine the radial stress distribution in the bar (Figure 8.63).

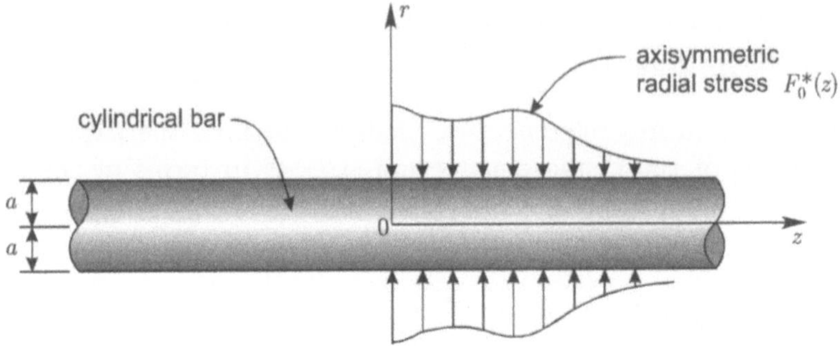

Figure 8.63: Uniform load radial loading of a cylindrical bar.

Solution

The analysis of the problem is reduced to the solution of (8.1056) subject to the traction boundary conditions

$$\sigma_{rr}(a, z) = 0 \qquad ; \qquad -\infty < z < 0 \tag{8.1073}$$

$$\sigma_{rr}(a, z) = -F_0^*(z) \qquad ; \qquad 0 < z < \infty \tag{8.1074}$$

$$\sigma_{rz}(a, z) = 0 \qquad ; \qquad -\infty < z < \infty \tag{8.1075}$$

In addition, the stress field should be finite in the cylindrical region; should decay to zero as $z \to -\infty$ and should decay to zero as $z \to \infty$ if $F_0^*(z) \to 0$. Since in the region of interest $r \in (0, a)$, the applicable form of the Fourier transform of Love's strain function is

$$\Phi^*(r, \xi) = C(\xi)I_0(\xi r) + \xi r D(\xi)I_1(\xi r) \tag{8.1076}$$

where $C(\xi)$ and $D(\xi)$ are arbitrary functions. In order to determine these unknown functions we make use of the boundary conditions (8.1073) to

(8.1075). We can transform the boundary conditions (8.1073) and (8.1074) and denote by $\overline{F}^*(\xi)$ the following;

$$\overline{F}^*(\xi) = \int_0^\infty F^*(z)e^{i\xi z}dz \tag{8.1077}$$

Noting that

$$\frac{d}{dr}\{I_0(\xi r)\} = \xi I_1(\xi r)$$

$$\frac{d}{dr}\{I_1(\xi r)\} = \xi\left[I_0(\xi r) - \frac{1}{\xi r}I_1(\xi r)\right] \tag{8.1078}$$

we can show that

$$\overline{\nabla}^2\overline{\Phi}^*(r,\xi) = 2\xi^2 B(\xi)I_0(\xi r) \tag{8.1079}$$

Using (8.1076) and (8.1079) in the equation (8.1072), we can obtain expressions for $\overline{\sigma}_{rr}(r,\xi)$ and $\overline{\sigma}_{zr}(r,\xi)$, in terms of $C(\xi)$ and $D(\xi)$. Using these expressions, the boundary conditions at $r = a$ give

$$C(\xi)\left[I_0(\xi a) - \frac{1}{\xi a}I_1(\xi a)\right]$$

$$+D(\xi)\left[(1-2\nu)I_0(\xi a) + \xi a\ I_1(\xi a)\right] = -\frac{i\overline{F}^*(\xi)}{\xi^3} \tag{8.1080}$$

$$C(\xi)\left[I_1(\xi a)\right] + D(\xi)\left[2(1-\nu)I_1(\xi a) + \xi a\ I_0(\xi a)\right] = 0$$

Solving (8.1080) we obtain

$$C(\xi) = \frac{ia}{\xi^3\Delta(\xi)}\left[2(1-\nu)I_1(\xi a) + \xi a\ I_0(\xi a)\right]\overline{F}^*(\xi)$$

$$\tag{8.1081}$$

$$D(\xi) = -\frac{ia}{\xi^3\Delta(\xi)}\left[I_1(\xi a)\right]\overline{F}^*(\xi)$$

where

$$\Delta(\xi a) = \left[(\xi a)^2 + 2(1 - \nu)\right] I_1^2(\xi a) - (\xi a)^2 I_0^2(\xi a) \tag{8.1082}$$

The expression for the transformed value of the radial stress $\overline{\sigma}_{rr}(r, \xi)$ is given by

$$\overline{\sigma}_{rr}(r, \xi) = \frac{a\overline{F}^*(\xi)}{r\Delta(\xi a)}[ar\xi^2 I_0(\xi a)I_0(\xi r) + \xi r I_0(\xi r)I_1(\xi a)$$

$$-\xi a I_0(\xi a)I_1(\xi r) - \left\{\xi^2 r^2 + 2(1 - \nu)\right\} I_1(\xi a)I_1(\xi r)] \tag{8.1083}$$

Taking the inverse transform of (8.1083) we can obtain an integral expressions for the radial stress; i.e.

$$\sigma_{rr}(r, z) = \frac{1}{2\pi} \int_{-\infty}^{\infty} \overline{\sigma}_{rr}(r, \xi)e^{-i\xi z} d\xi \tag{8.1084}$$

Formal integral expressions can be obtained for the variation of displacements and stresses within the solid bar, e.g.

$$2\mu u_r(r, z) = \frac{1}{2\pi} \int_{-\infty}^{\infty} \frac{a\overline{F}^*(\xi)}{\xi\Delta(\xi a)}\left[- \left\{2(1 - \nu)I_1(\xi a) + \xi a I_0(\xi a)\right\} I_1(\xi r)\right.$$

$$\left. +r\xi\, I_1(\xi a)I_0(\xi r)\right] e^{-i\xi z} d\xi \tag{8.1085}$$

etc. The integral expressions for displacements and stresses can be made specific by prescribing the variation of $F^*(z)$ along the surface of the bar. We can write integrals of the type (8.1084) as

$$\sigma_{rr}(r, z) = \frac{1}{2\pi} \int_{i\gamma-\infty}^{i\gamma+\infty} \overline{\sigma}_{rr}(r, \xi)e^{-i\xi z} d\xi \tag{8.1086}$$

where $\gamma > 0$. The integral of (8.1086) has no poles on the real axis, except at the origin. For inversion of (8.1080), $\Delta(\xi a)$ should be non-zero for all ξa. The contour integral can therefore be deformed into the real axis from $-\infty$ to ∞ with a clockwise contour semi-circle around the origin. Following procedures for contour integration, the integral can be evaluated as the sum of the integrals along the real axis less πi times the residue of the integrand at $\xi = 0$. The residues for the displacement and stress components can be obtained by using power series expansions for the various Bessel functions.

For σ_{rr}, the residue is $-i$ and the integral (8.1086) can be appropriately reduced to an integral from $\xi = 0$ to ∞.

Consider the particular case where the boundary conditions on the surface of the cylinder is subjected to the tractions

$$\sigma_{rr}(a, z) = -\sigma_0 \qquad\qquad 0 < z < \infty$$
$$\sigma_{rr}(a, z) = 0 \qquad\qquad -\infty < z < 0 \qquad\qquad (8.1087)$$
$$\sigma_{rz}(a, z) = 0 \qquad\qquad -\infty < z < \infty$$

To obtain $\overline{F}^*(\xi)$ we introduce the delta function due to Werner Heisenberg (1901-1976);

$$\delta_+(\xi) = \frac{1}{2}\delta(\xi) - \frac{1}{2\pi i \xi} \qquad\qquad (8.1088)$$

Consider the integral of $\delta_+(\xi)$

$$-\frac{\sigma_0}{2\pi} \int_{-\infty}^{\infty} \delta_+(\xi) e^{-i\xi z} d\xi = -\frac{\sigma_0}{2\pi}\left[\frac{1}{2} + \frac{1}{\pi}\int_0^{\infty} \frac{\sin(\xi z)}{\xi} d\xi\right]$$

$$= \begin{cases} -\dfrac{\sigma_0}{2\pi} \; ; \; 0 < z < \infty \\[2ex] 0 \quad ; -\infty < z < 0 \end{cases} \qquad\qquad (8.1089)$$

The result on the right hand side of (8.1089) is the radial stress defined by (8.1087) (except for the multiplicative constant $1/2\pi$). Therefore from the inversion of (8.1089) we have

$$\overline{F}^*(\xi) = 2\pi\sigma_0\delta_+(\xi) \qquad\qquad (8.1090)$$

The corresponding expression for σ_{rr} takes the form

$$\sigma_{rr}(r, z) = -\frac{\sigma_0}{2} + \frac{\sigma_0 a}{\pi r}\int_0^{\infty}\left[\frac{r}{a}\beta^2 I_0(\beta)I_0\left(\beta\frac{r}{a}\right) + \beta\frac{r}{a}I_1(\beta)I_0\left(\beta\frac{r}{a}\right)\right.$$
$$\left. - \left\{\left(\beta\frac{r}{a}\right)^2 + 2(1-\nu)\right\} I_1(\beta)I_1\left(\beta\frac{r}{a}\right)\right]$$

$$-I_0(\beta)I_1\left(\beta\frac{r}{a}\right)\Big] \frac{\sin\left(\beta\frac{z}{a}\right)}{\beta\Delta(\beta)}\,d\beta \tag{8.1091}$$

Expressions similar to (8.1091) can be derived for the displacements and remaining stress components. These integral expressions represent the formal mathematical solution to the problem of the discontinuous loading of the cylinder. Specific results can be obtained by numerical evaluation of the integrals.

8.10.13   Mixed boundary value problems

In the application of integral transform techniques to problems associated with both half-space and infinite space regions discussed previously attention has been restricted to cases where the surfaces $z = $ constant (or $r = $ constant), are subjected to either traction or displacement boundary conditions over the entire interval $r \in (0,\infty)$ (or $z \in (-\infty,\infty)$. With most practical situations involving stress analysis of half-space regions, for example, the surfaces $z = $ const. , can be subjected to *mixed* boundary conditions, where displacements are prescribed on a part of the surface and tractions are prescribed on the remainder of the surface. The analysis of this class of *mixed boundary value problems* in the classical theory of elasticity represents one of the most well researched areas of the mathematical theory of elasticity. No attempt will be made to cover this topic in any substantial detail. In the ensuing, however, we shall outline the application of Hankel transform techniques to the formulation and solution of a classical mixed boundary value problem involving the indentation of a half-space region by a rigid frictionless indentor with an axisymmetric base.

The specific problem deals with the indentation of the surface of a half-space region ($r \in (0,\infty)$; $z \in (0,\infty)$) by a rigid circular punch or indentor with a flat base.

This particular problem was first investigated by Boussinesq who obtained the solution from a consideration of the mathematical similarity between the elasticity problem and an analogous problem in potential theory. The application Hankel transforms to the solution of the problem is due to a celebrated paper by Harding and Sneddon who also obtained, through this procedure, the relationship between applied load P and the indentation w_0.

Love's strain function formulation of the mixed boundary value problem requires the solution of

$$\widehat{\nabla}^2\widehat{\nabla}^2\Phi(r,z) = 0 \tag{8.1092}$$

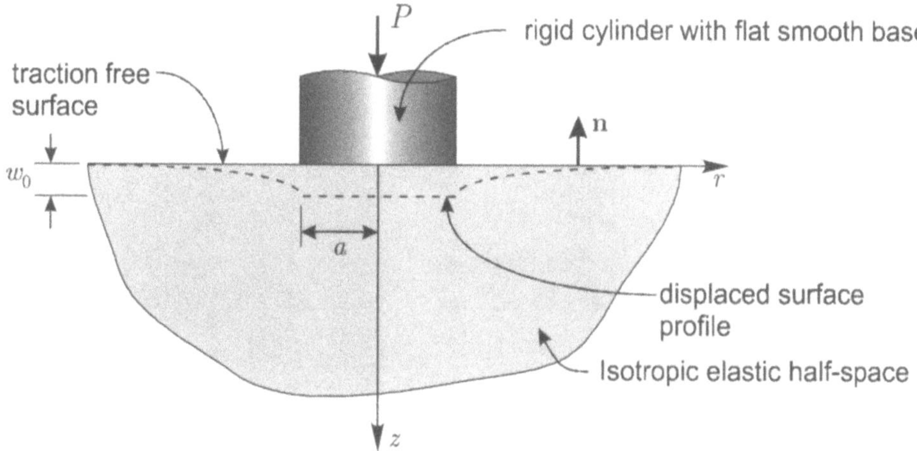

Figure 8.64: Frictionless indentation of a half-space by a rigid circular punch.

which is subject to the mixed boundary conditions

$$u_z(r,0) = w_0 \quad ; \quad 0 \leq r \leq a \tag{8.1093}$$

$$\sigma_{zz}(r,0) = 0 \quad ; \quad a < r < \infty \tag{8.1094}$$

$$\sigma_{rz}(r,0) = 0 \quad ; \quad 0 < r < \infty \tag{8.1095}$$

In addition, the displacements and stresses in the half-space region should reduce to zero as $r, z \to \infty$. To satisfy this regularity condition, the appropriate form of $\overline{\Phi}(\xi, z)$, the Hankel transform of Love's strain function should be of the form

$$\overline{\Phi}(\xi, z) = [A(\xi) + B(\xi)z]e^{-\xi z} \tag{8.1096}$$

From (8.1096) and (8.1024), to satisfy the boundary condition (8.1095) we require

$$A(\xi) = \frac{2\nu B(\xi)}{\xi} \tag{8.1097}$$

The mixed boundary conditions (8.1093) and (8.1094) are equivalent to the system of dual integral equations

$$\int_0^\infty -2\xi^2(1-\nu)B(\xi)J_0(\xi r)d\xi = 2\mu w_0 \quad ; \quad 0 \le r \le a \qquad (8.1098)$$

$$\int_0^\infty \xi^3 B(\xi)J_0(\xi r)d\xi = 0 \qquad ; \quad a < r < \infty \qquad (8.1099)$$

We can rewrite (8.1098) and (8.1099) as

$$\int_0^\infty M(\beta)J_0(\rho\beta)d\beta = w_0^* \qquad ; \qquad 0 \le \rho \le 1 \qquad (8.1100)$$

$$\int_0^\infty \beta M(\beta)J_0(\rho\beta)d\beta = 0 \qquad ; \qquad 1 < \rho < \infty \qquad (8.1101)$$

where

$$\beta = \xi a \quad ; \quad \rho = \frac{r}{a} \quad ; \quad M(\beta) = \beta^2 B(\xi) \quad ; \quad w_0^* = -\frac{\mu a^3 w_0}{(1-\nu)} \quad (8.1102)$$

For the solution of the system of dual integral equations defined by (8.1100) and (8.1101) we introduce the *finite Fourier cosine transform* of $M(\beta)$ defined by

$$M(\beta) = \int_0^1 \Omega(t)\cos(\beta t)dt \qquad (8.1103)$$

such that (8.1101) is identically satisfied. If we consider the transform

$$\mathcal{H}_0\left\{\xi^{-1}\mathcal{F}_c[g(t);\xi];r\right\} = \int_0^\infty J_0(\xi r)d\xi \sqrt{\frac{2}{\pi}}\int_0^\infty g(t)\cos(\xi t)dt$$

$$= \sqrt{\frac{2}{\pi}}\int_0^\infty g(t)dt \int_0^\infty J_0(\xi r)\cos(\xi t)d\xi \quad (8.1104)$$

The last integral of (8.1104) can be evaluated in closed form giving

$$\int_0^\infty J_0(\xi r) \cos(\xi t) d\xi = \frac{H(r-t)}{(r^2 - t^2)^{1/2}} \tag{8.1105}$$

where $H(r-t)$ is the Heaviside step function such that

$$H(r-t) = \begin{cases} 1 \; ; r > t \\ 0 \; ; r < t \end{cases} \tag{8.1106}$$

Hence

$$\frac{2}{\pi} \int_0^\infty J_0(\xi r) d\xi \int_0^\infty g(t) \cos(\xi t) dt = \sqrt{\frac{2}{\pi}} \int_0^\infty \frac{g(t) H(r-t) dt}{(r^2 - t^2)^{1/2}}$$

$$= \sqrt{\frac{2}{\pi}} \int_0^r \frac{g(t) dt}{(r^2 - t^2)^{1/2}} \tag{8.1107}$$

Similarly, noting that

$$\int_0^\infty J_0(\xi r) \sin(\xi t) d\xi = \frac{H(t-r)}{(t^2 - r^2)^{1/2}} \tag{8.1108}$$

where

$$H(t-r) = \begin{cases} 1 \; ; t > r \\ 0 \; ; t < r \end{cases} \tag{8.1109}$$

we can show that

$$\mathcal{H}_0 \left\{ \xi^{-1} \mathcal{F}_s[g(t); \xi]; r \right\} = \sqrt{\frac{2}{\pi}} \int_r^\infty \frac{g(t) dt}{(t^2 - r^2)^{1/2}} \tag{8.1110}$$

Also, we note that

$$\mathcal{F}_s\{g'(t); \xi\} = -\xi \mathcal{F}_c\{g(t); \xi\} \tag{8.1111}$$

$$\mathcal{F}_c\{g'(t); \xi\} = \xi \mathcal{F}_s\{g(t); \xi\} \tag{8.1112}$$

and (8.1112) is valid provided $g(0) = 0$. Hence we can write

$$\mathcal{H}_0\left\{\mathcal{F}_c[g(t);\xi];r\right\} = -\mathcal{H}_0\left\{\xi^{-1}\mathcal{F}_s[g'(t);\xi];r\right\}$$
$$= -\sqrt{\frac{2}{\pi}}\int_r^\infty \frac{g'(t)dt}{(t^2-r^2)^{1/2}} \tag{8.1113}$$

and

$$\mathcal{H}_0\left\{\mathcal{F}_s[g(t);\xi];r\right\} = \mathcal{H}_0\left\{\xi^{-1}\mathcal{F}_c[g'(t);\xi];r\right\}$$
$$= \sqrt{\frac{2}{\pi}}\int_0^r \frac{g'(t)dt}{(r^2-t^2)^{1/2}} \tag{8.1114}$$

with $g(0) = 0$.

If we assume that

$$g(t) = \Omega(t)H(1-t) \tag{8.1115}$$

then, we obtain from (8.1107) and (8.1113)

$$\mathcal{H}_0\left\{\beta^{-1}\int_0^1 \Omega(t)\cos(\beta t)dt \; ; \; \rho\right\} = \int_0^\rho \frac{\Omega(t)dt}{(\rho^2-t^2)^{\frac{1}{2}}} \quad ; \; 0 \le \rho \le 1 \tag{8.1116}$$

$$\mathcal{H}_0\left\{\int_0^1 \Omega(t)\cos(\beta t)dt \; ; \; \rho\right\} = 0 \qquad ; 1 < \rho < \infty \tag{8.1117}$$

Considering (8.1116) and (8.1117) with the system of dual integral equations (8.1100) and (8.1101) we note that (8.1101) is satisfied by the representation (8.1103), and the equation (8.1100) reduces to the integral equation

$$\int_0^\rho \frac{\Omega(t)dt}{(\rho^2-t^2)^{\frac{1}{2}}} = w_0^* \tag{8.1118}$$

This integral equation was first examined by Niels Henrik Abel (1802-1829). It occurs in the connection with the problem of obtaining the time of descent of a particle sliding down a smooth curve, starting from rest, purely under the influence of gravity. The problem originally posed by Johann Bernoulli (1667-1748) deals with the determination of the path of descent between two given points in the shortest time. The solution to the problem results in a path which is the *cycloid* between the given two points.

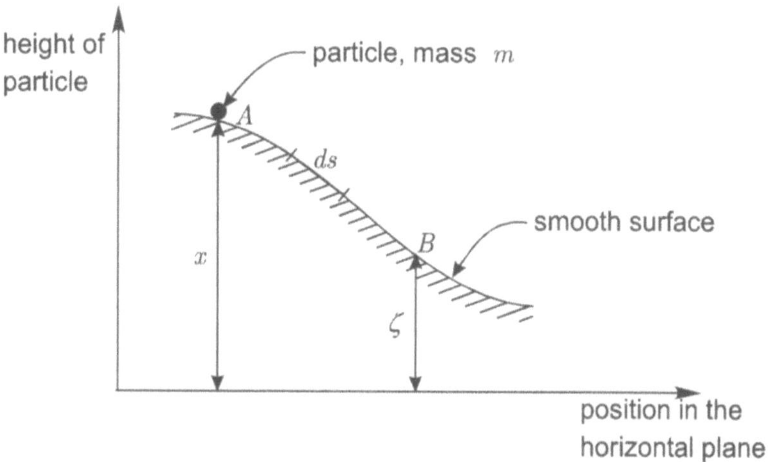

Figure 8.65: Descent of a particle along a smooth surface.

Briefly, if the particle descends under gravity along the *smooth surface*, all the potential energy lost during descent is converted to kinetic energy. If the velocity of the sliding particle at height ζ (Figure 8.65) is v, then

$$\frac{1}{2}mv^2 = mg(x - \zeta) \tag{8.1119}$$

or

$$v = \frac{ds}{dt} = \sqrt{2g(x - \zeta)} \tag{8.1120}$$

where ds is the incremental length along the surface. Time taken to reach the height ζ is

$$\int dt - \int \frac{ds}{\sqrt{2g(x - \zeta)}} \tag{8.1121}$$

Time to reach zero height is given by

$$t = -\int_0^x \frac{ds}{[2g(x-\zeta)]^{\frac{1}{2}}} = \int_0^x \frac{Q(\zeta)d\zeta}{(x-\zeta)^{\frac{1}{2}}} \qquad (8.1122)$$

where

$$Q(\zeta) = \frac{1}{\sqrt{2g}} \frac{ds}{d\zeta} \qquad (8.1123)$$

For a given path of descent, the time for descent can be evaluated. If, however the time for descent $f(x)$ is specified and the associated path of descent is unknown, (8.1122) gives rise to an equation of the Abel type:

$$f(x) = \int_0^x \frac{Q(\zeta)}{(x-\zeta)^{\frac{1}{2}}} d\zeta \qquad (8.1124)$$

If we multiply both sides of this equation by $x[t^2 - x^2]^{-\frac{1}{2}}$ and integrate between 0 to t we have

$$\int_0^t \frac{xf(x)dx}{(t^2 - x^2)^{\frac{1}{2}}} = \int_0^t \frac{x}{(t^2 - x^2)^{\frac{1}{2}}} \left[\int_0^x \frac{Q(\zeta)}{(x-\zeta)^{\frac{1}{2}}} d\zeta \right] dx \qquad (8.1125)$$

The order of the integration can be reversed by noting that

$$\int_0^t \int_0^x g(x,\zeta)d\zeta dx = \int_0^t \left[\int_\zeta^t g(x,\zeta)dx \right] d\zeta \qquad (8.1126)$$

The right hand side of the integral in (8.1125) can now be written as

$$\int_0^t Q(\zeta) \left[\int_\zeta^t \frac{x\,dx}{[(x-\zeta)(t^2 - x^2)]^{\frac{1}{2}}} \right] d\zeta = \frac{\pi}{2} \int_0^t Q(\zeta)d\zeta \qquad (8.1127)$$

Combining (8.1125) and (8.1127) we have

$$\int_0^t \frac{xf(x)dx}{(t^2 - x^2)^{\frac{1}{2}}} = \frac{\pi}{2} \int_0^t Q(\zeta)d\zeta \qquad (8.1128)$$

The solution of Abel's integral equation is obtained by differentiating both sides of (8.1128) which gives

$$Q(t) = \frac{2}{\pi} \frac{d}{dt} \int_0^t \frac{x f(x) dx}{(t^2 - x^2)^{\frac{1}{2}}} \tag{8.1129}$$

Using this relationship, the solution of (8.1118) is given by

$$\Omega(t) = \frac{2}{\pi} \frac{d}{dt} \int_0^t \frac{w_0^* \rho \, d\rho}{(t^2 - \rho^2)^{\frac{1}{2}}} = \frac{2 w_0^*}{\pi} \tag{8.1130}$$

Expressions for $B(\xi)$ can be derived by making use of the result for $\Omega(t)$ and (8.1103) and (8.1102). This would formally complete the solution to the mixed boundary value problem defined by (8.1093) to (8.1094). Results for the displacement and stress components can be derived by using the formal solution for $\Phi(r, z)$ and the results (8.869) and (8.870). A result of some engineering importance pertains to the evaluation of the force P required to induce the displacement w_0. By considering the equilibrium of normal tractions acting on the contact surface and the external force P we have

$$P = -2\pi \int_0^a r \sigma_{zz}(r, 0) dr \tag{8.1131}$$

Considering the expression for $\sigma_{zz}(r, 0)$, we have

$$\sigma_{zz}(r, 0) = \int_0^\infty \xi^3 B(\xi) J_0(\xi r) d\xi = \frac{1}{a^4} \int_0^\infty \beta M(\beta) J_0(\beta \rho) d\beta \tag{8.1132}$$

Using this expression in (8.1131) we obtain

$$\begin{aligned} P &= -\frac{2\pi}{a^2} \int_0^\infty \beta M(\beta) d\beta \int_0^1 \rho J_0(\beta \rho) d\rho \\ &= \frac{2\pi}{a^2} \int_0^\infty M(\beta) J_1(\beta) d\beta \end{aligned} \tag{8.1133}$$

Using the result (8.1103), the above equation gives

$$P = -\frac{2\pi}{a^2} \int_0^\infty \left[\int_0^1 \Omega(t) \cos(\beta t) dt \right] J_1(\beta) d\beta \tag{8.1134}$$

Since

$$\int_0^\infty \cos(\beta t) J_1(\beta) d\beta = 1 \quad \text{for} \quad 0 \le t \le 1 \tag{8.1135}$$

the result (8.1134) gives

$$P = -\frac{2\pi}{a^2} \int_0^1 \Omega(t) dt \tag{8.1136}$$

Substituting for $\Omega(t)$ from (8.1130) and for w_0^* from (8.1102), (8.1136) gives

$$P = \frac{4\mu a w_0}{(1-\nu)} \tag{8.1137}$$

which represents the load-displacement relationship for the smooth indentation of an elastic half-space by a rigid circular indentor with a flat base.

Also the normal contact stress at the indenting region is given by

$$\sigma_{zz}(r,0) = \frac{1}{a^4} \int_0^\infty \beta J_0(\beta\rho) d\beta \int_0^1 \Omega(t) \cos(\beta t) dt \tag{8.1138}$$

Using procedures outlined previously we can show that

$$\sigma_{zz}(\rho,0) = \frac{1}{a^4} \frac{\Omega(1)}{(1-\rho^2)^{\frac{1}{2}}} \tag{8.1139}$$

or

$$\sigma_{zz}(r,0) = -\frac{2\mu w_0}{\pi(1-\nu)(a^2-r^2)^{\frac{1}{2}}} \tag{8.1140}$$

The general procedure outlined here can be used to evaluate the load-displacement behaviour of smoothly indenting axisymmetric indentors with an arbitrary indentation profile within the region $0 \le r \le a$. The relationship between $\Omega(t)$ and the indentor profile $w(\rho)$ can be obtained from (8.1129) and (8.1130) provide complete contact is maintained over the entire region

$0 \leq r \leq a$. The mixed boundary value problem presented here is a classical problem in the theory of elasticity. Applications to other axisymmetric problems involving contact of elastic bodies in compression, crack problems, etc. have been extensively discussed in literature dealing with the application of the theory of elasticity to problems in mechanics.

8.11   Flexure of thin elastic plates

Any treatment of the biharmonic equation will not be complete without a discussion of its application to the study of the flexure of thin elastic plates which are subjected to transverse loads normal to is plane. A plate is usually regarded as a continuum region which is bounded by two surfaces of small or zero curvature. The distance between these surfaces is referred to as the thickness of the plate which, for thin plates, is assumed to be considerably smaller than its lateral dimensions. The definition of a *"thin elastic plate"* can also be approached by considering the various modes of energy stored in the plate during its deformation. In general, when a plate-shaped elastic structural element undergoes deformation, the stored energy is composed of *flexural strain energy* due to change in curvature, *shear strain energy* due to distortion and *extensional strain energy* due to stretching in the plane of the plate. In classical thin plate theory, the flexural energy is assumed to be the dominant component. This terminology is consistent with the definition of a thin plate introduced by Love (1863-1940). The study of the mechanics of thin plates has a long standing involvement of mathematical physicists, mechanicians and engineers. The classical studies dealing with the flexure of prismatic beams conducted by Jacques Bernoulli (1654-1705) and Euler (1707-1783) can be regarded as the precursor to the development of the theory of plates where notions of curvature of the beam due flexure, resultants of tractions acting on planes normal to the neutral axis and the equations of elasticity were combined to derive the differential equation governing the flexure of the beam.

The classical theory of thin plates, which results in the formulation of a non-homogeneous biharmonic equation for the transverse deflection of the plate was first proposed by Sophie Germain (1776-1831). Although the statement that the dynamic flexural behaviour of a plate can be characterized by a fourth-order partial differential equation is generally attributed to Lagrange (1736-1813), the fundamental work which led to the developments was conducted by Germain. Other important studies in this area were conducted by E.F.F. Chaladni (1756-1827) in connection with the identification of nodal curves of vibrating plates. A number of eminent mathematicians including Augustin Cauchy (1789-1857), Jakob Bernoulli (1759-1789), Simeon Poisson (1781-1840), L.M.H. Navier (1785-1836) and Clebsch (1833-1872) have made contributions to the development of the theory of plates. Navier, for example, developed a solution to the problem of a rectangular plate with simply supported edges. Poisson's studies dealt with the flexural vibrations of circular plates and provided the first complete proof of the statement attributed to Lagrange. M. Levy (1838-1910) also developed double Fourier series solutions to problems dealing with rectangular plates.

Although the partial differential equation governing the classical problem of the flexure of thin plates was firmly established, the formulation of the essential boundary conditions required for correctly formulating boundary value problems has been the subject of controversy. The three boundary conditions proposed by Poisson were related to the resultants of flexural stresses and shear tractions on the plate. Kirchhoff (1824-1887) showed that only two boundary conditions are necessary and sufficient to correctly define the traction boundary conditions along the edge of plate. The study of flexure of thin plates was extended by a number of eminent elasticians including Michell (1863-1940), Dougall (1867-1960), Love, Nadai (1883-1963) and Timoshenko (1878-1972) to include a variety of modifications including considerations of shear deformations, in-plane membrane effects, the influence of localized loading and clamped boundaries. The studies of Clebsch (1833-1872) and Föppl (1854-1924) dealt with the large deflections of thin plates undergoing flexure. Modern developments in the theory of plates are attributed to Korenev (1910-1998), Lekhnitski (1909-1981), Ufliand (1916-1991), Reissner (1913-1996), Mindlin (1906-1987) and others. The theory of plates forms one of the most well researched areas of the theory of elasticity and solid mechanics. The developments in theories of plate bending have found extensive applications in civil, mechanical, aerospace and materials engineering. In more recent work, the theories of flexural behaviour of laminates and fibre reinforced materials have formed the basis for the design of advanced composites, studies of flexural vibrations of piezoelectric and acoustic devices and in the study of advanced functionally graded materials.

The primary focus of this section is to examine the biharmonic equation resulting from the classical theory of plates. The developments presented in the ensuing sections, however, will also cover the theory of moderately thick plates where the effects of shear deformations are also included. This is a topic of considerable importance to modern developments in the study of advanced composites such as laminates. Its presentation is considered to be of some value not only to modelling these materials but also to extending the study of partial differential equations to non-classical topics. A secondary objective of this section is to develop the governing equations and the resulting partial differential equations in invariant forms which can be used to develop the relevant forms of the equations referred to any orthogonal curvilinear coordinate system. Such developments are relatively straightforward if, at the outset, attention is focussed on either fully three-dimensional or two-dimensional states. With plate problems, although the domain itself is three-dimensional, the geometry of the plate is such that, effectively, only two spatial coordinates are involved and spatial averaging is carried out in the third coordinate direction. This makes it awkward to use generalized formulations which are applicable for operational representations (e.g. gradient operator, divergence, etc.). In certain instances it is advantageous to either limit the invariant representations to two-dimensions and/or to employ indicial notation to obtain the desired results. The modifications to the operational notations will be presented as and when they are introduced.

8.11.1   Deformation of the plate region

We consider the geometry of deformation of a thin plate of thickness h which undergoes transverse deflections due to lateral loadings $p(x,y)$. To describe the deformations of the plate we select a rectangular Cartesian coordinate system (x, y, z) such that the $x-y$ plane coincides with the mid plane of the plate in its undisplaced or reference configuration (Figure 8.66). In discussing the kinematics of the plate deformation three important assumptions are involved. (i) We assume that all points located on the mid-plane of the plate have zero displacements in the x- and y- directions; as a consequence the mid-plane is also referred to as the neutral plane. (ii) It is assumed that no membrane forces or in-plane force resultants are induced in the plate during its flexure. (iii) Finally, we assume that straight material elements that are perpendicular to the mid-plane in the reference configuration remain straight but not necessarily perpendicular to the neutral plane in the deflected configuration of the plate. This allows for shear deformations of the plate which are characteristics of moderately thick plates. In the classical thin plate theory these elements not only remain straight in the deflected configuration but are also perpendicular to the neutral plane. The classical plate theory will therefore be recovered as a limiting case of the developments presented in this section.

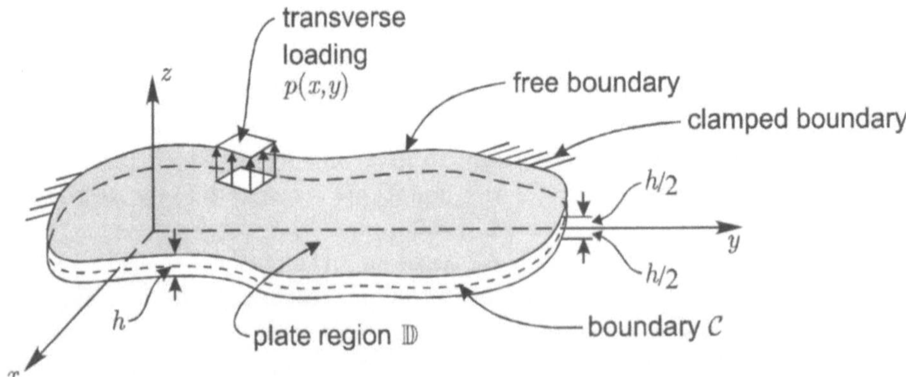

Figure 8.66: Configuration of the transversely loaded plate and boundary constraints.

We introduce a notation which will be used to signify operations and definitions which will involve only the in-plane coordinates x and y. For example, the two-dimensional equivalents of $\nabla, \mathfrak{S}, \mathfrak{E}$ defined in Section 8.2 are, respectively,

$$
\overset{\circ}{\nabla} = \mathbf{i}\frac{\partial}{\partial x} + \mathbf{j}\frac{\partial}{\partial y}
$$

$$
\overset{\circ}{\mathfrak{S}} = \mathbf{ii}\,\sigma_{xx} + \mathbf{ij}\,\sigma_{xy} + \mathbf{ji}\,\sigma_{yx} + \mathbf{jj}\,\sigma_{yy} = \overset{\circ}{\mathfrak{S}}{}^{c} \tag{8.1141}
$$

$$
\overset{\circ}{\mathfrak{E}} = \mathbf{ii}\,\epsilon_{xx} + \mathbf{ij}\,\epsilon_{xy} + \mathbf{ji}\,\epsilon_{yx} + \mathbf{jj}\,\epsilon_{yy} = \overset{\circ}{\mathfrak{E}}{}^{c}
$$

where $(\ )^{c}$ is the conjugate dyadic, and the stress and strain tensors $\boldsymbol{\sigma} = (\sigma_{ij})$ and $\boldsymbol{\epsilon}(= \epsilon_{ij})$ are interpreted appropriately. Also, other expressions with the notation $(\circ)$ will be interpreted in the same manner. To describe the kinematics of the plate deflections and the resulting internal deformations we consider a typical element bounded by the surfaces of the plate.

When the plate deflects, an element such as P_1P_2 in the reference configuration experiences both a displacement in the z-direction and a rotation. For ease of graphical illustration we shall consider only the region of the element above the neutral plane and denote by x',y' and z' the set of coordinates which are parallel to the reference coordinates x,y and z respectively where the origin P_1 of x',y',z' is now at the displaced position of the mid-plane.

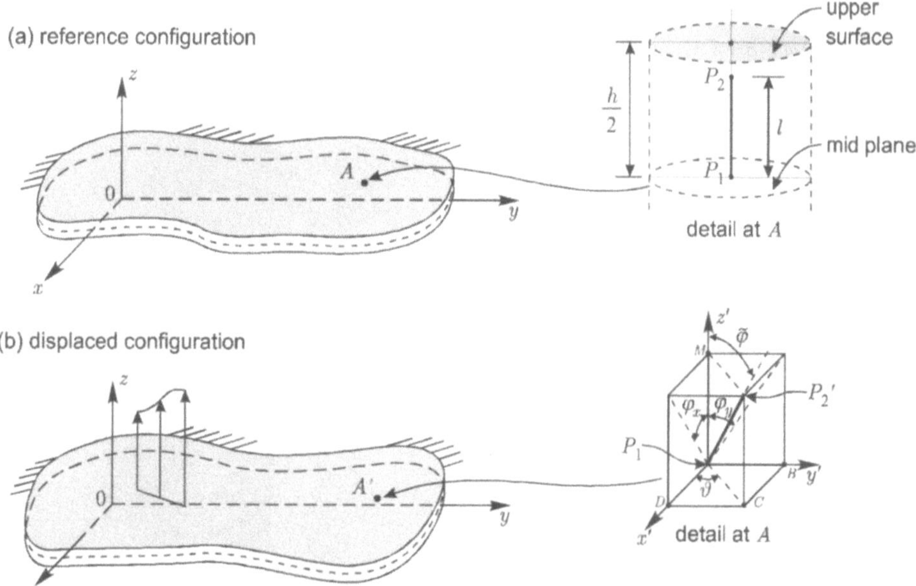

Figure 8.67: Rotations of a line element.

Since the thickness of the plate is assumed to remain constant during its deflection, the length of the line element $P_1\,P_2'$ in the displaced configuration is the same as it is in the reference configuration. Also in the displaced configuration, the angle between z' an/d line element $P_1\,P_2'$ is denoted by $\widetilde{\varphi}$. In the displaced configuration this line element subtends component angles φ_x and φ_y in the planes $x'-z'$ and $y'-z'$ respectively (see Figure 8.67(b)). From geometry

$$P_1 C = \ell \sin \widetilde{\varphi} \quad ; \quad P_1 M = \ell \cos \widetilde{\varphi}$$

$$P_1 D = \ell \sin \widetilde{\varphi} \cos \vartheta \quad ; \quad P_1 B = \ell \sin \widetilde{\varphi} \sin \vartheta \tag{8.1142}$$

For arbitrary finite rotations of the line element $P_1\,P_2$,

$$\tan \varphi_x = \frac{\ell \sin \widetilde{\varphi} \cos \vartheta}{\ell \cos \widetilde{\varphi}} \quad ; \quad \tan \varphi_y = \frac{\ell \sin \widetilde{\varphi} \sin \vartheta}{\ell \cos \widetilde{\varphi}} \tag{8.1143}$$

The assumptions of small deflections and small rotations allows us to linearize (8.1143) such that

$$\sin\widetilde{\varphi} \simeq \tan\widetilde{\varphi} \simeq \widetilde{\varphi} \quad ; \quad \cos\widetilde{\varphi} \simeq 1 \tag{8.1144}$$

Using (8.1144) we can reduce (8.1143) to the forms

$$\varphi_x = \widetilde{\varphi}\cos\vartheta \quad ; \quad \varphi_y = \widetilde{\varphi}\sin\vartheta \tag{8.1145}$$

Hence for small rotations, φ_x and φ_y are Cartesian components of a vector

$$\overset{\circ}{\varphi} = \mathbf{i}\,\varphi_x + \mathbf{j}\,\varphi_y \tag{8.1146}$$

referred to a two-dimensional space.

Let us consider the deflection of the plate and denote by $w(x,y)$ the deflection of the neutral plane from the reference $x-y$ plane. In addition to this, the elements within the plate will in general experience a displacement in the z-direction due to the rotation $\widetilde{\varphi}$. The Figure 8.68 illustrates this latter displacement for a plane with an orientation corresponding to the plane containing the rotated line element P_1P_2.

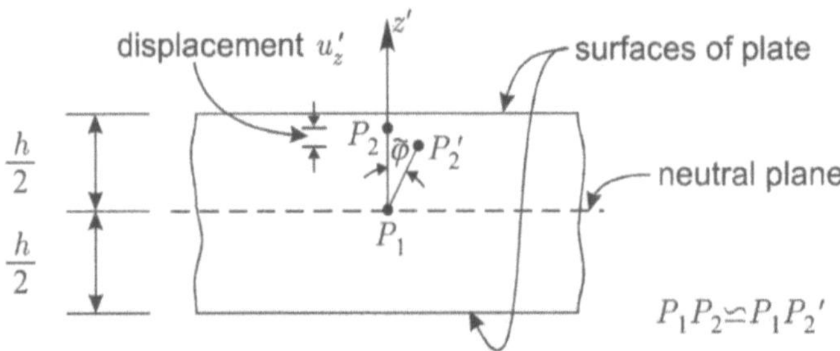

Figure 8.68: Displacements u'_z due to rotation of line element.

From geometry, for finite rotations, the displacement u'_z due to rotation of the line element is

$$u'_z = \ell - \ell\cos\widetilde{\varphi} \tag{8.1147}$$

If the rotations are small then $u_z' \equiv 0$. The displacement u_z of the plate is therefore identically equal to the *deflection of the neutral plane from the reference plane* which is denoted by $w(x,y)$. The remaining displacement components u_x and u_y can be determined by considering the component rotations in the respective planes. Referring to Figure 8.69, (note that the transverse displacements are suppressed)

$$u_x = z \tan \varphi_x \simeq z \sin \varphi_x = z\varphi_x$$

$$(8.1148)$$

$$u_y = z \tan \varphi_y \simeq z \sin \varphi_y = z\varphi_y$$

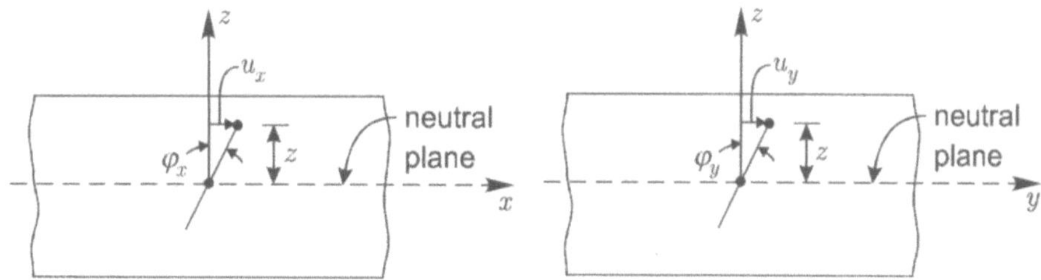

Figure 8.69: In-plane displacements within the plate element.

The displacement vector for a point within the plate can now be described in terms of the transverse deflection $w(x,y)$ and the rotations φ_x and φ_y such that

$$\mathbf{u} = z\, \varphi_x(x,y)\, \mathbf{i} + z\, \varphi_y(x,y)\, \mathbf{j} + w(x,y)\mathbf{k} \qquad (8.1149)$$

where $z \in (-h/2, h/2)$. Considering the plan configuration of an arbitrary plate region $\mathbb{D}$ with boundary $\mathcal{C}$ (Figure 8.70), it is useful to note the following vector identities.

The two-dimensional vectors $\overset{\circ}{\mathbf{n}}$ and $\overset{\circ}{\mathbf{t}}$ are defined by

$$\overset{\circ}{\mathbf{n}} = n_x\mathbf{i} + n_y\mathbf{j} \quad ; \quad \overset{\circ}{\mathbf{t}} = t_x\mathbf{i} + t_y\mathbf{j} \qquad (8.1150)$$

such that

$$\overset{\circ}{\mathbf{n}} \cdot \overset{\circ}{\mathbf{t}} = 0 \quad ; \quad \overset{\circ}{\mathbf{n}} \times \overset{\circ}{\mathbf{t}} = \mathbf{k} \qquad (8.1151)$$

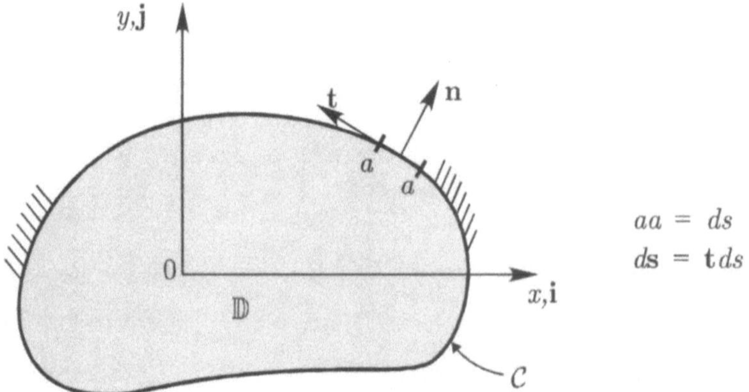

Figure 8.70: The plate region.

Also,

$$\overset{\circ}{\boldsymbol{\varphi}} = \varphi_x \mathbf{i} + \varphi_y \mathbf{j} = \varphi_n \overset{\circ}{\mathbf{n}} + \varphi \overset{\circ}{\mathbf{t}} \tag{8.1152}$$

where

$$\varphi_n = \overset{\circ}{\boldsymbol{\varphi}} \cdot \overset{\circ}{\mathbf{n}} \quad ; \quad \varphi_t = \overset{\circ}{\boldsymbol{\varphi}} \cdot \overset{\circ}{\mathbf{t}} \tag{8.1153}$$

Since the plate regions are essentially two-dimensional regions, where relevant; it is also convenient to adopt an indicial notation to denote certain relevant expressions. For example

$$\varphi_n = \varphi_\alpha n_\alpha = \varphi_x n_x + \varphi_y n_y$$

$$\varphi_t = \varphi_\alpha t_\alpha = \varphi_x t_x + \varphi_y t_y \tag{8.1154}$$

$$\varphi_\alpha = \varphi_n n_\alpha + \varphi_t t_\alpha$$

We can also define the two-dimensional alternating tensor

$$e_{\alpha\beta} = n_\alpha t_\beta - n_\beta t_\alpha \quad ; \quad \alpha, \beta = x, y \tag{8.1155}$$

The strain dyadic $\overset{\circ}{\boldsymbol{\varepsilon}}$ associated with the displacement field defined by (8.1149) can be obtained from the relationships (see e.g. (8.9))

$$\overset{\circ}{\boldsymbol{\mathfrak{E}}} = \frac{z}{2}\left[\overset{\circ}{\nabla}\overset{\circ}{\boldsymbol{\varphi}} + \left(\overset{\circ}{\nabla}\overset{\circ}{\boldsymbol{\varphi}}\right)^{T}\right] = \overset{\circ}{\boldsymbol{\mathfrak{E}}}{}^{c} \tag{8.1156}$$

where, (see e.g. (8.13))

$$\overset{\circ}{\nabla}\overset{\circ}{\boldsymbol{\varphi}} = \mathbf{ii}\frac{\partial\varphi_x}{\partial x} + \mathbf{ij}\frac{\partial\varphi_y}{\partial x} + \mathbf{ji}\frac{\partial\varphi_x}{\partial y} + \mathbf{jj}\frac{\partial\varphi_y}{\partial y}$$

$$\left(\overset{\circ}{\nabla}\overset{\circ}{\boldsymbol{\varphi}}\right)^{T} = \mathbf{ii}\frac{\partial\varphi_x}{\partial x} + \mathbf{ji}\frac{\partial\varphi_y}{\partial x} + \mathbf{ij}\frac{\partial\varphi_x}{\partial y} + \mathbf{jj}\frac{\partial\varphi_y}{\partial y} \tag{8.1157}$$

and

$$\epsilon_{iz} = \frac{1}{2}\left[\varphi_i + \frac{\partial w}{\partial x_i}\right] \quad ; \quad \epsilon_{zz} = 0 \quad ; \quad (i = x, y) \tag{8.1158}$$

In indicial notation (8.1157) and (8.1158) take the forms

$$\epsilon_{ij} = \frac{z}{2}\left[\varphi_{i,j} + \varphi_{j,i}\right] \quad ; \quad i = x, y \quad ; \quad j = x, y$$

$$\epsilon_{iz} = \frac{1}{2}\left[\varphi_i + w_{,i}\right] \quad ; \quad i = x, y \tag{8.1159}$$

$$\epsilon_{zz} = 0$$

When we are discussing the mechanics of beams attention is focussed on averages of strains, resultants of tractions etc. defined on surfaces normal to the neutral axis. Similarly, in discussing the mechanics of plates it is convenient to define the strain components in the plate by the expressions for the flexural moments $\overset{\circ}{\mathbf{m}}$ (or $\overset{\circ}{m}_{ij}$) and the shear strains $\overset{\circ}{\mathbf{q}}$ which are defined by

$$\overset{\circ}{\mathbf{m}} = \frac{12}{h^3}\int_{-\frac{h}{2}}^{\frac{h}{2}} z\,\overset{\circ}{\boldsymbol{\mathfrak{E}}}\,dz \tag{8.1160}$$

$$\overset{\circ}{\mathbf{q}} = \frac{2}{h}\int_{-\frac{h}{2}}^{\frac{h}{2}}\left[\overset{\circ}{\boldsymbol{\varphi}} + \overset{\circ}{\nabla}w\right]dz \tag{8.1161}$$

Since the dependency of $\overset{\circ}{\boldsymbol{\varphi}}$, $\overset{\circ}{\nabla}w$ and $\overset{\circ}{\boldsymbol{\epsilon}}$ on z is explicit (8.1160) and (8.1161) reduce to

$$\overset{\circ}{\mathbf{m}} = \frac{1}{2}\left[\overset{\circ}{\nabla}\overset{\circ}{\varphi} + \left(\overset{\circ}{\nabla}\overset{\circ}{\varphi}\right)^{T}\right]$$

$$(8.1162)$$

$$\overset{\circ}{\mathbf{q}} = \overset{\circ}{\varphi} + \nabla w$$

where $\overset{\circ}{\mathbf{m}}$ is the dyadic of flexural strains. In indicial notation

$$\overset{\circ}{\mathbf{m}} = \overset{\circ}{m}_{ij} = \frac{1}{2}\left[\varphi_{i,j} + \varphi_{j,i}\right] \quad ; \quad i,j = x,y$$

$$(8.1163)$$

$$q_i = \varphi_i + w_{,i}$$

and the relationships between $\overset{\circ}{\boldsymbol{\epsilon}}$, $\overset{\circ}{\mathbf{m}}$ and $\mathbf{q}$ are

$$\overset{\circ}{\epsilon}_{ij} = z\,\overset{\circ}{m}_{ij} \quad \text{or} \quad \overset{\circ}{\boldsymbol{\epsilon}} = z\,\overset{\circ}{\mathbf{m}} \quad ; \quad i,j = x,y$$

$$(8.1164)$$

$$\epsilon_{iz} = \frac{1}{2}q_i \quad ; \quad i = x,y$$

Since $\overset{\circ}{\mathbf{m}}$ is a tensor of order two the usual transformation rules discussed in relation to $\boldsymbol{\sigma}$ and $\boldsymbol{\epsilon}$ apply equally well to $\overset{\circ}{\mathbf{m}}$. We shall develop here the scalar components of $\overset{\circ}{\mathbf{m}}$ with respect to $\overset{\circ}{\mathbf{n}}\overset{\circ}{\mathbf{n}}$ and $\overset{\circ}{\mathbf{n}}\overset{\circ}{\mathbf{t}}$. Considering the increment of the vector $\overset{\circ}{\varphi}$ we have (see e.g. (1.21))

$$d\overset{\circ}{\varphi} = \overset{\circ}{\nabla}\overset{\circ}{\varphi}\cdot d\overset{\circ}{\mathbf{r}}$$

$$(8.1165)$$

where $d\overset{\circ}{\mathbf{r}} = \mathbf{i}dx + \mathbf{j}dy$. From this result we have

$$\frac{d\overset{\circ}{\varphi}}{\left|d\overset{\circ}{\mathbf{r}}\right|} = \frac{d\overset{\circ}{\mathbf{r}}}{\left|d\overset{\circ}{\mathbf{r}}\right|}\cdot\overset{\circ}{\nabla}\overset{\circ}{\varphi} = \overset{\circ}{\mathbf{n}}\cdot\overset{\circ}{\nabla}\overset{\circ}{\varphi} \equiv \overset{\circ}{\mathbf{D}}_n$$

$$(8.1166)$$

where $\overset{\circ}{\mathbf{n}} = d\overset{\circ}{\mathbf{r}}\,/\left|d\overset{\circ}{\mathbf{r}}\right|$ and $\overset{\circ}{\mathbf{D}}_n$ is identified as the directional derivative of $\overset{\circ}{\varphi}$ in the direction of the unit normal vector $\overset{\circ}{\mathbf{n}}$. Similarly considering the tangential direction

$$\frac{d\overset{\circ}{\varphi}}{|dt|} = \overset{\circ}{\mathbf{t}} \cdot \overset{\circ}{\nabla}\overset{\circ}{\varphi} \equiv \overset{\circ}{\mathbf{D}}_t \tag{8.1167}$$

and $\overset{\circ}{\mathbf{D}}_t$ is the directional derivative of $\overset{\circ}{\varphi}$ in the direction of the unit tangent vector $\overset{\circ}{\mathbf{t}}$. The scalar components of $\overset{\circ}{\mathbf{m}}$ with respect to $\overset{\circ}{\mathbf{n}}\overset{\circ}{\mathbf{n}}$ and $\overset{\circ}{\mathbf{n}}\overset{\circ}{\mathbf{t}}$ are given by the relationships

$$m_{nn} = \overset{\circ}{\mathbf{n}} \cdot \left[\left(\overset{\circ}{\mathbf{n}} \cdot \overset{\circ}{\nabla}\overset{\circ}{\varphi} \right) \right] = \overset{\circ}{\mathbf{n}} \cdot \overset{\circ}{\mathbf{D}}_n$$

$$\tag{8.1168}$$

$$2m_{nt} = \overset{\circ}{\mathbf{n}} \cdot \left[\overset{\circ}{\mathbf{t}} \cdot \overset{\circ}{\nabla}\overset{\circ}{\varphi} \right] + \overset{\circ}{\mathbf{t}} \cdot \left[\mathbf{n} \cdot \overset{\circ}{\nabla}\overset{\circ}{\varphi} \right] = \overset{\circ}{\mathbf{n}}\overset{\circ}{\mathbf{D}}_t + \overset{\circ}{\mathbf{t}}\overset{\circ}{\mathbf{D}}_n$$

In indicial notation

$$m_{nn} = m_{ij}n_i n_j = \frac{1}{2}\left[\varphi_{i,j} + \varphi_{j,i}\right] n_i n_j = \varphi_{i,j}n_i n_j \quad ; \quad (i,j = x,y)$$

$$\tag{8.1169}$$

$$m_{nt} = m_{ij}n_i t_j = \frac{1}{2}\left[\varphi_{i,j} + \varphi_{j,i}\right] n_i t_j \quad ; \quad (i,j - x,y)$$

The results given by (8.1168) are invariant representations of the components of the flexural strain components as such they can be used to calculate the strain components relative to any orthogonal curvilinear coordinate system.

The result (8.1165) can also be written in terms of the strain and rotation dyadics $\overset{\circ}{\mathbf{m}}$ and $\overset{\circ}{\mathbf{r}}$ i.e.

$$d\overset{\circ}{\varphi} = \overset{\circ}{\nabla}\overset{\circ}{\varphi} \cdot d\overset{\circ}{\mathbf{r}} = \left[\overset{\circ}{\mathbf{m}} + \overset{\circ}{\mathbf{r}}\right] \cdot d\overset{\circ}{\mathbf{r}} \tag{8.1170}$$

where the rotation dyadic is given by

$$\overset{\circ}{\mathbf{r}} = \frac{1}{2}\left[\overset{\circ}{\nabla}\overset{\circ}{\varphi} - \left(\overset{\circ}{\nabla}\overset{\circ}{\varphi}\right)^T\right] \tag{8.1171}$$

gives the components of the rotation due to flexure alone. In a similar fashion, we can determine the rotation vector $\overset{\circ}{\Omega}{}^*$ due to shear strains $\overset{\circ}{\mathbf{q}}$: i.e.

$$dw = \overset{\circ}{\nabla}w \cdot d\mathbf{r} = \frac{1}{2}\left[\overset{\circ}{\mathbf{q}} + \overset{\circ}{\Omega}{}^*\right] \cdot d\mathbf{r} \tag{8.1172}$$

where

$$\overset{\circ}{\Omega}{}^* = \overset{\circ}{\nabla}w - \overset{\circ}{\boldsymbol{\varphi}} \tag{8.1173}$$

In the case of rigid body motion, $\overset{\circ}{\mathbf{m}}$ and $\overset{\circ}{\mathbf{q}}$ vanish such that

$$\overset{\circ}{\nabla}\overset{\circ}{\boldsymbol{\varphi}} = -\left(\overset{\circ}{\nabla}\overset{\circ}{\boldsymbol{\varphi}}\right)^T \quad ; \quad \overset{\circ}{\boldsymbol{\varphi}} = -\overset{\circ}{\nabla}w \tag{8.1174}$$

which gives

$$\overset{\circ}{\mathbf{r}} = \overset{\circ}{\nabla}\overset{\circ}{\boldsymbol{\varphi}} \quad ; \quad \overset{\circ}{\Omega}{}^* = 2\,\overset{\circ}{\nabla}w \tag{8.1175}$$

and

$$w = w_0 + \frac{1}{2}tr\left(\overset{\circ}{\Omega}{}^*\,\mathbf{x}\right) \quad ; \quad \overset{\circ}{\boldsymbol{\varphi}} = -\frac{1}{2}\overset{\circ}{\Omega}{}^* \tag{8.1176}$$

where w_0 is a rigid body translation in the z direction and $\overset{\circ}{\Omega}{}^*$ characterize the infinitesimal rigid rotation of the mid-plane surface of the plate. From this analysis it is evident that a given strain field in the plate will not result in a unique displacement field unless additional independent conditions are specified to fix the value of w_0, Ω_x^* and Ω_y^*.

Finally, we note that the equations (8.1163) represent relationships between the displacement w, the rotations φ_x, φ_y and the five strain components of $\overset{\circ}{\mathbf{m}}$ and $\overset{\circ}{\mathbf{q}}$ designated by m_{xx}, m_{yy}, m_{xy}, q_x and q_y. As with the discussion related to classical compatibility equations of elasticity, (Section 8.2.6), the strains in the plate region can be derived explicitly if the displacement w and the rotations φ_x and φ_y are specified. If however, the five strain components are specified, the displacement w and the rotations φ_x and φ_y have to be determined be solving five first-order partial differential equations. In order to obtain unique expressions for w, φ_x and φ_y, the strains $\overset{\circ}{\mathbf{m}}$ and $\overset{\circ}{\mathbf{q}}$ must satisfy the integrability conditions characterized by the compatibility conditions. It can be shown that

$$
\begin{aligned}
S_{xy} &= m_{xx,yy} + m_{yy,xx} - 2m_{xy,xy} = 0 \\
R_x &= 2m_{xx,y} - 2m_{xy,x} + q_{y,xx} - q_{x,yx} = 0 \\
R_y &= 2m_{yy,x} - 2m_{yx,y} + q_{x,yy} - q_{y,xy} = 0
\end{aligned}
\tag{8.1177}
$$

It can be verified that the components of $\overset{\circ}{\mathbf{m}}$ and $\overset{\circ}{\mathbf{q}}$ identically satisfy (8.1177) and we can conclude that the compatibility equations are necessary to ensure single-valued expressions for w, φ_x and φ_y in a simply connected domain. For multiply connected domains (8.1177) are still necessary but an additional condition needs to be specified to satisfy certain conditions involving line integrals (a general discussion of such conditions is presented in Section 8.13). We can also show that (8.1177) are not independent since

$$
R_{x,y} + R_{y,x} + 2S_{xy} = 0
\tag{8.1178}
$$

8.11.2 Flexural stresses and stress resultants

When discussing mechanical actions causing flexural and shear deformations in a plate it is convenient to define the stress resultants

$$
\overset{\circ}{\mathbf{M}} = \int_{-h/2}^{h/2} \overset{\circ}{\boldsymbol{\sigma}}\, z dz
\tag{8.1179}
$$

$$
\overset{\circ}{\mathbf{Q}} = \int_{-h/2}^{h/2} \sigma_{iz}\, dz
\tag{8.1180}
$$

and $\overset{\circ}{\mathbf{M}}$ $(= M_{ij})$ is a second-order tensor. In view of the above definitions we note that $\overset{\circ}{\mathbf{M}}$ is a second-order tensor and $\overset{\circ}{\mathbf{Q}}$ vector in a two-dimensional space. By definition, the Cartesian components of $\overset{\circ}{\mathbf{M}}$ are

$$
\overset{\circ}{\mathbf{M}} = \begin{bmatrix} M_{xx} & M_{xy} \\ M_{yx} & M_{yy} \end{bmatrix} = \overset{\circ}{\mathbf{M}}^{\mathbf{T}}
\tag{8.1181}
$$

and the diagonal terms are the plate *bending moments* and the off-diagonal terms are the plate *twisting moments*. Both moments are measured per unit length. A sign convention needs to be defined for the elements of $\overset{\circ}{\mathbf{M}}$ and $\overset{\circ}{\mathbf{Q}}$ in relation to the sign conventions that were adopted for the stresses (see e.g. Section 8.3.5). Figure 8.71 illustrates the positive stresses acting on a plate undergoing flexure.

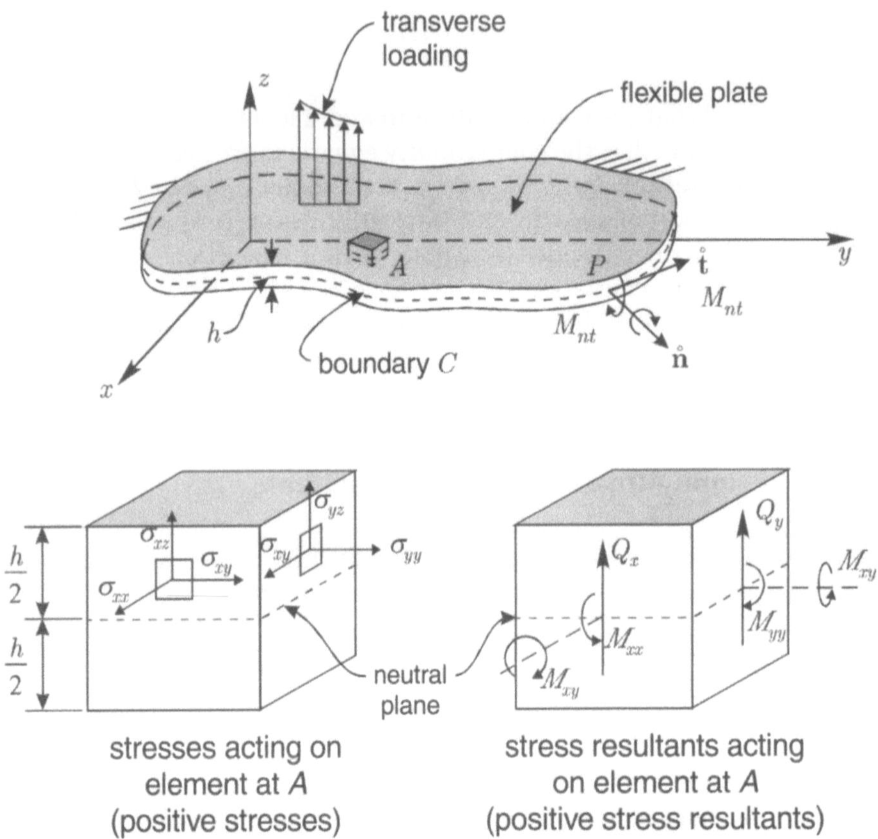

Figure 8.71: Positive stresses acting on a flexible plate.

Considering the traction vector acting at an arbitrary plane we can write

$$\overset{\circ}{\mathbf{T}} = \overset{\circ}{T_i} = \sigma_{ix} n_x + \sigma_{iy} n_y + \sigma_{iz} n_z \quad ; \quad i = x, y \tag{8.1182}$$

$$T_z = \sigma_{xz} n_x + \sigma_{yz} n_y + \sigma_{zz} n_z \tag{8.1183}$$

If we consider the tractions acting on an arbitrary point P on the boundary of the plate (Figure 8.71), $n_z \equiv 0$, and $\overset{\circ}{\mathbf{n}} \times \overset{\circ}{\mathbf{t}} = \mathbf{k}$. Hence on the boundary of the plate

$$\overset{\circ}{\mathbf{T}} = \overset{\circ}{\boldsymbol{\sigma}} \cdot \overset{\circ}{\mathbf{n}} \quad \text{or} \quad \overset{\circ}{T_i} = \overset{\circ}{\sigma}_{ij}\, n_j \quad ; \quad i = x, y$$

$$T_z = \overset{\circ}{\sigma}_{iz}\, n_i \tag{8.1184}$$

The tractions $\overset{\circ}{\mathbf{T}}$ defined by (8.1184) contributes to the moments at the boundary and T_z contributes to the shear. We evaluate the moment vector $\overset{\circ}{M}_i$ and the transverse shear force Q_n

$$\overset{\circ}{M}_i = \int_{-h/2}^{h/2} \overset{\circ}{T_i}\, z\,dz$$

$$\tag{8.1185}$$

$$Q_n = \int_{-h/2}^{h/2} T_z\, dz$$

Combining the equations (8.1179) to (8.1185) we obtain the expression for the moment vector $\overset{\circ}{M}_i$

$$\overset{\circ}{M}_i = \overset{\circ}{M}_{ij}\, n_j \quad ; \quad (i, j = x, y) \tag{8.1186}$$

and the transverse shear force Q_n as

$$Q_n = Q_i n_i = \overset{\circ}{\mathbf{Q}} \cdot \overset{\circ}{\mathbf{n}} \tag{8.1187}$$

The moment vector at P can be defined by

$$\overset{\circ}{\mathbf{M}} = M_{nn}\, \overset{\circ}{\mathbf{n}} + M_{nt}\, \overset{\circ}{\mathbf{t}} \tag{8.1188}$$

where

$$M_{nn} = \overset{\circ}{\mathbf{M}}_c \cdot \overset{\circ}{\mathbf{n}} = \overset{\circ}{M}_{ij}\, n_i n_j$$

$$\tag{8.1189}$$

$$M_{nt} = \overset{\circ}{\mathbf{M}}_c \cdot \overset{\circ}{\mathbf{t}} = \overset{\circ}{M}_{ij}\, t_i n_j$$

and $\overset{\circ}{\mathbf{M}}_c$ is the value of $\overset{\circ}{\mathbf{M}}$ on the boundary $\mathcal{C}$. In (8.1189) M_{nn} is a flexural moment and M_{nt} is a twisting moment.

8.11.3   Plate stress-strain relations

We assume isotropic elastic behaviour of the medium and consider the stress-strain relationship given by (8.242), i.e.

$$E\epsilon = (1+\nu)\boldsymbol{\sigma} - \nu tr(\boldsymbol{\sigma})\mathbf{I} \tag{8.1190}$$

For the plate problem we can write

$$E\,\overset{\circ}{\epsilon} = (1+\nu)\,\overset{\circ}{\boldsymbol{\sigma}} - \nu tr(\overset{\circ}{\boldsymbol{\sigma}})\,\overset{\circ}{\mathbf{I}} - \nu\sigma_{zz}\,\overset{\circ}{\mathbf{I}} \tag{8.1191}$$

and

$$\sigma_{zi} = 2\mu\epsilon_{zi} = \mu q_i \tag{8.1192}$$

where $\overset{\circ}{\epsilon}$ and $\overset{\circ}{\boldsymbol{\sigma}}$ are essentially two-dimensional matrices referred to the (x,y) system and $\overset{\circ}{\mathbf{I}}$ is the two-dimensional unit matrix.

From (8.1191) we have

$$E tr(\overset{\circ}{\epsilon}) = (1-\nu)tr(\overset{\circ}{\boldsymbol{\sigma}}) - 2\nu\sigma_{zz} \tag{8.1193}$$

Using (8.1193) in (8.1191) we have

$$\overset{\circ}{\boldsymbol{\sigma}} = \frac{E}{(1+\nu)}\,\overset{\circ}{\epsilon} + \frac{\nu E}{(1-\nu^2)} tr(\overset{\circ}{\epsilon})\,\overset{\circ}{\mathbf{I}} + \frac{\nu}{(1-\nu)}\sigma_{zz}\,\overset{\circ}{\mathbf{I}} \tag{8.1194}$$

Using the result $\overset{\circ}{\epsilon} = z\,\overset{\circ}{\mathbf{m}}$, (see e.g. (8.1164)) we can rewrite (8.1194) as

$$\overset{\circ}{\boldsymbol{\sigma}} = \frac{Ez}{(1-\nu^2)}\left[(1-\nu)\,\overset{\circ}{\mathbf{m}} +\nu tr(\overset{\circ}{\mathbf{m}})\,\overset{\circ}{\mathbf{I}}\right] + \frac{\nu}{(1-\nu)}\sigma_{zz}\,\overset{\circ}{\mathbf{I}} \tag{8.1195}$$

and the dilatation in the plate region is

$$tr(\overset{\circ}{\mathbf{m}}) = \overset{\circ}{\nabla}\cdot\overset{\circ}{\boldsymbol{\varphi}} = \frac{12}{h^3}\int_{-h/2}^{h/2} tr(\mathbf{u})dz \tag{8.1196}$$

Substituting (8.1195) into (8.1179) we have

$$\overset{\circ}{\mathbf{M}} = D\left[(1-\nu)\,\overset{\circ}{\mathbf{m}} -\nu tr(\overset{\circ}{\mathbf{m}})\,\overset{\circ}{\mathbf{I}}\right] + \frac{\nu\,\overset{\circ}{\mathbf{I}}}{(1-\nu)}\int_{-h/2}^{h/2} z\sigma_{zz}\,dz \tag{8.1197}$$

where

$$D = \frac{Eh^3}{12(1-\nu^2)} \tag{8.1198}$$

is defined as the flexural rigidity of the plate.

Similarly, by substituting (8.1192) into (8.1180) gives

$$\overset{\circ}{\mathbf{Q}} = \mu h\,\overset{\circ}{\mathbf{q}} \tag{8.1199}$$

At this point, it is convenient to introduce two simplifying assumptions which are attributed to the pioneering studies conducted by Mindlin (1906-1987) into the behaviour of moderately thick plates where shear deformations are included. In the first instance, we neglect the last term in the integral in (8.1197). According to this assumption, although the plate is subjected to transverse loadings they induce negligible normal stresses in the z-direction. Secondly, we assume that the shear modulus μ can be modified by a coefficient κ^2, which is referred to as the shear coefficient, and depends almost linearly with Poisson's ratio ν. Invoking these simplifying assumptions we can rewrite (8.1197) and (8.1199) in the forms

$$\overset{\circ}{\mathbf{M}} = D\left[(1-\nu)\,\overset{\circ}{\mathbf{m}} -\nu tr(\overset{\circ}{\mathbf{m}})\,\overset{\circ}{\mathbf{I}}\right] \tag{8.1200}$$

$$\overset{\circ}{\mathbf{Q}} = \kappa^2\mu h\,\overset{\circ}{\mathbf{q}} \tag{8.1201}$$

Also, using (8.1162), these equations can be rewritten in the invariant forms

$$\overset{\circ}{\mathfrak{M}} = \frac{D}{2}\left[(1-\nu)\left\{\overset{\circ}{\nabla}\overset{\circ}{\varphi} + \left(\overset{\circ}{\nabla}\overset{\circ}{\varphi}\right)^{T}\right\} + 2\nu\left(\overset{\circ}{\nabla}\cdot\overset{\circ}{\varphi}\right)\overset{\circ}{\mathfrak{I}}\right]$$

(8.1202)

$$\overset{\circ}{\mathbf{Q}} = \kappa^{2}\mu h\left[\overset{\circ}{\varphi} + \overset{\circ}{\nabla}w\right]$$

where $\overset{\circ}{\mathfrak{I}}$ is the two-dimensional unit dyadic. In indicial forms

$$\overset{\circ}{M}_{ij} = \frac{D}{2}\left[(1-\nu)\left\{\varphi_{i,j} + \varphi_{j,i}\right\} + 2\nu\varphi_{k,k}\,\delta_{ij}\right]$$

(8.1203)

$$\overset{\circ}{Q} = \kappa^{2}\mu h\left[\varphi_{i} + w_{,i}\right]$$

The Cartesian components of (8.1203) can be evaluated in the forms:

$$M_{xx} = D\left[\frac{\partial\varphi_{x}}{\partial x} + \nu\frac{\partial\varphi_{y}}{\partial y}\right]$$

$$M_{yy} = D\left[\frac{\partial\varphi_{y}}{\partial y} + \nu\frac{\partial\varphi_{x}}{\partial x}\right]$$

(8.1204)

$$M_{xy} = \frac{D}{2}(1-\nu)\left[\frac{\partial\varphi_{x}}{\partial y} + \frac{\partial\varphi_{y}}{\partial x}\right]$$

and

$$Q_{x} = \kappa^{2}\mu h\left[\varphi_{x} + \frac{\partial w}{\partial x}\right]$$

(8.1205)

$$Q_{y} = \kappa^{2}\mu h\left[\varphi_{y} + \frac{\partial w}{\partial y}\right]$$

Example 8.27

Derive the expressions for the components of $\overset{\circ}{\mathbf{M}}$ and $\overset{\circ}{\mathbf{Q}}$, referred to the (r, θ) coordinate system, in terms of the components of $\overset{\circ}{\varphi}$ referred to the same system.

Solution

In plane polar coordinates

$$\overset{\circ}{\nabla} = \mathbf{i}_r \frac{\partial}{\partial r} + \mathbf{i}_\theta \frac{1}{r} \frac{\partial}{\partial \theta} \tag{8.1206}$$

and

$$\overset{\circ}{\varphi} = \mathbf{i}_r \varphi_r + \mathbf{i}_\theta \varphi_\theta \tag{8.1207}$$

The matrices for $\overset{\circ}{\mathbf{M}}$ and $\overset{\circ}{\mathbf{Q}}$ take the forms

$$M_{ij} = \begin{bmatrix} M_{rr} & M_{r\theta} \\ M_{r\theta} & M_{\theta\theta} \end{bmatrix} \quad ; \quad Q_i = \begin{bmatrix} Q_r \\ Q_\theta \end{bmatrix} \tag{8.1208}$$

Considering developments presented in Sections 1.10 and 1.11,

$$h_r \mathbf{i}_r = \frac{\partial x}{\partial r} \mathbf{i} + \frac{\partial y}{\partial r} \mathbf{j} \quad ; \quad h_\theta \mathbf{i}_\theta = \frac{\partial x}{\partial \theta} \mathbf{i} + \frac{\partial y}{\partial \theta} \mathbf{j} \tag{8.1209}$$

where

$$h_r = 1 \quad ; \quad h_\theta = r \tag{8.1210}$$

The derivatives of the basis vectors $\mathbf{i}_r$ and $\mathbf{i}_\theta$ are linear combinations of the basis vectors: e.g.

$$\begin{bmatrix} \dfrac{\partial \mathbf{i}_r}{\partial r} & \dfrac{\partial \mathbf{i}_r}{\partial \theta} \\[4mm] \dfrac{\partial \mathbf{i}_\theta}{\partial r} & \dfrac{\partial \mathbf{i}_\theta}{\partial \theta} \end{bmatrix} = \begin{bmatrix} -\dfrac{1}{h_\theta}\dfrac{\partial h_r}{\partial \theta}\mathbf{i}_\theta & \dfrac{1}{h_r}\dfrac{\partial h_\theta}{\partial r}\mathbf{i}_\theta \\[4mm] \dfrac{1}{h_\theta}\dfrac{\partial h_r}{\partial \theta}\mathbf{i}_r & -\dfrac{1}{h_r}\dfrac{\partial h_\theta}{\partial r}\mathbf{i}_r \end{bmatrix} = \begin{bmatrix} 0 & \mathbf{i}_\theta \\[4mm] 0 & -\mathbf{i}_r \end{bmatrix} \tag{8.1211}$$

We can now perform the operations required to evaluate (8.1202): e.g.

$$\begin{aligned} \overset{\circ}{\nabla}\overset{\circ}{\varphi} &= \left[\mathbf{i}_r\frac{\partial}{\partial r} + \mathbf{i}_\theta\frac{1}{r}\frac{\partial}{\partial \theta}\right]\{\mathbf{i}_r\varphi_r + \mathbf{i}_\theta\varphi_\theta\} \\[2mm] &= \mathbf{i}_r\mathbf{i}_r\frac{\partial \varphi_r}{\partial r} + \mathbf{i}_r\mathbf{i}_\theta\left(\frac{\partial \varphi_\theta}{\partial r}\right) \\[2mm] &\quad + \mathbf{i}_\theta\mathbf{i}_r\left(\frac{1}{r}\frac{\partial \varphi_r}{\partial \theta} - \frac{\varphi_\theta}{r}\right) + \mathbf{i}_\theta\mathbf{i}_\theta\left(\frac{1}{r}\frac{\partial \varphi_\theta}{\partial \theta} + \frac{\varphi_r}{r}\right) \end{aligned} \tag{8.1212}$$

Also

$$\overset{\circ}{\nabla}\cdot\overset{\circ}{\varphi} = \frac{\partial \varphi_r}{\partial r} + \frac{\varphi_r}{r} + \frac{1}{r}\frac{\partial \varphi_\theta}{\partial \theta} \tag{8.1213}$$

Using the results (8.1212) and (8.1213) in the first equation of (8.1202) we obtain

$$M_{rr} = D\left[\frac{\partial \varphi_r}{\partial r} + \nu\left(\frac{1}{r}\frac{\partial \varphi_\theta}{\partial \theta} + \frac{\varphi_r}{r}\right)\right]$$

$$M_{\theta\theta} = D\left[\frac{1}{r}\frac{\partial \varphi_\theta}{\partial \theta} + \frac{\varphi_r}{r} + \nu\frac{\partial \varphi_r}{\partial r}\right] \tag{8.1214}$$

$$M_{r\theta} = \frac{D}{2}(1-\nu)\left[\frac{\partial \varphi_\theta}{\partial r} - \frac{\varphi_\theta}{r} + \frac{1}{r}\frac{\partial \varphi_r}{\partial \theta}\right]$$

Similarly the second equation of (8.1202) gives

$$Q_r = \kappa^2\mu h\left[\varphi_r + \frac{\partial w}{\partial r}\right] \quad ; \quad Q_\theta = \kappa^2\mu h\left[\varphi_\theta + \frac{1}{r}\frac{\partial w}{\partial \theta}\right] \tag{8.1215}$$

8.11.4 Equation of equilibrium for the plate

We consider the general equations of equilibrium for an elastic medium void of body forces (see e.g. (8.162) with $\mathbf{f}^b = \mathbf{0}$) and rewrite them in the following indicial forms:

$$\overset{\circ}{\sigma}_{ij,j} + \sigma_{iz,z} = 0 \tag{8.1216}$$

$$\overset{\circ}{\sigma}_{iz,i} + \sigma_{zz,z} = 0 \tag{8.1217}$$

Integrating (8.1217) over the thickness of the plate we can write

$$\int_{-h/2}^{h/2} \sigma_{iz,i} \, dz + \int_{-h/2}^{h/2} \sigma_{zz,z} \, dz = 0 \tag{8.1218}$$

The second integral of (8.1218) gives

$$\int_{-h/2}^{h/2} \sigma_{zz,z} \, dz = \sigma_{zz}\left(\frac{h}{2}\right) - \sigma_{zz}\left(-\frac{h}{2}\right) = p(x,y) \tag{8.1219}$$

where $p(x,y)$ is the intensity of the transverse loading acting on the plate (Figure 8.66). Using (8.1180) and (8.1219) we can reduce (8.1218) to the form

$$\overset{\circ}{\nabla} \cdot \overset{\circ}{\mathbf{Q}} + p(x,y) = 0 \tag{8.1220}$$

We can integrate (8.1216) in the form

$$\int_{-h/2}^{h/2} \overset{\circ}{\sigma}_{ij,j} \, zdz + \int_{-h/2}^{h/2} \sigma_{iz,z} \, zdz = 0 \tag{8.1221}$$

Integrating the second integral by parts we have

$$\int_{-h/2}^{h/2} \sigma_{iz,z} \, zdz = [z\sigma_{iz}]_{z=-h/2}^{z=h/2} - \int_{h/2}^{h/2} \sigma_{iz} \, dz = - \overset{\circ}{\mathbf{Q}} \tag{8.1222}$$

The transverse loading of the plate is usually applied in such a way that there are no shear stresses on the boundaries of the plate; i.e.

$$\sigma_{iz}(h/2) = \sigma_{iz}(-h/2) = 0 \tag{8.1223}$$

Using this result and (8.1180), (8.1222) gives

$$\int_{-h/2}^{h/2} \sigma_{iz,z}\, z dz = -\int_{h/2}^{h/2} \sigma_{iz} dz = -\overset{\circ}{\mathbf{Q}} \tag{8.1224}$$

Also

$$\int_{-h/2}^{h/2} \sigma_{ij,j}\, z dz = \overset{\circ}{M}_{ij,j} \equiv \overset{\circ}{\nabla} . \overset{\circ}{\mathfrak{M}} \tag{8.1225}$$

Using the results (8.1224) and (8.1225) in (8.1221) we obtain the moment equation of equilibrium

$$\overset{\circ}{\nabla} \cdot \overset{\circ}{\mathfrak{M}} - \overset{\circ}{\mathbf{Q}} = 0 \tag{8.1226}$$

Considering (8.1202) and (8.1203) we can write (8.1226) in the form

$$\begin{aligned}
\overset{\circ}{\mathbf{Q}} &= \frac{D}{2}\left[(1-\nu)\,\overset{\circ}{\nabla}^2\overset{\circ}{\varphi} +(1+\nu)\,\overset{\circ}{\nabla}\left(\overset{\circ}{\nabla}\cdot\overset{\circ}{\varphi}\right)\right] \\
&= \kappa^2\mu h\left[\overset{\circ}{\varphi}+\overset{\circ}{\nabla} w\right]
\end{aligned} \tag{8.1227}$$

Using the second equation of (8.1202) in (8.1220) we obtain

$$\overset{\circ}{\nabla}.\overset{\circ}{\mathbf{Q}} = \overset{\circ}{\nabla}.\left\{\kappa^2\mu h\left[\overset{\circ}{\varphi}+\overset{\circ}{\nabla} w\right]\right\} = -p(x,y) \tag{8.1228}$$

The equations (8.1227) and (8.1228) give the following:

$$\overset{\circ}{\nabla}^2 w = -\frac{\overset{\circ}{p}(x,y)}{\kappa^2\mu h} - \overset{\circ}{\nabla}.\overset{\circ}{\varphi} \tag{8.1229}$$

$$\frac{h^2}{12\kappa^2}\overset{\circ}{\nabla}^2\overset{\circ}{\varphi} - \overset{\circ}{\varphi} = \overset{\circ}{\nabla} w - \left(\frac{1+\nu}{1-\nu}\right)\frac{h^2}{12\kappa^2}\overset{\circ}{\nabla}\left(\overset{\circ}{\nabla}.\overset{\circ}{\varphi}\right) \tag{8.1230}$$

Considering (8.1196) and (8.1227) we have

$$\overset{\circ}{\nabla} \cdot \overset{\circ}{\mathbf{Q}} = \frac{D}{2}\left[(1-\nu)\,\overset{\circ}{\nabla}{}^2\left(\overset{\circ}{\nabla}\cdot\overset{\circ}{\boldsymbol{\varphi}}\right)+(1+\nu)\,\overset{\circ}{\nabla}\cdot\overset{\circ}{\nabla}\left(\overset{\circ}{\nabla}\cdot\overset{\circ}{\boldsymbol{\varphi}}\right)\right]$$
$$= D\,\overset{\circ}{\nabla}{}^2\left(\overset{\circ}{\nabla}\cdot\overset{\circ}{\boldsymbol{\varphi}}\right) \tag{8.1231}$$

Considering (8.1228) and (8.1231) we have

$$-p(x,y) = D\,\overset{\circ}{\nabla}{}^2\left(\overset{\circ}{\nabla}\cdot\overset{\circ}{\boldsymbol{\varphi}}\right) \tag{8.1232}$$

and eliminating $\overset{\circ}{\boldsymbol{\varphi}}$ between (8.1229) and (8.1232) we obtain

$$\overset{\circ}{\nabla}{}^4 w = \frac{p}{D} - \frac{\overset{\circ}{\nabla}{}^2 p}{\kappa^2 \mu h} \tag{8.1233}$$

Similarly by making use of (8.1196), (8.1229), (8.1230) and (8.1232) we can show that

$$\left(\frac{h^2}{12\kappa^2}\,\overset{\circ}{\nabla}{}^2 - 1\right)\left(D\,\overset{\circ}{\nabla}{}^4\overset{\circ}{\boldsymbol{\varphi}} - \overset{\circ}{\nabla} p\right) = \mathbf{0} \tag{8.1234}$$

It may be noted that the final *invariant* forms of the equilibrium equations (8.1233) and (8.1234) each contain the separate displacement variables $w(x,y)$ and $\overset{\circ}{\boldsymbol{\varphi}}(x,y)$. The equations are therefore completely *decoupled*.

8.11.5 Strain energy of a plate

In this section we shall derive an expression for the strain energy of a plate undergoing a general deformation. The measure of strain energy of a plate is utilized in topics related to the formulation of boundary conditions and for proof of uniqueness of the governing partial differential equation(s).

Considering the general expression for the strain energy density U in a continuum region (see e.g. (8.225)) we have

$$U = \frac{1}{2}tr(\boldsymbol{\sigma}.\boldsymbol{\epsilon}) = \frac{1}{2}\left[tr(\overset{\circ}{\boldsymbol{\sigma}}\cdot\overset{\circ}{\boldsymbol{\epsilon}}) + 2\sigma_{iz}\epsilon_{iz} + \sigma_{zz}\epsilon_{zz}\right] \tag{8.1235}$$

Considering the results given by (8.1159) and (8.1164), (8.1235) can be re-written in the form

$$U = \frac{1}{2}\left[z\,tr\left(\overset{\circ}{\boldsymbol{\sigma}} \cdot \overset{\circ}{\mathbf{m}}\right) + \overset{\circ}{\sigma}_{iz}\overset{\circ}{q}_i\right] \tag{8.1236}$$

Since the dependency on z is explicit, we can integrate (8.1236) to evaluate the strain energy density per unit area of the plate, U_A. Integrating (8.1236) over the thickness of the plate we obtain

$$U_A = \int_{-h/2}^{h/2} U\,dz = \frac{1}{2}\left[tr\left(\overset{\circ}{\mathbf{M}} \cdot \overset{\circ}{\mathbf{m}}\right) + tr\left(\overset{\circ}{\mathbf{Q}} \cdot \overset{\circ}{\mathbf{q}}\right)\right] \tag{8.1237}$$

Substituting the stress-strain relations (8.1200) and (8.1201) into (8.1237) we can represent U_A in terms of the strain components; i.e.

$$U_A = \frac{D}{2}\left[(1-\nu)tr\left(\overset{\circ}{\mathbf{m}}\right)^2 + \nu\left\{tr\left(\overset{\circ}{\mathbf{m}}\right)\right\}^2\right] + \frac{\kappa^2}{2}\mu h\,tr\left(\overset{\circ}{\mathbf{q}}\right)^2 \tag{8.1238}$$

Similarly we can express U_A in terms of the stress resultants $\overset{\circ}{\mathbf{M}}$ and $\overset{\circ}{\mathbf{Q}}$ of the plate: i.e.

$$U_A = \frac{1}{2(1-\nu^2)D}\left[(1+\nu)tr\left(\overset{\circ}{\mathbf{M}}\right)^2 - \nu\left\{tr\left(\overset{\circ}{\mathbf{M}}\right)\right\}^2\right]$$
$$+ \frac{tr\left(\overset{\circ}{\mathbf{Q}}\right)^2}{2\kappa^2\mu h} \tag{8.1239}$$

We can recover the appropriate forms of the stress resultants in terms of the strains (or vice versa) by taking the derivatives of U_A with respect to the appropriate arguments: i.e.

$$\overset{\circ}{\mathbf{M}} = \frac{\partial U_A}{\partial \overset{\circ}{\mathbf{m}}} = D\left[(1-\nu)\overset{\circ}{\mathbf{m}} + \nu tr\left(\overset{\circ}{\mathbf{m}}\right)\overset{\circ}{\mathbf{I}}\right]$$

$$\overset{\circ}{\mathbf{Q}} = \frac{\partial U_A}{\partial \overset{\circ}{\mathbf{q}}} = \kappa^2\mu h\,\overset{\circ}{\mathbf{q}} \tag{8.1240}$$

and

$$\overset{\circ}{\mathbf{m}} = \frac{\partial U_A}{\partial \overset{\circ}{\mathbf{M}}} = \frac{1}{(1-\nu^2)D} \left[(1+\nu)\,\overset{\circ}{\mathbf{M}} - \nu\, tr\left(\overset{\circ}{\mathbf{M}}\right)\overset{\circ}{\mathbf{I}} \right]$$

(8.1241)

$$\overset{\circ}{\mathbf{q}} = \frac{\partial U_A}{\partial \overset{\circ}{\mathbf{Q}}} = \frac{1}{\kappa^2 \mu h}\, \overset{\circ}{\mathbf{Q}}$$

8.11.6   The principle of virtual work

The principal of virtual work can be effectively utilized in the derivation of the boundary conditions applicable to the partial differential equation(s) governing flexure of plates. This principle relates the work done by external forces (δW) during a virtual displacement $\delta\mathbf{u}$, to the corresponding change in the strain energy (δU) for a solid in equilibrium i.e.

$$\delta W = \int\!\!\int\!\!\int_V \delta U\, dV$$

(8.1242)

and

$$\delta W = \int\!\!\int\!\!\int_V \mathbf{f}^b.\delta\mathbf{u}\, dV + \int\!\!\int_S \mathbf{T}^*.\delta\mathbf{u}\, ds$$

(8.1243)

where $\mathbf{T}^*$ is the surface traction vector prescribed on S_T the subset of S and $\mathbf{u}$ is the displacement vector prescribed on S_u, such that $S = S_u \cup S_T$. The symbol $\delta\mathbf{u}$ denotes a *virtual displacement field* that is arbitrary but subject to the condition that it does not violate the kinematic boundary conditions, i.e.

$$\delta\mathbf{u} = 0 \quad ; \quad \mathbf{x} \in S_u$$

(8.1244)

We can utilize the principle of virtual work to derive both the equations of equilibrium and the boundary conditions applicable to the solid region. Since $U = U(\boldsymbol{\epsilon})$ we can use (8.228) to write

$$\delta U = \frac{\partial U}{\partial \boldsymbol{\epsilon}} \cdot \delta\boldsymbol{\epsilon} = \boldsymbol{\sigma}.\delta\boldsymbol{\epsilon}$$

(8.1245)

Also from (8.9) and (8.10) we have

$$\delta(\boldsymbol{\epsilon}) + \delta(\boldsymbol{\omega}) = \delta(\nabla\mathbf{u}) = \nabla(\delta\mathbf{u}) \tag{8.1246}$$

and since $\boldsymbol{\sigma} = \boldsymbol{\sigma}^T$ and $\boldsymbol{\omega} = -\boldsymbol{\omega}^T$, we can write (8.1245) as

$$\delta U = \mathfrak{S}.\nabla(\delta\mathbf{u}) \tag{8.1247}$$

Considering (8.1242) we have

$$\delta W = \int\int\int_V \mathfrak{S}.\nabla(\delta\mathbf{u})dV \tag{8.1248}$$

Noting that

$$\nabla(\mathfrak{S}.\delta\mathbf{u}) = \mathfrak{S}.\nabla(\delta\mathbf{u}) + (\nabla.\mathfrak{S})\delta\mathbf{u} \tag{8.1249}$$

we have

$$\delta W = \int\int\int_V \left[(\nabla.\mathfrak{S})\delta\mathbf{u} - (\nabla.\mathfrak{S})\delta\mathbf{u}\right] dV \tag{8.1250}$$

Using the divergence theorem, we can rewrite (8.1250) as

$$\delta W = \int\int_{S_T} [\mathfrak{S}.\mathbf{n}]\,\delta\mathbf{u}\,dS - \int\int\int_V (\nabla.\mathfrak{S})\delta\mathbf{u}\,dV \tag{8.1251}$$

where the surface integral over S is now replaced by the surface integral over S_T in view of the condition (8.1244) imposed on $\delta\mathbf{u}$. Combining (8.1243) and (8.1251) we obtain

$$\int\int\int_V \left[\nabla.\mathfrak{S} + \mathbf{f}^b\right]\delta\mathbf{u}\,dV + \int\int_{S_T} [\mathbf{T}^* - \boldsymbol{\sigma}.\mathbf{n}]\,\delta\mathbf{u}\,dS = 0 \tag{8.1252}$$

Since the region V, its boundary S and the virtual displacement $\delta\mathbf{u}$ are arbitrary, (8.1252) is satisfied provided when each integral is separately zero. Then by appeal to the Dubois-Reymond Lemma, we obtain

$$\nabla.\mathfrak{S} + \mathbf{f}^b = \mathbf{0}$$

$$\mathbf{T}^* = \boldsymbol{\sigma}.\mathbf{n}$$

(8.1253)

As is evident, the application of the *principle of virtual work* gives not only the relevant *equations of equilibrium* but also furnishes the *admissible traction boundary conditions*.

8.11.7 Boundary conditions for plate problems

For a well posed boundary value problem associated with the partial differential equation(s) governing flexure of thin plates, it is necessary to prescribe boundary conditions which accurately reflect the physical constraints imposed on displacements, rotations and the traction resultants. With either simply supported or fixed boundaries, the relevant boundary conditions can be prescribed in a straightforward manner by appeal to deflections and rotations of the plate. The traction boundary conditions applicable to boundary regions of plates have, however, been the subject of considerable controversy since the inception of the theory of plates. For example, with reference to the classical theory of plates Poisson (1781-1840) proposed that at a boundary, the boundary conditions pertaining to the flexural moment, the twisting moment and the shearing force must be satisfied, *separately*, which made the problem over-determined except in situations involving axial symmetry when the twisting moments were identically zero throughout the plate. The correct form of the boundary conditions applicable to a thin plate was first proposed by Kirchhoff (1824-1887) several decades after the original developments, which led to the formulation of the biharmonic equation governing the deflection of the plate. Kirchhoff showed that the shearing forces and the twisting moments must collectively satisfy the appropriate traction boundary conditions. In this section we shall formulate the boundary conditions applicable to the theory of thin plates developed in Sections 8.11.1 to 8.11.5. The boundary conditions can be formulated in a consistent fashion by employing the principle of virtual work outlined in Section 8.11.6.

We consider a plate region $\mathbb{D}$ with boundary $\mathcal{C}$ which is subjected to a distributed transverse load $p(x,y)$ which acts on $\mathbb{D}$ and a vector of stress resultants $\overset{\circ}{\mathbf{M}}{}^*$ and Q_n which are prescribed on $\mathcal{C}$. The deformation of the plate is characterized by the transverse displacement $w(x,y)$ and the independent rotations $\overset{\circ}{\boldsymbol{\varphi}}$. We now impose a virtual deformation characterized by the virtual displacement field δw and rotations $\delta\overset{\circ}{\boldsymbol{\varphi}}$ such that the kinematic boundary conditions are not violated. The virtual work of all external "forces" acting on the plate and the virtual "deformation" field is given by

$$\delta W_P = \int\!\!\int_{\mathbb{D}} p(x,y)\delta w \; dA + \oint_C \left[\mathbf{\overset{\circ}{M}}{}^* . \delta\overset{\circ}{\varphi} + Q_n^*\delta w \right] dS \tag{8.1254}$$

Considering the variation in the strain energy of the plate δU_A, we have

$$\delta U_A = tr\left(\mathbf{\overset{\circ}{M}} . \delta\mathbf{\overset{\circ}{m}} \right) + \mathbf{\overset{\circ}{Q}} . \delta\mathbf{\overset{\circ}{q}} \tag{8.1255}$$

Using the results (8.217) and (8.218), by analogy we can show that

$$tr\left(\mathbf{\overset{\circ}{M}} . \delta\mathbf{\overset{\circ}{m}} \right) = \overset{\circ}{\mathfrak{M}} : \delta\left(\overset{\circ}{\nabla}\overset{\circ}{\varphi} \right) \tag{8.1256}$$

where (:) denotes the double scalar product and consequently

$$tr\left(\mathbf{\overset{\circ}{M}} . \delta\mathbf{\overset{\circ}{m}} \right) = \overset{\circ}{\nabla} . \left(\overset{\circ}{\mathfrak{M}} . \delta\overset{\circ}{\varphi} \right) - \left(\overset{\circ}{\nabla} . \overset{\circ}{\mathfrak{M}} \right) . \delta\overset{\circ}{\varphi} \tag{8.1257}$$

Integrating (8.1255) over the plate region we have

$$\int\!\!\int_{\mathbb{D}}\left[tr\left(\mathbf{\overset{\circ}{M}} . \delta\mathbf{\overset{\circ}{m}} \right) + \mathbf{\overset{\circ}{Q}} . \delta\mathbf{\overset{\circ}{q}} \right] dA$$

$$= \int\!\!\int_{\mathbb{D}}\left[\overset{\circ}{\nabla} . \left(\overset{\circ}{\mathfrak{M}} . \delta\overset{\circ}{\varphi} \right) - \left(\overset{\circ}{\nabla} . \overset{\circ}{\mathfrak{M}} \right) . \delta\overset{\circ}{\varphi} \right.$$

$$\left. + \mathbf{\overset{\circ}{Q}} . \left(\delta\overset{\circ}{\varphi} + \overset{\circ}{\nabla}\left(\delta w \right) \right) \right] dA \tag{8.1258}$$

Applying the divergence theorem to certain elements of the right hand side of (8.1258) we have

$$\int\!\!\int_{\mathbb{D}} \overset{\circ}{\nabla} . \left[\left(\overset{\circ}{\mathfrak{M}} . \delta\overset{\circ}{\varphi} \right) \right] dA = \oint_C \left[\overset{\circ}{\mathfrak{M}} . \mathbf{n} \right] . \delta\overset{\circ}{\varphi} \; dS$$

$$= \oint_C \left(\mathbf{\overset{\circ}{M}} \right) . \delta\overset{\circ}{\varphi} \; dS \tag{8.1259}$$

Also,

$$\int\int_{\mathbb{D}}\left[\overset{\circ}{\mathbf{Q}}\cdot\overset{\circ}{\nabla}\left(\delta w\right)\right]dA = \int\int_{\mathbb{D}}\left[\overset{\circ}{\nabla}\cdot\left(\overset{\circ}{\mathbf{Q}}\,\delta w\right) - \delta w\left(\overset{\circ}{\nabla}\cdot\overset{\circ}{\mathbf{Q}}\right)\right]dA$$

$$= \oint_{C}\left[\overset{\circ}{\mathbf{Q}}\cdot\mathbf{n}\right]\delta w\;dS$$

$$-\int\int_{\mathbb{D}}\left(\overset{\circ}{\nabla}\cdot\overset{\circ}{\mathbf{Q}}\right)\delta w\;dA \tag{8.1260}$$

where $\overset{\circ}{\mathbf{M}}$ is a vector of moments defined on C and from (8.1187)

$$\overset{\circ}{\mathbf{Q}}\cdot\overset{\circ}{\mathbf{n}} = Q_n \tag{8.1261}$$

where Q_n is the shear force vector on the boundary. From the principle of virtual work for a solid in equilibrium (8.1242) must be satisfied; combining (8.1254) and (8.1258) to (8.1261) we can write

$$\int\int_{\mathbb{D}} p(x,y)\delta w\;dA + \oint_{C}\left[\overset{\circ}{\mathbf{M}}^{*}\cdot\delta\overset{\circ}{\varphi} + Q_n^{*}\,\delta w\right]dS$$

$$= \int\int_{\mathbb{D}}\left[-\left(\overset{\circ}{\nabla}\cdot\overset{\circ}{\mathfrak{M}}\right)\cdot\delta\overset{\circ}{\varphi} + \overset{\circ}{\mathbf{Q}}\cdot\delta\overset{\circ}{\varphi} - \left(\overset{\circ}{\nabla}\cdot\overset{\circ}{\mathbf{Q}}\right)\delta w\right]dA$$

$$+ \oint_{C}\left[\overset{\circ}{\mathbf{M}}\cdot\delta\overset{\circ}{\varphi} + Q_n\,\delta w\right]dS \tag{8.1262}$$

We can rewrite (8.1262) in the form

$$\int\int_{\mathbb{D}}\left[\left(\overset{\circ}{\mathbf{Q}} - \overset{\circ}{\nabla}\cdot\overset{\circ}{\mathfrak{M}}\right)\cdot\delta\overset{\circ}{\varphi} - \left(\overset{\circ}{\nabla}\cdot\overset{\circ}{\mathbf{Q}} + p(x,y)\right)\delta w\right]dA$$

$$= \oint_{C}\left[\left(\overset{\circ}{\mathbf{M}}^{*} - \overset{\circ}{\mathbf{M}}\right)\cdot\delta\overset{\circ}{\varphi} + (Q_n^{*} - Q_n)\,\delta w\right]dS \tag{8.1263}$$

This equation is satisfied if and only if the integrands are zero independently; i.e.

$$\left(\overset{\circ}{\mathbf{Q}} - \overset{\circ}{\nabla}\cdot\overset{\circ}{\mathfrak{M}}\right)\cdot\delta\overset{\circ}{\varphi} - \left(\overset{\circ}{\nabla}\cdot\overset{\circ}{\mathbf{Q}} + p(x,y)\right)\delta w = 0 \quad;\quad \mathbf{x}\in\mathbb{D} \tag{8.1264}$$

$$\left(\overset{\circ}{\mathbf{M}}^{*} - \overset{\circ}{\mathbf{M}}\right)\cdot\delta\overset{\circ}{\varphi} + (Q_n^{*} - Q_n)\,\delta w = 0 \quad;\quad \mathbf{x}\in C \tag{8.1265}$$

Since the virtual deformation is arbitrary in $\mathbb{D}$ and $\mathcal{C}$ we require

$$\overset{\circ}{\nabla} \cdot \overset{\circ}{\mathfrak{M}} - \overset{\circ}{\mathbf{Q}} = 0 \quad ; \quad \overset{\circ}{\nabla} \cdot \overset{\circ}{\mathbf{Q}} + p(x,y) = 0 \quad ; \quad \mathbf{x} \in \mathbb{D} \tag{8.1266}$$

and

$$\overset{\circ}{\mathbf{M}}{}^{*} = \overset{\circ}{\mathbf{M}} \quad ; \quad Q_n^{*} = Q_n \quad ; \quad \mathbf{x} \in \mathcal{C} \tag{8.1267}$$

We have now recovered, through the application of the principle of virtual work the equations of equilibrium for the plate (8.1220) and (8.1226) which were derived through consideration of statics. Furthermore, the principle furnishes the set of statically admissible boundary conditions applicable to the plate boundary $\mathcal{C}$. These boundary conditions relate to the specification of *either*

$$\overset{\circ}{\mathbf{M}} \quad \text{or} \quad \overset{\circ}{\varphi}$$

with *either*

$$Q_n \quad \text{or} \quad w$$

being specified along $\mathcal{C}$. Considering the definition of rotation and moment vectors given by (8.1152) and (8.1188), we can specify the *three boundary conditions* in terms of *either*

$$M_{nn} \quad \text{or} \quad \varphi_n$$

with *either*

$$M_{nt} \quad \text{or} \quad \varphi_t$$

with *either*

$$Q_n \quad \text{or} \quad w$$

Once consistent boundary conditions are formulated, the boundary value problem (or an initial boundary value problem in the case of dynamics of plates) is well posed. We can now attempt to obtain solutions to specific problems by adopting particular variable separable forms and integral transform techniques discussed in the preceding section, in connection with the solution of two-dimensional plane problems in classical elasticity theory.

8.11.8 The classical theory of thin plates

The modified theory of thin plates presented in the previous sections is sufficiently general in that the classical theory of thin plates proposed by Germain, Lagrange and others can be recovered as a special case of this modified theory. A basic assumption in the classical theory of thin plates is that straight material elements normal to the mid-plane prior to its deformation remain straight and normal to the neutral plane of the deformed plate. As a consequence, the independent rotations $\overset{\circ}{\boldsymbol{\varphi}}$ are directly related to the spatial gradients of the displacement $w(x,y)$. Therefore, from (8.1174) we have

$$\overset{\circ}{\boldsymbol{\varphi}} = - \overset{\circ}{\nabla} w \tag{8.1268}$$

and the transverse shear deformations are now neglected, i.e.

$$\epsilon_{iz} = \frac{1}{2}\left[\varphi_i + w_{,i}\right] = 0 \tag{8.1269}$$

Expressions for $\overset{\circ}{\mathfrak{M}}$ and $\overset{\circ}{\mathbf{Q}}$ reduce to

$$\overset{\circ}{\mathfrak{M}} = -\frac{D}{2}\left[(1-\nu)\left\{\overset{\circ}{\nabla}\overset{\circ}{\nabla}w + \left(\overset{\circ}{\nabla}\overset{\circ}{\nabla}w\right)^T\right\} + 2\nu\,\overset{\circ}{\mathfrak{I}}\nabla^2 w\right] \tag{8.1270}$$

$$\overset{\circ}{\mathbf{Q}} = -D\left[\overset{\circ}{\nabla}\left(\overset{\circ}{\nabla}{}^2 w\right)\right]$$

where

$$\overset{\circ}{\mathfrak{I}} = \mathbf{ii} + \mathbf{jj} \tag{8.1271}$$

In indicial notation, (8.1270) can be written as

$$\overset{\circ}{M}_{ij} = -D\left[(1-\nu)w_{,ij} + \nu(\overset{\circ}{\nabla}^2 w)\delta_{ij}\right]$$

$$\tag{8.1272}$$

$$\overset{\circ}{Q}_i = \overset{\circ}{M}_{ij,j} = -D(\overset{\circ}{\nabla}^2 w)_{,i}$$

Since we are constraining the shear deformations of the plate to be zero,

$$\overset{\circ}{\mathbf{q}} = \underset{\kappa^2 \mu \to \infty}{\text{Lim}} \left(\frac{\overset{\circ}{\mathbf{Q}}}{\kappa^2 \mu h}\right) = 0 \tag{8.1273}$$

which is valid for arbitrary but finite values of $\overset{\circ}{\mathbf{Q}}$. Considering (8.1272) we have

$$\overset{\circ}{\nabla} \cdot \overset{\circ}{\mathbf{Q}} = -D\,\overset{\circ}{\nabla}^2\,\overset{\circ}{\nabla}^2 w \tag{8.1274}$$

and combining this with (8.1220), we obtain

$$D\,\overset{\circ}{\nabla}^2\,\overset{\circ}{\nabla}^2 w(x,y) = p(x,y) \tag{8.1275}$$

which is the *single biharmonic inhomogeneous partial differential equation* governing the deflection of a plate which exhibits only curvature due to bending and is void of any shear deformation.

The admissible boundary conditions governing the partial differential equation (8.1275) can also be obtained; again, by employing the principle of virtual work. For completeness, we shall record here the details, which are instrumental in reducing the *three* boundary conditions, effectively to *two*. The flexural strain energy in the plate can be computed by using the expression (8.1257) derived previously in connection with the modified theory of plates and invoke appropriate reductions obtained through (8.1268). The virtual work equation (8.1263) can now be written as

$$\int\int_{\mathbb{D}}\left[\left(\overset{\circ}{\nabla}\cdot\mathfrak{M} - \overset{\circ}{\mathbf{Q}}\right)\cdot\delta\left(\overset{\circ}{\nabla}w\right) - \left(\overset{\circ}{\nabla}\cdot\overset{\circ}{\mathbf{Q}} + p(x,y)\right)\delta w\right]dA$$

$$= \oint_C\left[\left(-\mathbf{M}^* + \overset{\circ}{\mathbf{M}}\right)\cdot\delta\left(\overset{\circ}{\nabla}w\right) + (Q_n^* - Q_n)\,\delta w\right]dS \tag{8.1276}$$

Considering the expression for $\overset{\circ}{\mathbf{M}} \cdot \delta\overset{\circ}{\varphi}$ in terms of the components of $(M_{nn}\,\delta\varphi_n)$ and $(M_{nt}\,\delta\varphi_t)$ we have

$$\overset{\circ}{\mathbf{M}} \cdot \delta\overset{\circ}{\varphi} = M_{nn}\,\delta\varphi_n + M_{nt}\,\delta\varphi_t \tag{8.1277}$$

If $(M_{nt}\,\delta w)$ is single-valued and continuous along the plate boundary, then

$$\oint_C \frac{\partial}{\partial S}(M_{nt}\,\delta w)\,dS = \oint_C M_{nt}\frac{\partial}{\partial S}(\delta w)dS + \oint_C \frac{\partial M_{nt}}{\partial S}\delta w\,dS$$
$$= 0 \tag{8.1278}$$

Hence

$$\oint_C M_{nt}\delta\varphi_t\,dS = -\oint_C M_{nt}\frac{\partial}{\partial S}(\delta w)\,dS = \oint_C \frac{\partial M_{nt}}{\partial S}\delta w\,dS \tag{8.1279}$$

Similarly

$$\oint_C M_{nn}\delta\varphi_n\,dS = -\oint_C M_{nn}\frac{\partial}{\partial n}(\delta w)\,dS$$
$$= -\oint_C M_{nn}\delta\left(\frac{\partial w}{\partial n}\right)dS \tag{8.1280}$$

Therefore for the classical plate theory, (8.1276) can be reduced to the form

$$\int\int_{\mathbb{D}}\left[\left(\overset{\circ}{\nabla}\cdot\overset{\circ}{\mathfrak{M}} - \overset{\circ}{\mathbf{Q}}\right)\cdot\nabla(\delta w) - \left(\overset{\circ}{\nabla}\cdot\overset{\circ}{\mathbf{Q}} + p(x,y)\right)\delta w\right]dA$$
$$= \oint_C\left[(V_n^* - V_n)\,\delta w + (M_{nn} - M_{nn}^*)\,\delta\left(\frac{\partial w}{\partial n}\right)\right]dS \tag{8.1281}$$

where

$$V_n = Q_n + \frac{\partial M_{nt}}{\partial S} \tag{8.1282}$$

is referred to as the "Kirchhoff shear" which combines the shear force Q_n and the twisting moment M_{nt} to represent the traction vector at the boundary. Again we observe that the equations of equilibrium recovered from (8.1281) are identical to those obtained previously (see e.g. (8.1266)) and the admissible set of boundary conditions applicable to the classical plate theory are *either*

$$V_n \text{ or } w$$

or

$$M_{nn} \text{ or } \frac{\partial w}{\partial n}$$

A physical interpretation of the Kirchhoff shear boundary condition is shown in Figure 8.72, where the twisting moment $M_{nt}dS$ acting over a length dS, is replaced by two parallel forces of magnitude M_{nt} separated by a distance dS. This gives rise to a net distribution of forces $(\partial M_{nt}/\partial S)\,dS$ acting over the length dS. This can be combined with the shear force Q_n to generate the replacement Kirchhoff shear force.

The admissible boundary conditions derived previously assume that each point on the boundary C has a unique normal $\mathbf{n}$ with a continuously turning tangent. The formulation therefore has to be modified if the plate region contains either exterior corners or re-entrant corners. In the ensuing we shall develop the corner conditions applicable to exterior corner regions by considering a virtual work approach. For the purposes of discussion, let us consider the case of a rectangular plate region where corner forces result from the discontinuous nature of $\mathbf{n}$ along C.

The virtual work of internal forces is given by

$$\delta U = \int\int_{\mathbb{D}} \delta U_A \, dA = \int\int_{\mathbb{D}} M_{ij} \, \delta w_{,ij} \, dA \tag{8.1283}$$

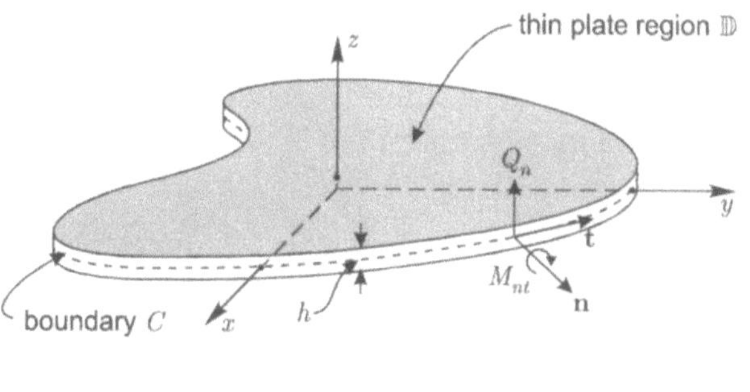

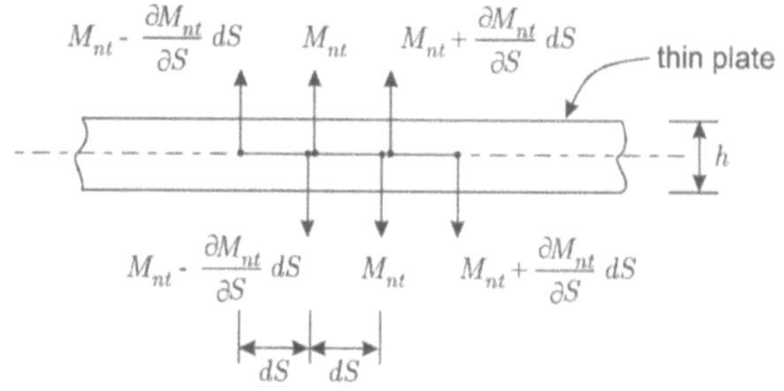

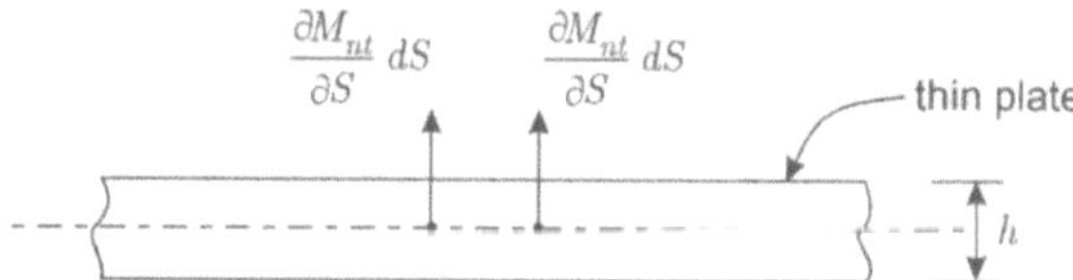

Figure 8.72: Kirchhoff shear forces at the boundary of a thin plate.

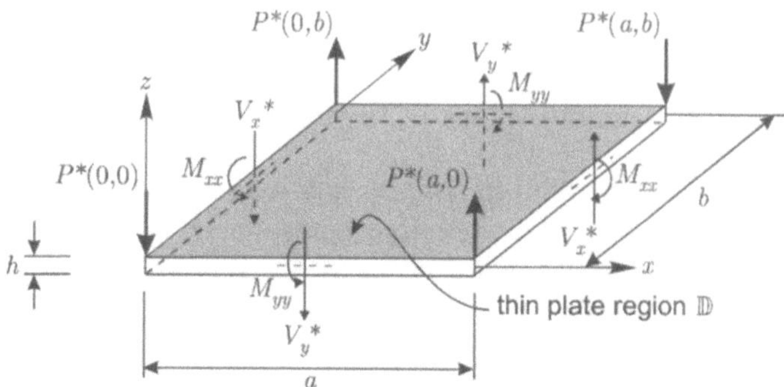

Figure 8.73: Corner forces acting on a rectangular plate.

We can perform integration by parts and rewrite (8.1283) in the form

$$\delta U = \int\!\!\int_{\mathbb{D}} M_{ij,ij} \, \delta w \, dA$$
$$+ \int_0^a \left[M_{yy} \, \delta\left(\frac{\partial w}{\partial y}\right) - V_y \, \delta w \right]_0^b dx$$
$$+ \int_0^b \left[M_{xx} \, \delta\left(\frac{\partial w}{\partial x}\right) - V_x \, \delta w \right]_0^a dy$$
$$+ \left[(M_{xy} \, \delta w)_0^a\right]_0^b + \left[(M_{yx} \, \delta w)_0^b\right]_0^a \tag{8.1284}$$

Considering the force resultants on the boundary of the rectangular plate, the virtual work of external forces is

$$\delta W = \int\!\!\int_{\mathbb{D}} p(x,y)\delta w \, dA - \oint_C M_{ij}^* n_j \, \delta\left(\frac{\partial w}{\partial x_i}\right) dS$$
$$+ \oint_C Q_i^* n_i \, \delta w \, dS \tag{8.1285}$$

Performing integration by parts (8.1285) gives

$$\delta W = \int\!\!\int_{\mathbb{D}} p(x,y)\delta w \, dA + \int_0^a \left[V_y^* \delta w - M_{yy}^* \, \delta\left(\frac{\partial w}{\partial y}\right) \right]_0^b dx$$
$$+ \int_0^b \left[V_x^* \, \delta w - M_{xx}^* \, \delta\left(\frac{\partial w}{\partial x}\right) \right]_0^a dy$$
$$- \left[(M_{xy}^* \, \delta w)_0^a\right]_0^b - \left[(M_{yx}^* \delta w)_0^b\right]_0^a \tag{8.1286}$$

where V_x^*, M_{xy}^*, ... etc., denote values at the boundary.

The results (8.1284) and (8.1286) in conjunction with the principle of virtual work will again furnish the plate bending equation (8.1275). The specific forms of the boundary conditions are given by the following:

(i)   on $x = 0$; $x = a$ we need to specify either

$$M_{xx} \text{ or } \frac{\partial w}{\partial x}, \text{ and either, } V_x = \left(Q_x + \frac{\partial M_{xy}}{\partial y} \right) \text{ or } w,$$

(ii) on $y = 0; y = b$ we need to specify either

$$M_{yy} \text{ or } \frac{\partial w}{\partial y}, \text{ and either, } V_y = \left(Q_y + \frac{\partial M_{xy}}{\partial x} \right) \text{ or } w,$$

(iii) at the corners $(x, y) = (0, 0), (a, 0), (0, b), (a, b)$ we need to specify either

$$M_{xy} \text{ or } w$$

The boundary conditions (i) and (ii) along the straight edges essentially correspond to those derived previously in connection with either $(V_n$ or $w)$ or $(M_{nn}$ or $\partial w / \partial n)$. The corner conditions (iii) can be interpreted along the lines presented previously in the illustration of the physical basis for the Kirchhoff shear (see e.g. Figure 8.74).

The twisting moments M_{xy} and M_{yx} are represented by their statical equivalent force intensities $(\partial M_{xy} / \partial y)$ and $(\partial M_{yx} / \partial x)$ respectively. At the corner or the plate these culminate in the generation of an equivalent force

$$M_{xy} + M_{yx} = 2M_{xy} = P \tag{8.1287}$$

This completes the formal presentation of the classical theory of thin plates as a well posed boundary value problem in the theory of partial differential equations. The classical theory of thin plates represents one of the most well researched and extensively documented theories in solid mechanics spanning nearly two centuries. It has appeal not only due to its mathematical tractability in the development of solutions but also due to the applicability of such solutions to problems of significant technological importance. The literature dealing with solutions to plate problems dealing with the classical theory is quite extensive and no attempt will be made to document a wider class of solutions. In the ensuing sections we shall solve certain typical boundary value problems dealing with the flexure of circular and rectangular plates governed by the classical thin plate theory.

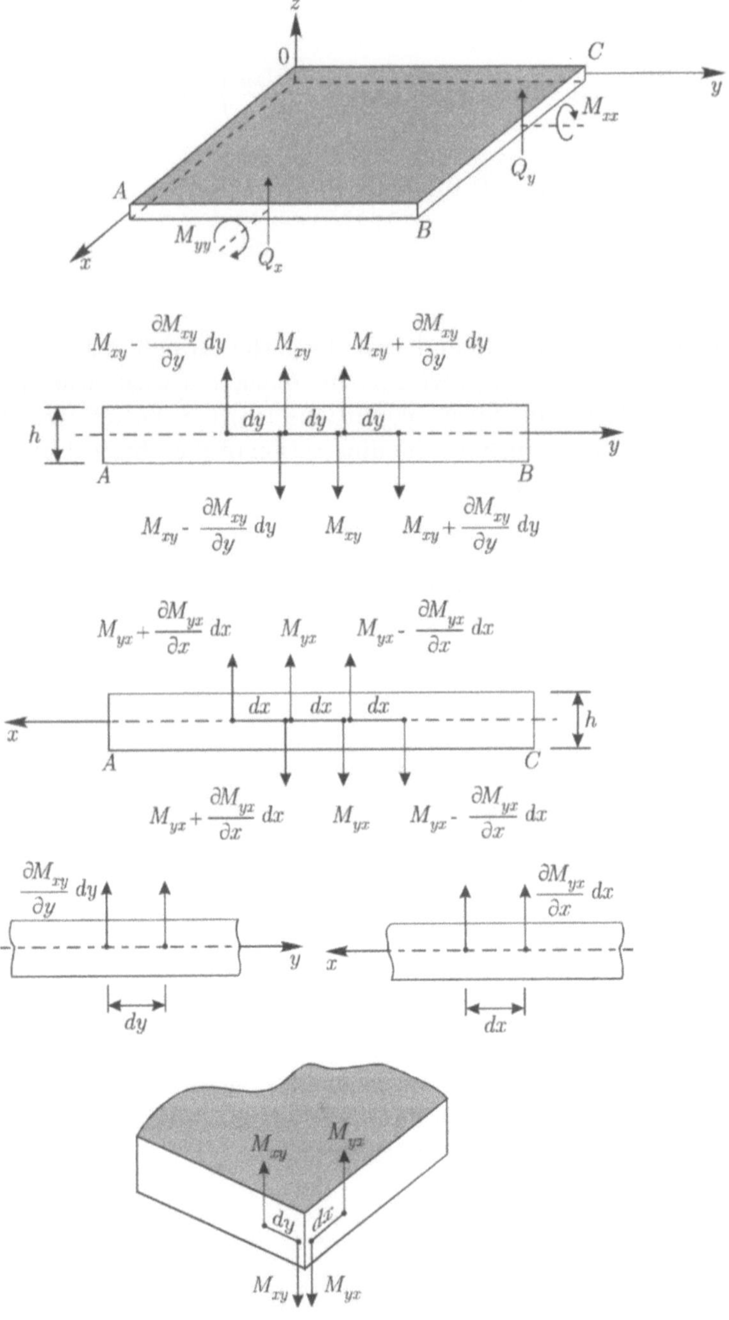

Figure 8.74: Corner forces acting on a thin plate.

8.11.9 Flexure of thin circular plates

We now apply the classical theory of thin elastic plates presented in Section 8.11.8 to develop solutions to certain selected problems dealing with flexure of circular plates. The range of problems dealing with the flexure of circular plates is so extensive that it is not possible to cover all aspects of circular plates relating to a wider class of non-symmetric loadings and boundary conditions. We shall consider here some specific problems related to the flexure of circular plates which employ elementary solutions of the governing partial differential equations. For completeness, we shall summarize here the relevant equations referred to the plane polar coordinate system (r, θ).

The partial differential equation governing flexure of the plate is

$$D \, \overset{\circ}{\nabla}{}^2 \, \overset{\circ}{\nabla}{}^2 w(r, \theta) = p(r, \theta) \tag{8.1288}$$

where

$$\overset{\circ}{\nabla}{}^2 = \frac{\partial^2}{\partial r^2} + \frac{1}{r}\frac{\partial}{\partial r} + \frac{1}{r^2}\frac{\partial^2}{\partial \theta^2} \tag{8.1289}$$

is Laplace's operator referred to the system of plane polar coordinates. Considering the results (8.1270) (see also 8.1214) we can show that

$$M_{rr}(r, \theta) = -D \left[\frac{\partial^2 w}{\partial r^2} + \nu \left(\frac{1}{r}\frac{\partial w}{\partial r} + \frac{1}{r^2}\frac{\partial^2 w}{\partial \theta^2} \right) \right]$$

$$M_{\theta\theta}(r, \theta) = -D \left[\frac{1}{r^2}\frac{\partial^2 w}{\partial \theta^2} + \frac{1}{r}\frac{\partial w}{\partial r} + \nu\frac{\partial^2 w}{\partial r^2} \right] \tag{8.1290}$$

$$M_{r\theta}(r, \theta) = -D(1 - \nu) \left[\frac{1}{r}\frac{\partial^2 w}{\partial r \partial \theta} - \frac{1}{r^2}\frac{\partial w}{\partial \theta} \right]$$

and

$$Q_r(r, \theta) = -D\frac{\partial}{\partial r}\left(\overset{\circ}{\nabla}{}^2 w \right)$$

$$\tag{8.1291}$$

$$Q_\theta(r, \theta) = -D\frac{1}{r}\frac{\partial}{\partial \theta}\left(\overset{\circ}{\nabla}{}^2 w \right)$$

as the complete expressions, respectively, for the flexural moments and shear forces in the thin plate. The components of the Kirchhoff shear applicable to a free boundary take the forms

$$
\begin{aligned}
V_r(r, \theta) &= Q_r + \frac{1}{r}\frac{\partial M_{r\theta}}{\partial \theta} \\
&= -D\left[\frac{\partial}{\partial r}\left(\overset{\circ}{\nabla}^2 w\right) + \frac{(1-\nu)}{r}\frac{\partial}{\partial \theta}\left(\frac{1}{r}\frac{\partial^2 w}{\partial r\partial\theta} - \frac{1}{r^2}\frac{\partial w}{\partial \theta}\right)\right]
\end{aligned}
$$

(8.1292)

$$
\begin{aligned}
V_\theta(r.\theta) &= Q_\theta + \frac{\partial M_{r\theta}}{\partial r} \\
&= -D\left[\frac{1}{r}\frac{\partial}{\partial \theta}\left(\overset{\circ}{\nabla}^2 w\right) + (1-\nu)\frac{\partial}{\partial r}\left(\frac{1}{r}\frac{\partial^2 w}{\partial r\partial\theta} - \frac{1}{r}\frac{\partial w}{\partial \theta}\right)\right]
\end{aligned}
$$

The boundary conditions applicable to a thin circular plate can be classified according to the following types:

(i) for a simply supported boundary $r = a$

$$
w(a, \theta) = 0 \quad ; \quad M_{rr}(a, \theta) = 0 \tag{8.1293}
$$

(ii) for a fixed boundary $r = a$

$$
w(a, \theta) = 0 \quad ; \quad \left[\frac{\partial w}{\partial r}\right]_{r=a} = 0 \tag{8.1294}
$$

(iii) for a free boundary $r = a$

$$
M_{rr}(a, \theta) = 0 \quad ; \quad V_r(a, \theta) = 0 \tag{8.1295}
$$

The general solution of the partial differential equation (8.1288) consists of a particular integral $w^P(r, \theta)$ and a homogeneous solution $w^H(r, \theta)$. In order to obtain a general solution of the homogeneous equation of (8.1288) we can follow the general procedures outlined in Section 8.9.7, in connection with the development of a general solution of the biharmonic equation for Airy's stress function in plane polar coordinates. We have

$$
w(r, \theta) = w^P(r, \theta) + w^H(r, \theta) \tag{8.1296}
$$

and the general solution (8.731) developed by Michell obviously qualifies as a solution

$$\overset{\circ}{\nabla}^2 \overset{\circ}{\nabla}^2 w^H(r,\theta) = 0 \tag{8.1297}$$

If the deflections of the plate are to be single-valued in $\mathbb{D}$, then the solutions

$$w^H(r,\theta) = D_0^* r^2 \theta + A'\theta - \frac{C_1}{2} r\theta \cos\theta \tag{8.1298}$$

are *inadmissible* for a complete plate region where $0 \le \theta \le 2\pi$. If we further impose the requirement that the gradient of the displacement should be single-valued in a complete plate region; then

$$w^H(r,\theta) = \frac{A_1}{2} r\theta \sin\theta \tag{8.1299}$$

is *inadmissible*. Hence, the most general form of the homogeneous solution applicable for a complete plate region reduces to

$$
\begin{aligned}
w^H(r,\theta) =\ & \left[A_0 \ln r + B_0 r^2 + C_0 r^2 \ln r + D_0 \right] \\
+\ & \left[A^* r + B_1 r^3 + \frac{A_1'}{r} + B_1' r \ln r \right] \cos\theta \\
+\ & \left[B^* r + D_1 r^3 + \frac{C_1'}{r} + D_1 r \ln r \right] \sin\theta \\
+\ & \sum_{n=2} \left[A_n r^n + B_n r^{2+n} + A_n' r^{-n} + B_n' r^{2-n} \right] \cos(n\theta) \\
+\ & \sum_{n=2} \left[C_n r^n + D_n r^{2+n} + C_n' r^{-n} + D_n' r^{2-n} \right] \sin(n\theta) \quad (8.1300)
\end{aligned}
$$

where $A_0, B_0, C_0, ...$ etc., are all constants.

This solution was also obtained by Clebsch (1833-1872) and Föppl (1854-1924) and forms the basis for examining a variety of problems dealing with the flexure of circular plates which are subjected to generalized loads. The first part of the homogeneous solution (8.1300) involving constants A_0, B_0, C_0 and D_0 is applicable to situations involving axisymmetric deflections of circular plates. For axial symmetry

$$w(r, \theta) = w(r) \quad ; \quad p(r, \theta) = p(r) \tag{8.1301}$$

and the partial differential equation (8.1288) reduces to

$$\frac{D}{r}\frac{d}{dr}\left\{ r\frac{d}{dr}\left[\frac{1}{r}\frac{d}{dr}\left(r\frac{dw}{dr} \right) \right] \right\} = p(r) \quad ; \quad (r, \theta) \in \mathbb{D} \tag{8.1302}$$

Equation (8.1302) can be directly integrated to obtain the general solution

$$w(r) = \int \frac{1}{r}\left\{ \int r\left[\int \frac{1}{r}\left\{ \int \frac{rp(r)dr}{D} \right\} dr \right] dr \right\} dr$$
$$+ \left\{ C_0 \ln r + C_1 r^2 + C_2 r^2 \ln r + C_3 \right\} \tag{8.1303}$$

where C_0, C_1, C_2 and C_3 are arbitrary constants.

Example 8.28

A circular plate of radius a is either (i) fixed at the boundary $r = a$ or (ii) simply supported at the boundary $r = a$. It is subjected to a uniform load of stress intensity p_0.

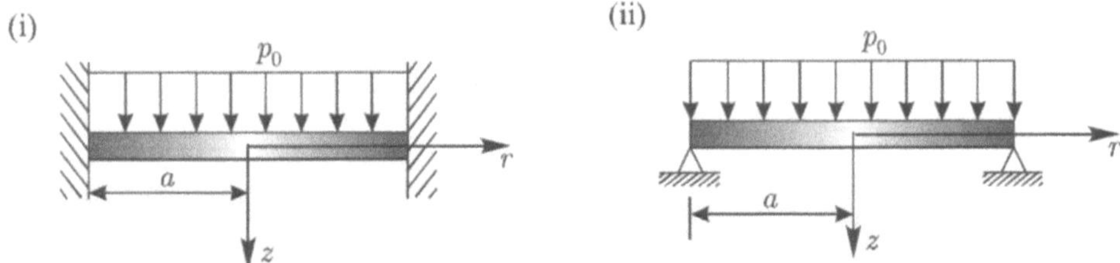

Figure 8.75: Uniform loading of a circular plate with (i) fixed and (ii) simply supported boundaries.

Derive the complete expressions for the deflection, the flexural moments and shear force in the plate.

Solution

We require a solution of

$$\overset{\circ}{\nabla}^2\,\overset{\circ}{\nabla}^2 w(r) = p(r) \quad ; \quad r \in (0, a) \tag{8.1304}$$

subject to the loading conditions

$$p(r) = p_0 \quad ; \quad r \in (0, a) \tag{8.1305}$$

and, in the case of a plate which is fixed at the boundary we require

$$w(a) = 0 \quad ; \quad \left[\frac{dw}{dr}\right]_{r=a} = 0 \tag{8.1306}$$

The particular integral of (8.1304) consistent with the constant loading obtained from (8.1303) is

$$w^P(r) = \frac{p_0 r^4}{64D} \tag{8.1307}$$

The homogeneous solution is chosen such that the deflections of the plate are bounded for $0 \le r \le a$. The appropriate complete solution indeterminate to within arbitrary constants C_1 and C_3 is

$$w(r) = \frac{p_0 r^4}{64D} + C_1 r^2 + C_3 \tag{8.1308}$$

These arbitrary constants can be determined from the boundary conditions (8.1306); i.e.

$$C_1 a^2 + C_3 + \frac{p_0 a^4}{64D} = 0 \quad ; \quad 2C_1 a + \frac{p_0 a^3}{16D} = 0 \tag{8.1309}$$

The solution for the deflection of the plate can be evaluated in the form

$$w(r) = \frac{p_0(a^2 - r^2)^2}{64D} \tag{8.1310}$$

and the non-zero flexural moments and shear are given by

$$M_{rr} = \frac{p_0}{16}\left[(1+\nu)a^2 - (3+\nu)r^2\right]$$

$$M_{\theta\theta} = \frac{p_0}{16}\left[(1+\nu)a^2 - (1+3\nu)r^2\right] \tag{8.1311}$$

$$Q_r = -\frac{p_0 r}{2}$$

In the instance when the boundary of the circular plate is simply supported, the boundary conditions are

$$w(a) = 0 \quad ; \quad M_{rr}(a) = 0 \tag{8.1312}$$

Using the general solution, the constants C_1 and C_3 are determined from the equations

$$C_1 a^2 + C_3 + \frac{p_0 a^4}{64D} = 0 \quad ; \quad 2(1+\nu)C_1 + \frac{p_0(3+\nu)a^2}{16} = 0 \tag{8.1313}$$

which gives

$$C_1 = -\frac{p_0}{32}\left(\frac{3+\nu}{1+\nu}\right)a^2 \quad ; \quad C_3 = \frac{p_0 a^4}{64D}\left(\frac{5+\nu}{1+\nu}\right) \tag{8.1314}$$

The corresponding results for $w(r)$, $M_{rr}(r)$ etc., take the forms

$$w(r) = \frac{p_0}{64D}\left[r^4 - 2\left(\frac{3+\nu}{1+\nu}\right)r^2 a^2 + \left(\frac{5+\nu}{1+\nu}\right)a^4\right]$$

$$M_{rr} = \frac{p_0}{16}(3+\nu)(a^2 - r^2)$$

$$\tag{8.1315}$$

$$M_{\theta\theta} = \frac{p_0}{16}\left[(3+\nu)a^2 - (1+3\nu)r^2\right]$$

$$Q_r = -\frac{p_0 r}{2}$$

Example 8.29

A circular plate of radius a is either fixed or simply supported at its boundary and subjected to a concentrated force of magnitude P_0 at the origin.

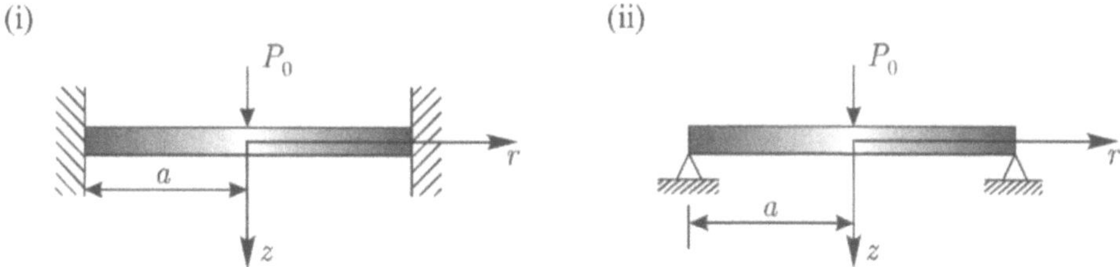

Figure 8.76: Concentrated loading of a circular plate with (i) fixed and (ii) simply supported boundaries.

Determine the deflection of the plate and the resulting moments and shear forces.

Solution

Due to the discontinuous nature of the transverse loading it is necessary to use a slightly different approach to solve this axisymmetric plate bending problem. This procedure involves the consideration of the vertical equilibrium of disc-shaped region of the plate of arbitrary radius r. Considering the equilibrium of the disc-shaped region we have

$$2\pi r Q_r + P_0 = 0 \tag{8.1316}$$

Using (8.1291) we can write

$$D\frac{d}{dr}\left(\overset{\circ}{\nabla}^2 w(r)\right) = \frac{P_0}{2\pi r} \tag{8.1317}$$

Integrating (8.1317) we have

$$w(r) = \frac{P}{2\pi D} \left[\frac{r^2}{4} \ln r \right] + C_1 r^2 + C_2 \ln r + C_3 \tag{8.1318}$$

where C_1, C_2 and C_3 are constants of integration. Since the deflection is to be finite within the plate region

$$C_2 \equiv 0 \tag{8.1319}$$

The constants C_1 and C_3 can be obtained by making use of the boundary conditions applicable to the fixed boundary; given by (8.1306). The final expression for the deflection of the plate is given by

$$w(r) = \frac{P_0 a^2}{16\pi D} \left[1 - \frac{r^2}{a^2} + 2\frac{r^2}{a^2} \ln \left(\frac{r}{a} \right) \right] \tag{8.1320}$$

The expressions for the flexural moments and shear force are given by

$$\begin{aligned}
M_{rr}(r) &= -\frac{P_0}{4\pi} \left[1 + (1+\nu) \ln \left(\frac{r}{a} \right) \right] \\
M_{\theta\theta}(r) &= -\frac{P_0}{4\pi} \left[\nu + (1+\nu) \ln \left(\frac{r}{a} \right) \right] \\
Q_r(r) &= -\frac{P_0}{2\pi r}
\end{aligned} \tag{8.1321}$$

Similarly, in the case of a plate which is simply supported at the boundary and subjected to the central force P_0, the boundary conditions (8.1312) can be used to determine C_1 and C_3. Avoiding details it can be shown that the expressions for the plate deflection, flexural moments and shear force take the forms

$$\begin{aligned}
w(r) &= \frac{P_0 a^2}{16\pi D} \left[\left(1 - \frac{r^2}{a^2} \right) \left(\frac{3+\nu}{1+\nu} \right) + 2\frac{r^2}{a^2} \ln \left(\frac{r}{a} \right) \right] \\
M_{rr}(r) &= -\frac{P_0(1+\nu)}{4\pi} \ln \left(\frac{r}{a} \right)
\end{aligned}$$

$$\tag{8.1322}$$

$$\begin{aligned}
M_{\theta\theta}(r) &= \frac{P_0}{4\pi} \left[(1-\nu) - (1+\nu) \ln \left(\frac{r}{a} \right) \right] \\
Q_r(r) &= -\frac{P_0}{2\pi r}
\end{aligned}$$

From these results it is evident that when the plate is subjected to a concentrated force, while the plate deflections remain finite, the flexural moment and the shear force exhibit singular behaviour at the point of application of the load.

Example 8.30

An annular thin plate is simply supported at the outer boundary $r = a$. It is subjected to uniformly distributed radial flexural moments M_A and M_B at the outer and inner boundaries of the plate, respectively.

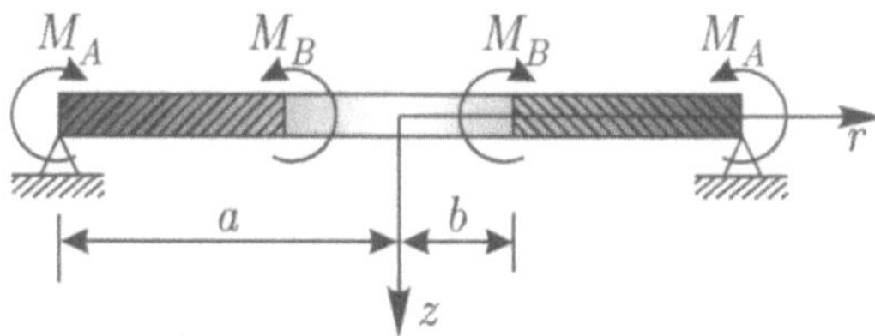

Figure 8.77: Loading of an annular plate by radial moments distributed along the boundaries.

Determine the deflected shape of the plate. Use this result to determine the amplification of the flexural moment in a plate containing a circular opening which is subjected to a far field pure bending.

Solution

Since the annular plate region is subjected to pure bending, the shear forces within the plate are zero; therefore considering the expression for the shear force we have

$$\frac{d}{dr}\left(\overset{\circ}{\nabla}^2 w(r) \right) = 0 \tag{8.1323}$$

Integrating (8.1323) we obtain

$$w(r) = C_1 \frac{r^2}{4} + C_2 \ln r + C_3 \quad ; \quad 0 < b \leq r \leq a \tag{8.1324}$$

The boundary conditions governing the problem are (refer to Figure 8.71 for the sign convention for the applied moments)

$$w(a) = 0 \quad ; \quad M_{rr}(a) = M_A \quad ; \quad M_{rr}(b) = M_B \tag{8.1325}$$

Considering these boundary conditions we can show that the deflection of the plate is given by

$$
w(r) = \left\{ \frac{M_B a^2 - M_A b^2}{2\,(1+\nu)\,D} \right\} \left(\frac{a^2 - r^2}{a^2 - b^2} \right)
$$
$$
+ \left\{ \frac{M_A - M_B}{(1-\nu)D} \right\} \left(\frac{a^2 b^2}{a^2 - b^2} \right) \ln\left(\frac{r}{a} \right) \tag{8.1326}
$$

The expressions for the flexural moments in the plate are

$$
M_{rr}(r) = \left\{ \frac{M_A a^2 - M_B b^2}{a^2 - b^2} \right\} + \frac{a^2 b^2}{r^2} \frac{(M_B - M_A)}{(a^2 - b^2)}
$$

$$
M_{\theta\theta}(r) = \left\{ \frac{M_A a^2 - M_B b^2}{a^2 - b^2} \right\} - \frac{a^2 b^2}{r^2} \frac{(M_B - M_A)}{(a^2 - b^2)}
$$

$$\tag{8.1327}$$

Consider the case when the plate is subjected only to the radial moment M_0 at $r = a$ and the inner boundary is free of any radial moment (i.e. $M_B = 0$). In this case, the expression for $M_{\theta\theta}$ given in (8.1327) reduces to

$$M_{\theta\theta}(r) = \frac{M_0 a^2}{(a^2 - b^2)} \left(1 + \frac{b^2}{r^2} \right) \tag{8.1328}$$

Evaluating $M_{\theta\theta}(r)$ at the inner radius we have

$$M_{\theta\theta}(b) = \frac{2M_0}{[1 - (b/a)^2]} \tag{8.1329}$$

If we let the inner radius b become small in comparison with a, (8.1329) reduces to

$$M_{\theta\theta}(b) \rightarrow 2M_0 \tag{8.1330}$$

indicating that there is a stress concentration around the inner boundary, approaching twice the applied far-field value.

This result can also be deduced by considering the problem of a plate of arbitrary shape with a smooth convex boundary containing a circular opening of radius b where the plate is subjected to a far-field state of pure bending M_0. The deflection of the plate should consist of a solution of the form

$$w(r) = C_1 r^2 + C_2 \ln r \qquad (8.1331)$$

Since the flexural moments contain derivatives of $w(r)$, the contribution to the flexural moments from the $C_2 \ln r$ term reduces at distances remote from the opening. The term $C_1 r^2$ gives rise to homogeneous bending. Considering the boundary conditions

$$M_{rr}(b) = 0 \quad ; \quad M_{rr}(\infty) = M_0 \qquad (8.1332)$$

we obtain the solution for the deflection of the plate

$$w(r) = -\frac{M_0}{D} \left[\frac{r^2}{2(1+\nu)} + \frac{b^2}{(1-\nu)} \ln r \right] \qquad (8.1333)$$

and the expression for $M_{\theta\theta}$ is given by

$$M_{\theta\theta}(r) = M_c \left(1 + \frac{b^2}{r^2} \right) \qquad (8.1334)$$

Again, at the boundary of the cavity $M_{\theta\theta}(b) = 2M_0$.

$$\bullet \ \bullet \ \bullet$$

The general results presented at the beginning of this section can also be used to examine solutions to plate bending problems where the state of deflection of the plate is no longer axisymmetric.

Example 8.31

The Figure 8.78 shows a section through a thin annular plate, the central region on which $(0 \leq r \leq b)$ contains a rigid stiffener in the form of a disc. This rigid stiffener is subjected to a small rotation Ω_0 and the outer boundary of the plate is held fixed. Derive an expression for the relationship between Ω_0 and the couple M_0 applied about the y-axis.

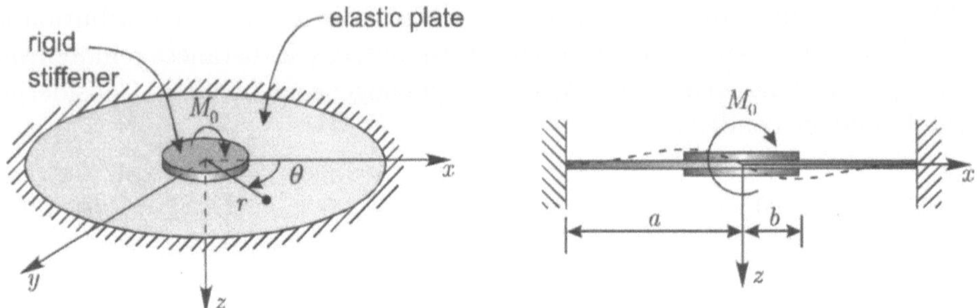

Figure 8.78: Asymmetric bending of a stiffened circular plate.

Solution

The boundary value problem requires the solution of the partial differential equation

$$D \overset{\circ}{\nabla}^2 \overset{\circ}{\nabla}^2 w(r, \theta) = 0 \tag{8.1335}$$

subject to the boundary conditions

$$w(b, \theta) = \Omega_0 b \cos \theta \quad ; \quad \left[\frac{\partial w}{\partial r}\right]_{r=b} = \left[\frac{w(r)}{r}\right]_{r=b}$$

$$\tag{8.1336}$$

$$w(a, \theta) = 0 \qquad ; \quad \left[\frac{\partial w}{\partial r}\right]_{r=a} = 0$$

Since the surface of the plate is unloaded, the particular integral (8.1288) is zero. The problem can be solved by employing purely the homogeneous solution. Considering the general form of the Fourier series solution for the plate deflection (8.1300) and the form of the inhomogeneous boundary condition of (8.1336) it is evident that the homogeneous solution for $w(r, \theta)$ should exhibit only a $\cos \theta$ dependence in θ. The appropriate solution for $w(r, \theta)$ takes the form

$$w(r, \theta) = \left[C_1 r + C_2 r^3 + \frac{C_3}{r} + C_4 r \ln r \right] \cos \theta \tag{8.1337}$$

The arbitrary constants C_1, C_2, C_3 and C_4 can be uniquely determined by satisfying the boundary conditions (8.1336). The final form of the deflection of the plate is given by

$$w(r, \theta) = \frac{\Omega_0}{\Delta^*} \left[(a^2 - b^2)r + 2r(a^2 + b^2) \ln \left(\frac{r}{a} \right) - r^3 + \frac{a^2 b^2}{r} \right] \cos \theta \tag{8.1338}$$

where

$$\Delta^* = \left\{ 2(a^2 - b^2) - 2(a^2 + b^2) \ln \left(\frac{a}{b} \right) \right\} \tag{8.1339}$$

The relationship between the rotation Ω_0 and the couple M_0 required to maintain this rotation can be obtained by considering the moment equilibrium at any location $r=$ const. within $b \leq r \leq a$, taken about the y-axis. It is convenient to choose $r = a$, to evaluate this resultant. The moment equilibrium of the plate about the y-axis is given by

$$2 \int_0^{\pi} \{ [M_{rr}(r, \theta)]_{r=a} \cos \theta + [M_{r\theta}(r, \theta)]_{r=a} \, a \sin \theta$$

$$+ [Q_r(r, \theta)]_{r=a} \, a \cos \theta \} \, a d\theta + M_0 = 0 \tag{8.1340}$$

In terms of (8.1337), the general expression for $M_{rr}, M_{r\theta}$ and Q_r take the forms

$$M_{rr}(r,\theta) = -D\left[2C_2(3+\nu)r + \frac{2C_3(1-\nu)}{r^3} + \frac{C_4(1-\nu)}{r}\right]\cos\theta$$

$$M_{r\theta}(r,\theta) = -D(1-\nu)\left[-2C_2 + \frac{2C_3}{r^3} - \frac{C_4}{r}\right]\sin\theta \tag{8.1341}$$

$$Q_r(r,\theta) = -D\left[8C_2 - \frac{2C_4}{r}\right]\cos\theta$$

Substituting the complete expressions for M_{rr}, $M_{r\theta}$ and Q_r in terms of Ω_o in (8.1340) and evaluating the result we obtain

$$\Omega_0 = \frac{M_0\Delta^*}{4\pi a^2 D(2+3\nu)} \tag{8.1342}$$

$$\bullet \ \bullet \ \bullet$$

The general form of the homogeneous solution presented in Equation (8.1300) can be employed to develop solutions to a variety of generalized loading conditions and edge support conditions. In the ensuing we shall consider a specific example related to the localized loading of a circular plate which is clamped along the boundary.

Example 8.32

A flexible circular plate of radius a is clamped along the edge and subjected to a concentrated force of magnitude P_0 at an arbitrary location (Figure 8.79). Develop an expression for the deflection of the plate.

Solution

At the outset it may be noted that the location of the concentrated force is quite arbitrary. We can without loss of generality take the reference coordinate system to be such that this force is located on the $x-$axis at the location $x = \xi$. For the solution of the problem we consider two regions of the plate; (i) an interior region $(\ )^{(i)}$ occupying $0 \leq r \leq \xi$; $0 < \theta < 2\pi$ and an exterior region $(\ )^{(e)}$ occupying the region $\xi \leq r \leq a$; $0 < \theta < 2\pi$. Since the loading is localized and located at the common boundary between the interior and exterior regions, $p(r,\theta)$ is zero in both regions. The boundary value problem is now reduced to obtaining solutions to the partial differential equations

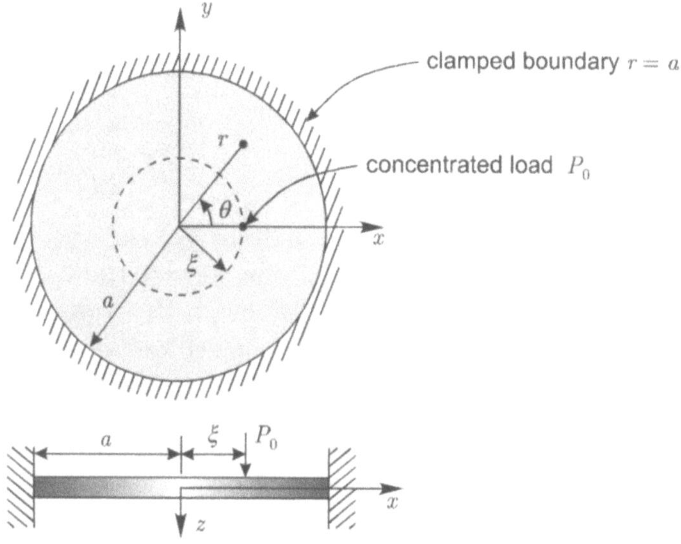

Figure 8.79: Localized loading of a clamped plate.

$$\overset{\circ}{\nabla}^2 \overset{\circ}{\nabla}^2 w^{(i)}(r,\theta) = 0 \quad ; \quad 0 \le r \le \xi \quad ; \quad 0 \le \theta \le 2\pi \tag{8.1343}$$

and

$$\overset{\circ}{\nabla}^2 \overset{\circ}{\nabla}^2 w^{(e)}(r,\theta) = 0 \quad ; \quad \xi \le r \le a \quad ; \quad 0 \le \theta \le 2\pi \tag{8.1344}$$

subject to the boundary conditions at $r = a$ continuity conditions at $r = \xi$ and regularity conditions at $r = 0$.

At the clamped boundary of the plate we require

$$w^{(e)}(a,\theta) = 0 \quad ; \quad \left[\frac{\partial w^{(e)}}{\partial r}\right]_{r=a} = 0 \tag{8.1345}$$

At the common boundary between the interior and exterior regions

$$w^{(i)}(\xi,\theta) = w^{(e)}(\xi,\theta);$$

$$\left[\frac{\partial w^{(i)}}{\partial r}\right]_{r=\xi} = \left[\frac{\partial w^{(e)}}{\partial r}\right]_{r=\xi} \; ; \quad \left[\frac{\partial w^{(i)}}{\partial \theta}\right]_{r=\xi} = \left[\frac{\partial w^{(e)}}{\partial \theta}\right]_{r=\xi} \tag{8.1346}$$

$$M_{rr}^{(i)}(\xi,\theta) = M_{rr}^{(e)}(\xi,\theta)$$

The remaining inhomogeneous boundary condition relates to the transverse shear obtained by making use of the condition that at the radius of application of the concentrated load, the difference in the transverse shears $Q_r^{(e)}(\xi,\theta)$ and $Q_r^{(i)}(\xi,\theta)$ must be equal to the applied force P_0; i.e.

$$-D\left[\frac{\partial}{\partial r}\,\overset{\circ}{\nabla}^2\left\{w^{(e)}(r,\theta) - w^{(i)}(r,\theta)\right\}\right]_{r=\xi,\theta=0} = P_0 \tag{8.1347}$$

In addition to these boundary conditions the deflections, derivatives of the deflection and moments at the centre of the plate ($r = 0$) must be finite and bounded. Furthermore, since the loading is symmetric about $\theta = 0$, we need to consider solutions of (8.1300) which involve only symmetric functions of θ. Considering the exterior region we have

$$w^{(e)}(r,\theta) = F_0(r) + \sum_{m=1}^{\infty} F_m(r)\cos m\theta \tag{8.1348}$$

where

$$F_0(r) = C_{10} + C_{20}r^2 + C_{30}\ln\left(\frac{r}{a}\right) + C_{40}r^2\ln\left(\frac{r}{a}\right)$$

$$F_1(r) = C_{11}r + C_{21}r^3 + C_{31}r^{-1} + C_{41}r\ln\left(\frac{r}{a}\right)$$

$$.... = ...$$

$$.... = ... \tag{8.1349}$$

$$F_m(r) = C_{1m}r^m + C_{2m}r^{2+m} + C_{3m}r^{-m} + C_{4m}r^{2-m} \; ; \quad (m > 1)$$

where $C_{10}, C_{20}, ..., C_{4m}$ are arbitrary constants. Similarly for the interior region of the plate

$$w^{(i)}(r,\theta) = F_0^*(r) + \sum_{m=1}^{\infty} F_m^*(r)\cos m\theta \tag{8.1350}$$

where

$$F_0^*(r) = C_{10}^* + C_{20}^* r^2 + C_{30}^* \ln\left(\frac{r}{a}\right) + C_{40}^* r^2 \ln\left(\frac{r}{a}\right)$$

$$F_1^*(r) = C_{11}^* r + C_{21}^* r^3 + C_{31}^* r^{-1} + C_{41}^* r \ln\left(\frac{r}{a}\right)$$

$$.... = ...$$

$$.... = ...$$

$$F_m^*(r) = C_{1m}^* r^m + C_{2m}^* r^{2+m} + C_{3m}^* r^{-m} + C_{4m}^* r^{2-m} \quad ; \quad (m > 1)$$

$$(8.1351)$$

where $C_{10}^*, C_{20}^*, ..., C_{4m}^*$ are arbitrary constants.

Since the deflection of the plate, its slope and flexural moments at the centre of the plate are finite, we require

$$C_{30}^* = C_{40}^* = 0$$

$$C_{31}^* = C_{41}^* = 0$$

$$.... = ...$$

$$.... = ...$$

$$C_{3m}^* = C_{4m}^* = 0$$

$$(8.1352)$$

Also due to the assumed form of the dependency of $w^{(i)}(r,\theta)$ and $w^{(e)}(r,\theta)$ on θ, the derivatives with respect to θ are continuous at $r = \xi$. The six remaining boundary conditions can be used to determine the six sets of constants $C_{1m}^*, C_{2m}^*, C_{1m}, ..., C_{4m}$. Considering the inhomogeneous boundary condition (8.1347), since the displacements are expressed in terms of Fourier expansions, it is necessary to represent the concentrated force also in terms of a Fourier series of the form

$$p(\theta) = \frac{A_0}{2} + \sum_{m=1}^{\infty} A_m \cos m\theta \tag{8.1353}$$

The function $p(\theta)$ can be visualized as the concentrated force P_0 which is symmetrically distributed over a distance $2(\Delta\chi)$ along the circumference of $r = \xi$ with periodicity 2π (Figure 8.80).

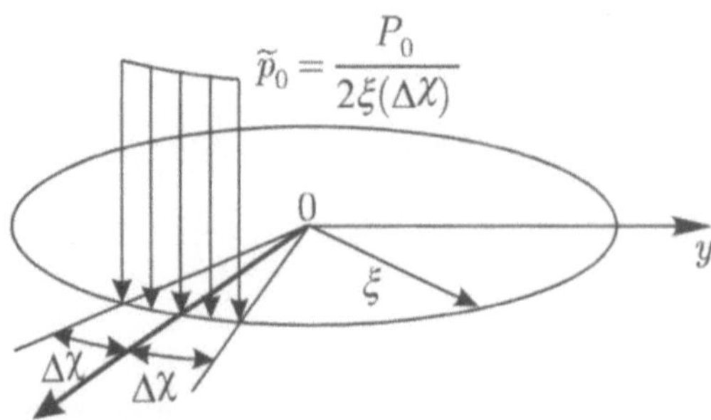

Figure 8.80: Equivalent periodic loading $p(\theta)$.

For the equivalent distributed loading

$$\widetilde{p}_0 = \frac{P_0}{2\xi(\Delta\chi)} \tag{8.1354}$$

The coefficients of the Fourier representation (8.1353) are obtained from the following:

$$A_0 = \frac{2}{\pi} \int_0^{\pi} \widetilde{p}_0 \; d\theta = \frac{2}{\pi} \int_0^{\Delta\chi} \frac{P_0 \; d\theta}{2\xi(\Delta\chi)} = \frac{P_0}{\pi\xi}$$

$$\tag{8.1355}$$

$$A_m = \frac{2}{\pi} \int_0^{\pi} \widetilde{p}_0 \cos m\theta \; d\theta = \frac{P_0}{\pi\xi} \frac{\sin(m\Delta\chi)}{(m\Delta\chi)}$$

We cam rewrite (8.1353) in the form

$$p(\theta) = \frac{P_0}{\pi\xi} \left[\frac{1}{2} + \sum_{m=1,}^{\infty} \frac{\sin(m\Delta\chi)}{(m\Delta\chi)} \cos(m\theta) \right] \tag{8.1356}$$

In the limit $(m\Delta\chi) \to 0$, (8.1356) gives

$$p(\theta) = \frac{P_0}{\pi\xi} \left[\frac{1}{2} + \sum_{m=1,}^{\infty} \cos(m\theta) \right] \tag{8.1357}$$

The boundary condition (8.1347) can now be written as

$$\left[\frac{\partial}{\partial r}\left\{\overset{\circ}{\nabla}^2 w^{(e)}(r,\theta)-\overset{\circ}{\nabla}^2 w^{(i)}(r,\theta)\right\}\right]_{r=\xi;\theta=0}$$
$$=-\frac{P_0}{\pi\xi D}\left[\frac{1}{2}+\sum_{m=1,}^{\infty}\cos(m\theta)\right] \tag{8.1358}$$

Expressions for the non-zero sets of functions $F_i(r)$ and $F_i^*(r)(i=0,1,...m)$ take the following forms:

$$F_0(r)=-\frac{P_0}{8\pi D}\left[(r^2+\xi^2)\ln\left(\frac{r}{a}\right)+\frac{(a^2+\xi^2)(a^2-r^2)}{2a^2}\right]$$

$$F_0^*(r)=-\frac{P_0}{8\pi D}\left[(r^2+\xi^2)\ln\left(\frac{\xi}{a}\right)+\frac{(a^2+r^2)(a^2-\xi^2)}{2a^2}\right]$$

$$F_1(r)=\frac{P_0\xi^3}{16\pi D}\left[\frac{1}{r}+\frac{2\left(a^2-\xi^2\right)r}{a^2\xi^2}-\frac{r^3\left(2a^2-\xi^2\right)}{a^4\xi^2}-\frac{4r}{\xi^2}\ln\left(\frac{a}{r}\right)\right]$$

$$F_1^*(r)-\frac{P_0\xi^3}{16\pi D}\left[\frac{2r\left(a^2-\xi^2\right)}{a^2\xi^2}+\frac{r^3\left(a^2-\xi^2\right)^2}{a^4\xi^2}-\frac{4r}{\xi^2}\ln\left(\frac{a}{\xi}\right)\right]$$

$$\tag{8.1359}$$

$$F_m(r)=-\frac{P_0\xi^m}{8m(m-1)\pi D}\left\{\frac{1}{r^m}\left(r^2-\frac{(m-1)}{(m+1)}\xi^2\right)\right.$$
$$\left.+\frac{r^m}{a^{2m}}\left[(m-1)\xi^2-ma^2+(m-1)r^2-\frac{m(m-1)}{(m+1)}\frac{\xi^2r^2}{a^2}\right]\right\}$$

$$F_m^*(r)=-\frac{P_0\xi^m}{8m(m-1)\pi D}\left\{\frac{r^m}{a^{2m}}\left[(m-1)\xi^2-ma^2+\frac{a^{2m}}{\xi^{2m-2}}\right]\right.$$
$$\left.+(m-1)\frac{r^{m+2}}{a^{2m}}\left[1-\frac{m}{(m+1)}\frac{\xi^2}{a^2}-\frac{1}{(m+1)}\frac{a^{2m}}{\xi^{2m}}\right]\right\}$$

As is evident, the solution to the problem of the localized loading of the clamped plate, although complete, is obtained in a form which involves the summation of a series. The evaluation of flexural moments and shear forces can be best attempted by using a symbolic mathematical manipulation scheme such as MATHEMATICA®or MAPLE®, particularly in view of the large number of terms in the series that need to be incorporated to accurately evaluate the variation of the moments and shear force in the vicinity of the concentrated load. The results for the flexural moments and shear forces will exhibit singular behaviour at the point of application of the loading, irrespective of the mode of support of the plate. A useful result of practical importance, namely the maximum deflection of the plate at the point of application of the load can be evaluated in exact closed form; i.e.

$$w(\xi, 0) = \frac{P_0}{16\pi D} \frac{(a^2 - \xi^2)^2}{a^2} \tag{8.1360}$$

In the special case when the concentrated load acts at the centre of the plate $\xi = 0$ and (8.1360) reduces to the result that can be obtained from (8.1320).

$$\bullet \ \bullet \ \bullet$$

In a general sense, the solution developed in this example represents the *Green's function* or *influence function* for the clamped plate, which can be utilized to develop solutions for problems involving generalized irregular loading of a clamping plate.

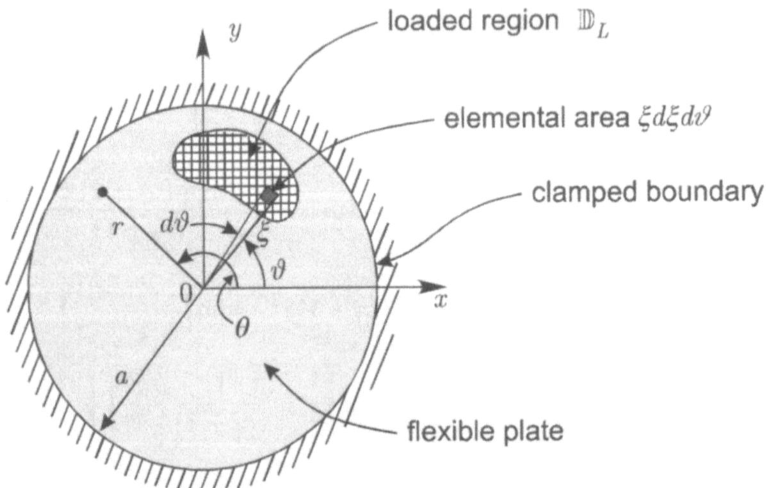

Figure 8.81: Arbitrary loading of a clamped plate.

By replacing θ in the expression for the plate deflection by $(\theta - \vartheta)$ we can obtain a solution to the deflection of the plate at (r, θ) when the concentrated loading acts at the location (ξ, ϑ). We can write this result in the form

$$w(r, \theta) = \frac{P_0 a^2}{8\pi D} W(r, \theta; \xi, \vartheta) \tag{8.1361}$$

where $W(r, \theta; \xi, \vartheta)$ is the series solution obtained previously. Considering the loaded area $\mathbb{D}_L$ shown in Figure 8.81, the deflection of the plate at an arbitrary location due to the load acting on the elemental area is

$$dw(r, \theta) = \frac{a^2 p(\xi, \vartheta) \xi \, d\xi \, d\vartheta}{8\pi D} W(r, \theta; \xi, \vartheta) \tag{8.1362}$$

where $p(\xi, \vartheta)$ is the intensity of loading over the region $\mathbb{D}_L$. The result (8.1362) can be integrated over the loaded region to obtain a general expression for the deflection of the plate; i.e.

$$w(r, \theta) = \frac{a^2}{8\pi D} \int \int_{\mathbb{D}_L} W(r, \theta; \xi, \vartheta) p(\xi, \vartheta) \xi d\xi d\vartheta \tag{8.1363}$$

The expression can be evaluated either numerically or analytically depending upon the nature of $p(\xi, \vartheta)$ and $\mathbb{D}_L$.

8.11.10 Complex variable method for circular plates

In view of the discussion relating to the solution of the biharmonic equation for the plane problem in classical elasticity by the use of complex variable methods it seems natural to enquire whether such a procedure extends to the solution of circular plate problems. Michell (1863-1940) applied a method based on a complex variable formulation to obtain the solution to the plate with a fixed clamped boundary. The solution was extended by Reissner (1913-1996) to include the case of a circular plate which is simply supported along the boundary, rotationally constrained along the circular boundary and subjected to a concentrated load at an arbitrary location.

In this section we shall present the problem of the localized loading of a clamped plate which is examined by the method of inversion proposed by Michell. Although the method is somewhat specialized in nature it results in the development of solutions in a compact form which is useful from the point of view of applications to a more generalized class of loading situations. The geometry, the loading and boundary conditions for the plate problem are as shown in Figure 8.79. The inversion technique employs a conformal transform defined by

$$zz' = k^2 \tag{8.1364}$$

where

$$z = x + iy = re^{i\theta} \quad ; \quad z' = x' + iy' = r'e^{i\theta'} \tag{8.1365}$$

and k is a real parameter referred to as the *radius of inversion*. The transformation consists of the inversion of the modulus $r' = k^2/r$ and a reflection about the real axis $\theta' = -\theta$. Additional properties of the transformation can be summarized. (i) If a function w is biharmonic in the $r - \theta$ plane then $(r')^2 w$ is biharmonic in the $r' - \theta'$ plane. (ii) The derivatives of the biharmonic function are given by

$$\frac{\partial}{\partial r'}\left\{(r')^2 w\right\} = \frac{2a^2}{r} w - a^2 \frac{\partial w}{\partial r}$$

$$\frac{1}{r'}\frac{\partial}{\partial \theta'}\left\{(r')^2 w\right\} = \frac{a^2}{r}\frac{\partial w}{\partial \theta} \tag{8.1366}$$

Hence for a clamped plate, where generally along the boundary the conditions

$$w = \frac{\partial w}{\partial r} = \frac{1}{r}\frac{\partial w}{\partial \theta} = 0 \tag{8.1367}$$

transforms into clamped boundary conditions in the $r'-\theta'$ plane. (iii) Circles in the $r-\theta$ plane will transform into circles in the $r'-\theta'$ plane. In particular, a circle with radius a and with its centre on the real axis at a distance h from the origin (Figure 8.82) will transform into a circle with

$$h' = \frac{k^2 h}{(h^2 - a^2)} \quad ; \quad a' = \frac{k^2 a}{(h^2 - a^2)} \tag{8.1368}$$

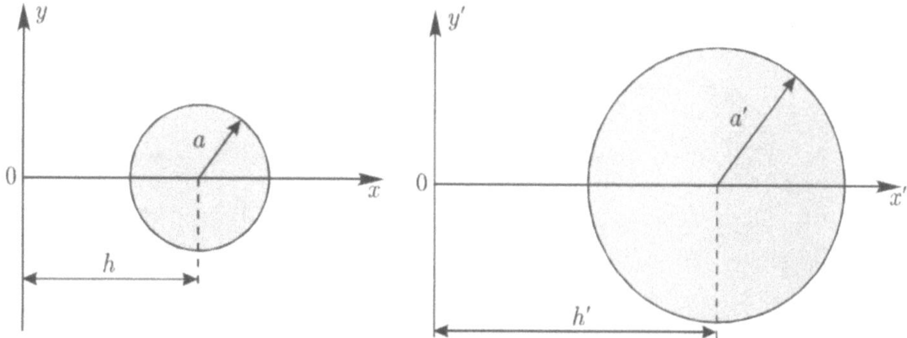

Figure 8.82: Transformation parameters for the method of inversion.

If $k^2 = h^2 - a^2$, the circle transforms into itself, and a circle which passes through the origin $(h = a)$ transforms into a straight line.

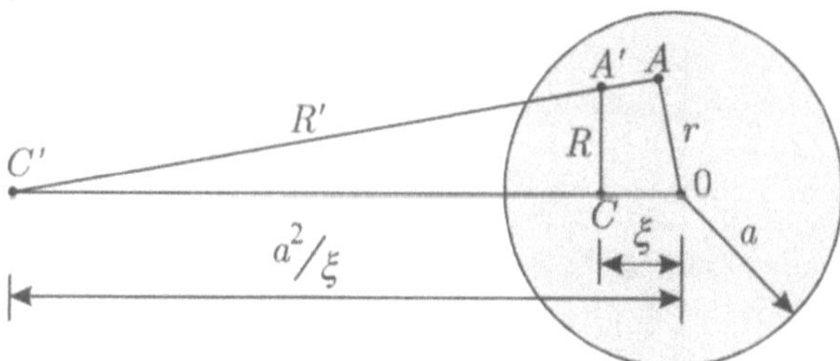

Figure 8.83: The inversion of a circle.

We shall utilize the inversion procedure to obtain the deflected mid-surface of a circular plate which is clamped on the boundary $r = a$ and subjected to a concentrated load P_0 at the location C (Figure 8.83). If we consider C' as the inverse of the point C with respect to a circle of radius a and ξ the distance between O and C. We start with a solution to the problem of a clamped circular plate which is subjected to a central load P_0 (see e.g. (8.1320))

$$w(r) = \frac{P_0 a^2}{16 \pi D} \left[\left(1 - \frac{r^2}{a^2} \right) + 2 \frac{r^2}{a^2} \ln \left(\frac{r}{a} \right) \right] \tag{8.1369}$$

where r is measured from O. We now apply the inversion procedure with
the constant of inversion.

$$k^2 = \frac{a^4}{\left(\xi^2 - a^2\right)} \tag{8.1370}$$

where the circle inverts into itself and the points O and C invert to C and
C' respectively. We denote

$$CA' = R \quad ; \quad C'A' = R' \tag{8.1371}$$

and from geometry (Figure 8.83)

$$\frac{R}{r} = \frac{R'}{a^2/\xi} \quad \text{or} \quad \frac{r}{a} = \left(\frac{a}{\xi}\right)\left(\frac{R}{R'}\right) \tag{8.1372}$$

Using these relationships we can rewrite (8.1369) in the form

$$(R')^2 w(r) = \frac{P_0 a^2}{16\pi D}\left[\left\{(R')^2 - \frac{a^2 R^2}{\xi^2}\right\} + 2\frac{a^2}{\xi^2}R^2 \ln\left(\frac{aR}{\xi R'}\right)\right] \tag{8.1373}$$

The function $(R')^2 w(r)$ is biharmonic and the boundary a is again clamped.
Hence if we set

$$w'(r) = \frac{(R')^2 w(r)}{a^2} \quad ; \quad P_0' = \frac{P_0 a^2}{\xi^2} \tag{8.1374}$$

we can rewrite (8.1373) as

$$w'(r) = \frac{P_0'}{16\pi D}\left[\left\{\frac{\xi^2 (R')^2}{a^2} - R^2\right\} + 2R^2 \ln\left(\frac{a}{\xi}\frac{R}{R'}\right)\right] \tag{8.1375}$$

which represents the mid-plane deflection of a clamped plate of radius a
under a concentrated load of magnitude P_0' applied at $r = \xi$.

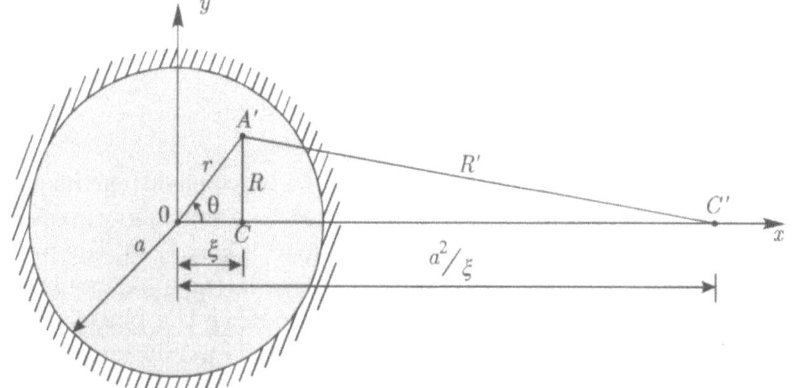

Figure 8.84: Localized loading of a clamped plate.

From the geómetry shown in Figure 8.84, we have

$$R^2 = r^2 + \xi^2 - 2r\xi\cos\theta \quad ; \quad (R')^2 = r^2 + \left(\frac{a^2}{\xi}\right)^2 - \frac{2ra^2}{\xi}\cos\theta \tag{8.1376}$$

Hence

$$\frac{\xi^2}{a^2}(R')^2 = R^2 = \frac{(a^2 - r^2)(a^2 - \xi^2)}{a^2} \tag{8.1377}$$

and the expression (8.1375) can be rewritten as

$$w'(r,\theta) = \frac{P_0'a^2}{16\pi D}\left[(1 - \bar{r}^2)(1 - \bar{\xi}^2)\right.$$

$$\left. + (\bar{r}^2 + \bar{\xi}^2 - 2\bar{r}\bar{\xi}\cos\theta)\ln\left\{\frac{\bar{r}^2 + \bar{\xi}^2 - 2\bar{r}\bar{\xi}\cos\theta}{1 + \bar{r}^2\bar{\xi}^2 - 2\bar{r}\bar{\xi}\cos\theta}\right\}\right] \tag{8.1378}$$

where

$$\bar{r} = \frac{r}{a} \quad ; \quad \bar{\xi} = \frac{\xi}{a} \tag{8.1379}$$

The result (8.1378) is valid throughout the plate region $0 \leq \theta \leq 2\pi$ and gives the exact result (8.1360) when $r = \xi$ and $\theta = 0$. Also, this result is a convenient form which can be used in the influence function technique summarized by (8.1361) to (8.1363).

8.11.11   Green's function for a thin plate

The Green's function for a thin plate is derived by considering the deflection of a plate of infinite extent, which is subjected to a unit concentrated force. Since the plate is of infinite extent and the load is localized, the problem is axisymmetric and it can be formulated in relation to the biharmonic equation where Laplace's operator is axisymmetric. Also since the plate is of infinite extent, no boundary conditions can be specified. The only known condition is that at any radius r from the point of application of the load, the shear force will be $1/2\pi r$. In Example 8.29, we determined the deflection of a plate of radius a which was simply supported at $r = a$, and subjected to a central load P_0. Setting $P_0 = 1$, this result can be rewritten as

$$w(r) = \frac{1}{8\pi D}r^2 \ln r$$
$$+\frac{1}{16\pi D}\left\{\left(\frac{3+\nu}{1+\nu}\right)a^2\left(1-\frac{r^2}{a^2}\right)-2r^2\ln a\right\} \tag{8.1380}$$

Therefore the only part of the solution relevant to a plate of infinite extent which is subjected to a concentrated force (at $r = 0$) is the first term of (8.1380) i.e.

$$w^G(r) = \frac{1}{8\pi D}r^2\ln r \tag{8.1381}$$

As discussed in the Example 8.32, Green's functions or influence functions can also be defined for particular situations involving plates of finite extent which are supported in a prescribed fashion. In such situations, the behaviour of the plate deflection, the flexural moments and shear force in the vicinity of the concentrated force should correspond to those that will be obtained by using the result (8.1381).

The result (8.1381) for the Green's function can be obtained in a variety of ways; in view of the axial symmetry of the problem and the infinite extent nature of the plate, it is also convenient to employ a Hankel integral transform technique. For the application of the Hankel transform technique we first consider a discontinuous uniform external load $p(r)$ such that

$$p(r) = \begin{cases} p_0 & 0 < r < a \\[2mm] 0 & a < r < \infty \end{cases} \tag{8.1382}$$

The zeroth-order Hankel transform of $p(r)$ is given by

$$\bar{p}^0(\xi) = \int_0^\infty r p(r) J_0(\xi r) dr \tag{8.1383}$$

Substituting (8.1382) in (8.1383) and performing the integration we have

$$\bar{p}^0(\xi) = p_0 \frac{a}{\xi} J_1(\xi a) \tag{8.1384}$$

We can recover the case of the concentrated *unit force* acting at the origin by setting $p_0 = 1/\pi a^2$ and taking the limit of (8.1384) as $a \to 0$ i.e.

$$\bar{p}_c^0(\xi) = \frac{1}{\pi} \operatorname*{Lim}_{a \to 0} \left(\frac{J_1(\xi a)}{\xi a} \right) = \frac{1}{2\pi} \tag{8.1385}$$

Hence for the concentrated force $p(r)$ must be of the form

$$p(r) = \frac{\delta(r)}{2\pi r} \tag{8.1386}$$

where $\delta(r)$ denotes the Dirac delta function. The Green's function is formally the particular integral of the partial differential equation

$$D \overset{\circ}{\nabla}^2 \overset{\circ}{\nabla}^2 w(r) = \frac{\delta(r)}{2\pi r} \tag{8.1387}$$

where $\overset{\circ}{\nabla}^2$ is the axisymmetric form of Laplace's operator. Operating on (8.1387) with the zeroth-order Hankel transform we have

$$D \xi^4 \bar{w}^0(\xi) = \frac{1}{2\pi} \tag{8.1388}$$

Inverting (8.1388) we obtain

$$w(r) = \frac{1}{2\pi D} \int_0^\infty \frac{J_0(\xi r)d\xi}{\xi^3} \tag{8.1389}$$

Since we are interested in the behaviour of this integral for small values of r we can introduce a length parameter a such that $\xi = \rho/a$ and expand $J_1(\rho r/a)$ in the form

$$J_0\left(\rho\frac{r}{a}\right) = 1 - \frac{1}{4}\left(\frac{r}{a}\right)^2 \rho^2 + 0\left(\frac{r^4}{a^4}\right) \tag{8.1390}$$

where $0(\;)$ is the Landau symbol (named after E. Landau (1877-1938). Substituting (8.1390) in (8.1389) and performing the integrations we can show that the deflected shape of the plate in the vicinity of the concentrated unit load, which also maintains the plate deflections finite at $r = 0$ is given by

$$w(r) = \frac{r^2}{8\pi D} \ln\left(\frac{r}{a}\right) \tag{8.1391}$$

The term $r^2 \ln a$ yields negligible displacements and stresses when the behaviour is local and (r/a) remains small.

8.11.12   Flexure of rectangular plates

The object of this and subsequent sections is to outline briefly the various mathematical procedures that can be applied to develop solutions to the class of problems dealing with the flexure of laterally loaded thin plates. As with problems dealing with circular plates, the literature in this area is quite extensive and no attempt will be made to develop solutions for wider class of loading and boundary conditions. The basic procedures presented here, however, are sufficiently general in that other similar problems can be treated as particular exercises. Problems involving rectangular domains are most easily formulated in reference a rectangular Cartesian coordinate system. For completeness and for ease of reference we shall summarize here the relevant equations governing a boundary value problem in classical plate theory. The partial differential equation governing the flexural deflection of the plate is

$$D \overset{\circ}{\nabla}^2 \overset{\circ}{\nabla}^2 w(x,y) = p(x,y) \tag{8.1392}$$

where $\overset{\circ}{\nabla}^2$ is Laplace's operator referred to the rectangular Cartesian coordinate system. Considering the results (8.1272) we can show that

$$M_{xx}(x,y) = -D\left[\frac{\partial^2 w}{\partial x^2} + \nu\frac{\partial^2 w}{\partial y^2}\right]$$

$$M_{yy}(x,y) = -D\left[\frac{\partial^2 w}{\partial y^2} + \nu\frac{\partial^2 w}{\partial x^2}\right] \tag{8.1393}$$

$$M_{xy}(x,y) = -D(1-\nu)\frac{\partial^2 w}{\partial x\partial y}$$

and

$$Q_x(x,y) = -D\frac{\partial}{\partial x}\left(\overset{\circ}{\nabla}^2 w\right) \quad ; \quad Q_y(x,y) = -D\frac{\partial}{\partial y}\left(\overset{\circ}{\nabla}^2 w\right) \tag{8.1394}$$

which form the complete expressions, respectively, for the flexural moment and shear forces in the thin plate. The components of the Kirchhoff shear applicable to a free boundary take the forms

$$V_x(x,y) = Q_x + \frac{\partial M_{xy}}{\partial y} = -D\left[\frac{\partial^3 w}{\partial x^3} + (2-\nu)\frac{\partial^3 w}{\partial x\partial y^2}\right]$$

$$\tag{8.1395}$$

$$V_y(x,y) = Q_x + \frac{\partial M_{xy}}{\partial x} = -D\left[\frac{\partial^3 w}{\partial y^3} + (2-\nu)\frac{\partial^3 w}{\partial y\partial x^2}\right]$$

The boundary conditions applicable to a thin plate can be classified according to the following types:

(i)   for a simply supported boundary $x = a$;

$$w(a,y) = 0 \quad ; \quad M_{xx}(a,y) = 0 \tag{8.1396}$$

(ii) for a fixed boundary $x = a$

$$w(a,y) = 0 \quad ; \quad \left[\frac{\partial w}{\partial x}\right]_{x=a} = 0 \tag{8.1397}$$

(iii) for a free boundary $x = a$

$$M_{xx}(a, y) = 0 \quad ; \quad V_x(a, y) = 0 \tag{8.1398}$$

Similar equations can be written down for the appropriate categories of boundary conditions for other edges of a rectangular plate e.g. $y = b$, etc.

The general solution of the partial differential equation (8.1392), again consists of a particular integral $w^P(x, y)$ and a homogeneous solution $w^H(x, y)$. The wider class of homogeneous solutions can be obtained via techniques similar to those discussed in connection with the development of solution associated with the biharmonic equation for Airy's stress function. For example the power series solutions defined by (8.448) to (8.452) are solutions of

$$\overset{\circ}{\nabla}^2 \overset{\circ}{\nabla}^2 w^H(x, y) = 0 \tag{8.1399}$$

Since they are solutions of the homogeneous equation, the bending of the plate must be maintained by flexural moments and shearing forces that are applied along the boundaries of the plate. Of course, not all such biharmonic solutions defined by (8.448) to (8.452) will lend themselves to realistic distributions of flexural moments and shearing forces which can be applied at the boundaries of a plate.

Example 8.33

A rectangular plate region is subjected to flexural moments M_A and M_B as showing in Figure 8.85. Determine the deflected shape of the plate assuming that the deflection and slope of the plate are zero at the origin.

Solution

Taken individually, the flexural moments M_A and M_B will induce cylindrical bending of the plate. The deflected shape of the plate should therefore be an additive variables separable solution of the form

$$w(x, y) = F(x) + G(y) \tag{8.1400}$$

Substituting (8.1400) into the first two equations of (8.1393) we obtain

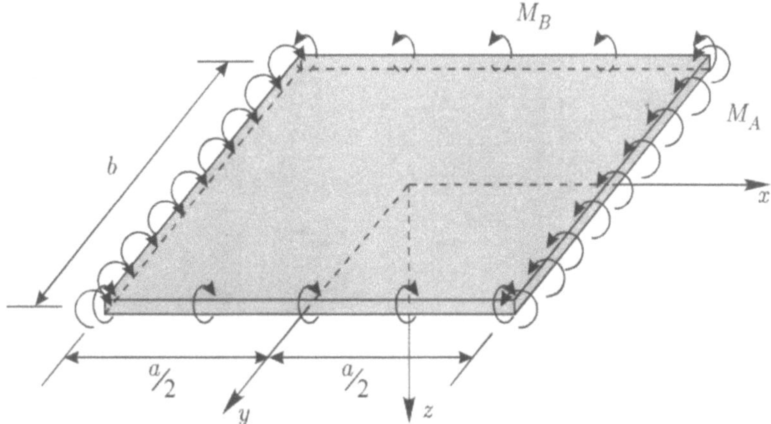

Figure 8.85: Pure bending of a rectangular plate.

$$\frac{d^2 F}{dx^2} + \nu \frac{d^2 G}{dy^2} = -\frac{M_A}{D} \quad ; \quad \frac{d^2 G}{dy^2} + \nu \frac{d^2 F}{dx^2} = -\frac{M_B}{D} \tag{8.1401}$$

Separately eliminating $G(y)$ and $F(x)$ between these equations we obtain, respectively

$$\frac{d^2 F}{dx^2} = -\left[\frac{M_A - \nu M_B}{D(1 - \nu^2)}\right] \quad ; \quad \frac{d^2 G}{dy^2} = -\left[\frac{M_B - \nu M_A}{D(1 - \nu^2)}\right] \tag{8.1402}$$

Integrating these equations we obtain

$$w(x, y) = -\left[\frac{M_A - \nu M_B}{2D(1 - \nu^2)}\right] x^2 - \left[\frac{M_B - \nu M_A}{2D(1 - \nu^2)}\right] y^2$$
$$+ C_1 x + C_2 y + C_3 \tag{8.1403}$$

where C_1, C_2 and C_3 are arbitrary constants. The conditions that can be applied to determine these constants are following:

$$w(0,0) = 0 \quad ; \quad \left[\frac{\partial w}{\partial x}\right]_{x=0, y=0} = 0 \quad ; \quad \left[\frac{\partial w}{\partial y}\right]_{x=0; y=0} = 0 \tag{8.1404}$$

These give $C_1 = C_2 = C_3 = 0$, and the complete solution for the deflected shape of the plate is given by

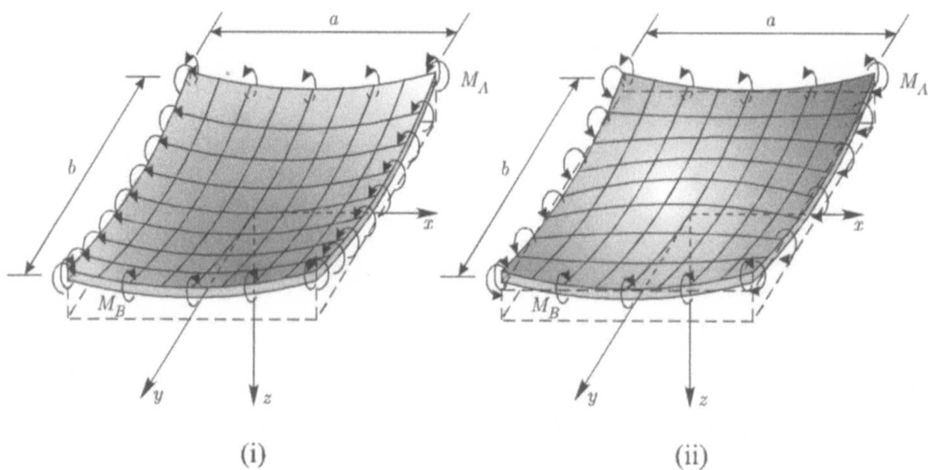

(i)   (ii)

Figure 8.86: Paraboloid and anticlastic bending of a plate under constant flexural moments along the boundary.

$$w(x,y) = -\left[\frac{M_A - \nu M_B}{2D(1-\nu^2)}\right] x^2 - \left[\frac{M_B - \nu M_A}{2D(1-\nu^2)}\right] y^2 \tag{8.1405}$$

Since the flexural moments are uniform, the solution developed here is applicable to any plate of finite dimensions. The problem of pure bending of a plate is exactly the analogue of the problem of uniform biaxial stressing discussed in connection with plane problems in elasticity discussed in this Chapter. The deflections are unbounded if $x, y \to \infty$; the bending moments, however, remain constant. The deflected shape of the plate, which is a *paraboloid of revolution,* is shown in Figure 8.86(i). When the direction of application of M_A is reversed, straight lines which are parallel to the x-axis become parabolic curves, convex downwards. the resulting deformed shape is shown in Figure 8.86(ii).

Example 8.34

A rectangular plate with dimensions $a \times b$ is subjected to twisting moments of constant value along its edges (Figure 8.87). Obtain the deflected shape of the plate by considering the asymmetry of the deflections and by assuming that at the origin the deflections are zero.

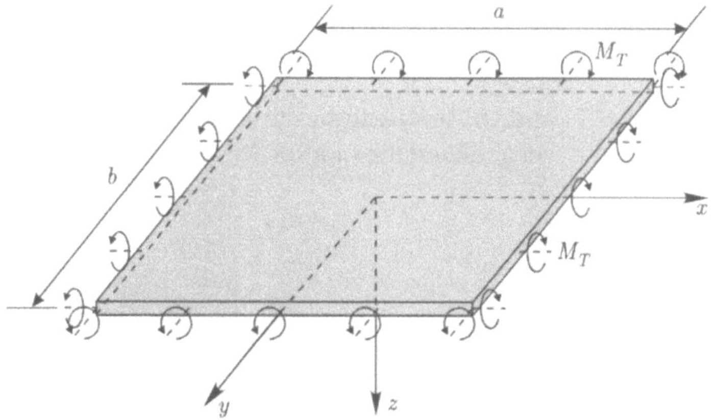

Figure 8.87: Pure twisting of a rectangular plate.

Solution

Since there are no transverse loads on the plate, the governing partial differential equation is

$$\overset{\circ}{\nabla}{}^2\, \overset{\circ}{\nabla}{}^2 w(x,y) = 0 \quad ; \quad x \in \left(-\frac{a}{2}, \frac{a}{2}\right) \quad ; \quad y \in \left(-\frac{b}{2}, \frac{b}{2}\right) \tag{8.1406}$$

which is subject to the imposed condition

$$w(0,0) = 0 \tag{8.1407}$$

and due to asymmetry of the deformation

$$w(x,0) = 0 \quad ; \quad w(y,0) = 0 \tag{8.1408}$$

On the boundaries, the moments should correspond to a constant twisting
moment M_T. Since the flexural moments M_{xx} and M_{yy} are zero at the boundaries of the plate we may assume, from (8.1393), that

$$\frac{\partial^2 w}{\partial x^2} = 0 \quad ; \quad \frac{\partial^2 w}{\partial y^2} = 0 \quad ; \quad \frac{\partial^2 w}{\partial x \partial y} = -\frac{M_T}{D(1-\nu)} \tag{8.1409}$$

Integrating these equations we can write

$$w(x,y) = -\frac{M_T xy}{D(1-\nu)} + xf(y) + yg(x) + h(x) + k(y) \qquad (8.1410)$$

The conditions (8.1407) and (8.1408) will be satisfied only if $g(x), f(y), h(x)$ and $k(y)$ are identically zero. The deflection of the plate is given by

$$w(x,y) = -\frac{M_T xy}{D(1-\nu)} \qquad (8.1411)$$

Therefore the application of a constant distribution of twisting moments along the boundaries of a plate, which maintains equilibrium, produces *anticlastic* bending where the plate deforms into a *hyperbolic paraboloid* surface. It can also be shown that this mode of pure twisting also occurs in a plate region during the application of constant flexural moments M_A and M_B of equal magnitude but opposite sense. If we set $M_{xx} = -M_0$ and $M_{yy} = M_0$ in the expression (8.1405) we obtain

$$w(x,y) = \frac{M_0}{2D(1-\nu)}(x^2 - y^2) \qquad (8.1412)$$

For ease of presentation we shall restrict attention to a *square plate region* and consider a rotation of the in-plane coordinate system as shown in Figure 8.87.

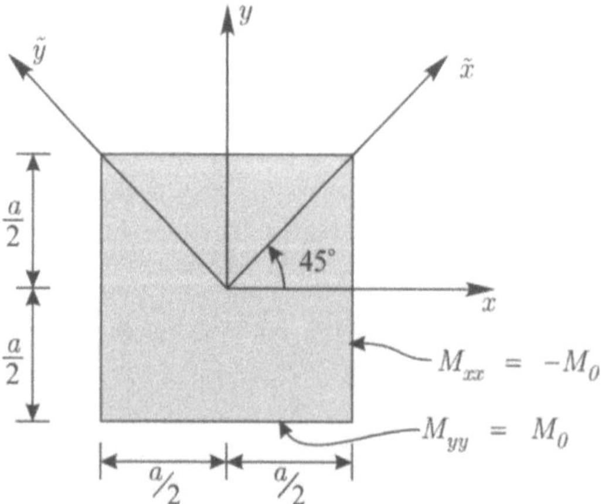

Figure 8.88: Rotation of coordinates.

Considering the rotation of coordinates we have

$$x = \widetilde{x}\cos 45^\circ - \widetilde{y}\sin 45^\circ = \frac{1}{\sqrt{2}}\left(\widetilde{x} - \widetilde{y}\right)$$

$$(8.1413)$$

$$y = \widetilde{x}\sin 45^\circ + \widetilde{y}\cos 45^\circ = \frac{1}{\sqrt{2}}\left(\widetilde{x} + \widetilde{y}\right)$$

Substituting (8.1413) in (8.1412) we obtain

$$w(\widetilde{x}, \widetilde{y}) = -\frac{M_0\widetilde{x}\widetilde{y}}{D(1-\nu)} \qquad (8.1414)$$

Therefore, anticlastic bending of the plate in relation to the $x-y$ axes induces a state of pure twisting with respect to the $\widetilde{x}, \widetilde{y}$ system of axes, obtained by a rotation of the original coordinates system by 45° in the counter-clockwise sense. This is analogous to the state of *pure shear* that would develop in *two-dimensional plane elements* which are subjected to principal stresses of equal magnitude but opposite sense.

8.11.13   Certain general solutions of the biharmonic equation

We now consider the variables separation technique to develop specific forms of solutions of the biharmonic equation. Unlike the second-order partial differential equations discussed in the previous Chapters, the biharmonic equation does not lend itself to a direct variables separation procedure. It will be shown that the separation procedure has to be applied repeatedly to extract solutions in a variables separable form. As a result, there are no assurances that the procedure will yield a complete set of solutions. We assume that $w(x,y)$ admits a variables separable solution of the form

$$w(x,y) = X(x)Y(y) \qquad (8.1415)$$

Substituting this in homogeneous biharmonic equation for $w(x,y)$ (see e.g. (8.1399))we obtain

$$Y\frac{d^4 X}{dx^4} + 2\frac{d^2 X}{dx^2}\frac{d^2 Y}{dy^2} + X\frac{d^4 Y}{dy^4} = 0 \qquad (8.1416)$$

Dividing throughout by XY we have

$$\frac{1}{X}\frac{d^4X}{dx^4} + 2\left(\frac{1}{X}\frac{d^2X}{dx^2}\right)\left(\frac{1}{Y}\frac{d^2Y}{dy^2}\right) + \frac{1}{Y}\frac{d^4Y}{dy^4} = 0 \tag{8.1417}$$

Therefore, there is no *direct* separation of the variables. We can, however, repeat the operation by introducing new dependent variables

$$F(x) = \frac{1}{X}\frac{d^4X}{dx^4} \quad ; \quad G(x) = \frac{1}{X}\frac{d^2X}{dx^2}$$

$$ \tag{8.1418}$$

$$H(y) = \frac{1}{Y}\frac{d^4Y}{dy^4} \quad ; \quad K(y) = \frac{1}{Y}\frac{d^2Y}{dy^2}$$

The equation (8.1417) can now be written as

$$F(x) + 2G(x)K(y) + H(y) = 0 \tag{8.1419}$$

Differentiating this equation with respect to x gives

$$\frac{dF}{dx} + 2K(y)\frac{dG}{dx} = 0 \tag{8.1420}$$

This equation is satisfied provided

$$\frac{dF/dx}{dG/dx} = -2K(y) \quad \text{or} \quad K(y) = -\lambda^2 = \text{const.} \tag{8.1421}$$

Similarly, by differentiating (8.1419) with respect to y we obtain

$$\frac{dH/dy}{dK/dy} = -2G(x) \quad \text{or} \quad G(x) = -\gamma^2 = \text{const.} \tag{8.1422}$$

We can now consider a class of variables separable solutions of the biharmonic equation depending upon whether

$$\lambda^2 > 0 \quad ; \quad \lambda^2 = 0 \quad ; \quad \lambda^2 < 0 \tag{8.1423}$$

and

$$\gamma^2 > 0 \quad ; \quad \gamma^2 = 0 \quad ; \quad \gamma^2 < 0 \tag{8.1424}$$

(i) If $\lambda^2 < 0$, then

$$\frac{d^2Y}{dy^2} + \lambda^2 Y = 0 \tag{8.1425}$$

which has solutions

$$Y(y) = A_0 \sin(\lambda y) + B_0 \cos(\lambda y) \tag{8.1426}$$

where A_0 and B_0 are arbitrary constants.

Using (8.1425) we can reduce (8.1417) to the following forms:

$$\frac{d^4X}{dx^4} - 2\lambda^2 \frac{d^2X}{dx^2} + \lambda^4 X = 0 \quad ; \quad (\lambda^2 > 0) \tag{8.1427}$$

The solution of the 4th order ordinary differential equation (8.1427)can be written in the form

$$X(x) = A_1 e^{\lambda x} + B_1 e^{-\lambda x} + C_1 x e^{\lambda x} + D_1 x e^{-\lambda x} \tag{8.1428}$$

where A_1, B_1, etc. are arbitrary constants. Therefore combining (8.1426) and (8.1428) we obtain the following set of solutions of the biharmonic equation for $w(x, y)$

$$w(x,y) = \left\{ A' e^{\lambda x} + B' e^{-\lambda x} + C' x e^{\lambda x} + D' x e^{-\lambda x} \right\} \sin(\lambda y)$$
$$+ \left\{ A^* e^{\lambda x} + B^* e^{-\lambda x} + C^* x e^{\lambda x} + D^* x e^{-\lambda x} \right\} \cos(\lambda y) \tag{8.1429}$$

where A', A^*, B', B^*, etc., are arbitrary constants, and with the constraint that $K(y) = -\lambda^2$ and $\lambda^2 > 0$. The result (8.1429) is identical to (8.536) and (8.537), obtained in connection with the solution of the biharmonic equation for the Airy stress function but, admittedly, by assuming at the outset the form of the stress function with an explicit dependency in y of the form $\sin(\lambda y)$ and $\cos(\lambda y)$ respectively.

(ii) If $\lambda^2 = 0$, then

$$\frac{d^2 Y}{dy^2} = 0 \quad \text{or} \quad Y(y) = A_0 + B_0 y \tag{8.1430}$$

where A_0 and B_0 are constants. Invoking the assumption $\lambda^2 = 0$, (8.1427) reduces to

$$\frac{d^4 X}{dx^4} = 0 \quad \text{or} \quad X(x) = A_1 x^3 + B_1 x^2 + C_1 x + D_1 \tag{8.1431}$$

where A_1, B_1, C_1 and D_1 are constants. The corresponding product form of the solution for $w(x, y)$ is given by

$$w(x, y) = \big\{ A_{31} x^3 y + A_{21} x^2 y + A_{30} x^3 + A_{11} xy$$
$$+ A_{10} x + A_{01} y + A_{00} \big\} \tag{8.1432}$$

where A_{ij} are constants. This polynomial solution up to one third degree in x forms a subset of the general polynomial solution of the Airy stress function given in (8.448).

(iii) If $\lambda^2 < 0$, then λ is imaginary and we have

$$\frac{d^2 Y}{dy^2} - (\lambda^*)^2 Y = 0 \quad ; \quad \lambda^* = -i\lambda > 0 \tag{8.1433}$$

where λ^* is real. The solutions of (8.1433) can be written as

$$Y(y) = A_0 \sinh(\lambda^* y) + B_0 \cosh(\lambda^* y) \tag{8.1434}$$

where A_0 and B_0 are constants. Again, using (8.1433), we can reduce (8.1417) to the following form:

$$\frac{d^4 X}{dx^2} + 2(\lambda^*)^2 \frac{d^2 X}{dx^2} + (\lambda^*)^4 X = 0 \tag{8.1435}$$

Solutions of (8.1435) (obtained by considering a form $X(x) = e^{\alpha x}$ with $\alpha = \pm i\lambda^*, \pm i\lambda^*$) can be written as

$$X(x) = \{A_1 \sin(\lambda^* x) + B_1 \cos(\lambda^* x)$$
$$+ C_1 x \sin(\lambda^* x) + D_1 x \cos(\lambda^* x)\} \tag{8.1436}$$

Combining (8.1434) and (8.1436) we can write the solution (8.1415) in the form

$$w(x, y) = \{A' \sin(\lambda^* x) + B' \cos(\lambda^* x))$$
$$+ C' x \sin(\lambda^* x) + D' x \cos(\lambda^* x)\} \sinh(\lambda^* y)$$
$$+ \left\{ \widetilde{A} \sin(\lambda^* x) + \widetilde{B} \cos(\lambda^* x) \right.$$
$$\left. + \widetilde{C} x \sin(\lambda^* x) + \widetilde{D} x \cos(\lambda^* x) \right\} \sinh(\lambda^* y) \tag{8.1437}$$

where $A', \widetilde{A}, B', \widetilde{B}$, etc. are real constants. if we replace λ^* by $-i\lambda$, consider the trigonometric identities

$$\cos(i\lambda x) = \cosh(\lambda x) = \frac{1}{2}\left(e^{\lambda x} + e^{-\lambda x}\right)$$

$$\sin(i\lambda x) = i\sinh(\lambda x) = \frac{i}{2}(e^{\lambda x} - e^{-\lambda x}) \tag{8.1438}$$

$$\cosh(i\lambda y) = \cos(\lambda y) \quad ; \quad \sinh(i\lambda y) = i\sin(\lambda y)$$

and assume $w(x, y)$ to be the *real part* of the resulting expression for (8.1437) we obtain

$$w(x, y) = \left\{ A e^{\lambda x} + B e^{-\lambda x} + C x e^{\lambda x} + D x e^{-\lambda x} \right\} \sin(\lambda y)$$
$$+ \left\{ \widehat{A} e^{\lambda x} + \widehat{B} e^{-\lambda x} + \widehat{C} x e^{\lambda x} + \widehat{D} x e^{-\lambda x} \right\} \cos(\lambda y) \tag{8.1439}$$

where $A', \widehat{A}, B', \widehat{B}$, etc. are constants; this result has the form of (8.1429) obtained by assuming at the outset $\lambda^2 > 0$.

We can use procedures similar to those presented in the preceding equations, to develop a class of variables separable solutions applicable for the case when

$G(x) = -\gamma^2$. In this instance it is sufficient to note the final forms of the solutions:

(i) for $\gamma^2 > 0$

$$w(x,y) = \left\{ A'e^{\gamma y} + B'e^{-\gamma y} + C'ye^{\gamma y} + D'ye^{-\gamma y} \right\} \sin(\gamma x)$$
$$+ \left\{ A^*e^{\gamma y} + B^*e^{-\gamma y} + C^*ye^{\gamma y} + D^*ye^{-\gamma y} \right\} \cos(\gamma x) \quad (8.1440)$$

where A', A^*, etc. are constants.

(ii) for $\gamma^2 = 0$

$$w(x,y) = \left\{ A_{13}xy^3 + A_{12}xy^2 + A_{03}y^3 + A_{11}xy \right.$$
$$\left. + A_{10}x + A_{01}y + A_{00} \right\} \quad (8.1441)$$

where A_{ij} are constants.

(iii) for $\gamma^2 < 0$ with $\gamma^* = -i\gamma$

$$w(x,y) = \left\{ A' \sin(\lambda^* y) + B' \cos(\gamma^* y) \right.$$
$$+ C'y \sin(\gamma^* y) + D'y \cos(\gamma^* y) \Big\} \sinh(\gamma^* x)$$
$$+ \left\{ \tilde{A} \sin(\gamma^* y) + \tilde{B} \cos(\gamma^* y) \right.$$
$$\left. + \tilde{C}y \sin(\gamma^* y) + \tilde{D}y \cos(\gamma^* y) \right\} \cosh(\gamma^* x) \quad (8.1442)$$

where A', $\tilde{A}$, etc. are constants (the constants occurring in equations (8.1426) to (8.1442) are generic and should be considered to be distinct when combinations of these solutions are used to solve a particular problem.

Of the categories of solutions described previously, the sets of solutions where (i) either $\lambda^2 = 0$, $\gamma^2 = 0$ or (ii) $\lambda^2 < 0, \gamma^2 < 0$, have limited general applicability. In (i) the plate deflections are maintained by boundary flexural moments and twisting moments which have prescribed variations which are very specific and perhaps somewhat artificial. In (ii), the plate deflections have a complex form which needs to be further reduced to obtain the real component. The class of solutions where $\lambda^2 > 0$ and $\gamma^2 > 0$ have the correct form and adaptability for the solution of problems related to flexure of rectangular plates. A general form of the series expression for the plate deflection which satisfies the homogeneous form of the partial differential equation for bending of a thin plate can be written as

$$
\begin{aligned}
w(x, y) = \sum_{m=1}^{\infty} & \left\{ A_m e^{\lambda_m x} + B_m e^{-\lambda_m x} + C_m x e^{\lambda_m x} \right. \\
& \left. + D_m x e^{-\lambda_m x} \right\} \sin(\lambda_m y) \\
+ \sum_{m=1}^{\infty} & \left\{ A_m^* e^{\lambda_m x} + B_m^* e^{-\lambda_m x} + C_m^* x e^{\lambda_m x} \right. \\
& \left. + D_m^* x e^{-\lambda_m x} \right\} \cos(\lambda_m y)
\end{aligned}
\tag{8.1443}
$$

where $A_m, B_m....D_m^*$ are either constants or functions of λ_m. A second solution is obtained by interchanging x and y in (8.1443). Examples of the application of these solutions to rectangular plate problems will be discussed in the ensuing sections.

8.11.14 Navier and Levy solutions for rectangular plates

The method of solution developed by Navier (1785-1836) and Levy (1838-1910) feature prominently in the literature dealing with the flexure of rectangular plates. Their approaches are distinct and for this reason the methods merit a complete exposition.

Navier's solution for the analysis of a *simply supported* rectangular plates is of historical interest, in the sense that it represents one of the earliest applications of the *double trigonometric Fourier series* of the solution of plate bending problems. The method assumes , *a priori*, that such a series solution can be developed to satisfy any variation of $p(x, y)$.

The boundary value problem for the simply supported plate can be posed as follows; the partial differential equation governing flexure of the transversely loaded simply supported plate is

$$
D \overset{\circ}{\nabla}^2 \overset{\circ}{\nabla}^2 w(x.y) = p(x, y) \quad ; \quad x \in (0, a) \quad ; \quad y \in (0, b)
\tag{8.1444}
$$

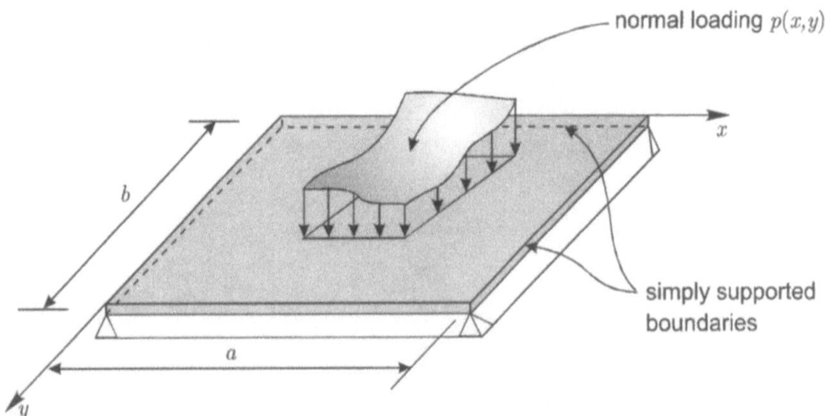

Figure 8.89: Rectangular plate simply supported along four edges.

which is subject to boundary conditions (see e.g. (8.1396)) applicable to simply supported edges:

$$w(0, y) = w(a, y) = 0$$
$$w(x, 0) = w(x, b) = 0$$
$$M_{xx}(0, y) = M_{xx}(a, y) = 0 \tag{8.1445}$$
$$M_{yy}(x, 0) = M_{yy}(x, b) = 0$$

Navier's solution assumes that the applied loading $p(x, y)$ and the deflection of the simply supported plate can be expanded as double *sine series* of the form

$$p(x, y) = \sum_{m=1,2,} \sum_{n=1,2,} p_{mn} \sin\left(\frac{m\pi x}{a}\right) \sin\left(\frac{n\pi y}{b}\right) \tag{8.1446}$$

$$w(x, y) = \sum_{m=1,2,} \sum_{n=1,2,} W_{mn} \sin\left(\frac{m\pi x}{a}\right) \sin\left(\frac{n\pi y}{b}\right) \tag{8.1447}$$

Substituting these representations in the governing partial differential equation we obtain

$$W_{mn} = \frac{p_{mn}}{D\pi^4 \left[\left(\frac{m}{a}\right)^2 + \left(\frac{n}{b}\right)^2\right]^2} \tag{8.1448}$$

The expression for the deflection of the plate can now be written as

$$w(x,y) = \frac{1}{D\pi^4} \sum_{m=1,2,} \sum_{n=1,2,} \frac{p_{mn} \sin\left(\frac{m\pi x}{a}\right) \sin\left(\frac{n\pi y}{b}\right)}{\left[\left(\frac{m}{a}\right)^2 + \left(\frac{n}{b}\right)^2\right]^2} \tag{8.1449}$$

and the corresponding expressions for the flexural and twisting moments take the forms

$$M_{xx} = \frac{1}{\pi^2} \sum_{m=1,2,} \sum_{n=1,2,} \frac{\left[\left(\frac{m}{a}\right)^2 + \nu\left(\frac{n}{b}\right)^2\right] p_{mn} \sin\left(\frac{m\pi x}{a}\right) \sin\left(\frac{n\pi y}{b}\right)}{\left[\left(\frac{m}{a}\right)^2 + \left(\frac{n}{b}\right)^2\right]^2}$$

$$M_{yy} = \frac{1}{\pi^4} \sum_{m=1,2,} \sum_{n=1,2,} \frac{\left[\left(\frac{n}{b}\right)^2 + \nu\left(\frac{m}{a}\right)^2\right] p_{mn} \sin\left(\frac{m\pi x}{a}\right) \sin\left(\frac{n\pi y}{b}\right)}{\left[\left(\frac{m}{a}\right)^2 + \left(\frac{n}{b}\right)^2\right]^2}$$

$$\tag{8.1450}$$

$$M_{xy} = \frac{(1+\nu)}{\pi^2} \sum_{m=1,2,} \sum_{n=1,2,} \frac{mn\, p_{mn} \cos\left(\frac{m\pi x}{a}\right) \cos\left(\frac{n\pi y}{b}\right)}{ab\left[\left(\frac{m}{a}\right)^2 + \left(\frac{n}{b}\right)^2\right]^2}$$

The expressions for the shear forces are given by

$$Q_x = \frac{1}{\pi} \sum_{m=1,2,} \sum_{n=1,2,} \frac{m p_{mn}}{a\left[\left(\frac{m}{a}\right)^2 + \left(\frac{n}{b}\right)^2\right]} \cos\left(\frac{m\pi x}{a}\right) \sin\left(\frac{n\pi y}{b}\right)$$

$$\tag{8.1451}$$

$$Q_x = \frac{1}{\pi} \sum_{m=1,2,} \sum_{n=1,2,} \frac{n p_{mn}}{b\left[\left(\frac{m}{a}\right)^2 + \left(\frac{n}{b}\right)^2\right]} \sin\left(\frac{m\pi x}{a}\right) \cos\left(\frac{n\pi y}{b}\right)$$

It may be noted that although the bending moments M_{xx} and M_{yy} are zero along the respective edges, the twisting moment, however, is non-zero along the edges of the plate and at the corners of the plate. This results in the alteration of the distribution of the reactions on the supports. The effects of these, by virtue of the principle of Saint-Venant (1797-1886), are felt only in the vicinity of the supports. In this type of simply supported plate it is assumed that the plate is adequately prevented from experiencing uplift or separation along the supports and at the corners. Such displacement constraints are necessary to generate the necessary corner reactions (see e.g.

(8.1287)). Navier's method is quite general in character, in that there are no constraints placed on the nature of $p(x, y)$. The external loading can display symmetry or asymmetry about the axes of symmetry of the plate.

Example 8.35

A rectangular plate is simply supported along its edges and subjected to a uniform loading over a finite rectangular area (Figure 8.90). Use Navier's method to develop an expression for the deflection of the plate and use this result to obtain the influence function or Green's function for a plate which is simply supported along its boundaries.

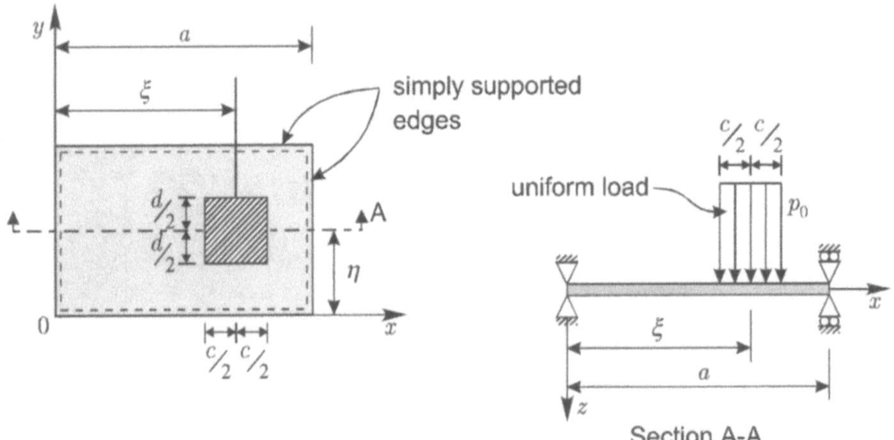

Figure 8.90: Loading of a simply supported plate over a rectangular patch.

Solution

We can apply Navier's analysis directly to obtain the solution to this problem. The general result for the deflection of plate given by (8.1449) is applicable provided p_{mn} can be evaluated. Considering the double Fourier series representation for $p(x, y)$ we have

$$p_{mn} = \frac{4p_0}{ab} \int_{\xi-c/2}^{\xi+c/2} \int_{\eta-d/2}^{\eta+d/2} \sin\left(\frac{m\pi x}{a}\right) \sin\left(\frac{n\pi y}{b}\right) dx\, dy \qquad (8.1452)$$

This gives

$$p_{mn} = \frac{16p_0}{\pi^2 mn} \sin\left(\frac{m\pi\xi}{a}\right) \sin\left(\frac{n\pi\eta}{b}\right) \sin\left(\frac{m\pi c}{2a}\right) \sin\left(\frac{m\pi d}{2b}\right) \quad (8.1453)$$

The expression for the deflected shape can now be obtained by directly substituting the result (8.1453) in (8.1449).

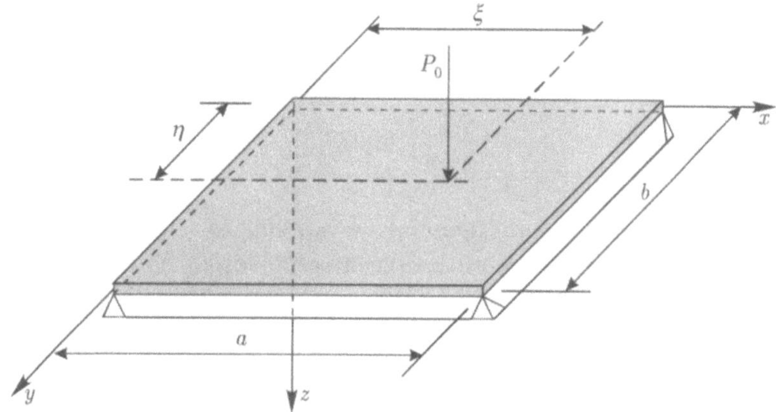

Figure 8.91: Concentrated force P_0 loading a simply supported plate.

To develop the solution to the problem of a concentrated load P_0 which acts at the location (ξ, η) of a simply supported plate (Figure 8.91) we assume that

$$p_0 = \frac{P_0}{cd} \quad (8.1454)$$

and take the limit of p_{mn} defined by (8.1453) as $c \to 0$ and $d \to 0$. if we denote the components of the Fourier expansion for P_0 by P_{mn}, then

$$P_{mn} = \lim_{d \to 0} \lim_{c \to 0} \left\{ \frac{4P_0}{ab} \sin\left(\frac{m\pi\xi}{a}\right) \sin\left(\frac{n\pi\eta}{b}\right) \right.$$
$$\left. \cdot \frac{\sin(m\pi c/2a)}{(m\pi c/2a)} \frac{\sin(n\pi d/2b)}{(n\pi d/2b)} \right\} \quad (8.1455)$$

or

$$P_{mn} = \frac{4P_0}{ab} \sin\left(\frac{m\pi\xi}{a}\right) \sin\left(\frac{n\pi\eta}{b}\right) \tag{8.1456}$$

and the *Green's function* for the simply supported rectangular plate subjected to a concentrated force (of magnitude P_0 rather than unity) takes the form

$$w(x,y;\xi,\eta) = \frac{4P_0}{\pi^4 abD} \sum_{m=1,2,} \sum_{n=1,2,}$$

$$\left\{ \frac{\sin\left(\frac{m\pi\xi}{a}\right) \sin\left(\frac{n\pi\eta}{b}\right) \sin\left(\frac{m\pi x}{a}\right) \sin\left(\frac{n\pi y}{b}\right)}{\left[\left(\frac{m}{a}\right)^2 + \left(\frac{n}{b}\right)^2\right]^2} \right\} \tag{8.1457}$$

The Navier series solution can therefore be applied to compute the deflections of the plate for the case of a concentrated force. For example, the deflection at the centre of a simply supported square plate when the load acts at the centre is given by

$$w(0,0) = \frac{4P_0 a^2}{\pi^4 D} \sum_{m=1,2,} \sum_{n=1,2,} \frac{1}{(m^2+n^2)^2} \simeq 0.01142 \frac{P_0 a^2}{D} \tag{8.1458}$$

This can be compared with the result

$$w(0,0) = 0.01160 \frac{P_0 a^2}{D} \tag{8.1459}$$

which is obtained by the *single* Fourier series method proposed by Levy, which will be discussed in the ensuing section. Although the displacements in the plate can be computed reasonably accurately using Navier's method, the convergence of the second derivative terms which use the double series technique is poor, particularly in the vicinity of localized loadings. This is not altogether unexpected, since, as we have observed in Sections 8.11.9 to 8.11.11 dealing with concentrated loading of plates, the flexural moments and shear forces are singular at the point of application of a concentrated load.

$$\bullet \; \bullet \; \bullet$$

Levy's method of solution was originally developed for the analysis of rectangular plates in which two opposite edges were simply supported. We consider the problem of a rectangular plate ($x \in (0, a); y \in (0, b)$) where the edges $x = 0$ and $x = a$ are simply supported and the remaining edges can be subjected to arbitrary boundary conditions. Levy's approach to the analysis of the rectangular plate problem assumes that the deflection of the plate can be expressed in a single Fourier series of the form

$$w(x, y) = \sum_{m=1,2,} F_m(y) \sin\left(\frac{m\pi x}{a}\right) \tag{8.1460}$$

This representation will automatically satisfy the simply supported boundary conditions along $x = 0$ and $x = a$, regardless of the choice of $F_m(y)$. It is now assumed that the general solution of the inhomogeneous form of the partial differential equation governing flexure of a rectangular plate (8.1444) can be represented in the form

$$w(x, y) = w^P(x, y) + w^H(x, y) \tag{8.1461}$$

Considering (8.1444) and the representation (8.1460) it is evident that the homogeneous solution should satisfy

$$\sum_{m=1,2,} \left[\frac{d^4 F_m^{(H)}}{dy^4} - 2\left(\frac{m\pi}{a}\right)^2 \frac{d^2 F_m^{(H)}}{dy^2} + \left(\frac{m\pi}{a}\right)^4 F_m^{(H)} \right] \sin\left(\frac{m\pi x}{a}\right)$$
$$= 0 \tag{8.1462}$$

where the superscript $^{(H)}$ refers to that part of the solution of $F_m(y)$ associated with the homogeneous partial differential equation governing $w^H(x, y)$. For (8.1462) to be valid for any x, we require

$$\frac{d^4 F_m^{(H)}}{dy^4} - 2\left(\frac{m\pi}{a}\right)^2 \frac{d^2 F_m^{(H)}}{dy^2} + \left(\frac{m\pi}{a}\right)^4 F_m^{(H)} = 0 \tag{8.1463}$$

The general solution for (8.1463) can be written in the form

$$F_m^{(H)}(y) = A_m e^{\frac{m\pi y}{a}} + B_m e^{-\frac{m\pi y}{a}} + C_m y e^{\frac{m\pi y}{a}} + D_m y e^{-\frac{m\pi y}{a}} \tag{8.1464}$$

where $A_m, B_m, ...$etc., are arbitrary constants. The homogeneous solution can therefore be written as

$$w^H(x,y) = \sum_{m=1,2,} F_m^{(H)}(y) \sin\left(\frac{m\pi x}{a}\right) \tag{8.1465}$$

This solution was obtained previously in Section 8.11.13 by considering a general variables separation technique. Levy's method however assumes, *a priori*, a specific trigonometric form in the dependence on one of spatial variables, dictated primarily by the nature of the rectangular plate bending problem involving simply supported edges.

Considering the particular solution $w^P(x,y)$ we observe that the simply supported boundary conditions are satisfied if we again choose a solution of the form

$$w^P(x,y) = \sum_{m=1,2,} F_m^{(P)}(y) \sin\left(\frac{m\pi x}{a}\right) \tag{8.1466}$$

We further assume that the external load can be represented in the form

$$p(x,y) = \sum_{m=1,2,} p_m(y) \sin\left(\frac{m\pi x}{a}\right) \tag{8.1467}$$

Considering the governing partial differential equation (8.1444) and the representations (8.1466) and (8.1467) we obtain

$$D\left\{\frac{d^4 F_m^{(P)}}{dy^4} - 2\left(\frac{m\pi}{a}\right)^2 \frac{d^2 F_m^{(P)}}{dy^2} + \left(\frac{m\pi}{a}\right)^4 F_m^{(P)}\right\} = p_m(y) \tag{8.1468}$$

where the particular solution corresponds to $F_m^{(P)}(y)$. Also from the Fourier series representation of $p(x,y)$ we have

$$p_m(y) = \frac{2}{a} \int_0^a p(x,y) \sin\left(\frac{m\pi x}{a}\right) dx \tag{8.1469}$$

The representation (8.1460) can now be re-written as

$$w(x,y) = \sum_{m=1,2,} \left\{ F_m^{(P)}(y) + F_m^{(H)}(y) \right\} \sin\left(\frac{m\pi x}{a}\right) \qquad (8.1470)$$

where $F_m^{(P)}(y)$ corresponds to the particular solution of (8.1468) and $F_m^{(H)}(y)$ is given by (8.1464) and the associated arbitrary constants A_m, B_m, etc. can be determined by satisfying the boundary conditions along the edges $y = 0$ and $y = b$.

The primary attraction of Levy's approach to the solution of rectangular plate problems is the advantage of a single series representation of the plate deflection which is convenient and efficient from the point of view of numerical computations. Secondly, the boundary conditions along the edges $y = 0$ and $y = b$ are unspecified in the general result (8.1470) and the arbitrary constants $A_m, B_m, ...$etc. can be adjusted to satisfy either simply supported, fixed, free or partially restrained edges, individually, along these edges. The method has the advantage that plates of finite width and both infinite and semi-infinite extent can be examined by selecting the appropriate forms of $F_m^{(H)}(y)$ applicable to such domains. Solutions developed via Levy's method can also be combined with other special solutions for rectangular plates to develop results for rectangular plates which are fully fixed along all boundaries.

Example 8.36

A thin elastic plate of finite width a and infinite length occupies the region $x \in (0, a)$; $y \in (-\infty, \infty)$, and simply supported along its edges $x = 0$ and $x = a$. The plate is subjected to a transverse line load as shown in Figure 8.92. Determine the deflection of the plate.

Solution

Since the loading is symmetric about the plane $y = 0$ we can restrict attention to a semi-infinite region $y \geq 0$, of the plate and adjust the boundary conditions on $y = 0$ to satisfy the loading symmetry conditions. Since the reduced domain of the plate is unloaded, the boundary value problem requires the solution of

$$\overset{\circ}{\nabla}^2 \overset{\circ}{\nabla}^2 w(x,y) = 0 \quad ; \quad x \in (0,a) \quad ; \quad y \in (0,\infty) \qquad (8.1471)$$

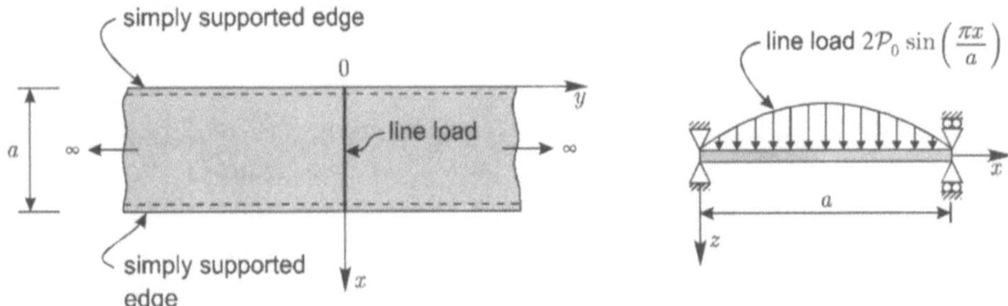

Figure 8.92: A simply supported infinite plate of finite width a with line loading.

which is subject to the boundary conditions

$$w(0, y) = w(a, y) = 0$$

$$(8.1472)$$

$$M_{xx}(0, y) = M_{xx}(a, y) = 0$$

$$\left[\frac{\partial w}{\partial y}\right]_{y=0} = 0 \tag{8.1473}$$

$$[Q_y]_{y=0} + \mathcal{P}_0 \sin\left(\frac{\pi x}{a}\right) = 0 \tag{8.1474}$$

In addition, the deflections of the plate should satisfy the regularity conditions

$$w(x, y) \to 0 \quad \text{as} \quad y \to \infty \tag{8.1475}$$

The appropriate form of the general solution for $w(x, y)$ which satisfies the boundary conditions (8.1472) and the regularity conditions (8.1475) is

$$w(x, y) = \sum_{m=1,2,} \{A_m + B_m y\}\, e^{-\frac{m\pi y}{a}} \sin\left(\frac{m\pi x}{a}\right) \tag{8.1476}$$

It can be shown that to satisfy the boundary condition (8.1473) for all x, we require

$$B_m = \frac{m\pi}{a} A_m \tag{8.1477}$$

The boundary condition (8.1474) can now be expressed as

$$\left[-D \sum_{m=1,2,} \frac{2m^3\pi^3}{a^3} A_m e^{-\frac{m\pi y}{a}} \sin\left(\frac{m\pi x}{a}\right) \right]_{y=0}$$
$$+ \mathcal{P}_0 \sin\left(\frac{\pi x}{a}\right) = 0 \tag{8.1478}$$

This equation is identically satisfied if $m = 1$ and

$$A_1 = \frac{\mathcal{P}_0 a^3}{2D\pi^3} \tag{8.1479}$$

The complete solution for the deflection of the plate occupying the region $x \in (0, a)$, $y \in (0, \infty)$ is

$$w(x, y) = \frac{\mathcal{P}_0 a^3}{2D\pi^3} \left\{ 1 + \frac{\pi y}{a} \right\} e^{-\frac{\pi y}{a}} \sin\left(\frac{\pi x}{a}\right) \tag{8.1480}$$

The procedure can be extended to include other forms of external line loads $\mathcal{P}(x)$ which are symmetric about the x-axis. In general, any line load can be represented in terms of its Fourier components, in the form

$$\mathcal{P}(x) = \sum_{n=1,2,} \mathcal{P}_m \sin\left(\frac{m\pi x}{a}\right) \tag{8.1481}$$

where

$$\mathcal{P}_m = \frac{2}{a} \int_0^a \mathcal{P}(x) \sin\left(\frac{m\pi x}{a}\right) dx \tag{8.1482}$$

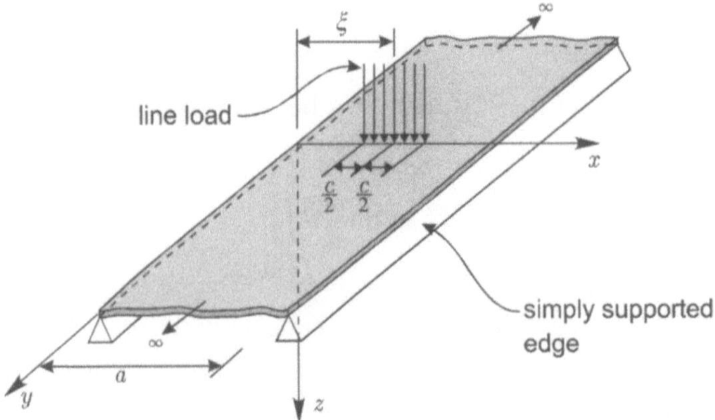

Figure 8.93: Simply supported plate of finite width subjected to a line load.

For example, when the line load has a constant intensity $\mathcal{P}_0$ and occupies the interval shown in Figure 8.93,

$$
\begin{aligned}
\mathcal{P}_m &= \frac{2\mathcal{P}_0}{\pi} \int_{\xi-\frac{c}{2}}^{\xi+\frac{c}{2}} \sin\left(\frac{m\pi x}{a}\right) dx \\
&= \frac{4\mathcal{P}_0}{\pi m} \sin\left(\frac{m\pi\xi}{a}\right) \sin\left(\frac{m\pi c}{2a}\right)
\end{aligned}
\tag{8.1483}
$$

The corresponding expression for A_m in (8.1478) now takes the form

$$
A_m = \frac{\mathcal{P}_0 a^3}{m^4 \pi^4 D} \sin\left(\frac{m\pi\xi}{a}\right) \sin\left(\frac{m\pi c}{2a}\right)
\tag{8.1484}
$$

The final expression for the deflection of the plate, applicable to the region $y \geq 0$ is given by

$$
w(x,y) = \frac{\mathcal{P}_0 a^3}{\pi^4 D} \sum_{m=1,2,} \left\{1 + \frac{m\pi y}{a}\right\} \frac{e^{-\frac{m\pi y}{a}}}{m^4}
$$
$$
\cdot \sin\left(\frac{m\pi\xi}{a}\right) \sin\left(\frac{m\pi c}{2a}\right) \sin\left(\frac{m\pi x}{a}\right)
\tag{8.1485}
$$

Taking the limit as $c \to 0$ with the conditions

$$\mathcal{P}_0 c = P_0 \quad ; \quad \lim_{c \to 0} \frac{\sin(m\pi c/2a)}{(m\pi c/2a)} = 1 \tag{8.1486}$$

we obtain from (8.1485), the solution to the problem of the localized loading of an infinite plate of finite width by a concentrated load which acts at a point $x = \xi$ along the axis of symmetry; i.e.

$$w(x,y) = \frac{P_0 a^2}{2\pi^3 D} \sum_{m=1,2,} \left\{ 1 + \frac{m\pi y}{a} \right\} \frac{e^{-\frac{m\pi y}{a}}}{m^3}$$
$$\cdot \sin\left(\frac{m\pi\xi}{a}\right) \sin\left(\frac{m\pi x}{a}\right) \tag{8.1487}$$

Example 8.37

A rectangular plate is simply supported along all four edges and subjected to a uniform normal load of stress intensity p_0 (Figure 8.94). Derive an expression for the deflection of the plate

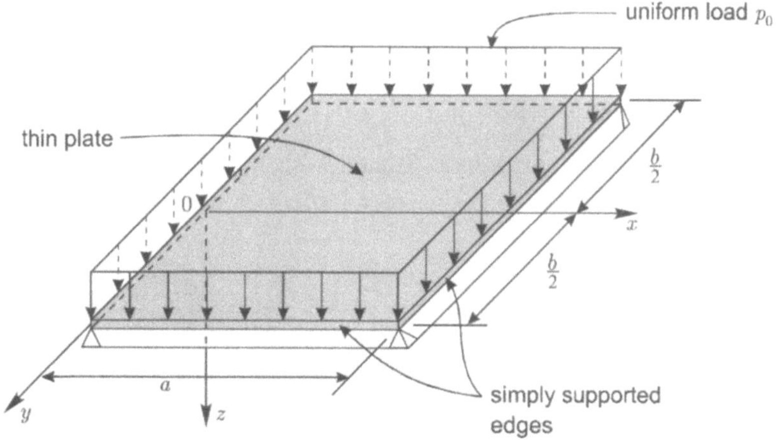

Figure 8.94: Uniform loading of a plate simply supported on all edges.

Solution

For the case of uniform loading (8.1469) gives

$$
p_m(y) = \begin{cases} \dfrac{4p_0}{m\pi} \ ; \ m = 1,3,5, \\[2em] 0 \ ; \ m = 2,4,6, \end{cases}
\tag{8.1488}
$$

The ordinary differential equation (8.1468) now reduces to

$$
D\left\{ \frac{d^4 F_m^{(P)}}{dy^4} - 2\left(\frac{m\pi}{a}\right)^2 \frac{d^2 F_m^{(P)}}{dy^2} + \left(\frac{m\pi}{a}\right)^4 F_m^{(P)} \right\} = \frac{4p_0}{m\pi}
\tag{8.1489}
$$

which gives

$$
F_m^{(P)}(y) = \frac{4p_0 a^4}{m^5 \pi^5 D}
\tag{8.1490}
$$

and (8.1466) can be written as

$$
w^P(x,y) = \frac{4p_0 a^4}{\pi^5 D} \sum_{m=1,3,5,} \frac{1}{m^5} \sin\left(\frac{m\pi x}{a}\right)
\tag{8.1491}
$$

Considering the homogeneous solution $w^H(x,y)$ we note that the deflected shape is symmetric about the x-axis. Hence we require

$$
F^{(m)}(y) = F^{(m)}(-y)
\tag{8.1492}
$$

This would be satisfied by selecting symmetric forms of $F^{(m)}(y)$; this will be more evident if (8.1464) is represented in terms of hyperbolic functions and the required expression for the complete deflection, indeterminate to within two sets of arbitrary constants, can be written as

$$
w(x,y) = \sum_{m=1,3,5,} \left\{ A_m \cosh\left(\frac{m\pi y}{a}\right) + B_m y \sinh\left(\frac{m\pi y}{a}\right) \right.
$$
$$
\left. + \frac{4p_0 a^4}{m^5 \pi^5 D} \right\} \sin\left(\frac{m\pi x}{a}\right)
\tag{8.1493}
$$

The remaining boundary conditions

$$w\left(x,\pm\frac{b}{2}\right)=0 \quad \text{and} \quad \left[\frac{\partial^2 w}{\partial y^2}\right]_{y=\pm\frac{b}{2}}=0 \tag{8.1494}$$

give

$$A_m\cosh(\zeta_m)+B_m\frac{b}{2}\sinh(\zeta_m)+\frac{4p_0a^4}{m^5\pi^5 D}=0$$

$$\tag{8.1495}$$

$$2\left(\frac{A_m\zeta_m}{b}+B_m\right)\cosh(\zeta_m)+B_m\zeta_m\sinh(\zeta_m)=0$$

where

$$\zeta_m=\frac{m\pi b}{2a} \tag{8.1496}$$

Solving the system of equations (8.1495) we obtain A_m and B_m. The final expression for the deflected shape of the plate is given by

$$w(x,y)=\frac{4p_0a^4}{\pi^5 D}\sum_{m=1,3,}^{\infty}\frac{1}{m^5}\left\{1-\left(\frac{\zeta_m\tanh\zeta_m+2}{2\cosh(\zeta_m)}\right)\cosh\left(\frac{m\pi y}{a}\right)\right.$$

$$\left.+\frac{1}{2\cosh(\zeta_m)}\left(\frac{m\pi y}{a}\right)\sinh\left(\frac{m\pi y}{a}\right)\right\}\sin\left(\frac{m\pi x}{a}\right) \tag{8.1497}$$

Expressions for the flexural moments and shear forces can be obtained by substituting (8.1497) in (8.1393) and (8.1394).

The deflection at the centre of the plate $x=a/2$, $y=0$ can be written as

$$w_c=w\left(\frac{a}{2},0\right)$$

$$=\frac{4p_0a^4}{\pi^5 D}\sum_{m=1,3,}^{\infty}\frac{1}{m^5}\left\{1-\left(\frac{\zeta_m\tanh\zeta_m+2}{2\cosh(\zeta_m)}\right)\right\}(-1)^{\left(\frac{m-1}{2}\right)} \tag{8.1498}$$

The first term in this series gives

$$\sum_{m=1,3,}^{\infty} \frac{(-1)^{\left(\frac{m-1}{2}\right)}}{m^5} = \frac{5\pi^5}{2^9(3)}$$

(8.1499)

and

$$w_c = \frac{5p_0 a^4}{384D} - \frac{4p_0 a^4}{\pi^5 D} \sum_{m=1,3,}^{\infty} \frac{(-1)^{\left(\frac{m-1}{2}\right)}}{m^5} \left\{ \frac{\zeta_m \tanh \zeta_m + 2}{2\cosh(\zeta_m)} \right\}$$

(8.1500)

As $b \to \infty$, $\zeta_m \to \infty$, and $\cosh \zeta_m \to \infty$ and $\tanh \zeta_m \to 1$ and the series term in (8.1500) reduces to zero and w_c reduces to the deflection of a simply supported plate of finite width a and infinite length. This result agrees with the solution obtained via an analysis which uses the Bernoulli-Euler beam theory where EI is replaced by D.

Example 8.38

A rectangular plate is simply supported along all its edges. The edge $x = 0$ is subjected to a distribution of moments of intensity $M_0^*(y)$ per unit length (Figure 8.95). Determine its deflected shape.

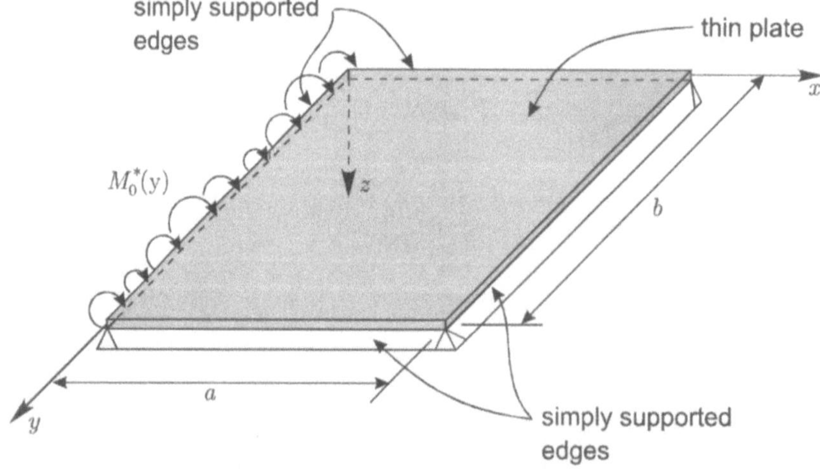

Figure 8.95: Edge loading of a simply supported plate by a distribution of moments $M_o^*(y)$ acting along one edge

Solution

This particular plate bending problem is of general interest since the resulting expression for the deflection of the plate can be used, in conjunction with other fundamental results, to examine the problem of a flexible plate which is clamped along all its boundaries. The boundary conditions applicable to the problem are

$$w(x,0) = w(x,b) = 0$$
$$w(0,y) = w(a,y) = 0$$
$$M_{yy}(x,0) = M_{yy}(x,b) = 0 \tag{8.1501}$$
$$M_{xx}(a,y) = 0$$

and the single inhomogeneous boundary condition is

$$M_{xx}(0,y) = M_0^*(y) \tag{8.1502}$$

Since there is no lateral loading of the plate and since the plate is simply supported we can select a solution of the form

$$w(x,y) = \sum_{m=1,2,} \left\{ A_m \sinh\left(\frac{m\pi x}{b}\right) + B_m \cosh\left(\frac{m\pi x}{b}\right) \right.$$
$$+ C_m x \sinh\left(\frac{m\pi x}{b}\right)$$
$$\left. + D_m x \cosh\left(\frac{m\pi x}{b}\right) \right\} \sin\left(\frac{m\pi y}{b}\right) \tag{8.1503}$$

which automatically satisfies the first and third boundary conditions of (8.1501) for any choice of $A_m, B_m, \ldots$. Considering the displacement boundary condition along the edge $x = 0$ we obtain the result

$$B_m = 0 \tag{8.1504}$$

The remaining boundary conditions can now be used to determine A_m, C_m and D_m, i.e.

$$-D \sum_{m=1,2,} \frac{2m\pi}{b} C_m \sin\left(\frac{m\pi y}{b}\right) = M_0^*(y)$$

$$A_m + C_m a + D_m a \coth(\chi) = 0 \tag{8.1505}$$

$$A_m + C_m \left\{1 + \frac{2}{(1-\nu)\chi} \coth(\chi)\right\}$$

$$+ D_m \left\{1 + \frac{2}{(1-\nu)\chi} \coth(\chi)\right\} = 0$$

where $\chi = (m\pi a/b)$. We can represent the distribution of flexural moments $M_0^*(y)$ as a Fourier series in the form

$$M_0^*(y) = \sum_{m=1,2,} M_m^* \sin\left(\frac{m\pi y}{b}\right) \tag{8.1506}$$

such that

$$M_m^* = \frac{2}{b} \int_0^b M_0^*(y) \sin\left(\frac{m\pi y}{b}\right) dy \tag{8.1507}$$

Expressions for A_m, C_m and D_m can now be obtained in terms of the coefficients M_m^* of the series (8.1506). The final expression for the deflection of the plate can be obtained in the form

$$w(x,y) = \frac{b}{2\pi D} \sum_{m=1,2,}^{\infty} \frac{M_m^* \sin(m\pi y/b)}{m \sinh(m\pi a/b)}$$

$$\cdot \left[x \cosh\left(\frac{m\pi(a-x)}{b}\right) \frac{a \sinh(m\pi x/b)}{\sinh(m\pi a/b)}\right] \tag{8.1508}$$

and the corresponding expressions for the flexural moments and shear forces can be obtained by substituting (8.1508) in the relevant expressions (8.1393) and (8.1394).

Example 8.39

The Figure 8.96 illustrates a rectangular plate which is simply supported along two opposite edges and clamped along the other edges. The plate is subjected to a uniform load of stress intensity p_0. Derive an expression for the deflection of the plate.

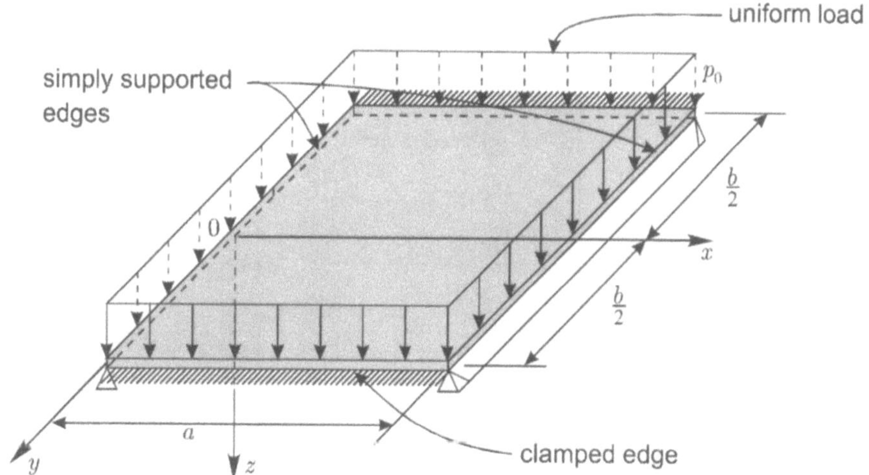

Figure 8.96: Uniformly loaded plate with clamped and simply supported edges.

Solution

The solution to this problem can be obtained in a straightforward manner by employing the single series solution procedure developed by Levy. In this example, however, we shall develop a solution to the plate problem by employing a superposition technique which utilizes aspects of the solutions developed in Examples 8.37 and 8.38. For the problem posed, we require the solution of the partial differential equation

$$D \, \overset{\circ}{\nabla}^2 \, \overset{\circ}{\nabla}^2 w(x, y) = p_0 \tag{8.1509}$$

which is subject to the boundary conditions

$$w\left(x, \frac{b}{2}\right) = w\left(x, -\frac{b}{2}\right) = 0 \quad ; \quad w(0, y) = w(a, y) = 0$$

$$\left[\frac{\partial w}{\partial y}\right]_{y=\frac{b}{2}} = \left[\frac{\partial w}{\partial y}\right]_{y=-\frac{b}{2}} \tag{8.1510}$$

$$M_{xx}(0, y) = M_{xx}(a, y) = 0$$

Considering the solution presented in Example 8.37 for the uniformly loaded plate simply supported along all its boundaries, we note that the solution satisfies all the boundary conditions (8.1510) applicable to the problem *except* that the zero moment boundary conditions for $M_{yy}\left(x, \pm\frac{b}{2}\right)$ is replaced by a zero slope boundary condition. If we denote the solution for the deflection of the uniformly loaded *simply supported plate* by $w^I(x, y)$ (see e.g. (8.1497)) we obtain

$$\left[\frac{\partial w^I}{\partial y}\right]_{y=\frac{b}{2}} = \frac{2p_0 a^3}{\pi^4 D} \sum_{m=1,3,} \frac{\sin\left(\frac{m\pi x}{a}\right)}{m^4}$$

$$\cdot \left[\zeta_m - \tanh \zeta_m (1 + \zeta_m \tanh \zeta_m)\right] \tag{8.1511}$$

where $\zeta_m = (m\pi b/2a)$.

We now refer to the problem where all the edges of the plate are simply supported and the edges $y = \pm b/2$ are subjected to symmetric distributions of moments $M_0^*(x)$ (see Figure 8.97).

Since the problem is symmetric about $y = 0$, we can select the solution for $w^{II}(x, y)$ in the form

$$w^{II}(x, y) = \sum_{m=1,3,} \left\{ \tilde{A}_m \cosh\left(\frac{m\pi y}{a}\right) \right.$$

$$\left. + \tilde{B}_m y \sinh\left(\frac{m\pi y}{a}\right) \right\} \sin\left(\frac{m\pi x}{a}\right) \tag{8.1512}$$

The solution (8.1512) satisfies the boundary conditions

$$w(0, y) = w(a, y) = 0$$

$$\tag{8.1513}$$

$$M_{xx}(0, y) = M_{xx}(a, y) = 0$$

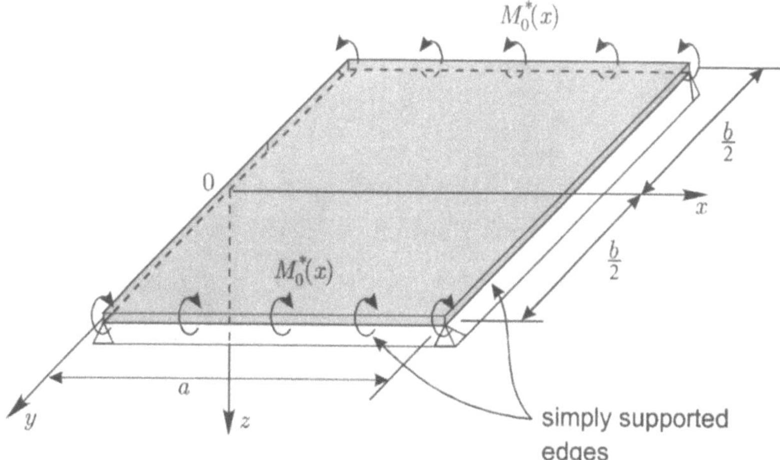

Figure 8.97: Simply supported plate subjected to symmetric distributions of edge moments.

and the remaining boundary conditions

$$w\left(x, \pm\frac{b}{2}\right) = 0 \quad ; \quad M_{yy}\left(x, +\frac{b}{2}\right) = M_0^*(x) \tag{8.1514}$$

can be satisfied by adopting the procedure described in Example 8.38. It can be shown that the deflection of the plate $w^{II}(x,y)$ can be obtained in the form

$$w^{II}(x,y) = \frac{a}{2\pi D} \sum_{m=1,3,} \frac{\widetilde{M}_m^*}{m} \left[\frac{b}{2} \tanh \zeta_m \cosh\left(\frac{m\pi y}{a}\right)\right.$$
$$\left. -y \sinh\left(\frac{m\pi y}{a}\right)\right] \sin\left(\frac{m\pi x}{a}\right) \tag{8.1515}$$

where

$$\widetilde{M}_m^* = \frac{2}{a} \int_0^a M_0^*(x) \sin\left(\frac{m\pi x}{a}\right) dx \tag{8.1516}$$

The slope of the deflected shape along $y = b/2$, is

$$\left[\frac{\partial w^{II}}{\partial y}\right]_{y=\frac{b}{2}} = \frac{a}{2\pi D} \sum_{m=1,3,} \frac{\widetilde{M_m^*}}{m} \left[\tanh \zeta_m \left(\zeta_m \tanh \zeta_m - 1\right)\right.$$

$$\left. -\zeta_m\right] \sin\left(\frac{m\pi x}{a}\right) \tag{8.1517}$$

The distributions of moments $M_0^*(x)$ along $y = \pm b/2$ is arbitrary. We can however determine a specific distribution of $M_0^*(x)$ such that

$$\left[\frac{\partial w^I}{\partial y}\right]_{y=\frac{b}{2}} + \left[\frac{\partial w^{II}}{\partial y}\right]_{y=\frac{b}{2}} = 0 \tag{8.1518}$$

Since the expressions for $w^I(x,y)$ and $w^{II}(x,y)$ are symmetric in y, the values of $(\partial w^I/\partial y)$ and $(\partial w^{II}/\partial y)$ at $y = -b/2$ will be of equal magnitude but opposite sign as those evaluated at $y = b/2$. Substituting (8.1511) and (8.1517) in (8.1518) we can obtain the following expression for $\widetilde{M_m^*}$;

$$\widetilde{M_m^*} = \frac{4p_0a^2}{m^3\pi^3} \left\{\frac{\zeta_m - \tanh \zeta_m(\zeta_m \tanh \zeta_m + 1)}{\zeta_m - \tanh \zeta_m(\zeta_m \tanh \zeta_m - 1)}\right\} \tag{8.1519}$$

The deflected shape of the thin plate which is simply supported along two opposite edges is given by the combination of $w^I(x,y)$ and $w^{II}(x,y)$ defined by (8.1497) and (8.1515) with $\widetilde{M_m^*}$ being defined by (8.1519).

8.11.15   Application of integral transform techniques

Integral transform techniques can, in principle, be applied to the analysis of plate bending problems. In a previous section dealing with the Airy stress function technique for solving two-dimensional plane problems in elasticity, integral transform methods were applied, quite effectively, to determine the state of stress in the elastic region. The method proved to be particularly effective, for example, when dealing with elastic regions of infinite and semi-infinite extent; e.g. a semi-infinite region where $x \in (-\infty, \infty)$; $y \in (0, \infty)$. For example, the fact that $|y|$ extends to infinity coupled with the requirement that the state of stress should satisfy some regularity condition as $y \to \infty$, resulted in the exclusion of solutions which gave divergent stress fields as $y \to \infty$. As a result, the solutions to the plane elasticity problem involving semi-infinite domains could be obtained in a relatively compact form. When the elastic region is finite in one dimension (say e.g. $x \in (-\infty, \infty)$; $y \in (0, H)$) all solutions of the biharmonic equation need to be utilized to satisfy the appropriate traction and/or displacement boundary conditions at $y = 0$ and $y = H$. The resulting integral results derived via the application of integral transform techniques have more complicated expressions in the integrands, in comparison to the analogous expressions that are obtained for the problem where H is infinite. The integral expressions derived for the case where H is finite can be evaluated only through some numerical scheme involving quadrature techniques.

When dealing with the biharmonic equation for bending of plates under transverse load, the formulation of the problem gives bounded plate deflections only if the domain is finite and the plate is adequately supported against unbounded deflections and rotations. Exceptions to this, can include highly specialized self equilibrating loadings which are applied along the boundary of a semi-infinite plate. For this reason, the solution of plate bending problems which utilizes an integral transform technique must, in general, include the complete solution of the biharmonic equation for plate bending (8.1275). As a result, the integral expressions for the deflections, moment and shear forces derived via an integral transform technique will invariably require numerical evaluation. In the ensuing we shall present an example of the application of a Fourier transform technique to the solution of a plate bending problem, involving an infinite plate of finite width.

Example 8.40

A flexible plate of infinite length and finite width is clamped along one boundary and simply supported along the other (Figure 8.98). The edge of the plate which is simply supported is subjected to a distribution of concentrated moments of intensity M_0 per unit length over a width $2c$. Use a Fourier cosine transform technique to determine the deflections of the plate.

Solution

Since the plate is subjected loadings which act along its boundary $p(x, y) = 0$. The boundary value problem requires the solution of the partial differential equation

$$\overset{\circ}{\nabla}{}^2 \, \overset{\circ}{\nabla}{}^2 w(x, y) = 0 \quad ; \quad x \in (-\infty, \infty) \; ; \; y \in (0, b) \tag{8.1520}$$

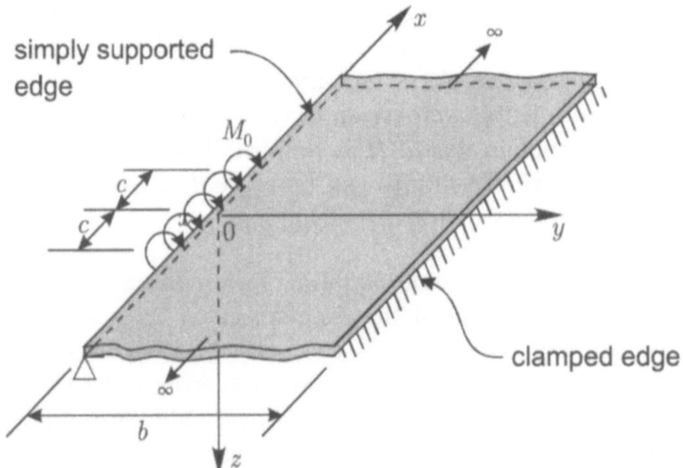

Figure 8.98: Edge loading of an infinite plate by a distribution of moments.

which is subject to boundary conditions

$$w(x,0) = 0 \quad ; \quad w(x,b) = 0$$

$$\left[\frac{\partial w}{\partial y}\right]_{y=b} = 0$$

$$(8.1521)$$

and

$$M_{yy}(x,0) = \begin{cases} M_0 \; ; \; |x| \le c \\[2ex] 0 \; ; \; |x| > c \end{cases} \tag{8.1522}$$

Since the problem is symmetric about the plane $x = 0$ and since $x \in (-\infty, \infty)$, we can formulate the problem by appeal to a Fourier cosine transform approach. We define the Fourier cosine transform of $w(x,y)$ as

$$\widetilde{w}^c(\xi, y) = \mathcal{F}_c\{w(x,y); \xi\} = \sqrt{\frac{2}{\pi}} \int_0^\infty w(x,y) \cos(\xi x)\,dx \tag{8.1523}$$

and the appropriate inversion theorem is

$$w(x, y) = \mathcal{F}_c^{-1} \left\{ \widetilde{w}^c(\xi, y); x \right\} = \sqrt{\frac{2}{\pi}} \int_0^\infty \widetilde{w}^c(\xi, y) \cos(\xi x) d\xi \qquad (8.1524)$$

Operating on (8.1520) with the Fourier cosine transform we obtain a fourth-order ordinary differential equation $\widetilde{w}^c(\xi, y)$;

$$\left(\frac{d^4}{dy^4} - \xi^2 \right)^2 \widetilde{w}^c(\xi, y) = 0 \qquad (8.1525)$$

Combining the solutions of (8.1525) and the inversion theorem (8.1524) we obtain

$$w(x, y) = \sqrt{\frac{2}{\pi}} \int_0^\infty \left\{ [A(\xi) + yB(\xi)] e^{-\xi y} \right.$$
$$\left. + [C(\xi) + yD(\xi)] e^{\xi y} \right\} \cos(\xi x) d\xi \qquad (8.1526)$$

Considering the boundary conditions (8.1521) we have

$$A(\xi) + C(\xi) = 0$$
$$\{A(\xi) + bB(\xi)\}e^{-\xi b} + \{C(\xi) + bD(\xi)\}e^{\xi b} = 0 \qquad (8.1527)$$
$$\{-\xi A(\xi) + (1 - \xi b)B(\xi)\}e^{-\xi b} + \{\xi C(\xi) + (1 + \xi b)D(\xi)\}e^{\xi b} = 0$$

and

$$M_{yy}(x, 0) = -D\sqrt{\frac{2}{\pi}} \int_0^\infty 2\xi\{D(\xi) - B(\xi)\} \cos(\xi x) d\xi \qquad (8.1528)$$

Considering the distribution of moments M_0 along the simply supported edge, we can write

$$M_{yy}(x, 0) = \sqrt{\frac{2}{\pi}} \int_0^\infty \widetilde{M}_0^c(\xi) \cos(\xi x) d\xi \qquad (8.1529)$$

where

$$\widetilde{M}_0^c(\xi) = \sqrt{\frac{2}{\pi}} \int_0^c M_0 \cos(\xi x)\, dx = \sqrt{\frac{2}{\pi}} M_0 \frac{\sin(\xi c)}{\xi} \tag{8.1530}$$

Comparing (8.1529) with (8.1528) and making use of the equations (8.1527) we can show that

$$D(\xi) = -\frac{M_0 \sin(\xi c)}{2\xi^2 D} \left\{ \frac{1 - e^\eta + \eta e^\eta}{1 + 2\eta e^\eta - e^{2\eta}} \right\} \tag{8.1531}$$

where $\eta = 2\xi b$. Similarly we can show that

$$A(\xi) = \left\{ \frac{b\eta e^\eta}{1 - e^\eta + \eta e^\eta} \right\} D(\xi) = -C(\xi)$$

$$\tag{8.1532}$$

$$B(\xi) = -\left\{ \frac{e^\eta(1 + \eta - e^\eta)}{1 - e^\eta + \eta e^\eta} \right\} D(\xi)$$

Since the functions $A(\xi), B(\xi), ...$etc. can be expressed in terms of M_0, the problem is formally solved. The integral expression for the deflection of the plate can be evaluated only by appeal to a numerical integration procedure.

Example 8.41

An infinite plate of width $2b$ is clamped along the edges $y = \pm b$ and subjected to a doubly symmetric loading $p(x, y)$ (Figure 8.99). Use an integral transform approach to develop a formal solution for the deflection of the plate.

Solution

For the formulation of the problem we consider the two-fold symmetry of both the loading and the plate geometry. The boundary value problem for the infinite plate can be reduced to the solution of

$$D \, \overset{\circ}{\nabla}^2 \overset{\circ}{\nabla}^2 w(x, y) = p(x, y) \quad ; \quad x \in (0, \infty) \; ; \; y \in (0, b) \tag{8.1533}$$

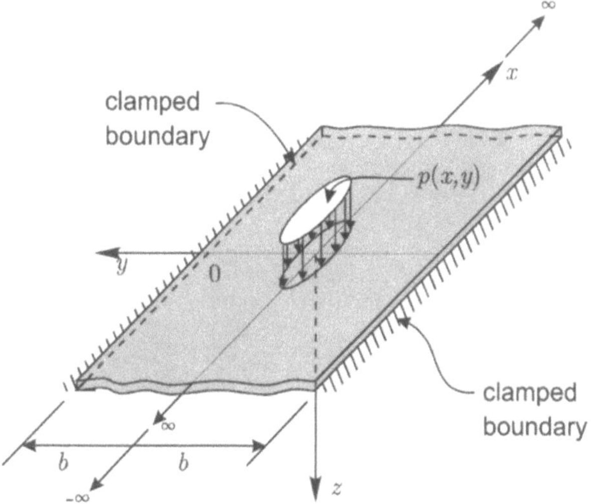

Figure 8.99: Loading of a clamped plate of infinite length by a doubly symmetric loading.

which should satisfy the symmetry conditions

$$\left(\frac{\partial w}{\partial x}\right)_{x=0} = 0 \quad ; \quad \left(\frac{\partial w}{\partial y}\right)_{y=0} = 0$$

$$M_{xy}(0, y) = 0 \quad ; M_{xy}(x, 0) = 0 \tag{8.1534}$$

and the boundary conditions

$$w(x, b) = 0 \quad ; \quad \left(\frac{\partial w}{\partial y}\right)_{y=b} = 0 \tag{8.1535}$$

Since the plate is subjected to a transverse load, the complete solution can be expressed in the form

$$w(x, y) = w^P(x, y) + w^H(x, y) \tag{8.1536}$$

where $w^P(x, y)$ is the particular solution corresponding to the external loading $p(x, y)$ and $w^H(x, y)$ is the homogeneous solution. To obtain the particular solution we assume that both $w^P(x, y)$ and $p(x, y)$ admit double

Fourier integral representations (see, e.g. Chapter 1) of the form

$$w^P(x,y) = \frac{2}{\pi} \int_0^\infty \int_0^\infty \widetilde{w}^P(\xi,\eta)\cos(\xi x)\cos(\eta y)d\xi d\eta \qquad (8.1537)$$

$$p(x,y) = \frac{2}{\pi} \int_0^\infty \int_0^\infty \widetilde{p}(\xi,\eta)\cos(\xi x)\cos(\eta y)d\xi d\eta \qquad (8.1538)$$

Using these representations in (8.1533) we obtain

$$\widetilde{w}^P(\xi,\eta) = \frac{\widetilde{p}(\xi,\eta)}{D(\xi^2+\eta^2)^2} \qquad (8.1539)$$

The Fourier integral expression for $w^P(x,y)$ can now be written in the form of a double Fourier series

$$w^P(x,y) = \frac{2}{\pi D} \int_0^\infty \int_0^\infty \frac{\widetilde{p}(\xi,\eta)}{(\xi^2+\eta^2)^2}\cos(\xi x)\cos(\eta y)d\xi d\eta \qquad (8.1540)$$

The choice of the double Fourier cosine integral representations of the form (8.1537) automatically satisfies the symmetry requirements of (8.1534) applicable to $x=0$. Also, this particular integral is obtained without specifying the boundary conditions (8.1535) applicable to the clamped edge. We can evaluate the displacement and rotation $(\partial w/\partial y)$ resulting from the particular solution as follows:

$$w^P(x,b) = \frac{2}{\pi D} \int_0^\infty \left[\int_0^\infty \frac{\widetilde{p}(\xi,\eta)\cos(\eta b)d\eta}{(\xi^2+\eta^2)^2} \right] \cos(\xi x)d\xi \qquad (8.1541)$$

$$\left[\frac{\partial w^P}{\partial y}\right]_{y=b} = \frac{2}{\pi D} \int_0^\infty \left[\int_0^\infty \frac{-\eta\widetilde{p}(\xi,\eta)\sin(\eta b)d\eta}{(\xi^2+\eta^2)^2} \right] \cos(\xi x)d\xi \qquad (8.1542)$$

Alternatively we can write these equations as

$$w^P(x,b) = \sqrt{\frac{2}{\pi}} \int_0^\infty \widetilde{\Omega}_d^c(\xi)\cos(\xi x)d\xi \qquad (8.1543)$$

$$\left[\frac{\partial w^P}{\partial y}\right]_{y=b} = \sqrt{\frac{2}{\pi}} \int_0^\infty \widetilde{\Omega}_r^c(\xi)\cos(\xi x)d\xi \qquad (8.1544)$$

where

$$\Omega_d^c(\xi) = \sqrt{\frac{2}{\pi}} \int_0^\infty \frac{\widetilde{p}(\xi, \eta)\cos(\eta b)d\eta}{(\xi^2 + \eta^2)^2} \tag{8.1545}$$

$$\Omega_r^c(\xi) = \sqrt{\frac{2}{\pi}} \int_0^\infty \frac{-\eta\widetilde{p}(\xi, \eta)\sin(\eta b)d\eta}{D(\xi^2 + \eta^2)^2} \tag{8.1546}$$

can be identified as the appropriate Fourier cosine transforms for the *deflection* (subscript d) and *rotation* (subscript r) due to $w^P(x, y)$ evaluated at $y = b$. Considering the general homogeneous solution of (8.1533) we can select Fourier integral solutions which will satisfy the symmetry conditions on $y = 0$, i.e.

$$w^H(x, y) = \sqrt{\frac{2}{\pi}} \int_0^\infty [A(\xi)\cosh(\xi y)$$
$$+ B(\xi)y\sinh(\xi y)]\cos(\xi x)d\xi \tag{8.1547}$$

where $A(\xi)$ and $B(\xi)$ are arbitrary functions.

If the boundary conditions (8.1535) are to be satisfied by the combination of the integral solutions (8.1540) and (8.1547), we require

$$A(\xi)\cosh(\xi b) + B(\xi)b\sinh(\xi b) + \widetilde{\Omega}_d^c(\xi) = 0$$

$$\tag{8.1548}$$

$$\xi A(\xi)\sinh(\xi b) + B(\xi)\left[\sinh(\xi b) + \xi b\cosh(\xi b)\right] + \widetilde{\Omega}_r^c(\xi) = 0$$

The arbitrary functions $A(\xi)$ and $B(\xi)$ can be expressed in terms of $\widetilde{\Omega}_d^c(\xi)$ and $\widetilde{\Omega}_r^c(\xi)$, i.e.

$$A(\xi) = \frac{1}{\Omega_0^*}\left[b\sinh(\xi b)\widetilde{\Omega}_r^c - \{\xi b\cosh(\xi b) + \sinh(\xi b)\}\widetilde{\Omega}_d^c\right]$$

$$B(\xi) = \frac{1}{\Omega_0^*}\left[\xi\sinh(\xi b)\widetilde{\Omega}_d^c - \cosh(\xi b)\widetilde{\Omega}_r^c\right] \tag{8.1549}$$

$$\Omega_0^* = \xi b + \sinh(\xi b)\cosh(\xi b)$$

The problem is now formally solved. Explicit integral expressions for the deflection of the plate and the flexural moments and forces can be obtained by combining (8.1540) and (8.1547).

$$\bullet \; \bullet \; \bullet$$

From the two examples presented above it is clear that although formal integral expressions can be obtained for the deflections, flexural moments, etc., these need to be numerically evaluated to generate results of interest to engineering applications. Nonetheless, the Fourier integral transform technique provides a further method for the analysis of plate problems of finite width and infinite length.

8.11.16 Complex variable methods for the solution of plate problems

The complex variable approach for the analysis of flexure of thin circular plates briefly discussed in Section 8.10.11 can also be adopted to examine problems involving thin plates, where the geometry is more conveniently represented by appeal to the rectangular Cartesian coordinates. Although the basic concepts concerning the application of complex variable methods to plate problems have been known for several decades, their routine application to the solution of specific plate problems is relatively scarce. The objective of this section is to briefly outline the basic elements of the method of solution of the biharmonic equation governing flexure of a thin plate and to provide formal results for certain problems involving *clamped edges*.

The complex variables $z = (x + iy)$ and $\bar{z} = (x - iy)$ are introduced as the independent variables. Again, as a matter of convention we shall retain this definition of the complex variables z and $\bar{z}$, and note that for plate problems, the deflection is independent of the coordinate variable z. The partial differential equation (8.1392) can be written as

$$D \overset{\circ}{\nabla}{}^2 \overset{\circ}{\nabla}{}^2 w(x,y) = 16D \frac{\partial^4 w}{\partial z^2 \partial \bar{z}^2} = p(z, \bar{z}) \tag{8.1550}$$

We assume that the solution of (8.1550) can be written in the form

$$w(z, \bar{z}) = w^P(z, \bar{z}) + w^H(z, \bar{z}) \tag{8.1551}$$

where $w^P(z, \bar{z})$ is the particular solution which satisfies

$$\frac{\partial^4 w^P}{\partial z^2 \partial \bar{z}^2} = \frac{p(z, \bar{z})}{16D} \tag{8.1552}$$

and $w^H(z, \bar{z})$ satisfies

$$\frac{\partial^4 w^H}{\partial z^2 \partial \bar{z}^2} = 0 \tag{8.1553}$$

We shall assume that the particular solution can be obtained by any one of the methods described in the preceding sections. It has also been shown that if $p(x, y)$ is an analytic function, then it is always possible to obtain $w^P(x, y)$ by using the result

$$w^P(z, \bar{z}) = \frac{1}{16D} \int_{z_0}^{z} (z - t) dt$$
$$\cdot \int_{\bar{z}_0}^{\bar{z}} (\bar{z} - \bar{t}) p \left[\left(\frac{t + \bar{t}}{2} \right), \left(\frac{t - \bar{t}}{2i} \right) \right] d\bar{t} \tag{8.1554}$$

For example, in the case of a uniformly loaded plate with constant stress intensity p_0,

$$w^P(z, \bar{z}) = \frac{p_0 z^2 \bar{z}^2}{64D} \tag{8.1555}$$

For a concentrated load P_0 which acts at the location $z_0 = x_0 + iy_0$,

$$w^P(z, \bar{z}) = \frac{P_0}{16\pi D} (z - z_0)(\bar{z} - \bar{z}_0) \ln \left[(z - z_0)(\bar{z} - \bar{z}_0) \right] \tag{8.1556}$$

Following the discussions presented in Section 8.7 in connection with the development of solutions for the Airy stress function in terms of complex valued analytic functions (see e.g. (8.405)), the general solution of (8.1553) can be expressed in the form

$$w^H(z, \bar{z}) = 2 \operatorname{Re} \{ z\Omega(z) + \omega(z) \} \tag{8.1557}$$

or, alternatively

$$w^H(z, \bar{z}) = \bar{z}\Omega(z) + z\overline{\Omega}(\bar{z}) + \omega(z) + \bar{\omega}(\bar{z}) \tag{8.1558}$$

where $\Omega(z)$ and $\omega(z)$ are analytic functions in the plate region $\mathbb{D}$ with boundary $\mathcal{C}$. Considering the expressions (8.1393) and (8.1394) we obtain the following expressions for M_{xx}, M_{yy}, etc. in terms of $\Omega(z)$, $\omega(z)$ and their conjugates: i.e.

$$M_{xx} + M_{yy} = -D(1 + \nu)\left[\Omega'(z) + \overline{\Omega}'(\bar{z})\right] \tag{8.1559}$$

$$M_{yy} - M_{xx} - 2iM_{xy} = 4D(1 - \nu)\left[\overline{\Omega}''(\bar{z}) + \bar{\omega}''(\bar{z})\right] \tag{8.1560}$$

$$Q_x + iQ_y = -8D\overline{\Omega}''(\bar{z}) \tag{8.1561}$$

....etc.

The unknown functions $\Omega(z)$ and $\omega(z)$ (and $\overline{\Omega}(\bar{z})$ and $\bar{\omega}(\bar{z})$) can be determined by making use of the boundary conditions of the problem.

In the ensuing we shall focus primarily on presentations where the *boundary of the plate is clamped*. The boundary conditions applicable to a clamped edge are

$$w(x, y) = 0 \quad ; \quad \frac{\partial w}{\partial n} = 0 \quad ; \quad (x, y) \in \mathcal{C} \tag{8.1562}$$

where $\partial/\partial n$ is the normal derivative, and $\mathbf{n}$ is the outward unit normal to $\mathcal{C}$. Due to these boundary condition we also require

$$\frac{\partial w}{\partial t} = 0 \quad ; \quad (x, y) \in \mathcal{C} \tag{8.1563}$$

where $\partial/\partial t$ is the tangential derivative. Considering these results we have

$$\left(\frac{\partial w}{\partial x} + i\frac{\partial w}{\partial y}\right) = \left(\frac{\partial w}{\partial n} + i\frac{\partial w}{\partial t}\right) e^{i\alpha} \tag{8.1564}$$

where α refers to the orientation of the unit normal to C at the point under consideration. Therefore for a clamped boundary we require

$$\left(\frac{\partial w}{\partial x} + i\frac{\partial w}{\partial y}\right) = \frac{\partial w}{\partial z} = 0 \quad ; \quad (x, y) \in C \tag{8.1565}$$

Considering (8.1551) and (8.1558), we can rewrite (8.1565) in the form

$$\left[\bar{z}\Omega'(z) + \overline{\Omega}(\bar{z}) + \omega'(z)\right] = -\frac{\partial w^P}{\partial z} \quad ; \quad (x, y) \in C \tag{8.1566}$$

For a clamped plate, the condition (8.1566) along with the condition

$$\left[\bar{z}\Omega(z) + z\overline{\Omega}(\bar{z}) + \omega(z) + \overline{\omega}(\bar{z})\right] = -w^P \quad ; \quad (x, y) \in C \tag{8.1567}$$

can be used to determine $\Omega(z)$ and $\omega(z)$. In the actual solution of a clamped plate problem, however, recourse must be made to a conformal mapping

$$z = f(\zeta) \quad ; \quad \zeta = \xi + i\eta \tag{8.1568}$$

which will map the region $\mathbb{D}$ on the regions such as a half-plane (say $\eta \geq 0$) or a unit circle $|\zeta| < 1$, where appropriate. For example if

$$\Omega(z) = \Omega_1(\zeta) \quad ; \quad \Omega'(z) = \frac{\Omega_1'(\zeta)}{f'(\zeta)}$$

$$\tag{8.1569}$$

$$\omega(z) = \omega_1(\zeta) \quad ; \quad \omega'(z) = \frac{\omega'(\zeta)}{f'(\zeta)}$$

then we can write (8.1566), for example, in the form

$$\overline{\Omega}_1(\bar{\zeta}) + \frac{\overline{f}(\bar{\zeta})}{f'(\zeta)}\Omega_1'(\zeta) + \frac{\omega_1'(\zeta)}{f'(\zeta)} = W^P(\zeta) \quad ; \quad (x, y) \in C \tag{8.1570}$$

where $W^P(\zeta)$ is obtained by expressing the right hand side of (8.1566) in terms of ζ.

Example 8.42

Examine the class of solutions of the partial differential equation (8.1553) obtained by assuming a variables separable form

$$w^H(z, \bar{z}) = F(z)G(\bar{z}) \tag{8.1571}$$

Solution

It is clear that a variables separable solution of an *additive form* yields a trivial result. Substituting (8.1571) in (8.1553) we obtain the following second order ordinary differential equations for $F(z)$ and $G(\bar{z})$:

$$\frac{d^2 F}{dz^2} = 0 \quad ; \quad \frac{d^2 G}{d\bar{z}^2} = 0 \tag{8.1572}$$

The solution for $w^H(z, \bar{z})$ can now be expressed in the form

$$w^H(z, \bar{z}) = Az\bar{z} + Bz + C\bar{z} + D \tag{8.1573}$$

where A, B, C and D are constants. By comparing (8.1573) with the general form for $w^H(z, \bar{z})$ given by (8.1558) we have

$$\Omega(z) = Az \quad ; \quad \overline{\Omega}(\bar{z}) = B$$
$$\tag{8.1574}$$
$$\omega(z) = D \quad ; \quad \overline{\omega}(\bar{z}) = C\bar{z}$$

Substituting these functions in (8.1559) and (8.1560) we have

$$M_{xx} + M_{yy} = -D(1 + \nu)A$$
$$\tag{8.1575}$$
$$M_{yy} - M_{xx} - 2iM_{xy} = 0$$

These are equivalent to

$$M_{xx} = M_{yy} = -\frac{D(1 + \nu)A}{2} \quad ; \quad M_{xy} = 0 \tag{8.1576}$$

The constant A can be suitably adjusted to give the uniform flexural moments of equal magnitude that can be applied to a plate region of arbitrary dimensions (see Example 8.33). The deflection of the plate is given by

$$w(x, y) = A(x^2 + y^2) \tag{8.1577}$$

Other forms of polynomial solutions for $w(z, \overline{z})$ can be used to determine the flexural response of plates which are subjected to boundary moments or equivalent loads. For example, the expression for the deflection

$$w(z, \overline{z}) = \frac{M_T i}{4(1 - \nu)D}(z^2 - \overline{z}^2) \tag{8.1578}$$

gives

$$\omega(z) = \frac{M_T i z^2}{4(1 - \nu)D} \quad ; \quad \overline{\omega}(z) = -\frac{M_T i \overline{z}^2}{4(1 - \nu)D} \tag{8.1579}$$

From (8.1559) and (8.1560) we obtain

$$M_{xx} = M_{yy} = 0 \quad ; \quad M_{xy} = M_T \tag{8.1580}$$

which agrees with the result developed in Example 8.34, for the pure twisting of a rectangular plate.

$$\bullet \ \bullet \ \bullet$$

The complex variable method can be more effectively applied to examine the class of plate problems where one or more of the boundaries are clamped. We first consider a semi-infinite plate occupying the region $x \in (-\infty, \infty)$; $y \in (0, \infty)$ and clamped along the edge $y = 0$. Taking the displacement of the plate as $w^H(z, \overline{z})$ defined by (8.1558), we have

$$\frac{\partial w^H}{\partial z} = \overline{z}\Omega'(z) + \overline{\Omega}(\overline{z}) + \omega'(z) \tag{8.1581}$$

We now introduce auxiliary potentials Ω_1 and ω_1 such that

$$\omega_1'(z) = -z\Omega_1'(z) + \Omega_1(z) \tag{8.1582}$$

The corresponding displacement $w_1^H(z,\bar z)$ is such that

$$4\,\mathrm{Re}\,\Omega_1(x) = \left(\frac{\partial w_1^H}{\partial x}\right)_{y=0} \equiv f(x) \quad ; \quad \left(\frac{\partial w_1^H}{\partial y}\right)_{y=0} = 0 \tag{8.1583}$$

Similarly the auxiliary potentials Ω_2 and ω_2 such that

$$\omega_2'(z) = -z\Omega_2'(z) - \Omega_2(z) \tag{8.1584}$$

give a displacement $w_2^H(z,\bar z)$ where

$$\left(\frac{\partial w_2^H}{\partial x}\right)_{y=0} = 0 \quad ; \quad 4\,\mathrm{Im}\,\Omega_2(x) = \left(\frac{\partial w_2^H}{\partial y}\right)_{y=0} \equiv g(x) \tag{8.1585}$$

Hence the problem of the half-plane region for a plate with

$$2\left(\frac{\partial w^H}{\partial z}\right)_{y=0} = f(x) + ig(x) = h(x) \tag{8.1586}$$

is essentially reduced to the determination of functions Ω_1 and Ω_2 which are analytic in $y \geq 0$ and have specified real and/or imaginary parts when $y = 0$. If we denote

$$\Omega \equiv \Omega_1 + \Omega_2 \quad ; \quad \omega \equiv \omega_1 + \omega_2 \tag{8.1587}$$

then

$$\omega'(z) = -z\Omega'(z) + \Omega_1(z) - \Omega_2(z) \tag{8.1588}$$

It can be shown that if $f(x)$ and $g(x)$ can be expressed as Fourier integrals then we can show that

$$4\Omega(z) = H(z) \quad ; \quad 4\omega'(z) = -zH'(z) + H^*(z) \tag{8.1589}$$

where

$$H(z) = \frac{1}{\pi}\int_0^\infty e^{izu}\,du \int_{-\infty}^\infty h(t)e^{-iut}\,dt \tag{8.1590}$$

or

$$H(z) = \frac{i}{\pi} \int_{-\infty}^{\infty} \frac{h(t)dt}{(z-t)} \tag{8.1591}$$

and $H^*(z)$ is the transform of $\overline{h}(t)$. The separate representations (8.1590) and (8.1591) will be convenient when clamped boundary conditions are specified over the entire or part of the boundary. This reduction is due to R. Tiffen who produced some seminal results related to the application of complex variable techniques to the analysis of plate bending problems.

Example 8.43

A semi-infinite plate region is clamped along the boundary $y = 0$, and subjected to a concentrated force P_0 which acts at the location (x_0, y_0) (Figure 8.100). Derive an expression for the deflection of the plate.

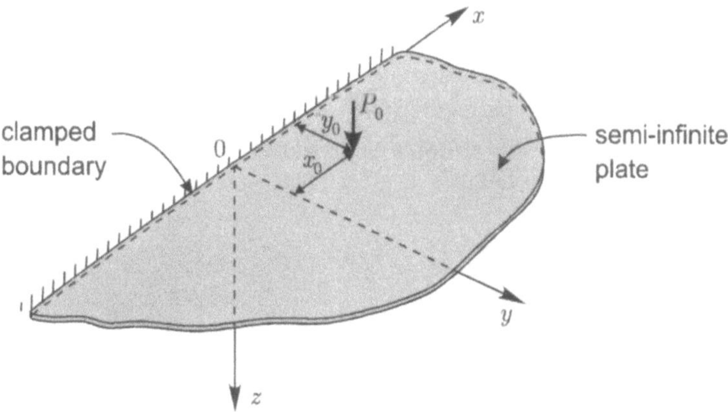

Figure 8.100: Loading of a clamped semi-infinite plate by a concentrated load.

Solution

It can be shown that the particular solution of the deflection $w^P(z, \overline{z})$ corresponding to (8.1556) is given by the potentials

$$\Omega_a(z) = \frac{P_0}{16\pi D}(z - z_0)\ln(z - z_0) \tag{8.1592}$$

$$\omega_a(z) = -\frac{P_0}{16\pi D}\overline{z}_0(z - z_0)\ln(z - z_0) \tag{8.1593}$$

and the potentials $\overline{\Omega}_a(\overline{z})$ and $\overline{\omega}_a(\overline{z})$ can be obtained by replacing z and z_0 in (8.1592) and (8.1593) by their respective conjugates.

By substituting $\Omega_a(z), \omega_a(z), \overline{\Omega}_a(\overline{z})$ and $\overline{\omega}_a(\overline{z})$ in the expression

$$w^P(z, \overline{z}) = \overline{z}\Omega_a(z) + z\overline{\Omega}_a(\overline{z}) + \omega_a(z) + \overline{\omega}_a(\overline{z}) \tag{8.1594}$$

we obtain the result (8.1556). As indicated in Figure 8.100, the plate occupies the region $y \geq 0$ and clamped along $y = 0$. The boundary conditions along the clamped edge are therefore

$$(w)_{y=0} = 0 \quad ; \quad \left(\frac{\partial w}{\partial z}\right)_{y=0} = 0 \tag{8.1595}$$

Since the deflection due to P_0 gives non-zero values along $y = 0$, we require additional potentials which are free of singularities in the plate region to satisfy the boundary conditions (8.1595). The logarithmic terms given by (8.1592) and (8.1593) in the boundary condition can be removed by adding "image potentials" of the form

$$\Omega_b(z) = -\frac{P_0}{16\pi D}(z - z_0)\ln(z - \overline{z}_0) \tag{8.1596}$$

$$\omega_b(z) = \frac{P_0}{16\pi D}\overline{z}(z - z_0)\ln(z - \overline{z}_0) \tag{8.1597}$$

Combining (8.1592), (8.1593) and (8.1596) and (8.1597) we obtain

$$\Omega_0(z) = \Omega_a + \Omega_b = \frac{P_0}{16\pi D}(z - z_0)\ln\left(\frac{z - z_0}{z - \overline{z}_0}\right) \tag{8.1598}$$

$$\omega_0(z) = \omega_a + \omega_b = -\frac{P_0}{16\pi D}\overline{z}_0(z - z_0)\ln\left(\frac{z - z_0}{z - \overline{z}_0}\right) \tag{8.1599}$$

If we define the displacement field obtained by $\Omega_0, \omega_0, ...,$ by $w_0(z, \overline{z})$ then

$$\frac{\partial w_0}{\partial z} = \frac{P}{16\pi D}\left[(\bar{z}-\bar{z}_0)\ln\left\{\frac{(z-z_0)(\bar{z}-\bar{z}_0)}{(z-\bar{z}_0)(\bar{z}-z_0)}\right\}+I(z,\bar{z})\right] \tag{8.1600}$$

where

$$I(z,\bar{z}) = \frac{P}{16\pi D}\frac{(z-\bar{z}_0)(z_0-\bar{z}_0)}{(z-\bar{z}_0)} \tag{8.1601}$$

Therefore

$$\left(\frac{\partial w_0}{\partial z}\right)_{y=0} = I(x,x) = J(x) = \frac{P}{16\pi D}(z_0-\bar{z}_0) \tag{8.1602}$$

Also

$$\left(\frac{\partial w_0}{\partial x}\right)_{y=0} = 2\operatorname{Re}J(x) = -f(x)$$

$$\left(\frac{\partial w_0}{\partial y}\right)_{y=0} = -2\operatorname{Im}J(x) = -g(x) \tag{8.1603}$$

Considering equations (8.1583) and (8.1585) we obtain

$$2\Omega_1(z) = -J(z) = -2\Omega_2(z) \tag{8.1604}$$

and from (8.1582), (8.1584) and (8.1587) we have

$$\Omega_C(z) = \Omega_1 + \Omega_2 = 0 \quad ; \quad \omega_C(z) = \omega_1 + \omega_2 = -\int J(z)dz \tag{8.1605}$$

These corrective solutions when combined with (8.1598) and (8.1599) give the complete potentials

$$\Omega(z) = \frac{P}{16\pi D}(z-z_0)\ln\left\{\frac{(z-z_0)}{(z-\bar{z}_0)}\right\} \tag{8.1606}$$

$$\omega(z) = \frac{P}{16\pi D}\left[-\bar{z}_0(z-z_0)\ln\left\{\frac{(z-z_0)}{(z-\bar{z}_0)}\right\}+z(\bar{z}_0-z_0)\right] \tag{8.1607}$$

which correctly accounts for the concentrated force that acts at the location (x_0, y_0) and satisfies the boundary conditions (8.1595) along the clamped edge $y = 0$. The method of solution, however, is certainly non-routine. Other methods of solution based on combining "*image solutions*" with integral transform techniques can be utilized to obtain equivalent solutions which involve infinite integrals.

Example 8.44

A square plate of edge length $2a$ is clamped along the edges and subjected to a uniform load of stress intensity p_0 over its entire surface area (Figure 8.101). Outline the essential steps in the complex variable technique which can be used to determine the deflection of the plate.

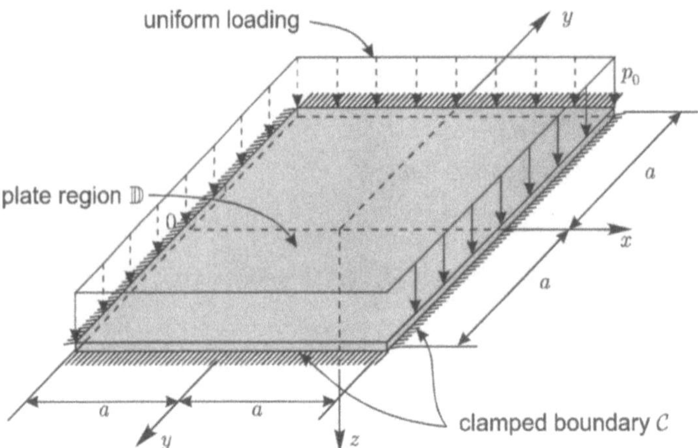

Figure 8.101: Uniform loading of a clamped plate.

Solution

Considering the uniform loading we can show that

$$w^P(x,y) = \frac{p_0}{64D}(x^2 + y^2)^2 \quad ; \quad (x,y) \in \mathbb{D} \tag{8.1608}$$

The boundary conditions for the clamped edges are

$$w(x, y) = 0 \quad ; \quad \frac{\partial w}{\partial n} = 0 \quad ; \quad (x, y) \in \mathcal{C} \tag{8.1609}$$

where $\mathcal{C}$ is the boundary of the clamped plate. Considering (8.1551), these boundary conditions give

$$w^H(x, y) = -w^P(x, y) \quad ; \quad (x, y) \in \mathcal{C} \tag{8.1610}$$

$$\frac{\partial w^H}{\partial n} = -\frac{\partial w^P}{\partial n} \quad ; \quad (x, y) \in \mathcal{C} \tag{8.1611}$$

The representation (8.1558) in its general form does not provide any insight into either the form or construction of the mapping function $f(\zeta)$ for the square plate. The mapping function can be obtained by considering Taylor series expansions of $\Omega(z)$ and $dw/dz = \psi(z)$, in terms of $f(\zeta)$; i.e.

$$\Omega[f(\zeta)] = \sum_{k=1}^{\infty} A_k \zeta^k \tag{8.1612}$$

$$\psi[f(\zeta)] = \sum_{k=1}^{\infty} A_k^* \zeta^k \tag{8.1613}$$

We also require the complex Fourier expansions (on the unit circle $\zeta = e^{i\theta}$) of the expressions

$$\frac{f(\zeta)}{\overline{f'}(\overline{\zeta})} = \sum_{k=-\infty}^{\infty} B_k e^{ik\theta} \tag{8.1614}$$

$$\frac{\partial w^H}{\partial x} + i \frac{\partial w^H}{\partial y} = \sum_{k=-\infty}^{\infty} B_k^* e^{ik\theta} \tag{8.1615}$$

The boundary conditions (8.1610) and (8.1611) will impose conditions on the unit circle $\zeta = 1$, which can be used to determine A_k and A_k^*. We obtain the following set of equations for the determination of A_k and A_k^*;

$$A_m + \sum_{k=1}^{\infty} k \overline{A}_k B_{m+k-1} = B_m^* \quad ; \quad (m = 1, 2, 3, \ldots) \tag{8.1616}$$

$$A_m^* + \sum_{k=1}^{\infty} k \overline{A}_k B_{-m+k-1} = B_{-m}^* \quad ; \quad (m = 0, 1, 2, , ...) \tag{8.1617}$$

The actual applications of the complex variable technique, the Taylor series of the mapping function (8.1568) is truncated at a finite number. The determination of $f(\zeta)$ is crucial to the application of the complex variable procedure. For the square plate region shown in Figure 8.101, the mapping function can be written in the form of the series

$$f(\zeta) = C^* \left[\zeta - \eta_1 \zeta^5 + \eta_2 \zeta^9 ...... \right] \tag{8.1618}$$

where η_1 and η_2 are real constants, and the constant C^* can be determined from the condition

$$f(1) = a \tag{8.1619}$$

8.11.17   Uniqueness of solution governing deflections of a plate

The mechanics of deformation of a thin elastic plate described in the preceding sections is a specialized state of deformation of an elastic continuum which satisfies, the compatibility conditions, the linear elastic constitutive relationships and the equations of equilibrium. This results in the development of the biharmonic equation which governs the deflection of the mid-plane of the elastic plate. The uniqueness of solution of the boundary value problem posed by the biharmonic equation is therefore assured by Kirchhoff's generalized theorem applicable to elastic media (see e.g. Section 8.5). It is, however, instructive to develop the uniqueness theorem by considering the specific flexural actions in the plate. For the sake of completeness we shall consider the plate behaviour characterized by the mid-plane deflection $w(x)$ and the rotations $\varphi(\mathbf{x})$. We first consider the strain energy density per unit area of the plate given by (8.1237)

$$2U_A = tr \left\{ \overset{\circ}{\mathbf{M}} . \overset{\circ}{\mathbf{m}} \right\} + tr \left\{ \overset{\circ}{\mathbf{Q}} . \overset{\circ}{\mathbf{q}} \right\} \tag{8.1620}$$

Considering the result (8.1254) it can be shown that

$$2U_A = \int \int_{\mathbb{D}} p(x, y) w \, dA + \oint_C \left[\overset{\circ}{\mathbf{M}}^* . \overset{\circ}{\varphi} + Q_n^* w \right] dS \tag{8.1621}$$

The equation (8.1621) indicates that the strain energy stored in the plates is composed of the work of the generalized 'forces' pdA, $\overset{\circ}{\mathbf{M}}{}^{*}dS$ and $Q_n^* dS$ and the generalized 'displacements' $w(\mathbf{x})$, $\overset{\circ}{\varphi}(S)$ and $w(S)$. The analogous expression for U_A applicable to the classical theory for thin plates is obtained by using the reduction (8.1268) in (8.1621). Referring to (8.1277) we have

$$\overset{\circ}{\mathbf{M}}{}^{*} \cdot \overset{\circ}{\varphi} = -M_{nn}^* \frac{\partial w}{\partial n} - M_{nt}^* \frac{\partial w}{\partial S} \tag{8.1622}$$

Since $M_{nt}^* w$ is continuous and single-valued on $\mathcal{C}$

$$\oint_{\mathcal{C}} \frac{\partial}{\partial S}\left(M_{nt}^* w\right) dS = \oint_{\mathcal{C}} \left\{ w \frac{\partial M_{nt}^*}{\partial S} + M_{nt}^* \frac{\partial w}{\partial S} \right\} dS = 0 \tag{8.1623}$$

Hence

$$\oint_{\mathcal{C}} \overset{\circ}{\mathbf{M}}{}^{*} \cdot \overset{\circ}{\varphi}\, dS = -\oint_{\mathcal{C}} M_{nn}^* \frac{\partial w}{\partial n} dS + \oint_{\mathcal{C}} w \frac{\partial M_{nt}^*}{\partial S} dS \tag{8.1624}$$

Also, considering (8.1282) we have

$$\frac{\partial M_{nt}^*}{\partial S} = V_n^* - Q_n^* \tag{8.1625}$$

Using these relationships we can show that

$$U_A = \frac{1}{2} \int\!\!\int_{\mathbb{D}} p(x,y) w\, dA + \frac{1}{2} \oint_{\mathcal{C}} M_{nn}^* \left(-\frac{\partial w}{\partial n}\right) dS$$
$$+ \frac{1}{2} \oint_{\mathcal{C}} V_n^* w\, dS \tag{8.1626}$$

THEOREM 8.12

The boundary value problem governing the flexure of the plate with deformations governed by $w(\mathbf{x})$ and $\overset{\circ}{\varphi}(x)$ requires the solution of the partial differential equations

$$\overset{\circ}{\nabla} \cdot \overset{\circ}{\mathfrak{M}} - \overset{\circ}{\mathbf{Q}} = 0 \quad ; \quad \overset{\circ}{\nabla} \cdot \overset{\circ}{\mathbf{Q}} + p(\mathbf{x}) = 0 \quad \mathbf{x} \in \mathbb{D} \tag{8.1627}$$

subject to boundary conditions, either

$$\overset{\circ}{\mathbf{M}} = \overset{\circ}{\mathbf{M}}^* \quad \text{or} \quad \overset{\circ}{\varphi} = \overset{\circ}{\varphi}^* \quad \text{on} \quad \mathbf{x} \in \mathcal{C}_1 \tag{8.1628}$$

and either

$$M_{nt} = M_{nt}^* \quad \text{or} \quad w = w^* \quad \text{on} \quad \mathbf{x} \in \mathcal{C}_2 \tag{8.1629}$$

and either

$$Q_n = Q_n^* \quad \text{or} \quad w = w^* \quad \text{on} \quad \mathbf{x} \in \mathcal{C}_3 \tag{8.1630}$$

where $\mathcal{C} = \mathcal{C}_1 \cup \mathcal{C}_2 \cup \mathcal{C}_3$. Prove that the solution to the above boundary value problem is unique.

PROOF

We assume that there exists two sets of solutions $\left\{ w^{(1)}(\mathbf{x}), \overset{\circ}{\varphi}^{(1)}(\mathbf{x}) \right\}$ and $\left\{ w^{(2)}(\mathbf{x}), \overset{\circ}{\varphi}^{(2)}(\mathbf{x}) \right\}$ which satisfy the governing partial differential equations (8.1627) and the appropriate choice of boundary conditions selected from (8.1628) to (8.1630). Since the governing partial differential equations are linear, the solutions

$$\widetilde{w}(\mathbf{x}) = w^{(1)}(\mathbf{x}) - w^{(2)}(\mathbf{x}) \quad ; \quad \overset{\widetilde{\circ}}{\varphi}(\mathbf{x}) = \overset{\circ}{\varphi}^{(1)}(\mathbf{x}) - \overset{\circ}{\varphi}^{(2)}(\mathbf{x}) \tag{8.1631}$$

will satisfy the $p(x,y) = 0$ and *null* prescribed conditions on $\mathcal{C}$. Therefore from (8.1626) we obtain

$$U_A = \frac{1}{2} \oint_{\mathcal{C}} \left(\overset{\widetilde{\circ}}{\mathbf{M}} \cdot \overset{\widetilde{\circ}}{\varphi} \right) dS + \frac{1}{2} \oint_{\mathcal{C}} \left(\widetilde{Q}_n \widetilde{w} \right) dS \tag{8.1632}$$

where

$$\overset{\widetilde{\circ}}{\mathbf{M}} = \overset{\circ}{\mathbf{M}}^{(1)} - \overset{\circ}{\mathbf{M}}^{(2)} \quad ; \quad \widetilde{Q}_n = Q_n^{(1)} - Q_n^{(2)} \tag{8.1633}$$

Since the two sets of solutions satisfy the same boundary conditions it is clear that

$$\overset{\approx}{\overset{\circ}{\mathbf{M}}} = \mathbf{0} \quad ; \quad \widetilde{Q}_n = 0 \quad ; \quad \widetilde{w} = 0 \quad \text{on} \quad \mathbf{x} \in \mathcal{C} \tag{8.1634}$$

Hence, the total strain energy of the plate

$$U = \int\int_{\mathbb{D}} U_A dA = 0 \tag{8.1635}$$

where U_A is the positive definite quadratic function in terms of the strain components. Since the volume under consideration is arbitrary (8.1635) implies that $U_A = 0$. However, by considering the result (8.1238), the positive definite quadratic form of U_A will be zero, only if

$$\overset{\approx}{\overset{\circ}{\mathbf{m}}} = \overset{\circ}{\mathbf{m}}^{(1)} - \overset{\circ}{\mathbf{m}}^{(2)} = \mathbf{0} \quad ; \quad \overset{\approx}{\overset{\circ}{\mathbf{q}}} = \overset{\circ}{\mathbf{q}}^{(1)} - \overset{\circ}{\mathbf{q}}^{(2)} = \mathbf{0} \tag{8.1636}$$

Since the strain components are identical, the plate stresses are also identical. The generalized displacements are determined by the integration of the strains, which will define the generalized displacements to within a rigid body displacement. These rigid body displacements can be made to vanish by fixing a part of the boundary $\mathcal{C}$. The procedures indicated here can also be adopted to prove the uniqueness of solutions governed by the boundary value problem for the plate deflection applicable to the classical thin plate theory; i.e.

$$D \overset{\circ}{\nabla}^2 \overset{\circ}{\nabla}^2 w(x,y) = p(x,y) \quad ; \quad \mathbf{x} \in \mathbb{D} \tag{8.1637}$$

subject to the sets of boundary conditions

$$w(x,y) = 0 \quad ; \quad M_{nn}(x,y) = 0 \quad \mathbf{x} \in C_S \tag{8.1638}$$

$$w(x,y) = 0 \quad ; \quad \frac{\partial w}{\partial n} = 0 \quad \quad \mathbf{x} \in C_f \tag{8.1639}$$

$$M_{nn}(x,y) = 0 \quad ; \quad V_n = 0 \quad \quad \mathbf{x} \in C_e \tag{8.1640}$$

8.11.18   Uniqueness of solution for plates with clamped boundaries

Other proofs of the uniqueness theorem can also be developed for specific situations involving plates with clamped boundaries. The ensuing proof for the uniqueness of solution for the clamped plate problem which utilizes the complex variable formulation of the problem was first presented by R. Tiffen.

THEOREM 8.13

We consider the flexure of a thin plate which is described by the partial differential equation

$$16D\frac{\partial^4 w}{\partial z^2 \partial \bar{z}^2} = p(z, \bar{z}) \quad ; \quad (z, \bar{z}) \in \mathbb{D} \tag{8.1641}$$

where $\mathbb{D}$ is a simply connected region with boundary $\mathcal{C}$. The plate is clamped along the *entire* boundary $\mathcal{C}$ giving rise to the boundary conditions

$$w = 0 \quad ; \quad \frac{\partial w}{\partial z} = 0 \quad ; \quad (z, \bar{z}) \in \mathcal{C} \tag{8.1642}$$

Prove that the solution to the boundary value problem described by (8.1641) and (8.1642) is unique.

PROOF

We assume that $w^{(1)}(z, \bar{z})$ and $w^{(2)}(z, \bar{z})$ represent two solutions, each of which satisfies the governing partial differential equation (8.1641), the clamped boundary conditions (8.1642) and the same surface loading, including concentrated forces, which give singularities. Since the governing partial differential equation is linear, the solution

$$\widetilde{w}(z, \bar{z}) = w^{(1)}(z, \bar{z}) - w^{(2)}(z, \bar{z}) \tag{8.1643}$$

satisfies

$$\frac{\partial^4 \widetilde{w}}{\partial z^2 \partial \bar{z}^2} = 0 \quad ; \quad (z, \bar{z}) \in \mathbb{D} \tag{8.1644}$$

and the boundary conditions

$$\widetilde{w} = 0 \quad ; \quad \frac{\partial \widetilde{w}}{\partial z} = 0 \quad ; \quad (z, \bar{z}) \in \mathcal{C} \tag{8.1645}$$

applicable to the clamped boundary $\mathcal{C}$. Also since $w^{(1)}(z, \bar{z})$ and $w^{(2)}(z, \bar{z})$ have the same specified isolated singularities in $\mathbb{D}$, are uniform and have uniform partial derivatives, except at these singularities, $\widetilde{w}(z, \bar{z})$ is free from singularities, is uniform and has uniform derivatives in the region $\mathbb{D}$. We now assume that the solution $\widetilde{w}(z, \bar{z})$ can be represented in terms of two analytic functions $\Omega_0(z)$ and $\omega_0(z)$ in the form (see e.g. (8.1558)).

$$\widetilde{w}(z, \bar{z}) = \bar{z}\Omega_0(z) + z\overline{\Omega}_0(\bar{z}) + \omega_0(z) + \overline{\omega}_0(\bar{z}) \tag{8.1646}$$

giving

$$\begin{aligned}
\frac{\partial \widetilde{w}}{\partial z} &= \bar{z}\Omega_0'(z) + \overline{\Omega}_0(\bar{z}) + \omega_0'(z) \\[2mm]
\frac{\partial^2 \widetilde{w}}{\partial z \partial \bar{z}} &= \Omega_0'(z) + \overline{\Omega}_0'(\bar{z}) \\[2mm]
\frac{\partial^3 \widetilde{w}}{\partial z \partial \bar{z}^2} &= \overline{\Omega}_0''(\bar{z}) \\[2mm]
\dots &= \dots
\end{aligned} \tag{8.1647}$$

etc., where the primes denote derivatives. Considering the boundary conditions (8.1645) on $\mathcal{C}$ and the uniformity of the derivatives given by (8.1647) we can state that the integral

$$I = \oint_{\mathcal{C}} \left[\frac{\partial \widetilde{w}}{\partial z} \frac{\partial^2 \widetilde{w}}{\partial z \partial \bar{z}} dz + \widetilde{w} \frac{\partial^3 \widetilde{w}}{\partial z \partial \bar{z}^2} d\bar{z} \right] = 0 \tag{8.1648}$$

Assuming continuity of the functions $\Omega_0, \Omega_0', \Omega_0'', \omega_0'$ and ω_0'' on $\mathbb{D}$, the integral I can be transformed into a surface integral

$$I = 2i \int\int_{\mathbb{D}} \left[\left(\frac{\partial^2 \widetilde{w}}{\partial z \partial \bar{z}} \right)^2 + \left(\frac{\partial \widetilde{w}}{\partial z} \frac{\partial^3 \widetilde{w}}{\partial z \partial \bar{z}^2} \right) \right.$$
$$\left. - \left(\frac{\partial \widetilde{w}}{\partial z} \frac{\partial^3 \widetilde{w}}{\partial z \partial \bar{z}^2} \right) - \left(\widetilde{w} \frac{\partial^4 \widetilde{w}}{\partial z^2 \partial \bar{z}^2} \right) \right] dA \tag{8.1649}$$

It can be shown that (8.1649) is equivalent to

$$\int\int_{\mathbb{D}} \left[\Omega_0'(z) + \overline{\Omega}_0'(\bar{z}) \right]^2 dA = 0 \tag{8.1650}$$

Since the region $\mathbb{D}$ is arbitrary, (8.1650) implies, by virtue of the Dubois-Reymond lemma, that

$$\Omega_0'(z) + \overline{\Omega}'(\bar{z}) = 0 \quad ; \quad (z, \bar{z}) \in \mathbb{D} \tag{8.1651}$$

From the theory of complex variables, it can be shown that, (8.1651) implies that

$$\Omega_0(z) = ciz + \alpha \tag{8.1652}$$

where c is a real constant and α is a complex constant. Also, from the second boundary condition of (8.1645), (8.1647) and (8.1652) we have

$$\omega_0'(z) = -\bar{z}ic + \bar{z}ic - \bar{\alpha} = -\bar{\alpha} \quad ; \quad (z, \bar{z}) \in \mathcal{C} \tag{8.1653}$$

If we consider any point in the interior of $\mathbb{D}$, then by virtue of the continuity and uniformity of $w_0'(z)$ and the applicability of Cauchy's integral formula (see e.g. (1.136),

$$\omega_0'(z) = \frac{1}{2\pi i} \oint_{\mathcal{C}} \frac{\bar{\alpha} d\zeta}{(z - \zeta)} = -\bar{\alpha} \tag{8.1654}$$

Hence

$$\omega_0(z) = -\bar{\alpha} z + \beta \quad ; \quad (z, \bar{z}) \in \mathbb{D} \tag{8.1655}$$

where β is a complex constant. From the first boundary condition of (8.1642)

$$\widetilde{w} = \overline{z}(ciz + \alpha) + z(-ci\overline{z} + \overline{\alpha}) + (-\overline{\alpha}z + \beta - \alpha\overline{z} + \overline{\beta}) = 0 \qquad (8.1656)$$

which gives

$$\beta = ik \qquad (8.1657)$$

where k is a real constant. From (8.1655) and (8.1657) we have

$$\omega_0(z) = -\overline{\alpha}z + ik \qquad (8.1658)$$

and using (8.1652) and (8.1658) in (8.1646) we can show that

$$\widetilde{w}(z,\overline{z}) \equiv 0 \qquad ; \qquad (z,\overline{z}) \in \mathbb{D} \qquad (8.1659)$$

and from the first boundary condition of (8.1642)

$$\widetilde{w}(z,\overline{z}) \equiv 0 \qquad ; \qquad (z,\overline{z}) \in \mathcal{C} \qquad (8.1660)$$

Hence $\widetilde{w}(z,\overline{z})$ is zero in the entire domain which implies that

$$w^{(1)}(z,\overline{z}) \equiv w^{(2)}(z,\overline{z}) \qquad ; \qquad (z,\overline{z}) \in \mathcal{C}, \mathbb{D} \qquad (8.1661)$$

and the solution to the boundary value problem posed by (8.1641) and (8.1642) is unique for any choice of $p(z,\overline{z})$.

8.12 Slow viscous flow

The biharmonic equation is also encountered in the analysis of problems dealing with fluid motions in viscous fluids where the convective inertial forces can be neglected. The formal study of slow viscous flows or creeping flows is generally attributed to Sir G.G. Stokes (1819-1903), who applied this approximation to the Navier-Stokes equations governing fluid motion to develop solutions to several problems of mathematical and technological importance. A notable result involves the evaluation of the frictional drag induced on a sphere moving at constant velocity within a Newtonian viscous fluid of infinite extent. The investigations of Lord Rayleigh (1842-1919) in this area include the fundamental study of unsteady motions induced in viscous fluids by impulsive motions inititiated at boundaries of viscous fluids. The theory of slow viscous flows can also be applied to describe flows which occur at low Reynolds numbers including fluid flow through porous media and other fluid mechanics problems dealing with lubrication theory. The celebrated work of H. Hele-Shaw (1854-1941) dealing with planar flow of viscous fluid in a narrow aperture forms the basis for the modelling of fluid flow in porous media through a viscous flow analogy. Two prominent fluid dynamicists of the past century, Sir G.I. Taylor (1886-1975) and Sir M.J. Lighthill (1924-1998) have also made fundamental contributions which deal with the application of slow viscous flows to the modelling of problems in biological fluid dynamics. The objective of this section is to present a brief account of the class of problems in the theory of slow viscous flows where the governing partial differential equation has a biharmonic form.

8.12.1   Kinematics of fluid flow

For the description of the kinematics of fluid flow we select the velocity vector $\mathbf{v}$ which is a function of position $\mathbf{x}$ and time. In the Lagrangian description, where each fluid particle is tracked throughout its movement through the flow, the identity of a particle of fluid is obtained by specifying its position $P_0(\mathbf{x_0})$ at time t_0 and its position $P(\mathbf{x})$ at time t, such that

$$\mathbf{x} = \mathbf{x}(\mathbf{x_0}, t) \tag{8.1662}$$

If $\mathbf{r}$ is the position vector characterizing the point P then

$$\mathbf{v} = \frac{d\mathbf{r}}{dt} \tag{8.1663}$$

and the acceleration $\mathbf{a}$ is given by

$$\mathbf{a} = \frac{d\mathbf{v}}{dt} \tag{8.1664}$$

In the Eulerian description, we observe the velocity field at each point of the fluid domain, regardless of the particle identity. At every point $P(\mathbf{x})$ there is a velocity $\mathbf{v}(\mathbf{x}, t)$ and a fluid pressure $p(\mathbf{x}, t)$ and $\mathbf{x}$ and t are regarded as independent variables. The Eulerian approach is more convenient for describing steady motions when all dependent variables become independent of time. As shown in Section 7.8, the acceleration vector $\mathbf{a}$ can be obtained by considering it to be a function of the four variables $\mathbf{x}$ and t. (see e.g. (7.588)) i.e.

$$\mathbf{a} = \frac{d\mathbf{v}}{dt} + (\mathbf{v}.\nabla)\mathbf{v} \tag{8.1665}$$

and the operator

$$\frac{D}{Dt} = \frac{\partial}{\partial t} + \mathbf{v}.\nabla \tag{8.1666}$$

is the *material time derivative* and is also referred to as the *total* and *substantial derivative*.

The definition of fluid velocity $\mathbf{v}$ in the Eulerian description of the flow of a fluid also allows the definition of a *rate of deformation* (or strain rate) which is analogous to the definition of the strain in a solid undergoing deformation. If the fluid continuum has velocities $\mathbf{v}'(\mathbf{x})$ and $\mathbf{v}(\mathbf{x})$ at two neighbouring points such that

$$\mathbf{v}' = \mathbf{v} + (\nabla\mathbf{v})\,d\mathbf{x} \tag{8.1667}$$

Then we can write (8.1667) in the form

$$\mathbf{v}' = \mathbf{v} + \frac{1}{2}\left[\nabla\mathbf{v} + \mathbf{v}\nabla\right]d\mathbf{x} + \frac{1}{2}\left[\nabla\mathbf{v} - \mathbf{v}\nabla\right]d\mathbf{x} \tag{8.1668}$$

In keeping with the definitions of strain and rotation dyadics (see e.g. (8.21) and (8.22)) we can define the *rate of deformation dyadic* as

$$\mathfrak{F} = \frac{1}{2}\left[\nabla\mathbf{v} + \mathbf{v}\nabla\right] \tag{8.1669}$$

and the *vorticity dyadic* as

$$\mathfrak{V} = \frac{1}{2}\left[\nabla\mathbf{v} - \mathbf{v}\nabla\right] \tag{8.1670}$$

The 'rate of deformation' is derived from the fact that the various components of (8.1668) can be identified with extensional and shear strain rates synonymous with the deformations associated with deformable solids. The *vorticity dyadic* is the analogue of the dyadic for the rotation of a deformable solid.

Viscous fluids can be classified as either compressible or incompressible with regard to their volume change characteristics. In this section, however, attention will be restricted to the behaviour of viscous fluids which are *incompressible*. For such fluids the equation of continuity or the equation of mass conservation is an added constraint on the deformation. (This aspect was discussed in several of the preceding Chapters, particularly in Chapters 5 and 6). The equation of mass conservation can be obtained by considering a fixed control volume dV within a fluid domain V bounded by a fixed control surface S. From the observation that the rate of change of mass within V must be balanced by the fluid movement through the control surface S. Assuming that there are no fluid *sources* or *sinks* we obtain

$$\frac{d}{dt}\int\int\int_V \rho(\mathbf{x},t)dV = -\int\int_S \rho(\mathbf{x},t)\mathbf{v}.\mathbf{n}\,dS \tag{8.1671}$$

where $\rho(\mathbf{x},t)$ is the mass density of the viscous fluid and $\mathbf{n}$ is the *outward* unit normal to the surface S. Considering the divergence theorem and the Dubois-Reymond Lemma, we can express (8.1671) in the form

$$\frac{\partial\rho}{\partial t} + \nabla.(\rho\mathbf{v}) = 0 \tag{8.1672}$$

and using (8.1666) we can write

$$\frac{D\rho}{Dt} + \rho\nabla.\mathbf{v} = 0 \quad ; \quad \text{or} \quad \nabla.\mathbf{v} = \frac{D}{Dt}\ln\left(\frac{1}{\rho}\right) \tag{8.1673}$$

If the viscous fluid is incompressible, then we obtain

$$\nabla.\mathbf{v} = 0 \tag{8.1674}$$

8.12.2 Equation of motion

The state of stress at a point in the viscous fluid at any time t is defined in relation to the stress vectors similar to those defined in Section 8.3, in connection with the definition of the state of stress in a solid. We can consider the dynamic equilibrium of a region V with surface S under the action of tractions $\mathbf{T}$ on S, body forces $\mathbf{f}^b$ in V and acceleration effects of $D\mathbf{v}/Dt$ in V, i.e.

$$\int\int_S \mathbf{T}\,dS + \int\int\int_V \mathbf{f}^b\,dV = \int\int\int_V \rho\frac{D\mathbf{v}}{Dt}\,dV \tag{8.1675}$$

Considering the procedures outlined in equations (8.158) and (8.159) we can show that the equation of motion can be expressed in terms of the stress dyadic $\mathfrak{S}$ in the form

$$\nabla.\mathfrak{S} + \mathbf{f}^b = \rho\frac{D\mathbf{v}}{Dt} \tag{8.1676}$$

In indicial notation (8.1676) takes the form

$$\sigma_{ij,j} + f_i^b = \rho\frac{Dv_i}{Dt} \tag{8.1677}$$

where

$$\frac{Dv_i}{Dt} = \frac{\partial v_i}{\partial t} + v_j v_{i,j} \tag{8.1678}$$

now represents a *non-linear* term involving the velocity components and velocity gradients.

8.12.3 Constitutive equations for a viscous fluid

We shall restrict attention to linearly viscous fluids or Newtonian fluids and assume that the stress matrix $\boldsymbol{\sigma}$ defined in relation to a system of orthogonal coordinates (see e.g. (8.144)) can separated into isotropic and deviatoric parts such that

$$\mathfrak{S} = -p\mathfrak{I} + \mathfrak{S}' \tag{8.1679}$$

or

$$\boldsymbol{\sigma} = -p\mathbf{I} + \boldsymbol{\sigma}' \tag{8.1680}$$

where p is a scalar pressure, $\mathfrak{I}$ is the unit dyadic and $\mathbf{I}$ is the unit matrix. Also, $\mathfrak{S}'$ and $\boldsymbol{\sigma}'$ represent, respectively, the deviatoric components of the stress dyadic and the stress matrix, and $-p\mathfrak{I}$ or $-p\mathbf{I}$ is an inviscid component. We assume that for a Newtonian viscous fluid the deviator stress is a linear function of the linearized strain matrix dyadic $\mathfrak{F}$ or the linearized strain rate matrix $\mathbf{s}$ can be written as (see e.g.(8.33))

$$\mathfrak{F} = \frac{1}{2} \left[\nabla \mathbf{v} + \mathbf{v}\nabla \right]$$
$$= \mathbf{ii}\frac{\partial v_x}{\partial x} + \mathbf{ij}\frac{1}{2}\left(\frac{\partial v_x}{\partial y} + \frac{\partial v_y}{\partial x}\right) + \mathbf{ik}\frac{1}{2}\left(\frac{\partial v_x}{\partial z} + \frac{\partial v_z}{\partial x}\right)$$
$$+ \mathbf{ji}\frac{1}{2}\left(\frac{\partial v_x}{\partial y} + \frac{\partial v_y}{\partial x}\right) + \mathbf{jj}\frac{\partial v_y}{\partial y} + \mathbf{jk}\frac{1}{2}\left(\frac{\partial v_y}{\partial z} + \frac{\partial v_z}{\partial y}\right)$$
$$+ \mathbf{ki}\frac{1}{2}\left(\frac{\partial v_x}{\partial z} + \frac{\partial v_z}{\partial x}\right) + \mathbf{kj}\frac{1}{2}\left(\frac{\partial v_y}{\partial z} + \frac{\partial v_z}{\partial y}\right) + \mathbf{kk}\frac{\partial v_z}{\partial z} \tag{8.1681}$$

or in indicial notation

$$\mathbf{s} = \frac{1}{2}\left[v_{i,j} + v_{j,i}\right] \tag{8.1682}$$

Furthermore, we restrict attention to isotropic viscous fluids where the viscosity properties are *direction independent*. (Certain polymeric materials and viscous fluids containing dispersion of particulates can exhibit effective viscosities which are direction dependent.) The constitutive relationship governing $\boldsymbol{\sigma}$ and $\mathbf{s}$ can be derived by considering procedures similar to those that were employed in connection with the development of a constitutive relationship for an isotropic linear elastic solid. The most general constitutive relationship for $\mathfrak{S}'$ applicable to a Newtonian viscous fluid can be written in the form

$$\mathfrak{S}' = \kappa(\nabla \cdot \mathbf{v})\mathfrak{I} + 2\eta\mathfrak{F} \tag{8.1683}$$

where κ is the *bulk viscosity* and η is the *dynamic shear viscosity*. In terms of the stress matrix $\boldsymbol{\sigma}'$, (8.1683) takes the form

$$\boldsymbol{\sigma}' = \kappa\,(\nabla.\mathbf{v})\mathbf{I} + 2\eta\mathbf{s} \tag{8.1684}$$

or in indicial form

$$\sigma_{ij} = \kappa s_0 \delta_{ij} + 2\eta s_{ij} \tag{8.1685}$$

where

$$s_0 = tr(s_{ij}) \tag{8.1686}$$

The viscosity parameters should satisfy the thermodynamic constraints and these will be discussed in the ensuing section. For a Newtonian viscous fluid, the dynamic shear viscosity η is a measure of the response of the fluid to shearing and extensional deformation rates. The bulk viscosity κ is relevant to *Newtonian viscous fluids* which are compressible. Therefore, for Newtonian viscous fluids which are *incompressible, isotropic* and *void of bulk viscosity*, the constitutive equation can be written in the forms

$$\mathfrak{S} = -p\mathfrak{I} + 2\eta\mathfrak{F} \quad \text{or} \quad \boldsymbol{\sigma} = -p\mathbf{I} + 2\eta\mathbf{s} \tag{8.1687}$$

In indicial form,(8.1687) can be written as

$$\sigma_{ij} = -p\delta_{ij} + 2\eta s_{ij} \tag{8.1688}$$

8.12.4 Thermodynamic constraints on the viscosity coefficients

The constitutive relationships for Newtonian viscous fluids presented in the previous section are based on purely mathematical arguments and assumptions pertaining to the linear dependence of the stresses on the strain rates. It is therefore important to establish the thermodynamic constraints that can be imposed on the values of the viscosity coefficients κ and η. An essential concept to the formulation of such constraints is the notion of *internal energy dissipation* in the viscous fluid during its deformation. Consider a control volume V of a Newtonian viscous fluid which always encloses the same fluid particles. The kinetic energy K and the internal energy E of the region are given by

$$K = \int\!\!\int\!\!\int_V \frac{1}{2}\rho v^2 dV \quad ; \quad E = \int\!\!\int\!\!\int_V \rho e\, dV \tag{8.1689}$$

where $\mathbf{v}$ is the velocity of a point within V and e is the internal energy per unit mass. The time rates of increase of K and E are

$$\frac{DK}{Dt} = \int\int\int_V \rho\mathbf{a}\mathbf{v}\,dV$$

$$\frac{DE}{Dt} = \int\int\int_V \rho\frac{De}{Dt}dV \tag{8.1690}$$

where $\mathbf{a}$ is the acceleration vector. Considering the equation of motion (8.1676) we can write

$$\nabla.\mathfrak{S} + \mathbf{f}^b = \rho\mathbf{a} \tag{8.1691}$$

The energy balance equation is obtained by the following: the rate of increase of kinetic energy and internal energy *equals* the rate of work of the tractions on S the boundary of V, + the rate of working of the body forces $\mathbf{f}^b$ + rate at which heat is supplied to V; i.e.

$$\frac{D}{Dt}(K + E) = \int\int_S -(\mathbf{n}.\mathfrak{S})\mathbf{v}dS$$
$$+ \int\int\int_V \mathbf{v}\mathbf{f}^b dV + \int\int\int_V QdV \tag{8.1692}$$

where Q is the rate per unit volume of the heat that is supplied to the control volume V. This heat can be supplied through, for example, by conduction through S or within V through some exothermic processes. Combining (8.1690) to (8.1692) and using the divergence theorem we obtain

$$\int\int\int_V \left\{\rho\mathbf{a}\mathbf{v} - \mathbf{f}^b\mathbf{v} - \nabla.(\mathfrak{S}.\mathbf{v}) + \rho\frac{De}{Dt} - Q\right\}dV = 0 \tag{8.1693}$$

Noting that (see e.g. (8.217))

$$\nabla.(\mathfrak{S}.\mathbf{v}) = (\nabla.\mathfrak{S}).\mathbf{v} + \mathfrak{S}:\nabla\mathbf{v} \tag{8.1694}$$

where (:) denotes the double scalar product. Since the volume of integration is arbitrary, using (8.1691) and (8.1694) in (8.1693) and considering the Dubois-Reymond Lemma we obtain

$$\rho \frac{De}{Dt} = Q + \mathfrak{S} : \nabla \mathbf{v} \tag{8.1695}$$

The constitutive equation for the incompressible Newtonian viscous fluid can be written as

$$\mathfrak{S} = -p\mathfrak{I} + \kappa(\nabla . \mathbf{v})\mathfrak{I} + 2\eta\mathfrak{F} \quad \text{or} \quad \boldsymbol{\sigma} = -p\mathbf{I} + \kappa(\nabla . \mathbf{v})\mathbf{I} + 2\eta\mathbf{s} \tag{8.1696}$$

From (8.1669) and (8.1670)

$$\nabla \mathbf{v} = \mathfrak{F} + \mathfrak{V} \tag{8.1697}$$

and since $\mathfrak{S} = \mathfrak{S}^C$ (see e.g. (8.174)) and $\mathfrak{V} = -\mathfrak{V}^C$

$$\mathfrak{S} : \nabla \mathbf{v} = tr\,[\boldsymbol{\sigma}.\mathbf{s}] \tag{8.1698}$$

Hence (8.1695) can be written as

$$\rho \frac{De}{Dt} = tr\,[\boldsymbol{\sigma}.\mathbf{s}] + Q \tag{8.1699}$$

From the first law of thermodynamics it is assured that heat is a form of energy. Considering the viscous fluid we assume that there exists an internal energy function e which depends on the state variables p, ρ and T where p is a pressure, ρ is the density and T is the absolute temperature. When the fluid is supplied with a small quantity of heat q we have

$$q = de + pd\vartheta \tag{8.1700}$$

where $d\vartheta$ is the incremental change in volume, de is the excess of energy supplied over the mechanical work done by the pressure. Also

$$q = T\,dS^* \tag{8.1701}$$

where dS^* is the differential of a function S^* referred to as the entropy. hence for a unit mass

$$T\,dS^* = de + p\,d\left(\frac{1}{\rho}\right) \tag{8.1702}$$

Considering the *convected time derivative* of the heat gain per unit volume given by (8.1702) we have

$$\rho T \frac{DS^*}{Dt} = \rho \frac{De}{Dt} + p\rho \frac{D}{Dt}\left(\frac{1}{\rho}\right)$$

(8.1703)

Using (8.1673) we can rewrite (8.1703) as

$$\rho T \frac{DS^*}{Dt} = \rho \frac{De}{Dt} + p(\nabla .\mathbf{v})$$

(8.1704)

Considering (8.1699) and (8.1704) we have

$$\rho T \frac{DS^*}{Dt} = tr\,[\boldsymbol{\sigma}.\mathbf{s}] + p(\nabla .\mathbf{v}) + Q$$

(8.1705)

By definition, Q is the rate at which heat is being supplied by conduction and other external causes. Therefore the rate of heat generation by other dissipative phenomena, such as internal friction within the viscous fluid, denoted by $\mathcal{D}^*$ is given by

$$\mathcal{D}^* = tr\,\{\boldsymbol{\sigma}.\mathbf{s}\} + p(\nabla .\mathbf{v})$$

(8.1706)

Substituting (8.1696) in (8.1706) and noting that $tr\{\mathbf{s}.\mathbf{I}\} = \nabla .\mathbf{v}$, we obtain

$$\mathcal{D}^* = 2\eta \, tr(\mathbf{s}^2\} + \kappa\{\nabla .\mathbf{v}\}^2$$

(8.1707)

Therefore for the dissipation to be non-negative during any prescribed history of deformation we require

$$\eta > 0 \quad ; \quad \kappa > 0$$

(8.1708)

These thermodynamic bounds obtained for the Newtonian viscous fluid through consideration of the positive definiteness of the internal energy dissipation is analogous to the thermodynamic bounds obtained for the elastic constants through consideration of the positive definiteness of the stored energy function (see e.g. Section 8.4.5).

8.12.5 Navier-Stokes equation

The constitutive equation for a Newtonian viscous fluid with bulk viscosity κ and shear viscosity η can be written, respectively, in the dyadic form or the matrix forms as follows:

$$\mathfrak{S} = \{-p + \kappa(\nabla.\mathbf{v})\}\mathfrak{I} + 2\eta\mathfrak{F}$$

$$\boldsymbol{\sigma} = \{-p + \kappa(\nabla.\mathbf{v})\}\mathbf{I} + 2\eta\mathbf{s} \tag{8.1709}$$

or in indicial form

$$\sigma_{ij} = -p\delta_{ij} + \kappa(v_{k,k})\delta_{ij} + \eta(v_{i,j} + v_{j,i}) \tag{8.1710}$$

These equations can be combined with the equations of motion (8.1676) or (8.1677) to give the following expressions:

$$\eta\nabla^2\mathbf{v} + \left(\kappa + \frac{\eta}{3}\right)\nabla(\nabla.\mathbf{v}) - \nabla p + \mathbf{f}^b = \rho\frac{D\mathbf{v}}{Dt} \tag{8.1711}$$

or in indicial form

$$\eta v_{i,jj} + \left(\kappa + \frac{\eta}{3}\right)(v_{j,j})_{,i} - p_{,i} + f_i^b = \rho\left[\frac{\partial v_i}{\partial t} + v_j v_{i,j}\right] \tag{8.1712}$$

In the case of an *incompressible Newtonian viscous fluid,* void of bulk viscosity the equivalent forms for (8.1711) and (8.1712) are given by

$$\eta\nabla^2\mathbf{v} - \nabla p + \mathbf{f}^b = \rho\frac{D\mathbf{v}}{Dt} \tag{8.1713}$$

and

$$\eta v_{i,jj} - p_{,i} + f_i^b = \rho\left[\frac{\partial v_i}{\partial t} + v_j(v_{i,j})\right] \tag{8.1714}$$

respectively. The equations (8.1711) to (8.1714) were obtained by C.L.M.H. Navier (1785-1836), S.D. Poisson (1781-1840), B. de Saint-Venant (1797-1886) and Sir G.G. Stokes (1819-1903) in different forms. These are usually referred to as the *Navier-Stokes equations*. They are non-linear partial differential equations for the velocity components $\mathbf{v}(\mathbf{x}, t)$, the non-linearity arising from the non-linear form of the material time derivative (see e.g. (8.1676) and (8.1677)). To date only a limited number of analytical solutions have been obtained for situations where the non-linear terms are non-zero.

8.12.6 Biharmonic formulations of problems in slow viscous flow

The solution of the non-linear Navier-Stokes equations forms an important aspect of both theoretical and computational fluid dynamics. A limited number of solutions of these non-linear partial differential equations mostly involving spatially one-dimensional problems are given in the literature. Solutions of practical interest have been obtained for cases where, with suitable approximations, the equations are reduced to linear partial differential equations. The theory of slow viscous flows or creeping flows is one such approximation where the flows occur with a very low Reynolds number. With this constraint the non-linear acceleration terms in the flow domain are assumed to be negligible in comparison with the viscous terms.

In the special case when steady motion occurs in an incompressible Newtonian viscous fluid with constant viscosity, the convective terms in (8.1713) (and (8.1714)) can be neglected and the partial differential equations governing the problem are the equations of motion

$$\eta \nabla^2 \mathbf{v} - \nabla p + \mathbf{f}^b = 0 \tag{8.1715}$$

and the continuity equation

$$\nabla . \mathbf{v} = 0 \tag{8.1716}$$

Eliminating $\mathbf{v}$ between (8.1715) and (8.1716) we have

$$\nabla^2 p - \nabla . \mathbf{f}^b = 0 \tag{8.1717}$$

and if the body forces are negligible in comparison to the viscous forces (8.1717) reduces to

$$\nabla^2 p = 0 \tag{8.1718}$$

Therefore for slow viscous motions of Newtonian fluids, the fluid pressure field is *harmonic*.

The operations which are used to arrive at a biharmonic formulation of the slow viscous flow problem are very similar to those employed in connection with the Galerkin vector formulation in classical elasticity (see e.g. 8.10.2). We first assume that the velocity vector $\mathbf{v}$ can be expressed in terms of a function Ψ such that

$$\mathbf{v} = \nabla \times \Psi \tag{8.1719}$$

where Ψ is a divergence free field i.e.

$$\nabla.\Psi = 0 \tag{8.1720}$$

If the condition (8.1720) is satisfied then Ψ itself could be derived from another vector $\mathbf{M}$ such that

$$\Psi = \nabla \times \mathbf{M} \tag{8.1721}$$

Combining (8.1719) to (8.1721) (compare with (8.812))

$$\mathbf{v} = \nabla \times \Psi = \nabla \times (\nabla \times \mathbf{M}) = \nabla(\nabla.\mathbf{M}) - \nabla^2\mathbf{M} \tag{8.1722}$$

Now applying the curl operator on (8.1715) gives

$$\eta\nabla \times (\nabla^2\mathbf{v}) + \nabla \times \mathbf{f}^b = 0 \tag{8.1723}$$

If we assume that $\mathbf{f}^b$ is either zero or derived from a separate potential, then (8.1723) reduces to

$$\nabla \times (\nabla^2\mathbf{v}) = 0 \tag{8.1724}$$

Also (8.1722) gives

$$\nabla^2\nabla^2(\nabla \times \mathbf{v}) - \nabla \times (\nabla^2\nabla^2\mathbf{M}) = 0 \tag{8.1725}$$

Example 8.45

Consider the vector-valued function

$$\mathbf{M} = \mathbf{k}\, F(x, y, z) \tag{8.1726}$$

where $F(x, y, z)$ is an arbitrary function. Using the developments presented in earlier show that F is biharmonic.

Solution

From (8.1721)

$$\Psi = \nabla \times \mathbf{M} = \begin{vmatrix} \mathbf{i} & \mathbf{j} & \mathbf{k} \\ \dfrac{\partial}{\partial x} & \dfrac{\partial}{\partial y} & \dfrac{\partial}{\partial z} \\ 0 & 0 & F \end{vmatrix} = \left\{ \frac{\partial F}{\partial y}\mathbf{i} - \frac{\partial F}{\partial x}\mathbf{j} + 0\mathbf{k} \right\} \tag{8.1727}$$

and

$$\mathbf{v} = \nabla \times \Psi = \begin{vmatrix} \mathbf{i} & \mathbf{j} & \mathbf{k} \\ \dfrac{\partial}{\partial x} & \dfrac{\partial}{\partial y} & \dfrac{\partial}{\partial z} \\ \dfrac{\partial F}{\partial y} & -\dfrac{\partial F}{\partial x} & 0 \end{vmatrix}$$

$$= \left\{ \frac{\partial^2 F}{\partial x \partial z}\mathbf{i} + \frac{\partial^2 F}{\partial y \partial z}\mathbf{j} + \left(-\frac{\partial^2 F}{\partial x^2} - \frac{\partial^2 F}{\partial y^2} \right)\mathbf{k} \right\} \tag{8.1728}$$

Also, $F(x, y, z)$ can be specified to within an arbitrary F_0; i.e.

$$F_0(x, y, z) = G(x, y) + H(z) \tag{8.1729}$$

The form of (8.1729) is such that all the velocity components derived from (8.1728) will be zero only if

$$\frac{\partial^2 G}{\partial x^2} + \frac{\partial^2 G}{\partial y^2} = 0 \tag{8.1730}$$

Also

$$\nabla^2 \mathbf{v} = \left\{ \frac{\partial^2}{\partial x \partial z}(\nabla^2 F)\mathbf{i} + \frac{\partial^2}{\partial y \partial z}(\nabla^2 F)\mathbf{j} \right. $$
$$\left. + \left[-\frac{\partial^2}{\partial x^2}(\nabla^2 F) - \frac{\partial^2}{\partial y^2}(\nabla^2 F) \right] \mathbf{k} \right\} \tag{8.1731}$$

substituting (8.1731) in (8.1724) we have

$$\nabla \times (\nabla^2 \mathbf{v}) = \left\{ -\frac{\partial}{\partial y}(\nabla^4 F)\mathbf{i} + \frac{\partial}{\partial x}(\nabla^4 F)\mathbf{j} + 0\mathbf{k} \right\} = 0 \tag{8.1732}$$

This gives

$$\frac{\partial}{\partial x}(\nabla^4 F) = 0 \quad ; \quad \frac{\partial}{\partial y}(\nabla^4 F) = 0 \tag{8.1733}$$

or, by virtue of (8.1729)

$$\nabla^4 F(x, y, z) = 0 \tag{8.1734}$$

Therefore any biharmonic function $F(x, y, z)$ is a solution of a particular problem involving slow viscous flow. The velocity field is given by (8.1728) and from (8.1715), the pressure is given by

$$p = \eta \frac{\partial}{\partial z}(\nabla^2 F) + p_0 \tag{8.1735}$$

where p_0 is a constant.

8.12.7 Planar problems in slow viscous flow

We now examine the category of two-dimensional slow viscous flow problems involving incompressible Newtonian viscous fluids. The steady planar fluid flow occurs in the $x - y$ plane. Accordingly the velocity components v_x and v_y and the scalar pressure p are independent of the z-coordinate. Hence

$$\mathbf{v} = \{v_x(x,y)\mathbf{i} + v_y(x,y)\mathbf{j}\}$$

$$p = p(x,y) \tag{8.1736}$$

The basic equations (8.1716) and (8.1715) (with $\mathbf{f}^b = \mathbf{0}$), give

$$\frac{\partial v_x}{\partial x} + \frac{\partial v_y}{\partial y} = 0 \tag{8.1737}$$

and

$$\frac{\partial p}{\partial x} = \eta \,\overset{\circ}{\nabla}{}^2 v_x \quad ; \quad \frac{\partial p}{\partial y} = \eta \,\overset{\circ}{\nabla}{}^2 v_y \tag{8.1738}$$

where

$$\overset{\circ}{\nabla}{}^2 = \frac{\partial^2}{\partial x^2} + \frac{\partial^2}{\partial y^2} \tag{8.1739}$$

The formulation of the problem it is convenient to introduce a "stream fun-ction" $\Psi^*(x,y)$ such that

$$v_x = -\frac{\partial \Psi^*}{\partial y} \quad ; \quad v_y = -\frac{\partial \Psi^*}{\partial x} \tag{8.1740}$$

The representation satisfies (8.1737) for any choice of Ψ^*. The equations (8.1738) now give

$$\frac{\partial p}{\partial x} = \eta \frac{\partial}{\partial y}\left(\overset{\circ}{\nabla}{}^2 \Psi^*\right) ; \frac{\partial p}{\partial y} = -\eta \frac{\partial}{\partial x}\left(\overset{\circ}{\nabla}{}^2 \Psi^*\right) \tag{8.1741}$$

Eliminating p and Ψ^* successively we obtain

$$\overset{\circ}{\nabla}{}^2 \overset{\circ}{\nabla}{}^2 \Psi^*(x,y) = 0 \quad ; \quad \overset{\circ}{\nabla}{}^2 p(x,y) = 0 \tag{8.1742}$$

The single vorticity component is given by

$$\Omega_z = \frac{\partial v_x}{\partial y} - \frac{\partial v_y}{\partial x} = -\,\overset{\circ}{\nabla}{}^2 \Psi^* \tag{8.1743}$$

Hence it is also possible to use Ω_z instead of Ψ as the dependent variable. The equations (8.1741) now can be written as

$$\frac{\partial p}{\partial x} = -\eta \frac{\partial \Omega_z}{\partial y} \quad ; \quad \frac{\partial p}{\partial y} = \eta \frac{\partial \Omega_z}{\partial x} \tag{8.1744}$$

We can identify these as the Cauchy-Riemann equations (see e.g. (8.398)) indicating that the solution can be expressed in the form of a function with a complex argument

$$f(x + iy) = p(x, y) + i\eta \, \overset{\circ}{\nabla}^2 \Psi^* \tag{8.1745}$$

As expected, the theory of the complex variable formulations discussed in connection with the plane problem in elasticity theory can be adapted for the analysis of the plane problem in slow viscous flow.

8.12.8 Axisymmetric problems in slow viscous flow

Axisymmetric problems in the theory of slow viscous flows can also be reduced to the solution of a fourth order partial differential equation where the operator is $\left\{ \widehat{\nabla}^2 - 2(\partial/\partial r) \right\}^2$ instead of being $(\widehat{\nabla}^2)^2$. The ensuing developments are restricted to the system of cylindrical polar coordinates (r, θ, z); the formulations can, however, be extended to other situations involving axisymmetric problems which can be described more conveniently in relation to a system of orthogonal curvilinear coordinates. The velocity vector is given by

$$\mathbf{v} = \{v_r(r, z)\mathbf{i}_r + v_z(r, z)\mathbf{i}_z\} \tag{8.1746}$$

where $v_r(r, z)$ and $v_z(r, z)$ are the velocity components in the r and z directions respectively, and owing to the assumed symmetry of the motion about the z-axis, the velocity component $v_\theta \equiv 0$. The incompressibility condition (8.1716) gives

$$\frac{\partial v_r}{\partial r} + \frac{v_r}{r} + \frac{\partial v_z}{\partial z} = 0 \tag{8.1747}$$

In the absence of body forces $\mathbf{f}^b$, the equations of motion for steady flows in a Newtonian viscous fluid reduce to

$$\frac{\partial p}{\partial r} = \eta \left[\widehat{\nabla}^2 v_r - \frac{v_r}{r^2} \right]$$

$$\frac{\partial p}{\partial z} = \eta \left[\widehat{\nabla}^2 v_z \right]$$

(8.1748)

where $\widehat{\nabla}^2$ is the axisymmetric form of Laplace's operator referred to the cylindrical polar coordinate system; i.e.

$$\widehat{\nabla}^2 = \frac{\partial^2}{\partial r^2} + \frac{1}{r}\frac{\partial}{\partial r} + \frac{\partial^2}{\partial z^2} \tag{8.1749}$$

Again, for the formulation of a governing partial differential equation, it is convenient to introduce a "stream function" $\chi(r,z)$ such that

$$v_r = -\frac{1}{r}\frac{\partial \chi^*}{\partial z} \quad ; \quad v_z = \frac{1}{r}\frac{\partial \chi^*}{\partial r} \tag{8.1750}$$

which satisfies (8.1747) for any choice of $\chi^*(r,z)$. The equations of motion (8.1748) can now be written in the forms

$$\frac{\partial p}{\partial r} = -\frac{\eta}{r}\frac{\partial}{\partial z}(E^2\chi^*) \quad ; \quad \frac{\partial p}{\partial z} = \frac{\eta}{r}\frac{\partial}{\partial r}(E^2\chi^*) \tag{8.1751}$$

where E^2 is Stokes' operator given by

$$E^2 = \widehat{\nabla}^2 - \frac{2}{r}\frac{\partial}{\partial r} = \frac{\partial^2}{\partial r^2} - \frac{1}{r}\frac{\partial}{\partial r} + \frac{\partial^2}{\partial z^2} \tag{8.1752}$$

Eliminating p from (8.1751), we obtain

$$E^2 E^2 \chi^*(r,z) = 0 \tag{8.1753}$$

and by eliminating χ^* from (8.1751) we have

$$\widehat{\nabla}^2 p(r,z) = 0 \tag{8.1754}$$

Although the axisymmetric problem dealing with steady viscous flow does not give rise to a partial differential equation which is strictly "biharmonic", many of the techniques applicable to the solution of problems involving the biharmonic partial differential equation (expressed in terms of the operator $\widehat{\nabla}^2$) can also be adopted for the solution of (8.1753).

8.12.9 Boundary conditions

In order to complete the formulation of a well-posed boundary value problem associated with the modelling of slow viscous flow, appropriate boundary conditions need to be specified. Consider a rigid boundary S_1 which is in contact with a viscous fluid (Figure 8.102a). This boundary moves with a velocity $\mathbf{V}$. The unit normal to the surface is denoted by $\mathbf{n}$.

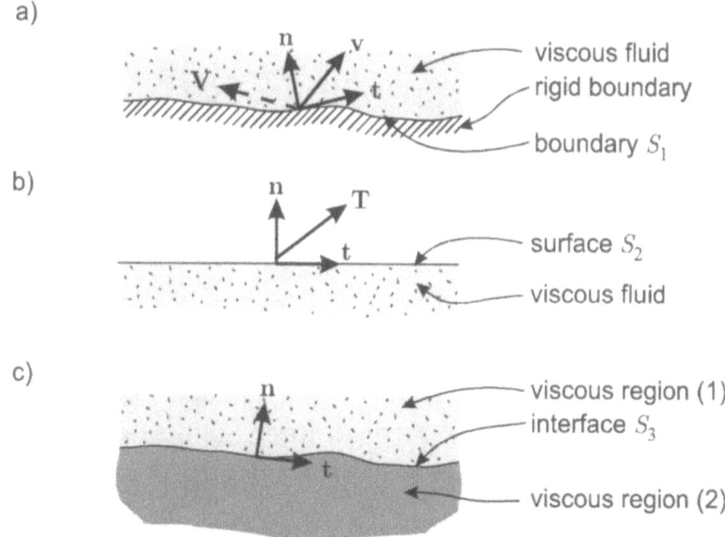

Figure 8.102: Boundary conditions for problems in slow viscous flow.

(i) The kinematic condition that the normal velocity in the viscous fluid in contact with the boundary moving with velocity $\mathbf{V}$ is equal to the normal component of the fluid velocity $\mathbf{v}$ at the boundary, gives

$$\mathbf{v}.\mathbf{n} = \mathbf{V}.\mathbf{n} \quad ; \quad \mathbf{x} \in S_1 \tag{8.1755}$$

this result is identical to the kinematic condition (5.43) developed in connection with potential flow problems associated with *inviscid fluids*. The second kinematic boundary condition at S_1 is based on an *"adherence condition"* or a *"no-slip condition"* which is derived largely on the physical insight and experimental observations of processes that can exist at a surface in contact with a viscous fluid. The assumption states that the tangential velocity along S_1 is the same for both the Newtonian viscous fluid and the boundary, i.e.

$$\mathbf{v.t} = \mathbf{V.t} \qquad \mathbf{x} \in S_1 \tag{8.1756}$$

where $\mathbf{t}$ is the tangent vector to S_1.

(ii) At the surface of a viscous fluid, the boundary conditions can be formulated by considering the equilibrium of tractions at the surface. Consider the surface of a Newtonian viscous fluid which is subjected to tractions $\mathbf{T}$ (Figure 8.102b).

For equilibrium at the boundary S_2

$$\mathbf{T} = \{-p\mathbf{I} + \kappa(\nabla.\mathbf{v})\mathbf{I} + 2\eta\mathbf{s}\}.\mathbf{n} \quad ; \quad \mathbf{x} \in S_2 \tag{8.1757}$$

or in indicial notation

$$T_i = \{-p\delta_{ij} + \kappa v_{k,k}\delta_{ij} + 2\eta s_{ij}\}n_j \quad ; \quad \mathbf{x} \in S_2 \tag{8.1758}$$

(iii) Consider an interface S_3 separating two immiscible viscous fluids with differing viscosity characteristics (Figure 8.102c). At such an interface, the normal tractions and the viscous stresses are continuous provided the surface tension effects are negligible, i.e.

$$\boldsymbol{\sigma}^{(1)}.\mathbf{n} = \boldsymbol{\sigma}^{(2)}.\mathbf{n} \quad ; \quad \mathbf{x} \in S_3 \tag{8.1759}$$

In addition to these mechanical constraints the velocity fields in the two regions should satisfy the kinematic continuity constraints applicable to the continuity of velocity components at the interface, i.e.

$$\mathbf{v}^{(1)} = \mathbf{v}^{(2)} \quad ; \quad \mathbf{x} \in S_3 \tag{8.1760}$$

Example 8.46

The two-dimensional "Hele-Shaw Model" consists of a thin layer of viscous fluid of uniform thickness $2h$ which is in steady motion. The analogue can be used to examine the flow pattern around obstacles as shown in Figure 8.103.

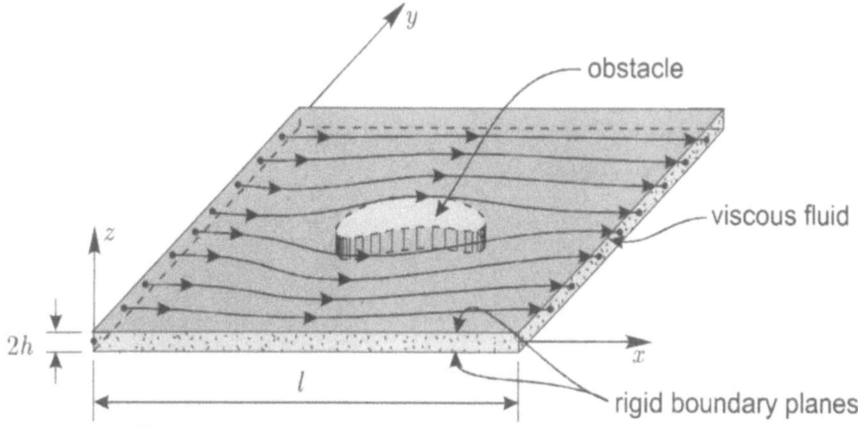

Figure 8.103: Two-dimensional Hele-Shaw Model.

Derive an expression for the velocity distribution in the thin layer as a function of the gradients of the fluid pressure p.

Solution

Since the thickness of the fluid layer $(2h)$ is small in comparison with the in-plane dimensions (ℓ) we can assume that the derivatives of v_x and v_y with respect to x and y are much smaller than the derivatives with respect to z. Since the geometry of the region is unspecified, the only boundary conditions applicable to the problem are the no-slip boundary conditions at the rigid boundary planes $z = \pm h$ (but not at the boundary of the obstacle, Figure 8.103) i.e.

$$v_x(x, y, \pm h) = 0 \quad ; \quad v_y(x, y \pm h) = 0 \tag{8.1761}$$

Since we are interested only in a solution which undeterminate to within a pressure gradient, we can utilize the equation of equilibrium (8.1715) (with $\mathbf{f}^b = \mathbf{0}$). For steady flow with zero velocity in the z-direction, (8.1715) gives

$$\frac{\partial p}{\partial x} = \eta \frac{\partial^2 v_x}{\partial z^2} \quad ; \quad \frac{\partial p}{\partial y} = \eta \frac{\partial^2 v_y}{\partial z^2} \tag{8.1762}$$

and

$$\frac{\partial p}{\partial z} = 0 \tag{8.1763}$$

The equation (8.1763) implies that

$$p = p(x, y) \tag{8.1764}$$

Integrating (8.1762) we obtain

$$\frac{\partial v_x}{\partial z} = \frac{z}{\eta} \frac{\partial p}{\partial x} + \zeta_1(x, y)$$

$$\frac{\partial v_y}{\partial z} = \frac{z}{\eta} \frac{\partial p}{\partial y} + \zeta_2(x, y) \tag{8.1765}$$

where $\zeta_1(x, y)$ and $\zeta_2(x, y)$ are arbitrary functions. Since, from symmetry

$$\left(\frac{\partial v_x}{\partial z} \right)_{z=0} = 0 \quad ; \quad \left(\frac{\partial v_y}{\partial z} \right)_{z=0} = 0 \tag{8.1766}$$

we require

$$\zeta_1(x, y) = \zeta_2(x, y) = 0 \tag{8.1767}$$

Using this result in (8.1765), performing the integrations and making use of the boundary conditions (8.1761) we obtain

$$v_x(x, y, z) = -\frac{(h^2 - z^2)}{2\eta} \frac{\partial p}{\partial x}$$

$$v_y(x, y, z) = -\frac{(h^2 - z^2)}{2\eta} \frac{\partial p}{\partial y} \tag{8.1768}$$

We can define the *average fluid velocities* within the aperture of thickness $2h$ as

$$\bar{v}_x(x,y,z) = \frac{1}{2h} \int_{-h}^{h} v_x(x,y,z)dz$$

$$\bar{v}_y(x,y,z) = \frac{1}{2h} \int_{-h}^{h} v_y(x,y,z)dz$$

(8.1769)

Substituting (8.1768) in (8.1769) we have

$$\bar{v}_x = -\frac{h^2}{3\eta}\frac{\partial p}{\partial x} \quad ; \quad \bar{v}_y = -\frac{h^2}{3\eta}\frac{\partial p}{\partial y}$$

(8.1770)

In vector form,

$$\overset{\circ}{\mathbf{v}} = -\left(\frac{h^2}{3\eta}\right)\overset{\circ}{\nabla}\, p$$

(8.1771)

where $(\circ)$ denotes the two-dimensional form. We can assume that the fluid pressure p is a component of a total potential $\Phi(x,y)$ such that

$$\Phi(x,y) = \Phi_P(x,y) + \Phi_D(x,y)$$

(8.1772)

where Φ_P and Φ_D are pressure and datum potentials respectively, with

$$\Phi_P = \frac{p}{\gamma}$$

and γ is the unit weight of the Newtonian viscous fluid. If we assume that the total potential is measured in relation to the plane $z = 0$, then (8.1771) can be written as

$$\overset{\circ}{\mathbf{v}} = -\frac{(2h)^2}{12\eta}\gamma\,\overset{\circ}{\nabla}\,\Phi$$

(8.1773)

This result can be identified as a form of "*Darcy's Law for a fluid conducting fracture*", where the Darcy constant k is

$$k = \frac{K\gamma}{\eta} \tag{8.1774}$$

where

$$K = \frac{(2h)^2}{12} \tag{8.1775}$$

which has units of (length)^2 is referred to as the "Permeability" of the aperture. The *parallel plate* or *Hele-Shaw* model of an aperture is used quite extensively in the modelling of fluid transport in narrow apertures such as joints and fissures in geological media. Such simplified planar approximations of flow processes in fissures form an important aspect of the models that are used to examine transport problems in geo-environmental engineering.

Example 8.47

The Figure 8.104 shows a cross-section through a thin-film lubrication zone between a stationary plate and a sliding plate. The aperture (h) of the thin film is smaller than the width of the stationary plate (L). The fluid is contained in a chamber such that the excess fluid pressure at the locations $x = 0$ and $x = L$ are zero (the excess fluid pressure is in relation to the hydrostatic fluid pressure). Considering the mechanics of the thin film and the viscous stresses in the fluid show that the "drag force" generated due to a steady movement of the sliding plate at a velocity V_0 (in the x-direction) is given by $\eta V_0 L/2$.

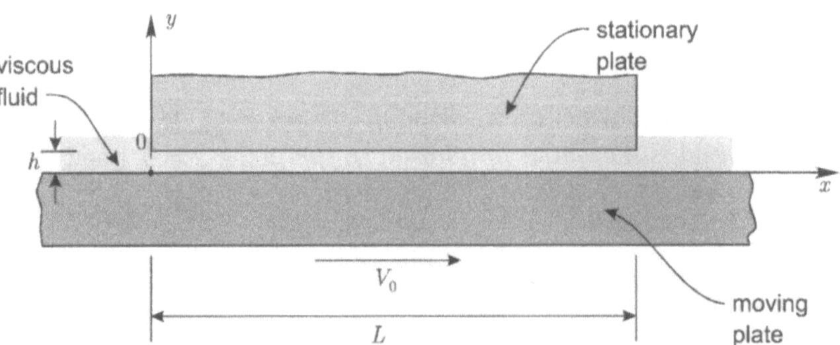

Figure 8.104: Mechanics of a thin lubrication zone.

Solution

In principle, this problem can be formulated in relation to the "stream function" approach which gives rise to the biharmonic equation (see e.g. Section 8.12.7). If one employs such a procedure then the appropriate form of the stream function needs to be selected from many possible alternatives, including polynomial solutions and generalized variables separable solutions. When dealing with Newtonian viscous flow problems which involve thin films or small apertures, it is more convenient to directly introduce the relevant approximations into the equations of motion and fluid continuity equations and to solve them directly.

From considerations of the geometry of the thin film lubrication zone, we can assume that if $(h/L) \ll 1$, then $(y/L) \ll 1$ and if the steady movement of the plate is such that $(V_0 L/\eta)(y/L) \ll 1$, then we can assume that

$$v_x = v_x(y) \quad ; \quad v_y = 0$$

$$p = p(x, y) \tag{8.1776}$$

Also, in view of the two-dimensional nature of the problem $v_z \equiv 0$ and the velocity and pressure fields are independent of z. The equations of motion (8.1738) now give

$$\frac{\partial p}{\partial x} = \eta \frac{\partial^2 v_x}{\partial y^2} \quad ; \quad \frac{\partial p}{\partial y} = 0 \tag{8.1777}$$

The second equation of (8.1777) implies that $p = p(x)$; using this reduction and integrating the first of these equations twice we obtain

$$v_x = \frac{y^2}{2\eta} \left(\frac{dp}{dx} \right) + y f_1(x) + f_2(x) \tag{8.1778}$$

where $f_1(x)$ and $f_2(x)$ are arbitrary functions. To determine these functions we employ the boundary conditions applicable to the planes $y = 0$ and $y = h$. Since the viscous fluid is adhering to the surface during the motion of the fluid we require

$$v_x(x, 0) = V_0 \quad ; \quad v_x(x, h) = 0 \tag{8.1779}$$

which gives

$$f_1(x) = -\frac{1}{h}\left[\frac{h^2}{2\eta}\frac{dp}{dx} + V_0\right] \quad ; \quad f_2(x) = V_0 \tag{8.1780}$$

The result for the velocity v_x can now be written in the form

$$v_x(x,y) = V_0\left(1 - \frac{y}{h}\right) - \frac{yh}{2\eta}\left(1 - \frac{y}{h}\right)\frac{dp}{dx} \tag{8.1781}$$

which is indeterminate to within the pressure gradient (dp/dx). The remaining available equation is the equation of continuity which needs to be used to determine $p(x,y)$. Considering the fluid discharge at any location $x = constant$, we have

$$Q = \int_0^h v_x dy \tag{8.1782}$$

Substituting (8.1781) in (8.1782) and performing the integrations we obtain

$$\frac{dp}{dx} = \frac{6\eta V_0}{h^2} - \frac{12\eta Q}{h^3} \tag{8.1783}$$

Since the right hand side of (8.1783) is constant we can integrate this equation to obtain

$$p(x) = \left(\frac{6\eta V_0}{h^2} - \frac{12\eta Q}{h^3}\right)x + C \tag{8.1784}$$

where C is a constant of integration. Since $p(x)$ refers to the excess fluid pressure we have

$$p(0) = 0 \quad ; \quad p(L) = 0 \tag{8.1785}$$

which give

$$C = 0 \quad ; \quad Q = \frac{V_0 h}{2} \tag{8.1786}$$

Hence $p(x) \equiv 0$. It must be emphasized that this result is a direct consequence of the assumption of a constant value for the thin film thickness. If the film thickness varies over $x \in (0, L)$, (in most practical situations it does vary), the value of $p(x)$ will be finite.

The non-zero shear stress in the viscous fluid layer is given by

$$\sigma_{xy} = \eta \left[\frac{\partial v_x}{\partial y} \right] = -\eta \frac{V_0}{h} \tag{8.1787}$$

The "drag force" D_0 exerted on the moving plate can be obtained by considering the equilibrium of the plate; i.e.

$$D_0 + \int_0^L [\sigma_{xy}]_{y=0} \, dx = 0 \tag{8.1788}$$

or

$$D_0 = \frac{\eta V_0 L}{h} \tag{8.1789}$$

8.12.10 Stokes' paradox and related problems

We consider the problem of the slow flow of a viscous fluid past a cylinder of diameter $2a$ and infinite length, with velocity V_∞ in the x-direction. The boundary value problem can be formulated in relation to Stokes' stream function $\psi(r, \theta)$, referred to the plane polar coordinate system. The partial differential equation governing the stream function is

$$\overset{\circ}{\nabla}^2 \overset{\circ}{\nabla}^2 \psi(r, \theta) = 0 \tag{8.1790}$$

where the components of the velocity vector

$$\mathbf{v} = \{v_r(r, \theta)\mathbf{i}_r + v_\theta(r, \theta)\mathbf{i}_\theta + 0\mathbf{i}_z\} \tag{8.1791}$$

are given by

$$v_r = \frac{1}{r}\frac{\partial \psi}{\partial \theta} \quad ; \quad v_\theta = -\frac{\partial \psi}{\partial r} \tag{8.1792}$$

such that the incompressibility condition

$$\nabla \cdot \mathbf{v} = 0 \tag{8.1793}$$

is satisfied for any choice of $\psi(r, \theta)$. The boundary conditions governing the problem relate to the non-slip conditions on the surface of the cylinder; i.e.

$$v_r(a, \theta) = 0 \quad ; \quad v_\theta(a, \theta) = 0 \tag{8.1794}$$

and the uniform far-field velocity conditions

$$v_r(r, \theta) \to V_\infty \cos\theta \quad \text{as} \quad r \to \infty$$

$$\tag{8.1795}$$

$$v_\theta(r, \theta) \to -V_\infty \sin\theta \quad \text{as} \quad r \to \infty$$

Considering the boundary conditions (8.1794), the far-field conditions (8.1795) and the stream function representations of $\mathbf{v}$, we seek a solution of (8.1790) of the form

$$\psi(r, \theta) = F(r)\sin\theta \tag{8.1796}$$

where $F(r)$ is an arbitrary function. Substituting (8.1796) into (8.1790) gives rise to an Euler-type ordinary differential equation for $F(r)$; which has the solution (see e.g. (8.731))

$$F(r) = \frac{A_1}{r} + A_2 r + A_3 r \ln r + A_4 r^3 \tag{8.1797}$$

where $A_n(n = 1, 2, ..., 4)$ are arbitrary constants. In order to satisfy the far-field conditions (8.1795) we require

$$A_3 = A_4 = 0 \quad ; \quad A_2 \equiv V_\infty \tag{8.1798}$$

The remaining single solution cannot satisfy both boundary conditions on $r = a$, unless $(\mathbf{v})_{x\to\infty} = 0$. This insolvability of the two-dimensional slow viscous flow problem past a cylinder gives rise to "Stokes' paradox" (sometimes, also referred to as the "Whitehead paradox"). It can also be shown that no solution can be found for two-dimensional uniform slow viscous flow past a body located in an unbounded domain, which will satisfy *all* the appropriate boundary and regularity conditions. In other words, steady *uniform flow past cylindrical objects is unattainable at low Reynolds numbers.* A more general proof of the *non-existence* of solutions for creeping flow past a cylinder located in an infinite fluid domain is given in Section 8.14. Certain solutions, however, do exist for unbounded viscous fluid domains; these include slow viscous shear flow in an unbounded domain containing elliptical (or circular) bodies and slow viscous flow past boundaries containing protuberances.

Example 8.48

A rigid cylinder is located in a viscous fluid domain of infinite extent. The fluid is subjected to uniform shearing motion with Cartesian velocity vector

$$\mathbf{v} = (\Omega_0 y\mathbf{i} + 0\mathbf{j} + 0\mathbf{k}) \tag{8.1799}$$

and the cylinder is held stationary (Figure 8.105). If the fluid flow is governed by equations of slow viscous flow, derive an expression for the torque that should be applied to the cylinder to prevent its rotation.

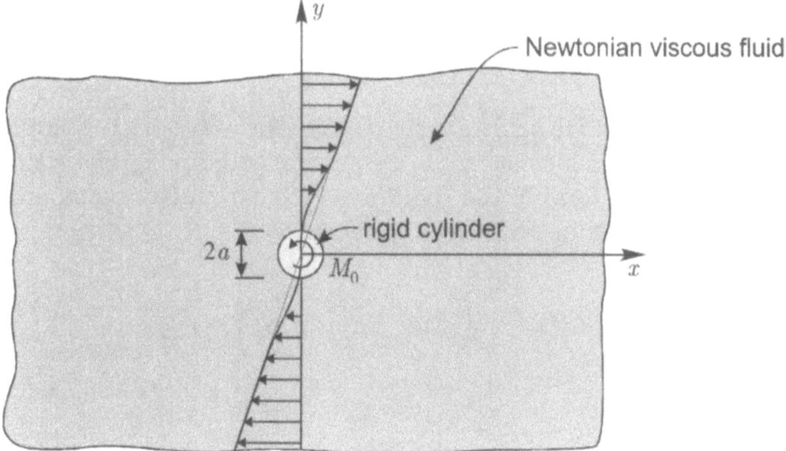

Figure 8.105: Rigid stationary cylinder in a viscous fluid region subjected to uniform shear.

Solution

The boundary value problem governing the slow viscous shearing flow past the stationary cylinder can be formulated in relation to Stokes' stream function governed by

$$\overset{\circ}{\nabla}^2\,\overset{\circ}{\nabla}^2\psi(r,\theta) = 0 \tag{8.1800}$$

The velocity components derived from $\psi(r,\theta)$ should satisfy the boundary conditions

$$v_r(a,\theta) = 0 \quad ; \quad v_\theta(r,\theta) = 0 \tag{8.1801}$$

and the far field regularity conditions

$$v_r(r,\theta) \to \frac{\Omega_0}{2} r \sin 2\theta \quad ; \quad r \to \infty$$

$$\tag{8.1802}$$

$$v_\theta(r,\theta) \to -\frac{\Omega_0}{2} r\left(1 - \cos 2\theta\right) \quad ; \quad r \to \infty$$

Considering the boundary conditions (8.1801) and the far-field regularity conditions (8.1802) we seek solutions of (8.1800) of the form

$$\psi(r,\theta) = F_1(r) + F_2(r)\cos 2\theta \tag{8.1803}$$

where $F_1(r)$ and $F_2(r)$ are arbitrary functions. The appropriate solutions for $F_1(r)$ and $F_2(r)$ can be gleaned from the general solution for the Airy stress function (8.731). The form of the solution which can satisfy the appropriate far field conditions can be written in the form

$$\psi(r,\theta) = A_0 \ln r + B_0 r^2 + \left(A_2 r^2 + \frac{C_2}{r^2} + D_2 \right) \cos 2\theta \tag{8.1804}$$

Using (8.1804) in (8.1792) we obtain

$$v_r(r, \theta) = -\left\{2A_2 r + 2\frac{C_2}{r^3} + 2\frac{D_2}{r}\right\} \sin 2\theta$$

$$v_\theta(r, \theta) = -\frac{A_0}{r} - 2B_0 r + \left\{-2A_2 r + 2\frac{C_2}{r^3}\right\} \cos 2\theta$$

$$(8.1805)$$

Considering the boundary conditions and the far field (8.1801) and (81802) respectively we can determine the arbitrary constants A_0, B_0, A_2, C_2 and D_2 explicitly; we have

$$A_0 = -\frac{\Omega_0 a^2}{2} \quad ; \quad B_0 = -A_2 = \frac{\Omega_0}{4}$$

$$C_2 = -\frac{\Omega_0 a^4}{4} \quad ; \quad D_2 = \frac{\Omega_0 a^2}{2}$$

$$(8.1806)$$

The velocity components are obtained in the form

$$v_r(r, \theta) = \frac{\Omega_0 a}{2} \left[\left(\frac{r}{a}\right) + \left(\frac{a^3}{r^3}\right) - 2\left(\frac{a}{r}\right)\right] \sin 2\theta$$

$$v_\theta(r, \theta) = \frac{\Omega_0 a}{2} \left[\left(\frac{a}{r} - \frac{r}{a}\right) + \left(\frac{r}{a} - \frac{a^3}{r^3}\right) \cos 2\theta\right]$$

$$(8.1807)$$

The shear stress in the viscous fluid $\sigma_{r\theta}$ is given by

$$\sigma_{r\theta} = 2\eta \epsilon_{r\theta} = \eta \left[\frac{\partial v_\theta}{\partial r} - \frac{v_\theta}{r} + \frac{1}{r}\frac{\partial v_r}{\partial \theta}\right] \qquad (8.1808)$$

Using (8.1807) in the above we have

$$\sigma_{r\theta} = \Omega_0 \eta \left(-\frac{a^2}{r^2} + \left\{1 - 2\frac{a^2}{r^2} + 3\frac{a^4}{r^4}\right\} \cos 2\theta\right) \qquad (8.1809)$$

Only the shear tractions can contribute to a moment resultant at the centre of the cylinder. Hence, the couple M_0 required to maintain the cylinder in a stationary position without rotation is given by

$$M_0 + \int_0^{2\pi} \sigma_{r\theta}(a,\theta)a^2 d\theta = 0 \tag{8.1810}$$

Using (8.1809) in (8.1810) and evaluating the result we obtain

$$M_0 = 2\pi a^2 \Omega_0 \eta \tag{8.1811}$$

8.12.11   Slow viscous flow past a sphere

We now examine the problem of a rigid sphere of radius a which is held stationary in a region of slow viscous flow which has a uniform far field velocity

$$\mathbf{v} = V_0 \mathbf{i}_z \tag{8.1812}$$

where V_0 is a constant (Figure 8.106). This axisymmetric three-dimensional problem, unlike the two-dimensional problem presented in connection with Stokes' paradox, is well-posed and was first investigated by Stokes. It represents one of the most celebrated and extensively used solutions associated with the theory of slow viscous flows or creeping flows. The versatility of the solution stems not only from the simplicity of the mathematical analysis but also from the range of extensive applications of the exact closed form result for the drag force generated on the stationary sphere due to the steady viscous flow. This result has important applications in the study of sedimentation of particles encountered in water treatment processes, fluvial processes, mine tailings disposal and in chemical engineering applications of particulate mixtures. Other applications of particular relevance to environmental applications include the study of airborne fine particulates such as dust and mist droplets. Stokes' solution also forms the basis for many of the standard techniques that are used for the measurement of fluid viscosity.

In view of the geometry of problem, it is convenient to formulate the boundary value problem governing slow viscous flow in relation to the system of spherical polar coordinates (R, ϑ, Θ). The basic partial differential equations governing the velocity vector

$$\mathbf{v} = \{v_R(R,\Theta)\mathbf{i}_R + 0\mathbf{i}_\vartheta + v_\Theta(R,\Theta)\mathbf{i}_\Theta\} \tag{8.1813}$$

and the pressure $p(R,\Theta)$ are

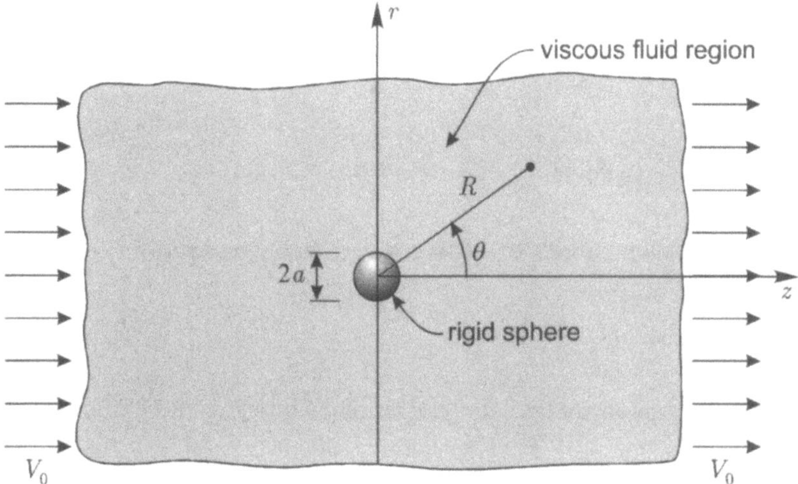

Figure 8.106: Stationary rigid sphere located in a slow viscous flow region of uniform velocity V_0.

$$\widehat{\nabla}_S^2 \mathbf{v} = \nabla p \tag{8.1814}$$

and

$$\nabla \cdot \mathbf{v} = 0 \tag{8.1815}$$

where

$$\widehat{\nabla}_S^2 = \frac{\partial^2}{\partial R^2} + \frac{2}{R}\frac{\partial}{\partial R} + \frac{1}{R^2}\frac{\partial^2}{\partial \Theta^2} + \frac{\cot\Theta}{R^2}\frac{\partial}{\partial \Theta} \tag{8.1816}$$

is the axisymmetric form of Laplace's operator referred to the spherical polar coordinate system.

The boundary conditions governing the sphere problem are the following; on the surface of the sphere

$$v_R(a, \Theta) = 0 \quad ; \quad v_\Theta(a, \Theta) = 0 \tag{8.1817}$$

and at distances remote from the sphere

$$v_R(R,\Theta) \to V_0 \cos\Theta \quad , \quad \text{as} \quad r \to \infty$$

$$v_\Theta(R,\Theta) \to -V_0 \sin\Theta \quad , \quad \text{as} \quad r \to \infty \tag{8.1818}$$

Also the fluid pressure needs to satisfy a regularity condition

$$p(R,\Theta) \to p_\infty \quad \text{as} \quad R \to \infty \tag{8.1819}$$

Since the flow is axisymmetric, the continuity equation (8.1815) reduces to

$$\frac{\partial v_R}{\partial R} + 2\frac{v_R}{R} + \frac{1}{R}\frac{\partial v_\Theta}{\partial \Theta} + \frac{\cot\Theta}{R}v_\Theta = 0 \tag{8.1820}$$

We can introduce Stokes' stream function $\psi(R,\Theta)$ such that

$$v_R = \frac{1}{R^2\sin\Theta}\frac{\partial\psi}{\partial\Theta} \quad ; \quad v_\Theta = -\frac{1}{R\sin\Theta}\frac{\partial\psi}{\partial R} \tag{8.1821}$$

and the continuity equation (8.1820) is satisfied for all choices of $\psi(R,\Theta)$ and the equations of motion (8.1814) can be reduced to the partial differential equations

$$E_S^2 E_S^2 \psi(R,\Theta) = 0 \tag{8.1822}$$

$$\widehat{\nabla}_S^2 p(R.\Theta) = 0 \tag{8.1823}$$

where E_S^2 is the axisymmetric form of Stokes' operator given by

$$E_S^2 = \frac{\partial^2}{\partial R^2} + \frac{1}{R^2}\frac{\partial^2}{\partial\Theta^2} - \frac{\cot\Theta}{R^2}\frac{\partial}{\partial\Theta} \tag{8.1824}$$

The boundary conditions (8.1817) on the surface of the sphere can now be written as

$$\frac{\partial\psi}{\partial R} = 0 \quad ; \quad \frac{\partial\psi}{\partial\Theta} = 0 \quad \text{on} \quad R = a \tag{8.1825}$$

and the far field uniform velocity boundary conditions can be replaced by the requirement

$$\psi(R,\Theta) \to \frac{V_0 R^2}{2} \sin^2\Theta \quad \text{as} \quad R \to \infty \tag{8.1826}$$

Considering the form of the revised boundary conditions (8.1825) on the surface of the sphere and the far field boundary condition (8.1826), we seek a variables separable solution of $\psi(R,\Theta)$ of the form

$$\psi(R,\Theta) = \frac{V_0}{2} F(R) \sin^2\Theta \tag{8.1827}$$

where $F(R)$ is an arbitrary function. Substituting (8.1827) in (8.1822) gives the fourth-order ordinary differential equation

$$\left(\frac{d^2}{dR^2} - \frac{2}{R^2} \right)^2 F(R) = 0 \tag{8.1828}$$

This equation can be satisfied by a polynomial solution of the form

$$F(R) = C_n R^n \tag{8.1829}$$

(where C_n are arbitrary constants) provided n satisfies the algebraic equation

$$[(n-2)(n-3) - 2]\,[n(n-1) - 2] = 0 \tag{8.1830}$$

giving roots

$$n = -1, 1, 2 \text{ and } 4 \tag{8.1831}$$

The relevant general solution for $\psi(R,\Theta)$ can now be written as

$$\psi(R,\Theta) = \frac{V_0}{2} \left[\frac{A}{R} + BR + CR^2 + DR^4 \right] \sin^2\Theta \tag{8.1832}$$

where A, B, C and D are arbitrary constants.

Considering the revised far field regularity conditions (8.1826), it is evident that for the far field conditions to be satisfied we require

$$C = 1 \quad ; \quad D = 0 \tag{8.1833}$$

The boundary conditions on the surface of the sphere give

$$A = \frac{a^3}{2} \quad ; \quad B = -\frac{3a}{2} \tag{8.1834}$$

The complete solution for Stokes' stream function which satisfies the boundary conditions on the surface of the sphere and the far field uniform velocity boundary conditions has the form

$$\psi(R, \Theta) = \frac{V_0}{2} \left[R^2 - \frac{3aR}{2} + \frac{a^3}{2R} \right] \sin^2 \Theta \tag{8.1835}$$

The velocity components derived from (8.1821) are

$$v_R(R, \Theta) = V_0 \left[\frac{1}{2} \left(\frac{a}{R} \right)^3 - \frac{3}{2} \left(\frac{a}{R} \right) + 1 \right] \cos \Theta$$

$$\tag{8.1836}$$

$$v_\Theta(R, \Theta) = V_0 \left[\frac{1}{4} \left(\frac{a}{R} \right)^3 + \frac{3}{4} \left(\frac{a}{R} \right) - 1 \right] \sin \Theta$$

The remaining quantity that needs to be determined is the scalar pressure field $p(R, \Theta)$, which is given by (8.1814) gives

$$\frac{\partial p}{\partial R} = \eta \left[\widehat{\nabla}_S^2 v_R - \frac{2v_R}{R^2} - \frac{2}{R^2} \frac{\partial v_\Theta}{\partial \Theta} - \frac{2v_\Theta}{R^2} \cot \Theta \right] \tag{8.1837}$$

$$\frac{1}{R} \frac{\partial p}{\partial \Theta} = \eta \left[\widehat{\nabla}_S^2 v_\Theta - \frac{v_\Theta}{R^2 \sin^2 \Theta} + \frac{2}{R^2} \frac{\partial v_R}{\partial \Theta} \cot \Theta \right] \tag{8.1838}$$

Substituting (8.1836) into (8.1837) and (8.1838) and integrating the resulting expressions we can show that

$$p(R, \Theta) = p_\infty - \frac{3\eta a V_0 \cos \Theta}{2R^2} \tag{8.1839}$$

The problem is now formally solved. A result of importance to engineering applications of the sphere problem relates to the evaluation of the drag force (F_D) exerted by the viscous fluid, moving with a far field velocity V_0, on the stationary rigid sphere. In order to compute the drag force, we need to evaluate the resultant of tractions acting on the surface of the rigid sphere $R = a$, along the z-direction. The components of the tractions due to stresses σ_{RR} and $\sigma_{R\Theta}$ take the forms

$$\sigma_{RR} = -p + 2\eta \frac{\partial v_R}{\partial R}$$

$$\sigma_{R\Theta} = \eta \left(\frac{\partial v_\Theta}{\partial R} - \frac{v_\Theta}{R} + \frac{1}{R} \frac{\partial v_R}{\partial \Theta} \right)$$

$$(8.1840)$$

Using (8.1836) and (8.1839), these expressions give

$$\sigma_{RR} = -p_\infty + \eta \frac{V_0}{a} \left[\frac{9}{2} \left(\frac{a}{R} \right)^2 - 3 \left(\frac{a}{R} \right)^4 \right] \cos \Theta$$

$$(8.1841)$$

$$\sigma_{R\Theta} = \eta \frac{V_0}{a} \left[-\frac{3}{2} \left(\frac{a}{R} \right)^4 \right] \sin \Theta$$

The action of these stress components on the surface of the sphere are shown in Figure 8.107.

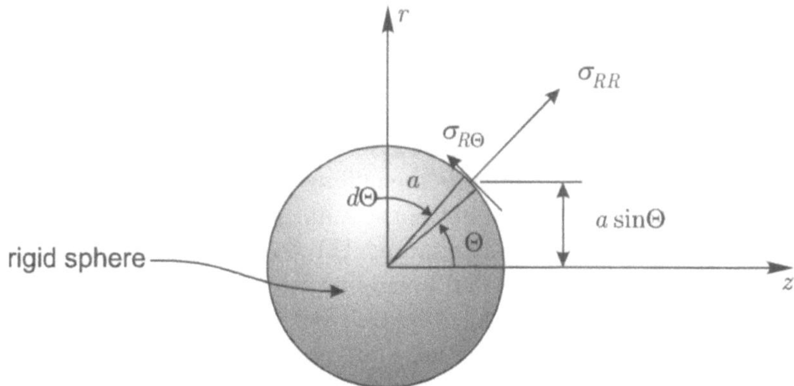

Figure 8.107: Tractions acting on the surface of the rigid sphere.

Considering the tractions acting on the surface of the rigid sphere we can show that

$$F_D - \int_0^\pi \{\sigma_{RR}\cos\Theta - \sigma_{R\Theta}\sin\Theta\}_{R=a}\, 2\pi a^2 \sin\Theta\, d\Theta = 0 \qquad (8.1842)$$

Substituting the expression (8.1841) into (8.1842) and performing the integrations we obtain

$$F_D = 6\pi\eta a V_0 \qquad (8.1843)$$

which is the classical result obtained by Stokes for the viscous drag on the rigid sphere located in a slow viscous flow region with uniform far field velocity V_0.

8.12.12   Application of integral transform techniques

Integral transform techniques offer alternative procedures for examining certain types of problems involving slow viscous flow. With reference to infinite fluid domains, integral transform techniques become more effective when the geometry of the embedded object corresponds to either a plane or a circular disc shaped region, where the plane of the object can be made to coincide with a coordinate plane. When dealing with two-dimensional problems involving slow viscous flow at uniform velocity, by virtue of Stokes' paradox, there is no solution to the problem which deals with a cylinder with an elliptical cross-section which is located in a viscous flow region of infinite extent (figure 8.108). As such, Stokes' paradox extends to the limiting case of a flat plate which is located in an infinite viscous flow region at uniform velocity.

We have shown, however, that solutions can be developed for two-dimensional problems where the symmetric object is located in a fluid domain of infinite extent which is subjected to steady shear flow. This would suggest that the solution to the problem of shear flow past a flat plate could be developed by adopting an integral transform technique. With axisymmetric bodies such as oblate and prolate spheroids and their limiting shapes (e.g. disc, sphere and elongated needle shaped objects) the slow viscous flow problem is well posed and solvable. Therefore the axisymmetric problem related to uniform slow viscous flow past a flat circular disc can be formulated by appeal to an integral transform technique involving Hankel transforms. In this section we shall outline two examples where integral transform techniques can be used to formulate certain slow viscous flow problems for infinite fluid domains, as mixed boundary value problems. These approaches however result in systems of dual integral equations, the solutions of which are only provided by appeal to certain standard procedures.

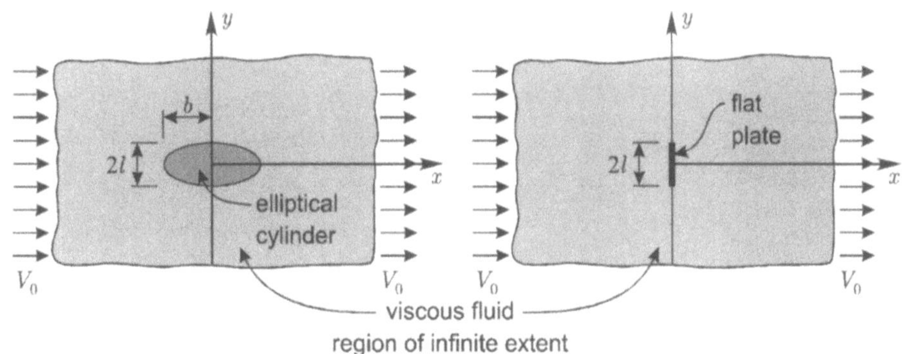

Figure 8.108: Examples of two-dimensional slow viscous flow problems for which no solutions exist by virtue of Stokes' paradox.

Example 8.49

A flat rigid plate of width 2ℓ is located in a viscous fluid region of infinite extent. The flat plate is subjected to a steady rotation Ω_0 about its centre . (Figure 8.109). Use an integral transform technique to develop a relationship between Ω_0 and the moment M_0 that should be applied at the centre of the plate to maintain the steady rotation. Show that this solution when combined with an appropriate far field velocity distribution gives the solution to the problem of a stationary rigid plate which is placed in a region of uniform shear flow.

Solution

The boundary value problem dealing with the steady rotation of the flat plate requires the solution of the biharmonic equation for the stream function

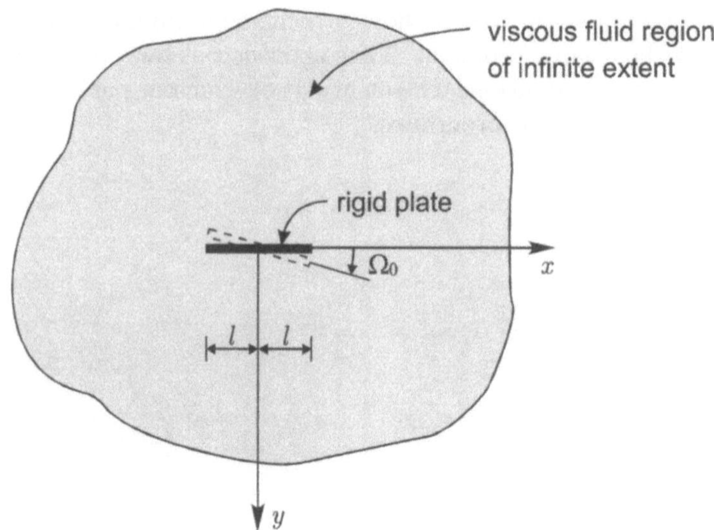

Figure 8.109: Steady rotation of a flat plate contained in a viscous fluid region of infinite extent.

$$\overset{\circ}{\nabla}^2 \, \overset{\circ}{\nabla}^2 \psi(x,y) = 0 \tag{8.1844}$$

which is subject to certain boundary conditions and regularity conditions. Considering the infinite viscous fluid domain, we note that the rotation induces a velocity field which exhibits asymmetry about the plane $y = 0$. Therefore, we can restrict attention to a half-plane region of the viscous fluid occupying $x \in (-\infty, \infty)$, $y \in (0, \infty)$ where the plane $y = 0$ is now subjected to mixed boundary conditions determined by the state of asymmetry and the 'no slip' boundary conditions at the plate viscous fluid interface. These correspond to

$$v_y(x,0) = \begin{cases} \Omega_0 x & 0 \le x \le \ell \\ \\ -\Omega_0 x & -\ell \le x \le \ell \end{cases} \tag{8.1845}$$

$$v_x(x,0) = 0 \qquad -\ell \le x \le \ell \tag{8.1846}$$

which are determined by the rotation and no slip conditions in the plate region and

$$v_x(x,0) = 0 \quad ; \quad -\infty \leq x \leq -\ell \quad ; \quad \ell \leq x \leq \infty \tag{8.1847}$$

$$\sigma_{yy}(x,0) = 0 \quad ; \quad -\infty < x < -\ell \quad ; \quad \ell < x < \infty \tag{8.1848}$$

which are determined by virtue of the *asymmetry* of the problem about $y = 0$. In addition, the velocity and stresses determined from the stream function $\psi(x, y)$ should reduce to zero as $|x| \to \infty$. The velocity components can be expressed in terms of $\psi(x, y)$ in the form

$$v_x = \frac{\partial \psi}{\partial y} \quad ; \quad v_y = -\frac{\partial \psi}{\partial x} \tag{8.1849}$$

The stress component relevant to the boundary condition (8.1848) is given by

$$\sigma_{yy} = -p - 2\eta \frac{\partial^2 \psi}{\partial x \partial y} \tag{8.1850}$$

where $p(x, y)$ is determined from the equations

$$\frac{\partial p}{\partial x} = \eta \frac{\partial}{\partial y}\left(\overset{\circ}{\nabla}^2 \psi\right) \quad ; \quad \frac{\partial p}{\partial y} = -\eta \frac{\partial}{\partial x}\left(\overset{\circ}{\nabla}^2 \psi\right) \tag{8.1851}$$

In (8.1844) and (8.1851), $\overset{\circ}{\nabla}^2$ is the two-dimensional form of Laplace's operator referred to the $x - y$ system. Considering a Fourier transform development of (8.1844) it can be shown that (see e.g. Section 8.8.3) an appropriate integral solution of (8.1844) is

$$\psi(x, y) = \sqrt{\frac{2}{\pi}} \int_0^\infty \left[A(\xi) + B(\xi)y\right] e^{-\xi y} \cos(\xi x) d\xi \tag{8.1852}$$

gives the appropriate asymmetry of the velocity component v_y about $x = 0$ and satisfies the regularity conditions as $y \to \infty$. Considering the boundary conditions (8.1846) and (8.1847) we note that

$$v_x(x,0) = 0 \quad ; \quad x \in (-\infty, \infty) \tag{8.1853}$$

From (8.1849) and (8.1852) we note that in order to satisfy (8.1853) we require

$$B(\xi) = \xi A(\xi) \tag{8.1854}$$

The reduced form for $\psi(x, y)$ is

$$\psi(x, y) = \sqrt{\frac{2}{\pi}} \int_0^\infty A(\xi) \left[1 + \xi y\right] e^{-\xi y} \cos(\xi x) d\xi \tag{8.1855}$$

Using this result in (8.1851) and integrating the equations we can show that

$$p(x, y) = \eta \sqrt{\frac{2}{\pi}} \int_0^\infty 2\xi^2 A(\xi) e^{-\xi y} \sin(\xi x) d\xi \tag{8.1856}$$

Since the appropriate asymmetry about $x = 0$ is already incorporated we can restrict attention to the interval $x \in (0, \infty)$ and pose the mixed boundary conditions governing the rotation of the flat plate as follows:

$$v_y(x, 0) = \Omega_0 x \quad ; \quad 0 \le x \le \ell \tag{8.1857}$$

$$\sigma_{yy}(x, 0) = 0 \quad ; \quad \ell < x < \infty \tag{8.1858}$$

Considering the solutions (8.1855) and (8.1856) we can show that the mixed boundary conditions (8.1857) and (8.1858) are equivalent to the system of dual integral equations

$$\int_0^\infty \xi A(\xi) \sin(\xi x) d\xi = \sqrt{\frac{\pi}{2}} \Omega_0 x \quad ; \quad 0 \le x \le \ell \tag{8.1859}$$

$$\int_0^\infty \xi^2 A(\xi) \sin(\xi x) d\xi = 0 \quad ; \quad \ell < x < \infty \tag{8.1860}$$

The solution of the dual system defined by (8.1859) and (8.1860) is given in several texts and papers on the theory of dual integral equations. A comprehensive treatment of the various method of solution of dual integral equations is given in the authoritative works by L.A. Galin (1912-1981), Y.Ia. Ufliand (1916-1991) and I.N. Sneddon (1919-). In this we shall present the final solution for the unknown function $A(\xi)$. Using the relationship

$$J_{1/2}(\xi x) = \left(\frac{2}{\pi\xi x}\right)^{1/2} \sin(\xi x) \tag{8.1861}$$

we can rewrite (8.1859) and (8.1860) in the forms

$$\int_0^\infty \left[\xi^{5/2}A(\xi)\right]\xi^{-1}J_{1/2}(\xi x)d\xi = \Omega_0 x^{1/2} \quad ; \quad 0 \le x \le \ell$$

$$\int_0^\infty \left[\xi^{5/2}A(\xi)\right]J_{1/2}(\xi x)d\xi = 0 \quad ; \quad \ell < x < \infty \tag{8.1862}$$

Using procedures outlined by Sneddon (Fourier Transforms, Chapter 9). The solution of (8.1862) for $A(\xi)$ takes the form

$$A(\xi) = \sqrt{\frac{\pi}{2}}\Omega_0\ell\frac{J_1(\xi\ell)}{\xi^2} \tag{8.1863}$$

The normal stress acting on the plane $y = 0$ is given by

$$\sigma_{yy}(x,0) = -2\eta\sqrt{\frac{2}{\pi}}\int_0^\infty \xi^2 A(\xi)\sin(\xi x)d\xi \tag{8.1864}$$

Substituting (8.1863) in (8.1864) and performing the integrations we obtain

$$\sigma_{yy}(x,0) = \begin{cases} -\dfrac{2\eta\Omega_0 x}{(\ell^2 - x^2)^{1/2}} & |x| < \ell \\[2ex] 0 & |x| > \ell \end{cases} \tag{8.1865}$$

The moment (M_0) rotation (Ω_0) relationship for the flat plate can be obtained by considering the equilibrium of the plate. We note that the contact stresses acting on the face of the plate in contact with the half-plane region $y \ge 0$ will contribute to only one half of the applied moment M_0. Therefore the equilibrium equation for the flat plate is given by

$$\frac{M_0}{2} + \int_{-\ell}^{\ell} x\sigma_{yy}(x,0)dy = 0 \tag{8.1866}$$

Substituting (8.1865) in (8.1866) we obtain the result

$$M_0 = 2\pi \ell^2 \Omega_0 \eta \tag{8.1867}$$

Example 8.50

A flat circular plate is located in a viscous fluid region of infinite extent. The disc is subjected to a uniform axial translation with velocity U_0. (Figure 8.110) If the surfaces of the disc exhibit '*no-slip*' boundary conditions, derive an expression for the axial force P_0 required to maintain the uniform velocity U_0.

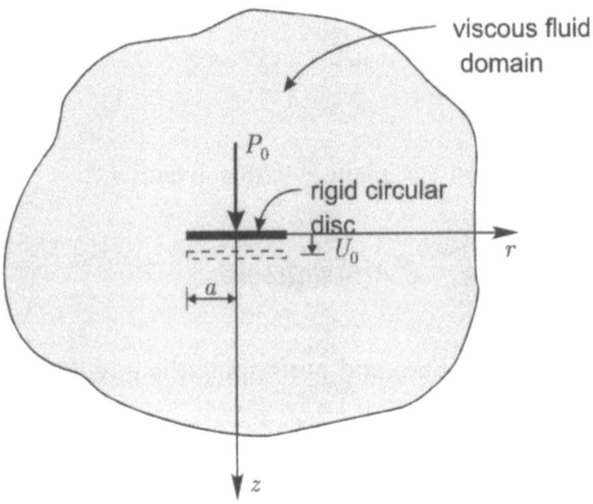

Figure 8.110: Axial translation of a rigid disc located in a viscous fluid domain of infinite extent.

Solution

The axisymmetric boundary value problem governing the translation of the disc can be formulated with reference to Stokes' stream function which satisfies the partial differential equation

$$E^2 E^2 \chi^*(r, z) = 0 \tag{8.1868}$$

The boundary conditions governing the problem can be formulated by considering the asymmetry of the motion about the plane $z = 0$. due to this asymmetry, we can formulate the boundary conditions in relation to the viscous half-space region occupying $z \in (0, \infty)$.

The boundary conditions applicable to the exterior region $r \in (a, \infty)$ of the plane $z = 0$ are

$$v_r(r, 0) = 0 \quad ; \quad a \le r < \infty \tag{8.1869}$$

$$\sigma_{zz}(r, 0) = 0 \quad ; \quad a < r < \infty \tag{8.1870}$$

In the disc region, the imposed velocity and the assumed non-slip conditions give rise to the following boundary conditions:

$$v_r(r, 0) = 0 \quad ; \quad 0 \le r \le a \tag{8.1871}$$

$$v_z(r, 0) = U_0 \quad ; \quad 0 \le r \le a \tag{8.1872}$$

In addition to these boundary conditions, the velocity and stress fields determined from $\chi^*(r, z)$ should reduce to zero as $r, z \to \infty$. In view of the axial symmetry of the boundary value problem, we can seek variables separable solutions. The fourth-order operator $E^2 E^2$, however, does not lend itself to the development of solutions based on a generalized variables separation scheme (similar limitations were observed in connection with the development of solutions for the biharmonic equation). The choice of the form of a solution has to be narrowed down by selecting a specific dependence in r. Since the velocity components v_r and v_z are determined via (8.1750) in terms of $\chi^*(r, z)$, the velocities should be finite as $r \to 0$ and should decay to zero as $r \to \infty$. Considering the representations in (8.1750) we can select a specific form for $\chi^*(r, z)$ as

$$\chi^*(r, z) = G(z) r J_1(\xi r) \tag{8.1873}$$

which gives bounded velocity components at $r = 0$. Substituting (8.1873) in (8.1868) gives the ordinary differential equation

$$\left(\frac{d^2}{dz^2} - \xi^2 \right)^2 G(z) = 0 \tag{8.1874}$$

for the unknown function $G(z)$. Of the four independent solutions of (8.1874) the ones that are applicable to the half-space region $z \in (0, \infty)$ are

$$G(z) = [A(\xi) + zB(\xi)]\, e^{-\xi z} \tag{8.1875}$$

where $A(\xi)$ and $B(\xi)$ are arbitrary functions. We can construct an integral solution of $\chi^*(r, z)$ in the form

$$\chi^*(r, z) = \int_0^\infty \xi\,[A(\xi) + zB(\xi)]\, e^{-\xi z} r J_1(\xi r)\, d\xi \tag{8.1876}$$

Combining the boundary conditions (8.1869) and (8.1871) applicable to the radial velocity we note that

$$v_r(r, 0) = 0 \quad ; \quad r \in (0, \infty) \tag{8.1877}$$

To satisfy (8.1877) we require

$$B(\xi) = \xi A(\xi) \tag{8.1878}$$

The reduced expression for $\chi^*(r, z)$ given by

$$\chi^*(r, z) = \int_0^\infty \xi A(\xi)\,[1 + \xi z]\, e^{-\xi z} r J_1(\xi r)\, d\xi \tag{8.1879}$$

can be used in conjunction with (8.1751) to determine the hydrostatic pressure $p(r, z)$; i.e.

$$p(r, z) = 2\eta \int_0^\infty \xi^3 A(\xi) e^{-\xi z} J_0(\xi r)\, d\xi \tag{8.1880}$$

The expressions for $v_z(r, z)$ and $\sigma_{zz}(r, z)$ required to formulate the remaining mixed boundary conditions can be evaluated in the forms

$$v_z(r, z) = \int_0^\infty \xi^2 A(\xi)\,[1 + \xi z]\, e^{-\xi z} J_0(\xi r)\, d\xi \tag{8.1881}$$

$$\sigma_{zz}(r, z) = -2\eta \int_0^\infty \xi^3 A(\xi)\,[1 + \xi z]\, e^{-\xi z} J_0(\xi r)\, d\xi \tag{8.1882}$$

Using the expressions in (8.1870) and (8.1872) we obtain the following system of dual integral equations for the unknown function $A(\xi)$:

$$\int_0^\infty \xi^2 A(\xi) J_0(\xi r)d\xi = U_0 \quad ; \quad 0 \leq r \leq a$$

$$\int_0^\infty \xi^3 A(\xi) J_0(\xi r)d\xi = 0 \quad ; \quad a < r < \infty$$

$$(8.1883)$$

This system of dual integral equations is identical to that discussed in Section 8.10.13 in connection with the indentation of an isotropic elastic half-space by a rigid smooth punch with a flat base. The details of the method of solution is given in Section 8.10.13 and will not be repeated here. It is sufficient to note that the solution of the dual system gives the following expression for the stress $\sigma_{zz}(r,z)$ in the region of the disc;

$$\sigma_{zz}(r,0) = -\frac{4\eta U_0}{\pi(a^2-r^2)^{1/2}} \quad ; \quad 0 < r < a \qquad (8.1884)$$

The magnitude of the force P_0 required to maintain the steady velocity U_0 can be obtained by considering the equilibrium of the rigid circular disc. We also note that the stress $\sigma_{zz}(r,0^-)$ acting on the surface of the disc in contact with the viscous half-space region $z \in (-\infty,0)$ has the same distribution as given by (8.1884) except for a change in sign to reflect the tensile nature of the stress (Figure 8.111)

Considering the equilibrium of the disc, we have

$$P_0 - 2\pi \int_0^a \left[\sigma_{zz}(r,0^+) - \sigma_{zz}(r,0^-)\right] rdr = 0 \qquad (8.1885)$$

where

$$\sigma_{zz}(r,0^+) = -\sigma_{zz}(r,0^-) = -\frac{4\eta U_0}{\pi(a^2-r^2)^{1/2}} \qquad (8.1886)$$

Evaluating (8.1885) we obtain

$$P_0 = 16\eta U_0 a \qquad (8.1887)$$

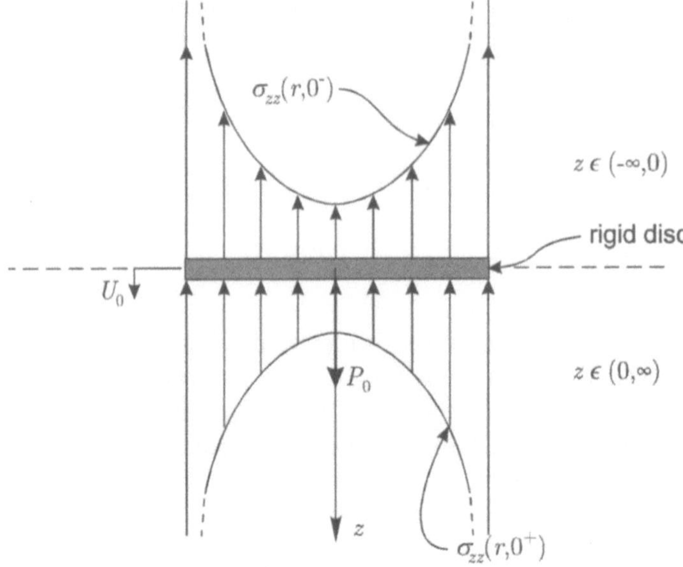

Figure 8.111: Free body diagram for the rigid disc contained within a viscous fluid.

which agrees with the classical result obtained as a limiting case of a rigid oblate spheroid which translates with uniform velocity U_0, along the minor axis.

8.12.13   Diffusive motions in viscous fluids

Although this topic does not directly relate to the biharmonic equation, it is pertinent to include a discussion of the application of the diffusion equation to the study of motions in Newtonian viscous fluids. This is particularly facilitated by the more general treatment of the equations governing slow viscous flow presented in the preceding sections. As discussed briefly in Section 6.1.6, unsteady fluid motion can be initiated at boundaries of viscous fluid domains, when such boundaries are subjected to sudden motions. This type of unsteady motion in viscous fluids was investigated by Sir G.G. Stokes (1819-1903) and Lord Rayleigh (1842-1919). The former examined the problem of an oscillating flat plate in contact with a viscous fluid domain of semi-infinite extent and the latter examined the transient effects which occur in a viscous half-space region due to impulsive shearing motion of a plate resting on its surface. In this section we shall examine certain basic solutions for spatially one-dimensional diffusive motions in viscous fluids.

Example 8.51

A viscous fluid domain occupies the region $x \in (-\infty, \infty)$, $y \in (-\infty, \infty)$ and $z \in (0, \infty)$. The surface of the viscous is in contact with a rigid flat plate of infinite extent. This plate is set in motion with a velocity V_0 in the x-direction, at time $t = 0$. The partial differential equation governing the velocity $v_x(z, t)$ in the viscous half-space region is given by

$$\nu^* \frac{\partial^2 v_x}{\partial z^2} = \frac{\partial v_x}{\partial t} \tag{8.1888}$$

where $\nu^*(= \eta/\rho)$ is the kinematic viscosity, η is the viscosity and ρ is the mass density. Use a Laplace transform technique to determine the velocity distribution in the half-space region.

Solution

The initial condition governing the problem is

$$v_x(z, 0) = 0 \tag{8.1889}$$

and the solution should satisfy the following boundary and regularity conditions:

$$v_x(0, t) = V_0 \; ; t > 0$$

$$\tag{8.1890}$$

$$v_x(z, t) \to 0 \; ; z \to \infty \; ; \; t > 0$$

In view of the homogeneous initial condition, it is also convenient to adopt a Laplace transform technique for the solution of the initial boundary value problem. We denote the Laplace transform of $v_x(z, t)$ by

$$\bar{v}_x(z, s) = \mathcal{L}\{v_x(z, t)\} \tag{8.1891}$$

and

$$\mathcal{L}\left\{\frac{\partial v_x}{\partial t}\right\} = s\bar{v}_x(z, s) - v_x(z, 0) \tag{8.1892}$$

Applying the Laplace transform, (8.1888) gives

$$\nu^* \frac{d^2 \bar{v}_x}{dz^2} - s\bar{v}_x = 0 \tag{8.1893}$$

The solution of (8.1893) can be written as

$$\bar{v}_x(z,s) = Ae^{-z\sqrt{s/\nu^*}} + Be^{z\sqrt{s/\nu^*}} \tag{8.1894}$$

where A and B are either constants or functions of the transform parameter s. They can be determined by satisfying the boundary conditions and regularity conditions given by (8.1890). To satisfy the regularity condition as $z \to \infty$, we require $B \equiv 0$, and since

$$\bar{v}_x(0,s) = \mathcal{L}\{v_x(0,t)\} = \frac{V_0}{s} \tag{8.1895}$$

the relevant solution of (8.1894) can be reduced to

$$\bar{v}_x(z,s) = \frac{V_0}{s} e^{-z\sqrt{s/\nu^*}} \tag{8.1896}$$

The inverse Laplace transform of (8.1896) can be obtained via a complex inversion procedure, i.e.

$$v_x(z,t) = \frac{1}{2\pi i} \int_{\gamma-i\infty}^{\gamma+i\infty} \frac{V_0 e^{st-z\sqrt{s/\nu^*}}}{s} ds \tag{8.1897}$$

where the real number γ is chosen such that the line $x = \gamma$ in the complex plane lies to the right of all the singularities of the integrand. The inversion is left as an exercise (see e.g. Problem 1.13). The final result for the velocity distribution can be obtained in terms of the error function:

$$v_x(z,t) = V_0 \left[1 - \mathrm{erf}\left(\frac{z}{2\sqrt{\nu^* t}} \right) \right] \tag{8.1898}$$

where

$$\mathrm{erf}(\xi) = \frac{2}{\sqrt{\pi}} \int_0^{\xi} e^{-u^2}\, du \tag{8.1899}$$

and tabulated values for $\mathrm{erf}(\xi)$ are given the literature.

Example 8.52

A Newtonian viscous fluid of viscosity η and density ρ is contained in a long horizontal tube of length ℓ and diameter $2a(\ell \gg 2a)$. Initially the fluid is at rest and at time $t = 0$, a pressure gradient $(p_0 - p_\ell)/\ell$ is induced in the system (Figure 8.112). If unsteady laminar flow is induced in the fluid region, determine the transient velocity distribution induced in the fluid prior to the attainment of a steady state.

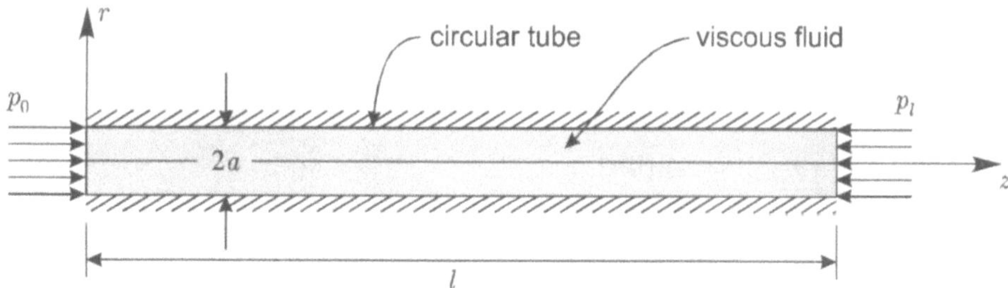

Figure 8.112: Unsteady laminar flow in a circular tube.

Solution

Since the length of the tube is much longer than its diameter, we can neglect the "*end effects*" at the locations $z = 0$ and $z = \ell$ and assume that the unsteady laminar flow is spatially one-dimensional with velocity vector

$$\mathbf{v} = 0\,\mathbf{i}_r + 0\,\mathbf{i}_\theta + v_z(r,t)\,\mathbf{i}_z \tag{8.1900}$$

The equation of slow viscous flow (8.1713) now reduce to

$$\eta\left(\frac{\partial^2 v_z}{\partial r^2} + \frac{1}{r}\frac{\partial v_z}{\partial r}\right) - \frac{\partial p}{\partial z} = \rho\frac{\partial v_z}{\partial t}$$

$$\frac{\partial p}{\partial r} = \frac{1}{r}\frac{\partial p}{\partial \theta} = 0 \tag{8.1901}$$

Considering the one-dimensional pressure gradient along the length of the tube we can rewrite the first equation of (8.1901) as

$$\nu^* \left\{ \frac{\partial^2 v_z}{\partial r^2} + \frac{1}{r}\frac{\partial v_z}{\partial r} \right\} + \frac{p_o - p_\ell}{\rho \ell} = \frac{\partial v_z}{\partial t} \tag{8.1902}$$

which is the governing partial differential equation. The boundary conditions applicable to the flow region are

$$v_z(a,t) = 0 \qquad ; t > 0$$
$$v_z(r,t) \to \text{finite} \; ; r \in (0,a); \; t > 0 \tag{8.1903}$$

and the initial condition is

$$v_z(r,0) = 0 \quad ; \quad r \in (0,a) \tag{8.1904}$$

Since the initial condition (8.1904) is homogeneous, we can introduce a change dependent variable such that

$$v_z(r,t) = V_z(r,t) + \widehat{v}_z(r) \tag{8.1905}$$

The initial boundary value problem defined by (8.1902) to (8.1904) can now be reduced to the following modified initial boundary value problem for $V_z(r,t)$:

$$\nu^* \left\{ \frac{\partial^2 V_z}{\partial r^2} + \frac{1}{r}\frac{\partial V_z}{\partial r} \right\} = \frac{\partial V_z}{\partial t} \tag{8.1906}$$

with boundary conditions

$$V_z(a,t) = 0 \qquad ; t > 0$$
$$V_z(r,t) \to \text{finite} \; ; r \in (0,a) \; ; \; t > 0 \tag{8.1907}$$

and inhomogeneous initial condition

$$V_z(r,0) = -\widehat{v}_z(r) \tag{8.1908}$$

The boundary value problem for $\widehat{v}_z(r)$ is governed by the ordinary differential equation

$$\nu^* \left\{ \frac{d^2\widehat{v}_z}{dr^2} + \frac{1}{r}\frac{d\widehat{v}_z}{dr} \right\} + \frac{(p_0 - p_\ell)}{\rho\ell} = 0 \tag{8.1909}$$

with boundary conditions

$$\widehat{v}_z(a) = 0$$
$$\tag{8.1910}$$
$$\widehat{v}_z(r) \to \text{finite} \; ; \; r \in (0,a)$$

The solution of the boundary value problem for $\widehat{v}_z(r)$ is elementary and we have

$$\widehat{v}_z(r) = \frac{(p_0 - p_\ell)}{4\nu^*\rho\ell} \left\{ a^2 - r^2 \right\} \tag{8.1911}$$

This result represents the classical steady state solution for laminar viscous flow in a circular tube developed independently by G. Hagen (1797-1884) and J. Poiseulle (1799-1869).

For the solution of the initial boundary value problem $V_z(r,t)$ we employ a variables separation technique and follow the general procedures outlined in Examples 6.6 and 6.7. Avoiding details, we can show that the general form of the solution of (8.1906) which gives finite velocity fields in the region $r \in (0,a)$ can be represented in the form of an infinite series

$$V_z(r,t) = \sum_{n=1,2,} A_n e^{-\xi_n^2 \frac{\nu^* t}{a^2}} J_0\left(\frac{\xi_n r}{a} \right) \tag{8.1912}$$

where ξ_n correspond the roots of the equation

$$J_0(\xi) = 0 \tag{8.1913}$$

The constants A_n are determined by making use of the initial condition (8.1908).

This gives

$$-\frac{(p_0 - p_\ell)}{4\nu^* \rho \ell}(a^2 - r^2) = \sum_{n=1,2,} A_n J_0\left(\frac{\xi_n r}{a}\right) \tag{8.1914}$$

To obtain A_n we use a Bessel series representaion for $\widehat{v}_z(r)$ and the orthogonality properties of Bessel functions. Using such a procedure we have

$$A_n = -\frac{2}{J_1^2(\xi_n)}\left[\frac{(p_0 - p_\ell)\,a^2}{4\nu^*\rho\ell}\right]\int_0^a \frac{r}{a^2}\left(1 - \frac{r^2}{a^2}\right) J_0\left(\frac{\xi_n r}{a}\right) dr \tag{8.1915}$$

Performing the integrations we obtain

$$A_n = -4\left\{\frac{(p_0 - p_\ell)}{4\nu^*\rho\ell}\right\}\frac{J_2(\xi_n)}{\xi_n^2 J_1^2(\xi_n)} \tag{8.1916}$$

The final result for the velocity distribution can be obtained in the form

$$v_z(r,t) =$$
$$\frac{(p_0 - p_\ell)a^2}{4\eta\ell}\left[\left(1 - \frac{r^2}{a^2}\right) - 4\sum_{n=1,2,} \frac{J_2(\xi_n)J_0(\frac{\xi_n r}{a})}{\xi_n^2 J_1^2(\xi_n)} e^{\frac{-\nu^* \xi_n^2 t}{a^2}}\right] \tag{8.1917}$$

$$\bullet\ \bullet\ \bullet$$

A further category of unsteady flow problem deals with circular motions in viscous fluid regions which are symmetric about an axis. The velocity field is characterized by

$$\mathbf{v} = 0\ \mathbf{i}_r + v_\theta(r,t)\ \mathbf{i}_\theta + 0\ \mathbf{i}_z \tag{8.1918}$$

This category of unsteady flow problems have been examined in connection with the study of decay of line and circular vortices in viscous fluid regions. To obtain the relevant partial differential equation governing the unsteady motion we make use of the identity

$$\nabla \times (\nabla \times \mathbf{v}) = \nabla \nabla . \mathbf{v} - \nabla^2 \mathbf{v} \tag{8.1919}$$

where

$$\nabla \nabla . \mathbf{v} = \left\{ \mathbf{i}_r \frac{\partial}{\partial r} + \mathbf{i}_\theta \frac{1}{r} \frac{\partial}{\partial \theta} + \mathbf{i}_z \frac{\partial}{\partial z} \right\} \left(\frac{\partial v_r}{\partial r} + \frac{v_r}{r} + \frac{1}{r} \frac{\partial v_\theta}{\partial \theta} + \frac{\partial v_z}{\partial z} \right) \tag{8.1920}$$

$$\nabla \times (\nabla \times \mathbf{v})$$

$$= \frac{1}{r} \begin{vmatrix} \mathbf{i}_r & r\mathbf{i}_\theta & \mathbf{i}_z \\[2mm] \dfrac{\partial}{\partial r} & \dfrac{\partial}{\partial \theta} & \dfrac{\partial}{\partial z} \\[3mm] \dfrac{1}{r}\dfrac{\partial v_z}{\partial \theta} - \dfrac{\partial v_\theta}{\partial z} & r\left(\dfrac{\partial v_r}{\partial z} - \dfrac{\partial v_z}{\partial r} \right) & \dfrac{1}{r}\dfrac{\partial}{\partial r}(r v_\theta) - \dfrac{1}{r}\dfrac{\partial v_r}{\partial \theta} \end{vmatrix} \tag{8.1921}$$

which gives

$$\nabla^2 \mathbf{v} = \mathbf{i}_r \left\{ \nabla^2 v_r - \frac{v_r}{r^2} - \frac{2}{r^2}\frac{\partial v_\theta}{\partial \theta} \right\} + \mathbf{i}_\theta \left\{ \nabla^2 v_\theta + \frac{2}{r^2}\frac{\partial v_r}{\partial \theta} - \frac{v_\theta}{r^2} \right\}$$
$$+ \mathbf{i}_z \nabla^2 v_z \tag{8.1922}$$

Considering (8.1918) and (8.1922), the equation (8.1713) gives

$$\nu^* \left\{ \frac{\partial^2 v_\theta}{\partial r^2} + \frac{1}{r}\frac{\partial v_\theta}{\partial r} - \frac{v_\theta}{r^2} \right\} = \frac{\partial v_\theta}{\partial t} \tag{8.1923}$$

Alternatively, the equation of motion can be obtained by observing that for the state of rotational symmetry defined by the velocity (8.1918), the single non-zero stress component $\sigma_{r\theta}$ is given by

$$\sigma_{r\theta} = 2\eta \epsilon_{r\theta} = \eta \left(\frac{\partial v_\theta}{\partial r} - \frac{v_\theta}{r} \right) \tag{8.1924}$$

and the corresponding equation of equilibrium in terms of $\sigma_{r\theta}$ is

$$\frac{\partial \sigma_{rr}}{\partial r} + 2\frac{\sigma_{r\theta}}{r} = \rho\frac{\partial v_\theta}{\partial t} \tag{8.1925}$$

Substituting of (8.1924) in (8.1925) gives (8.1923). The analysis of this class of radially symmetric unsteady viscous flow problem is now reduced to the solution of the partial differential equation (8.1923) subject to appropriate boundary, regularity and initial conditions.

Example 8.53

A Newtonian viscous fluid is contained within a rigid circular boundary of radius a (Figure 8.113). Two-dimensional rotational flow is induced by suddenly rotating the outer boundary with a constant velocity $\Omega_0 a$. Determine the unsteady motion induced in the viscous fluid region.

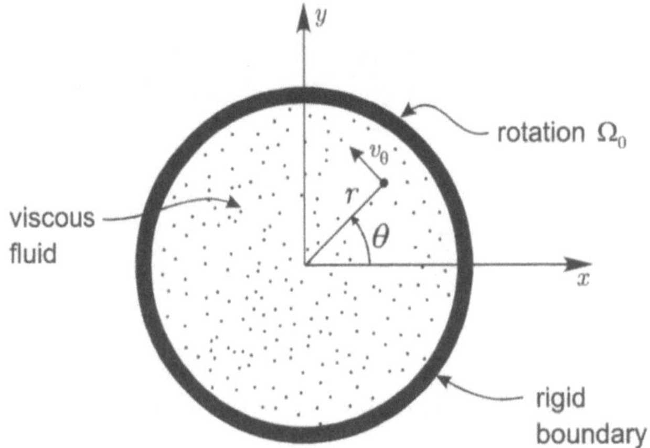

Figure 8.113: Unsteady flow in a circular fluid domain.

Solution

As indicated in connection with Examples 8.51 and 8.52, initial boundary value problems of this type can be examined by adopting either a variables separation technique which yields solutions in the form of an infinite series or a Laplace transform technique which requires Laplace transform inversion usually via a complex inversion procedure. Here we present a series solution and the solution based on Laplace transform techniques is left as an exercise.

The initial boundary value problem requires the solution of the partial differential equation

$$\nu^* \left\{ \frac{\partial^2 v_\theta}{\partial r^2} + \frac{1}{r} \frac{\partial v_\theta}{\partial r} - \frac{v_\theta}{r^2} \right\} = \frac{\partial v_\theta}{\partial t} \quad ; \quad r \in (0, a) \quad ; \quad t > 0 \qquad (8.1926)$$

subject to the boundary and regularity conditions

$$v_\theta(a,t) = \Omega_0 a \quad ; t > 0$$

$$v_\theta(r,t) \to \text{finite} \; ; r \to 0 \quad ; \quad t > 0 \tag{8.1927}$$

and the initial condition

$$v_\theta(r,0) = 0 \quad ; \quad r \in (0,a) \tag{8.1928}$$

In view of the homogeneous nature of the initial condition (8.1928) we consider a change in the dependent variable such that

$$v_\theta(r,t) = V_\theta(r,t) + \widehat{v}_\theta(r) \tag{8.1929}$$

This yields a modified initial boundary value problem for $V_\theta(r,t)$ with the governing partial differential equation

$$\nu^* \left\{ \frac{\partial^2 V_\theta}{\partial r^2} + \frac{1}{r}\frac{\partial V_\theta}{\partial r} - \frac{V_\theta}{r^2} \right\} = \frac{\partial V_\theta}{\partial t} \quad ; \quad r \in (0,a) \quad ; \quad t > 0 \tag{8.1930}$$

boundary and regularity conditions

$$V_\theta(a,t) = 0 \qquad ; t > 0$$

$$V_\theta(r,t) \to \text{finite} \; ; r \to 0 \quad ; \quad t > 0 \tag{8.1931}$$

and initial condition

$$V_\theta(r,0) = -\widehat{v}_\theta(r) \quad ; \quad r \in (0,a) \tag{8.1932}$$

The boundary value problem governing $\widehat{v}_\theta(r)$ requires the solution of the ordinary differential equation

$$\frac{d^2\widehat{v}_\theta}{dr^2} + \frac{1}{r}\frac{d\widehat{v}_\theta}{dr} - \frac{\widehat{v}_\theta}{r^2} = 0 \quad ; \quad r \in (0,a) \tag{8.1933}$$

with boundary and regularity conditions

$$\widehat{v}_\theta(a) = \Omega_0 a$$

$$\widehat{v}_\theta(r) \to \text{finite} \; ; r \to 0 \tag{8.1934}$$

The solution of (8.1933) subject to conditions (8.1934) gives

$$\widehat{v}_\theta(r) = \Omega_0 r \tag{8.1935}$$

For the solution of the modified initial boundary value problem defined by (8.1930) to (8.1932) we assume that $V_\theta(r,t)$ admits a solution of the form

$$V_\theta(r,t) = F(r)T(t) \tag{8.1936}$$

The ordinary differential equations governing $F(r)$ and $T(t)$ are

$$\frac{d^2 F}{dr^2} + \frac{1}{r}\frac{dF}{dr} + \left(\frac{\xi^2}{a^2} - \frac{1}{r^2}\right) F = 0 \tag{8.1937}$$

and

$$\frac{dT}{dt} = -\nu^* \frac{\xi^2}{a^2} T \tag{8.1938}$$

respectively where (ξ^2/a^2) is a separation constant which is assumed to have real roots. The general solution for $V_\theta(r,t)$ can be written as

$$V_\theta(r,t) = \left[A J_1\left(\frac{\xi r}{a}\right) + B Y_1\left(\frac{\xi r}{a}\right) \right] e^{-\nu^* \frac{\xi^2 t}{a^2}} \tag{8.1939}$$

where A and B can either be constants or functions of ξ. Since $V_\theta(r,t)$ is to be bounded for $r \in (0,a)$ we require

$$B = 0 \tag{8.1940}$$

and for the solution to be non trivial, the boundary condition in (8.1931) gives

$$J_1(\xi) = 0 \tag{8.1941}$$

If the roots of (8.1941) are denoted by ξ_n, the series form of the solution for $V_\theta(r,t)$ can be written as

$$V_\theta(r,t) = \sum_{n=1,2,} A_n e^{-\frac{\nu^* \xi_n^2 t}{a^2}} J_1\left(\frac{\xi_n r}{a}\right) \tag{8.1942}$$

The constants A_n are determined by considering the initial condition (8.1932) with $\widehat{v}_\theta(r)$ defined by (8.1935); i.e.

$$-\Omega_0 r = \sum_{n=1,2,} A_n J_1\left(\frac{\xi_n r}{a}\right) \tag{8.1943}$$

Again we make use of a normalizing integral approach to determine A_n; the final solution for $V_\theta(r,t)$ can be combined with $\widehat{v}_\theta(r)$ to obtain $v_\theta(r,t)$; we have

$$v_\theta(r,t) = \Omega_0 a \left[\frac{r}{a} - 2 \sum_{n=1,2,} e^{-\frac{\nu^* \xi_n^2 t}{a^2}} \frac{J_1\left(\frac{\xi_n r}{a}\right)}{\xi_n J_2(\xi_n)}\right] \tag{8.1944}$$

The series converges relatively rapidly for large or moderate value of $(\nu^* t/a^2)$. For small values of $(\nu^* t/a^2)$ a large number of terms needs to be considered.

$$\bullet \ \bullet \ \bullet$$

When examining diffusive motions in viscous fluids, it is also instructive to examine the formulation of the problem in terms of the time-dependent decay of "*vorticity*" in the fluid. The vorticity vector $\boldsymbol{\omega}$ can be defined in terms of the curl of the velocity vector $\mathbf{v}$; i.e.

$$\boldsymbol{\omega} = \boldsymbol{\nabla} \times \mathbf{v} \tag{8.1945}$$

The equations governing motion of an incompressible Newtonian viscous fluid are the continuity equation (8.1674) and the Navier-Stokes equation (8.1713) which can be re-written in the form

$$\eta \nabla^2 \mathbf{v} - \nabla(p + \rho F) = \rho \frac{D\mathbf{v}}{Dt} \tag{8.1946}$$

where it is assumed that the body force vector $\mathbf{f}^b$ can be derived from a potential F. Consider the vector identity

$$\mathbf{v}.\nabla\mathbf{v} = \boldsymbol{\omega} \times \mathbf{v} + \nabla\left(\frac{1}{2}q^2\right) \tag{8.1947}$$

where q is referred to as the 'speed' of flow and

$$q^2 = \mathbf{v}.\mathbf{v} \tag{8.1948}$$

The Navier-Stokes equation can now be written as

$$\frac{\partial\mathbf{v}}{\partial t} + \boldsymbol{\omega} \times \mathbf{v} = -\nabla H - \nu^*\nabla \times \boldsymbol{\omega} \tag{8.1949}$$

where

$$H = \frac{1}{2}q^2 + \frac{p}{\rho} + F \tag{8.1950}$$

can be identified as the "*Bernoulli Potential*" of the fluid. Taking the curl of equation (8.1949) we obtain

$$\frac{D\boldsymbol{\omega}}{Dt} = \boldsymbol{\omega}.\nabla\mathbf{v} + \nu^*\nabla^2\boldsymbol{\omega} \tag{8.1951}$$

Since for slow viscous flows the higher order terms can be neglected we obtain from (8.1951) the following partial differential equation governing $\boldsymbol{\omega}$; i.e.

$$\nu^*\nabla^2\boldsymbol{\omega} = \frac{\partial\boldsymbol{\omega}}{\partial t} \tag{8.1952}$$

which is the classical diffusion equation for the vorticity vector $\boldsymbol{\omega}$. Hence vorticity variations in the flow field generally results in the diffusion of the same vorticity. Furthermore for an incompressible Newtonian viscous fluid, vorticity cannot be generated at the interior of the fluid; vorticity must be diffused inwards from boundary motions.

Example 8.54

Consider the problem of a line vortex of "strength" Ω^* which is initiated at $t = 0$ in an incompressible fluid domain of infinite extent. The cylindrical polar coordinate system (r, θ, z) is chosen such that the z-axis coincides with the line vortex. Assuming that the velocity vector $\mathbf{v}$ associated with the motion in the viscous fluid is given by

$$\mathbf{v} = 0 \; \mathbf{i}_r + v_\theta(r, t) \; \mathbf{i}_\theta + 0 \; \mathbf{i}_z \tag{8.1953}$$

determine the time-dependent velocity $v_\theta(r, t)$.

Solution

The vorticity vector $\boldsymbol{\omega}$ which corresponds to the velocity (8.1953) can be obtained from (8.1945) i.e.

$$\boldsymbol{\omega} = 0 \; \mathbf{i}_r + 0 \; \mathbf{i}_\theta + \omega_z \; \mathbf{i}_z \tag{8.1954}$$

where

$$\omega_z = \frac{1}{r} \frac{\partial}{\partial r} \{ r v_\theta(r, t) \} \tag{8.1955}$$

Since $\omega_z = \omega_z(r, t)$, the result (8.1952) reduces to

$$\nu^* \left\{ \frac{\partial^2 \omega_z}{\partial r^2} + \frac{1}{r} \frac{\partial \omega_z}{\partial r} \right\} = \frac{\partial \omega_z}{\partial t} \tag{8.1956}$$

Again, this partial differential equation can be solved by "similarity transformation" which combines the independent variables r, t to form a single variable (see e.g. Problem 8.61) defined by

$$\chi = \frac{r^2}{4\nu^* t} \tag{8.1957}$$

which gives

$$\omega_z = \frac{\Omega^*}{4\pi\nu^* t}e^{-\chi} \tag{8.1958}$$

Using this expression for ω_z in (8.1955) and integrating the result we obtain

$$v_\theta(r,t) = \frac{\Omega^*}{2\pi r}(1 - e^{-\chi}) + \frac{g(t)}{r} \tag{8.1959}$$

where $g(t)$ is an arbitrary function. To ensure that $v_\theta(r,t)$ is finite as $r \to 0$ we set $g(t) = 0$. (Note that by using the rule of L'Hospital (1661-1704), the first term of (8.1959) gives $v_\theta(0,t) \equiv 0$). The velocity field therefore reduces to

$$v_\theta(r,t) = \frac{\Omega^*}{2\pi r}(1 - e^{-\chi}) \tag{8.1960}$$

Other types of motions can be obtained by simply differentiating (8.1958) with respect to t; for example, it can be verified that

$$\omega_z(r,t) = \frac{\widetilde{\Omega}}{2\pi\nu^* t^2}\left\{1 - \chi\right\}e^{-\chi} \tag{8.1961}$$

is a solution of (8.1956) and corresponds to a line vortex of angular momentum of $\widetilde{\Omega}$ and velocity

$$v_\theta(r,t) = \frac{r\widetilde{\Omega}e^{-\chi}}{4\pi\nu^* t^2} \tag{8.1962}$$

8.13   The compatibility conditions

The concept of compatibility of strains in a deforming continuum was introduced in Section 8.2.6. In this section we shall demonstrate that the compatibility conditions (8.102) applicable to infinitesimal strains are both necessary and sufficient for their integrability to generate a single-valued and unique displacement field in a simply connected domain. Although the developments can be presented with reference to the dyadic notation representations of $\boldsymbol{\epsilon}$, it is more convenient to adopt an exposition in terms of an indicial notation.

THEOREM 8.14

The strain-displacement relationships for a continuum region undergoing infinitesimal strains are given by

$$\epsilon_{ij} = \frac{1}{2}(u_{i,j} + u_{j,i}) \tag{8.1963}$$

where ϵ_{ij} are the infinitesimal strains, u_i are the displacement components and the comma denotes partial differentiation with respect to the Cartesian coordinate variables x_i. Prove that the conditions necessary and sufficient for the integrability of (8.1963) to generate a single-valued and unique displacement field in a simply connected domain correspond to

$$\epsilon_{ik,j\ell} + \epsilon_{j\ell,ik} = \epsilon_{i\ell,jk} + \epsilon_{jk,i\ell} \tag{8.1964}$$

PROOF

If the displacement field u_i is given, it is then a relatively easy matter to compute the strain matrix simply by using (8.1963). The converse problem of determining the displacement field from the strain field is not so straightforward; we are presented with the six linear partial differential equations (8.1963) for the determination of three displacement components u_i. Such a system is over-determined and for the existence of a single-valued and continuous solutions for u_i (i.e. the continuum should deform without the occurrence of cracks, discontinuities, dislocations, material overlapping, etc.) it becomes necessary to impose certain restrictions on ϵ_{ij}. These conditions are referred to as the *compatibility conditions*. In mathematical terms, the compatibility conditions are a statement of the conditions both *necessary* and *sufficient* for ensuring the single-valuedness of the displacement field u_i derived by considering the strain field ϵ_{ij}.

The necessary compatibility conditions can be obtained by eliminating u_i from the strain displacement relations (8.1963). From the definition of the displacement gradient (8.62) we can write

$$u_{i,j} = \epsilon_{ij} + \omega_{ij} \tag{8.1965}$$

where ω_{ij} in the rotation matrix. Differentiating (8.1965) we can write

$$(u_{i,j})_{,k} = \epsilon_{ij,k} + \omega_{ij,k} \tag{8.1966}$$

Also from (8.1966) we can interchange the order of integration and write

$$(u_{i,k})_{,j} = \epsilon_{ik,j} + \omega_{ik,j} \tag{8.1967}$$

From (8.1966) and (8.1967) we have

$$\epsilon_{ij,k} - \epsilon_{ik,j} = \omega_{ik,j} - \omega_{ij,k} \tag{8.1968}$$

Also from the definition of the rotation matrix we have

$$\omega_{ik,j} - \omega_{ij,k} = \omega_{jk,i} \tag{8.1969}$$

Substituting (8.1969) in (8.1968) and differentiating the result with respect to x_ℓ we have

$$\epsilon_{ij,k\ell} - \epsilon_{ik,j\ell} = \omega_{jk,i\ell} \tag{8.1970}$$

Similarly we can show that

$$\epsilon_{\ell j,ki} - \epsilon_{\ell k,ji} = \omega_{jk,\ell i} \tag{8.1971}$$

Since the differentiations on the right hand side of (8.1970) and (8.1971) commute we have

$$\epsilon_{ij,k\ell} + \epsilon_{k\ell,ij} = \epsilon_{ik,j\ell} + \epsilon_{j\ell,ik} \tag{8.1972}$$

These are the compatibility equations which *must* be satisfied by the strain components ϵ_{ij} if they are to be related to the displacement u_i through (8.1963), which in turn implies that u_i are single valued and continuous. They are, consequently, the *necessary conditions*.

We can now focus on the proof of sufficiency of the compatibility equations, for generating a single-valued continuous displacement field u_i in a simply connected domain (In a simply connected domain, every closed curve can be continuously reduced to a point without crossing the boundary of the domain). Consider a point $P(x_i)$ within the continuum at which the displacement field u_i^P and the rotation ω_{ij}^P are known. The displacement at any other location Q in the continuum can be represented by a *line integral* along a continuous curve from P to Q, i.e.

$$u_i^Q = u_i^P + \oint_P^Q du_i \tag{8.1973}$$

we note that

$$du_i = \frac{\partial u_i}{\partial x_j} dx_j = u_{i,j}\, dx_j \tag{8.1974}$$

Using (8.1965) and (8.1974) in (8.1973) we obtain

$$u_i^Q = u_i^P + \oint_P^Q \epsilon_{ij} dx_j + \oint_P^Q \omega_{ij}\, dx_j \tag{8.1975}$$

Since x_j^Q is a fixed location we can replace dx_j by $d\left(x_j - x_j^Q\right)$. Therefore, the last integral in (8.1975) can be integrated by parts as follows:

$$\oint_P^Q \omega_{ij} d\left(x_j - x_j^Q\right) = \left[\omega_{ij}\left(x_j - x_j^Q\right)\right]_P^Q$$
$$- \oint_P^Q \left(x_j - x_j^Q\right) d\omega_{ij} \tag{8.1976}$$

Noting that

$$d\omega_{ij} = (\omega_{ij,k}) dx_k \tag{8.1977}$$

the result (8.1976) can be re-written as

$$\oint_P^Q \omega_{ij} d\left(x_j - x_j^Q\right) = -\,\omega_{ij}^P\left(x_j^P - x_j^Q\right)$$
$$- \oint_P^Q \left(x_j - x_j^Q\right) \omega_{ij,k}\, dx_k \tag{8.1978}$$

From the definition of the rotation matrix ω_{ij} we have

$$\omega_{ij,k} = \frac{1}{2}\left[(u_{i,j})_{,k} - (u_{j,i})_{,k}\right] \tag{8.1979}$$

By adding and subtracting the term $(1/2)u_{k,ij}$ to (8.1979) and re-arranging terms we can write

$$\omega_{ij,k} = \epsilon_{ik,j} - \epsilon_{jk,i} \tag{8.1980}$$

Combining (8.1975), (8.1978) and (8.1980) we have

$$u_i^Q = u_i^P - \omega_{ij}^P\left(x_j^P - x_j^Q\right) + \oint_P^Q \Phi_{ik}\,dx_k \tag{8.1981}$$

where

$$\Phi_{ik} = \epsilon_{ik} - \left(x_j - x_j^Q\right)\left(\epsilon_{ik,j} - \epsilon_{jk,i}\right) \tag{8.1982}$$

For u_i^Q to be continuous and single-valued, the integral in (8.1981) must be *path-independent*. It implies that the integrand of the last term of (8.1981) must be an *exact differential*. From the theory of line integrals applicable to a simple connected domain, the necessary and sufficient condition for $\Phi_{ik}dx_k$ to be an exact differential is that

$$\Phi_{ik,\ell} = \Phi_{i\ell,k} \tag{8.1983}$$

From (8.1982) and (8.1983) we have

$$\begin{aligned}
&\epsilon_{ik,\ell} - x_{j,\ell}(\epsilon_{ik,j} - \epsilon_{jk,i}) - (x_j - x_j^Q)(\epsilon_{ik,j\ell} - \epsilon_{jk,i\ell}) \\
&= \epsilon_{i\ell,k} - x_{j,k}(\epsilon_{i\ell,j} - \epsilon_{j\ell,i}) - (x_j - x_j^Q)(\epsilon_{i\ell,jk} - \epsilon_{j\ell,ik})
\end{aligned} \tag{8.1984}$$

Noting that $x_{i,j} = \delta_{ij}$, (8.1984) can be reduced to the result

$$\epsilon_{ik,j\ell} + \epsilon_{j\ell,ik} = \epsilon_{i\ell,jk} + \epsilon_{jk,i\ell} \tag{8.1985}$$

A simple interchange of indices will reveal that (8.1985) is identical to (8.1972). Consequently, we have proved that the conditions (8.1972) are also *sufficient* to ensure integrability of the strain-displacement relations (8.1963), to generate a single-valued and continuous displacement field u_i. The mathematical aspects of the proof that the compatibility requirements applicable to elastic continua were considered by a number of eminent elasticians including Saint-Venant (1797-1886) Beltrami (1835-1900), Morera (1856-1909), Cesaro (1859-1906), Boussinesq (1842-1929) and Michell(1863-1940).

8.14 A proof of Stokes' paradox

In Section 8.12.10 we presented a discussion which showed, by considerations of elementary forms of solutions for $\psi(r,\theta)$, that slow creeping flow past a circular cylinder is not possible. Here, we give a slightly more generalized theorem to prove Stokes' paradox.

THEOREM 8.15

The partial differential equation governing "*slow viscous flow*" or "*creeping flow*" in two-dimensions is given by

$$\overset{\circ}{\nabla}^2 \, \overset{\circ}{\nabla}^2 \psi(r,\theta) = 0 \tag{8.1986}$$

where (r,θ) are plane polar coordinates, $\psi(r,\theta)$ is Stokes' stream function and the velocity vector is given by

$$\mathbf{v} = \left\{ \frac{1}{r}\frac{\partial\psi}{\partial\theta}\,\mathbf{i}_r - \frac{\partial\psi}{\partial r}\,\mathbf{i}_\theta \right\} \tag{8.1987}$$

Prove that steady creeping flow past a fixed cylinder is not possible.

PROOF

If $\psi(r,\theta)$ is *biharmonic*, it implies that $\psi(r,\theta)$ is also *analytic*, and for any circular annular region, $\psi(r,\theta)$ can be expanded in Fourier series in the form

$$\psi(r,\theta) = \sum_{n=1,2,} \{A_n(r)\cos(n\theta) + B_n(r)\sin(n\theta)\} \tag{8.1988}$$

In (8.1988), $A_n(r)$ and $B_n(r)$ are solutions of the ordinary differential equations

$$\widetilde{D}_n^2 \widetilde{D}_n^2 A_n(r) = 0 \quad ; \quad \widetilde{D}_n^2 \widetilde{D}_n^2 B_n(r) = 0 \tag{8.1989}$$

and the operator $\widetilde{D}_n^2$ is given by

$$\widetilde{D}_n^2 = \frac{d^2}{dr^2} + \frac{1}{r}\frac{d}{dr} - \frac{n^2}{r^2} \tag{8.1990}$$

If the velocity field is determined from (8.1987) then for the velocity vector to be bounded at infinity we can show that

$$A_1(r) = A_1 r + \frac{A_{-1}}{r} \tag{8.1991}$$

and

$$A_n(r) = \frac{A_n'}{r^{n-2}} + \frac{A_n^*}{r^n} \quad ; \quad n \geq 2 \tag{8.1992}$$

Hence the general solution of $\psi(r,\theta)$ can be written as

$$\psi(r,\theta) = V_\infty r \sin\theta + \psi_0 + A_0 \ln r + \frac{A_1}{r}\cos\theta + \frac{B_1}{r}\sin\theta \tag{8.1993}$$

$$+ \sum_{n=2}^{\infty}\left\{ \left(\frac{A_n'}{r^{n-2}} + \frac{A_n^*}{r^n}\right)\cos n\theta + \left(\frac{B_n'}{r^{n-2}} + \frac{B_n^*}{r^n}\right)\sin n\theta\right\}$$

where V_∞ and ψ_0 are, respectively, far-field and near-field measures of the velocity field.

Considering Green's second identity for a plane region $\mathbb{D}$ with boundary $\mathcal{C}$, we have

$$\int\int_{\mathbb{D}} \left\{ \Psi \overset{\circ}{\nabla}^2\Phi - \Phi \overset{\circ}{\nabla}^2\Psi \right\} dA = \int_{\mathcal{C}} \left\{ \Psi \frac{\partial\Phi}{\partial n} - \Phi \frac{\partial\Psi}{\partial n} \right\} d\mathcal{C} \tag{8.1994}$$

where Ψ and Φ are arbitrary functions. If we set

$$\Psi = \overset{\circ}{\nabla}^2 \psi \quad ; \quad \Phi = \psi \tag{8.1995}$$

and note that since $\overset{\circ}{\nabla}^2 \overset{\circ}{\nabla}^2 \psi = 0$, (8.1994) reduces to

$$\int\!\!\int_{\mathbb{D}} \left(\overset{\circ}{\nabla}^2 \psi\right)^2 dA = \int_{\mathcal{C}} \left\{ \overset{\circ}{\nabla}^2 \psi \frac{\partial \psi}{\partial n} - \psi \frac{\partial}{\partial n}\left(\overset{\circ}{\nabla}^2 \psi\right) \right\} d\mathcal{C} \tag{8.1996}$$

Let $\mathbb{D}$ be the domain between the fixed cylinder with boundary $\mathcal{C}_i$ and a larger circle of radius r with boundary $\mathcal{C}_e$. From the *no-slip* velocity boundary conditions

$$\mathbf{v} = 0 \quad ; \quad (r, \theta) \in \mathcal{C}_i \tag{8.1997}$$

and from the representation (8.1987) we obtain

$$\psi = \frac{\partial \psi}{\partial n} = 0 \quad ; \quad (r, \theta) \in \mathcal{C}_i \tag{8.1998}$$

Hence the integral on the right hand side of (8.1996) vanishes on $\mathcal{C}_i \in \mathcal{C}$. On the outer circle, since

$$\psi = 0(r) \qquad ; \overset{\circ}{\nabla} \psi = 0(1)$$
$$\overset{\circ}{\nabla}^2 \psi = 0\left(\frac{1}{r}\right) ; \overset{\circ}{\nabla}\left(\overset{\circ}{\nabla}^2 \psi\right) = 0\left(\frac{1}{r^2}\right) \tag{8.1999}$$

we have

$$\int_{\mathcal{C}_e} \left\{ \overset{\circ}{\nabla}^2 \psi \frac{\partial \psi}{\partial n} - \psi \frac{\partial}{\partial n}\left(\overset{\circ}{\nabla}^2 \psi\right) \right\} dS = 0\left(\frac{1}{r}\right) \tag{8.2000}$$

This integral will vanish as $r \to \infty$. Hence the result (8.1996) is equivalent to

$$\int\!\!\int_{\mathbb{D}} \left(\overset{\circ}{\nabla}^2 \psi\right)^2 d\Lambda = 0 \tag{8.2001}$$

Since the integrand of (8.2001) is positive definite, it follows from the Dubois-Reymond lemma that (8.2001) is satisfied if and only if

$$\overset{\circ}{\nabla}^2 \psi(r,\theta) \equiv 0 \tag{8.2002}$$

The equation (8.2002) implies that the stream function must be harmonic. Hence $A'_n = B'_n = 0$ in (8.1993). Also from the condition of *no-slip* on the fixed cylinder

$$\frac{\partial \psi}{\partial r} = 0 \quad ; \quad (r,\theta) \in C_i \tag{8.2003}$$

we require

$$\psi = \psi_0 \quad ; \quad V_\infty = 0 \tag{8.2004}$$

Therefore the no-slip boundary conditions on the cylinder are satisfied only when $V_\infty \equiv 0$.

8.15   A uniqueness theorem for viscous flows

The formulation of the two-dimensional problem in slow viscous flow or creeping flow as a problem related to the biharmonic equation was presented in Section 8.12.6. In section 8.12.8 the discussion was extended to the class of axisymmetric slow viscous flow problems governed by Stokes' operator. In this section we shall outline a general theorem which establishes the uniqueness of the slow viscous flow problem.

THEOREM 8.16

Consider an incompressible viscous fluid within a *bounded region* $V = V(t)$ whose boundary S consists of as many finite rigid surfaces moving in a prescribed manner. The viscous fluid satisfies the incompressibility condition

$$\nabla.\mathbf{v} = 0 \quad ; \quad \mathbf{x} \in V \tag{8.2005}$$

the equations of motion

$$\rho\frac{D\mathbf{v}}{Dt} = \eta\nabla^2\mathbf{v} - \nabla p + \mathbf{f}^b \quad ; \quad \mathbf{x} \in V \tag{8.2006}$$

and by virtue of the adherence condition (see Section 8.12.9), the velocity of the fluid on S is the velocity of S itself. If two flows in the bounded region $V = V(t)$ have the same velocity distribution at $t = 0$ in $V(t)$ and S, then the velocity fields are identical.

PROOF

Let $\mathbf{v}^{(1)}(\mathbf{x}, t)$ and $\mathbf{v}^{(1)}(\mathbf{x}, t)$ be two velocity fields which will satisfy the governing equations (8.2005) and (8.2006) and the identical boundary conditions on S. Let

$$\mathbf{v}^* = \mathbf{v}^{(1)}(\mathbf{x}, t) - \mathbf{v}^{(2)}(\mathbf{x}, t) \quad \mathbf{x} \in V \tag{8.2007}$$

which satisfies

$$\mathbf{v}^*(\mathbf{x}, 0) = 0 \quad ; \quad \mathbf{x} \in V \tag{8.2008}$$

$$\mathbf{v}^*(\mathbf{x}, 0) = 0 \quad ; \quad \mathbf{x} \in S \tag{8.2009}$$

The kinetic energy associated with the difference in motion $\mathbf{v}^*(\mathbf{x}, t)$ is

$$\mathcal{K} = \frac{\rho}{2}\int\int\int_V (\mathbf{v}^* . \mathbf{v}^*)dV \tag{8.2010}$$

Then, the convected time derivative of $\mathcal{K}$ is

$$\frac{D\mathcal{K}}{Dt} = -\int\int\int_V [\eta\nabla\mathbf{v}^* : \nabla\mathbf{v}^* + \rho\mathbf{v}^* . \mathbf{s} . \mathbf{v}^*]\, dV \tag{8.2011}$$

where (:) denotes the double scalar product and $\mathbf{s}$ is strain rate tensor for motion $\mathbf{v}^{(2)}$ defined by (8.1682). In order to prove (8.2011) we note that both $\mathbf{v}^{(1)}(\mathbf{x}, t)$ and $\mathbf{v}^{(2)}(\mathbf{x}, t)$ satisfy the Navier-Stokes equations (8.2006) and by subtracting the resulting equations gives

$$\frac{\partial\mathbf{v}^*}{\partial t} + \mathbf{v}^{(1)} . \nabla\mathbf{v}^* + \mathbf{v}^* . \nabla\mathbf{v}^{(2)} = -\nabla\left\{\frac{p^{(1)} - p^{(2)}}{\rho}\right\} + \frac{\eta}{\rho}\nabla^2\mathbf{v}^* \tag{8.2012}$$

Taking the scalar product of (8.2012) and $\mathbf{v}^*$ and using the incompressibility condition

$$\nabla.\mathbf{v}^* = \nabla.\mathbf{v}^{(2)} = \nabla\mathbf{v}^{(1)} = 0 \tag{8.2013}$$

gives

$$\frac{\partial}{\partial t}\left[\frac{1}{2}(\mathbf{v}^*)^2\right] = \nabla \cdot \left[\frac{\eta}{\rho}\nabla\frac{1}{2}(\mathbf{v}^*.\mathbf{v}^*) - \left(\frac{p^{(1)} - p^{(2)}}{\rho}\right)\mathbf{v}^* - \frac{1}{2}(\mathbf{v}^*.\mathbf{v}^*)\mathbf{v}^{(1)}\right]$$
$$- \frac{\eta}{\rho}\nabla\mathbf{v}^* : \nabla\mathbf{v}^* - \mathbf{v}^*.\mathbf{s}.\mathbf{v}^* \tag{8.2014}$$

The integral (8.2011) is obtained by integration of (8.2014) over V and through the application of the condition (8.2009) on S.

Let $-m$ be a lower bound of the characteristic values of the tensor $\mathbf{s}$ in the time interval $0 < t < \tau$. Since $tr\ \mathbf{s} = \nabla.\mathbf{v}^* = 0$ we observe that $m \geq 0$. From the definition of m and the properties of characteristic values it follows that for $\mathbf{x} \in V$ and for all $t \in (0,\tau)$

$$\mathbf{v}^*\cdot\mathbf{s}\cdot\mathbf{v}^* \geq -m(\mathbf{v}^*.\mathbf{v}^*) \tag{8.2015}$$

Hence from (8.2011)

$$\frac{D\mathcal{K}}{Dt} \leq m\rho \int\int\int_V (\mathbf{v}^*.\mathbf{v}^*)dV = 2m\mathcal{K} \tag{8.2016}$$

which can be written as

$$\frac{D}{Dt}\left(\mathcal{K}e^{-2mt}\right) \leq 0 \tag{8.2017}$$

Integrating this result from $t = 0$ to $t = \tau$, gives

$$\mathcal{K}(\tau)e^{-2m\tau} \leq 0 \tag{8.2018}$$

Since τ was an arbitrary time it follows that $\mathcal{K}$ must be identically zero. Hence $\mathbf{v}^* = 0$ or $\mathbf{v}^{(1)} \equiv \mathbf{v}^{(2)}$.

As assumed in the derivation of the uniqueness theorem, the region V is considered to be bounded. The theorem will not hold if the region is unbounded, since the result (8.2010) and the divergence theorem are applicable only to bounded regions.

$$\odot \; \odot \; \odot$$

8.16 PROBLEM SET 8

8.1 The deformation of a continuum region is defined by the displacements

$$\mathbf{u} = \left\{ \frac{xz}{r^3}\mathbf{i} + \frac{yz}{r^3}\mathbf{j} + \left[\frac{z^2}{r^3} + \left(\frac{\lambda + 3\mu}{\lambda + \mu} \right) \frac{1}{r} \right] \mathbf{k} \right\}$$

where u is the displacement vector referred to the Cartesian coordinate system (x, y, z); $r^2 = x^2 + y^2 + z^2$ and λ and μ are Lamé's constants.

(i) Determine the strain matrix $\boldsymbol{\epsilon}$ and the rotation matrix $\boldsymbol{\omega}$.

(ii) Referring to an infinite region bounded internally by a spherical cavity of radius a, find the deformed shape of the cavity.

8.2 A two-dimensional region of a continuum with a square shape is subjected to the deformation shown in Figure 8.114, where $\xi \ll 1$.

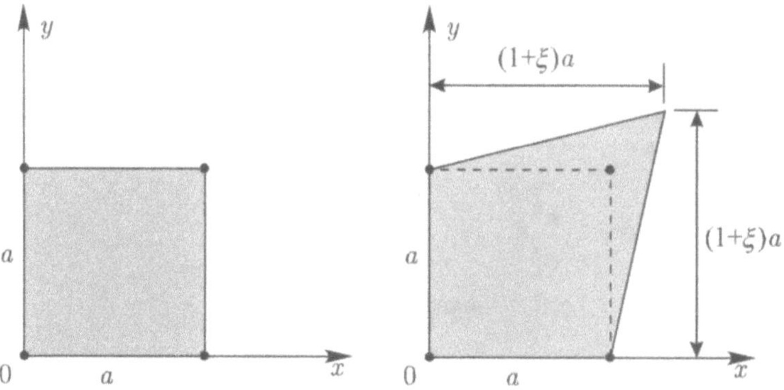

Figure 8.114: Deformation of a continuum region.

Determine the following

(i)   The deformation gradient matrix

(ii) The strain and rotation matrices

(iii) The principal strains

(iv) The change in area of the square

8.3   The Figure 8.115 shows the cross-section through a cylindrical borehole drilled into a geological medium with gravitational stresses.

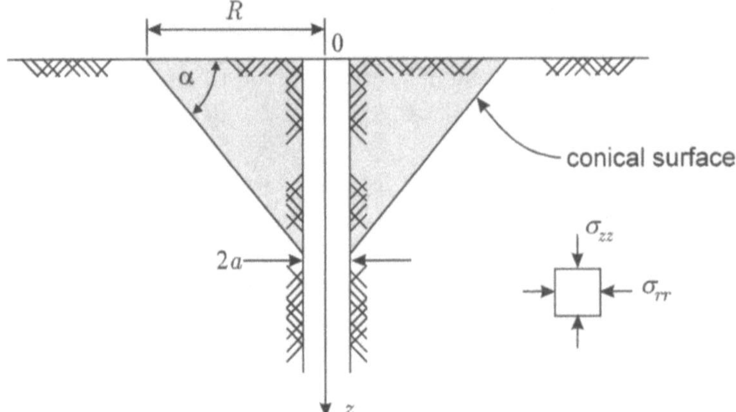

Figure 8.115: Gravity stresses around a borehole.

The state of stresses in the geomaterial, referred to the cylindrical polar coordinate system (r, θ, z) is given by

$$\boldsymbol{\sigma} = \begin{bmatrix} \rho g z(1 - \dfrac{a^2}{r^2}) & 0 & 0 \\ 0 & \rho g z(1 + \dfrac{a^2}{r^2}) & 0 \\ 0 & 0 & \rho g z \end{bmatrix}$$

where ρg is the unit weight of the geomaterial. Derive expressions for the variation of normal and shear tractions on the conical plane shown in Figure 8.115.

8.4 Boussinesq's solution for the loading of the surface of a half-space region
by a concentrated normal force P_B gives rise to the state of stress given
in Section 8.10.10. Consider the spherical surface of radius a which has
its centre O' at a distance a from the origin O along the z-axis (Figure
8.116). By evaluating the resultant of tractions on this surface in the
z-direction show that the internal stresses are in equilibrium with the
applied normal force P_B.

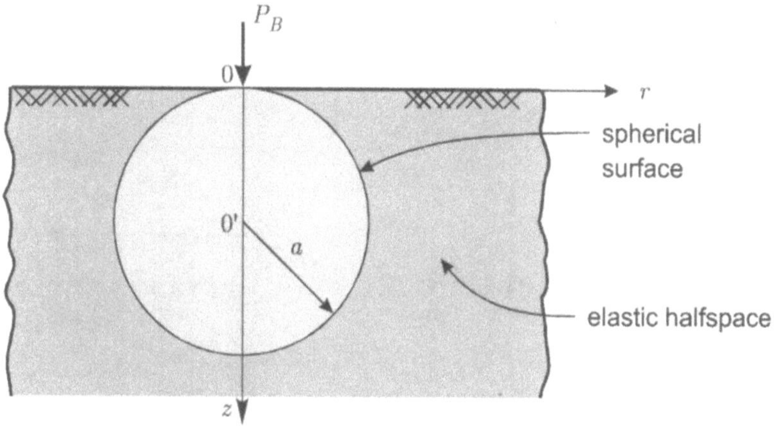

Figure 8.116: Boussinesq's problem.

8.5 The state of stress at a point P in an elastic continuum is given by

$$\sigma_{ij} = C \begin{bmatrix} \bar{x}^2\bar{y} & \bar{x}(1-\bar{y}^2) & \bar{y}^2 \\ \bar{x}(1-\bar{y}^2) & \dfrac{1}{3}(\bar{y}^3 - 3\bar{y}) & \bar{x}^2 \\ \bar{y}^2 & \bar{x}^2 & 2\bar{z}^2 \end{bmatrix}$$

where C is a constant and $\bar{x}, \bar{y},$ and $\bar{z}$ are non-dimensional coordinates.

(i) Determine the *body force distribution* $\mathbf{f}^b$ if the equations of static
equilibrium are satisfied at the field point P.

(ii) The coordinates of P are now specified as $\bar{x} = 1, \bar{y} = 1,$ and $\bar{z} = 2$.
Calculate the traction vectors at P on a cylindrical surface which
passes through P and the axis of which coincides with the $\bar{z}$-axis.

8.6   A prismatic elastic bar of negligible weight and cross-sectional dimensions $2a \times 2a$ is *just immersed* in an inviscid fluid region (of unit weight γ) by the application of an external normal stress on one surface (Figure 8.117). The plane ends of the bar are smoothly restrained in such a way that the displacements in the z-direction are zero.

(i) Use an appropriate polynomial form of an Airy stress function to determine state of stress in the prismatic elastic bar

(ii) Determine the change in length of the edge AB.

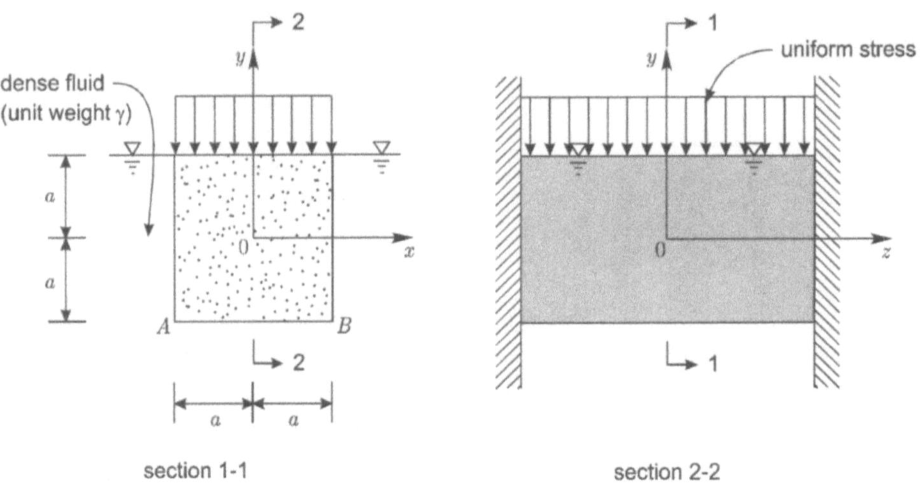

Figure 8.117: Stress analysis of a submerged elastic prism.

8.7   The Figure 8.118 shows an elastic mound with a boundary $y = ax^2$ where a is an arbitrary constant. The elastic material has a unit weight γ. Derive expressions for the stress components in the elastic mound by considering an Airy stress function of the form

$$\varphi(x, y) = Ax^2 y + By^2$$

where A and B are constants.

Also, by assuming that the length of the mound in the z-direction is infinite develop expressions for $u(x,y)$ and $v(x,y)$ by imposing the conditions $u(0,0) = v(0,0) = 0$.

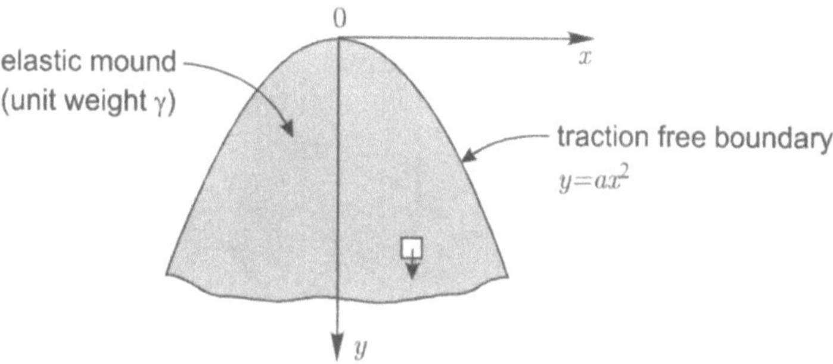

Figure 8.118: Gravity stresses in an elastic mound.

8.8   A hollow elastic pipe of length l and external diameter $2a$, where $l >> 2a$ is just immersed in a dense fluid of unit weight γ (Figure 8.119). The ends of the pipe are restrained from movement and sealed. Any excess pressures developed in the interior region are vented. Develop an expression for the ovalization of the interior of the pipe.

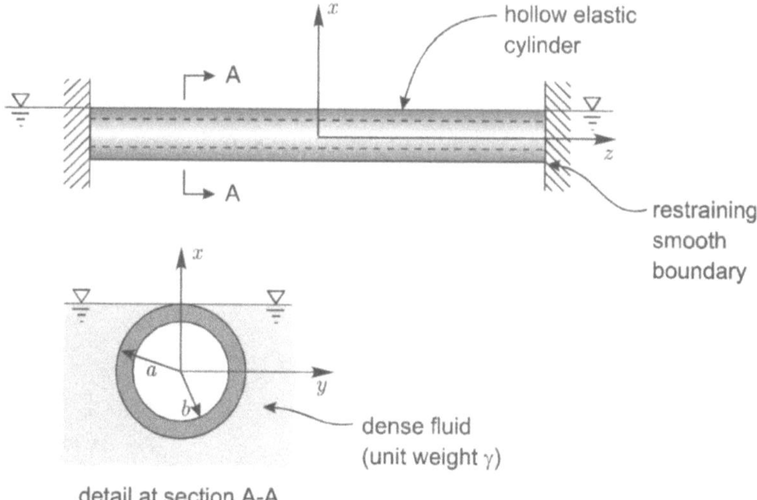

Figure 8.119: Hollow cylinder immersed in a dense fluid.

8.9 A sheet of an isotropic elastic material is stretched by an all round tensile stress σ_o and then rigidly clamped at the radius b. A small circular hole of radius a is made at the centre of the clamped sheet (Figure 8.120). The ratio (a/b) is such that terms of the order $(a/b)^2$ can be neglected. The boundary of the hole is traction free. When the hole is introduced in the stretched sheet, some energy is released. Derive an expression for the released energy.

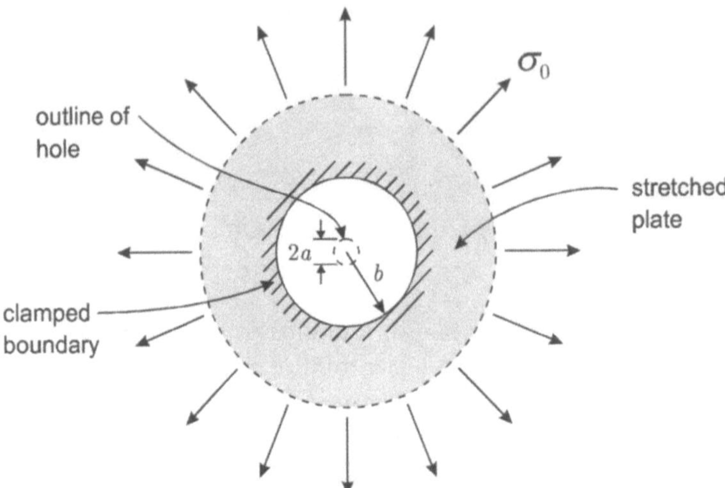

Figure 8.120: Energy release from a stretched-clamped plate due to introduction of hole.

8.10 A cross-section through a cylindrical cavity in a geological medium is shown in Figure 8.121. The cavity conveys a dense inviscid fluid of unit weight γ which is purely under hydrostatic pressure . The dimensions of the cavity are such that the problem can be modelled as a plane strain problem. Use an Airy stress function approach to determine the state of stresses $\sigma_{\theta\theta}(a, \theta)$ along the boundary of the cavity.

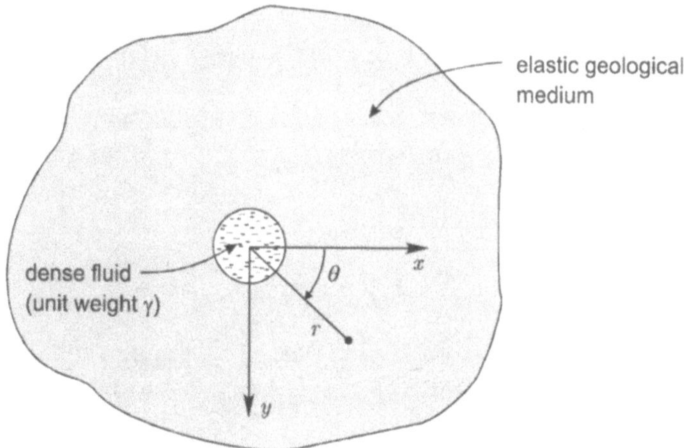

Figure 8.121: Fluid filled cylindrical cavity in an elastic geological medium.

8.11 The Figure 8.122 shows an elastic *spring mount* which consists of an annular elastic region which is bonded to a rigid cylindrical outer boundary and an inner rigid cylinder. The *spring mount* operates by providing an elastic resistance to the displacement of the inner cylinder while the outer boundary is restrained from movement. The application of a central force P_0^* per unit length (in the z-direction) induces a displacement Δ_0 of the inner cylinder. Assuming plane strain behaviour of the *spring mount*, develop an expression relating P_0^* and Δ_0.

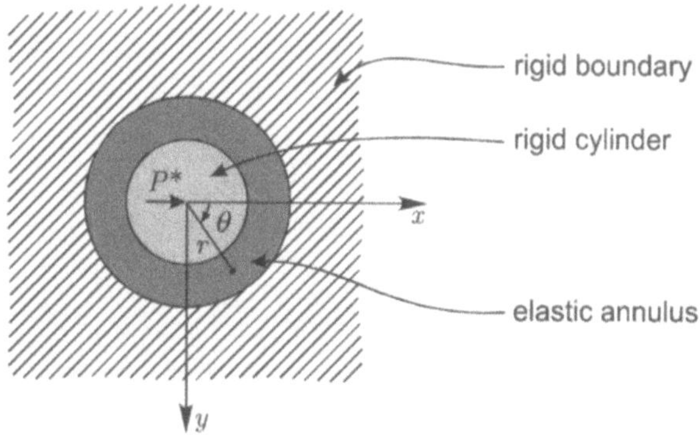

Figure 8.122: The elastic *spring mount* problem

8.12 The Figure 8.123 shows a square plate which is subjected to a generalized two-dimensional state of plane stress. If the relevant Airy stress function is

$$\varphi(r, \theta) = \sigma r^2 \{1 - \sin\theta\cos\theta\}$$

where σ is a constant, determine

(i) the stress components σ_{xx}, σ_{yy} and σ_{xy} and

(ii) the displacements $u(a, a)$ and $v(a, a)$ assuming that the point $(0,0)$ experiences zero displacement and zero rotation.

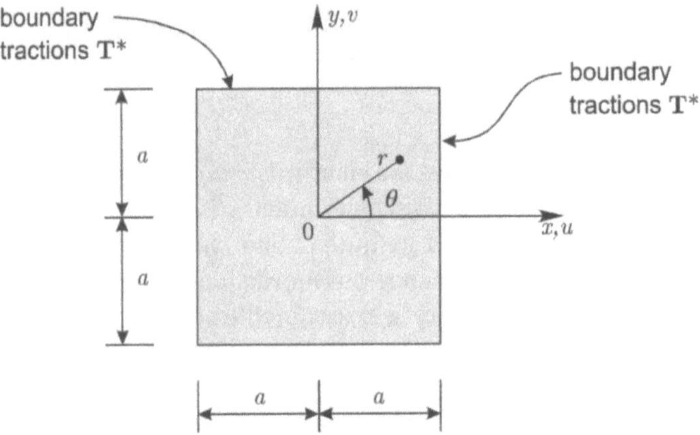

Figure 8.123: Boundary loading of a square plate.

8.13 The Figure 8.124 shows a rectangular elastic plate region which is subjected to boundary tractions. The Airy stress function proposed for the analysis of the state of stress in the plate is

$$\varphi(x, y) = \frac{\sigma_0}{6}\left[y^2(y - 3x\tan\alpha) + x^2\tan^2\alpha(3y - x\tan\alpha)\right]$$

where σ_0 is a constant.

Interpret the stress distribution obtained by $\varphi(x, y)$ in relation to the rectangular region ABCD. Also determine the stresses acting on the surfaces of the region ABCD.

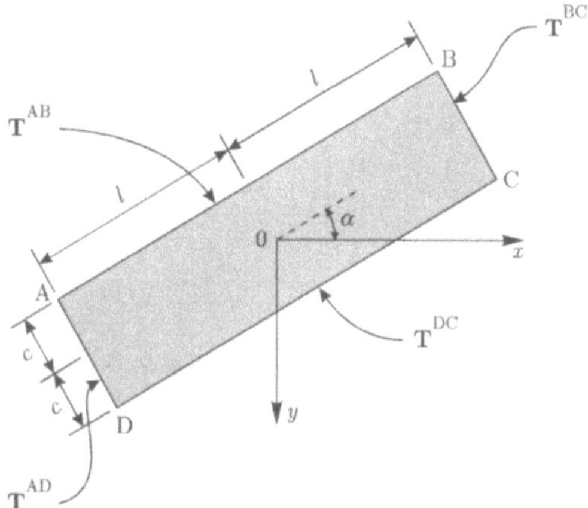

Figure 8.124: Boundary tractions on an elastic plate.

8.14 The cantilever beam shown in Figure 8.125 is subjected to a distributed shear stress on the upper boundary $y = h$. Using a polynomial Airy stress function of the form

$$\varphi(x, y) = C_{02}y^2 + C_{03}y^3 + C_{04}y^4 + C_{05}y^5 + C_{20}x^2 + C_{21}x^2y$$

$$+ C_{22}x^2y^2 + C_{23}x^2y^3$$

where C_{ij} are constants, determine the stress distribution in the beam which satisfies explicitly the traction boundary conditions on $y = \pm h$ and gives zero force and moment resultants at the free end.

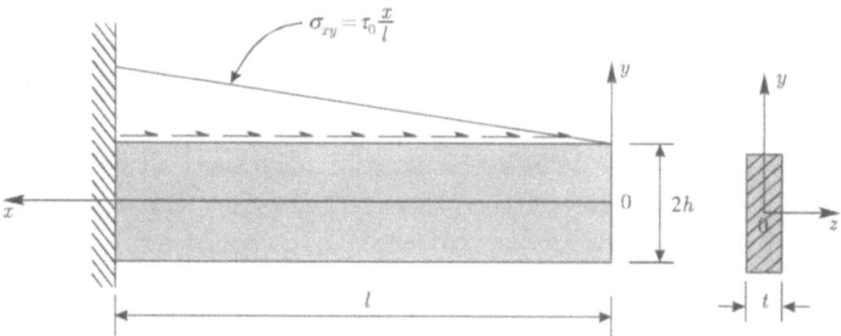

Figure 8.125: Boundary shear loading of a cantilever plate.

8.15 An infinite strip of an elastic material is subjected to shear stress distributions

$$\sigma_{xy}(x, h) = -\tau_0 \sin\left(\frac{2\pi x}{l}\right)$$

$$\sigma_{xy}(x, -h) = \tau_0 \sin\left(\frac{2\pi x}{l}\right)$$

No other tractions are applied on the planes $y = \pm h$ (Figure 8.126).

(i) Using a variables separable solution of the biharmonic equation for the Airy stress function, derive the expressions for σ_{xx}, σ_{yy} and σ_{xy} throughout the strip, assuming that the deformations of the strip conforms to a state of generalized plane stress.

(ii) Evaluate the resultant of tractions in the x-direction at the locations $x = 0$ and $x = l/2$. Use these results to check the validity of the solution.

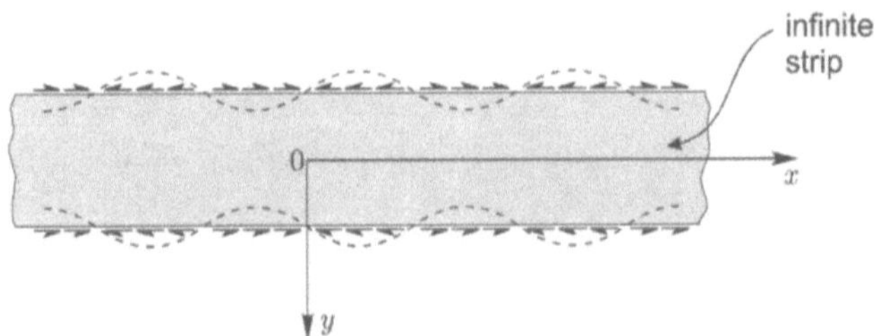

Figure 8.126: Shear loading of an infinite strip.

8.16 An elastic layer of finite thickness $2h$ rests on a *Winkler-Foundation* which is equivalent to a continuous distribution of linearly deformable spring elements with spring constant k (Figure 8.127). This particular form of *elastic support* is associated with the name of E. Winkler (1835-1888)). The Winkler foundation imposes a displacement dependent normal traction at the base of the layer. The surface of the elastically supported layer is subjected to a compressive normal traction

$$\sigma_{yy}(x, h) = -\sigma_0 \cos\left(\frac{\lambda x}{h}\right)$$

where σ_0 has units of stress and the state of deformation in the layer corresponds to state plane strain.

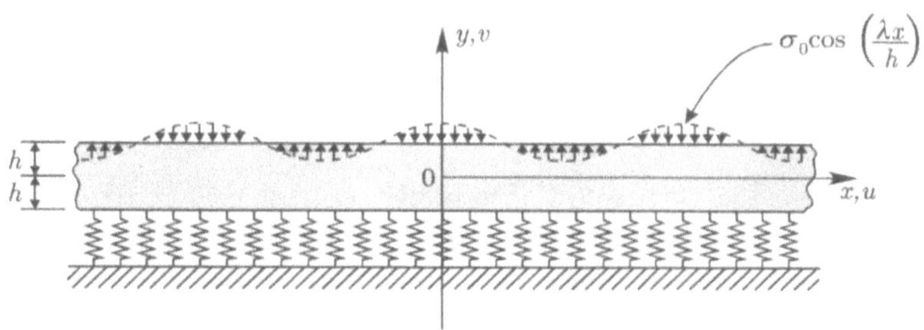

Figure 8.127: Flexure of a layer on a Winkler foundation.

(i) Derive an expression for the deflection $v(x,0)$ at the mid-plane of the layer.

(ii) Compare this result with the deflection of an elastically supported thin plate, which for plane strain deformations satisfies the ordinary differential equation

$$D\frac{d^4 w}{dx^4} + kw = p(x)$$

8.17 The Figure 8.128 shows an infinite plate which is bounded by two edges OA and OB. The plate is subjected to a localized load through a knife edge which exerts a force of magnitude P_0/h where h is the thickness of the plate. Assuming that a suitable equilibrating force to P_0/h is applied at the remote boundary, determine the resultant of tractions in the x-and y-directions transmitted across the circular arc CD.

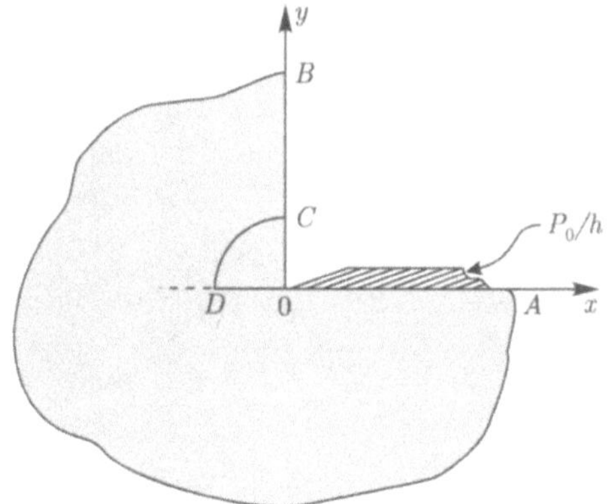

Figure 8.128: Loading of an elastic plate at a re-entrant corner.

8.18 A wedge shaped plate region with $r \in (0, \infty)$ and $\theta \in (-\alpha, \alpha)$ is subjected to an inclined load P^* at its vertex (Figure 8.129). Use Airy stress functions of the form

$$\varphi(r, \theta) = Ar\theta \sin\theta + Br\theta \cos\theta$$

where A and B are arbitrary constants, to determine the stress state in the plate.

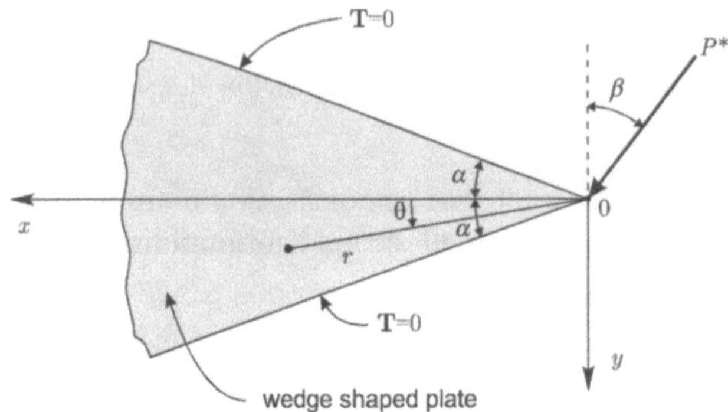

Figure 8.129: Loading of a wedge shaped plate at the apex.

8.19 A circular cantilever beam shown Figure 8.130 is subjected to a shear force T_0 at the boundary $\theta = 0$ and zero tractions on the boundaries $r = a$ and $r = b$. Considering an Airy stress function of the form

$$\varphi(r, \theta) = f(r) \sin\theta$$

show that the boundary conditions on $r = a$ can be satisfied exactly and the boundary conditions at $\theta = 0$ and $\theta = \pi/2$ can be satisfied in an integral sense.

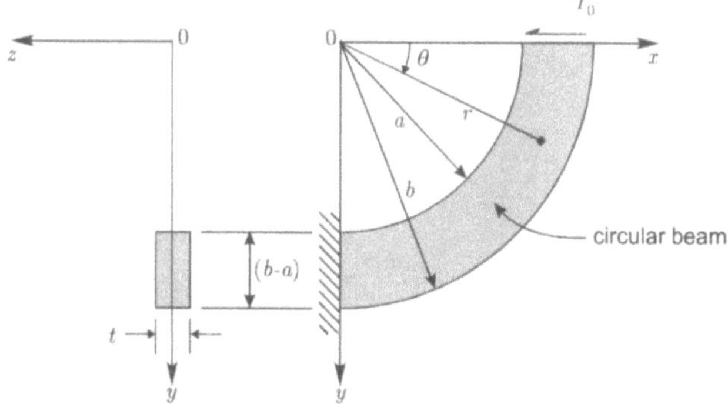

Figure 8.130: Bending of a cantilever beam.

8.20 The Figure 8.131 shows the cross-section through a cylindrical cavity located in an elastic block, contained between smooth rigid boundaries. The diameter of the cavity $2a$ is much smaller than the dimensions of the block (L). If the block is subjected to a uniform compressive stress σ_0 in the x-direction, determine the maximum value of $\sigma_{\theta\theta}$ at the boundary of the cylindrical cavity.

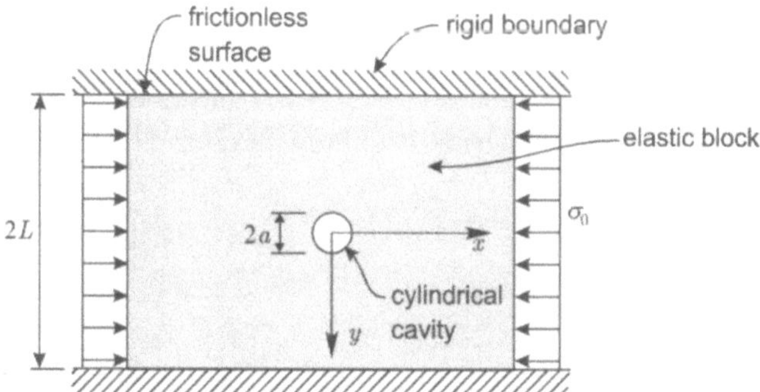

Figure 8.131: Cylindrical cavity in a constrained elastic block subjected to uniform compression.

8.21 An elastic plate of infinite extent is bounded internally by a circular cavity. A rigid disc inclusion is bonded to the boundary of the circular cavity such that the displacements at the bonded boundary are zero. (Figure 8.132). Determine the maximum radial stress that is generated at the bonded interface due to the action of shear stress field τ_0.

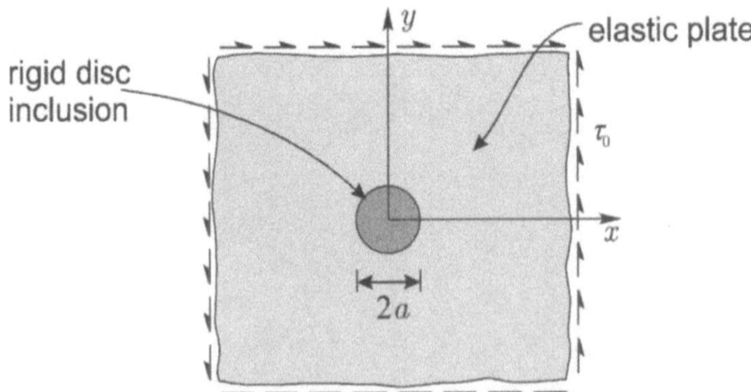

Figure 8.132: Embedded rigid disc inclusion in a shear stress field.

8.22 (i) If $V(x,y)$ is a harmonic function then show that $xV(x,y)$, $yV(x,y)$ and $(x^2+y^2)V(x,y)$ are biharmonic.

(ii) Prove that the Airy stress function can be represented in the form

$$\varphi(z,\bar{z}) = \mathrm{Re}\{\bar{z}\psi(z) + \chi(z)\}$$

where $z = (x+iy)$, $\bar{z} = (x-iy)$, Re refers to the real part, and $\psi(z)$ and $\chi(z)$ are suitably chosen analytic functions.

8.23 Starting from the basic equations of equilibrium and stress-strain relations, show that displacement and stress components can be represented in terms of the analytical functions $\psi(z)$ and $\chi(z)$ such that

$$\sigma_{xx} + \sigma_{yy} = 2\left[\psi'(z) + \overline{\psi}'(\bar{z})\right]$$

$$\sigma_{yy} - \sigma_{xx} + 2i\sigma_{xy} = 2\left[\bar{z}\psi''(z) + \chi''(z)\right]$$

$$2\mu(u+iv) = \kappa\psi(z) - z\overline{\psi}'(\bar{z}) - \overline{\chi}'(\bar{z})$$

where for plane strain and plane stress problems, respectively,

$$\kappa = (3 - 4\nu)$$

and

$$\kappa = \left(\frac{3-\nu}{1+\nu}\right)$$

8.24 Show that the problem of a hollow elastic cylinder which is subjected to radially symmetric tractions.

$$\sigma_{rr}(a,\theta) = -p_a \quad ; \quad \sigma_{r\theta}(a,\theta) = 0$$

$$\sigma_{rr}(b,\theta) = -p_b \quad ; \quad \sigma_{r\theta}(b,\theta) = 0$$

can be solved by selecting complex potentials

$$\psi(z) = Az \quad ; \quad \chi(z) = \frac{B}{z}$$

Derive the complete expressions for the stresses and displacements in the hollow cylinder.

8.25 In relation to a half-plane region $x \in (0,\infty), y \in (-\infty,\infty)$ examine the traction boundary value problem which is solved by the Airy stress function

$$\varphi(r,\theta) = Cr^2\theta$$

where C is a constant. Determine the displacement field (u_r, u_θ) associated with the stress state defined by this stress function. (neglect any rigid body components)

Use these expressions for the displacements to determine the displacements $v(x,0)$ along the edge of a semi-infinite plate due to a uniform normal stress of finite width (see Figure 8.133).

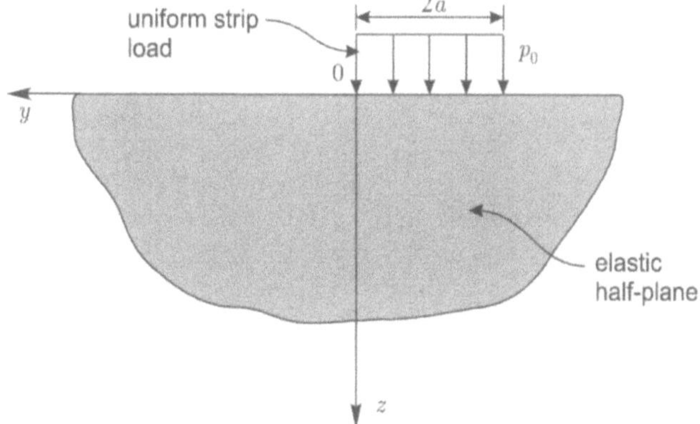

Figure 8.133: Strip loading of the boundary of a half-plane.

8.26 The Figure 8.134 shows an elastic half-plane which is *continuously elastically supported* along its plane boundary and subjected to a concentrated load P_0 at the surface of the half-plane. Use a Fourier transform development of the Airy stress function approach to determine the stress distribution in the elastic support.

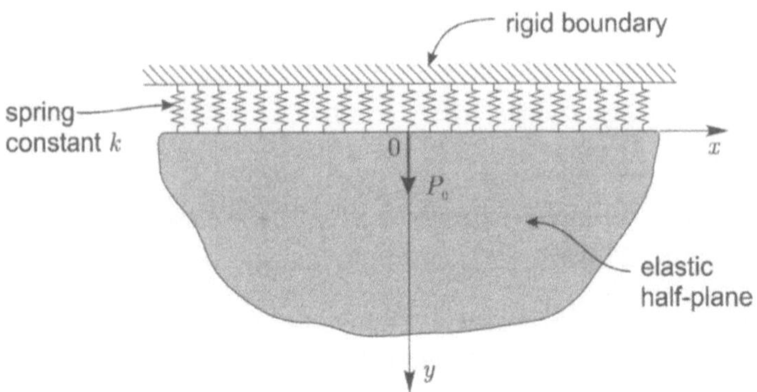

Figure 8.134: Concentrated loading of an isotropic elastic half-plane with an elastically supported plane boundary.

8.27 The Figure 8.135 shows an infinite plane region which is made up by bonding two half-plane regions of differing elastic properties. The interface region is subjected to a concentrated load of magnitude P_0 which acts along the y-direction. The interface exhibits complete continuity of displacement and tractions. Use a Fourier transform development of the Airy stress function technique to develop a solution for the stress states in the half-plane regions.

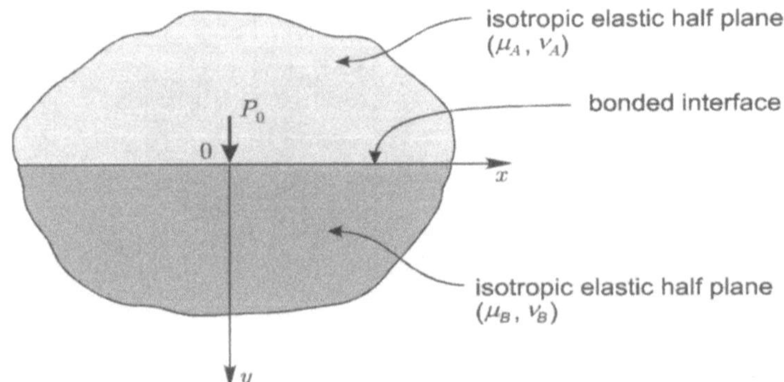

Figure 8.135: Concentrated loading of a bi-material elastic infinite plane.

8.28 A conical elastic solid occupies the region $R \in (0, \infty), \Theta \in (0, \Theta_0)$. The cone is subjected to a concentrated force P_0 which acts along the axis $\Theta = 0$. Use a Love strain function formulation to determine the displacements and stresses in the conical region (Figure 8.136).

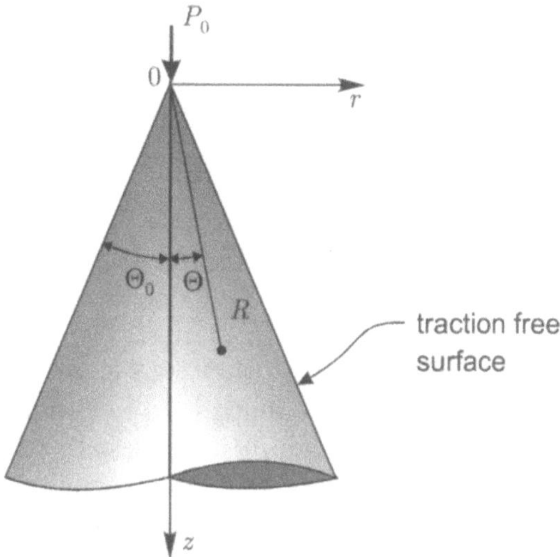

Figure 8.136: Axisymmetric loading of a conical elastic region.

8.29 An elastic cylinder of length $2H$ and diameter $2a$ is subjected to a radial stress distribution

$$\sigma_{rr}(a, z) = \sigma_0 \left(1 - \frac{z^2}{H^2} \right) \quad ; \quad z \in (H, -H)$$

over the cylindrical boundary (Figure 8.137) All other tractions acting on the cylinder are zero. Determine the state of stress in the cylinder by satisfying the traction boundary conditions on $r = a$ exactly and the tractions on $z = \pm H$ in an average sense.

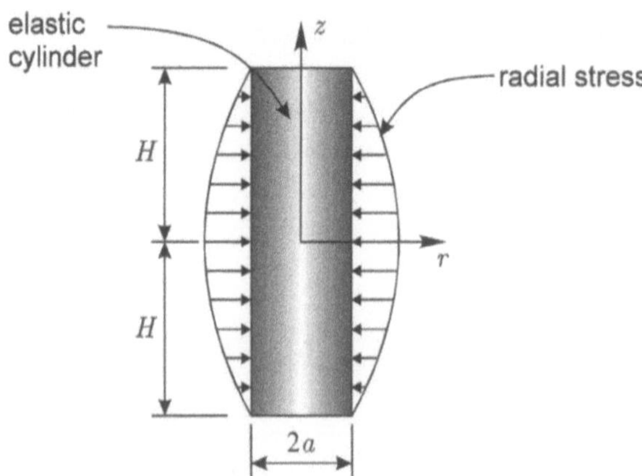

Figure 8.137: Radial loading of an isotropic elastic cylinder.

8.30 An isotropic elastic infinite space region is bounded internally by a spherical rigid inclusion which is embedded in *bonded contact* (Figure 8.138) The rigid spherical inclusion is subjected to a central force P_0 which is directed along the z-axis. Use Love's strain function approach to develop a force displacement relationship for the spherical inclusion.

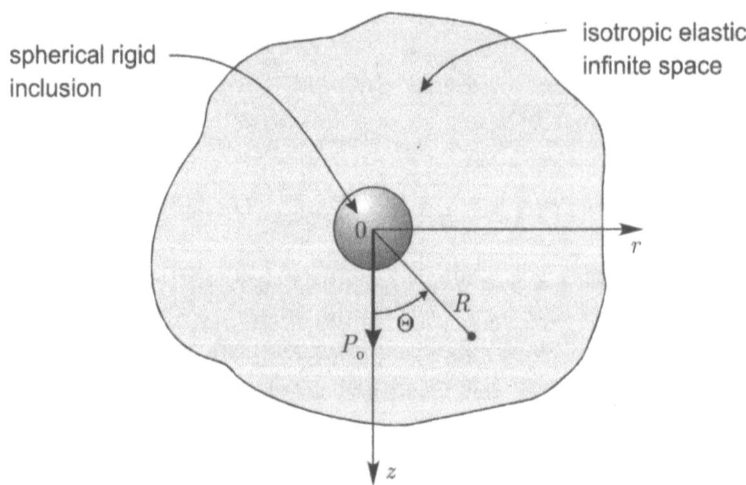

Figure 8.138: Axial loading of a spherical rigid inclusion embedded in bonded contact in an elastic infinite space.

8.31 The Figure 8.139 shows an isotropic elastic solid which is bounded internally by a spherical cavity which is filled with an inviscid fluid of compressibility C_w. The elastic solid is subjected to the far field stress state which corresponds to a compressive axial stress σ_1 and a compressive radial stress σ_3. Derive an expression for the fluid pressure generated in the cavity as a result of this external stress state.

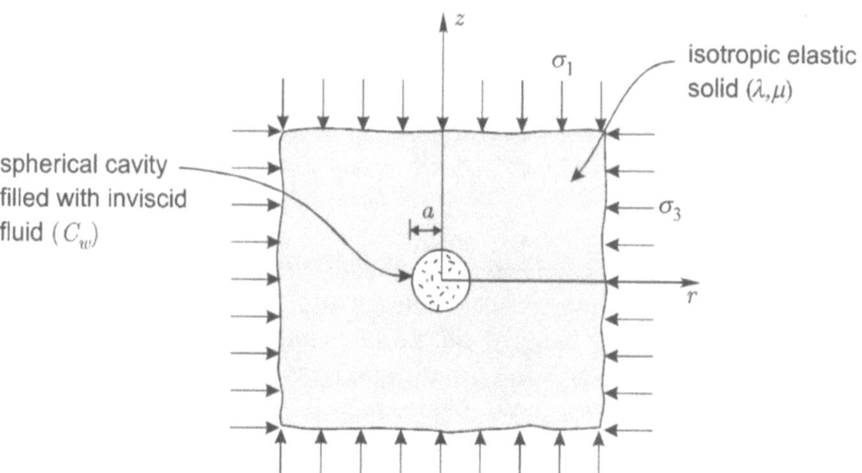

Figure 8.139: Triaxial loading of an elastic solid containing a compressible fluid inclusion.

8.32 The Figure 8.140 shows an isotropic elastic infinite space which is bounded internally by a spherical cavity. The surface of the cavity is subjected to an *"inextensibility constraint"* which prevents displacements in θ-direction. The elastic solid is subjected to an uniaxial far field stress σ_0 in the z-direction. Use the Love strain function approach to determine the amplification of the applied stress at the boundary of the cavity. (Assume that the *inextensibility constraint* is stress independent.)

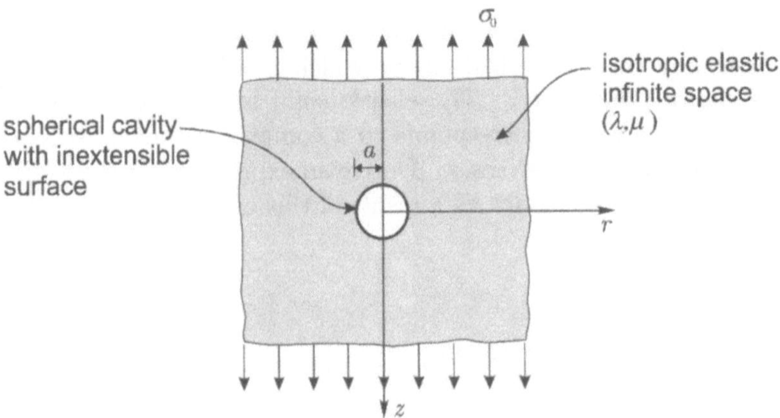

Figure 8.140: Uniform loading of an elastic solid containing a
spherical cavity with an inextensibility constraint.

8.33 A spherical rigid inclusion is embedded in smooth contact in an elastic
medium of infinite extent (Figure 8.141). The inclusion experiences an
environmentally induced dilatation which changes its radius from a
to $(a + \epsilon a)$ where ϵ is a small quantity. In this state the inclusion is
subjected to a magnetic field which induces a central force resultant
P_0. Evaluate the displacement Δ_0 resulting from the application of the
force P_0, which will just induce separation at the smooth interface.
Derive an expression for this result in terms of the parameter ϵ and the
elasticity characteristics of the solid.

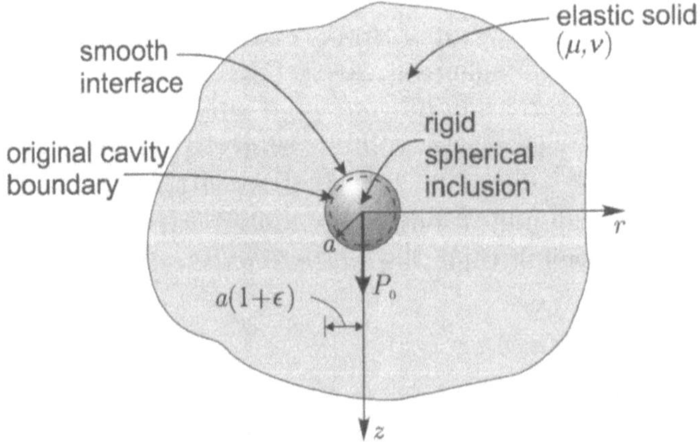

Figure 8.141: Displacement of a smoothly embedded rigid spherical
inclusion.

8.34 An elastic layer of finite thickness and infinite extent is subjected to circular normal stresses of constant stress intensity as shown in Figure 8.142. No other forces or tractions are applied either on the surfaces $z = \pm H$ or within the interior of the layer. Use a Hankel transform development of Love's strain function to obtain an integral expression for the normal stress distribution at the mid-plane of the layer.

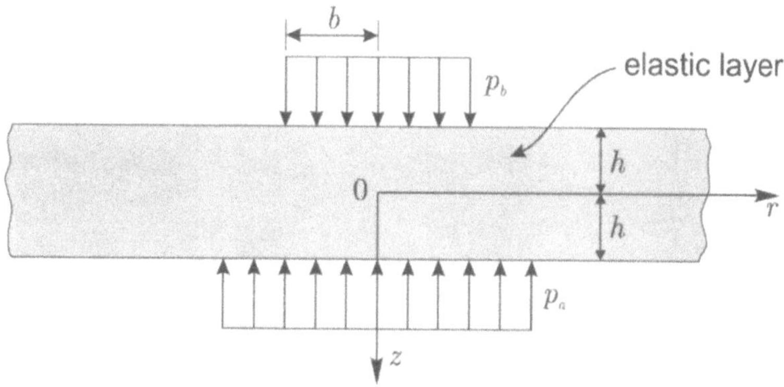

Figure 8.142: Axisymmetric compression of an elastic layer of infinite lateral extent.

8.35 The surface of an isotropic elastic half-space region $(z \geq 0)$ is bonded to a rigid boundary. The half-space region is subjected to a concentrated force of magnitude P_0 which is located at a distance H_0 from the bonded surface (Figure 8.143) Formulate the boundary value problem by appeal to Love's strain function approach and a Hankel transform development of the resulting biharmonic equation. Determine the integral expressions for the displacements and stresses.

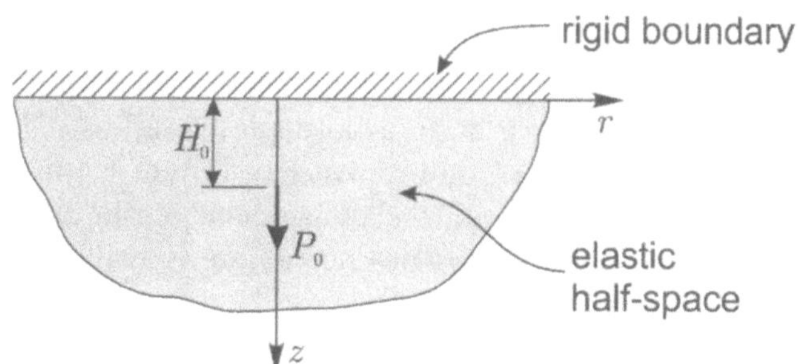

Figure 8.143: Localized loading of an isotropic elastic half-space with a bonded boundary.

8.36 A rigid circular disc inclusion is embedded at the boundary between two dissimilar elastic half-space regions. The plane containing the disc inclusion is considered to be *inextensible* (Figure 8.144). Use a Hankel transform development of the biharmonic equation for Love's strain function to develop the axial displacement Δ_0 of the rigid disc inclusion which is subjected to an axial load P_0. Show that when the half-space regions are identical this result reduces to the exact solution for the axial displacement of a rigid disc inclusion embedded in an elastic infinite space.

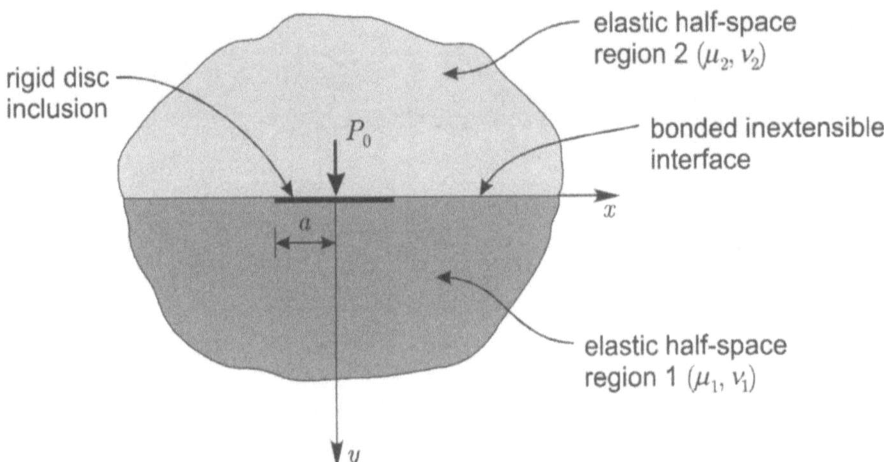

Figure 8.144: Disc inclusion embedded at a constrained bi-material interface.

8.37 The *pressuremeter problem* in geomechanics involves the application of a band of pressure (p_0) to a borehole located in a geological medium of infinite extent. Use the Love strain function approach to formulate the elasticity problem dealing with the application of a uniform pressure band of pressure of length $2l$ to a borehole of radius a located in an isotropic elastic medium of infinite extent (Figure 8.145). Develop an integral expression for the radial displacement at the boundary of the cylindrical cavity. Use a quadrature technique to evaluate $u_r(a, 0)$.

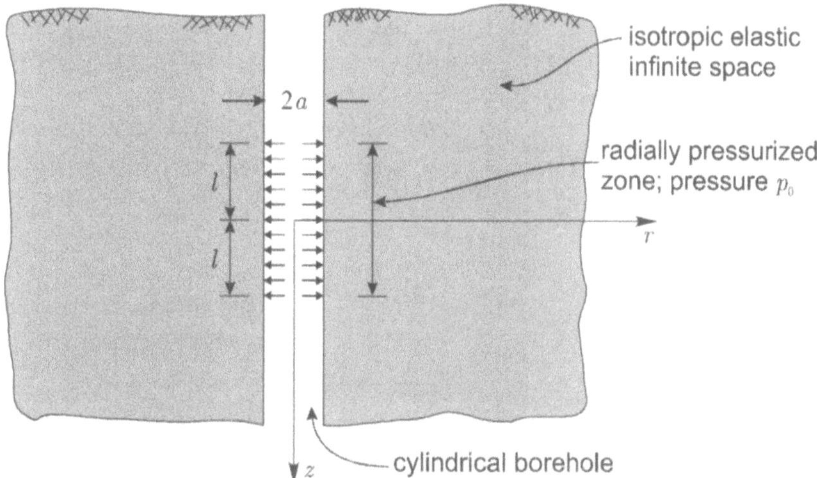

Figure 8.145: Radially pressurized zone of a borehole of infinite length.

8.38 The Figure 8.146 shows a flexible circular plate which is clamped at the boundary $r = a$ and simply supported at the location $r = b$. The plate is loaded by a concentrated force which acts at its centre. Develop a solution for the deflection of the plate by considering a superposition of solutions for the circular plate with a clamped boundary.

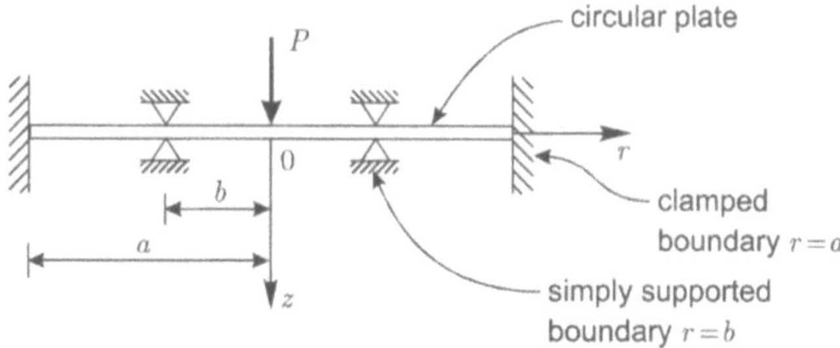

Figure 8.146: Flexure of a clamped circular plate with internal displacement constraint.

8.39 Find the maximum deflection and maximum radial flexural moment in a *circular plate* which is either (i) clamped or (ii) simply supported at the boundary $r = a$ and subjected to a loading

$$p(x, y) = p_0 xy.$$

8.40 The flexural behaviour of a heavy elastic plate can be modelled by the Germain-Poisson-Kirchhoff theory for thin plates. A heavy plate of infinite extent rests on a rigid frictionless base. It is lifted by a concentrated force P_0 (Figure 8.147). If the displacement of the plate at the point of application of the load is Δ_0, develop a relationship between P_0 and Δ_0.

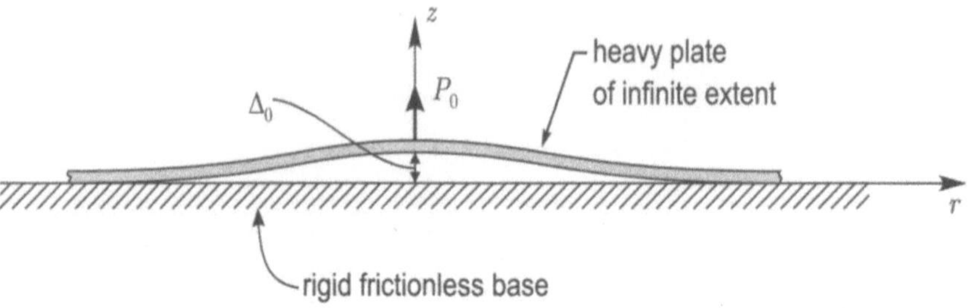

Figure 8.147: Lifting of a heavy plate by a localized load.

8.41 A baffle in a circular pipe of radius a consists of a flexible plate which has a circular aperture of radius b (Figure 8.148). Assuming that the boundary $r = a$ of the baffle is fixed, determine expressions for the deflections and flexural moments in the plate for an arbitrary axisymmetric distribution of differential fluid pressure $p(r)$.

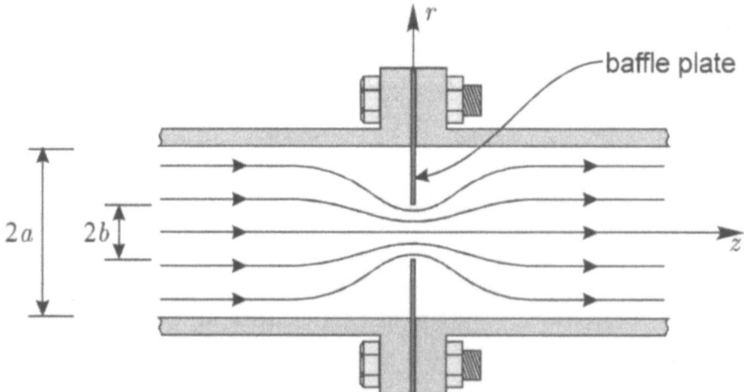

Figure 8.148: Plate with circular aperture located in an axisymmetric flow field.

8.42 A uniformly loaded thin plate is fixed along the boundary $r = a$ and simply supported along the x-axis over the entire diameter (Figure 8.149). Develop an expression for the deflection of the plate.

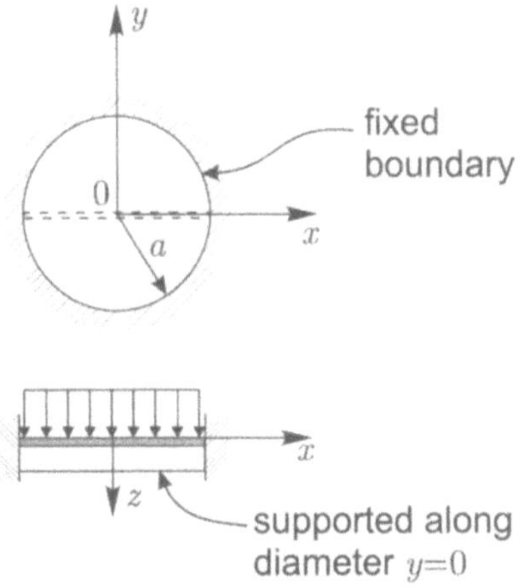

Figure 8.149: Bending of a flexible circular plate with a diametral displacement constraint.

8.43 The Figure 8.150 shows the section through the plane rigid wall of a container which retains a dense fluid of unit weight γ. This plane boundary contains a circular opening to which a flexible plate is attached. The boundary of the circular plate is restrained from displacement and rotation. Develop an expression for the deflection of the flexible plate.

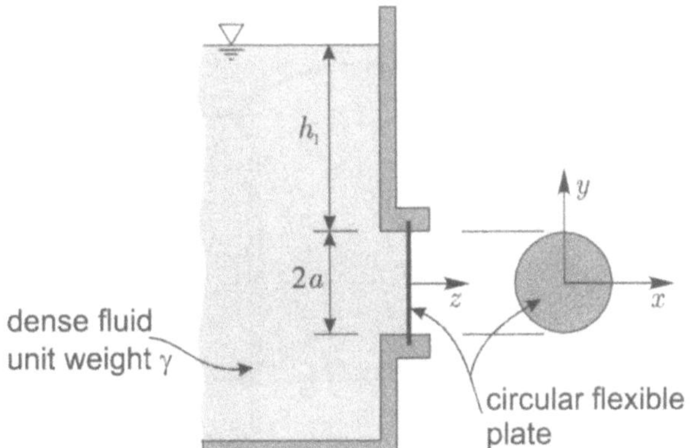

Figure 8.150: Circular plate loaded by non-uniform fluid pressures.

8.44 A circular plate which is welded to a rigid cylinder is simply supported along the boundary (Figure 8.151). If the rigid cylinder is subjected to an eccentric axial load, derive an expression for the deflection of the plate.

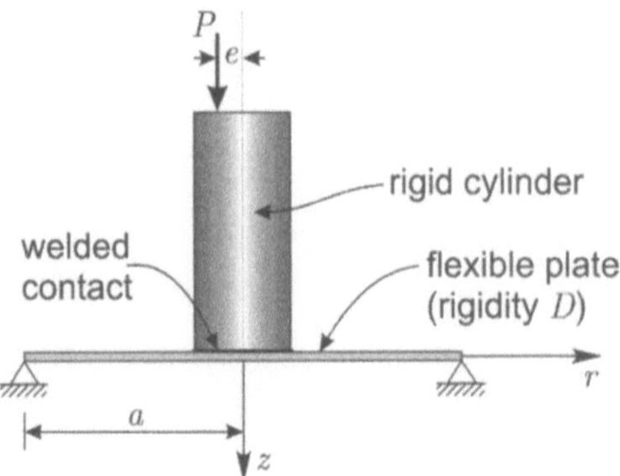

Figure 8.151: Loading of flexible plate through bonded rigid cylinder.

8.45 The circular base plate of a cylindrical grain storage bin consists of a flexible elastic plate of rigidity D which is restrained against deflection along its boundary $r = a$. The release mechanism for the plate is based on a specified displacement limit at the centre of the plate (Figure 8.152). Derive a relationship between the central deflection of the plate w_0, in terms of the given loading geometry and the flexural characteristics of the plate which can be used to control the release mechanism for the base plate.

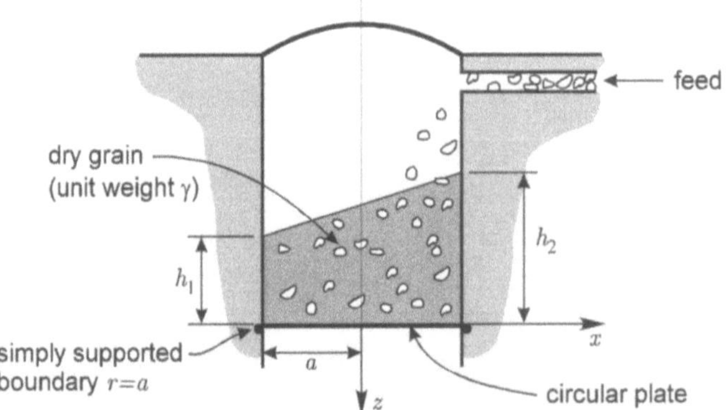

Figure 8.152: Non-uniform loading of the base plate of a silo.

8.46 The flexural behaviour structural plate in the elastic range can be idealized by the Germain-Poisson-Kirchhoff thin plate theory. The Figure 8.153 illustrates a typical connection between a rigid circular column and the structural slab. The plate is subjected to a *far field* state of pure bending by flexural moments $M_{xx} = M_o$. Assume a perfect welded connection between the rigid circular column and the structural slab. Determine the distribution of flexural stresses M_{rr}, $M_{\theta\theta}$ and $M_{r\theta}$ at the interface between the slab and the column.

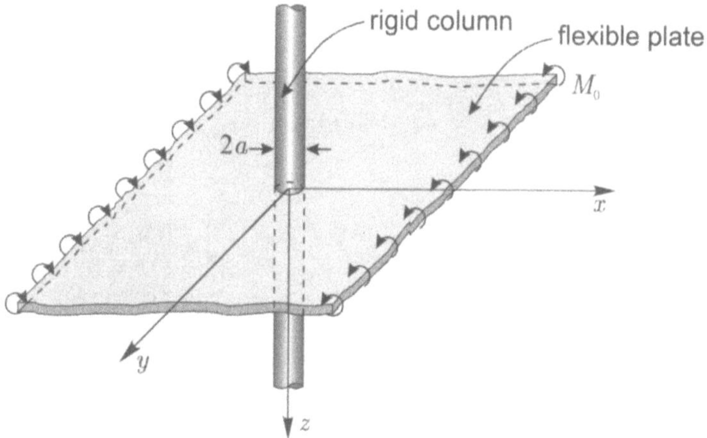

Figure 8.153: Flexure of a plate-column connection.

8.47 The Figure 8.154 shows a semi-infinite plate, the boundary of which is subjected to a periodic distribution of moments which have the variation

$$M_{yy}(x, 0) = M_0 \sin\left(\frac{\pi x}{a}\right)$$

Assuming that the displacement of the plate at the origin is zero, determine an expression for the deflected shape of the plate.

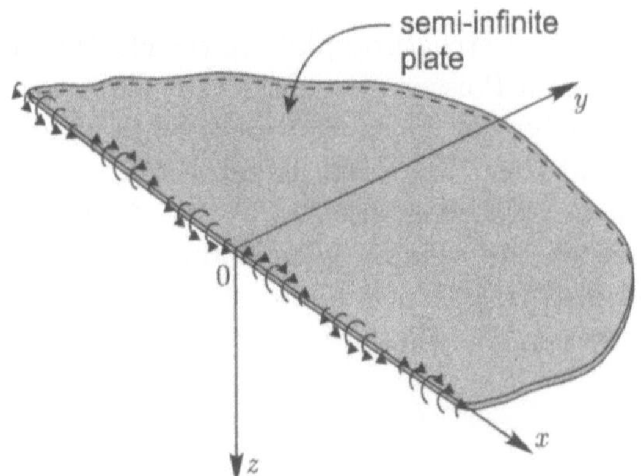

Figure 8.154: Distributed flexural moments along boundary of semi-infinite plate.

8.48 A thin plate of infinite extent is constrained between two rigid rollers which correspond to a *line support* (Figure 8.155). The plate is subjected to localized loads as shown. Use a Fourier transform technique to develop a formal integral expression for the deflection of the plate in the region $x \in (-\infty, \infty)$, $y \in (0, \infty)$.

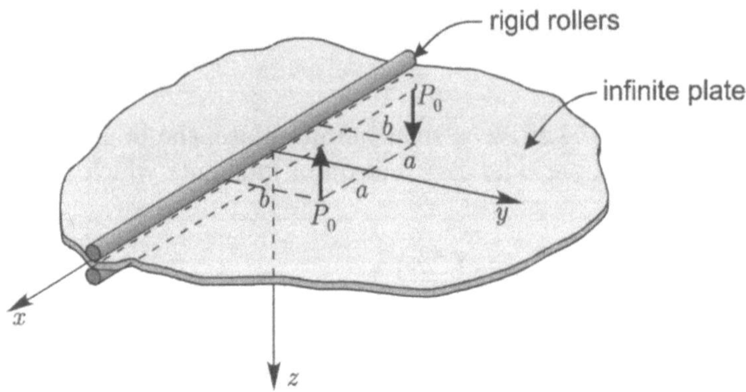

Figure 8.155: Flexure of an elastic plate of infinite extent with line support.

8.49 An infinite plate of finite width b is clamped along one of the boundaries and subjected to a uniform line load of constant intensity P_0 and finite width $2c$, at the free end (Figure 8.156). Use an integral transform technique to obtain formal integral expressions for the deflection and flexural moments in the plate.

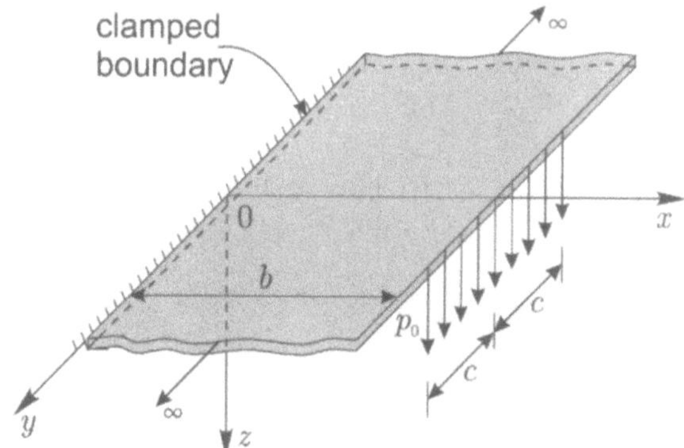

Figure 8.156: Edge loading of a clamped infinite plate.

8.50 A plate of infinite length and finite width $2b$ is simply supported along the edges $y = b$ and $y = -b$ (Figure 8.157). Both these edges are subjected to distributions of moments of constant magnitude over the intervals $x \in (0, \infty)$ and $x \in (-\infty, 0)$ respectively. Use a Fourier transform technique to obtain formal integral expressions for the deflections and flexural moments in the plate.

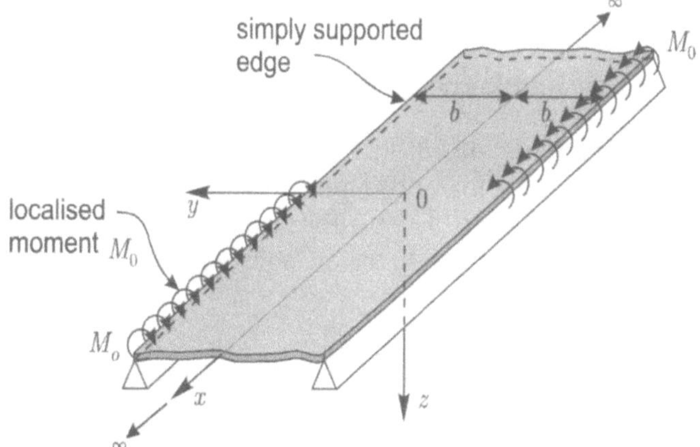

Figure 8.157: Localized edge loading of an infinite plate by uniform distributions of edge moments.

8.51 Two semi-infinite plates of differing elastic materials are fully (i.e., continuously) joined at the interface $y = 0$, as shown in Figure 8.158. The plate is subjected to a concentrated load P_0 at the location (ξ, η). Derive expressions for the deflections of the plates in the separate regions and the flexural moments M_{yy} along the boundary $y = 0$.

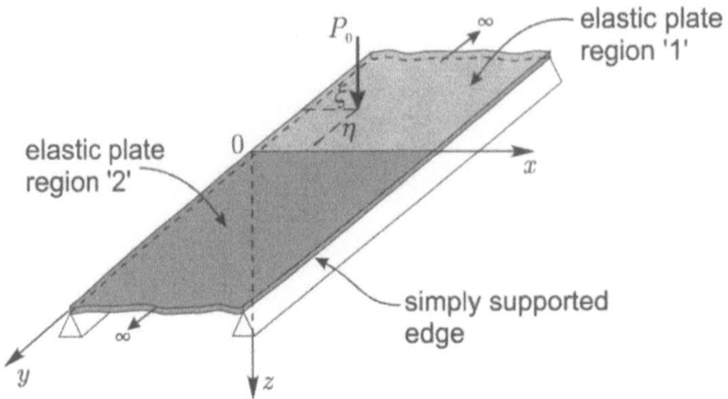

Figure 8.158: Concentrated loading of a simply supported two material plate region.

8.52 A water retaining structure for a hydroelectric project consists of a series of flat panels which can be moved along guided tracks located in a system of piers (Figure 8.159). The guided tracks prevent displacements along the vertical edges and these edges are capable of experiencing rotation. The base of the panel is fully restrained from displacement and rotation. Examine the flexural behaviour of the panel.

8.53 The thin elastic plate shown in Figure 8.160 is simply supported along all its four edges and loads of equal magnitude P_0 are applied at the locations shown. If there is no separation between the plate and the supports, derive a series expansion solution for the deflection of the plate.

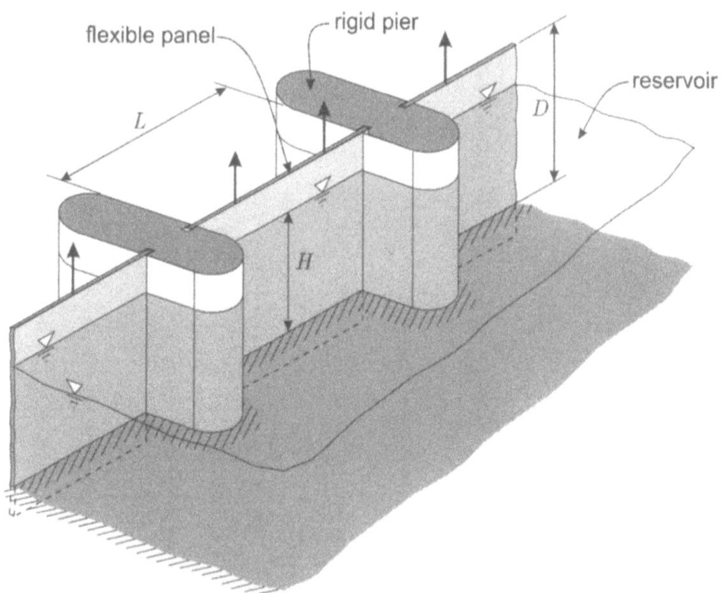

Figure 8.159: Flexure of a panel subjected to fluid pressure.

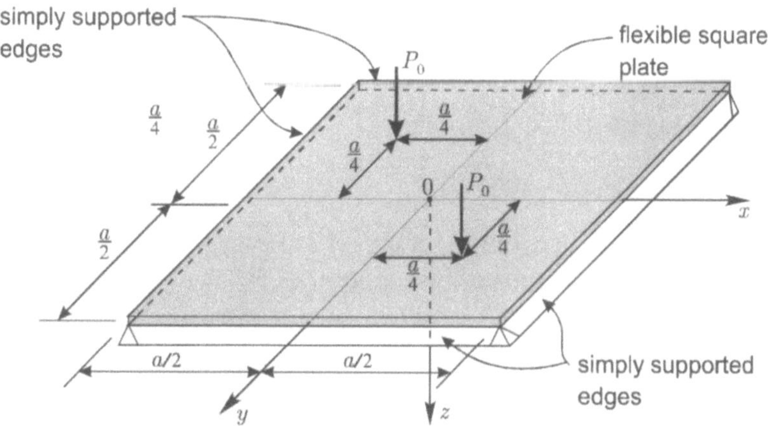

Figure 8.160: Axisymmetric loading of a square plate simply
supported along its edges.

8.54 The Figure 8.161 shows a cross-section through a slider bearing where
the aperture has sinusoidal variation. The aperture contains a viscous
fluid and the motion of the lower plate with constant velocity V_0 induces
slow viscous flow in the region. Assuming that the parameters h/L and

δ/L are small in comparison with unity, derive an expression for the frictional drag exerted on the moving plate. ($L = 2nl$; n is an integer; $n \gg 1$).

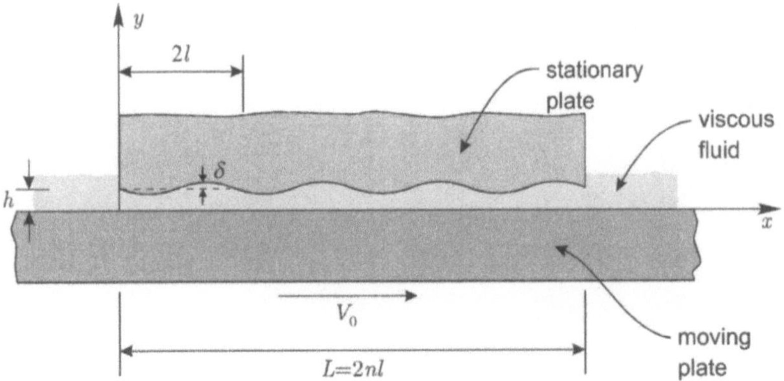

Figure 8.161: Slider bearing with non-uniform aperture.

8.55 The Figure 8.162 shows the cross-section through a cylindrical shaft-bearing system where the cylinder moves with constant velocity V_0. Formulate the problem related to the mechanics of viscous fluid in the narrow non-uniform aperture by considering axial symmetry of the problem with velocity vector

$$\mathbf{v} = \{0\mathbf{i}_r + 0\mathbf{i}_\theta + v(r)\mathbf{i}_z\}$$

Determine the distribution of excess fluid pressure induced in the aperture region due to the uniform motion of the shaft. Derive an expression for the frictional drag on the shaft and compare this result with the elementary result derived in Example 8.47, as $\frac{R}{h_o} \to$ large.

8.56 A rigid cylinder with a deformable surface coating is placed in a slow flow region with a velocity field which corresponds to uniform shear (Figure 8.163). Derive expressions for the state of stress in the surface coating if the deformable material is a rubber-like incompressible elastic solid.

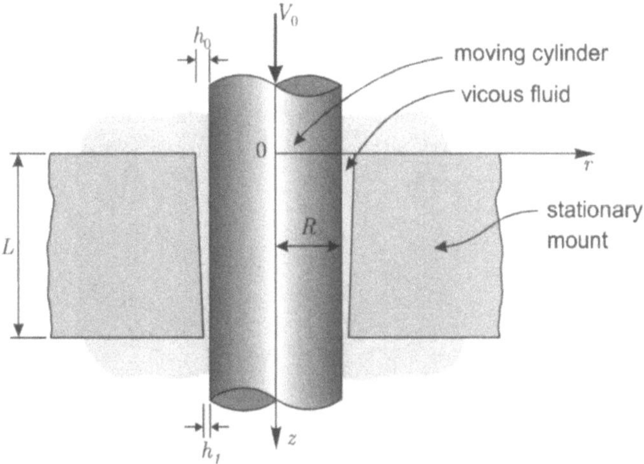

Figure 8.162: Cylindrical slider bearing.

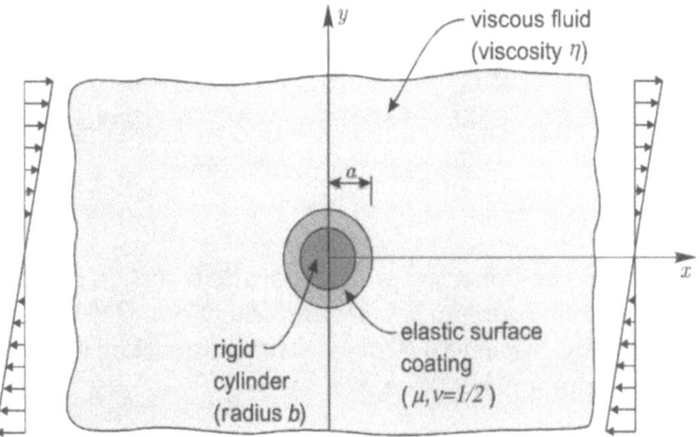

Figure 8.163: Cylinder with an elastic surface coating in a viscous
flow region with shear flow.

8.57 An infinitely long cylinder of radius a is contained in a *double shear
flow region* with a far field velocity vector defined by the Cartesian
components

$$\mathbf{v} = \{\Omega_1 y \mathbf{i} + \Omega_2 x \mathbf{j} + 0\mathbf{k}\}$$

where Ω_1 and Ω_2 are arbitrary rotations with units of radians/unit
time (Figure 8.164). Derive an expression for the torque T_0 that should

be applied to maintain the cylinder in a stationary configuration. Use this result to show that when $\Omega_1 = \Omega_2 = \Omega_0$, the expression for the torque reduces to that obtained for the "Couette Flow" (named after M.F.A Couette (1858-1943)) type problem where a cylinder of radius a, which is located in a viscous fluid region of infinite extent, rotates with angular velocity Ω_0.

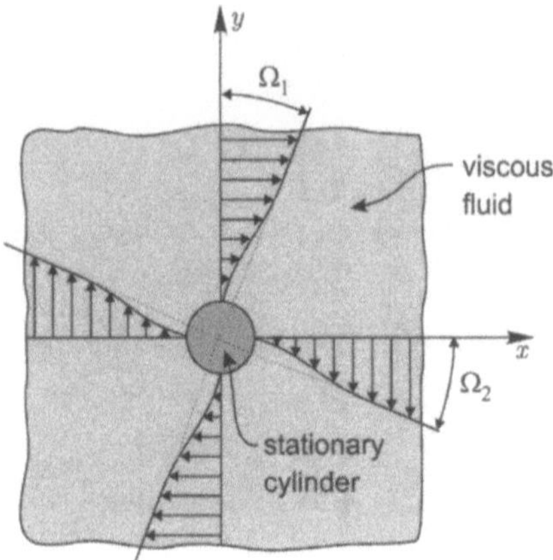

Figure 8.164: Rigid cylinder located in a viscous fluid domain undergoing double shear.

8.58 Show that for the spherical polar coordinate (R, ϑ, Θ) formulation of the axisymmetric problem (i.e. symmetric about $\Theta = 0$) governing slow viscous flow, by appeal to Stokes' stream function $\psi(R, \Theta)$ such that the velocity components are given by

$$\mathbf{v} = \nabla \times \boldsymbol{\Psi}^*$$

where

$$\boldsymbol{\Psi}^* = \frac{\psi(R, \Theta)}{R \sin \Theta} \mathbf{i}_\vartheta$$

satisfies the continuity equation

$$\nabla \cdot \mathbf{v} = 0$$

Also, show that the equations of motion can be reduced to the operator equations

$$E_s^2 E_s^2 \psi(R,\Theta) = 0 \;\; ; \;\; \widehat{\nabla}_s^2 p(R,\Theta) = 0$$

where

$$E_s^2 = \frac{\partial^2}{\partial R^2} + \frac{1}{R^2}\frac{\partial^2}{\partial \Theta^2} - \frac{\cot\Theta}{R^2}\frac{\partial}{\partial\Theta}$$

is Stokes' operator and

$$\widehat{\nabla}_s^2 = \frac{\partial^2}{\partial R^2} + \frac{2}{R}\frac{\partial}{\partial R} + \frac{1}{R^2}\frac{\partial^2}{\partial\Theta^2} + \frac{\cot\Theta}{R^2}\frac{\partial}{\partial\Theta}$$

is Laplace's operator.

8.59 The figure 8.165 shows a viscous spherical inclusion which is held stationary in a slow viscous flow region of uniform velocity V_0. The viscosity of inclusion material is η_i and that of the exterior region is η_0 and $\eta_i >> \eta_0$. Derive an expression for the frictional drag on the viscous inclusion region assuming that the spherical shape is nearly maintained during attainment of steady flow.

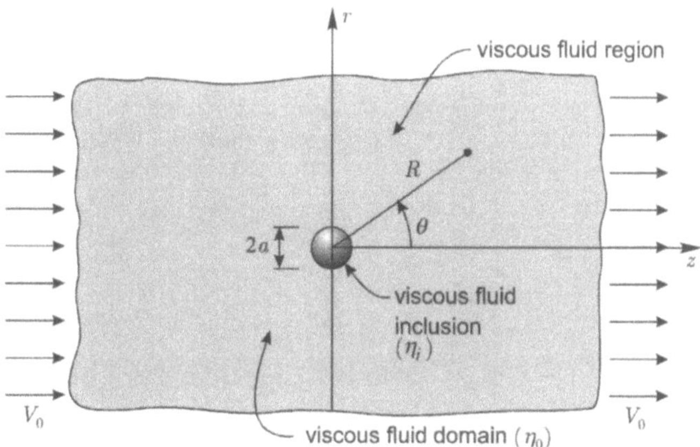

Figure 8.165: Spherical viscous fluid inclusion in a slow viscous flow region.

8.60 A rigid sphere of radius a is located in a slow viscous flow region with uniform far field velocity V_0 (Figure 8.166). The surface of sphere exhibits sliding friction such that the tangential traction on $R = a$ is directly proportional to the tangential displacement $u_\theta(a,\theta)$ and the constant of proportionality is a viscous friction β which has the same units as η. Derive an expression for the frictional drag on the rigid sphere.

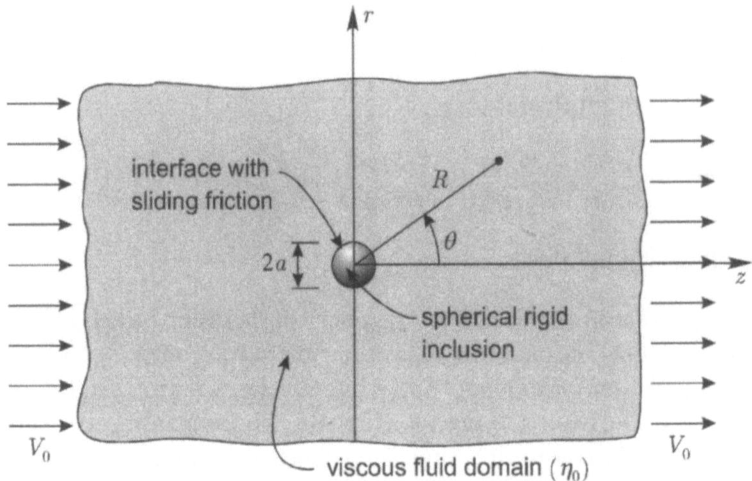

Figure 8.166: Spherical rigid inclusion with sliding viscous frictional interface located in a slow viscous flow region.

8.61 A viscous fluid domain of the infinite extent contains a thin flat plate of large extent. The fluid region is initially at rest. The plate is subjected to a sudden motion such that for all times $t > 0$, the plate has a velocity of magnitude V_0 in the plane of the plate (Figure 8.167).

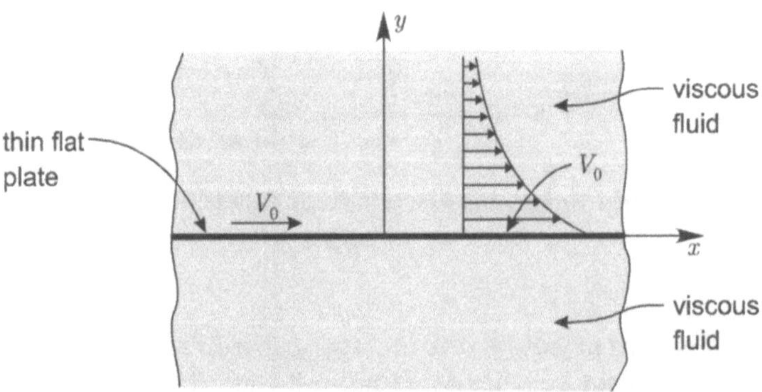

Figure 8.167: Stokes' problem for a flat plate.

Derive the partial differential equation governing the velocity $v_x(y,t)$ assuming Newtonian viscous behaviour of the fluid. Formulate the initial boundary value problem governing $v_x(y,t)$. Show that by introducing a single independent variable $\xi = y/2(\nu t)^{\frac{1}{2}}$ where $\nu = \eta/\rho$ is the kinematic viscosity, the partial differential equation for $v_x = V_0 F(\xi)$, can be reduced to a single ordinary differential equation for $F(\xi)$. (The change of independent variable is obtained through considerations of non-dimensional groups. As such, the procedure results in a *similarity* solution.) Develop an expression for the time-dependent variation of $v_x(y,t)$.

8.62 Consider the infinite plate problem discussed in Problem 8.61. Develop the solution for the case when the infinite plate is stationary and the one-dimensional velocity at infinity corresponds to

$$v_x(y,t) \to V_0 + V^* \cos \omega t \quad \text{as} \quad y \to \infty$$

8.63 The surface of a viscous fluid layer of thickness h (Figure 8.168) is impulsively subjected to a constant shear stress τ_0 over its entire surface. Formulate the initial boundary value problem governing the variation of velocity $v_x(y,t)$ over the depth of the layer. Use a variables separation technique to develop a series expression for $v_x(y,t)$.

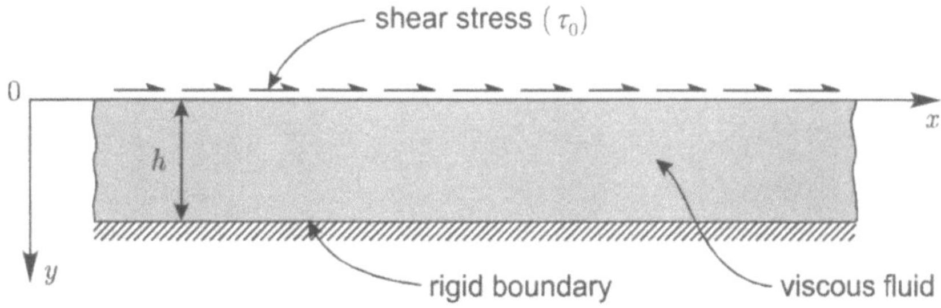

Figure 8.168: Surface loading of a viscous fluid layer by an impulsively applied surface shear stress.

8.64 The region exterior to a rigid circular cylinder of infinite length (in the z-direction) contains a Newtonian viscous fluid adheres to the cylinder thereby providing a no-slip boundary condition (Figure 8.169). The cylinder is impulsively subjected to a rotation Ω_0 Assuming that the resulting unsteady motion in the viscous fluid exhibits circular symmetry, use a Laplace transform technique to develop an integral expression for the time-dependent distribution of velocity $v_\theta(r, t)$ in the viscous fluid.

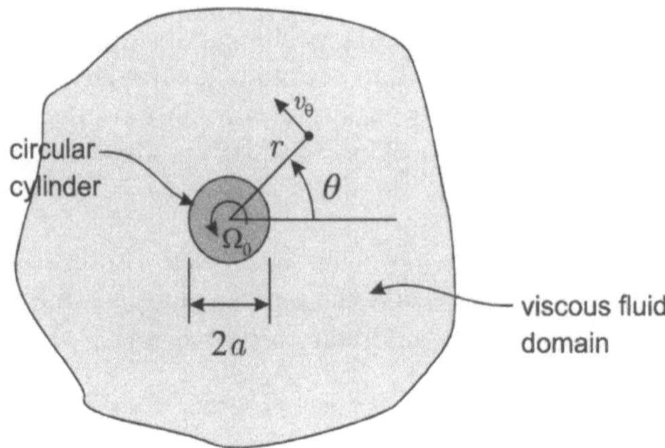

Figure 8.169: Impulsively rotated cylinder contained within a viscous fluid region of infinite extent.

Chapter 9

Poisson's equation

Poisson's equation is the inhomogeneous equivalent of Laplace's equation. It is encountered in the modelling of a variety of problems in mechanics and physics, ranging from the study of fluid flows in porous media to the theory of gravitation. In Chapter 5 dealing with Laplace's equation, we have briefly encountered Poisson's equation in connection with the development of the Green's function for Laplace's equation. In this Chapter, however, the emphasis is on the examination of specific types of problems where the governing partial differential equation corresponds to the inhomogeneous form of Laplace's equation. Furthermore, the function which contributes to the inhomogeneous form results from the incorporation of specific aspects of the mechanics and physics of a problem. The origins of Poisson's equation is generally associated with the study of problems in field theory, such as electrostatics, mechanics and physics of conducting media and in the description of gravitational potential in regions with mass density. In these instances, the variation of the field characterized by either an electric potential or a gravitational potential in the presence of a spatial distribution (of either an electronic charge density or a mass density exterior to a point), gives rise to Poisson's equation. The range of applicability of Poisson's equation also extends to the study of steady fluid flow through porous media which are internally subjected to either fluid influx or withdrawal. Also, in the area of fluid mechanics, the steady axial motion of viscous fluids in conduits and steady gravity flows of viscous fluids along inclined planes can also be modelled by appeal to Poisson's equation. In the area of applied mechanics, Poisson's equation is encountered in the modelling of the transverse small deflection behaviour of stretched membranes which are subjected to transverse loads. In the study of mechanics of deformable elastic solids, Poisson's equation governs the axial displacement field of prismatic solids of arbitrary cross-section which are subject to a state of torsion. The celebrated work of Barre de Saint Venant (1797-1886) on elastic torsion represents an elegant application of Poisson's equation to a problem not only of considerable engineering importance but also of fundamental mathematical interest. The objective of this chapter is to outline briefly the modelling of certain problems in mechanics which gives rise to a partial differential equation of the Poisson-type. In many presentations of Poisson's equation, attention is usually restricted to examples involving classical potential theory. In this chapter, however, advantage is taken of developments in flow through porous media given in Chapter 5 and developments in solid mechanics and mechanics of viscous fluids given in Chapter 8 to present a more comprehensive range of application of Poisson's equation to problems in mechanics.

9.1 Flow through porous media with internal sources

A complete discussion of the modelling associated with the flow of a fluid through a porous medium was presented in Section 5.1.2 of Chapter 5. In the modelling of the flow process it is assumed that the flow takes place through the void phase of volumetric porosity n^* (defined by $n^* = V_v/V$; where V_v is the volume of voids with a total volume V), with average fluid velocity $\overline{\mathbf{v}}$. The volume flux or the fluid volume rate per unit cross-sectional area $\widehat{\mathbf{v}}$ (which takes into account both the solid and void phases of the porous medium) can be obtained by a mass conservation equation through any area A normal to the flow direction; i.e.

$$\widehat{\mathbf{v}} = n^*\overline{\mathbf{v}} \tag{9.1}$$

This represents the "smeared" average velocity flow vector which is much smaller than $\overline{\mathbf{v}}$, since for porous media with low porosity, $n^* << 1$. We now consider a fixed control volume V of the porous medium where the entire pore space is saturated. The average outward fluid velocity at S, the boundary of V is denoted by $\widehat{\mathbf{v}}$. We also assume that fluid is generated within the volume V at a constant volume rate of q per unit total volume of the porous medium. In the case of a three-dimensional problem involving fluid recharge, this assumption is highly artificial if q is distributed within the volume. One can visualize a set of well points with entry locations at the interior of V which can recharge at a constant rate throughout V.

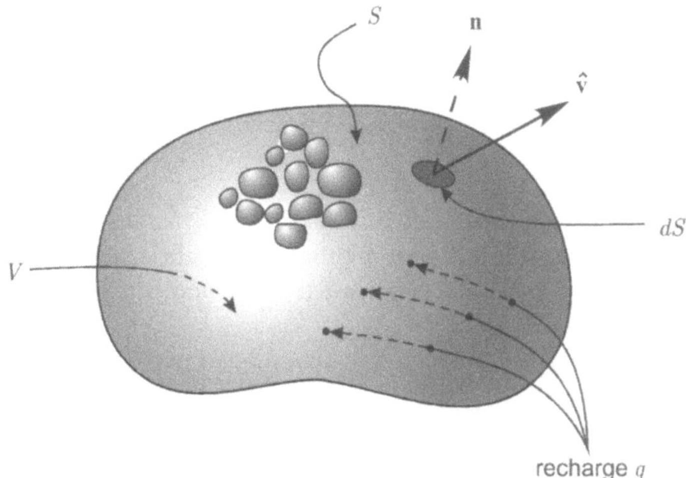

Figure 9.1: Porous medium with internal recharge of pore fluid.

If fluid mass is conserved then the rate at which fluid mass *accumulates* within V must be identically equal to the rate at which fluid mass is *generated* by sources within V less the rate at which fluid mass *leaves* the control surface S; i.e.

$$\frac{d}{dt}\iiint_V \rho n^* dV = \iiint_V \rho q dV - \iint_S \rho \hat{\mathbf{v}}.\mathbf{n} dS \tag{9.2}$$

where ρ is the mass density of the fluid. Using Green's theorem we can rewrite (9.2) in the form

$$\iiint_V \left[\frac{\partial}{\partial t}(\rho n^*) - \rho q + \nabla.(\rho\hat{\mathbf{v}})\right] dV = 0 \tag{9.3}$$

Since the volume region of the porous medium V which is fixed in time, is arbitrary, applying the Dubois-Reymond lemma to (9.3) we obtain

$$\frac{\partial}{\partial t}(\rho n^*) - \rho q + \nabla.(\rho\hat{\mathbf{v}}) = 0 \tag{9.4}$$

Introducing the concept of a reference density of the fluid ρ_0 such that

$$D_R = \log_e\left(\frac{\rho}{\rho_0}\right) \tag{9.5}$$

we can rewrite (9.3) in the form

$$\nabla.\widehat{\mathbf{v}} - q = -\frac{\partial n^*}{\partial t} - \left(n^*\frac{\partial}{\partial t} + \widehat{\mathbf{v}}.\nabla\right)D_R \tag{9.6}$$

For most applications involving steady flow through porous geological media, the fabric of the medium is assumed to remain unchanged during steady groundwater flow and the fluid is assumed to be incompressible. These assumptions reduce (9.6) to

$$\nabla.\widehat{\mathbf{v}} - q = 0 \tag{9.7}$$

We now employ Darcy's law which relates the average velocity of flow $\widehat{\mathbf{v}}$ at any point within the porous medium to the gradient of the hydraulic potential $\Phi(\mathbf{x})$ defined by D. Bernoulli (1700-1783). The reduced form of the Bernoulli potential applicable for the study of flow in porous media takes into consideration only the datum potential $\Phi_D(\mathbf{x})$ and the pressure potential $\Phi_P(\mathbf{x})$ and neglects the velocity potential $\Phi_v(\mathbf{x})$, i.e.

$$\Phi(\mathbf{x}) = \Phi_D(\mathbf{x}) + \Phi_P(\mathbf{x}) \tag{9.8}$$

From Darcy's observations, the average fluid velocity at a point within the porous medium is proportional to the gradient of the Bernoulli potential $\Phi(\mathbf{x})$. If the porous medium is *hydraulically isotropic*, Darcy's law can be written as

$$\widehat{\mathbf{v}} = -K\nabla\Phi \tag{9.9}$$

where K is the hydraulic conductivity (with units of L/T) of the porous medium. The negative sign is indicative of the fact the gradient decreases with increase in $\mathbf{x}$ if flow velocities are considered positive in the positive coordinate directions. Substituting (9.9) in (9.7) we obtain

$$\nabla^2\Phi = -\frac{q}{K} \tag{9.10}$$

which is a partial differential equation of the Poisson-type. It should be noted that q represents the fluid *influx* to the porous medium as a result of a distributed set of *sources*. In the event fluid *efflux* takes place from the porous medium due to a distributed set of *sinks*, then q is simply replaced by $-q$ and the resulting partial differential equation is still of the Poisson-type.

9.2 Heat conduction in the presence of heat sources

The topic of heat conduction in solids was discussed in detail in Section 6.1.1 of Chapter 6. The steady state distribution of temperature in a heat conducting solid which contain either *heat sources* or *heat sinks* is of interest to the thermal modelling of solids which can produce radiogenic heating or materials which can experience *exothermic* or *endothermic* reactions due to chemical processes. A common example of an exothermic reaction is the heating of fresh concrete during hydration of a cement particle. Freshly mixed concrete will emit heat during the solidification process. The heat generated during decay of organic material also represents an example of a distributed heat source. Examples of these include heat generation in decaying landfills and in stored agricultural products which mature with time.

We consider an arbitrary region V of a heat conducting solid with surface S, and examine the total heat flux or the quantity of heat leaving the surface per unit area. If $\mathbf{q}$ is the vector of heat flow, then from Fourier's law

$$\mathbf{q} = -k\nabla T \tag{9.11}$$

where T is the temperature and k is the thermal conductivity of the solid. The quantity of heat *leaving* the surface S per unit time is

$$\iint_S \mathbf{q}.\mathbf{n}\, dS = \iint_S (-k\nabla T).\mathbf{n}\, dS = \iiint_V \nabla.(-k\nabla T)dV \tag{9.12}$$

We assume that heat is generated within the body by endogenous processes such as radioactive decay or exothermic reactions. The rate of heat generation per unit volume of the solid is denoted by $g(\mathbf{x}, t)$. The rate of total heat generation within the volume V is

$$G = \iiint_V g\, dV \tag{9.13}$$

At steady state, there is no heat accumulation within the body and the rate at which heat is leaving the surface S must balance the rate of heat generation within V. Combining (9.12) and (9.13) and assuming that the heat conducting solid is thermally homogeneous we obtain

$$\iiint_V [-k\nabla.\nabla T - g]\, dV = 0 \qquad (9.14)$$

Since the volume region of the heat conducting solid is arbitrary (9.14), by virtue of the Dubois-Reymond lemma, gives

$$k\nabla^2 T = -g \qquad (9.15)$$

which is again a partial differential equation of the Poisson-type for the temperature distribution within the solid.

9.3 Transverse deflections of a stretched transversely loaded membrane

The equation of motion of a stretched membrane is subjected to transverse loading $p(x,y)$ in the direction of the z-axis was derived in Section 7.4.1 of Chapter 7. The details of the derivation will not be pursued here. The partial differential equation governing the small static deflection $w(x,y)$ of a membrane which is subjected to uniform tension T_m and loaded by a transverse load $p(x,y)$ in the direction of $w(x,y)$ is given by

$$\nabla^2 w(x,y) = -\frac{p(x,y)}{T_m} \qquad (9.16)$$

which is again a partial differential equation of the Poisson-type.

9.4 Boundary conditions

Poisson's equation is a second-order partial differential equation of the *elliptic* type. As noted in connection with the solution of Laplace's equation, in order to develop a well posed boundary value problem, it is necessary to impose certain boundary conditions applicable to the various applications of Poisson's equation discussed in the previous sections. These boundary conditions are generally of the Dirichlet and Neumann types and other mixed (Robin) type boundary conditions are also encountered.

9.4.1   Boundary conditions for flow in porous media

The Dirichlet type boundary conditions governing the Bernoulli potential $\Phi(\mathbf{x})$ in its reduced form are encountered along boundaries where the potential is prescribed a priori. Referring to Figure 9.2, which represents a porous domain V with boundaries S_1, S_2, S_3 and S_4, the boundaries S_1 and S_2 are of the Dirichlet type. Similarly, on the boundaries S_3 and S_4, the flux normal to the surfaces is zero. By virtue of Darcy's law, these latter boundary conditions are of the Neumann type. Referring to Figure 9.2 we have

$$\Phi = \Phi_1 \; ; \; \mathbf{x} \in S_1$$

$$\Phi = \Phi_2 \; ; \; \mathbf{x} \in S_2 \tag{9.17}$$

which represent the Dirichlet boundary conditions, and

$$\frac{\partial \Phi}{\partial n} = 0 \; ; \; \mathbf{x} \in S_3 \; ; \; \mathbf{x} \in S_4 \tag{9.18}$$

represent the Neumann boundary conditions.

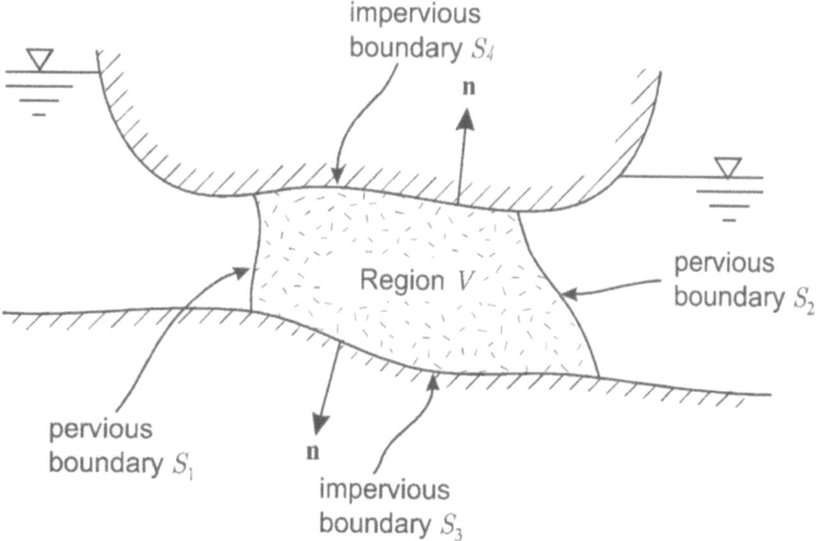

Figure 9.2: Fluid flow in porous media.

9.4.2 Boundary conditions for heat conduction

The Dirichlet and Neumann type boundary conditions applicable to steady
state heat conduction in solids due to boundary and internal heating can be
expressed in terms of the value of the temperature along prescribed boundaries. Referring to Figure 9.3 we can identify three types of boundary conditions. Dirichlet boundary conditions are applicable to surfaces on which
the temperature is prescribed, i.e.

$$T = T_0(\mathbf{x}) \quad ; \quad \mathbf{x} \in S_1 \tag{9.19}$$

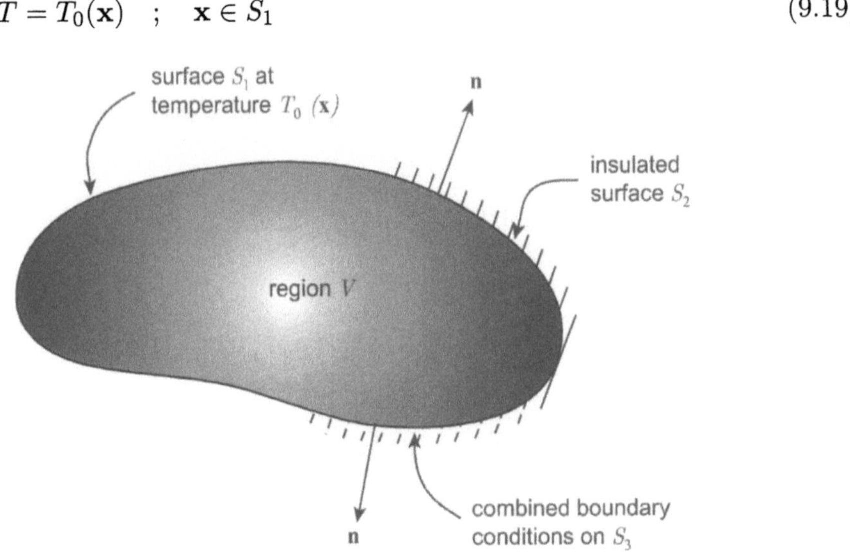

Figure 9.3: Boundary conditions for a steady state heat conduction.

Neumann boundary conditions are applicable to the surface S_2 which is
perfectly insulated (i.e. the normal component of the heat flux vector is zero
on S_2). Hence

$$\frac{\partial T}{\partial n} = 0 \quad ; \quad \mathbf{x} \in S_2 \tag{9.20}$$

The Robin-type combined boundary condition is applicable to boundaries
which are subjected to Newton-type cooling. This type of boundary condition can be written in the generalized form

$$\frac{\partial T}{\partial n} + \lambda_1 T = \lambda_2 T_0^*(\mathbf{x}) \quad ; \quad \mathbf{x} \in S_3 \tag{9.21}$$

where λ_1 and λ_2 are physical parameters which can be expressed in terms of the thermal conductivity and heat transfer properties at the boundary S_3.

9.4.3 Boundary conditions for loaded membranes

Physically realistic boundary conditions applicable to membranes which are subjected to a state of isotropic tension T_m are usually formulated by specifying the transverse displacements along the boundary. Along the segment of a boundary $\mathcal{C}$, such a boundary condition takes the form

$$w(\mathbf{x}) = W_0(\mathbf{x}) \quad ; \quad \mathbf{x} \in \mathcal{C} \tag{9.22}$$

where $\mathcal{C}$ is the boundary of the membrane (Figure 9.4a). Homogeneous Neumann type boundary conditions are applicable to boundaries which correspond to planes of symmetry involving sub-domains of a stretched membrane under the action of pressure $p(x,y)$. Referring to Figure 9.4b, the boundary conditions applicable to the sub-domain of the stretched membrane are

$$w(\mathbf{x}) = 0 \quad ; \quad \mathbf{x} \in \mathcal{C}_1 \tag{9.23}$$

and

$$\frac{\partial w}{\partial n} = 0 \quad ; \quad \mathbf{x} \in \mathcal{C}_2 \tag{9.24}$$

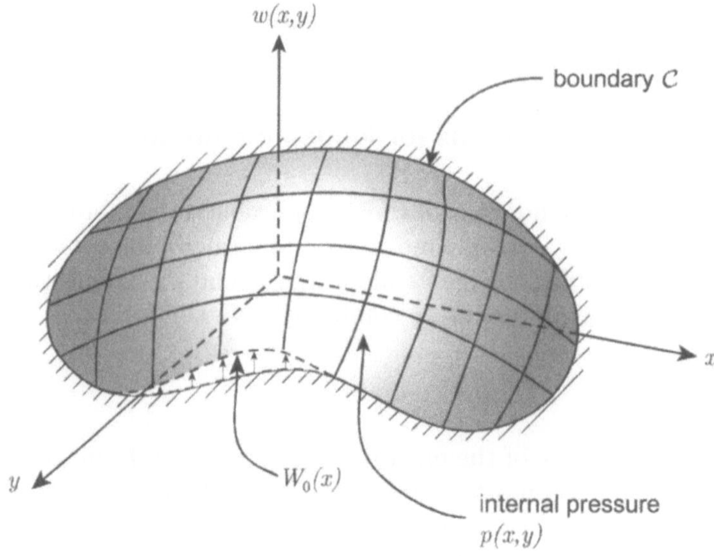

(a) Stretched membrane with boundary displacement.

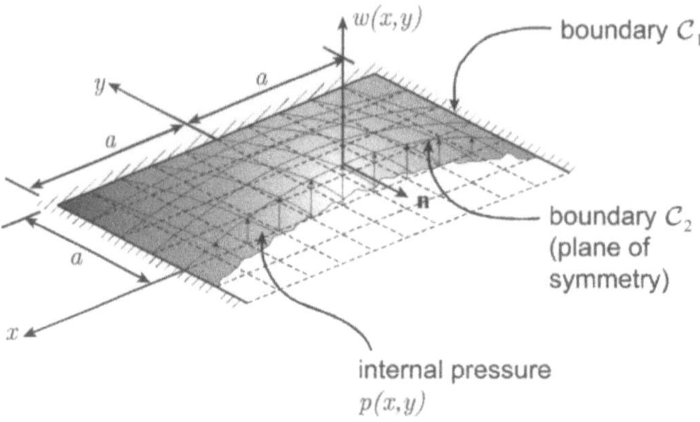

(b) Internally pressurized membrane with boundary restraint.

Figure 9.4: Boundary conditions for the pressurized membrane problem.

In situations where a pressurized stretched membrane interacts with a rigid profile of known shape, only the displacement at the separation point between the membrane and the rigid object can be specified (Figure 9.5).

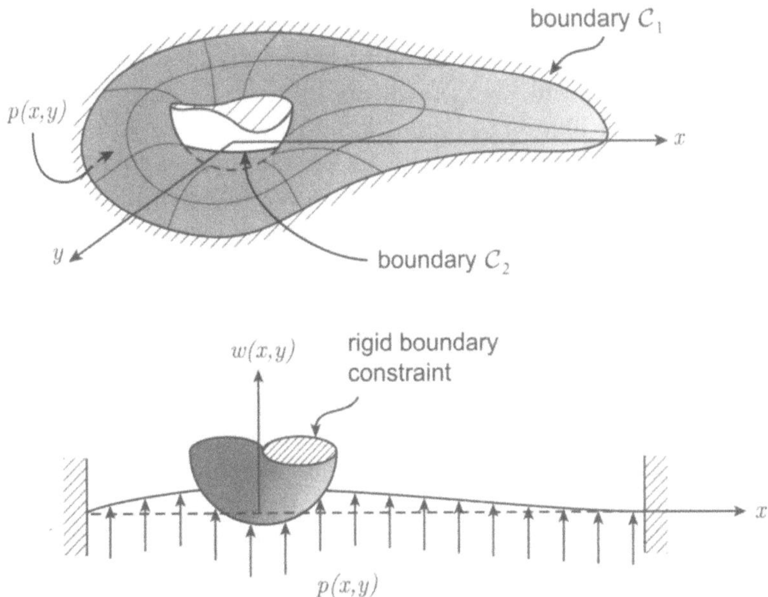

Figure 9.5: Interaction of pressurized membrane with a stationary rigid boundary.

Referring to Figure 9.5, the boundary conditions applicable to the stretched member problem are

$$w(\mathbf{x}) = 0 \quad ; \quad \mathbf{x} \in \mathcal{C}_1 \tag{9.25}$$

and

$$w(\mathbf{x}) = w_c(\mathbf{x}) \quad ; \quad \mathbf{x} \in \mathcal{C}_2 \tag{9.26}$$

where $w_c(\mathbf{x})$ is derived from the geometry of the rigid obstacle up to the separation line between the rigid boundary and the membrane. The boundary $\mathcal{C}_2$ itself is, of course, an unknown in the problem.

9.5 Generalized results

In connection with Laplace's equation, we were able to develop certain generalized results which address qualitative aspects of solutions of Laplace's equation for well posed problems involving Dirichlet, Neumann and Robin type boundary conditions (see e.g. Section 5.3 of Chapter 5). The discussions were restricted to domains of finite extent and the development of a wider class of general results was facilitated by the homogeneous nature of the governing partial differential equation. In the case of Poisson's equation, the inhomogeneous nature of the partial differential equation does not lend itself to the development of extensive sets of generalized results covering Dirichlet, Neumann and Robin type boundary conditions applicable to a finite domain. To develop such generalized results it becomes necessary to invoke certain *maximum-minimum principles* similar to those considered in connection with Laplace's equation. In this section we shall present certain generalized results which can be developed by appeal to concepts derived largely from maximum-minimum principles. We shall establish a *mean value theorem* and a maximum principle which will be of assistance in generating additional theorems involving *subharmonic functions*.

A function $\varphi(\mathbf{x})$ satisfying

$$\nabla^2 \varphi \geq 0 \quad ; \quad \mathbf{x} \in V \tag{9.27}$$

where V is a bounded domain is said to be *subharmonic* in V or simply subharmonic. Similarly, if $\varphi(\mathbf{x})$ satisfies

$$\nabla^2 \varphi \leq 0 \quad ; \quad \mathbf{x} \in V \tag{9.28}$$

such that $-\varphi(\mathbf{x})$ is subharmonic, we define $\varphi(\mathbf{x})$ as a *superharmonic function*.

Let $\varphi(\mathbf{x})$ be a subharmonic function and $\widetilde{\varphi}(\mathbf{x})$ a harmonic function in V with boundary S and $\Omega = V \cup S$, its closure. The function

$$\varphi^*(\mathbf{x}) = \varphi(\mathbf{x}) - \widetilde{\varphi}(\mathbf{x}) \quad ; \quad \mathbf{x} \in V \tag{9.29}$$

will be subharmonic in V. If $\varphi(\mathbf{x})$ and $\widetilde{\varphi}(\mathbf{x})$ coincide on S, then $\varphi^*(\mathbf{x})$ will vanish on S and by the maximum principle, will be negative throughout V. Thus if

$$\varphi^*(\mathbf{x}) \leq 0 \quad ; \quad \mathbf{x} \in V \tag{9.30}$$

then

$$\varphi(\mathbf{x}) \le \widetilde{\varphi}(\mathbf{x}) \quad ; \quad \mathbf{x} \in V \tag{9.31}$$

The term subharmonic is derived from this particular property. The values of a subharmonic function $\varphi(\mathbf{x})$ in $\mathbf{x} \in \Omega$ where $\Omega = V \cup S$, are always lower than the values of the harmonic function which coincides with $\varphi(\mathbf{x})$ on S the boundary of V.

THEOREM 9.1

Let $\varphi(\mathbf{x})$ be a subharmonic function in V which satisfies

$$\nabla^2 \varphi \ge 0 \quad ; \quad \mathbf{x} \in V \tag{9.32}$$

Then the *mean value* of $\varphi(\mathbf{x})$ at the point $\overline{\mathbf{x}}$ defined by $\varphi(\overline{\mathbf{x}})$ satisfies the inequality

$$\varphi(\overline{\mathbf{x}}) \le \frac{1}{4\pi R_0^2} \iint_{S_{S_0}} \varphi(\mathbf{x}) dS \tag{9.33}$$

where S_{S_0} is the spherical surface of radius R_0 centered at $\overline{\mathbf{x}}$, but contained within V.

PROOF

Let V_S and S_S, respectively, be the volume and surface area of a sphere centered at $\overline{\mathbf{x}}$ and with radius R.

From the divergence theorem applicable to a bounded domain V_S with smooth boundary S_S and (9.32) we have

$$\iiint_{V_S} \nabla^2 \varphi \, dV = \iiint_{V_S} \nabla \cdot (\nabla \varphi) \, dV = \iint_{S_S} \frac{\partial \varphi}{\partial n} dS \ge 0 \tag{9.34}$$

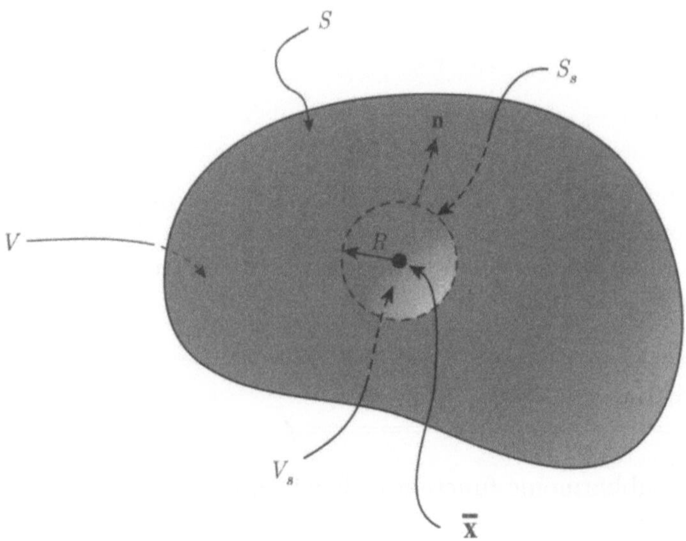

Figure 9.6: Spherical subregion within V.

Since the outward unit normal to the spherical surface centered at $\overline{\mathbf{x}}$ is the radial direction, we can use spherical coordinates to express the surface integral in (9.34) as

$$\iint_{S_s} \frac{\partial \varphi}{\partial n} dS = \int_0^\pi \int_0^{2\pi} R^2 \frac{\partial \varphi}{\partial R} \sin \Theta d\Theta d\vartheta \tag{9.35}$$

where (R, ϑ, Θ) are the spherical coordinates. We now allow R to vary between 0 and R_0, each V_S now being a sphere centered at $\overline{\mathbf{x}}$. As stated previously, the value of R_0 is taken sufficiently small such that the resulting spherical region V_{S_0} with surface S_{S_0} is entirely within V. Integrating (9.35) from 0 to R_0 and interchanging the order of integration we can show that

$$\int_0^\pi \int_0^{2\pi} \varphi(R, \vartheta, \Theta) \sin \Theta d\Theta d\vartheta - 4\pi R_0^2 \varphi(\overline{\mathbf{x}}) \geq 0 \tag{9.36}$$

or

$$\varphi(\overline{\mathbf{x}}) \leq \frac{1}{4\pi R_0^2} \iint_{S_{S_0}} \varphi(\mathbf{x}) dS \tag{9.37}$$

The right side of (9.37) is the mean value of $\varphi(\mathbf{x})$ taken over S_{S_0}, the surface of the sphere with centre $\overline{\mathbf{x}}$. Hence (9.37) assures us that the value of any subharmonic function at any point $\mathbf{x} \in V$ is bounded by the mean value over any spherical region with V having its centre at $\overline{\mathbf{x}}$.

In the case when $\varphi(\mathbf{x})$ is *harmonic*, the inequality (9.37) is valid for both $\varphi(\mathbf{x})$ and $-\varphi(\mathbf{x})$. Consequently we can state the *mean value theorem* for *harmonic functions* without a formal proof.

THEOREM 9.2

If $\varphi(\mathbf{x})$ satisfies

$$\nabla^2\varphi = 0 \quad ; \quad \mathbf{x} \in V \tag{9.38}$$

then $\varphi(\overline{\mathbf{x}})$ is equal to its mean value taken over any sphere in V with radius R_0 and surface S_{S_0} and centre located at $\overline{\mathbf{x}}$, i.e.

$$\varphi(\overline{\mathbf{x}}) = \frac{1}{4\pi R_0^2} \iint_{S_{S_0}} \varphi(\mathbf{x})dS \tag{9.39}$$

PROOF

Proof follows from Theorem 9.1.

THEOREM 9.3

Consider a region V of surface S in which $\varphi(\mathbf{x})$ is a solution of Poisson's equation

$$\nabla^2\varphi = -f(\mathbf{x}) \quad ; \quad \mathbf{x} \in V \tag{9.40}$$

If

$$f(\mathbf{x}) < 0 \quad ; \quad \mathbf{x} \in V \tag{9.41}$$

then φ_M, the maximum value of $\varphi(\mathbf{x})$ occurs in S.

[We note here that in the development of Poisson's Equation with applications to groundwater movement in porous media, heat conduction in solids, deflections of stretched membranes (see e.g. Sections 9.1 to 9.3) and in forthcoming sections dealing with axial flow of viscous fluids (Section 9.8) and torsion of prismatic elastic solids (Section 9.9), the general form of Poisson's equation occurs in the form indicated by (9.40) with $-f(\mathbf{x})$ on the right side of the equation. It must be clearly understood that the sign accompanying $f(\mathbf{x})$ is indicative of a physical process utilized in the formulation , choice of a coordinate direction, etc. (e.g. groundwater recharge as opposed to groundwater depletion; direction of measurement of the deflection of a membrane in relation to the direction of application of the pressure causing the membrane deflection, etc.). The form of Poisson's equation given by (9.40) in more commonly used in mathematical literature. In conection with the ensuing developments, when the function $f(\mathbf{x})$ is interpreted as either being positive or negative it is important to relate the resulting conclusions in relation to the envisaged physical application of Poisson's equation.]

PROOF

We assume that $\varphi(\mathbf{x})$ is continuous in $\Omega = V \cup S$, the closure of V and S, and attains it maximum in V. More specifically, we assume that the maximum occurs at a specific location $\mathbf{x}_0$. At the location $\mathbf{x}_0$ where the maximum occurs

$$\nabla\varphi = 0 \quad ; \quad \mathbf{x}_0 \in V \tag{9.42}$$

and

$$\frac{\partial^2\varphi}{\partial x^2} \leq 0 \quad ; \quad \frac{\partial^2\varphi}{\partial y^2} \leq 0 \quad ; \quad \frac{\partial^2\varphi}{\partial z^2} \leq 0 \quad ; \quad \mathbf{x}_0 \in V \tag{9.43}$$

The result (9.43) requires that

$$\nabla^2\varphi \leq 0 \quad ; \quad \mathbf{x}_0 \in V \tag{9.44}$$

This contradicts the fact that $f(\mathbf{x}) < 0$, therefore if $f(\mathbf{x}) < 0$, the maximum must occur on S. Thus if $\varphi(\mathbf{x})$ satisfies (9.40) with $f(\mathbf{x}) < 0$ for $\mathbf{x} \in V$ and if $\varphi(\mathbf{x}) \leq \varphi_M$ in $\mathbf{x} \in S$, we find that $\varphi(\mathbf{x}) \leq \varphi_M$ in $\mathbf{x}_0 \in V$.

◎ ◎ ◎

A physical interpretation of this maximum principle can be obtained by considering, for example, the problem of a stretched membrane which is subjected to a transverse load $p(x, y)$, where the small transverse deflection of the membrane $w(x, y)$ satisfies (9.16) with the understanding that $w(x, y)$ is positive for $p(x, y) > 0$ (i.e. $p(x, y)$ acts in the positive direction of $w(x, y)$; Figure 9.5). If $p(x, y)$ acts in the opposite direction, the displacements will be negative, and the maximum displacement will occur at the boundary of the membrane. This theorem also confirms the fact that a strictly upward force distribution at every point of the stretched membrane cannot produce a downward deflection.

A similar general conclusion can be drawn in applications related to groundwater flow. Consider, for example, the process of groundwater flow with recharge, where the form of Poisson's equation governing the Bernoulli potential Φ is identical to (9.40). If we consider recharge to be such that in (9.10) $q > 0$, then Φ_M, the maximum of $\Phi(\mathbf{x})$, must occur at $\mathbf{x} = \mathbf{x_0}$ where $\mathbf{x_0} \in V$. Conversely, if there is groundwater depletion within $\mathbf{x} \in V$, the Φ_M will occur in $\mathbf{x_0} \in S$.

THEOREM 9.4

A maximum principle for a subharmonic function was presented in Chapter 5 (Theorem 5.7) in connection with Laplace's equation. We can also develop this proof by a slightly different approach which makes use of the mean value concepts presented previously.

Consider a subharmonic function $\varphi(\mathbf{x})$ which satisfies

$$\nabla^2 \varphi \geq 0 \quad ; \quad \mathbf{x} \in V \tag{9.45}$$

and attains its maximum φ_M at some point $\mathbf{x}_0 \in V$. Then

$$\varphi(\mathbf{x}) \equiv \varphi_M \quad ; \quad \mathbf{x} \in V \tag{9.46}$$

PROOF

Assume that $\varphi(\mathbf{x})$ satisfies the inequality (9.45) in $\mathbf{x} \in V$ and attains its maximum at some location $\mathbf{x}_0 \in V$. Since

$$\varphi(\mathbf{x}) \leq \varphi_M \quad ; \quad \mathbf{x} \in V \tag{9.47}$$

and since at $\mathbf{x}_0$

$$\varphi(\mathbf{x}_0) \equiv \varphi_M \quad ; \quad \mathbf{x}_0 \in V \tag{9.48}$$

we can conclude from (9.37), that (9.48) is satisfied on every sphere centered at $\mathbf{x}_0$ and situated in V. Let us now assume that there is a point $\mathbf{x}_N \in V$ where

$$\varphi(\mathbf{x}_N) \equiv \varphi_M \quad ; \quad \mathbf{x}_N \in V \tag{9.49}$$

Then, the same is true in the neighbourhood of $\mathbf{x}_N$. We now connect $\mathbf{x}_0$ and $\mathbf{x}_N$ by a curve in V and let $\mathbf{x}_P$ be the first point on this curve (Figure 9.7) where

$$\varphi(\mathbf{x}_P) = \varphi_M \quad ; \quad \mathbf{x}_P \in V \tag{9.50}$$

It implies that $\varphi(\mathbf{x})$ is *not* identically equal to φ_M on any sufficiently small circle centered at $\mathbf{x}_P$, i.e.

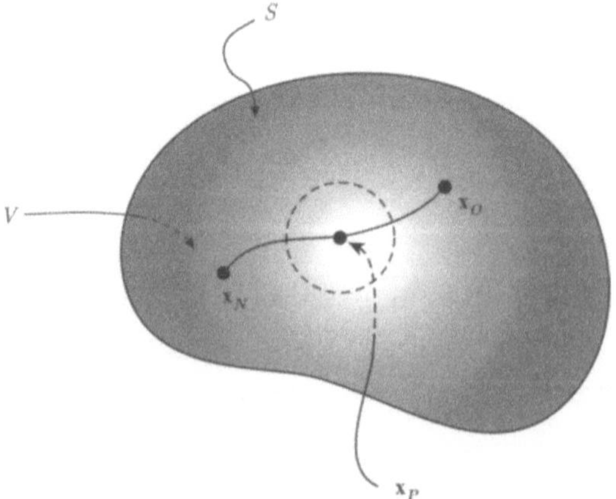

Figure 9.7: Maximum principle for subharmonic functions.

$$\varphi(\mathbf{x}_P) \neq \varphi_M \tag{9.51}$$

This contradicts the inequality (9.37) and the inequality is established when (9.46) is satisfied.

⊚ ⊚ ⊚

We can now focus attention on the development of a uniqueness theorem for Poisson's equation by considering the three general classes of boundary conditions.

THEOREM 9.5

Consider a bounded region V with boundary S in which the function $\varphi(\mathbf{x})$ is continuously differentiable. Let $\Omega = V \cup S$ define the closure. Then there exists at least one solution to Poisson's equation

$$\nabla^2 \varphi = -f(\mathbf{x}) \quad ; \quad \mathbf{x} \in V \tag{9.52}$$

subject to the Dirichlet boundary conditions

$$\varphi(\mathbf{x}) = f_D(\mathbf{x}) \quad ; \quad \mathbf{x} \in S_D \tag{9.53}$$

and subject to the Robin boundary conditions

$$\frac{\partial \varphi}{\partial n} + \lambda(\mathbf{x})\varphi(\mathbf{x}) = f_M(\mathbf{x}) \quad ; \quad \mathbf{x} \in S_M \tag{9.54}$$

where $f(\mathbf{x})$, $f_D(\mathbf{x})$ and $f_M(\mathbf{x})$ are given functions; $\lambda(\mathbf{x}) \geq 0$ is bounded, continuous and not identically equal to zero and $S = S_D \cup S_M$.

PROOF

Let us assume that there exists two solutions $\varphi_1(\mathbf{x})$ and $\varphi_2(\mathbf{x})$ to the boundary value problem posed by (9.52) to (9.54). Then

$$\varphi^*(\mathbf{x}) - \varphi_1(\mathbf{x}) - \varphi_2(\mathbf{x}) \tag{9.55}$$

constitutes a solution of the boundary value problem governed by

$$\nabla^2\varphi^*(\mathbf{x}) = 0 \quad ; \quad \mathbf{x} \in V \tag{9.56}$$

with boundary conditions

$$\varphi^*(\mathbf{x}) = 0 \quad ; \quad \mathbf{x} \in S_D \tag{9.57}$$

and

$$\frac{\partial\varphi^*}{\partial n} + \lambda(\mathbf{x})\varphi^*(\mathbf{x}) = 0 \quad ; \quad \mathbf{x} \in S_M \tag{9.58}$$

Considering Green's first identity (e.g. (1.45) of Chapter 1 and setting $\psi = \varphi = \varphi^*$) we obtain

$$\iint_{S_D} \varphi^*\frac{\partial\varphi^*}{\partial n}\,dS + \iint_{S_M} \varphi^*\frac{\partial\varphi^*}{\partial n}\,dS = \iiint_V (\nabla\varphi^*)^2 dV \tag{9.59}$$

Using (9.57) and (9.58) we can reduce (9.59) to

$$-\iint_{S_M} \lambda(\mathbf{x})\,[\varphi^*(\mathbf{x})]^2\,dS = \iiint_V (\nabla\varphi^*)^2 dV \tag{9.60}$$

Since $\lambda(\mathbf{x}) \geq 0$, the integrands of these integrals are non-negative, and we require

$$\nabla\varphi^* = 0 \quad ; \quad \mathbf{x} \in V \tag{9.61}$$

The equation (9.61) implies that

$$\varphi^*(\mathbf{x}) = \text{constant} \quad ; \quad \mathbf{x} \in \Omega \tag{9.62}$$

which (since $\varphi^*(\mathbf{x})$ is continuously differentiable in Ω) gives

$$\frac{\partial\varphi^*}{\partial n} = 0 \quad ; \quad \mathbf{x} \in S \tag{9.63}$$

The Robin condition (9.58) and the result (9.63) give

$$\lambda(\mathbf{x})\varphi^*(\mathbf{x}) = 0 \quad ; \quad \mathbf{x} \in S \tag{9.64}$$

Hence $\varphi^*(\mathbf{x}) = 0$ on some point of S; which implies that

$$\varphi^*(\mathbf{x}) = 0 \quad ; \quad \mathbf{x} \in \Omega \tag{9.65}$$

or

$$\varphi_1(\mathbf{x}) = \varphi_2(\mathbf{x}) \quad ; \quad \mathbf{x} \in \Omega \tag{9.66}$$

which proves uniqueness and the existence of at most one solution for the boundary value problem described by (9.52) to (9.54).

We can extend the discussion to the consideration of purely Neumann boundary conditions which also result in a *compatibility condition* for the existence of a solution.

THEOREM 9.6

Consider the bounded domain V with boundary S. The function $\varphi(\mathbf{x})$ is twice continuously differentiable in V and with continuous normal derivative in S. If $\varphi(\mathbf{x})$ satisfies Poisson's equation

$$\nabla^2 \varphi = -f(\mathbf{x}) \quad ; \quad \mathbf{x} \in V \tag{9.67}$$

with Neumann boundary condition

$$\frac{\partial \varphi}{\partial n} = f_N(\mathbf{x}) \quad ; \quad \mathbf{x} \in S \tag{9.68}$$

then any two solutions to this boundary value problem can differ only by a constant. Also a necessary *compatibility condition* for the existence of a solution is that $f(\mathbf{x})$ and $f_N(\mathbf{x})$ satisfy the condition

$$\iiint_V f(\mathbf{x})dV + \iint_S f_N(\mathbf{x})dS = 0 \tag{9.69}$$

PROOF

The procedures used in connection with the Proof of Theorem 9.5 can be used to prove the uniqueness of solution for the Neumann problem.

Considering Green's first identity for scalar functions $\varphi(\mathbf{x})$ and $\psi(\mathbf{x})$

$$\iint_S \psi \frac{\partial \varphi}{\partial n} dS = \iiint_V \left[\psi \nabla^2 \varphi + (\nabla \varphi).(\nabla \psi) \right] dV \tag{9.70}$$

We set $\psi = 1$, giving

$$\iint_S \frac{\partial \varphi}{\partial n} dS = \iiint_V \nabla^2 \varphi \, dV \tag{9.71}$$

Substituting (9.67) and (9.68) in (9.71) immediately yields the compatibility condition (9.69).

$$\odot \; \odot \; \odot$$

A maximum principle governing Poisson's equation was presented in Section 5.8 of Chapter 5. The principle can be extended to include both subharmonic and superharmonic functions.

THEOREM 9.7

Consider a function $\varphi(\mathbf{x})$, defined in a bounded region V with surface S and continuously differentiable in $\Omega(= V \cup S)$, its closure.

If

$$\nabla^2 \varphi \leq 0 \quad ; \quad \mathbf{x} \in V \; , \; \varphi(\mathbf{x}) \geq 0 \quad ; \quad \mathbf{x} \in S \tag{9.72}$$

then

$$\varphi(\mathbf{x}) \geq 0 \, ; \, \mathbf{x} \in V \tag{9.73}$$

Similarly if

$$\nabla^2 \varphi \geq 0 \quad ; \quad \mathbf{x} \in V \ , \ \varphi(\mathbf{x}) \leq 0 \quad ; \quad \mathbf{x} \in S \tag{9.74}$$

then

$$\varphi(\mathbf{x}) \leq 0 \, ; \, \mathbf{x} \in V \tag{9.75}$$

PROOF

If we consider the set of conditions defined by (9.72) and assume that there exists values of $\mathbf{x} \in V$ with $\varphi(\mathbf{x}) < 0$, then $\varphi(\mathbf{x})$ achieves a negative minimum value at some location $\mathbf{x}_0 \in V$. We introduce an auxiliary function

$$\chi(\mathbf{x}) = \varphi(\mathbf{x}) - \epsilon \, |\mathbf{x} - \mathbf{x}_0|^2 \tag{9.76}$$

Since $\varphi(\mathbf{x}) > 0$ on $\mathbf{x} \in S$, we can adjust the value of ϵ such that

$$\chi(\mathbf{x}) > -\epsilon \, |\mathbf{x} - \mathbf{x}_0|^2 \geq -\epsilon R^2 > \varphi(\mathbf{x}_0) = \chi(\mathbf{x}_0) \, ; \, \mathbf{x} \in S \tag{9.77}$$

where $R = \max\{|\mathbf{x} - \mathbf{x}^*| \ ; \ \mathbf{x}, \ \mathbf{x}^* \in \Omega\}$ is the diameter of V. The inequality (9.77) shows that $\chi(\mathbf{x})$ assumes its minimum at some point $\widehat{\mathbf{x}} \in V$. Since $\chi(\widehat{\mathbf{x}})$ is a minimum, the second derivative test gives $0 \leq \left[\nabla^2 \chi\right]_{\mathbf{x}=\widehat{\mathbf{x}}}$ and this leads to the contradiction

$$0 \leq \left[\nabla^2 \chi\right]_{\mathbf{x}=\widehat{\mathbf{x}}} = \left[\nabla^2 \varphi\right]_{\mathbf{x}=\widehat{\mathbf{x}}} - 6\epsilon \ < 0 \tag{9.78}$$

The result (9.78) implies that the original assumption, that $\varphi(\mathbf{x})$ could assume negative values, is false. Hence,

$$\varphi(\mathbf{x}) \geq 0 \quad ; \quad \mathbf{x} \in V \tag{9.79}$$

The Proof of the Theorem associated with statements (9.74) and (9.75) can be obtained by applying the above arguments to the function $-\varphi(\mathbf{x})$.

9.6   Green's function for Poisson's equation

In the chapter dealing with Laplace's equation, we have discussed the development of a fundamental solution for the two-dimensional problem involving a Poisson-type partial differential equation where the inhomogeneous function corresponded to a source in the form of a Dirac delta function, situated at an arbitrary location. Here, we shall generalize this approach to develop a three-dimensional equivalent for a Green's function. At the outset we should note that the Green's function is a solution to a specific boundary value problem (or an initial boundary value problem). Therefore it will depend not only on the domain of applicability of the differential operator but also on the boundary conditions applicable to the domain. Consequently, we can expect each separate type of boundary condition, namely, Dirichlet, Neumann and Robin type to yield separate forms of Green's functions applicable in general to finite domains, but with modifications, to infinite domains. In the following we shall obtain an important representation formula applicable to functions which are twice continuously differentiable in a three-dimensional region.

Consider first, the spherically symmetric form of Laplace's equation referred to the spherical polar coordinate system (R, ϑ, Φ); i.e.

$$\nabla^2 \varphi = \frac{d^2 \varphi}{dR^2} + \frac{2}{R} \frac{d\varphi}{dR} = 0 \tag{9.80}$$

which yields solutions

$$\varphi(R) = \frac{C_1}{R} + C_2 \tag{9.81}$$

where C_1 and C_2 are constants. We can normalize $\varphi(R)$ to be zero at infinity such that $C_2 = 0$ and set $C_1 = 1/4\pi$. This gives

$$\varphi(R) = \frac{1}{4\pi R} \tag{9.82}$$

If we select two locations $P(x, y, z)(= P(\mathbf{x}))$ and $Q(x^*, y^*, z^*)(= Q(\mathbf{x}^*))$ such that

$$R^2 = (x - x^*)^2 + (y - y^*)^2 + (z - z^*)^2 \tag{9.83}$$

and we can write (9.82) as

$$\varphi(R) = \frac{1}{4\pi \, |\mathbf{x} - \mathbf{x}^*|} \tag{9.84}$$

The function $\varphi(R)$ is referred to as the *fundamental solution* to Laplace's equation. It is harmonic in $\mathbf{x}$ or in $\mathbf{x}^*$ for $\mathbf{x} \neq \mathbf{x}^*$ and has a singularity at $\mathbf{x} = \mathbf{x}^*$.

Consider the region V with boundary S where $\varphi(\mathbf{x})$ is twice continuously differentiable in V and continuously differentiable in the closure $\Omega = V \cup S$. We wish to apply Green's second identity

$$\iint_S \left[\varphi \frac{\partial \psi}{\partial n} - \psi \frac{\partial \varphi}{\partial n} \right] dS = \iiint_V \left[\varphi \nabla^2 \psi - \psi \nabla^2 \varphi \right] dV \tag{9.85}$$

with

$$\psi = \frac{1}{4\pi R} \tag{9.86}$$

where $\mathbf{x} \in V$ is fixed and integration is with respect to $\mathbf{x}^*$. Green's second identity cannot be applied directly due to the presence of singularity of ψ at $\mathbf{x} = \mathbf{x}^*$. Therefore, we exclude from V a small sphere centered at Q and with volume V^* and surface area S^* (Figure 9.8).

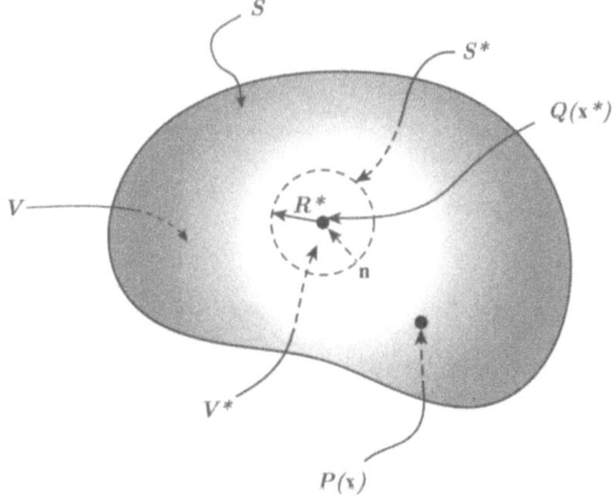

Figure 9.8: Region with isolated boundaries.

The radius of the spherical region centered at Q is denoted by R^*. We fix $R^* > 0$ such that $V^* \subset V$. We define the domain $\widehat{V} = V - V^*$ with boundary $\widehat{S} = S \cup S^*$. We can now apply Green's second identity (9.85) with ψ defined by (9.86) to obtain

$$\iiint_{\widehat{V}} \left\{ \frac{1}{4\pi R} \nabla^2 \varphi - \varphi \nabla^2 \left(\frac{1}{4\pi R} \right) \right\} dV$$
$$= \iiint_{\widehat{V}} \left\{ \frac{1}{4\pi R} \nabla^2 \varphi \right\} dV$$
$$= \iint_{\widehat{S}} \left\{ \frac{1}{4\pi R} \frac{\partial \varphi}{\partial n} - \varphi \frac{\partial}{\partial n} \left(\frac{1}{4\pi R} \right) \right\} dS \tag{9.87}$$

The integral over $\widehat{S}$ consist the integral over S the boundary of V which is independent of R^* and the integral over S^*. We consider the first integral on the right side of (9.87). Since φ has continuous partial derivatives in V

$$\left| \frac{\partial \varphi}{\partial n} \right| = |\nabla \varphi . \mathbf{n}| \leq M \tag{9.88}$$

for some constant M and for all points $\mathbf{x}^*$ in the proximity of $\mathbf{x}$. Then

$$\left| \iint_{S^*} \frac{1}{4\pi R} \frac{\partial \varphi}{\partial n} dS \right| \leq \frac{M}{4\pi R^*} \iint_{S^*} dS = \frac{M}{4\pi R^*} 4\pi (R^*)^2 \tag{9.89}$$

Hence

$$\underset{R^* \to 0}{\text{Lim}} \left| \iint_{S^*} \frac{1}{4\pi R} \frac{\partial \varphi}{\partial n} dS \right| \to 0 \tag{9.90}$$

Similarly, we can show that as $R^* \to 0$

$$\iiint_{\widehat{V}} \frac{1}{4\pi R} \nabla^2 \varphi \, dV \to \iiint_{V} \frac{1}{4\pi R} \nabla^2 \varphi \, dV \tag{9.91}$$

Considering the results given by (9.90) and (9.91) we can rewrite (9.87) as

$$\iiint_{\widehat{V}} \frac{1}{4\pi R} \nabla^2 \varphi \, dV = \iint_{\widehat{S}} \left\{ \frac{1}{4\pi R} \frac{\partial \varphi}{\partial n} - \varphi \frac{\partial}{\partial n} \left(\frac{1}{4\pi R} \right) \right\} dS \tag{9.92}$$

On S^* the surface of the sphere V^*, the outward unit normal $\partial/\partial n = -\partial/\partial R$. Using spherical polar coordinates with origin at $\mathbf{x}$ to evaluate the second integral of (9.92), i.e.

$$\iint_{S^*} \varphi \frac{\partial}{\partial n}\left(\frac{1}{4\pi R}\right) dS$$
$$= \int_0^{2\pi} \int_0^{\pi} \varphi(\mathbf{x} + R^*\mathbf{d}) \frac{1}{4\pi(R^*)^2}(R^*)^2 \sin\Theta d\Theta d\vartheta \tag{9.93}$$

where

$$\mathbf{d} = \{\cos\vartheta \sin\Theta \mathbf{i} + \sin\vartheta \sin\Theta \mathbf{j} + \cos\Theta \mathbf{k}\} \tag{9.94}$$

In the limit as $R^* \to 0$, (9.93) gives

$$\frac{1}{4\pi} \int_0^{2\pi} \int_0^{\pi} \varphi(\mathbf{x}) \sin\Theta d\Theta d\vartheta = \varphi(\mathbf{x}) \tag{9.95}$$

Considering the reduction derived in (9.95) we can obtain the limiting value of (9.92) as $R^* \to 0$ in the form

$$\varphi(\mathbf{x}) = \iint_S \frac{1}{4\pi R}\frac{\partial\varphi}{\partial n} dS - \iint_S \varphi\frac{\partial}{\partial n}\left(\frac{1}{4\pi R}\right) dS$$
$$- \iiint_V \frac{1}{4\pi R}\nabla^2\varphi \, dV \tag{9.96}$$

This equation indicates that the solution to the potential function $\varphi(\mathbf{x})$ which is twice continuously differentiable in V and continuously differentiable in Ω, the closure $V \cup S$, can be expressed in terms of integrals over V and S. The first integral over S is commonly referred to as a *single layer potential* with density $\partial\varphi/\partial n$. The second integral over S is referred to as a *double layer potential*. The integral over V is referred to as a *volume potential*. We can state that, in general, every function which satisfies the differentiability criteria $(C^2(V) \; ; \; C^1(\Omega))$ can be expressed as the sum of a single layer potential, a double layer potential and a volume potential. Let us now apply this result to the development of a solution to Poisson's equation defined by the boundary value problem

$$\nabla^2 \varphi = -f(\mathbf{x}) \quad ; \quad \mathbf{x} \in V \tag{9.97}$$

subject to Dirichlet boundary conditions

$$\varphi = f_D(\mathbf{x}) \quad ; \quad \mathbf{x} \in S \tag{9.98}$$

Since $f(\mathbf{x})$ and $f_D(\mathbf{x})$ are specified, the second and third integrals of (9.96) can be evaluated. The first integral of (9.96) which involves $\partial\varphi/\partial n$ cannot be evaluated since only φ is specified. Specifying both $\varphi(\mathbf{x})$ and $\partial\varphi/\partial n$, on S will make the problem over-determined. To avoid this difficulty we can generalize (9.96). We define a harmonic function $g(\mathbf{x})$ and any function $\varphi(\mathbf{x})$ in $\mathbf{x} \in V$, both of which are twice continuously differentiable in V and continuously differentiable in the closure $\Omega = V \cup S$. Applying Green's second identity to these two functions (with $\psi = g(\mathbf{x})$) we obtain

$$-\iiint_V g\nabla^2\varphi \, dV + \iint_S \left[g\frac{\partial\varphi}{\partial n} - \varphi\frac{\partial g}{\partial n} \right] dS = 0 \tag{9.99}$$

Subtracting (9.99) from (9.96) we obtain

$$\varphi(\mathbf{x}) = \iint_S \left\{ \frac{1}{4\pi R} - g \right\} \frac{\partial\varphi}{\partial n} dS - \iint_S \varphi \frac{\partial}{\partial n} \left[\frac{1}{4\pi R} - g \right] dS$$
$$- \iiint_V \nabla^2\varphi \left\{ \frac{1}{4\pi R} - g \right\} dV \tag{9.100}$$

We can choose the harmonic function $g(\mathbf{x})$ such that the first integral in (9.100) vanishes. In other words, for every $\mathbf{x} \in V$ we select a function $g(\mathbf{x}, \mathbf{x}^*)$ such that the function satisfies the differentiability criteria in V and Ω and constitutes a solution to the following Dirichlet boundary value problem

$$(\nabla^*)^2 g(\mathbf{x}, \mathbf{x}^*) = 0 \; ; \; \mathbf{x}^* \in V \tag{9.101}$$

$$g(\mathbf{x}^*) = \frac{1}{4\pi \, |\mathbf{x} - \mathbf{x}^*|} \quad ; \quad \mathbf{x}^* \in S \tag{9.102}$$

The result (9.100) now can be written as

$$\varphi(\mathbf{x}) = -\iint_S \left\{ \frac{\partial}{\partial n^*} G(\mathbf{x}, \mathbf{x}^*) \right\} f_D(\mathbf{x}^*) dS$$

$$+ \iiint_V G(\mathbf{x}, \mathbf{x}^*) f(\mathbf{x}^*) dV \tag{9.103}$$

which constitutes the solution to the Dirichlet boundary value problem by (9.97) and (9.98) where

$$G(\mathbf{x}, \mathbf{x}^*) = \frac{1}{4\pi |\mathbf{x} - \mathbf{x}^*|} - g(\mathbf{x}, \mathbf{x}^*) \tag{9.104}$$

is referred to as the Green's function for the Dirichlet problem in V.

The result (9.103) is an important result for both Poisson's equation and Laplace's equation in that, once the Green's function $G(\mathbf{x}, \mathbf{x}^*)$ for the domain in known, the general solution for any arbitrary choices of $f(\mathbf{x})$ and $f_D(\mathbf{x})$ can be obtained by simply performing the surface and domain integrals. The development of the Green's function for various types of domains and for various boundary conditions of the Dirichlet, Neumann and Robin types therefore forms an important part of these applications. In the ensuing we shall present certain general results and techniques for generating Green's functions for specific domains of interest. Integral transform techniques represent the most convenient approach for developing such Green's functions. They are particularly effective when dealing with domains of infinite extent in one or more directions. Other techniques based on variables separation techniques, method of eigenfunctions, complex variable techniques, method of images, etc., can also be successfully applied to develop Green's functions for partial differential equations of the Laplace or Poisson types.

We shall focus attention on the Dirichlet problem for Poisson's equation defined by (9.97) and (9.98), the solution of which in terms of the Green's function $G(\mathbf{x}, \mathbf{x}^*)$ is given by (9.103). Although the basic procedures are illustrated with reference to the Dirichlet problem, they can be extended to include Neumann and Robin-type boundary conditions.

9.6.1   Integral transform techniques

Integral transform techniques can, in principle, be applied to determine solutions of Poisson's equation for bounded domains. As has been observed in sections of pervious chapters dealing with applications of integral transforms, the procedures become more effective when the domain of interest is unbounded in at least one coordinate direction. Examples of application of integral transform techniques for the development of Green's functions for Laplace's equation, the diffusion equation and for the wave equation are discussed in previous Chapters. As an example, we consider the following two-dimensional Dirichlet problem for an infinite strip subjected to concentrated source at its interior.

Example 9.1

Develop the Green's function $G(x, y \; ; \; 0, y^*)$ for Poisson's equation for an infinite strip of finite width (Figure 9.9) governed by

$$\nabla^2 G(\mathbf{x}; \mathbf{x}^*) = -\delta(x)\delta(y - y^*) \; ; \; x \in (-\infty, \infty) \; ; \quad y \in (0, a) \quad (9.105)$$

with $0 < y^* < a$, and subject to Dirichlet boundary conditions

$$G(x, 0 \; ; \; 0, y^*) = 0 \; ; \quad x \in (-\infty, \infty)$$

$$(9.106)$$

$$G(x, a \; ; \; 0, y^*) = 0 \; ; \quad x \in (-\infty, \infty)$$

In addition, the Green's function should satisfy the regularity condition

$$\mathop{\mathrm{Lim}}_{|x| \to \infty} G(\mathbf{x}; \mathbf{x}^*) = 0 \; ; \quad y \in (0, a) \tag{9.107}$$

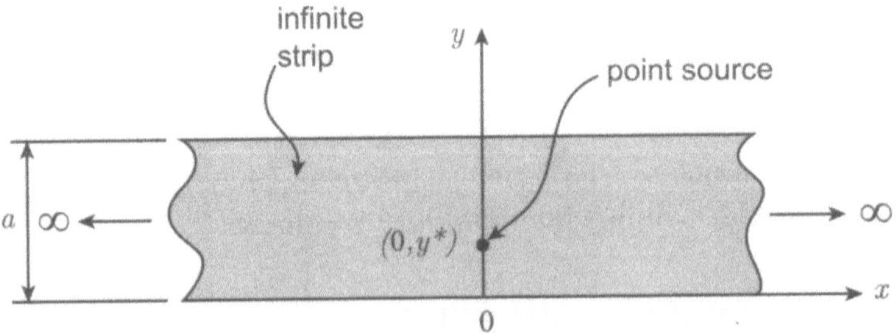

Figure 9.9: Potential problem for an infinite strip.

Solution

Consider the Fourier transform of $G(\mathbf{x}; \mathbf{x}^*)$ with respect to x and for the moment let us suppress the dependency of the solution on $\mathbf{x}^*$. The exponential Fourier transform of $G(x, y)$ is defined by

$$g(\xi, y) = \frac{1}{\sqrt{2\pi}} \int_{-\infty}^{\infty} e^{i\xi x} G(x, y)\, dx \tag{9.108}$$

The inversion theorem is

$$G(x, y) = \frac{1}{\sqrt{2\pi}} \int_{-\infty}^{\infty} e^{-i\xi x} g(\xi, y)\, d\xi \tag{9.109}$$

and we shall assume that both the exponential Fourier transform of $G(\mathbf{x}; \mathbf{x}^*)$ and its inverse exist, where ξ is the transform parameter. Multiplying both sides of the equation (9.105) by $e^{i\xi x}$ and integrating with respect to x between $-\infty$ and $+\infty$. We have

$$\int_{-\infty}^{\infty} e^{i\xi x} \frac{\partial^2 G}{\partial x^2}\, dx + \int_{-\infty}^{\infty} e^{i\xi x} \frac{\partial^2 G}{\partial y^2}\, dx$$
$$= -\int_{-\infty}^{\infty} e^{i\xi x} \delta(x)\delta(y - y^*)\, dx \tag{9.110}$$

In the second integral on the left we interchange the order of integration and differentiation and the first integral is integrated by parts twice. Noting that the integral on the right is identically equal to $-\delta(y - y^*)$, the equation (9.110) can be reduced to the form

$$\frac{d^2 g}{dy^2} - \xi^2 g = -\delta(y - y^*) \tag{9.111}$$

since the boundary conditions (9.106) require $G(x, 0)$ and $G(x, a)$ to be zero, from (9.108), it is evident that

$$g(\xi, 0) = 0 \quad ; \quad g(\xi, a) = 0 \tag{9.112}$$

The appropriate solution of (9.111) can be obtained in the form

$$g(\xi) = \frac{1}{\xi \sinh(\xi a)} \sinh \xi [y]_{\min} \sinh \xi(a - [y]_{\max}) \tag{9.113}$$

where

$$[y]_{\min} = \text{lesser of } (y, y^*)$$

$$[y]_{\max} = \text{greater of } (y, y^*) \tag{9.114}$$

The Fourier inversion theorem (9.109) can be used, in conjunction with (9.113) to obtain an integral expression for the Green's function, i.e.

$$G(x, y \; ; \; 0, y^*) = \frac{1}{2\pi} \int_{-\infty}^{\infty} g(\xi) e^{-i\xi x} d\xi \tag{9.115}$$

We also note that if the source is placed at (x^*, y^*) instead of $(0, y^*)$, the appropriate form of the Green's function for the boundary value problem defined by the partial differential equation

$$\nabla^2 G(\mathbf{x}; \mathbf{x}^*) = -\delta(x - x^*)\delta(y - y^*) \quad ; \quad x \in (-\infty, \infty)$$
$$; \quad y \in (0, a) \tag{9.116}$$

with boundary conditions

$$G(x, 0 \; ; \; x^*, y^*) = 0 \quad ; \quad x \in (-\infty, \infty)$$

$$G(x, a \; ; \; x^*, y^*) = 0 \quad ; \quad x \in (-\infty, \infty) \tag{9.117}$$

and the regularity condition (9.107), can be evaluated in the form

$$G(x, y \; ; \; x^*, y^*)$$
$$= \frac{1}{2\pi} \int_{-\infty}^{\infty} \frac{e^{-i\xi(x-x^*)}}{\xi \sinh(\xi a)} \sinh \xi [y]_{\min} \sinh \xi(a - [y]_{\max}) d\xi \tag{9.118}$$

9.6.2 Series solution techniques

In this section we shall briefly discuss the application of a variables separation technique to the development of a series solution representation for the Green's function. The procedure is more suitable for situations where the domain is bounded. We shall focus on the development of a Green's function which can be used for solving the Dirichlet problem for Laplace's equation applicable to a rectangular region.

Example 9.2

Consider a rectangular region $x \in (0, a), y \in (0, b)$ in which the Green's function for the point source satisfies the Poisson's equation

$$\nabla^2 \varphi = -\delta(x - x^*)\delta(y - y^*) \tag{9.119}$$

and the homogenous boundary conditions

$$\varphi(0, y) = \varphi(a, y) = \varphi(x, 0) = \varphi(x, b) = 0 \tag{9.120}$$

Solution

At regions close to the point source (x^*, y^*) (i.e. $(\widetilde{r}/a) \ll 1; (\widetilde{r}/b) \ll 1$;

$\widetilde{r} = [(x - x^*)^2 + (y - y^*)^2]^{1/2})$ the influences of the boundaries will be negligible and the solution for an unbounded domain is

$$\varphi(x, y) = -\ln[(x - x^*)^2 + (y - y^*)^2] \tag{9.121}$$

We can therefore expect the solution of (9.119) for a finite domain to approach infinity $-2\ln\widetilde{r}$ as $\widetilde{r}$ becomes smaller in relation to the distance between (x^*, y^*) and any nearest boundary point.

The most direct way of solving (9.119) is to observe that the boundary conditions are identically satisfied by choosing eigen-function solutions of the form

$$\varphi(x, y) = \sum_{m=1,2,} \sum_{n=1,2,} A_{mn} \sin\left(\frac{m\pi x}{a}\right) \sin\left(\frac{n\pi y}{b}\right) \tag{9.122}$$

where m, n take integer values. We note here that a double series of the form (9.122) is notoriously slow in convergence and a significant number of terms in the series is required to obtain computational accuracy. It should be noted that (9.122) is not a harmonic solution, but to satisfy (9.119) it should be harmonic everywhere except at (x^*, y^*).

We now assume that $f(x, y)$, the general form of the function applicable to Poisson's equation (see e.g. (9.97)) admits a double series expansion of the form

$$f(x, y) = \sum_{m=1,2} \sum_{n=1,2} F_{mn} \sin\left(\frac{m\pi x}{a}\right) \sin\left(\frac{n\pi y}{b}\right) \tag{9.123}$$

such that

$$F_{mn} = \frac{4}{ab} \int_0^a \int_0^b f(x, y) \sin\left(\frac{m\pi x}{a}\right) \sin\left(\frac{n\pi y}{b}\right) dx dy \tag{9.124}$$

For the specific case of $f(x, y)$ defined by (9.119)

$$F_{mn} = \frac{4}{ab} \sin\left(\frac{m\pi x^*}{a}\right) \sin\left(\frac{n\pi y^*}{b}\right) \tag{9.125}$$

Considering (9.119), (9.122), and (9.125) we obtain

$$A_{mn} = \frac{4 \sin\left(\dfrac{m\pi x^*}{a}\right) \sin\left(\dfrac{n\pi y^*}{b}\right)}{ab \left[\dfrac{m^2\pi^2}{a^2} + \dfrac{n^2\pi^2}{b^2}\right]} \tag{9.126}$$

The solution of (9.119) can now be written as

$$\varphi(x, y \; ; \; x^*, y^*) = \frac{4}{ab} \sum_{m=1,2,} \sum_{n=1,2,} \tag{9.127}$$

$$\frac{\left[\sin\left(\dfrac{m\pi x^*}{a}\right) \sin\left(\dfrac{n\pi y^*}{b}\right) \sin\left(\dfrac{m\pi x}{a}\right) \sin\left(\dfrac{n\pi y}{b}\right)\right]}{\left[\dfrac{m^2\pi^2}{a^2} + \dfrac{n^2\pi^2}{b^2}\right]}$$

The poor convergence of the solution (9.127) results from the fact that the solution must reproduce the logarithmically singular behaviour as $\mathbf{x} \to \mathbf{x}^*$. In general the double power series solution for $\varphi(x, y)$ yields reasonably efficiently convergent results for values of $f(x, y)$ which do not contain singularities.

$$\bullet \ \bullet \ \bullet$$

In the ensuing we shall outline an alternative series solution for the solution of the boundary value problem for the Poisson's equation defined in Example 9.2.

Example 9.3

Develop an alternative solution to the Dirichlet boundary value problem for Poisson's equation defined by (9.119) and (9.120) by considering a series solution of the form

$$\varphi(\mathbf{x}, \mathbf{x}^*) = \sum_{n=1,2,} F_n(y) \sin\left(\frac{n\pi x}{a}\right) \tag{9.128}$$

Solution

Substituting this form of the solution in the governing partial differential equation (9.119) we have

$$\sum_{n=1,2,} \sin\left(\frac{n\pi x}{a}\right) \left\{ \frac{d^2 F_n}{dy^2} - \frac{n^2 \pi^2}{a^2} F_n \right\} = -\delta(x - x^*)\delta(y - y^*) \tag{9.129}$$

We now assume that the delta function $\delta(x - x^*)$ admits a representation

$$\delta(x - x^*) = \sum_{n=1,2} C_n \sin\left(\frac{n\pi x}{a}\right) \tag{9.130}$$

The coefficients C_n are determined from the result

$$C_n = \frac{2}{a} \int_0^a \delta(x - x^*) \sin\left(\frac{n\pi x}{a}\right) dx = \frac{2}{a} \sin\left(\frac{n\pi x^*}{a}\right) \tag{9.131}$$

Combining (9.129), (9.130) and (9.131) we obtain the following ordinary differential equation for $F_n(y)$; i.e.

$$\frac{d^2 F_n}{dy^2} - \frac{n^2 \pi^2}{a^2} F_n = -\frac{2}{a} \sin\left(\frac{n\pi x^*}{a}\right) \delta(y - y^*) \qquad (9.132)$$

The integral with respect to y of this delta function over any internal is non-zero when the interval contains $y = y^*$ and otherwise it is zero. Hence at any point

$$\frac{dF_n}{dy} = \frac{n^2 \pi^2}{a^2} \int_0^y F_n(\zeta) d\zeta + C_1 - \frac{2}{a} \sin\left(\frac{n\pi x^*}{a}\right) H(y - y^*) \qquad (9.133)$$

where $H(y - y^*)$ is the Heaviside step function defined by

$$H(y - y^*) = 1 \quad ; \quad (y - y^*) \geq 0$$

$$\qquad (9.134)$$

$$H(y - y^*) = 0 \quad ; \quad (y - y^*) < 0$$

and C_1 is a constant of integration.

From the result (9.133) we have

$$\left[\frac{dF_n}{dy}\right]_{y=y^*+0} - \left[\frac{dF_n}{dy}\right]_{y=y^*-0} = -\frac{2}{a} \sin\left(\frac{n\pi x^*}{a}\right) \qquad (9.135)$$

which implies that dF_n/dy is discontinuous at $y = y^*$. If we integrate (9.133), we note that F_n is continuous in the domain and F_n satisfies the homogenous equation

$$\frac{d^2 F_n}{dy^2} - \frac{n^2 \pi^2}{a^2} F_n = 0 \qquad (9.136)$$

for the entire range $y \in (0, b)$, except at $y = y^*$. Invoking the boundary conditions at $y = 0$ and $y = b$, we can show that the solution of (9.136) which satisfies the boundary conditions (9.120) take the forms

$$F_n(y) = D^* \frac{\sinh\left[\dfrac{n\pi}{a}(y-b)\right]}{\sinh\left[\dfrac{n\pi}{a}(y^*-b)\right]} \quad ; \quad y \geq y^* \tag{9.137}$$

$$F_n(y) = D^* \frac{\sinh\left[\dfrac{n\pi y}{a}\right]}{\sinh\left[\dfrac{n\pi y^*}{a}\right]} \quad ; \quad y \leq y^* \tag{9.138}$$

where D^* is a constant. This constant can be evaluated by using the condition (9.135). This gives

$$D^* = -\left(\frac{2}{n\pi}\right) \frac{\sinh\left(\dfrac{n\pi y^*}{a}\right) \sinh\left[\dfrac{n\pi(y^*-b)}{a}\right]}{\sinh\left(\dfrac{n\pi b}{a}\right)} \sin\left(\frac{n\pi x^*}{a}\right) \tag{9.139}$$

The solution for $\varphi(\mathbf{x},\mathbf{x}^*)$ can be written in the following form

$$\begin{aligned}
\varphi(\mathbf{x};\mathbf{x}^*) = &-\frac{2}{\pi} \sum_{n=1,2,} \frac{1}{n} \operatorname{cosech}\left(\frac{n\pi b}{a}\right) \sinh\left[\frac{n\pi}{a}(y-b)\right] \\
&\cdot \sinh\left[\frac{n\pi y^*}{a}\right] \sin\left(\frac{n\pi x^*}{a}\right) \sin\left(\frac{n\pi x}{a}\right) \quad ; \quad y \geq y^*
\end{aligned} \tag{9.140}$$

$$\begin{aligned}
\varphi(\mathbf{x};\mathbf{x}^*) = &-\frac{2}{\pi} \sum_{n=1,} \frac{1}{n} \operatorname{cosech}\left(\frac{n\pi b}{a}\right) \sinh\left[\frac{n\pi}{a}(y^*-b)\right] \\
&\cdot \sinh\left[\frac{n\pi y}{a}\right] \sin\left(\frac{n\pi x}{a}\right) \sin\left(\frac{n\pi x^*}{a}\right) \quad ; \quad y \leq y^*
\end{aligned} \tag{9.141}$$

Again, the solutions (9.140) and (9.141) have the drawback that in (9.130), the delta function is represented as a divergent Fourier series. The convergence of the series is thus affected by this representation.

9.6.3 Method of eigenfunctions

The method of eigenfunctions for the solution of Poisson's equation can be viewed as a generalization of the separation of variables technique discussed in Section 9.6.2. Consider the boundary value problem for Poisson's equation

$$\nabla^2 \varphi = -f(\mathbf{x}) \quad ; \quad \mathbf{x} \in V \tag{9.142}$$

subject to Dirichlet boundary conditions

$$\varphi = f_D(\mathbf{x}) \quad ; \quad \mathbf{x} \in S \tag{9.143}$$

We have shown that if $G(\mathbf{x}, \mathbf{x}^*)$ is the Green's function which satisfies

$$\nabla^2 G = -\delta(\mathbf{x}, \mathbf{x}^*) \quad ; \quad \mathbf{x} \in V \tag{9.144}$$

$$G = 0 \quad ; \quad \mathbf{x} \in S \tag{9.145}$$

then the solution to the boundary value problem can be written as

$$\varphi(\mathbf{x}) = \iiint_V G(\mathbf{x}, \mathbf{x}^*) f(\mathbf{x}^*) dV - \iint_S f_D \frac{\partial G}{\partial n} dS \tag{9.146}$$

The method of eigen-functions seeks solutions of the eigen-value problem

$$\nabla^2 \varphi + \lambda \varphi = 0 \quad ; \quad \mathbf{x} \in V$$
$$\varphi = 0 \quad ; \quad \mathbf{x} \in S \tag{9.147}$$

where λ is a complex parameter. For most values of λ, (9.147) admits a trivial solution $\varphi(\mathbf{x}) \equiv 0 \, ; \mathbf{x} \in V$. For other values of λ for which (9.147) admits a non-trivial solution φ are called the eigenvalues of ∇^2 in V, and the corresponding non-trivial solution of φ is an eigenfunction associated with the eigenvalue λ. The convention $\nabla^2 \varphi + \lambda \varphi = 0$ is chosen, instead of $\nabla^2 \varphi = \lambda \varphi$, so that all eigenvalues λ will be positive.

It should be remarked that the solutions of (9.147) are developed for Dirichlet boundary conditions on S. Therefore, more appropriately, the solutions of (9.147) should be referred to as Dirichlet eigenvalues and eigenfunctions, in view of the relevant boundary conditions on S. It should be clear that appropriate Neumann and/or Robin eigenvalues and eigenfunctions can be determined by altering the boundary condition in (9.147), respectively, to the Neumann or Robin type.

At this point it is useful to record some general properties applicable to eigenvalues and eigenfunctions with specific reference to the eigenvalue problems defined previously. To aid the presentation of the proof it is instructive to introduce certain definitions and concepts. The *inner product* of two functions $\varphi(x)$ and $\psi(x)$ which are finite-valued and integrable over the interval $0 \le x \le \ell$ is defined by

$$\langle \varphi, \psi \rangle = \int_0^\ell w(x)\varphi(x)\psi(x)dx \tag{9.148}$$

where $w(x)$ is the weight function for $x \in (0, \ell)$, which is often set to unity. The inner product is symmetric; i.e.

$$\langle \varphi, \psi \rangle = \langle \psi, \varphi \rangle \tag{9.149}$$

The *norm* of the function $\varphi(x)$ is defined in terms of the inner product; i.e.

$$\|\varphi\| = \langle \varphi, \varphi \rangle^{\frac{1}{2}} = \left[\int_0^\ell w(x)[\varphi(x)]^2 dx \right]^{\frac{1}{2}} \tag{9.150}$$

It is evident that the *norm* is non-negative and, is indicative of the magnitude of $\varphi(x)$ over the interval of interest. If the function $\varphi(x)$ is continuous in the interval then $\|\varphi\| = 0$ implies that $\varphi(x) = 0$. Any two functions $\varphi(x)$ and $\psi(x)$ are said to be orthogonal over the integral $x \in (0, \ell)$ if

$$\langle \varphi, \psi \rangle = 0 \tag{9.151}$$

If the functions φ and ψ are complex-valued, the inner product (9.148) is replaced by the Hermitian inner product (named after Charles Hermite (1822-1901))

$$\langle \varphi, \psi \rangle = \int_0^\ell w(x)\varphi(x)\overline{\psi(x)}dx \tag{9.152}$$

where $\overline{\psi(x)}$ is the complex conjugate. This gives

$$\langle \varphi, \psi \rangle = \langle \overline{\psi, \varphi} \rangle \tag{9.153}$$

which defines the Hermitian symmetry of the inner product. The relevant norm is defined by

$$\|\varphi\| = \left[\overline{\langle \varphi, \varphi \rangle}\right]^{\frac{1}{2}} = \left[\int_0^\ell w(x)\{|\varphi(x)|\}^2 dx\right]^{\frac{1}{2}} \tag{9.154}$$

THEOREM 9.8

Consider the eigenvalue problem for the bounded domain V with boundary S, defined by the partial differential equation with Dirichlet boundary conditions, as stated by (9.147). Prove that the eigenvalues are real and positive. Also show that if $\lambda = 0$, the corresponding solution of (9.147) is not an eigenfunction.

PROOF

Consider the complex-valued eigenfunctions $\varphi_1(\mathbf{x})$ and $\varphi_2(\mathbf{x})$ in $\mathbf{x} \in V$. The inner product of these eigenfunctions is defined by

$$\langle \varphi_1, \varphi_2 \rangle = \iiint_V \varphi_1(\mathbf{x})\overline{\varphi_2(\mathbf{x})}dV \tag{9.155}$$

where the bar denotes the complex conjugate. The norm in V is defined by

$$\|\varphi\| = \langle \varphi, \varphi \rangle^{\frac{1}{2}} = \left\{\iiint_V |\varphi|^2\, dV\right\}^{\frac{1}{2}} \tag{9.156}$$

Also, if $\varphi_1(\mathbf{x})$ and $\varphi_2(\mathbf{x})$ are orthogonal then

$$\langle \varphi_1, \varphi_2 \rangle = 0 \tag{9.157}$$

We multiply the partial differential equation in (9.147) by $\overline{\varphi(x)}$ and integrate over the region V. This gives

$$\lambda \left\| \varphi \right\|^2 = - \iiint_V \overline{\varphi} \nabla^2 \varphi \, dV \qquad (9.158)$$

Considering Green's first identity for functions $\overline{\varphi}$ and φ we can write

$$\iiint_V \left[\overline{\varphi} \nabla^2 \varphi + \nabla \varphi . \nabla \overline{\varphi} \right] dV = \iint_S \overline{\varphi} \frac{\partial \varphi}{\partial n} dS \qquad (9.159)$$

Considering the Dirichlet boundary condition in (9.147) we obtain from (9.158) and (9.159) the result

$$\lambda \left\| \varphi \right\|^2 = \iiint_V \left[\nabla \varphi . \nabla \overline{\varphi} \right] dV = \iiint_V \left| \nabla \varphi \right|^2 dV \qquad (9.160)$$

Since the integrand of the right side is positive definite, $\left\| \varphi \right\| \neq 0$ if φ is an eigenfunction. Hence λ is real and non-negative. Furthermore, if $\lambda = 0$, this requires $\nabla \varphi = 0$ in $\mathbf{x} \in V$ or $\varphi = $const., in $\mathbf{x} \in V$. By virtue of the Dirichlet boundary condition in (9.147), this implies $\varphi = 0$; $\mathbf{x} \in V$. Thus the solution which corresponds to $\lambda = 0$ is not an eigenfunction.

◉ ◉ ◉

A further useful theorem relates to the orthogonality of eigenfunctions. We shall present here a brief exposition of the proof of orthogonality of eigenfunctions by appeal to the concept of norms described previously.

THEOREM 9.9

Consider the eigenfunctions φ and ψ, with real eigenvalues λ and μ, respectively, such that $\lambda \neq \mu$. Prove that the eigenfunctions φ and ψ are orthogonal.

PROOF

Since the eigenvalues λ and μ are real, the Dirichlet eigenvalue problems correspond, respectively, to

$$\nabla^2 \varphi + \lambda \varphi = 0 \quad ; \quad \mathbf{x} \in V$$

$$\varphi = 0 \quad ; \quad \mathbf{x} \in S \tag{9.161}$$

and either

$$\nabla^2 \psi + \mu \psi = 0 \quad ; \quad \mathbf{x} \in V$$

$$\psi = 0 \quad ; \quad \mathbf{x} \in S \tag{9.162}$$

or

$$\nabla^2 \overline{\psi} + \mu \overline{\psi} = \nabla^2 \overline{\psi} + \overline{\mu} \overline{\psi} = 0 \quad ; \quad \mathbf{x} \in V$$

$$\overline{\psi} = 0 \quad ; \quad \mathbf{x} \in S \tag{9.163}$$

We now multiply the partial differential equation in (9.161) by $\overline{\psi}$ and the partial differential equation in (9.163) by φ, subtract one from the other and integrate the result over V. We obtain

$$\lambda \iiint_V \overline{\psi} \varphi \, dV - \mu \iiint_V \varphi \overline{\psi} dV$$

$$= \iiint_V \left[\varphi \nabla^2 \overline{\psi} - \overline{\psi} \nabla^2 \varphi \right] dV \tag{9.164}$$

Considering (9.152) and the symmetric nature of the inner product, the left side can be written as

$$(\lambda - \mu) \langle \varphi, \psi \rangle = \iiint_V \left[\varphi \nabla^2 \overline{\psi} - \overline{\psi} \nabla^2 \varphi \right] dV \tag{9.165}$$

In view of the homogeneous Dirichlet boundary conditions applicable to φ and $\overline{\psi}$, Green's second identity gives the right side of (9.165) to be identically zero; hence

$$(\lambda - \mu) \langle \varphi, \psi \rangle = 0 \tag{9.166}$$

Since $\lambda \neq \mu$ we require

$$\langle \varphi, \psi \rangle = 0 \tag{9.167}$$

Therefore, from the result (9.157) the eigenfunctions φ and ψ corresponding to the different eigenvalues λ and μ, respectively, are orthogonal.

$$\odot \; \odot \; \odot$$

Returning to the eigenvalue problem defined by (9.147), we restrict attention to the following two-dimensional eigenvalue problem.

Example 9.4

Consider the two-dimensional eigenvalue problem

$$\nabla^2\varphi + \lambda\varphi = 0 \quad ; \quad x \in (0,a) \ , \ y \in (0,b) \tag{9.168}$$

with Dirichlet boundary conditions

$$\varphi = 0 \quad ; \quad x = 0, a \quad ; \quad y = 0, b \tag{9.169}$$

Obtain an eigenfunction expansion solution for the Green's function $G(x,y;x^*,y^*)$ which satisfies

$$\nabla^2 G = -\delta(x - x^*)\delta(y - y^*) \quad ; \quad x \in (0,a) \ , \ y \in (0,b) \tag{9.170}$$

with

$$G = 0 \quad ; \quad x = 0, a \quad ; \quad y = 0, b \tag{9.171}$$

Solution

Let φ_{mn} be the eigenfunctions and λ_{mn} the corresponding eigenvalues. We expand $G(x,y;x^*,y^*)$ and $\delta(x,y;x^*,y^*)$ in terms of the eigenfunctions φ_{mn} in the forms

$$G(x,y;x^*,y^*) = \sum_{m=1,\,n=1,} \sum A_{mn}(x,y)\varphi_{mn}(x^*,y^*) \tag{9.172}$$

$$\delta(x,y;x^*,y^*) = \sum_{m=1,\,n=1,} \sum B_{mn}(x,y)\varphi_{mn}(x^*,y^*) \tag{9.173}$$

where

$$B_{mn} \doteq \frac{1}{\|\varphi_{mn}\|^2} \int_0^a \int_0^b \delta(x^* - x)\delta(y^* - y)\varphi_{mn}(x^*, y^*)dx^*dy^*$$
$$= \frac{\varphi_{mn}(x, y)}{\|\varphi_{mn}\|^2} \tag{9.174}$$

and

$$\|\varphi_{mn}\|^2 = \int_0^a \int_0^b \varphi_{mn}^2(x^*, y^*)dx^*dy^* \tag{9.175}$$

Substituting (9.172) and (9.173) into (9.170) and noting that

$$\nabla^2 \varphi_{mn} + \lambda_{mn}\varphi_{mn} = 0 \tag{9.176}$$

we can write

$$\sum_{m=1, \, n=1,} \sum \lambda_{mn} A_{mn}(x, y)\varphi_{mn}(x^*, y^*)$$
$$= \sum_{m=1, \, n=1,} \sum \frac{\varphi_{mn}(x, y)\varphi_{mn}(x^*, y^*)}{\|\varphi_{mn}\|^2} \tag{9.177}$$

Equation (9.177) gives

$$A_{mn}(x, y) = \frac{\varphi_{mn}(x, y)}{\lambda_{mn}\|\varphi_{mn}\|^2} \tag{9.178}$$

and the Green's function can be written as

$$G(x, y; x^*, y^*) = \sum_{m=1, \, n=1,} \sum \frac{\varphi_{mn}(x, y)\varphi_{mn}(x^*, y^*)}{\lambda_{mn}\|\varphi_{mn}\|^2} \tag{9.179}$$

Example 9.5

A thin porous seam of hydraulic conductivity K is contained between horizontally bedded geological strata of differing hydraulic conductivity (Figure 9.10). The rock above the seam can be considered to be impermeable and the highly permeable fractured rock below the seam is under artesian pressure. When only the boundaries of this porous seam are exposed during an excavation, in-plane flow takes place in the porous seam which is uniformly fed by the underlying fractured rock, at steady recharge rate of q (with units, e.g. m^3/sec per square metre of the plan area of the porous seam).

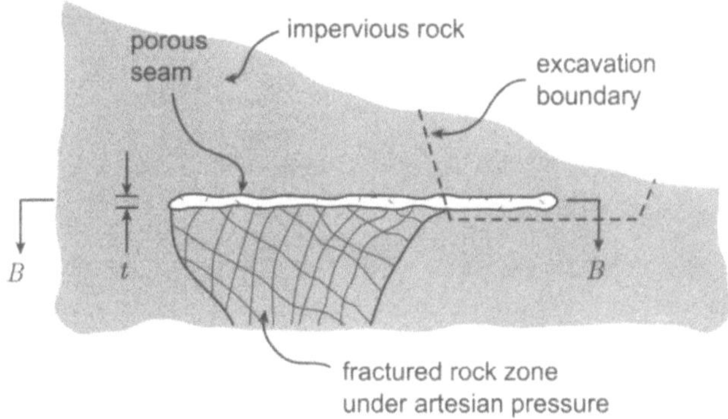

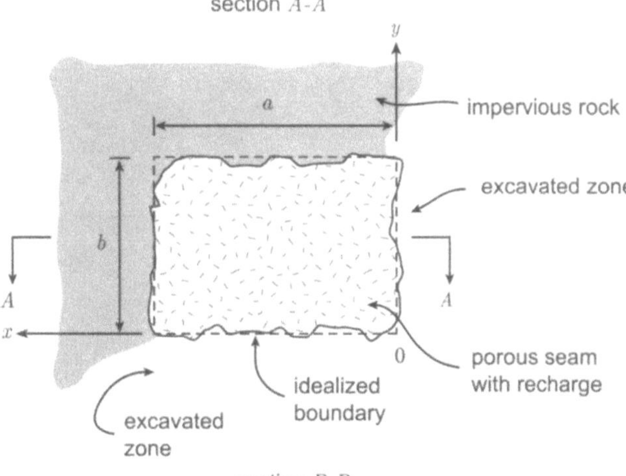

Figure 9.10: In-plane seepage in a porous seam subjected to distributed recharge.

Formulate the boundary value problem governing the distribution of pressure head $H(x,y)$ in the porous seam and derive an eigenfunction expansion solution for $H(x,y)$. Show that the in-plane flow in the porous seam results in a total discharge of qab at the exposed boundaries of the porous seam.

Solution

The partial differential equation governing in-plane flow in the porous seam can be derived by considering an elemental area as shown in Figure 9.11.

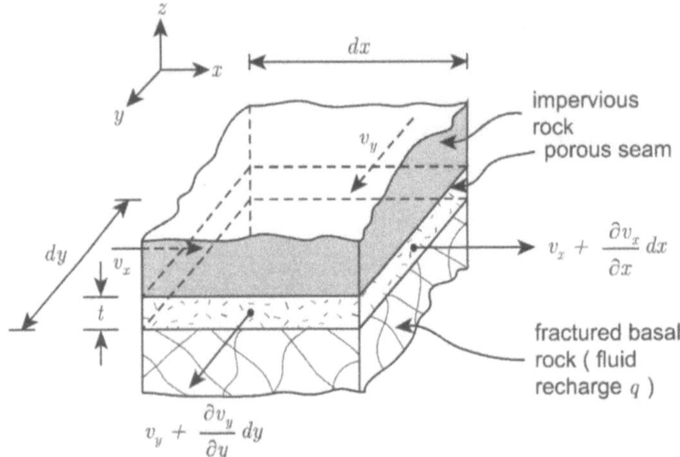

Figure 9.11: Elemental region of porous seam.

Assuming in-plane flow within the porous seam (i.e. $v_z = 0$), the equation of continuity of flow gives

$$\left\{ v_x + \frac{\partial v_x}{\partial x} dx \right\} t\,dy + \left\{ v_y + \frac{\partial v_y}{\partial y} dy \right\} t\,dx$$
$$- v_x t\,dy - v_y t\,dx - q\,dx\,dy = 0 \tag{9.180}$$

This reduces to

$$\frac{\partial v_x}{\partial x} + \frac{\partial v_y}{\partial y} = \overset{\circ}{\nabla} . \mathbf{v} = \frac{q}{t} \tag{9.181}$$

Assuming the datum to be at the level of the porous seam, Darcy's law can be written as

$$\mathbf{v} = -K\nabla H \tag{9.182}$$

where $H(x,y)$ is the pressure head. Combining (9.181) and (9.182) we obtain

$$\nabla^2 H = -\frac{q}{Kt} \tag{9.183}$$

Considering the idealized flow domain which has a rectangular planform (Figure 9.10) the boundary conditions governing the problem are

$$H(x,0) = 0 \quad ; \quad H(0,y) = 0 \tag{9.184}$$

$$\frac{\partial H}{\partial x} = 0 \quad ; \quad x = a \tag{9.185}$$

$$\frac{\partial H}{\partial y} = 0 \quad ; \quad y = b \tag{9.186}$$

In the solution procedure involving eigenfunction expansions we seek solutions of

$$\frac{\partial^2 H}{\partial x^2} + \frac{\partial^2 H}{\partial y^2} + \lambda H = 0 \tag{9.187}$$

where λ is a constant. The governing boundary conditions are given by (9.184) to (9.186). The solution of (9.187) obtained via a separation of variables technique, which satisfies these boundary conditions can be written as

$$H(x,y) = A_{mn} \sin\left(\frac{m\pi x}{2a}\right) \sin\left(\frac{n\pi y}{2b}\right) \tag{9.188}$$

This solution satisfies the boundary conditions (9.184) explicitly and will satisfy the boundary conditions (9.185) and (9.186) provided we set $m, n = 1, 3, 5, \ldots$ The value of the constant λ takes the form

$$\lambda = \pi^2 \left[\left(\frac{m\pi}{2a}\right)^2 + \left(\frac{n\pi}{2b}\right)^2\right] \tag{9.189}$$

and the eigenfunction expansion solution takes the form

$$H(x,y) = \sum_{m=1,3,\,n=1,3,} \sum A_{mn} \sin\left(\frac{m\pi x}{2a}\right)\sin\left(\frac{n\pi y}{2b}\right) \tag{9.190}$$

We now assume that the right side of (9.183) can be represented as a double Fourier series in the form

$$\sum_{m=1,3}\sum_{n=1,3} C_{mn} \sin\left(\frac{m\pi x}{2a}\right)\sin\left(\frac{n\pi y}{2b}\right) = \frac{q}{Kt} \tag{9.191}$$

and the constants C_{mn} are given by

$$C_{mn} = \frac{4}{ab}\int_0^a\int_0^b \left(\frac{q}{Kt}\right)\sin\left(\frac{m\pi x}{2a}\right)\sin\left(\frac{n\pi y}{2b}\right)dxdy \tag{9.192}$$

which gives

$$C_{mn} = \frac{16q}{\pi^2 K t mn} \tag{9.193}$$

Substituting (9.190) and (9.191) in (9.183) we obtain

$$A_{mn} = \frac{64q}{\pi^4 K t mn \left\{\dfrac{m^2}{a^2} + \dfrac{n^2}{b^2}\right\}} \tag{9.194}$$

and the variation of pressure head within the porous seam is given by

$$H(x,y) = \frac{64q}{\pi^4 K t}\sum_{m=1,3,\,n=1,3,}\sum \frac{\sin\left(\dfrac{m\pi x}{2a}\right)\sin\left(\dfrac{n\pi y}{2b}\right)}{mn\left\{\dfrac{m^2}{a^2}+\dfrac{n^2}{b^2}\right\}} \tag{9.195}$$

which completes the solution of the problem. We can establish the accuracy of the solution by evaluating the steady discharge on the exposed boundaries, $x = 0$ and $y = 0$, of the porous seam.

The total steady discharge along the exposed boundaries is given by

$$Q = \int_0^b Kt\left[\frac{\partial H}{\partial x}\right]_{x=0} dy + \int_0^a Kt\left[\frac{\partial H}{\partial y}\right]_{y=0} dx \tag{9.196}$$

where the sign of Q has been adjusted to account for the efflux nature of the discharge. Using the expression for $H(x,y)$ in (9.196) we obtain

$$Q = \frac{64qab}{\pi^4} \sum_{m=1,3,} \sum_{n=1,3,} \frac{1}{m^2 n^2} = qab \tag{9.197}$$

9.6.4 Symmetry of the Green's function

The symmetry of the Green's function is an important property particularly when it is employed to obtain further solutions via superposition techniques such as the method of images, which will be discussed in the ensuing section. We shall present here a proof of the symmetry of the Green's function, restricting attention to the *Dirichlet boundary condition*.

THEOREM 9.10

Prove that the Green's function $G(\mathbf{x}; \mathbf{x}^*)$ which satisfies

$$\nabla^2 G(\mathbf{x}; \mathbf{x}^*) = -\delta(\mathbf{x} - \mathbf{x}^*) \quad ; \quad \mathbf{x} \in V \tag{9.198}$$

and by definition

$$G(\mathbf{x}; \mathbf{x}^*) = 0 \quad ; \quad \mathbf{x} \in S \tag{9.199}$$

satisfies the symmetry condition

$$G(\mathbf{x}; \mathbf{x}^*) = G(\mathbf{x}^*; \mathbf{x}) \tag{9.200}$$

PROOF

Let $G(\mathbf{x}^*; \mathbf{x})$ and $G(\mathbf{x}; \widehat{\mathbf{x}})$ be Green's functions for $\mathbf{x} \in V$, corresponding to sources located at $\mathbf{x}^*$ and $\widehat{\mathbf{x}}$ respectively. Hence

$$\nabla^2 G(\mathbf{x};\mathbf{x}^*) = -\delta(\mathbf{x} - \mathbf{x}^*) \quad ; \quad (\mathbf{x},\mathbf{x}^*) \in V$$

$$G(\mathbf{x};\mathbf{x}^*) = 0 \quad\quad\quad\quad ; \quad \mathbf{x} \in S \tag{9.201}$$

and

$$\nabla^2 G(\mathbf{x};\widehat{\mathbf{x}}) = -\delta(\mathbf{x} - \widehat{\mathbf{x}}) \quad ; \quad (\mathbf{x},\widehat{\mathbf{x}}) \in V$$

$$G(\mathbf{x};\widehat{\mathbf{x}}) = 0 \quad\quad\quad\quad ; \quad \mathbf{x} \in S \tag{9.202}$$

We now multiply the partial differential equation of (9.201) by $G(\mathbf{x};\widehat{\mathbf{x}})$ and the partial differential equation of (9.202) by $G(\mathbf{x};\mathbf{x}^*)$, subtract one from the other and carry out the integrations over V. This gives

$$\iiint_V \left[G(\mathbf{x};\mathbf{x}^*)\nabla^2 G(\mathbf{x};\widehat{\mathbf{x}}) - G(\mathbf{x};\widehat{\mathbf{x}})\nabla^2 G(\mathbf{x};\mathbf{x}^*) \right] dV$$

$$= \iiint_V \left[G(\mathbf{x};\widehat{\mathbf{x}})\delta(\mathbf{x} - \mathbf{x}^*) - G(\mathbf{x};\mathbf{x}^*)\delta(\mathbf{x} - \widehat{\mathbf{x}}) \right] dV$$

$$= G(\mathbf{x}^*;\widehat{\mathbf{x}}) - G(\widehat{\mathbf{x}};\mathbf{x}^*) \tag{9.203}$$

Using Green's second identity, we can convert the left side of (9.203) to an area integral, i.e.

$$\iiint_V \left[G(\mathbf{x};\mathbf{x}^*)\nabla^2 G(\mathbf{x};\widehat{\mathbf{x}}) - G(\mathbf{x};\widehat{\mathbf{x}})\nabla^2 G(\mathbf{x};\mathbf{x}^*) \right] dV$$

$$= \iint_S \left[G(\mathbf{x};\mathbf{x}^*)\frac{\partial}{\partial n}G(\mathbf{x};\widehat{\mathbf{x}}) - G(\mathbf{x};\widehat{\mathbf{x}})\frac{\partial}{\partial n}G(\mathbf{x};\mathbf{x}^*) \right] dS \tag{9.204}$$

where $\mathbf{n}$ is the outward unit normal to S. Since $G(\mathbf{x};\mathbf{x}^*)$ and $G(\mathbf{x};\widehat{\mathbf{x}})$ are zero on the boundary S, we obtain from (9.203) and (9.204) the *reciprocity principle* for the Green's function

$$G(\widehat{\mathbf{x}};\mathbf{x}^*) = G(\mathbf{x}^*;\widehat{\mathbf{x}}) \tag{9.205}$$

The physical interpretation of the reciprocity principle is that the potential induced at $\widehat{\mathbf{x}}$ due to a source at $\mathbf{x}^*$ is identically equal to the potential induced at $\mathbf{x}^*$ due to the source at $\widehat{\mathbf{x}}$.

9.6.5   The method of images

The method of images is essentially a superposition technique that can be used to generate the Green's function for specific domains of finite or semi-infinite extent. We shall illustrate the application of the method of images by considering a specific region, such as a half-space. Consider Poisson's equation for the Green's function $G(\mathbf{x}; \mathbf{x}^*)$ related to the domain V where

$$\nabla^2 G = -\delta(\mathbf{x} - \mathbf{x}^*) \quad ; \quad \mathbf{x} \in V \tag{9.206}$$

subject to Dirichlet boundary conditions

$$G = 0 \quad ; \quad \mathbf{x} \in S \tag{9.207}$$

The Green's function can be represented as the sum of the solution $\widetilde{G}(\mathbf{x}; \mathbf{x}^*)$ which is the Green's function for an *infinite space* and a harmonic solution $F(\mathbf{x}, \mathbf{x}^*)$ which satisfies the following

$$\nabla^2 \widetilde{G} = -\delta(\mathbf{x} - \mathbf{x}^*) \quad ; \quad \mathbf{x} \in V \tag{9.208}$$

and

$$\nabla^2 F = 0 \qquad ; \quad \mathbf{x} \in V \tag{9.209}$$

$$F = -\widetilde{G}(\mathbf{x}; \mathbf{x}^*) \quad ; \quad \mathbf{x} \in S \tag{9.210}$$

Thus, once the Green's function $\widetilde{G}(\mathbf{x}; \mathbf{x}^*)$ for the infinite space is known, the problem of determining the Green's function $G(\mathbf{x}; \mathbf{x}^*)$ for a finite domain is reduced to solving the Laplace boundary value problem defined by (9.209) and (9.210). It may be noted that the fundamental solutions satisfying (9.208) take the following forms:

In two dimensions:

$$\widetilde{G}(\mathbf{x}; \mathbf{x}^*) = \frac{1}{2\pi} \log \frac{1}{|\mathbf{x} - \mathbf{x}^*|} \tag{9.211}$$

In three dimensions:

$$\tilde{G}(\mathbf{x};\mathbf{x}^*) = \frac{1}{4\pi} \frac{1}{|\mathbf{x}-\mathbf{x}^*|} \tag{9.212}$$

The formal procedure described above can be very effectively applied to determine the Green's function applicable to finite domains. The use of *"image solutions"* feature prominently in many of the procedures.

Example 9.6

Develop the Green's function for a half-space V_0 occupying the region $x \in (-\infty, \infty)$, $y \in (-\infty, \infty)$ and $z \in (0, \infty)$, where a unit source is placed at $\mathbf{x}^*$ and the boundary $z = 0$ is maintained at zero potential.

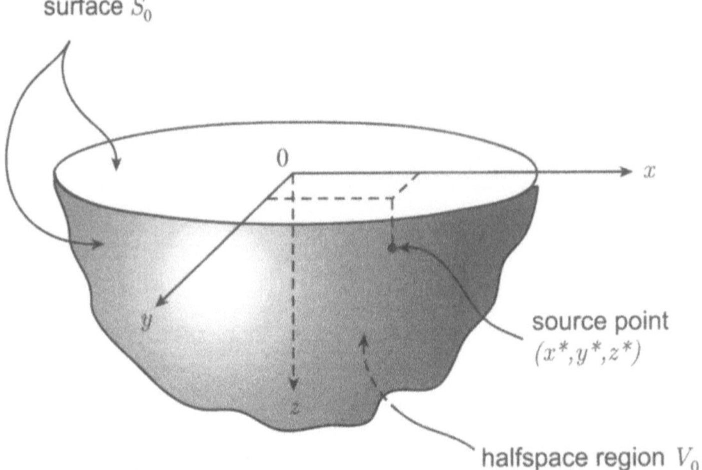

Figure 9.12: Green's function for a half-space region.

Solution

The boundary value problem to be solved for $G(\mathbf{x};\mathbf{x}^*)$ is given by

$$\nabla^2 G = -\delta(\mathbf{x}-\mathbf{x}^*) \quad ; \quad \mathbf{x} \in V_0 \tag{9.213}$$

and the boundary condition

$$G = 0 \quad ; \quad z = 0 \tag{9.214}$$

Also to ensure that the solution to the problem is unique, we specify that $G \to \infty$ as $|\mathbf{x}| \to \infty$ with $z \in (0, \infty)$. The solution to the problem can be written in the form

$$G(\mathbf{x}; \mathbf{x}^*) = \widetilde{G}(\mathbf{x}; \mathbf{x}^*) + F(\mathbf{x}; \mathbf{x}^*) \tag{9.215}$$

The surface S_0 of the region V_0 consists of the bounding plane $z = 0$ and the boundary surface at infinity. Since the Dirichlet Green's function must satisfy

$$G(\mathbf{x}; \mathbf{x}^*) \to 0 \quad \text{as } |\mathbf{x}| \to \infty \tag{9.216}$$

and

$$G(\mathbf{x}, \mathbf{x}^*) = 0 \quad \text{on } z = 0 \tag{9.217}$$

it is evident that we require an image source at a position $\mathbf{x}'$ which is the reflection of $\mathbf{x}^*$ about the plane $z = 0$ (see e.g. Figure 9.13), so that the image source is located in $z < 0$, a domain outside V_0.

Therefore the function $F(\mathbf{x}; \mathbf{x}^*)$ in (9.215) is derived from (9.212) with the source location at $\mathbf{x}' = (x^*\mathbf{i} + y^*\mathbf{j} - z^*\mathbf{k})$ and

$$G(\mathbf{x}; \mathbf{x}^*) = \frac{1}{4\pi} \left[\frac{1}{|\mathbf{x} - \mathbf{x}^*|} - \frac{1}{|\mathbf{x} - \mathbf{x}'|} \right] \tag{9.218}$$

Now consider the Dirichlet problem for the half-space, defined by

$$\nabla^2 \varphi = 0 \quad ; \quad \mathbf{x} \in V_0$$

$$\varphi = f_D(\mathbf{x}) \quad ; \quad \mathbf{x} \in S_0 \tag{9.219}$$

where $f_D(x, y)$ is an arbitrary function. Considering the general result for the Dirichlet problem for the Poisson's equation given by (9.146), we obtain, for $f(\mathbf{x}) = 0$

$$\varphi(\mathbf{x}) = - \iint_{S_0} f_D(\mathbf{x}^*) \frac{\partial G}{\partial n} dS \tag{9.220}$$

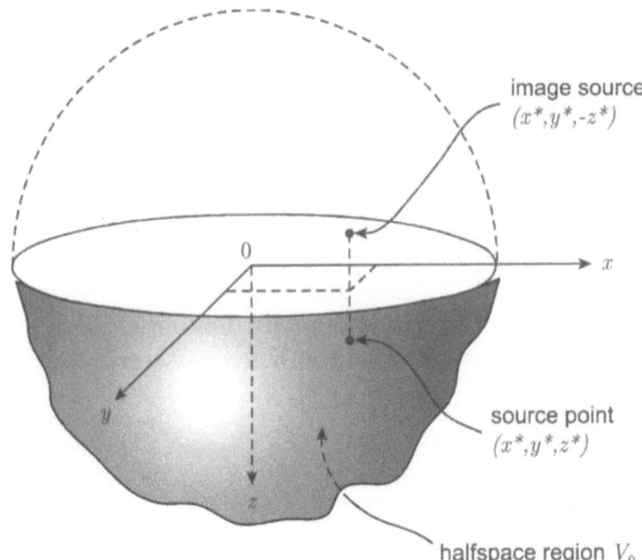

Figure 9.13: Location of image source for development of the Green's function for a half-space.

and the integrations are interpreted with respect to the variables x^* and y^*. We note that

$$\frac{\partial G}{\partial n} = \mathbf{n}.\nabla G(\mathbf{x}; \mathbf{x}^*) \tag{9.221}$$

and on the surface of the half-space $\mathbf{n} = -\mathbf{k}$, and

$$\frac{\partial G}{\partial n} = -\mathbf{k}.\nabla G(\mathbf{x}; \mathbf{x}^*) \tag{9.222}$$

Evaluating the result on $z = 0$ we have

$$\left[\frac{\partial G}{\partial n}\right]_{z=0} = -\frac{z^*}{2\pi\left[(x-x^*)^2 + (y-y^*)^2 + (z^*)^2\right]^{\frac{3}{2}}} \tag{9.223}$$

Considering the symmetry of $G(\mathbf{x}; \mathbf{x}^*)$ we can write

$$\left[\frac{\partial G}{\partial n^*}\right]_{z^*=0} = -\frac{z}{2\pi\left[(x-x^*)^2 + (y-y^*)^2 + z^2\right]^{\frac{3}{2}}} \tag{9.224}$$

The solution to the Dirichlet boundary value problem for the half-space region defined by (9.219) can be written as

$$\varphi(\mathbf{x}) = \frac{z}{2\pi} \int_{-\infty}^{\infty} \int_{-\infty}^{\infty} \frac{f_D(x^*, y^*)dx^* dy^*}{[(x - x^*)^2 + (y - y^*)^2 + z^2]^{\frac{3}{2}}} \tag{9.225}$$

Similarly, the Dirichlet boundary value problem for Poisson's equation applicable to a half-space region can be obtained by including the volume integral in (9.103) with $f(\mathbf{x}^*)$ defined through the general form of the partial differential equation (see e.g. (9.97)).

The method of images can also be used to obtain Green's function for other types of domains where superposition of *image* solutions can be used to satisfy both Dirichlet and Neumann boundary conditions applicable to quarter-space and quarter-plane regions. Also, an integration of the fundamental solution in three dimensions, (9.212), can be used to develop further solutions applicable to both two-dimensional and three-dimensional problems.

9.7 A monotonicity result for Poisson's equation

In section 9.5 we presented certain generalized results and theorems which are applicable to both superharmonic and subharmonic functions. We can utilize these results to establish a monotonicity result for Poisson's equation. Let us consider a bounded region V having a smooth boundary S with closure $\Omega = V \cup S$. Let $\varphi(\mathbf{x})$ be a continuous function in Ω such that $\nabla^2 \varphi$ is continuous in V. We consider the solution $\varphi_{\min}$ and $\varphi_{\max}$ such that

$$\varphi_{\min} = \frac{\min}{\mathbf{x} \in S}\varphi(\mathbf{x}) \ ; \ \varphi_{\max} = \frac{\max}{\mathbf{x} \in S}\varphi(\mathbf{x}) \tag{9.226}$$

Then, from the results presented in Section 9.5 we can conclude the following:

(i) $\nabla^2 \varphi \geq 0$; $\mathbf{x} \in V$ implies that $\varphi \leq \varphi_{\max}$; $\mathbf{x} \in \Omega$ $\quad$ (9.227)

(ii) $\nabla^2 \varphi = 0$; $\mathbf{x} \in V$ implies that $\varphi_{\min} \leq \varphi \leq \varphi_{\max}$
$$; \quad \mathbf{x} \in \Omega \tag{9.228}$$

(iii) $\nabla^2 \varphi \leq 0$; $\mathbf{x} \in V$ implies that $\varphi \geq \varphi_{\min}$; $\mathbf{x} \in \Omega$ $\quad$ (9.229)

Since $\varphi_{\min}$ and $\varphi_{\max}$ are specified on the boundary S, the statements (9.227) to (9.229) constitute a "*weak maximum principle*", since there are no assurances concerning the possibility of a maximum value occurring within the interior of the region.

The preceding results can be extended to constitute a "*strong maximum principle*". Consider a bounded region V with boundary S and closure $\Omega = V \cup S$, and let $\varphi(\mathbf{x})$ be continuous in V. The maximum and minimum values of $\varphi(\mathbf{x})$ are defined according to (9.226).

As established in Theorem 9.4, if $\nabla^2 \varphi \geq 0$, then $\varphi = \varphi_{\max}$ at some location within V, implies that $\varphi = \varphi_{\max}$ at all points of Ω. Similarly if $\nabla^2 \varphi \leq 0$ then $\varphi = \varphi_{\min}$ at some location within V, which implies that $\varphi = \varphi_{\min}$ at all points of Ω. These remarks add to the observations made previously in Section 9.5 that unless φ is a constant in the region of interest, the maximum and minimum values of φ occur on S. A monotonicity result for Poisson's equation be developed by making use of the preceeding results.

THEOREM 9.11

Consider a bounded region V with a smooth boundary S and $\Omega = V \cup S$ the closure. Let $\varphi(\mathbf{x})$ be a function which is continuous in V, satisfies

$$\nabla^2 \varphi = -f(\mathbf{x}) \quad ; \quad \mathbf{x} \in V \tag{9.230}$$

where $f(\mathbf{x})$ is a forcing function, and the Dirichlet boundary condition

$$\varphi = f_D(\mathbf{x}) \quad ; \quad \mathbf{x} \in S \tag{9.231}$$

Let $\varphi_1(\mathbf{x})$ and $\varphi_2(\mathbf{x})$ be solutions of (9.230) which correspond to forcing functions $f_1(\mathbf{x})$ and $f_2(\mathbf{x})$ respectively. Then

$$f_1(\mathbf{x}) \leq f_2(\mathbf{x}) \quad ; \quad \mathbf{x} \in V \tag{9.232}$$

implies that

$$\varphi_1(\mathbf{x}) \leq \varphi_2(\mathbf{x}) \quad ; \quad \mathbf{x} \in \Omega \tag{9.233}$$

PROOF

If we consider

$$\varphi^*(\mathbf{x}) = \varphi_1(\mathbf{x}) - \varphi_2(\mathbf{x}) \tag{9.234}$$

then $\varphi^*(\mathbf{x})$ satisfies

$$\nabla^2\varphi^*(\mathbf{x}) = -f_1(\mathbf{x}) + f_2(\mathbf{x}) \geq 0 \quad ; \quad \mathbf{x} \in V \tag{9.235}$$

and

$$\varphi^*(\mathbf{x}) = 0 \quad ; \quad \mathbf{x} \in S \tag{9.236}$$

From the previous discussion, we have

$$\varphi^*_{\max} = \mathop{\max}_{\mathbf{x} \in V} \varphi^* = 0 \tag{9.237}$$

which implies that $\varphi^*(\mathbf{x})$ is non-positive in Ω. Also, the strong maximum principle implies that $\varphi_1(\mathbf{x}) - \varphi_2(\mathbf{x})$ is strictly negative in V if $f_1(\mathbf{x}) < f_2(\mathbf{x})$. Alternatively, (9.233) is proved, leading to the monotonicity argument for Poisson's equation.

9.8 Viscous flow in conduits

Poisson's equation also occurs in the analysis of steady unidirectional viscous flows in conduits of arbitrary shape. We assume that the constitutive equation governing slow viscous flows can be defined in relation to an incompressible Newtonian viscous fluid. The equations of motion for steady flow, expressed in Cartesian coordinates take the forms (see e.g. (8.1714) with change in direction of $\mathbf{f}$):

$$\frac{\partial p}{\partial x} + f_x = \eta\nabla^2 v_x$$

$$\frac{\partial p}{\partial y} + f_y = \eta\nabla^2 v_y \tag{9.238}$$

$$\frac{\partial p}{\partial z} + f_z = \eta\nabla^2 v_z$$

where

$$\nabla^2 = \frac{\partial^2}{\partial x^2} + \frac{\partial^2}{\partial y^2} + \frac{\partial^2}{\partial z^2} \tag{9.239}$$

η is the dynamic shear viscosity, p is the excess pressure, $\mathbf{f} = \{f_x\mathbf{i}+f_y\mathbf{j}+f_z\mathbf{k}\}$ is the body force vector and $\mathbf{v} = \{v_x\mathbf{i} + v_y\mathbf{j} + v_z\mathbf{k}\}$ is the velocity vector.

We now restrict attention to a very special class of slow viscous flow problems where the body force vector $\mathbf{f}$ is derived from a single potential Ω such that

$$\mathbf{f} = \nabla\Omega \tag{9.240}$$

In Chapter 8 we considered flow of a viscous fluid between parallel plates in connection with the modelling of Hele-Shaw type flow problems. We refer to the flow between two parallel plates at a distance h apart due to motion of one plate relative to the other with velocity V_0 in the z-direction.

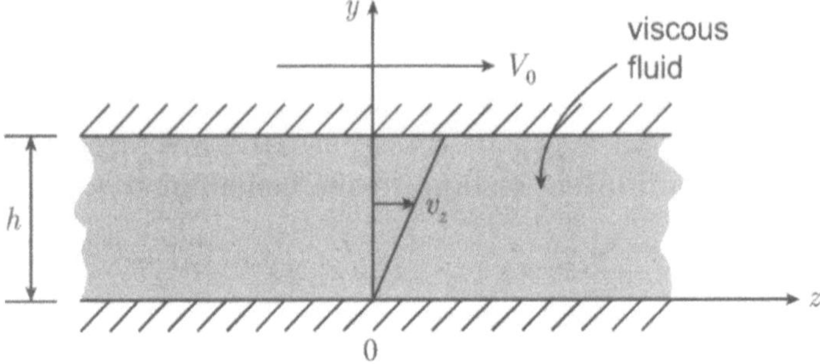

Figure 9.14: Plane Couette flow.

For plane Couette flow $v_x = v_y = 0$ and $v_z = v_z(y)$. Considering the influence of gravity effects and the choice of the direction of increasing y we have

$$\Omega = \rho g y \tag{9.241}$$

The non-trivial equations of motion give

$$\frac{\partial p}{\partial x} = 0 \quad ; \quad \frac{\partial p}{\partial y} = -\rho g \tag{9.242}$$

which implies that the pressure variation with y is hydrostatic. Differentiating the second equation of (9.242) with respect to z we have

$$\frac{\partial^2 p}{\partial y \partial z} = 0 \tag{9.243}$$

which implies that $\partial p/\partial z$ is independent of y. It is also independent of x since the flow is planar or two-dimensional. The equation of motion in the z-direction gives

$$\eta \frac{d^2 v_z}{dy^2} = \frac{\partial p}{\partial z} \tag{9.244}$$

Since from (9.244), $\partial p/\partial z$ can only be a function of y and from (9.243) it is independent of y, we conclude that $\partial p/\partial z$ can only be a constant. If there is no pressure gradient in the z-direction, the integration of (9.244) along with the no-slip boundary condition at $y = 0$ and the prescribed velocity boundary condition at $y = h$ gives

$$v_z(x, y, z) = \frac{V_0 y}{h} \tag{9.245}$$

Let us now consider the situation where *axial flow* takes place in a confined region of finite cross-section A and bounded by parallel planes which are inclined to the horizontal plane.

For steady axial flow of the viscous fluid, $v_x = v_y = 0$ and $v_z = v_z(x, y)$. The equations of motion (9.238) now give

$$\frac{\partial p}{\partial x} + f_x = 0 \quad ; \quad \frac{\partial p}{\partial y} + f_y = 0 \tag{9.246}$$

$$\eta \overset{\circ}{\nabla}{}^2 v_z = \frac{\partial p}{\partial z} + f_z \tag{9.247}$$

where

$$\overset{\circ}{\nabla}{}^2 = \frac{\partial^2}{\partial x^2} + \frac{\partial^2}{\partial y^2} \tag{9.248}$$

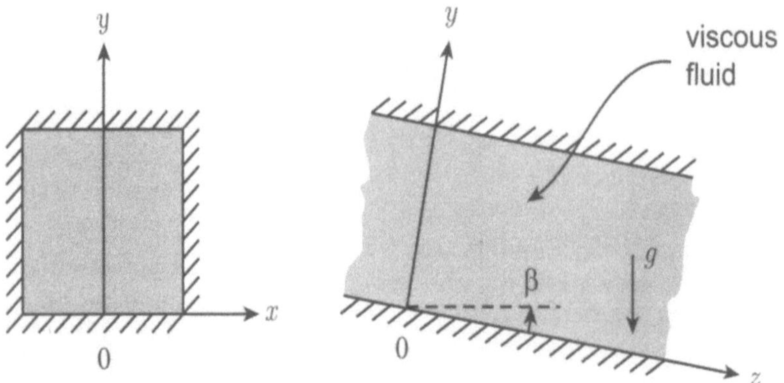

Figure 9.15: Axial flow of a viscous fluid in an inclined duct.

is Laplace's operator in two-dimensions. Considering the gravity effects in the inclined configuration of the duct and noting the choice of the direction of increasing y we obtain

$$\frac{\partial p}{\partial x} = 0 \quad ; \quad \frac{\partial p}{\partial y} = -\rho g \cos \beta \tag{9.249}$$

$$\eta \overset{\circ}{\nabla}^2 v_z = \frac{\partial p}{\partial z} - \rho g \sin \beta \tag{9.250}$$

We infer from (9.249) and (9.250) that $\partial p/\partial z$ is a constant and the Poisson-type partial differential equation governing v_z is given by

$$\eta \overset{\circ}{\nabla}^2 v_z = -C - \rho g \sin \beta \tag{9.251}$$

where

$$C = -\frac{\partial p}{\partial z} \tag{9.252}$$

is a constant.

Example 9.7

A long duct with a rectangular cross-section contains a viscous fluid of viscosity η. Axial flow is initiated in the viscous fluid under a constant pressure gradient in the z-direction (Figure 9.16). If no-slip conditions exist at the boundary of the duct determine the axial velocity distribution over the cross-section of the duct.

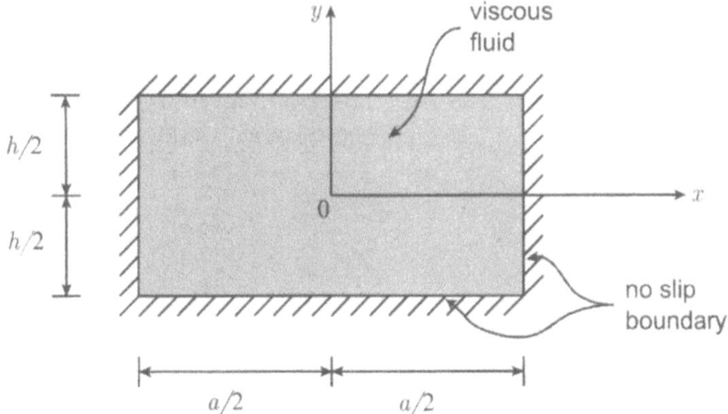

Figure 9.16: Axial flow of a viscous fluid in a duct.

Solution

The partial differential equation governing the axial velocity field $v_z(x, y)$ is given by

$$\overset{\circ}{\nabla}^2 v_z(x, y) = -\frac{C}{\eta} \tag{9.253}$$

where $C \, (= -\partial p / \partial z)$ is the constant pressure gradient. The boundary conditions governing v_z are the following

$$v_z(x, \pm h/2) = 0 \quad ; \quad v_z(\pm a/2, y) = 0 \tag{9.254}$$

The solution of (9.253) can be represented in the form

$$v_z(x, y) = v_z^H(x, y) + v_z^P(x, y) \tag{9.255}$$

where the superscripts $(\ )^H$ and $(\ )^P$ refer to the homogenous and particular solutions respectively. As can be observed, the particular solution can be expressed in a number of ways;

$$v_z^P(x,y) = -\frac{C}{2\eta}y^2 + C_1 y + C_2$$

$$(9.256)$$

$$v_z^P(x,y) = -\frac{C}{2\eta}x^2 + C_1^* x + C_2^*$$

where C_1, C_2, C_1^* and C_2^* are arbitrary constants, are possible particular solutions of (9.253), as are any linear combinations of the solutions (9.256). We shall restrict attention to the *first* solution of (9.256) and determine the constants C_1 and C_2 by considering the no-slip boundary conditions by $y = \pm\, h/2$; i.e.

$$v_z(x, \pm h/2) = 0 \qquad\qquad (9.257)$$

We have

$$v_z^P(x,y) = \frac{C}{2\eta}\left(\frac{h^2}{4} - y^2\right) \qquad\qquad (9.258)$$

The homogenous solutions are essentially solutions of Laplace's equation which can be represented in the series form

$$v_z^H(x,y) = \sum_{n=1,}^{\infty} A_n \cosh\left\{\frac{(2n-1)\pi x}{h}\right\}\cos\left\{\frac{(2n-1)\pi y}{h}\right\} \qquad (9.259)$$

where the solution is symmetric with respect to both x and y and satisfies the boundary conditions (9.254) on $y = \pm\, h/2$ for all integer values of n. The constants A_n in (9.259) are determined by satisfying the boundary conditions on $x = \pm\, a/2$. Avoiding details of calculations we have

$$A_n = \frac{2h^2 C}{\eta\lambda_n^3}\frac{\left\{\lambda_n \cos\left(\frac{\lambda_n}{2}\right) - 2\sin\left(\frac{\lambda_n}{2}\right)\right\}}{\cosh\left\{\frac{\lambda_n a}{2h}\right\}} \qquad\qquad (9.260)$$

where

$$\lambda_n = (2n - 1)\pi \tag{9.261}$$

The complete solution to the axial fluid flow problem is obtained from (9.255) where $v_z^P(x, y)$ and $v_z^H(x, y)$ are completely defined through (9.258) and (9.259).

9.9 Torsion of prismatic elastic solids

The theory of torsion of prismatic elastic solids represents an important topic in continuum solid mechanics not only for its intrinsic mathematical interest but also for the extensive range of applicability of the results to practical problems in engineering. The formal study of the torsion of elastic solids dates back to the classical work of Coulomb (1736-1806), who examined the response of prismatic bars of circular cross-section which were subjected to torsional loading. Coulomb's solution, which assumes that cross-sections of a circular prismatic elastic bar remain plane and rotate without any distortion, led to the development of an indirect solution of this classical torsion problem. The necessity for the cross-sections to remain plane without distortion or warping can be observed by considerations of symmetry of the deformation field in the circular bar. Refering to Figure 9.17a, let us assume that the cross-section undergoes an axisymmetric warping displacement in the direction towards the end A. We can now rigidly rotate the bar about the y-axis to the configuration shown in Figure 9.17b, with the warped configuration as indicated. The torsional loadings of the two configurations are, however, identical, which requires the plane $x - y$ of the circular bar to remain plane and undistorted.

The theory developed by Coulomb was applied by Navier (1785-1836) to examine the torsional behaviour of prismatic bars with non-circular cross-sections. Navier's solution erroneously predicts that for a given torque, the angle of twist of the non-circular prismatic bar is inversely proportional to the centroidal polar moment of inertia of the cross-section, and that the greatest shear stress occurs at points furthest away from the centroid of the cross-section. Furthermore, it can be shown that Navier's assumption leads to a contradiction in the boundary conditions governing the torsion problem. Consider, for example, the case of a prismatic bar with a rectangular cross-section which is subjected to torsion. The torsional loads are applied in the plane of the cross-section and the prismatic surface of the bar is free of trac-

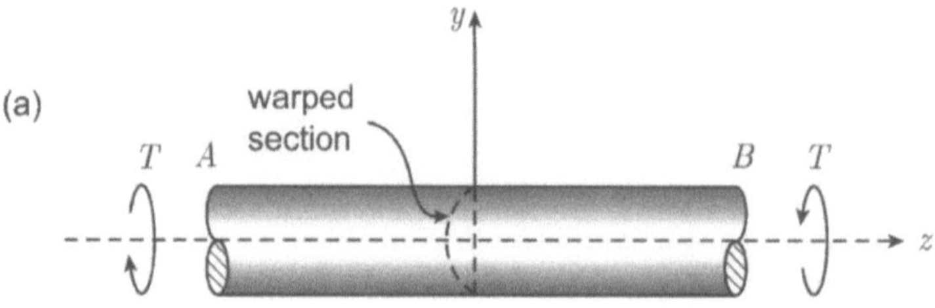

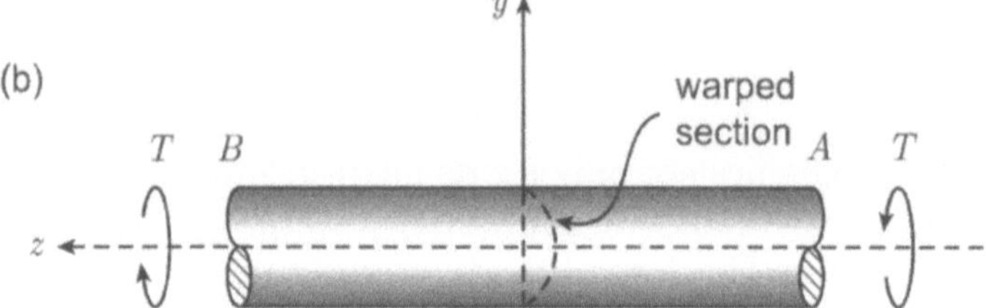

Figure 9.17: Torsion of a prismatic bar with circular bar.

tions. According to Navier's assumption, on any radius vector from the centroid, the resultant of tractions should act normal to the radius vector. (Figure 9.18). This is plausible at points within the interior of the section, but will be inadmissible at the boundary of the cross-section.

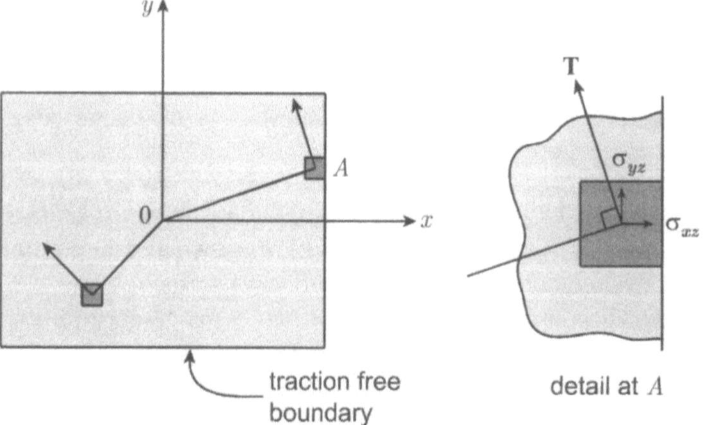

Figure 9.18: Torsional stresses in a prismatic bar with a rectangular cross-section – Navier's assumption.

Referring to the location A in Figure 9.18 it is evident that the resultant shear traction on the $x-y$ plane can be represented by components σ_{xz} and σ_{yz} as indicated. The shear stress σ_{yz} will induce, within the prismatic bar, a complementary shear stress which will not influence the traction free requirement on the boundary of the prismatic surface. The complementary shear stress $\sigma_{zx}(=\sigma_{xz})$ will, however, require a non-zero distribution of tractions on the prismatic surface of the bar, which is contrary to the requirement for a *traction free* prismatic surface.

The problem of the torsion of non-circular prismatic bars by end couples was correctly solved by Barre de Saint-Venant (1797-1886) in a celebrated memoir published in 1855. Saint-Venant employed a procedure which is now referred to as a semi-inverse method of analysis. The procedure first requires an ingenious guess as to the possible deformation of the twisted prismatic bar. Through such an assumption, Saint-Venant showed that by selecting a compatible deformation field, he could satisfy the constitutive equations of elasticity, the equations of equilibrium and boundary conditions of the torsion problem, explicitly. The correctness of the solution is established by appeal to the uniqueness theorem for classical elasticity. Saint-Venant's solution represents one of the most important contributions in theoretical solid mechanics. The solution procedure also addresses a further principle which is now attributed to Saint-Venant. This involves the manner in which the prismatic bar is subjected to torsional loadings. Saint-Venant also postulated a principle which established the fact that for prismatic solids composed of isotropic elastic materials, the effects of the mode of application of the torsional loadings (see e.g. Figure 9.19) is felt only within a region restricted to the ends of the prismatic bar (the dimensions of the region being comparable to the cross-sectional dimensions of the bar). Saint-Venant's solution becomes applicable to a greater region of the prismatic bar provided $(a/\ell) \ll 1$; $(b/\ell) \ll 1$, where ℓ is the length of the prismatic bar. The study of the decay of the perturbation in the classical Saint-Venant solution, which is defined in terms of the resultant torque T_0 acting at the ends of the prismatic bar, continues to be a topic of continuing interest to both applied mathematicians and engineers.

The purpose of this section is to present the study of the classical torsion problem for a prismatic elastic body which culminates in the development of a partial differential equation of the Poisson-type.

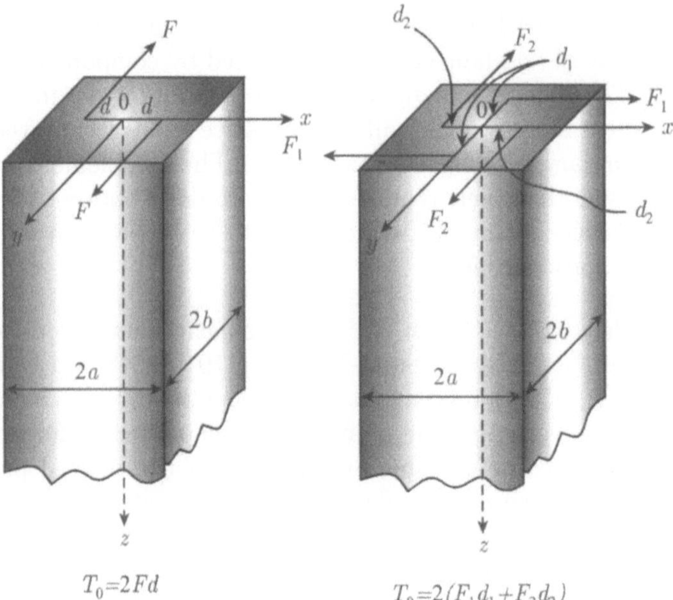

Figure 9.19: Equivalent torsional loading of prismatic bar.

9.9.1 General formulation of the torsion problem

Prismatic elastic bodies are solids of uniform cross-section, the dimensions of the cross-section being small in comparison with its length. The equations governing the torsion of prismatic elastic solids is most conveniently formulated in relation to a Cartesian coordinate system. Consider a prismatic isotropic elastic bar with an arbitrary but simply connected cross-section and a smooth convex boundary $\mathcal{C}$, which is twisted by couples applied at the plane ends.

The formulation of the torsion problem and the development of the relevant partial differential equation can be approached in a variety of ways. In this section we shall consider an engineering approach to the development of the governing equation. We select a rectangular Cartesian coordinate system such that the z-axis is parallel to the generators of the boundary surface S of the prismatic cylinder. The torque T_0 on a plane end is considered to be positive when the vector representing T_0 acts in the direction of the outward unit normal to the plane end region $\mathcal{A}$ (i.e. the right hand rule). Consider a point P with a line vector OP in the initial configuration which rotates about the z-axis by a small angle α during the twisting of the bar.

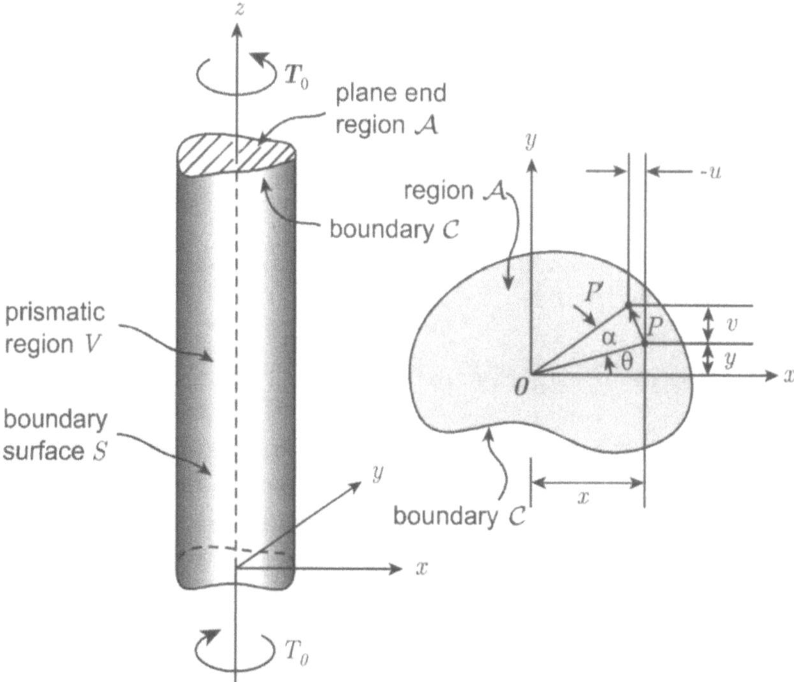

Figure 9.20: Prismatic bar twisted by end torques T_0.

The displacement vector $\mathbf{u}$ referred to the rectangular Cartesian coordinate system is given by

$$\mathbf{u} = \{u\mathbf{i} + v\mathbf{j} + w\mathbf{k}\} \tag{9.262}$$

The location of the axis of rotation, O, is referred to as the centre of twist (where $u = v = 0$) and depends on the shape of the cross-section. Since the angle of twist α is small, the arc PP' (Figure 9.20) is assumed to be the straight line normal to OP. The components of the displacement vector in the $x - y$ plane are given by

$$u = -\alpha r \sin\theta = -\alpha y$$

$$\tag{9.263}$$

$$v = \alpha r \cos\theta = \alpha x$$

In addition to these in-plane displacements, the cross-section can experience a warping displacement in the z-direction. Since warping of each cross-section is assumed to be the same, the axial displacement w is only a function of the spatial variables x and y. If the z-axis is located at the centre of twist, the angle of twist α is given by

$$\alpha = \Omega_0 z \tag{9.264}$$

where Ω_0 is the twist per unit length in the z-direction. The displacement vector can now be written as

$$\mathbf{u} = \{-\Omega_0 y z \mathbf{i} + \Omega_0 x z \mathbf{j} + w(x,y)\mathbf{k}\} \tag{9.265}$$

The assumed form of the displacement field (9.265) is only a partial solution in the *semi-inverse* procedure. The governing partial differential equation(s) need to be determined by appeal to the stress-strain relations and the equations of equilibrium. For the displacement field defined by (9.265), the strain matrix (see e.g. (8.68)) is given by

$$\epsilon = \begin{bmatrix} 0 & 0 & \frac{1}{2}\left(\dfrac{\partial w}{\partial x} - \Omega_0 y\right) \\[2ex] 0 & 0 & \frac{1}{2}\left(\dfrac{\partial w}{\partial y} + \Omega_0 x\right) \\[2ex] \frac{1}{2}\left(\dfrac{\partial w}{\partial x} - \Omega_0 y\right) & \frac{1}{2}\left(\dfrac{\partial w}{\partial y} + \Omega_0 x\right) & 0 \end{bmatrix} \tag{9.266}$$

and the non-zero components of stresses derived from (9.266) and (8.236) are

$$\sigma_{xz} = 2\mu\epsilon_{xz} = \mu\left(\frac{\partial w}{\partial x} - \Omega_0 y\right)$$

$$\tag{9.267}$$

$$\sigma_{yz} = 2\mu\epsilon_{yz} = \mu\left(\frac{\partial w}{\partial y} + \Omega_0 x\right)$$

where μ is the linear elastic shear modulus. We can now eliminate $w(x,y)$ from (9.267) by differentiating the first equation with respect to y, the second equation with respect to x and subtracting one from the other; this gives

$$\frac{\partial \sigma_{xz}}{\partial y} - \frac{\partial \sigma_{yz}}{\partial x} = -2\mu\Omega_0 \tag{9.268}$$

This result can also be obtained by integrating the appropriate compatibility equation for strains (see e.g. (8.106) and by making use of the stress-strain relations (9.267). This gives

$$\frac{\partial \epsilon_{xz}}{\partial y} - \frac{\partial \epsilon_{yz}}{\partial x} = -\Omega_0 \tag{9.269}$$

Considering the equations of equilibrium (see e.g. (8.162)), in the absence of body forces, the single equation governing static equilibrium is

$$\frac{\partial \sigma_{xz}}{\partial x} + \frac{\partial \sigma_{yz}}{\partial y} = 0 \tag{9.270}$$

We now introduce a "*warping function*" $\phi(x,y)$ such that

$$w(x,y) = \Omega_0 \phi(x,y) \tag{9.271}$$

Substituting this in the stress-strain relations (9.267) and using the results in the equation of equilibrium (9.270) we obtain

$$\overset{\circ}{\nabla}^2 \phi = \frac{\partial^2 \phi}{\partial x^2} + \frac{\partial^2 \phi}{\partial y^2} = 0 \quad ; \quad (x,y) \in \mathcal{A} \tag{9.272}$$

Hence, in the theory of torsion of prismatic elastic solids the warping function is harmonic. To determine the boundary conditions governing ϕ, we consider the boundary $\mathcal{C}$ of the region $\mathcal{A}$, and the stresses acting on an element located at the boundary .

The components of the outward unit normal to $\mathcal{C}$ are given by

$$\mathbf{n} = \{n_x\mathbf{i} + n_y\mathbf{j} + 0\mathbf{k}\} \tag{9.273}$$

For the prismatic surface S to be traction free

$$\mathbf{T} = \boldsymbol{\sigma}.\mathbf{n} = 0 \tag{9.274}$$

Two of the conditions in (9.274) are identically satisfied and the remaining boundary condition gives

$$\sigma_{xz} n_x + \sigma_{yz} n_y = 0 \quad ; \quad (x, y) \in C \tag{9.275}$$

This requirement implies that the resultant of shear traction at a point along C is directed along the tangent to the boundary. Considering the infinitesimal element shown in Figure 9.21, with s increasing with θ along C we have

$$n_x = \frac{dy}{ds} \quad ; \quad n_y = -\frac{dx}{ds} \tag{9.276}$$

The traction boundary condition (9.275) can now be written as

$$\left(\frac{\partial \phi}{\partial x} - y\right) \frac{dy}{ds} - \left(\frac{\partial \phi}{\partial y} + x\right) \frac{dx}{ds} = 0 \quad ; \quad (x, y) \in C \tag{9.277}$$

Therefore, in terms of the warping function ϕ, the torsion problem is reduced to the solution of Laplace's equation for ϕ subject to the Neumann-type boundary condition (9.277).

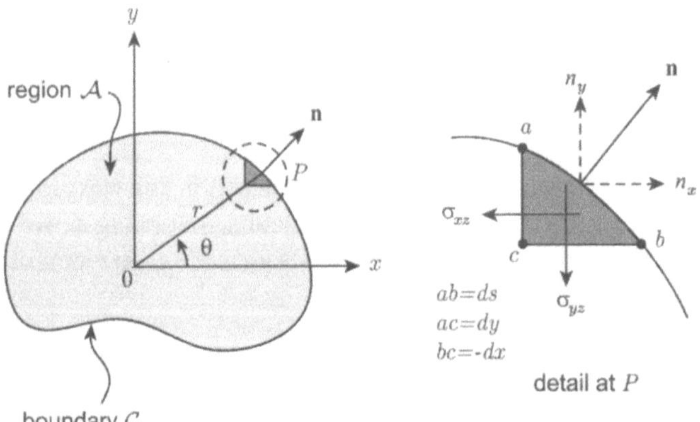

Figure 9.21: Traction boundary conditions for the torsion problem.

An alternative formulation of the torsion problem is due to Ludwig Prandtl (1875-1953), who introduced a 'stress function' $\chi(x, y)$ which considerably reduces the complicated form of the boundary condition (9.277). The stress function formulation assumes that the stress components σ_{xz} and σ_{yz} can be expressed in terms of $\chi(x, y)$ such that

$$\sigma_{xz} = \frac{\partial \chi}{\partial y} \quad ; \quad \sigma_{yz} = -\frac{\partial \chi}{\partial x} \tag{9.278}$$

and the equation of equilibrium (9.270) is satisfied for all choices of $\chi(x, y)$. Using the stress-strain relations (9.267) expressed in terms of $\phi(x, y)$ and (9.278) we have

$$\frac{\partial \chi}{\partial y} = \mu \Omega_0 \left(\frac{\partial \phi}{\partial x} - y \right)$$

$$\tag{9.279}$$

$$-\frac{\partial \chi}{\partial x} = \mu \Omega_0 \left(\frac{\partial \phi}{\partial y} + x \right)$$

Eliminating ϕ between these two equations we obtain

$$\overset{\circ}{\nabla}^2 \chi = \frac{\partial^2 \chi}{\partial x^2} + \frac{\partial^2 \chi}{\partial y^2} = -2\mu \Omega_0 \quad ; \quad (x, y) \in \mathcal{A} \tag{9.280}$$

which is a partial differential equation of the Poisson's type for the stress function $\chi(x, y)$.

The single boundary condition (9.275) for the traction free condition on S (or equivalently on $\mathcal{C}$) can now be written in terms of the stress function $\chi(x, y)$ as

$$\frac{\partial \chi}{\partial y} \frac{dy}{ds} + \left(-\frac{\partial \chi}{\partial x} \right) \left(-\frac{dx}{ds} \right) = \frac{d\chi}{ds} = 0 \quad ; \quad (x, y) \in \mathcal{C} \tag{9.281}$$

The result (9.281) indicates that χ must be constant along the boundary of the cross-section. In the case of a prismatic bar with a simply connected cross-section $\mathcal{A}$ (e.g. a solid bar) this constant (arbitrary) can be set to zero without loss of generality, since it has no effect on the stress components and the displacement field is influenced only to within a rigid displacement.

We can now evaluate the resultant of tractions that need to be applied on the plane ends of the prismatic bar in order to maintain the state of torsion defined by the displacement field (9.265). The outward unit normal to the plane end $z = 0$ has direction cosines

$$\mathbf{n} = (0\mathbf{i} + 0\mathbf{j} - 1\mathbf{k}) \tag{9.282}$$

and at a location $z = \ell$, the outward unit normal has direction cosines

$$\mathbf{n} = (0\mathbf{i} + 0\mathbf{j} + 1\mathbf{k}) \tag{9.283}$$

Consider the tractions acting on elements located within the region $\mathcal{A}$ as indicated on Figure 9.22. The resultant of tractions arising from the stress components σ_{xz} and σ_{yz} are denoted by F_x, F_y and T_0.

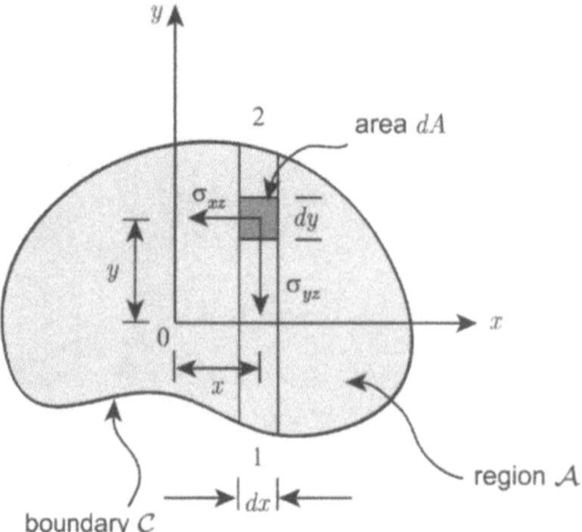

Figure 9.22: Tractions acting on elemental area located at plane $z = 0$.

The x-component of the resultant force on the plane regions $\mathcal{A}$ located at $z = 0, \ell$ are given by

$$F_x = \mp \iint_{\mathcal{A}} \left[\frac{\partial \phi}{\partial x} - y \right] dA \tag{9.284}$$

If we consider the representation for σ_{xz} in terms of the warping function given by (9.267) we have

$$\iint_{\mathcal{A}} \sigma_{xz} dA = \mu \Omega_0 \iint_{\mathcal{A}} \left[\frac{\partial \phi}{\partial x} - y \right] dA \tag{9.285}$$

The right side of (9.285) can be written as

$$\iint_{\mathcal{A}} \left[\frac{\partial \phi}{\partial x} - y \right] dA$$
$$= \iint_{\mathcal{A}} \left\{ \frac{\partial}{\partial x} \left[x \left(\frac{\partial \phi}{\partial x} - y \right) \right] + \frac{\partial}{\partial y} \left[x \left(\frac{\partial \phi}{\partial y} + x \right) \right] \right\} dA \quad (9.286)$$

From Green's theorem in two dimensions

$$\iint_{\mathcal{A}} \left\{ \frac{\partial F_1}{\partial x} + \frac{\partial F_2}{\partial y} \right\} dA = \oint_{C} [F_1 dy - F_2 dx] \quad (9.287)$$

Also, using the results (9.276), and the above results we obtain

$$\iint_{\mathcal{A}} \sigma_{xz} dA = \mu \Omega_0 \oint_{C} x \left[\frac{\partial \phi}{\partial n} - y n_x + x n_y \right] ds \quad (9.288)$$

The boundary condition (9.277) can be written as

$$\frac{\partial \phi}{\partial x} n_x + \frac{\partial \phi}{\partial y} n_y = \frac{\partial \phi}{\partial n} = y n_x - x n_y \quad (9.289)$$

Considering (9.289), we obtain from (9.288) the requirement

$$\iint_{\mathcal{A}} \sigma_{xz} dA = 0 \quad (9.290)$$

In a similar manner we can show that

$$\iint_{\mathcal{A}} \sigma_{yz} dA = 0 \quad (9.291)$$

Finally, considering the moments of tractions acting on region $\mathcal{A}$, about the axis of torsion we have

$$T_0 = \iint_{\mathcal{A}} [x \sigma_{yz} - y \sigma_{xz}] dA \quad (9.292)$$

Substituting the expressions for σ_{xz} and σ_{yz} in terms of the warping function we obtain

$$T_0 = \mu\Omega_0 \iint_A \left[x\frac{\partial\phi}{\partial y} - y\frac{\partial\phi}{\partial x} + x^2 + y^2 \right] dA \tag{9.293}$$

The equation (9.293) introduces the *Torsional Rigidity* of the prismatic solid J, defined by

$$J = \iint_A \left[x\frac{\partial\phi}{\partial y} - y\frac{\partial\phi}{\partial x} + x^2 + y^2 \right] dA \tag{9.294}$$

The results (9.290), (9.291) and (9.293) can also be derived by adopting the Prandtl stress function formulation discussed previously. The resultant of tractions in the x-direction is given by

$$F_x = \iint_A \sigma_{xz} dA = \iint_A \frac{\partial\chi}{\partial y} dA \tag{9.295}$$

We note that

$$\iint_A \frac{\partial\chi}{\partial y} dA = \iint_A \nabla\cdot[\mathbf{j}\chi] \, dA \tag{9.296}$$

From Green's theorem in two dimensions

$$\iint_A \nabla\cdot[\mathbf{j}\chi] \, dA = \oint_C \mathbf{j}\cdot\mathbf{n}\chi ds = \oint_C \chi n_y ds \tag{9.297}$$

Since χ is specified to be either a constant or zero on the boundary, the integral over the boundary is zero and we obtain the result (9.290). Using a similar procedure we can also establish the result (9.291).

Considering the moments of tractions acting of the region $\mathcal{A}$ about the axis of torsion we have

$$T_0 = \iint_A [x\sigma_{yz} - y\sigma_{xz}] \, dA = -\iint_A \left[x\frac{\partial\chi}{\partial x} + y\frac{\partial\chi}{\partial y} \right] dA \tag{9.298}$$

The second integral in (9.298) can be written in the form

$$-\iint_{\mathcal{A}} \left[x\frac{\partial \chi}{\partial x} + y\frac{\partial \chi}{\partial y} \right] dA$$

$$= 2\iint_{\mathcal{A}} \chi dA - \iint_{\mathcal{A}} \left[\frac{\partial}{\partial x}(x\chi) + \frac{\partial}{\partial y}(y\chi) \right] dA \tag{9.299}$$

Using Green's theorem, we have

$$\iint_{\mathcal{A}} \left[\frac{\partial}{\partial x}(x\chi) + \frac{\partial}{\partial y}(y\chi) \right] dA = \oint_{C} \{x\chi n_x + y\chi n_y\}\, ds \tag{9.300}$$

and if χ vanishes on the boundary C of the simply connected region $\mathcal{A}$, (9.298) gives

$$T_0 = 2\iint_{\mathcal{A}} \chi dA \tag{9.301}$$

The salient results for the torsion of a prismatic bar can thus be expressed in terms of either the warping function $\phi(x,y)$ or the Prandtl stress function $\chi(x,y)$. The formulation in terms of the warping function gives rise to Laplace's equation for $\phi(x,y)$, with Neumann type boundary conditions, which can be solved by a variety of ways as indicated in Chapter 5 of Volume I. The formulation in terms of the Prandtl stress function gives rise to Poisson's equation for $\chi(x,y)$ with much simpler Dirichlet boundary conditions. This latter approach represents a relatively convenient formulation of the torsion problem. Many of the methods of analyses and general theorems have been successfully employed to develop specific and general results for the torsion of prismatic solids.

9.9.2 Torsion of prismatic elements with multiply connected cross-sections

In the previous section we have shown that the torsional rigidity of a prismatic element is dependent on the polar moment of inertia of the cross-section about the centre of twist. To obtain the maximum value of the torsional rigidity for a given cross-sectional area $\mathcal{A}$, it becomes necessary to re-distribute the solid material, by introducing prismatic forms of voids within the prismatic member. This invariably leads to cross-sections which are multiply connected.

Consider the multiply connected cross-section of a prismatic bar shown in Figure 9.23 where $\mathcal{A}$ is the total area of the region bounded by $\mathcal{C}$. The material region is denoted by $\mathcal{A}^*$ and $\mathcal{A}_i(i = 1, 2....)$ correspond to the void regions within $\mathcal{A}$, i.e.

$$\mathcal{A} = \mathcal{A}^* + \sum_{i=1,2...} \mathcal{A}_i \tag{9.302}$$

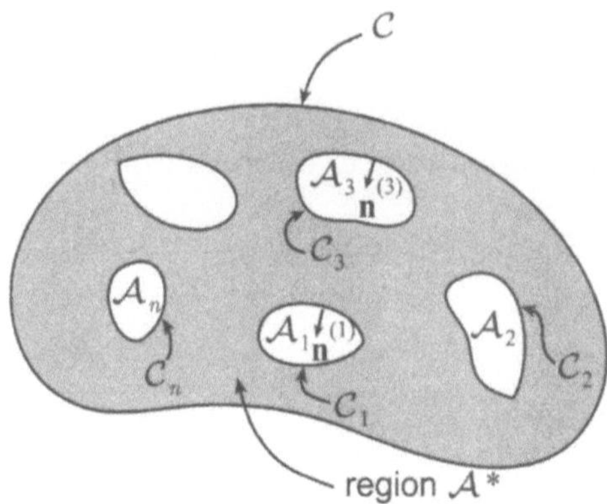

Figure 9.23: Multiply connected cross-section.

Let the boundaries of the void domains $\mathcal{A}_i$ be denoted by $\mathcal{C}_i$. The boundary value problem governing torsion of a prismatic bar can be formulated either in terms of the warping function $\phi(x, y)$ or the Prandtl stress function $\chi(x, y)$. For a multiply-connected cross-section, the boundary value problem in terms of $\phi(x, y)$ requires the solution of

$$\overset{\circ}{\nabla}^2\phi = 0 \quad ; \quad (x, y) \in \mathcal{A}^* \tag{9.303}$$

subject to the Neumann boundary conditions

$$\mathbf{n}^{(i)}.\nabla\phi = yn_x^{(i)} - xn_y^{(i)} \quad ; \quad (x, y) \in \mathcal{C}_i \tag{9.304}$$

where $\mathbf{n}^{(i)}(= n_x^{(i)}\mathbf{i} + n_y^{(i)}\mathbf{j})$ are the outward unit normals to $\mathcal{C}_i$.

Similarly, the torsion boundary value problem formulated in terms of the stress function $\chi(x, y)$ applicable to a multiply connected region $\mathcal{A}$ requires the solution of the Poisson equation

$$\overset{\circ}{\nabla}^2 \chi = -2\mu\Omega_0 \quad ; \quad (x,y) \in \mathcal{A}^* \tag{9.305}$$

subject to boundary conditions

$$\chi = \text{const} \quad ; \quad (x,y) \in \mathcal{C}, \mathcal{C}_i \tag{9.306}$$

Let us now consider the zero lateral force requirement over the material region $\mathcal{A}^*$ of the multiply connected cross-section of the prismatic bar. For example, the total lateral force in the positive y-direction (see Figure 9.22) is given by

$$F_y = \iint_{\mathcal{A}^*} \frac{\partial \chi}{\partial x} dA = \iint_{\mathcal{A}} \frac{\partial \chi}{\partial x} dA - \sum_{i=1,2,} \iint_{\mathcal{A}_i} \frac{\partial \chi}{\partial x} dA \tag{9.307}$$

Using the Green's theorem, the area integrals can be reduced to contour integrals as follows, i.e.

$$F_y = \oint_{\mathcal{C}} \chi n_x ds - \sum_{i=1,2,} \oint_{\mathcal{C}_i} \chi n_x^{(i)} ds \tag{9.308}$$

Since $\chi(x,y)$ is either constant (or zero) on any boundary $\mathcal{C}$ and $\mathcal{C}_i$

$$\oint_{\mathcal{C}} n_x ds = \oint_{\mathcal{C}_i} n_x^{(i)} ds \tag{9.309}$$

and the value of $F_y = 0$. Similarly we can show that $F_x = 0$ in the material region $\mathcal{A}^*$.

Next, consider the expression for the torque T_0 required to maintain the rotation Ω_0. Considering the results (9.298) and (9.300) we have

$$T_0 = -\oint_{\mathcal{C}} \left\{ x\chi n_x + y\chi n_y \right\} ds$$
$$+ \sum_{i=1,2,} \oint_{\mathcal{C}_i} \left\{ x\chi n_x^{(i)} + y\chi n_y^{(i)} \right\} ds + 2\iint_{\mathcal{A}^*} \chi dA \tag{9.310}$$

We note that

$$\oint [x n_y + y n_x]\, ds = \oint [x\, dy - y\, dx] \tag{9.311}$$

From Green's theorem

$$\oint_C [x\, dy - y\, dx] = \iint_{A^*} \left[\frac{\partial x}{\partial x} + \frac{\partial y}{\partial y} \right] dx\, dy = 2 A^* \tag{9.312}$$

The result (9.310) can now be written as

$$T_0 = \iint_{A^*} 2\chi\, dA + 2 \sum_{i=1,2,} K_i A_i - 2 K^* A^* \tag{9.313}$$

where K^* and K_i are arbitrary constants. We can, without loss of generality, set $K^* = 0$, and (9.313) reduces to

$$T_0 = \iint_{A^*} 2\chi\, dA + 2 \sum_{i=1,2,} K_i A_i \tag{9.314}$$

The need for maintaining K_i unspecified can be better understood by imposing the requirement that $u_z(x, y)$ be single-valued in $(x, y) \in A^*$. When the warping function $\phi(x, y)$ is single-valued, the displacement and stress components in A^* will also be single-valued; thereby preserving the continuum character of the prismatic bar before and after application of the torque. For a multiply-connected region, we need to examine $\phi(x, y)$, by considering a circuit around an interior contour, to ensure that the warping function is single-valued, i.e.

$$\oint_{C_i} d\phi = \oint_{C_i} \left[\frac{\partial \phi}{\partial x} dx + \frac{\partial \phi}{\partial y} dy \right] = 0 \tag{9.315}$$

Considering the relationship between the warping function $\phi(x, y)$ and the stress function $\chi(x, y)$ obtained through the relationships (9.279) we can rewrite (9.315) as

$$\oint_{C_i} \left[\frac{\partial \chi}{\partial y} dx - \frac{\partial \chi}{\partial x} dy - \mu \Omega_0 (x\, dy - y\, dx) \right] = 0 \tag{9.316}$$

Considering Green's theorem we have

$$\oint_{C_i} (x\,dy - y\,dx) = 2\mathcal{A}_i \tag{9.317}$$

where $\mathcal{A}_i$ is the *area inside* the contour C_i. Since

$$dy = n_x ds \quad ; \quad dx = -n_y ds \tag{9.318}$$

we can now rewrite (9.316) as

$$\oint_{C_i} \left[\frac{\partial \chi}{\partial x} n_x + \frac{\partial \chi}{\partial y} n_y \right] ds = -2\mu\Omega_0\mathcal{A}_i \tag{9.319}$$

Alternatively,

$$\oint_{C_i} \frac{\partial \chi}{\partial n} ds = -2\mu\Omega_0\mathcal{A}_i \tag{9.320}$$

Therefore, the stress function should satisfy this additional constraint (9.320) to render the displacements single-valued in $\mathcal{A}^*$. The arbitrary constants K_i are therefore uniquely determined through this additional set of constraints.

9.9.3 Solutions for torsion problems derived from equations of boundaries

In the previous section we have shown that the boundary value problem governing torsion of prismatic elastic solids can be formulated either in terms of the warping function $\phi(x, y)$ which satisfies Laplace's equation with associated Neumann boundary conditions of the inhomogeneous type or in terms of the Prandtl stress function $\chi(x, y)$ which satisfies Poisson's equation with Dirichlet boundary conditions where $\chi(x, y)$ is prescribed to be zero or a constant value. Although the resulting boundary value problems are well-posed, the solution of specific problems related to torsion of prismatic solids with cross-sections of an arbitrary shape cannot be easily achieved in view of the difficulties in satisfying the boundary conditions explicitly. Solutions to torsion problems involving cross-sections with certain types of geometries can be obtained by considering the two approaches described previously.

Prior to examining certain specific cases, we shall briefly outline the relationship between the two separate approaches to solving torsion problems by appeal to the theory of complex variables. It can be shown that if $z_c = (x+iy)$ (we shall use the lower case z_c to denote a complex number, in order to avoid confusion with z which is now used to denote the axial coordinate) then any function $f(z_c)$ satisfies Laplace's equation. We can select a polynomial in z_c of degree n to generate polynomials in x and y, which could be used to solve the harmonic boundary value problem. Not all polynomial solutions have applications to problems of practical interest. We note that solutions of the type, $xy, (x^2 - y^2), (x^3 - 3xy^2), (3x^2y - y^3)$, etc. are all possible solutions of Laplace's equation.

Consider now the warping function $\phi(x, y)$ which satisfies

$$\overset{\circ}{\nabla}{}^2\phi = 0 \quad ; \quad (x, y) \in \mathcal{A} \tag{9.321}$$

where $\mathcal{A}$ is the cross-section region of the prismatic elastic bar with smooth, continuous boundary $\mathcal{C}$. Since ϕ is harmonic, there exists a conjugate harmonic function $\psi(x, y)$ such that ϕ and ψ satisfy the Cauchy-Riemann conditions

$$\frac{\partial \phi}{\partial x} = \frac{\partial \psi}{\partial y} \quad ; \quad \frac{\partial \phi}{\partial y} = -\frac{\partial \psi}{\partial x} \tag{9.322}$$

Using (9.322) and the representation for stresses in terms of the Prandtl stress function (9.278) we can rewrite the stress-strain relations (9.267) in the form

$$\sigma_{xz} = \frac{\partial \chi}{\partial y} = \Omega_0\mu\left(\frac{\partial \psi}{\partial y} - y\right)$$

$$\sigma_{yz} = -\frac{\partial \chi}{\partial x} = -\Omega_0\mu\left(\frac{\partial \psi}{\partial x} - x\right)$$

$$\tag{9.323}$$

Integrating (9.323) we have

$$\chi(x,y) = \mu\Omega_0 \left\{ \psi - \frac{1}{2}(x^2 + y^2) + C \right\} \tag{9.324}$$

where C is constant of integration which can be chosen arbitrarily. We note that, for the case of a simply connected domain $\mathcal{A}$, the torsion problem can be formulated as follows in terms of the conjugate harmonic function $\psi(x,y)$; i.e.

$$\overset{\circ}{\nabla}^2 \psi = 0 \quad ; \quad (x,y) \in \mathcal{A} \tag{9.325}$$

subject to the boundary conditions

$$\psi = \tfrac{1}{2}(x^2 + y^2) + \widehat{c} \quad ; \quad (x,y) \in \mathcal{C} \tag{9.326}$$

and when there is a single boundary, we usually consider a solution for which the constant $\widehat{c}$ is zero.

Example 9.8

Develop a solution for the torsion of a cylindrical elastic bar with cross-sectional radius 'a'.

Solution

Considering the form of the equation for the circular cross-section, we note that a Prandtl stress function of the form

$$\chi(x,y) = C_0 \left\{ a^2 - x^2 - y^2 \right\} \tag{9.327}$$

will satisfy the boundary condition

$$\chi(x,y) = 0 \ \text{ on } \ (x,y) \in \mathcal{C} \tag{9.328}$$

where on $\mathcal{C}$, $r = a$. The value of the constant C_0 can be obtained by considering the partial differential equation governing χ; i.e.

$$\overset{\circ}{\nabla}^2 \chi = -4C_0 = -2\Omega_0\mu \tag{9.329}$$

which gives

$$C_0 = \frac{\Omega_0 \mu}{2} \tag{9.330}$$

The stress components are given by

$$\sigma_{xz} = \frac{\partial \chi}{\partial y} = -\Omega_0 \mu y \quad ; \quad \sigma_{yz} = -\frac{\partial \chi}{\partial x} = \Omega_0 \mu x \tag{9.331}$$

From (9.301), the torque T_0 that needs to be applied to maintain the rotation Ω_0 is given by

$$\begin{aligned}
T_0 &= 2 \iint_A \frac{\Omega_0 \mu}{2} \left[a^2 - (x^2 + y^2) \right] dA \\
&= 2\pi \Omega_0 \mu \int_0^a (a^2 - r^2) r\, dr
\end{aligned} \tag{9.332}$$

Evaluating (9.332) we obtain

$$T_0 = \frac{\pi \Omega_0 \mu a^4}{2} \tag{9.333}$$

which is the result that can be obtained by appeal to a derivation based on elementary solid mechanics calculations, which assumes a state of pure shear in the cylindrical bar. Also, we note that there is no warping of the cross-section since $\phi(x, y) \equiv 0$.

Example 9.9

Show that the conjugate harmonic function

$$\psi(x, y) = C_1(x^2 - y^2) + C_2 \tag{9.334}$$

where C_1 and C_2 are arbitrary functions can be used to examine the torsional response of a prismatic elastic bar with a cross-section in the form of an ellipse.

Solution

From the given form of the conjugate harmonic function and (9.324), we obtain the Prandtl stress function $\chi(x, y)$ as

$$\chi(x, y) = \mu\Omega_0\left[\left(C_1 - \frac{1}{2}\right)x^2 - \left(C_1 + \frac{1}{2}\right)y^2 + C_2\right] \tag{9.335}$$

The boundary of the elliptical cross-section is

$$\frac{x^2}{a^2} + \frac{y^2}{b^2} = 1 \tag{9.336}$$

Therefore $\chi(x, y)$ can be made to vanish along the boundary of the cross-section if we set

$$C_1 = \frac{a^2 - b^2}{2(a^2 + b^2)} \quad ; \quad C_2 = \frac{a^2 b^2}{(a^2 + b^2)} \tag{9.337}$$

Hence we can rewrite (9.335) as

$$\chi(x, y) = \frac{\mu\Omega_0 a^2 b^2}{(a^2 + b^2)}\left[1 - \frac{x^2}{a^2} - \frac{y^2}{b^2}\right] \tag{9.338}$$

which satisfies Poisson's equation (9.280) within the elliptical region and the boundary condition $\chi(x, y) = 0$ on the boundary of the region. The result (9.338) therefore represents the complete closed form solution for the Prandtl stress function governing torsion of a prismatic bar with an elliptical cross-section.

The torque T_0 that should be applied to maintain the rotation Ω_0 is given by

$$T_0 = \frac{2a^2 b^2 \mu\Omega_0}{(a^2 + b^2)}\left[\iint dx\,dy - \frac{1}{a^2}\iint x^2\,dx\,dy - \frac{1}{b^2}\iint y^2\,dx\,dy\right] \tag{9.339}$$

we note that

$$\iint x^2\, dx\, dy = I_{yy} = \frac{\pi b a^3}{4}$$
$$\iint y^2\, dx\, dy = I_{xx} = \frac{\pi a b^3}{4} \tag{9.340}$$
$$\iint dx\, dy = A_0 = \pi a b$$

where I_{xx} and I_{yy} are the moments of inertia of the elliptical cross-section about the x and y axes respectively and A_0 in the area of the section. Substituting (9.340) in (9.339) we have

$$T_0 = \frac{\pi a^3 b^3 \mu \Omega_0}{(a^2 + b^2)} \tag{9.341}$$

From (9.278) and (9.338), the stresses in the elliptical bar are given by

$$\sigma_{xz} = -\frac{2a^2 \mu \Omega_0 y}{(a^2 + b^2)} \quad ; \quad \sigma_{yz} = \frac{2b^2 \mu \Omega_0 x}{(a^2 + b^2)} \tag{9.342}$$

since the stress components act on planes $z =$ const., the shear stress vector τ at any point is the vector sum of these stress components; i.e.

$$\tau^2 = \sigma_{xz}^2 + \sigma_{yz}^2 \tag{9.343}$$

Considering the representation in terms of Prandtl's stress function

$$\tau^2 = \left(\frac{\partial \chi}{\partial x}\right)^2 + \left(\frac{\partial \chi}{\partial y}\right)^2 \tag{9.344}$$

Therefore

$$\overset{\circ}{\nabla}{}^2(\tau^2) = 2\frac{\partial \chi}{\partial x}\frac{\partial}{\partial x}\left(\overset{\circ}{\nabla}{}^2\chi\right) + 2\frac{\partial \chi}{\partial y}\frac{\partial}{\partial y}\left(\overset{\circ}{\nabla}{}^2\chi\right)$$
$$+ 2\left(\frac{\partial^2 \chi}{\partial x^2}\right)^2 + 2\left(\frac{\partial^2 \chi}{\partial y^2}\right)^2 + 4\left(\frac{\partial^2 \chi}{\partial x \partial y}\right)^2 \tag{9.345}$$

Since

$$\overset{\circ}{\nabla}^2 \chi = -2\mu\Omega_0 = \text{const.} \tag{9.346}$$

The expression (9.345) reduces to

$$\overset{\circ}{\nabla}^2(\tau^2) = 2\left(\frac{\partial^2\chi}{\partial x^2}\right)^2 + 2\left(\frac{\partial^2\chi}{\partial y^2}\right)^2 + 4\left(\frac{\partial^2\chi}{\partial x\partial y}\right)^2 \geq 0 \tag{9.347}$$

Hence τ^2 is subharmonic, and the maximum value of τ occurs on the boundary of the region (see e.g. Section 9.5). For the elliptical cross-section, these are the two points along the minor axis which are located closest to the centre of the twist. (i.e. $y = \pm b$, if $a > b$). The maximum shear stress is given by

$$|\tau_{\text{max}}| = \frac{2T_0}{\pi a b^2} = \frac{2a^2 b\mu\Omega_0}{(a^2 + b^2)} \tag{9.348}$$

9.9.4   Variables separable solutions for the torsion problem

The method of separation of variables offers a further procedure which can be used to develop solutions to problems involving torsion of prismatic elastic solids. The variable separation procedure can be formulated either in relation to rectangular Cartesian coordinates or plane polar coordinates depending upon the specific form of the cross-section of the prismatic bar. In view of the relatively simple form of homogeneous Dirichlet boundary conditions which result from the Prandtl stress function formulation, it is convenient to adopt the stress function formulation as opposed to the formulation in terms of the warping function. The basic features of the separation of variables procedure can be demonstrated by appeal to certain specific examples.

Example 9.10

The cross-section of a prismatic elastic bar has a rectangular shape as shown in Figure 9.24. The partial differential equation governing the torsional response of the prismatic bar is defined in relation to the Prandtl stress function $\chi(x,y)$ which satisfies Poisson's equation. Formulate the boundary value problem and obtain an expression for the resultant torque T_0 which needs to be applied to the prismatic bar to maintain the rotation Ω_0.

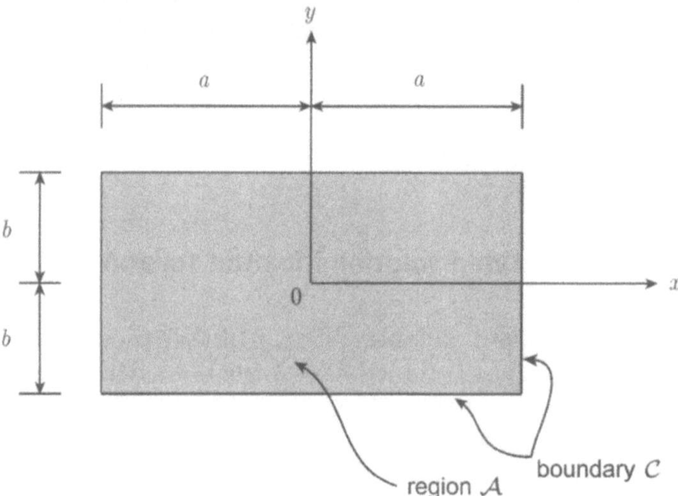

Figure 9.24: Cross-section of prismatic elastic bar.

Solution

The boundary value problem governing the Prandtl stress function $\chi(x,y)$ requires the solution of Poisson's equation

$$\overset{\circ}{\nabla}^2 \chi(x,y) = -2\Omega_0\mu \quad ; \quad (x,y) \in \mathcal{A} \tag{9.349}$$

subject to the boundary conditions

$$\chi(\pm a, y) = 0 \quad ; \quad \chi(x, \pm b) = 0 \tag{9.350}$$

As discussed in connection with the previous exposition of the application of the variables separation technique to Poisson's equation, we assume that the solution of (9.349) can be represented in the form

$$\chi(x,y) = \chi^P(x,y) + \chi^H(x,y) \tag{9.351}$$

where the superscripts $(\ )^P$ and $(\ )^H$ refer, respectively, to the particular solution of (9.349) and the general solution of the homogeneous partial differential equation, i.e. Laplace's equation

$$\overset{\circ}{\nabla}^2 \chi^H(x,y) = 0 \tag{9.352}$$

This general solution can be obtained by the method of separation of variables discussed in Chapter 5. Avoiding details it can be shown that

$$\chi^H(x,y) = \sum_{n=1,2,} \{A_n^* \sinh(\lambda_n y) + B_n^* \cosh(\lambda_n y)\}$$
$$\cdot \{C_n \sin(\lambda_n x) + D_n \cos(\lambda_n x)\} \tag{9.353}$$

where $\lambda_n (> 0)$ are arbitrary parameters and A_n^*, B_n^*, C_n and D_n are arbitrary constants. Alternatively, we can obtain the solution in the form where x and y in (9.353) are interchanged. For the present discussion, we restrict attention to the form of $\chi^H(x,y)$ defined by (9.353). Considering the symmetry of $\chi^H(x,y)$ with respect to the y-axis, the homogeneous Dirichlet boundary conditions along the sides $x = \pm a$ can be satisfied by selecting $\lambda_n = n\pi/2a$, and setting $C_n = 0$. The resulting homogeneous solution has the form

$$\chi^H(x,y) = \sum_{n=1,3,} \left\{ A_n \sinh\left(\frac{n\pi y}{2a}\right) + B_n \cosh\left(\frac{n\pi y}{2a}\right) \right\} \cos\left(\frac{n\pi x}{2a}\right) \tag{9.354}$$

A variety of particular solutions can be constructed to satisfy Poisson's equation. In view of the fact that the boundary conditions on $x = \pm a$ can be satisfied with dependency in $\cos(n\pi x/2a)$ we can expand the right side of (9.349) as a Fourier cosine series, i.e.,

$$\Gamma(x) = \frac{L_0}{2} + \sum_{n=1,} L_n \cos\left(\frac{n\pi x}{2a}\right) = -2\mu\Omega_0 \quad ; \quad -a < x < a \tag{9.355}$$

Evaluating the Fourier coefficients we obtain

$$L_0 = 0 \quad ; \quad L_n = -\frac{8\mu\Omega_0(-1)^{(n-1)/2}}{n\pi} \tag{9.356}$$

The particular solution can be represented in the form

$$\chi^P(x,y) = \sum_{n=1,3,} \frac{32\mu\Omega_0 a^2}{n^3\pi^3}(-1)^{(n-1)/2}\cos\left(\frac{n\pi x}{2a}\right) \tag{9.357}$$

The complete solution can now be written as

$$\chi(x,y) = \sum_{n=1,3,} \left[A_n \sinh\left(\frac{n\pi y}{2a}\right) + B_n \cosh\left(\frac{n\pi y}{2a}\right) \right.$$
$$\left. + \frac{32\mu\Omega_0 a^2}{n^3\pi^3}(-1)^{(n-1)/2} \right] \cos\left(\frac{n\pi x}{2a}\right) \tag{9.358}$$

The arbitrary sets of constants A_n and B_n can be determined by satisfying the boundary conditions on $y = \pm b$. This gives

$$A_n = 0 \quad ; \quad B_n = \frac{-32\mu\Omega_0 a^2(-1)^{(n-1)/2}}{n^3\pi^3 \cosh(n\pi b/2a)} \tag{9.359}$$

The final result for the Prandtl stress function $\chi(x,y)$ which satisfies Poisson's equation (9.349) and the boundary conditions (9.350) takes the form

$$\chi(x,y) = \frac{32\mu\Omega_0 a^2}{\pi^3} \sum_{n=1,3,} \frac{(-1)^{(n-1)/2}}{n^3}$$
$$\cdot \left[1 - \frac{\cosh(n\pi y/2a)}{\cosh(n\pi b/2a)} \right] \cos\left(\frac{n\pi x}{2a}\right) \tag{9.360}$$

We note that the alternative approach, which selects the particular solution in terms of the independent variable y, can be deduced by observing the similarity between the torsion problem and the axial viscous flow problem for a rectangular duct discussed in Example 9.7. The stress components are given by

$$\sigma_{xz} = \frac{\partial \chi}{\partial y} \tag{9.361}$$

$$= -\frac{16\mu\Omega_0 a}{\pi^2} \sum_{n=1,3,} \frac{(-1)^{(n-1)/2}}{n^2} \frac{\sinh(n\pi y/2a)}{\cosh(n\pi b/2a)} \cos\left(\frac{n\pi x}{2a}\right)$$

$$\sigma_{yz} = -\frac{\partial \chi}{\partial x} \tag{9.362}$$

$$= \frac{16\mu\Omega_0 a}{\pi^2} \sum_{n=1,3,} \frac{(-1)^{(n-1)/2}}{n^2} \left[1 - \frac{\cosh(n\pi y/2a)}{\cosh(n\pi b/2a)}\right] \sin\left(\frac{n\pi x}{2a}\right)$$

The torque T_0 required to maintain the rotation Ω_0 is given by (9.301). Substituting (9.360) in (9.301) we obtain

$$T_0 = \frac{64\mu\Omega_0 a^2}{\pi^3} \int_{-a}^{a} \int_{-b}^{b} \sum_{n=1,3,} \frac{(-1)^{(n-1)/2}}{n^3} \left[1 - \frac{\cosh(n\pi y/2a)}{\cosh(n\pi b/2a)}\right]$$
$$\cdot \cos\left(\frac{n\pi x}{2a}\right) dx dy \tag{9.363}$$

Carrying out the integrations and noting that

$$\sum_{n=1,3,} \frac{1}{n^4} = \frac{\pi^4}{96} \tag{9.364}$$

we obtain

$$T_0 = \frac{16\mu\Omega_0 a^3 b}{3} - \frac{1024\mu\Omega_0 a^4}{\pi^5} \sum_{n=1,3,} \frac{1}{n^5} \tanh\left(\frac{n\pi b}{2a}\right) \tag{9.365}$$

The torque (T_0) vs. twist (Ω_0) relationship can be obtained by summation of the series term in (9.365) for specific choices of the aspect ratio (a/b). In the specific case when $a = b$, (i.e. a square cross-section) the result (9.365) reduces to

$$T_0 \simeq \frac{9}{4}\mu\Omega_0 a^4 \tag{9.366}$$

In comparison, for the cylindrical bar of radius a (see e.g. (9.333))

$$T_0 = \frac{\pi}{2}\mu\Omega_0 a^4 \tag{9.367}$$

Also if we consider a circular bar of radius c such that the area of the circular bar is equal to the area of the section, $c = (2/\sqrt{\pi})a$ and the torsional stiffness of the prismatic bar with radius c is given by

$$T_0 = \frac{8}{\pi}\mu\Omega_0 a^4 \tag{9.368}$$

These results provide the inequalities

$$\frac{3}{2\pi} < \frac{9\mu\Omega_0 a^4}{4T_0} < \frac{9\pi}{32} \tag{9.369}$$

and suggests the conclusion that the torsional stiffness of prismatic elastic bodies can be bounded by considering equivalent results for cross-sectional regions with continuous shapes the dimensions of which are ordered through consideration of equivalence in area. This topic was extensively studied by mathematician G. Pólya (1887-1985) and co-workers.

Example 9.11

Develop an approximate result for the torque (T_0) vs twist (Ω_0) relationship for a *plate* with a rectangular cross-section where $(a/b) \ll 1$.

Solution

In this case an approximate result for the torque-twist relationship can be derived by assuming a Prandtl stress function of the form

$$\chi(x,y) = A\left[1 - \left(\frac{x}{a}\right)^2\right] \tag{9.370}$$

where A is an arbitrary constant. The solution (9.370) satisfies the boundary condition

$$\chi(\pm a, y) = 0 \tag{9.371}$$

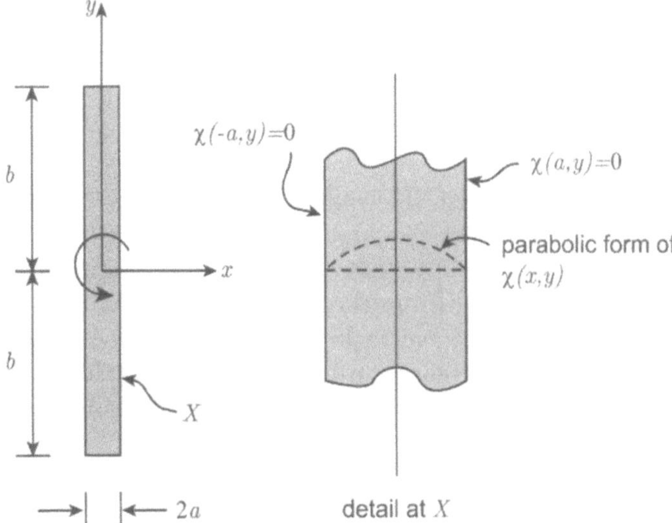

Figure 9.25: Torsion of a flat plate.

but violates the boundary condition

$$\chi(x, \pm b) = 0 \tag{9.372}$$

Since we are interested in developing an *approximate solution* to the problem we can assume that the solution (9.370) is valid for a large part of the cross-section, when $b \gg a$. The arbitrary constant A can be determined by substituting the solution (9.370) in Poisson's equation (9.280). This gives

$$A = \mu\Omega_0 a^2 \tag{9.373}$$

and the non-zero stress component σ_{yz} is given by

$$\sigma_{yz}(x, y) = -\frac{\partial \chi}{\partial x} = 2\mu\Omega_0 x \tag{9.374}$$

The torque T_0 required to maintain the rotation Ω_0 can be obtained from the result (9.301);

$$T_0 = 2\int_{-a}^{a}\int_{-b}^{b} \chi(x, y)dxdy \tag{9.375}$$

or

$$T_0 = \frac{16\mu\Omega_0 a^3 b}{3} \tag{9.376}$$

This result can be compared with the leading term of (9.365) for the torque-twist relationship derived for the rectangular cross-section where no restrictions are imposed on the aspect ratio a/b. As $(a/b) \to 0$, $\tanh(n\pi b/2a) \to 1$, and if the $a^3 b$ term is factorized from the result (9.365), the term involving the summation will be proportional to (a/b), and will reduce to zero much faster than the first term. The solution to the problem of the torsion of a flat prismatic plate is therefore contained in the complete solution.

$$\bullet\ \bullet\ \bullet$$

The procedures discussed in the Example 9.10 can also be extended to include plane polar coordinates which are better suited to examine torsion of prismatic members with curved boundaries. We first note that for torsional loading of prismatic elastic solids, the non-zero components of the stress matrix $\boldsymbol{\sigma}$ (see e.g. (8.857)) referred to the cylindrical polar coordinates (r, θ, z) are σ_{rz} and $\sigma_{\theta z}$, i.e.

$$\boldsymbol{\sigma} = \begin{bmatrix} \sigma_{rr} & \sigma_{r\theta} & \sigma_{rz} \\ \sigma_{r\theta} & \sigma_{\theta\theta} & \sigma_{\theta z} \\ \sigma_{rz} & \sigma_{\theta z} & \sigma_{zz} \end{bmatrix} = \begin{bmatrix} 0 & 0 & \sigma_{rz} \\ 0 & 0 & \sigma_{\theta z} \\ \sigma_{rz} & \sigma_{\theta z} & 0 \end{bmatrix} \tag{9.377}$$

This fact can also be established by using the transformation rule (8.182) applicable to stresses.

For torsion of prismatic bodies, the single equation of equilbrium reduces to

$$\frac{\partial \sigma_{rz}}{\partial r} + \frac{1}{r}\frac{\partial \sigma_{\theta z}}{\partial \theta} + \frac{\sigma_{rz}}{r} = 0 \tag{9.378}$$

We can introduce the Prandtl stress function $\chi(r, \theta)$ such that

$$\sigma_{rz} = \frac{1}{r}\frac{\partial \chi}{\partial \theta} \ ; \ \sigma_{\theta z} = -\frac{\partial \chi}{\partial r} \tag{9.379}$$

such that the equation of equilibrium (9.378) is identically satisfied for all choices of $\chi(r, \theta)$.

Referred to the cylindrical polar coordinate system, the displacement vector has the form

$$\mathbf{u} = \{0\mathbf{i}_r + \Omega_0 rz\mathbf{i}_\theta + \Omega_0\phi(r, \theta)\mathbf{i}_z\} \tag{9.380}$$

where Ω_0 is the twist per unit length and $\phi(r, \theta)$ is the warping function. The fact that the radial displacement is zero can be established by considering the "*second-order*" nature of such a displacement. The shear strains ϵ_{rz} and $\epsilon_{\theta z}$ can be expressed in terms of the derivatives of the components of (9.380): i.e.

$$\epsilon_{rz} = \frac{\Omega_0}{2}\frac{\partial\phi}{\partial r} \quad ; \quad \epsilon_{\theta z} = \frac{\Omega_0}{2}\left\{r + \frac{1}{r}\frac{\partial\phi}{\partial\theta}\right\} \tag{9.381}$$

The corresponding expressions for the stress components expressed in terms of $\phi(r, \theta)$ can be obtained from the results

$$\sigma_{rz} = 2\mu\epsilon_{rz} \quad ; \quad \sigma_{\theta z} = 2\mu\epsilon_{\theta z} \tag{9.382}$$

Substituting (9.381) into (9.382) and the resulting expressions in (9.378) we obtain

$$\overset{\circ}{\nabla}^2\phi(r, \theta) = 0 \tag{9.383}$$

where

$$\overset{\circ}{\nabla}^2 = \frac{\partial^2}{\partial r^2} + \frac{1}{r}\frac{\partial}{\partial r} + \frac{1}{r^2}\frac{\partial^2}{\partial\theta^2} \tag{9.384}$$

is Laplace's operator referred to the plane coordinate system. Considering (9.379) and (9.382) we obtain

$$\sigma_{rz} = \Omega_0\mu\left\{\frac{\partial\phi}{\partial r}\right\} = \frac{1}{r}\frac{\partial\chi}{\partial\theta} \tag{9.385}$$

$$\sigma_{\theta z} = \Omega_0\mu\left\{r + \frac{1}{r}\frac{\partial\phi}{\partial\theta}\right\} = -\frac{\partial\chi}{\partial r} \tag{9.386}$$

Eliminating $\phi(r,\theta)$ between (9.385) and (9.386) we obtain

$$\overset{\circ}{\nabla}^2 \chi(r,\theta) = -2\mu\Omega_0 \tag{9.387}$$

The results (9.383) and (9.387) are a confirmation of the results (9.272) and (9.280) respectively, and establishes their *invariance property*. The boundary conditions governing $\phi(r,\theta)$ and $\chi(r,\theta)$ follow from results of the form (9.277) and (9.281), which can be transformed to their equivalents in the (r,θ) coordinate system.

We also note that $\psi(r,\theta)$, the conjugate harmonic function to $\phi(r,\theta)$ satisfies

$$\overset{\circ}{\nabla}^2 \psi(r,\theta) = 0 \quad ; \quad (r,\theta) \in \mathcal{A} \tag{9.388}$$

and the Cauchy-Riemann equations

$$\frac{\partial \phi}{\partial r} = \frac{1}{r}\frac{\partial \psi}{\partial \theta} \quad ; \quad \frac{1}{r}\frac{\partial \phi}{\partial \theta} = -\frac{1}{r}\frac{\partial \psi}{\partial r} \tag{9.389}$$

The boundary condition for $\psi(r,\theta)$, corresponding to (9.326) takes the form

$$\psi(r,\theta) = \frac{r^2}{2} + \text{const.}, \ (r,\theta) \in \mathcal{C} \tag{9.390}$$

where $\mathcal{C}$ is the boundary of the region $\mathcal{A}$. The boundary condition for the Prandtl stress function takes the form

$$\chi(r,\theta) = \text{const.}, \ (r,\theta) \in \mathcal{C} \tag{9.391}$$

and the torque T_0 that must be applied to the prismatic bar to maintain the rotation Ω_0 can be evaluated from the result

$$T_0 = 2 \iint_{\mathcal{A}} \chi(r,\theta)dA \tag{9.392}$$

Extension of these results to cover multiply connected domains is left as an exercise.

Example 9.12

Obtain the solution to the problem of the torsional loading of a prismatic elastic bar with a cross-section in the form of a semi-circle of a radius a (Figure 9.26). This problem is a special case of the general problem relating to the torsion of a prismatic elastic bar with a cross-section in the form of a sector of a circle, investigated by Saint-Venant (1797-1886) using the semi-inverse method of analysis.

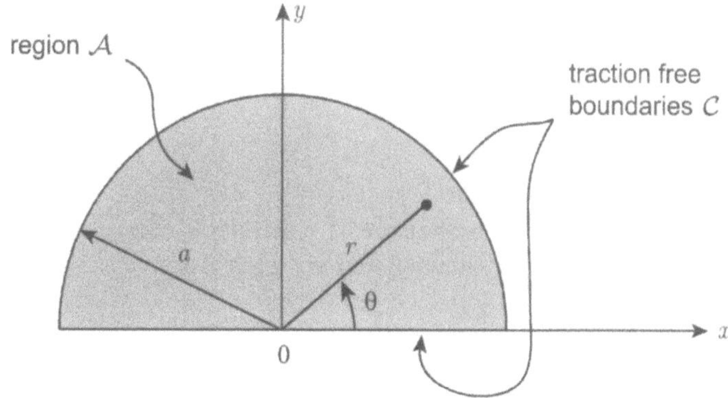

Figure 9.26: Torsion of a prismatic bar with a semi-circular cross-section.

Solution

For the solution of this problem we shall employ the Prandtl stress function approach where $\chi(r,\theta)$ satisfies Poisson's equation

$$\overset{\circ}{\nabla}^2\chi(r,\theta) = -2\mu\Omega_0 \quad ; \quad (r,\theta) \in \mathcal{A} \tag{9.393}$$

and the boundary conditions

$$\chi(a,\theta) = 0 \quad ; \quad \chi(r,0) = 0 \quad ; \quad \chi(r,\pi) = 0 \tag{9.394}$$

Now consider a Fourier *sine series* representation of the right side of (9.393) in the form

$$-2\mu\Omega_0 = \sum C_n^* \sin(n\theta) \tag{9.395}$$

The constants C_n^* can be evaluated in the form

$$C_n^* = -\frac{4\mu\Omega_0}{\pi}\left[\frac{1+(-1)^{n+1}}{n}\right] \tag{9.396}$$

and (9.393) can be written as

$$\overset{\circ}{\nabla}^2\chi(r,\theta) = -\frac{4\mu\Omega_0}{\pi}\sum_{n=1,3,}\left[\frac{1+(-1)^{n+1}}{n}\right]\sin(n\theta) \tag{9.397}$$

We now assume that $\chi(r,\theta)$ admits a solution of the form

$$\chi(r,\theta) = \sum_{n=1,3,} F_n(r)\sin(n\theta) \tag{9.398}$$

Substituting (9.398) in (9.397) we obtain the following set of ordinary differential equations which should be satisfied by $F_n(r)$: i.e.

$$\frac{d^2 F_n}{dr^2} + \frac{1}{r}\frac{dF_n}{dr} - \frac{n^2}{r^2}F_n = -\frac{4\mu\Omega_0}{\pi}\left[\frac{1+(-1)^{n+1}}{n}\right] \quad ; \; n=1,3, \tag{9.399}$$

The particular solution of (9.399) takes the form

$$F_n^P(r) = -\frac{4\mu\Omega_0\left[1+(-1)^{n+1}\right]r^2}{\pi n(4-n^2)} \tag{9.400}$$

The solution of the homogeneous equation of (9.399) gives

$$F_n^H(r) = A_n r^n + \frac{B_n}{r^n} \tag{9.401}$$

where A_n and B_n are arbitrary constants.

Since the cross-section of the prismatic bar is a complete semi-circle, the domain $\mathcal{A}$ includes the origin and the stresses and warping displacements should be bounded at $r=0$. For the homogeneous solution (9.401) to satisfy this requirement we set $B \equiv 0$. The complete solution of $\chi(r,\theta)$ indeterminate to within the arbitrary constants A_n can be written in the form

$$\chi(r,\theta) = \sum_{n=1,3,} \left\{ A_n r^n - \frac{4\mu\Omega_0[1 + (-1)^{n+1}]r^2}{\pi n(4 - n^2)} \right\} \sin(n\theta) \qquad (9.402)$$

The constants A_n can be determined by satisfying the boundary condition on $r = a$; i.e.

$$A_n = \frac{4\mu\Omega_0[1 + (-1)^{n+1}]}{\pi n(4 - n^2)a^{n-2}} \qquad (9.403)$$

The final result for the Prandtl stress function is

$$\chi(r,\theta) = \frac{4\mu\Omega_0 a^2}{\pi} \sum_{n=1,3,} \frac{[1 + (-1)^{n+1}]}{n(4 - n^2)} \left(\frac{r}{a}\right)^2$$
$$\cdot \left[\left(\frac{r}{a}\right)^{n-2} - 1 \right] \sin(n\theta) \qquad (9.404)$$

The expressions for the stresses σ_{rz} and $\sigma_{\theta z}$ can be determined by substituting (9.404) in (9.379). The maximum values of these stresses can be obtained by a numerical evaluation of the resulting expressions. A result of some general interest relates to the evaluation of the torque T_0 required to maintain the rotation Ω_0. Using the result (9.392), and noting that $\chi(r,\theta)$ is in a variables separated form, we can write

$$T_0 = 2 \int_0^a \int_0^\pi \chi(r,\theta) r\, dr\, d\theta \qquad (9.405)$$

Performing the integrations in (9.405) we obtain

$$T_0 = \frac{2\mu\Omega_0 a^4}{\pi} \sum_{n=1,3,} \frac{[1 + (-1)^{n+1}]^2}{n^2(2 + n)^2} \qquad (9.406)$$

The series can be evaluated to any required accuracy, and an approximate result can be obtained as

$$T_0 \simeq 0.296\mu\Omega_0 a^4 \qquad (9.407)$$

Using Polya's approach, a complete circular cylinder with a cross-sectional area equal to that of the semi-circular cylinder gives

$$T_0 \simeq 0.392\mu\Omega_0 a^4 \tag{9.408}$$

Similarly, the cylinder with a maximum diameter inscribed within the semicircular region gives

$$T_0 \simeq 0.10\mu\Omega_0 a^4 \tag{9.409}$$

The results (9.407) to (9.409) confirm Polya's Theorem for the stiffness bounds, similar to those developed in (9.369).

9.9.5   A complex variable formulation of the torsion problem

The complex variable method offer a further technique for the formulation and solution of the torsion problem related to prismatic elastic solids. We restrict attention to simply connected regions and when the traction free condition associated with the prismatic surface is expressed in terms of the Prandtl stress function as

$$\chi(x,y) = \text{const.} \quad ; \quad (x,y) \in \mathcal{C} \tag{9.410}$$

where $\mathcal{C}$ is the boundary of the cross-section. The stress function can be expressed in the form

$$\chi(x,y) = \mu\Omega_0 \left[\psi(x,y) - \frac{1}{2}(x^2 + y^2) \right] + \text{const.} \tag{9.411}$$

where $\psi(x,y)$ is the conjugate harmonic function to the warping function $\phi(x,y)$ and on the boundary $\mathcal{C}$

$$\psi(x,y) = \frac{1}{2}(x^2 + y^2) + \text{const.} = \frac{r^2}{2} + \text{const.} \tag{9.412}$$

Since ϕ and ψ are related to each other by the Cauchy-Riemann equations we can define an analytic function

$$\omega(z_c) = \phi(x,y) + i\psi(x,y) \tag{9.413}$$

where z_c and $\bar{z}_c$ are, respectively, the complex variable and its conjugate defined by

$$z_c = (x + iy) \quad ; \quad \overline{z}_c = (x - iy) \tag{9.414}$$

(We note again that it is necessary to introduce this distinction when using the complex variable notation since z, the coordinate variable occurs in the developments associated with the torsion problem.)

Since the function $\omega(z_c)$ could be complex, in addition to containing z_c, a bar is place above it to define the complex conjugate (e.g. if $\omega(z_c) = iz_c$, then $\overline{\omega}(\overline{z}_c) = i\overline{z}_c$). It is convenient at this point to represent the salient results for the displacements and stress components in the following forms. We define

$$D = (u + iv) \quad ; \quad \Psi_S = (\sigma_{xz} + i\sigma_{yz}) \tag{9.415}$$

where D and Ψ_S can be complex-valued functions. From (9.265) and (9.415) we have

$$D = i\Omega_0 z z_c \tag{9.416}$$

Also, taking the derivative of $\omega(z_c)$ and using the Cauchy-Riemann conditions we obtain

$$\omega'(z_c) = \frac{\partial \psi}{\partial y} + i\frac{\partial \psi}{\partial x} \tag{9.417}$$

and we can rewrite the second equation of (9.415) as

$$\Psi_S = \mu\Omega_0 \left\{ iz_c + \overline{\omega}'(\overline{z}_c) \right\} \tag{9.418}$$

The torque T_0 required to maintain the rotation Ω_0 can be obtained from (9.293), the Cauchy-Riemann equation (9.322) and (9.417); i.e.

$$T_0 = \mu\Omega_0 \, \mathrm{Re}\left\{ \iint_A \left\{ iz_c\omega'(z_c) + z_c\overline{z}_c \right\} dA \right\} \tag{9.419}$$

The two-dimensional form of Green's theorem (see e.g. (1.51))

$$\iint_A \left[\frac{\partial L}{\partial x} - \frac{\partial M}{\partial y} \right] dx\,dy = \oint_C [L\,dx + M\,dy] \tag{9.420}$$

can be expressed in terms of the complex variable z_c and its conjugate $\bar{z}_c$ in the form

$$\iint_{\mathcal{A}} \left[\frac{\partial}{\partial z_c}(L + iM) + \frac{\partial}{\partial \bar{z}_c}(L - iM) \right] dA$$
$$= \frac{1}{2} \oint_C \left[(-M - iL)dz_c + (-M + iL) \right] d\bar{z}_c \tag{9.421}$$

By setting $M = iL = -f(z_c, \bar{z}_c)$, the result (9.421) gives

$$2i \iint_{\mathcal{A}} \frac{\partial f}{\partial \bar{z}_c} dA = \oint_C f(z_c, \bar{z}_c) dz_c \tag{9.422}$$

Using this result, we can express (9.419) in the form

$$T_0 = \frac{\mu \Omega_0}{4} \operatorname{Re} \left\{ \oint_C z_c \bar{z}_c \left[2\omega'(z_c) - i\bar{z}_c \right] dz_c \right\} \tag{9.423}$$

Finally, we note that the boundary condition (9.410) or its equivalent

$$\psi(x, y) = \frac{1}{2}(x^2 + y^2) \tag{9.424}$$

can be expressed in terms of the complex function $\omega(z_c)$ as

$$\omega(z_c) - \overline{\omega}(\bar{z}_c) = iz_c \bar{z}_c + \text{ const.}, \quad (z_c, \bar{z}_c) \in C \tag{9.425}$$

The method for obtaining solutions for the torsion of prismatic elastic solids developed via a complex variable approach is now reduced to the evaluation of the complex potential $\omega(z_c)$. As indicated previously, for certain regular cross-sectional shapes, the function $\omega(z_c)$ can be obtained through consideration of the equation of the boundary curve expressed in terms of the complex variables z_c and $\bar{z}_c$.

Example 9.13

Obtain the complex function $\omega(z_c, \bar{z}_c)$ that would be applicable for the solution of the torsion of a prismatic bar with an elliptical cross-section (length of the major axis in the x-direction is $2a$ and the length of the minor axis is $2b$).

Solution

Considering the definitions (9.414) we can write

$$x = \frac{1}{2}(z_c + \bar{z}_c) \quad ; \quad y = -\frac{i}{2}(z_c - \bar{z}_c) \tag{9.426}$$

Considering the form of the boundary condition (9.425) we assume that, the relationship between z_c and $\bar{z}_c$ at all points on the boundary can be put in the form

$$z_c \bar{z}_c = f(z_c) + \overline{f}(\bar{z}_c) \tag{9.427}$$

The requirement on the boundary C given by (9.425) will be identically satisfied by the relationship

$$\omega(z_c) = i f(z_c) + \text{const.} \tag{9.428}$$

For the prismatic bar with an elliptical cross-section

$$\frac{x^2}{a^2} + \frac{y^2}{b^2} = 1 \tag{9.429}$$

Substituting (9.426) in (9.429) and separating terms we obtain

$$z_c \bar{z}_c = \frac{2a^2 b^2}{(a^2 + b^2)} + \frac{(a^2 - b^2)}{2(a^2 + b^2)}(z_c^2 + \bar{z}_c^2) \tag{9.430}$$

Comparing (9.430) with (9.428) and noting that $(z_c^2 + \bar{z}_c^2) = i z_c^2$ we obtain

$$\omega(z_c) = \frac{(a^2 - b^2)}{2(a^2 + b^2)} i z_c^2 + \text{const.}, \tag{9.431}$$

Substituting (9.431) in either (9.419) or (9.423) gives the expression (9.341) for the relationship between the torque T_0 and the resulting rotation Ω_0. For example, we can consider the result (9.419) that, involves the derivative of $\omega(z_c)$ which removes the arbitrary constant in (9.431); we obtain from (9.419)

$$T_0 = \mu\Omega_0 \operatorname{Re}\left\{\iint_{\mathcal{A}}\left[-\frac{z_c^2(a^2-b^2)}{(a^2+b^2)}+(x^2+y^2)\right]\right\}dA \qquad (9.432)$$

This reduces to

$$T_0 = \frac{2\mu\Omega_0 a^2 b^2}{(a^2+b^2)}\iint_{\mathcal{A}}\left[\frac{x^2}{a^2}+\frac{y^2}{b^2}\right]dA = \frac{\pi\mu\Omega_0 a^3 b^3}{(a^2+b^2)} \qquad (9.433)$$

which is the agreement with the solution obtained previously (Example 9.9), by considering the Prandtl stress function in the form of the boundary equation of the ellipse.

$$\bullet\ \bullet\ \bullet$$

In instances where the boundary $\mathcal{C}$ of the cross-section of the prismatic bar posesses a shape which cannot be described by a simple continuous function, recourse must be made to the selection of a *conformal transformation* or a *mapping function*, which will map the region $\mathcal{A}$ to the interior of a unit circle in the ζ-plane (Figure 9.27) such that $\zeta = \xi + i\eta$.

We restrict attention to simply connected regions and to situations where the boundary condition governing the Prandtl stress function is homogeneous,

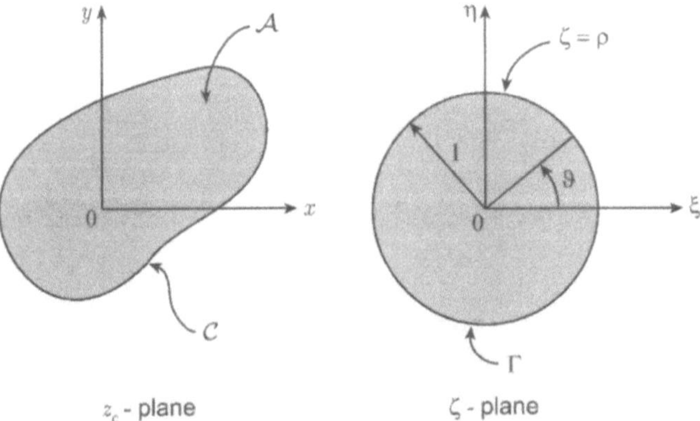

Figure 9.27: Conformal transformation.

i.e. $\chi = 0$; $(x, y) \in \mathcal{C}$. For this case, the stress function can be written as

$$\chi(x, y) = \mu \Omega_0 \left[\psi(x, y) - \frac{1}{2}(x^2 + y^2) \right] = 0 \tag{9.434}$$

where $\psi(x, y)$ is the conjugate harmonic function to the warping function $\phi(x, y)$, and

$$\psi(x, y) = \frac{1}{2}(x^2 + y^2) = \frac{r^2}{2} \tag{9.435}$$

Since ψ and ϕ are related by the Cauchy-Riemann equations, we can define an analytic function $\omega(z_c)$ such that ϕ and ψ are defined as, respectively, the real and imaginary parts of $\omega(z_c)$ (see e.g. (9.413)). We define a mapping $W(\zeta)$ which allows a simply connected cross-section in the z_c-plane to be determined from a mapping from the interior of the unit circle in the $\zeta-$plane in the form

$$z_c = W(\zeta) \tag{9.436}$$

At the boundary of the region in the z_c plane, the function ψ should satisfy the condition

$$\psi(x, y) = \frac{1}{2}(x^2 + y^2) = \frac{1}{2}(x + iy)(x - iy) = \frac{z_c \bar{z}_c}{2} \tag{9.437}$$

In the ζ-plane we specify the problem as

$$\omega(\zeta) = F(z_c) = F[W(\zeta)] \tag{9.438}$$

and using (9.413) we can rewrite the expression (9.438) in the form

$$-\frac{1}{2} i [\omega(z_c) - \overline{\omega(z_c)}] = \frac{1}{2} z_c \bar{z}_c \tag{9.439}$$

On the boundary of the unit circle $\zeta = \rho$

$$\omega(\rho) - \overline{\omega(\rho)} = i W(\zeta) \overline{W(\zeta)} \tag{9.440}$$

Multiplying (9.440) by $(1/2\pi i)\,[d\rho/(\rho-\zeta)]$ where ζ is a point at the interior of Γ (Figure 9.27), and integrating around Γ, we have

$$\frac{1}{2\pi i}\oint_\Gamma \frac{\omega(\rho)}{(\rho-\zeta)}d\rho - \frac{1}{2\pi i}\oint_\Gamma \frac{\overline{\omega(\rho)}}{(\rho-\zeta)}d\rho$$
$$= \frac{1}{2\pi}\oint_\Gamma \frac{W(\rho)\overline{W(\rho)}}{(\rho-\zeta)}d\rho \tag{9.441}$$

Using Cauchy's integral formula

$$\frac{1}{2\pi i}\oint_\Gamma \frac{\omega(\rho)d\rho}{(\rho-\zeta)} = \omega(\zeta) \tag{9.442}$$

the first integral of (9.441) reduces to $\omega(\zeta)$. To evaluate the second integral we note that $\overline{\rho} = 1/\rho$ and write

$$\omega(\zeta) = \omega(0) + \omega'(0)\zeta + \omega''(0)\frac{\zeta^2}{2!} + \dots \tag{9.443}$$

such that

$$\overline{\omega}\left(\frac{1}{\rho}\right) = \overline{\omega}(0) + \overline{\omega}'(0)\frac{1}{\rho} + \overline{\omega}''(0)\frac{1}{\rho^2 2!} + \dots \tag{9.444}$$

This gives

$$\frac{1}{2\pi i}\oint_\Gamma \frac{\overline{\omega(\rho)}}{(\rho-\zeta)}d\rho = \frac{1}{2\pi i}\oint_\Gamma \overline{\omega}\left(\frac{1}{\rho}\right)\sum_{n=0}^{\infty}\left(\frac{\zeta}{\rho}\right)^n \frac{d\rho}{\rho} \tag{9.445}$$

From the residue theorem, (9.445) gives

$$\frac{1}{2\pi i}\oint_\Gamma \frac{\overline{\omega(\rho)}}{(\rho-\zeta)}d\rho = \overline{\omega(0)} \tag{9.446}$$

Using these reductions, (9.441) can be reduced to the result

$$\omega(\zeta) - \overline{\omega(0)} = \frac{1}{2\pi}\oint_\Gamma \frac{W(\rho)\overline{W(\rho)}}{(\rho-\zeta)}d\rho \tag{9.447}$$

The complex potential $\omega(\zeta)$ can be written as

$$\omega(\zeta) = \frac{1}{2\pi} \oint_\Gamma \frac{W(\rho)\overline{W(\rho)}}{(\rho - \zeta)} d\rho + \text{const.} \tag{9.448}$$

The complex potential $\omega(\zeta)$ is now expressed in terms of the conformal transformation of the mapping function $W(\zeta)$. Once the function $\omega(\zeta)$ is determined the torsional response of the prismatic elastic bar is fully defined. For example, the stresses can be calculated from the second equation of (9.415) and (9.418) as follows:

$$\Psi_S = \mu\Omega_0 \left\{ iW(\zeta) + \frac{\overline{\omega}'(\overline{\zeta})}{\overline{W}'(\overline{\zeta})} \right\} \tag{9.449}$$

Similarly, the torque T_0 that needs to be applied to the prismatic bar to maintain the rotation Ω_0 is given by

$$T_0 = \frac{\mu\Omega_0}{4} \text{Re} \left\{ \oint_\Gamma W(\rho)\overline{W}(\overline{\rho}) \left[2\omega'(\rho) - i\overline{W}(\overline{\rho})W'(\rho)d\rho \right] \right\} \tag{9.450}$$

Since the expressions for the stresses and the torsional response, given by (9.449) and (9.450) respectively, both involve only derivatives of $\omega(\rho)$, the arbitrary constant (9.448) has no influence on these results. Although the formal developments governing the analysis of torsion problem by appeal to a conformal transformation technique are quite straightforward, the adaptability of the procedure largely rests on the availability of mapping functions in convenient forms. In particular, the requirement for mapping functions which transform the interior of the cross-section of the prismatic bar to the interior of a unit circle makes the simple implementation of the procedure a less than convenient exercise.

Example 9.14

Use a conformal transformation technique to develop a solution to the torsional response of a prismatic elastic bar with a circular section of radius a.

Solution

Since the circular domain is simply connected the mapping function $W(\zeta)$ corresponds to

$$W(\zeta) = a\zeta \tag{9.451}$$

The complex potential $\omega(\zeta)$ is obtained from (9.448) as follows:

$$\omega(\zeta) = \frac{1}{2\pi} \oint_\Gamma \frac{(a\rho)(a/\rho)}{(\rho - \zeta)} \, d\rho = a^2 i \tag{9.452}$$

The torque T_0 required to maintain a twist Ω_0 is obtained by substituting (9.451) and (9.452) in (9.450) i.e.

$$T_0 = \frac{\mu\Omega_0}{4} \, \text{Re} \left\{ \oint_\Gamma (a\rho) \frac{a}{\rho} \left[-i \left(\frac{a}{\rho} \right) a \right] d\rho \right\} \tag{9.453}$$

Evaluating (9.453) we have

$$T_0 = \frac{\mu\Omega_0 a^4}{4} \, \text{Re} \left\{ -i \int_0^{2\pi} \frac{ie^{i\theta} d\theta}{e^{i\theta}} \right\} = \frac{\pi\mu\Omega_0 a^4}{2} \tag{9.454}$$

which is in agreement with the solution (9.333) obtained by considering the equation of the boundary of the circular cross-section.

9.10 An alternative derivation of the elastic torsion problem

In this section we shall present an alternative derivation of the partial differential equation governing the torsional behaviour of a prismatic elastic bar. We consider a prismatic elastic region V with surface region S which is subjected to a torque T_0 about the z-axis. The cross-section of the prismatic bar is arbitrary with area $\mathcal{A}$ and boundary $\mathcal{C}$. The continuous boundary $\mathcal{C}$ is assumed to have a continuous derivative. The prismatic bar occupies the interval $z \in (-L, L)$ (Figure 9.28).

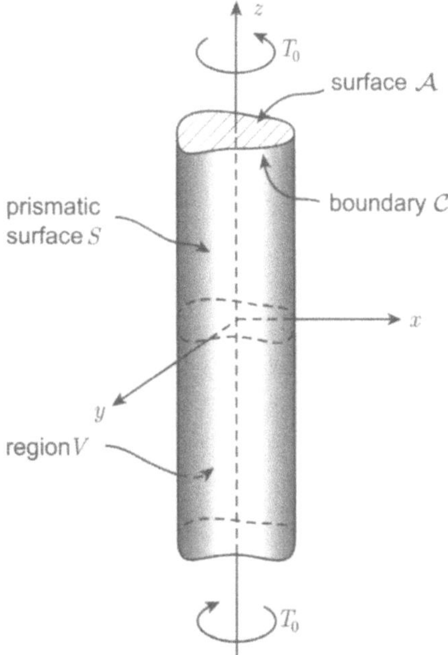

Figure 9.28: Torsion of a prismatic elastic bar.

The state of deformation in the prismatic bar is defined by the displace-
ment vector $\mathbf{u}$ and the state of stress is defined by the *stress dyadic* $\mathfrak{S}$ and
the state of strain by the *strain dyadic* $\mathfrak{E}$. In the absence of body forces,
the elastostatic boundary value problem governing the torsion problem requires the determination of a solution which satisfies the equations of static
equilibrium (see e.g. (8.161))

$$\nabla.\mathfrak{S} = 0 \quad ; \quad \mathbf{x} \in V \tag{9.455}$$

the compatibility conditions (see e.g. (8.102))

$$\nabla \times \mathfrak{E} \times \nabla = 0 \quad ; \quad \mathbf{x} \in V \tag{9.456}$$

in the region of the prismatic bar, and traction boundary conditions

$$\boldsymbol{\sigma}.\mathbf{n} = 0 \quad ; \quad \mathbf{x} \in S \tag{9.457}$$

on the surface of the prismatic bar.

Consider a cylindrical polar coordinate system (r, θ, z) with the axis z located about the centre of the twist and $\mathbf{r} = r\mathbf{i}_r$ is a two-dimensional position vector. The traction resultants should satisfy

$$\mathbf{i}_z.\mathfrak{S}.\mathbf{i}_z = 0 \quad ; \quad (x, y) \in \mathcal{A} \tag{9.458}$$

and

$$\iint_A \mathbf{i}_z.\mathfrak{S}\, dA = 0 \quad ; \quad (x, y) \in \mathcal{A} \tag{9.459}$$

$$\iint_A r\mathbf{i}_r \times (\mathbf{i}_z.\mathfrak{S})dA = \pm T_0\mathbf{i}_z \tag{9.460}$$

The equations (9.459) and (9.460) correspond to the resultants of tractions acting on the end planes of the prismatic bar.

In the indirect method of analysis of the torsion problem we postulate the form of the deformation field and assume that it can be separated into two parts. The first consists of a rigid rotation of any cross-section about the z-axis. The angle of rotation being dependent on the distance of the section from $z = 0$, and the second consisting of the uniform warping of every section along the z-axis. The displacement vector $\mathbf{u}$ can then be represented in the form

$$\mathbf{u} = 0\mathbf{i}_r + \Omega_0 f(z)r\mathbf{i}_\theta + \Psi^*(r, \theta)\mathbf{i}_z \tag{9.461}$$

where Ψ^* is a function independent of z and Ω_0 is the rotation per unit length of the prismatic bar. We choose the function $f(z)$ such that $f(0) = 0$ and $f(L) = L$. The solution of this problem is thus reduced to the determination of Ψ^* and f to satisfy the governing equations (9.455) to (9.460). Since the formulation of the problem is directly in terms of the displacement vector, the compatibility conditions are satisfied. Considering (9.461), the assumed form of the displacement vector referred to the cylindrical polar coordinate system (r, θ, z), we have

$$\nabla\mathbf{u} = \nabla\Psi^*\mathbf{i}_z + \Omega_0 \left[r\frac{df}{dz}\mathbf{i}_z\mathbf{i}_\theta + f(\mathbf{i}_r\mathbf{i}_\theta - \mathbf{i}_\theta\mathbf{i}_r) \right] \tag{9.462}$$

The strain dyadic $\mathfrak{E}$ is given by

$$\mathfrak{E} = \frac{1}{2}\left(\nabla\mathbf{u} + \mathbf{u}\nabla\right) = \frac{1}{2}\left[\nabla\Psi^*\mathbf{i}_z + \mathbf{i}_z\nabla\Psi^*\right]$$
$$+\frac{\Omega_0 r}{2}\frac{df}{dz}\{\mathbf{i}_z\mathbf{i}_\theta + \mathbf{i}_\theta\mathbf{i}_z\} \tag{9.463}$$

We note that

$$|\mathfrak{E}| = 1 \tag{9.464}$$

indicating that the torsional deformational field (9.461) results in an *isochoric* or *volume preserving* deformations of the elastic solid.

Considering the linear elastic constitutive relationship for isotropic elastic materials (see e.g. (8.238)) we have

$$\mathfrak{S} = \lambda\mathfrak{I}(\nabla.\mathbf{u}) + 2\mu\mathfrak{E} \tag{9.465}$$

Since $\nabla.\mathbf{u} = 0$, using (9.463) we can rewrite (9.465) in the form

$$\mathfrak{S} = \mu\left[\nabla\Psi^*\mathbf{i}_z + \mathbf{i}_z\nabla\Psi^* + \Omega_0 r\frac{df}{dz}(\mathbf{i}_z\mathbf{i}_\theta + \mathbf{i}_\theta\mathbf{i}_z)\right] \tag{9.466}$$

The stress components derived from (9.466) satisfies the condition (9.458). Substituting (9.466) in the equation of equilibrium (9.455) we have

$$\nabla.\nabla\Psi^*\mathbf{i}_z + \Omega_0 r\frac{d^2 f}{dz^2}\mathbf{i}_\theta = 0 \tag{9.467}$$

Therefore, for the equations of equilibrium to be identically satisfied we require

$$\nabla.\nabla\Psi^* = \overset{\circ}{\nabla}^2\Psi^* = 0 \; ; \; \mathbf{x} \in \mathcal{A} \tag{9.468}$$

and

$$\frac{d^2 f}{dz^2} = 0 \; ; \quad \mathbf{x} \in S \tag{9.469}$$

To satisfy the boundary conditions at $z = 0$ and $z = L$, we require

$$f(z) = z \tag{9.470}$$

Therefore, to have uniform torsion, the rigid rotation of the cross-section at any distance z is proportional to the distance z from the mid-plane. The result (9.466) for $\mathfrak{S}$ can now be written as

$$\mathfrak{S} = \mu \left[\nabla \Psi^* \mathbf{i}_z + \mathbf{i}_z \nabla \Psi^* + \Omega_0 r (\mathbf{i}_z \mathbf{i}_\theta + \mathbf{i}_\theta \mathbf{i}_z) \right] \tag{9.471}$$

The boundary condition on $\mathcal{C}$, the boundary of the cross-section with domain $\mathcal{A}$, is given by

$$\mathbf{n} . \left[\Omega_0 r \mathbf{i}_\theta + \nabla \Psi^* \right] = 0 \quad ; \quad \mathbf{x} \in \mathcal{C} \tag{9.472}$$

The results (9.468) and (9.472) now define the Neumann-type boundary value problem for the unknown function Ψ^*. We can introduce the conjugate of Ψ^* defined by Φ^*, which is related to Ψ^* through the Cauchy-Riemann equations, to convert the Neumann problem to an equivalent Dirichlet problem for the plane harmonic function Φ^*; i.e.

$$\nabla \Psi^* = -\mathbf{i}_z \times \nabla \Phi^* \tag{9.473}$$

Substituting (9.473) in (9.468) we obtain

$$\nabla . \nabla \Psi^* = \overset{\circ}{\nabla}^2 \Phi^* = 0 \; ; \; \mathbf{x} \in \mathcal{A} \tag{9.474}$$

with boundary conditions

$$\mathbf{n} . \mathbf{i}_z \times \nabla \left(\Phi^* - \frac{\Omega_0 r^2}{2} \right) = \mathbf{i}_z . \mathbf{n} \times \nabla \left(\Phi^* - \frac{\Omega_0 r^2}{2} \right)$$
$$= 0 \quad ; \quad \mathbf{x} \in \mathcal{C} \tag{9.475}$$

The result (9.475) implies that

$$\Phi^* - \frac{\Omega_0 r^2}{2} = \text{const.}, \quad ; \quad \mathbf{x} \in \mathcal{C} \tag{9.476}$$

The equations (9.474) and (9.476) represent a Dirichlet boundary value problem for Φ^*.

We now introduce the Prandtl stress function $\chi(x, y)$ which is related to $\Phi^*(x, y)$ according to

$$\chi(x, y) = \mu \left[\Phi^* - \frac{\Omega_0 r^2}{2} \right] \tag{9.477}$$

Substituting (9.477) in (9.474) we obtain

$$\overset{\circ}{\nabla}^2 \chi = -2\Omega_0 \mu \quad ; \quad (x, y) \in \mathcal{A} \tag{9.478}$$

with boundary condition

$$\chi(x, y) = \text{const.}, \quad ; \quad (x, y) \in \mathcal{C} \tag{9.479}$$

which is the Dirichlet boundary value problem for Prandtl's stress function $\chi(x, y)$ governed by Poisson's equation. Again, if the cross-section $\mathcal{A}$ is simply connected, the constant term in (9.479) can be set in zero.

The stress dyadic $\mathfrak{S}$ can be expressed in terms of the Prandtl stress function as follows:

$$\mathfrak{S} = \mathbf{i}_z \nabla \chi \times \mathbf{i}_z + \nabla \chi \times \mathbf{i}_z \mathbf{i}_z \tag{9.480}$$

The resultant of tractions on any cross-section can be evaluated using (9.459); i.e.

$$\iint_{\mathcal{A}} \mathbf{i}_z \mathfrak{S} \, dA = - \iint_{\mathcal{A}} \mathbf{i}_z \times \nabla \chi dA \tag{9.481}$$

We can use Green's theorem to transform the surface integral in (9.481) to a line integral, i.e.

$$- \iint_{\mathcal{A}} \mathbf{i}_z \times \nabla \chi dA = - \oint_{\mathcal{C}} \chi d\mathbf{r} \tag{9.482}$$

Since $\chi = 0$ on the boundary $\mathcal{C}$, the boundary condition (9.459) is identically satisfied. Note that the boundary condition (9.459) would be identically satisfied even if $\chi(x, y) = \text{const.}$, $(x, y) \in \mathcal{C}$; since on any closed contour $\mathcal{C}$

$$\oint_{\mathcal{C}} d\mathbf{r} = 0 \tag{9.483}$$

Considering the expression (9.460) for the total torque, we have

$$\iint_{\mathcal{A}} r\mathbf{i}_r \times (\mathbf{i}_z.\mathfrak{S})dA = \iint_{\mathcal{A}} (\mathbf{i}_\theta \times \mathbf{i}_z) \times (\nabla\chi \times \mathbf{i}_z)rdA \tag{9.484}$$

$$= 2\mathbf{i}_z \iint_{\mathcal{A}} \chi dA - \mathbf{i}_z \iint_{\mathcal{A}} \mathbf{i}_z.\nabla \times (r\mathbf{i}_\theta\chi)dA$$

Using Green's theorem for the plane to convert the last integral of (9.484) to a line integral, i.e.

$$\iint_{\mathcal{A}} \mathbf{i}_z.\nabla \times (r\mathbf{i}_\theta\chi)dA = \oint_{\mathcal{C}} r\chi d\mathbf{r}.\mathbf{i}_\theta \tag{9.485}$$

The right side of the equation (9.485) is zero if $\chi = 0$ on $\mathcal{C}$. If χ is non-zero on $\mathcal{C}$ (say χ^*) then

$$\oint_{\mathcal{C}} \chi^* r d\mathbf{r}.\mathbf{i}_\theta = 2\chi^* \iint_{\mathcal{A}} dA \tag{9.486}$$

Combining (9.484) and (9.486), the magnitude of the torque T_0 is given by

$$T_0 = 2 \iint_{\mathcal{A}} [\chi - \chi^*] dA \tag{9.487}$$

which is in agreement with the result (9.301) derived through a different procedure.

9.11   The gravitational potential

As a concluding application of Poisson's equation, we consider the classical problem in potential theory related to the theory of gravitation. First let us consider the interaction of two particles P and Q of masses m_P and m_Q respectively. According to the law of gravitation proposed by Isaac Newton (1642-1727), the magnitude of the force of attraction between the two particles is given by

$$F = \frac{Gm_P m_Q}{r^2} \tag{9.488}$$

where G is referred to as the universal gravitational constant and r is the distance between the particles. We can choose the unit of mass as that of a particle which, when placed at a unit distance from a particle of equal mass, exerts a force of attraction of unity. Then (9.488) can be written as

$$F = \frac{m_P m_Q}{r^2} \tag{9.489}$$

Also, if we denote the vector $\overrightarrow{PQ}$ by $\mathbf{r}$, we can express the force per unit mass at P attracting a mass m at Q by a vector relationship

$$\mathbf{F} = -\frac{m\mathbf{r}}{r^2} = \nabla\left(\frac{m}{r}\right) \tag{9.490}$$

In (9.490), $\mathbf{F}$ is referred to as the *intensity of force* or the *gravitational field* of force at the location P, and it is expressed as the gradient of the scalar m/r. Furthermore, if $\mathbf{F}$ can be expressed as a gradient of a scalar function, then the field $\mathbf{F}$ is said to be *conservative*, and a requirement for conservative fields is

$$\nabla \times \mathbf{F} = \nabla \times \nabla\left(\frac{m}{r}\right) = 0 \tag{9.491}$$

For example, in relation to mechanics, we can interpret conservative force fields such as $\mathbf{F}$ by considering the work done by $\mathbf{F}$ when it moves along a specified path. In general, the work done by $\mathbf{F}$ can be expressed in the form

$$W = \int \mathbf{F}.d\mathbf{r} \tag{9.492}$$

where $d\mathbf{r}$ refers to an element of the path along which $\mathbf{F}$ moves. A force field for which (9.492) depends on the path of integration as well as the end points of the line is referred to as a *non-conservative force field*. The term *non-conservative* implies that not all the energy is conserved during motion and either a part or all of the energy can be dissipated by processes such as *friction* or other manifestations of *frictional phenomena*. There are force fields for which (9.492) depends only on the end points of the integration and independent of the path. Such fields are referred to as *conservative force fields*. In such cases all energy is conserved. The only *necessary* and *sufficient* condition for (9.492) to be independent of path or alternatively for $\mathbf{F}$ to be conservative is the requirement

$$\nabla \times \mathbf{F} = 0 \tag{9.493}$$

and

$$\nabla \times \mathbf{F} \neq 0 \tag{9.494}$$

is a requirement for a *non-conservative* force field. If we assume that $\mathbf{F}$ is derived from a function $W(\mathbf{x})$ such that

$$\mathbf{F} = F_x \mathbf{i} + F_y \mathbf{j} + F_z \mathbf{k} = \nabla W \tag{9.495}$$

Then

$$F_x = \frac{\partial W}{\partial x} \quad ; \quad F_y = \frac{\partial W}{\partial y} \quad ; \quad F_z = \frac{\partial W}{\partial z} \tag{9.496}$$

and

$$\nabla \times \mathbf{F} = \begin{vmatrix} \mathbf{i} & \mathbf{j} & \mathbf{k} \\ \dfrac{\partial}{\partial x} & \dfrac{\partial}{\partial y} & \dfrac{\partial}{\partial z} \\ \dfrac{\partial W}{\partial x} & \dfrac{\partial W}{\partial y} & \dfrac{\partial W}{\partial y} \end{vmatrix} \tag{9.497}$$

Since the differentiations commute, the three components of (9.497) are zero. Hence any conservative force field can be expressed as the gradient of a scalar potential. Alternatively, any force field which is derived from a scalar potential is conservative. Let us consider the total work done by the unit particle as it moves from ∞ to r_P. We have

$$\int_\infty^{r_P} \mathbf{F}.d\mathbf{r} = \int_\infty^{r_P} \nabla\left(\frac{m}{r}\right) d\mathbf{r} = \left[\frac{m}{r}\right]_\infty^{r_P} = \frac{m}{r_P} \tag{9.498}$$

Since the force field is conservative, this is independent of the path followed by the particle as it arrives at r_P and is referred to as the *potential* at P due to the particle of identical mass at Q. We shall denote this potential by $\mathcal{V}_P$ and write (9.490) as

$$\mathbf{F} = \nabla\left(\frac{m}{r_P}\right) = \nabla\mathcal{V}_P \tag{9.499}$$

The equation (9.499) implies that the intensity of force at a point is equal to the gradient of the potential at P. The result (9.490) can be extended to an assembly of particles m_1, m_2, ..., m_n, whose spatial positions in relation to P are denoted by r_1, r_2, ..., r_n, respectively. The resulting forces of attraction per unit mass at P due to the system masses is the vector sum of the intensities due to each mass, i.e.

$$\mathbf{F}^* = \nabla\left(\frac{m_1}{r_1}\right) + \nabla\left(\frac{m_2}{r_2}\right) + \ldots + \nabla\left(\frac{m_n}{r_n}\right) = \nabla\sum_{i=1}^{n}\left(\frac{m_i}{r_i}\right) \tag{9.500}$$

We can apply the procedure described previously and define the potential $\mathcal{V}^*$ as the work done by the attracting forces on a particle of unit mass, as it moves from infinity to P, by

$$\int_{\infty}^{r_P} \mathbf{F}^*.d\mathbf{r} = \sum_{i=1}^{n}\frac{m_i}{r_i} = \mathcal{V}^* \tag{9.501}$$

Hence the potential at P due to the assembly of particles is the sum of the potentials due to the individual particles. The potential $\mathcal{V}^*$ is a scalar function of position an except at positions Q_i where the masses m_i are located, it satisfies

$$\nabla^2\mathcal{V}^* = \nabla^2\sum_{i=1}^{n}\frac{m_i}{r_i} = \sum_{i=1}^{n}\nabla^2\left(\frac{m_i}{r_i}\right) = 0 \tag{9.502}$$

Hence $\mathcal{V}^*$ satisfies Laplace's equation. Alternatively, if

$$\mathbf{F}^* = \nabla\mathcal{V}^* \tag{9.503}$$

then

$$\nabla.\mathbf{F}^* = 0 \qquad (9.504)$$

at all locations except those referring to Q_i.

Example 9.15

Consider the two force fields

$$\mathbf{F}^{(1)} = 2xy\mathbf{i} + x^2\mathbf{j} + 0\mathbf{k} \qquad (9.505)$$

and

$$\mathbf{F}^{(2)} = -y^2\mathbf{i} + xy\mathbf{j} + 0\mathbf{k} \qquad (9.506)$$

Apply the criterion $\nabla \times \mathbf{F} = 0$ to identify the conservative and non-conservative force fields. Verify the result by evaluating the work done by the forces as they move from $(0,0,0)$ to $(1,1,1)$ along the paths $y = x$ and $y = x^2$.

Solution

Considering $\mathbf{F}^{(1)}$, we have

$$\nabla \times \mathbf{F}^{(1)} = \begin{vmatrix} \mathbf{i} & \mathbf{j} & \mathbf{k} \\ \dfrac{\partial}{\partial x} & \dfrac{\partial}{\partial y} & \dfrac{\partial}{\partial z} \\ 2xy & x^2 & 0 \end{vmatrix} = 0\mathbf{i} + 0\mathbf{j} + 0\mathbf{k} = \mathbf{0} \qquad (9.507)$$

Hence the force field $\mathbf{F}^{(1)}$ is conservative. Similarly

$$\nabla \times \mathbf{F}^{(2)} = \begin{vmatrix} \mathbf{i} & \mathbf{j} & \mathbf{k} \\ \dfrac{\partial}{\partial x} & \dfrac{\partial}{\partial y} & \dfrac{\partial}{\partial z} \\ -y^2 & xy & 0 \end{vmatrix} = 0\mathbf{i} + 0\mathbf{j} + 3y\mathbf{k} \neq \mathbf{0} \qquad (9.508)$$

Hence the force field $\mathbf{F}^{(2)}$ is *non-conservative.*

The vector $d\mathbf{r}$ used in the computation of W is given by

$$d\mathbf{r} = \mathbf{i}dx + \mathbf{j}dy + \mathbf{k}dz \tag{9.509}$$

and the general expression for the work done by $\mathbf{F}$ can be written as

$$W = \int [F_x\mathbf{i} + F_y\mathbf{j} + F_z\mathbf{k}][\mathbf{i}dx + \mathbf{j}dy + \mathbf{k}dz] \tag{9.510}$$

Considering the force field $\mathbf{F}^{(1)}$, we can write

$$W^{(1)} = \int [2xy\ dx + x^2 dy] \tag{9.511}$$

For the path of integration $y = x$; $dy = dx$, and

$$W_1^{(1)} = \int_0^1 [2x^2 dx + x^2 dx] = [x^3]_0^1 = 1 \tag{9.512}$$

For the path of integration $y = x^2$; $dy = 2x dx$, and

$$W_2^{(1)} = \int_0^1 \{2x^3 dx + 2x^3 dx] = [x^4]_0^1 = 1 \tag{9.513}$$

Hence $W_1^{(1)} = W_2^{(1)}$, which re-affirms the conservative nature of $\mathbf{F}^{(1)}$. Considering the force field $\mathbf{F}^{(2)}$ we have, in general

$$W^{(2)} = \int [-y^2 dx + xy\ dy] \tag{9.514}$$

For the path of integration $y = x$; $dy = dx$, and

$$W_1^{(2)} = \int_0^1 [-x^2 dx + x^2 dx] = 0 \tag{9.515}$$

For the path of integration $y = x^2$; $dy = 2x\ dx$, and

$$W_2^{(2)} = \int_0^1 [-x^4 dx + 2x^4 dx] = \left[\frac{x^5}{5}\right] = \frac{1}{5} \tag{9.516}$$

Hence $W_1^{(2)} \neq W_2^{(2)}$, which re-affirms the non-conservative nature of $\mathbf{F}^{(2)}$.

9.11.1   Spatial distribution of matter

We now focus attention on the problem where the attracting force field results from the distribution of matter, in a continuous fashion, within the region V bounded by surface S. The continuous distribution mass is such that we can define the density of matter at a point as ρ. Considering an elemental volume dV (Figure 9.29), the mass in the element is ρdV.

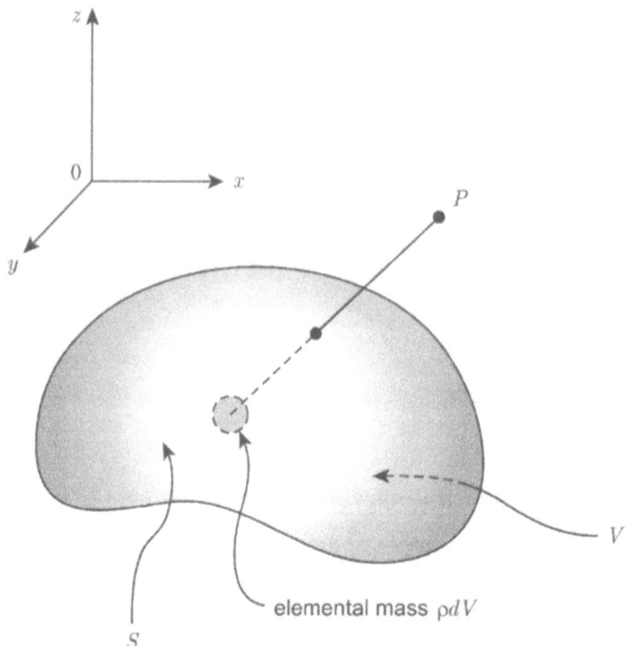

Figure 9.29: Force field due to distributed matter.

If the point P is located outside the region V, then the discrete sum (9.501) can be represented as an integral relationship of the form

$$V^* = \iiint_V \frac{\rho dV}{r} \tag{9.517}$$

where r is the distance of the elemental volume dV from the location P. The integral (9.517) gives the potential at P due to the entire body. If the point P is located inside the body, the integrand is infinite. In this case, the potential can be calculated by isolating the point P by a closed spherical surface V_s (Figure 9.30) and by considering the potential at P due to all matter surrounding P but exterior to V_s.

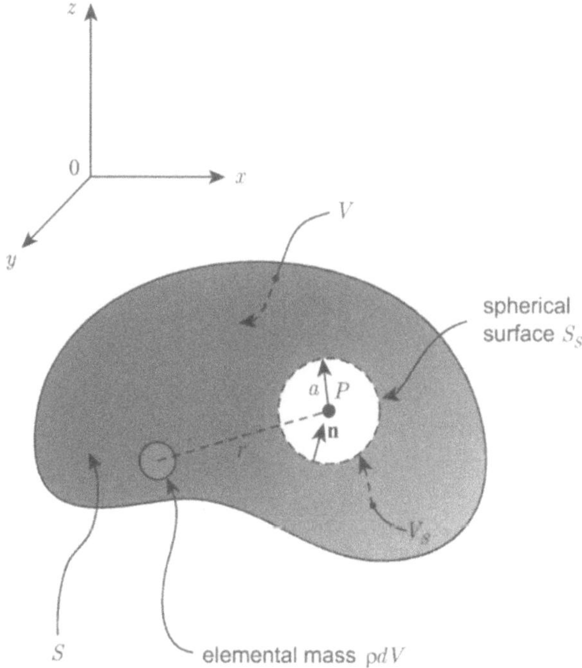

Figure 9.30: Force field due to distributed matter; interior location of field point.

The corresponding integrand in (9.517) is finite everywhere since P is outside $(V - V_s)$. We can let the region V_s decrease indefinitely and in the limit converge to the point P. The potential $\mathcal{V}_P$ is now defined by

$$\mathcal{V}_P = \operatorname*{Lim}_{V_s \to 0} \iiint_{V_s}^{V} \frac{\rho dV}{r} \tag{9.518}$$

Noting that the volume of V_s is of order r^3, where r is the radius of the sphere of volume V_s, and that the integrand varies as $1/r$, the integral in (9.518) will converge to a definite limit provided ρ is finite. The integral

(9.518) represents the potential at P due to the entire region V. Let us now consider the material region bounded externally by S and internally by the spherical surface S_s for radius a. The point P is located at the centre of this spherical region. The gravitational field at P consists of the vector sum of the gravitational field $\mathbf{F}_s$ that is produced by the matter inside S_s, and $\mathbf{F}_0$, the gravitational field of matter outside S_s; i.e.

$$\mathbf{F}_P = \mathbf{F}_0 + \mathbf{F}_s \tag{9.519}$$

Taking the divergence of (9.519) we can write

$$\nabla.\mathbf{F}_P = \nabla.\mathbf{F}_0 + \nabla.\mathbf{F}_s \tag{9.520}$$

Since $\mathbf{F}_0$ is produced by matter outside S_s we have, from (9.504),

$$\nabla.\mathbf{F}_0 = 0 \tag{9.521}$$

Consequently

$$\nabla.\mathbf{F}_P = \nabla.\mathbf{F}_s \tag{9.522}$$

The mass of the spherical region V_s of radius a is

$$m_s = \frac{4}{3}\pi\rho a^3 \tag{9.523}$$

As $a \to 0$, the intensity of the gravitational field on the surface S_s has a magnitude

$$F_s = \frac{m_s}{a^2} = \frac{4}{3}\pi\rho a \tag{9.524}$$

and acts in a direction of the inward unit normal to S_s. From the divergence theorem we have

$$\iiint_{V_s} \nabla.\mathbf{F}_s dV = \iint_{S_s} \mathbf{F}_s.\mathbf{n}dS = -\frac{4}{3}\pi\rho a \iint_{S_s} dS \tag{9.525}$$

In the limit $a \to 0$, (9.525) can be written as

$$\mathop{\text{Lim}}_{a \to 0} (\nabla.\mathbf{F}_s) \frac{4}{3}\pi a^3 = \mathop{\text{Lim}}_{a \to 0} \left(-\frac{4}{3}\pi\rho a\right)\left(4\pi a^2\right) \tag{9.526}$$

and from (9.522) we obtain

$$\nabla.\mathbf{F}_s = \nabla.\mathbf{F}_P = -4\pi\rho \tag{9.527}$$

As in the case of (9.503) we can assume that since $\mathbf{F}_P$ is a conservative field, which can be represented in terms of a potential $\mathcal{V}^P$ such that

$$\mathbf{F}_P = \nabla\mathcal{V}^P \tag{9.528}$$

Combining (9.527) and (9.528) we find that the gravitational potential in a region containing matter satisfies Poisson's equation

$$\nabla^2\mathcal{V}^P = -4\pi\rho \tag{9.529}$$

where ρ is the mass density. This result was first obtained by Poisson in 1833.

The procedures employed in developing Poisson's equation for the gravitational potential can be easily extended to include other types of phenomena where basic law governing the forces of attraction can be described by equation of the type (9.488). An example of such an application concerns electrostatics where, according to Coulomb's Law, two point charges of electricity in the same homogeneous medium exert forces on one another the magnitude of which is given by (9.488) where the masses are replaced by the charge and with the *proviso* that like charges repel and opposite charges attract. The constant of proportionality will be determined by the units used and the properties of the medium in which the charges are located. In view of this mathematical equivalence, any problem in attraction which hinges on the *inverse square law* can be interpreted either in terms of gravitation or in terms of electrostatics.

Returning to the result (9.517) and (9.529) it is evident that the solution to the potential $\mathcal{V}^P$ can be represented in the form

$$\mathcal{V}^P(\mathbf{x}) = \frac{1}{4\pi} \iiint_V \frac{f(\mathbf{x}, \mathbf{x}^*)}{r} dV^* \tag{9.530}$$

is a solution of

$$\nabla^2 \mathcal{V}^P(\mathbf{x}) = -f(\mathbf{x}) \tag{9.531}$$

In a general treatment, the result (9.530) which can be written in the form

$$\mathcal{V}^P(x, y, z) = \frac{1}{4\pi} \iiint_V \frac{f(x^*, y^*, z^*)dx^*dy^*dz^*}{[(x - x^*)^2 + (y - y^*)^2 + (z - z^*)^2]^{1/2}} \tag{9.532}$$

is a solution of (9.531) and $\mathbf{x}$ is the field point location and $\mathbf{x}^*$ is the source point location. In the development of potential $\mathcal{V}^P$ we have explicitly assumed that the gravitational potential energy at infinity is zero. The solution of Poisson's equation determined from (9.532) therefore reduces to zero at infinity. If a solution to Poisson's equation where the potential reduces to zero at a finite location is required, then solutions of Laplace's equation

$$\nabla^2 \mathcal{V}_H^P = 0 \tag{9.533}$$

have to be employed to satisfy the conditions at the boundary. This procedure was extensively utilized in many of the problems discussed in this Chapter and in Chapter 5 of Volume I.

Example 9.16

A homogenous circular disc of radius a is located on the $y - z$ plane with the centre of the disc at the origin (Figure 9.31).

If the area mass density of the disc is ρ^*, determine the attraction due to the disc on a particle located at a point P along the x-axis.

Solution

Since the problem is symmetric about the x-axis, the force of attraction will have only a single component, F_x.

The attraction between particle P and elementary mass $\rho^* \eta \, d\eta \, d\vartheta$ is given by

$$d\widehat{F} = \frac{\rho^* \eta \, d\eta \, d\vartheta}{[\eta^2 + x^2]} \tag{9.534}$$

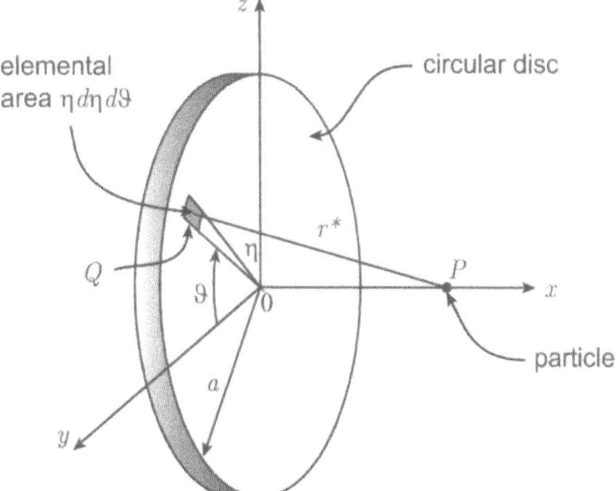

Figure 9.31: Gravitational attraction between disc and particle.

This force is directed along PQ, and has a component

$$dF_x = \frac{\rho^*\eta\,d\eta\,d\vartheta}{[\eta^2 + x^2]}\left(-\frac{x}{[\eta^2 + x^2]^{1/2}}\right) \tag{9.535}$$

The total force of attraction is given by

$$F_x = -\int_0^{2\pi}\int_0^a \frac{\rho^*x\eta\,d\eta\,d\vartheta}{[\eta^2 + x^2]^{3/2}} \tag{9.536}$$

Evaluating (9.536) we obtain

$$F_x = -2\pi\rho^*x\left[\frac{1}{|x|} - \frac{1}{\sqrt{a^2 + x^2}}\right] \tag{9.537}$$

Example 9.17

A line charge of intensity q per unit length is placed at the location $(\ell, 0)$ of a conducting medium. Determine the distribution of potential in the medium if a cylindrical cavity of radius a $(a < \ell)$ is kept at zero potential.

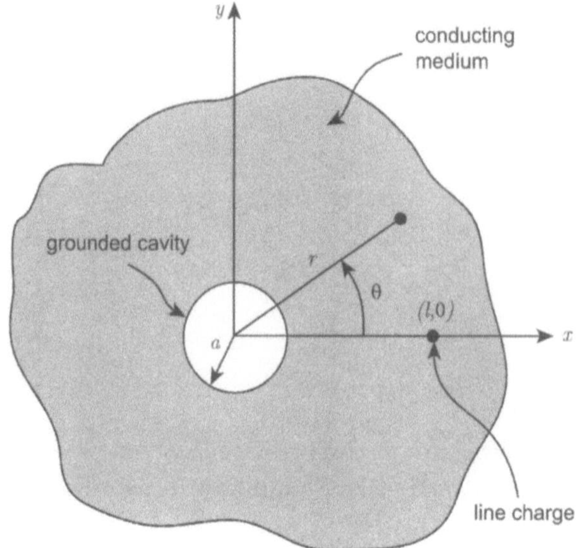

Figure 9.32: Potential problem for a conducting medium.

Solution

We first consider the problem of a complete conducting medium of infinite extent (i.e. without a cavity) which contains the concentrated line charge of unit intensity per unit length located at $\tilde{x}_o$. Since the medium is of infinite extent, there are no boundaries and we can formulate the problem through the introduction of the Green's function $\mathcal{V}_G(\mathbf{x}, \mathbf{x}_o)$ which satisfies

$$\nabla^2 \mathcal{V}_G(\mathbf{x}, \mathbf{x}_o) = -4\pi\delta(\mathbf{x} - \mathbf{x}_o) \tag{9.538}$$

Since the problem represents a steady state process with a concentrated line source located at $\mathbf{x} = \mathbf{x}_o$, the solution is symmetric about the axis of the line source. If we denote the radius vector $\mathbf{r} = (\mathbf{x} - \mathbf{x}_o)$ with $|\mathbf{r}| = [(x - x_o)^2 + (y - y_o)^2]^{1/2}$ and note that source function is zero everywhere except at $\mathbf{x} = \mathbf{x}_o$, we have

$$\nabla^2 \mathcal{V}_G = \frac{1}{r}\frac{d}{dr}\left(r\frac{d\mathcal{V}_G}{dr}\right) = 0 \quad ; \quad r \neq 0 \tag{9.539}$$

The general solution of (9.539) is

$$\mathcal{V}_G(r) = C_1 \ln r + C_2 \tag{9.540}$$

where C_1 and C_2 are arbitrary constants. These constants, which account for the singularity, can be determined by integrating around a circle containing the line charge of unit intensity, i.e.,

$$\int\int_{\mathcal{A}} \nabla^2 \mathcal{V}_G \, dA = -4\pi \tag{9.541}$$

Using the divergence theorem, (9.541) can be written as

$$\int\int_{\mathcal{A}} \nabla \cdot \nabla \mathcal{V}_G \, dA = \int_{\mathcal{C}} \nabla \mathcal{V}_G \cdot \mathbf{n} \, dC = -4\pi \tag{9.542}$$

where $\mathcal{C}$ corresponds to the boundary of $\mathcal{A}$. Since $\mathbf{n}$ to the contour $\mathcal{C}$ is radially directed, and since

$$\nabla \mathcal{V} = \frac{d\mathcal{V}_G}{dr}\mathbf{i}_r + 0\mathbf{i}_\theta + 0\mathbf{i}_z \tag{9.543}$$

is constant along any radius $r = const.$, (9.542) gives

$$2\pi r \frac{d\mathcal{V}_G}{dr} = -4\pi \tag{9.544}$$

Using the solution (9.540) in (9.544) we obtain

$$C_1 = -2 \tag{9.545}$$

The second arbitrary constant C_2 can be assigned any arbitrary value. We can set this to zero without loss of generality. The Green's function for distribution of potential in an infinite conducting medium containing a line source at $\mathbf{x}_o$ can be written as

$$\mathcal{V}_G(r, \theta; r_o, \theta_o) = -2\ln(\mathbf{r} - \mathbf{r}_o) \tag{9.546}$$

Considering a series expansion for the logarithmic term, and a line source of intensity q (per unit length along the axis) which acts at (r_o, θ_o), we can express the Green's function (9.546) as

$$\mathcal{V}_G(r,\theta;r_o,\theta_o) = -2q\ln(r) \tag{9.547}$$
$$+2q\sum_{n=1,}^{\infty}\frac{1}{n}\left(\frac{r_o}{r}\right)^n\cos[n(\theta-\theta_o)] \quad ; \quad r>r_o$$

$$\mathcal{V}_G(r,\theta;r_o,\theta_o) = -2q\ln(r_o) \tag{9.548}$$
$$+2q\sum_{n=1,}^{\infty}\frac{1}{n}\left(\frac{r}{r_o}\right)^n\cos[n(\theta-\theta_o)] \quad ; \quad r<r_o$$

We now consider the special case where the line source of intensity q per unit length along its axis, is located at $(\ell,0)$ and since the radius of the cavity $a < \ell$, the potential due to the the line source if the grounded cylinder was absent would be

$$\mathcal{V}_G(r,\theta;\ell,\theta_o) = -2q\ln(r) \tag{9.549}$$
$$+2q\sum_{n=1}^{\infty}\frac{1}{n}\left(\frac{\ell}{r}\right)^n\cos(n\theta) \quad ; \quad r>\ell$$

$$\mathcal{V}_G(r,\theta;\ell,\theta_o) = -2q_o\ln(\ell) \tag{9.550}$$
$$+2q\sum_{n=1}^{\infty}\frac{1}{n}\left(\frac{r}{\ell}\right)^n\cos(n\theta) \quad ; \quad r<\ell$$

To achieve zero potential boundary conditions on $r = a$, we require solutions of Laplace's equation which (by introducing suitable constant terms) can be written as:

$$\mathcal{V}_H(r,\theta) = 2q\ln(\frac{\ell r}{a}) - 2q\sum_{n=1}^{\infty}\frac{1}{n}\left(\frac{a^2}{\ell r}\right)^n\cos(n\theta) \tag{9.551}$$

Combining (9.550) and (9.551) we obtain the complete solution

$$\tilde{\mathcal{V}}_T(r,\theta;\ell,0) = 2q\ln(\frac{\ell}{a}) \tag{9.552}$$
$$-2q\sum_{n=1,2,}^{\infty}\frac{1}{n}\left[\left(\frac{a^2}{\ell r}\right)^n - \left(\frac{\ell}{r}\right)^n\right]\cos(n\theta) \quad ; \quad r>\ell$$

and

$$\tilde{\mathcal{V}}_T(r, \theta; \ell, 0) = 2q \ln\left(\frac{r}{a}\right) \tag{9.553}$$

$$-2q \sum_{n=1,2,} \frac{1}{n}\left[\left(\frac{a^2}{\ell r}\right)^n - \left(\frac{r}{\ell}\right)^n\right]\cos(n\theta) \quad ; \quad a \leq r < \ell$$

The zero potential boundary condition at $r = a$, will be indentically satisfied by (9.553).

Example 9.18

A point charge of strength Q_o is located at a finite distance $z = h$ from the grounded surface of a conducting half-space region occupying $r \in (0, \infty)$, $\theta \in (0, 2\pi)$, and $z \in (0, \infty)$. Considering the axisymmetric nature of the problem determine the distribution of charge within the half-space region.

Solution

We first consider the problem where the point charge is located within an infinite space region occupying $r \in (0, \infty)$, $\theta \in (0, 2\pi)$, and $z \in (-\infty, \infty)$. The partial differential equation governing the distribution of charge in the infinite space region is given by

$$\nabla^2 \mathcal{V}^P = -Q_0 \delta(\mathbf{x} - \mathbf{x}^*) \tag{9.554}$$

where $\mathbf{x}^*$ is the source location. The solution for $\mathcal{V}^P(\mathbf{x})$ now represents the Green's function corresponding to a point charge of strength Q_0. Since the problem is axisymmetric and the source is located at $z = h$, the solution of (9.554) is given by (see e.g. (9.212))

$$\mathcal{V}^P(r, z) = \frac{Q_0}{4\pi \left[r^2 + (z - h)^2\right]^{1/2}} \tag{9.555}$$

which possesses the singularity at $r = 0$, $z = h$. The potential at $z = 0$ due to the point charge at $z = h$ is given by

$$\mathcal{V}^P(r, 0) = \frac{Q_0}{4\pi \left[r^2 + h^2\right]^{1/2}} \quad ; \quad r \in (0, \infty) \tag{9.556}$$

The boundary condition that needs to be satisfed by the complete solution $\mathcal{V}(r, z)$ at the grounded plane is

$$\mathcal{V}(r, 0) = 0 \quad ; \quad r \in (0, \infty) \tag{9.557}$$

To satisfy the boundary condition (9.557) we seek solutions of Laplace's equation by adapting a Hankel transform approach. The general solution of Laplace's equation based on a Hankel transform approach can be expressed in the form

$$\mathcal{V}^H(r, z) = \int_0^\infty \left[A(\xi)e^{-\xi z} + B(\xi)e^{\xi z} \right] J_0(\xi r)d\xi \tag{9.558}$$

where $A(\xi)$ and $B(\xi)$ are arbitrary functions of the transformation parameter ξ and $J_0(\xi r)$ is the zeroth-order Bessel function of the first-kind.

Since the charge distribution in the half-space must decay to zero as $z \to \infty$, we require

$$B(\xi) = 0 \tag{9.559}$$

The remaining solution of (9.558) can be used to satisfy the zero potential boundary condition at $z = 0$, i.e.

$$\int_0^\infty A(\xi) J_0(\xi r)d\xi + \frac{Q_0}{4\pi \left[r^2 + h^2\right]^{1/2}} = 0 \tag{9.560}$$

Representing the second term of (9.560) as a Hankel integral we can show that

$$A(\xi) = - \frac{Q_0}{4\pi} e^{-\xi h} \tag{9.561}$$

Hence the corrective solution of Laplace's equation $\mathcal{V}^H(r, z)$ that should be added to the solution for the infinite space problem, takes the form

$$\mathcal{V}^H\left(r,z\right) = -\frac{Q_0}{4\pi}\int_0^\infty e^{-(\xi-h)z}J_0(\xi r)d\xi \tag{9.562}$$

$$= -\frac{Q_0}{4\pi\left[r^2+(z+h)^2\right]^{1/2}}$$

The final solution for the distribution of charge in the half-space region which has a grounded plane surface $z = 0$ is obtained by combined (9.555) and (9.562): i.e.

$$\mathcal{V}\left(r,z\right) = \frac{Q_0}{4\pi}\left\{\frac{1}{\left[r^2+(z-h)^2\right]^{1/2}} - \frac{1}{\left[r^2+(z+h)^2\right]^{1/2}}\right\} \tag{9.563}$$

It is evident that the solution (9.563) corresponds to a combination of a point charge Q_0 located at $(0,h)$ and its negative *mirror image* located at $(0,-h)$. (See e.g. Example 9.6.)

Example 9.19

The point charges of equal strength Q_0 are located in a conducting medium of infinite extent and spaced at a distance $2h$. A thin circular conducting disc of radius a is placed midway between the charges with its axis collinear with the line connecting the point charges. Obtain an expression for the density of charge in the disc if it is maintained at zero potential.

Solution

Consider the problem of the conducting medium which contains *only* the two point charges of strength Q_0. The distribution of potential due to these charges can be obtained by superposing solutions obtained for the Green's function for a conducting medium of infinite extent. The spatial variation of potential due to the two charges placed at $z = \pm h$ is given by

$$\mathcal{V}^P(r,z) = \frac{Q_0}{4\pi}\left[\frac{1}{\left[r^2+(z-h)^2\right]^{1/2}} + \frac{1}{\left[r^2+(z+h)^2\right]^{1/2}}\right] \tag{9.564}$$

The potential on the plane of symmetry is given by

$$\mathcal{V}^P(r, 0) = \frac{Q_0}{2\pi \left[r^2 + h^2\right]^{1/2}} = f(r) \qquad (9.565)$$

We now assume that the potential $\mathcal{V}(r, z)$ can be represented in the form

$$\mathcal{V}(r, z) = \mathcal{V}^P(r, z) + \mathcal{V}^H(r, z) \qquad (9.566)$$

where $\mathcal{V}^H(r, z)$ is a solution of Laplace's equation.

The boundary conditions applicable to the problem are such that the final solution for the potential $\mathcal{V}(r, z)$ should satisfy the *mixed boundary* conditions

$$\mathcal{V}(r, 0) = 0 \qquad ; \qquad 0 \leq r \leq a \qquad (9.567)$$

$$\left[\frac{\partial \mathcal{V}}{\partial z}\right]_{z=0} = 0 \qquad ; \qquad a < r < \infty \qquad (9.568)$$

The boundary condition (9.568) implies that the potential field is symmetric about the plane of a disc. We can therefore focus attention on a single half-space region occupying $r \in (0, \infty)$ and $z \in (0, \infty)$. Since the influence of the concentrated charges are accounted for by the solutions based on Poisson's equation, the mixed boundary conditions (9.567) and (9.568) can be satisfied by selecting solutions of Laplace's equation based on Hankel transforms. The relevant solution of $\mathcal{V}^H(r, z)$, which also satisfies the regularity condition $\mathcal{V}^H(r, z) \to 0$ as $z \to \infty$, can be written as

$$\mathcal{V}^H(r, z) = \int_0^\infty A(\xi) e^{-\xi z} J_0(\xi r) d\xi \qquad (9.569)$$

where $A(\xi)$ is an unknown function. To satisfy the mixed boundary conditions, we require

$$\int_0^\infty A(\xi) J_0(\xi r) d\xi = -f(r) \qquad ; \qquad 0 \leq r \leq a \qquad (9.570)$$

$$\int_0^\infty \xi A(\xi) J_0(\xi r) d\xi = 0 \qquad ; \qquad a < r < \infty \qquad (9.571)$$

The equations (9.570) and (9.571) form a system of *dual integral equations* which have been studied extensively in literature dealing with potential theory, hydrodynamics, and contact and crack problems in the theory of elasticity. The procedure for the reduction of the system of dual integral equations encountered in Boussinesq's classical contact problem was presented in Section 8.10.13. The method of solution involves the introduction of a finite Fourier cosine transform such that integral equation for the exterior region $r \in (a, \infty)$ is identically satisfied. The remaining integral equation can be reduced to an integral equation of the Abel type. By introducing a function $\psi^* (t)$ such that

$$A (\xi) = \int_0^a \psi^* (t) \cos (\xi t) \, dt \tag{9.572}$$

we can identically satisfy the equation (9.571) and the equation (9.570) can be reduced to an Abel integral equation, the solution of which gives

$$\psi^* (t) = \frac{2}{\pi} \frac{d}{dt} \int_0^t \frac{r f (r)}{(t^2 - r^2)^{1/2}} dr \tag{9.573}$$

Considering (9.570) and (9.573) we obtain

$$\psi^* (t) = \frac{2 Q_0 h}{\pi (h^2 + t^2)} \tag{9.574}$$

The result of interest, namely the distribution of charge within the circular disc can be obtained by evaluating (9.571) within the region $0 < r < a$. The details of the procedures can be found in the texts on elasticity and potential theory cited in the bibliography. The charge distribution on the disc region is given by

$$\left[\frac{\partial \mathcal{V}}{\partial z} \right]_{z=0} = \left[\frac{\partial \mathcal{V}^P}{\partial z} \right]_{z=0} + \left[\frac{\partial \mathcal{V}^H}{\partial z} \right]_{z=0} \quad ; \quad r \in (0, a) \tag{9.575}$$

From (9.564) we note that

$$\left[\frac{\partial \mathcal{V}^P}{\partial z} \right]_{z=0} \equiv 0 \tag{9.576}$$

Hence

$$\left[\frac{\partial \mathcal{V}}{\partial z}\right]_{z=0} = \left[\frac{\partial \mathcal{V}^H}{\partial z}\right]_{z=0} = \frac{1}{4\pi r}\frac{d}{dr}\int_r^a \frac{s\psi^*(s)}{(s^2 - r^2)^{1/2}}\,ds \tag{9.577}$$

Evaluating (9.577) we obtain

$$\left[\frac{\partial \mathcal{V}^H}{\partial z}\right]_{z=0} = \frac{-Q_0 h}{2\pi^2 \left(h^2 + r^2\right)^{3/2}}\left[\tan^{-1}\sqrt{\frac{a^2 - r^2}{h^2 + r^2}} + \sqrt{\frac{h^2 + r^2}{a^2 - r^2}}\right] \tag{9.578}$$

The final expressions for the potential in the half-space region $z \in (0, \infty)$ consists of $\mathcal{V}^P(r, z)$ due to the combination of point charges defined by (9.564) and the solution for $\mathcal{V}^H(r, z)$ defined by (9.569). Although the results for the charge distribution on the disc can be evaluated in closed form, general expressions for both the potential and charge distribution within the entire region can only be obtained in integral form.

9.12   PROBLEM SET 9

9.1   Develop the Green's function $G(\mathbf{x}; \mathbf{x}^*)$ for the Poisson-type equation for an infinite strip of finite width, defined by

$$\nabla^2 G = -\delta(x - x^*) \quad ; \quad x \in (-\infty, \infty); y \in (0, a)$$

subject to the Dirichlet boundary conditions

$$G(x, 0; x^*, 0) = 0 \quad ; \quad x \in (-\infty, \infty)$$

$$G(x, a; x^*, 0) = 0 \quad ; \quad x \in (-\infty, \infty)$$

and the regularity condition

$$\operatorname*{Lim}_{|x| \to \infty} G(\mathbf{x}; \mathbf{x}^*) = 0 \quad ; \quad y \in (0, a)$$

9.2   Consider the problem of the in-plane seepage in a porous layer of thickness t and plan dimensions such that $x \in (0, a)$ and $y \in (0, b)$. Also $t \ll a, b$. The region is subjected to a fluid influx rate $q\ m^3/sec$ per square metre of the plan area (see e.g. Figure 9.10). Develop a solution to the distribution of pressure head $H(x, y)$ in the porous layer which is governed by the partial differential equation of the Poisson type

$$\nabla^2 H = -\frac{q}{Kt}$$

by assuming that $H(x, y)$ admits a solution of the form

$$H(x, y) = -\frac{qx^2}{2Kt} + H^*(x, y)$$

where H^* satisfies Laplace's equation.

9.3   A concrete column has a square cross-section. During curing of the fresh concrete, the exothermic reactions in the cement paste produces heat at a steady rate of g (Watts/m^3) (Figure 9.33) Two adjacent sides of the curing concrete have imperfect insulation which allows the temperatures at these surfaces to correspond to the ambient value T_0. Develop an expression for the steady state distribution of temperatures within the section of the column.

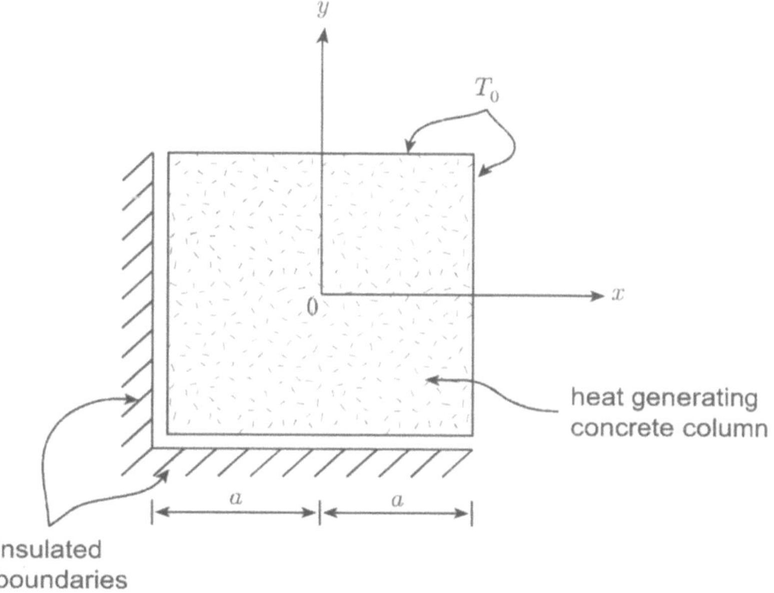

Figure 9.33: Partially insulated column.

9.4   The Figure 9.34 illustrates the end region of a borosilicate plate containing distributed quantities of irradiated nuclear fuel waste. The rate of heat generation is g (Watts/m^3). The outer boundaries of the strip are kept at a constant temperature by immersion in a water pool with water circulation. The steady state distribution of temperature $T(x, y)$ in the strip satisfies the partial differential equation

$$\nabla^2 T = -\frac{g}{k}$$

where k is the thermal conductivity of the dosed borosilicate. Assuming that the plate can be idealized as a semi-infinite strip occupying the region $x \in (0, \infty)$ and $y \in (0, a)$, determine the steady state distribution of temperature in the end zone.

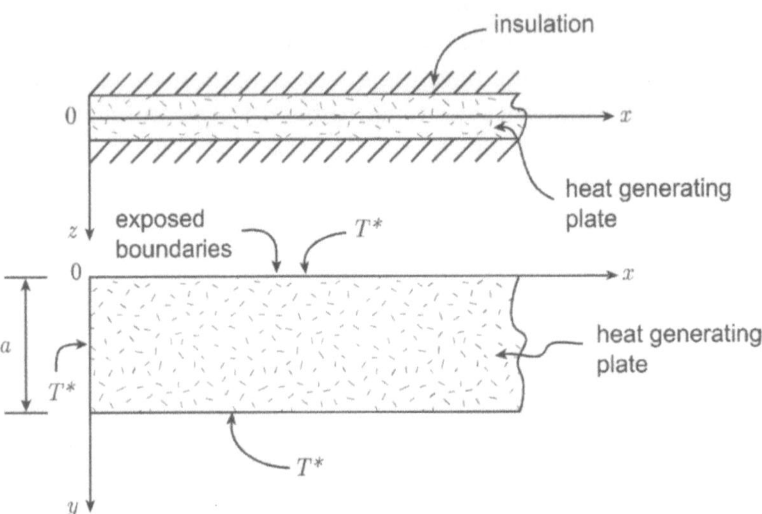

Figure 9.34: Internal heating of an infinite strip.

9.5 Consider the idealized problem of in-plane fluid flow in a thin porous seam of thickness t which is small in comparison to the plan dimensions of the porous seam. The porous seam is subjected to a steady recharge q m^3/sec per square metre of the plan area over a limited region (Figure 9.35). A line of "well points" are installed along two boundaries of the porous seam and pumping is carried out to reduce the pressure head along these boundaries to the datum level. Derive expressions for the total pumping rate at the respective boundaries necessary to maintain this condition.

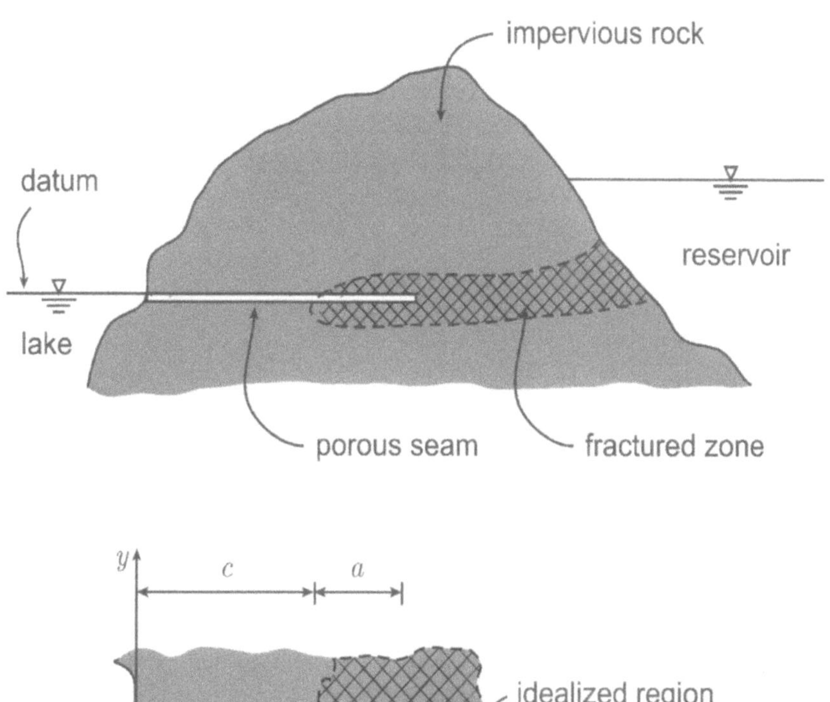

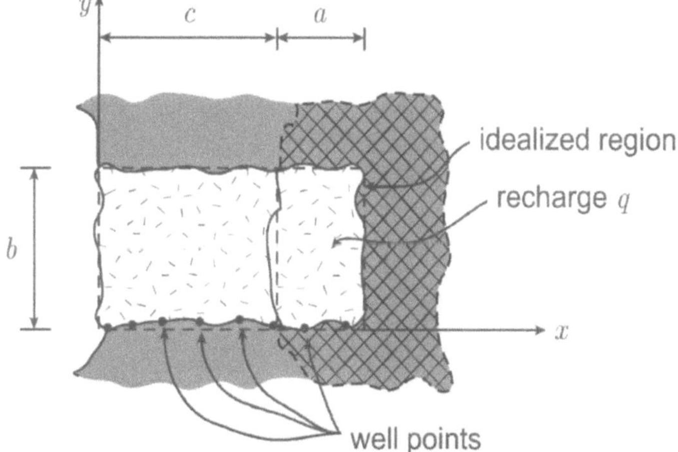

Figure 9.35: Non-uniform recharge of a porous aquifer.

9.6   A membrane which is stretched with uniform tension T is fixed to a rectangular rigid frame (Figure 9.36). If a part of this membrane is loaded by a uniform transverse pressure derive an expression for the transverse displacement along the axis of symmetry.

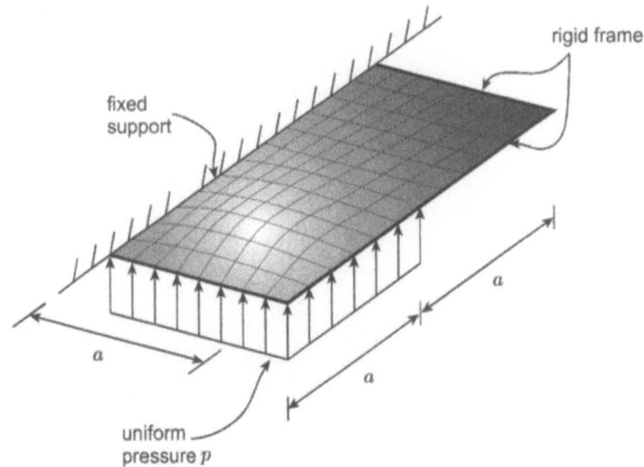

Figure 9.36: Partially loaded stretched membrane.

9.7   A stretched membrane is attached to the end of a hollow cylindrical tube of radius a. The tube is lowered into a water container at an inclination ω as shown in Figure 9.37. Obtain an expression for the deflected shape of the membrane induced by the action of hydrostatic pressure.

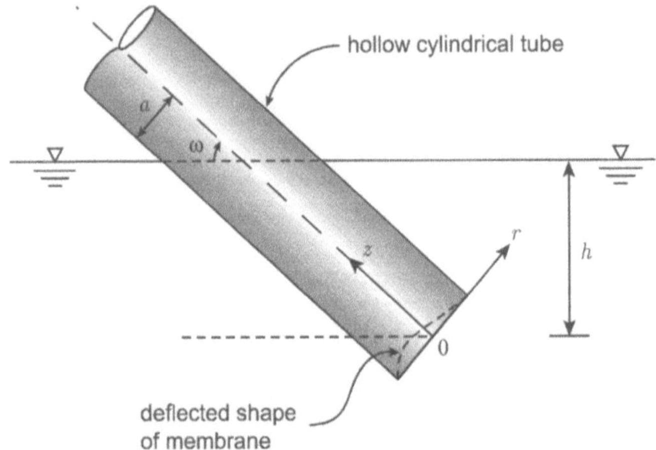

Figure 9.37: Deflection of stetched membrane due to fluid pressure.

9.8   Figure 9.38 shows the general view and plan of a membrane stretched to a unform tension T and subjected to a spatially constant pressure p_0. Develop a series solution of the deflected shape of the membrane and calculate the deflection at the location $(0,0)$.

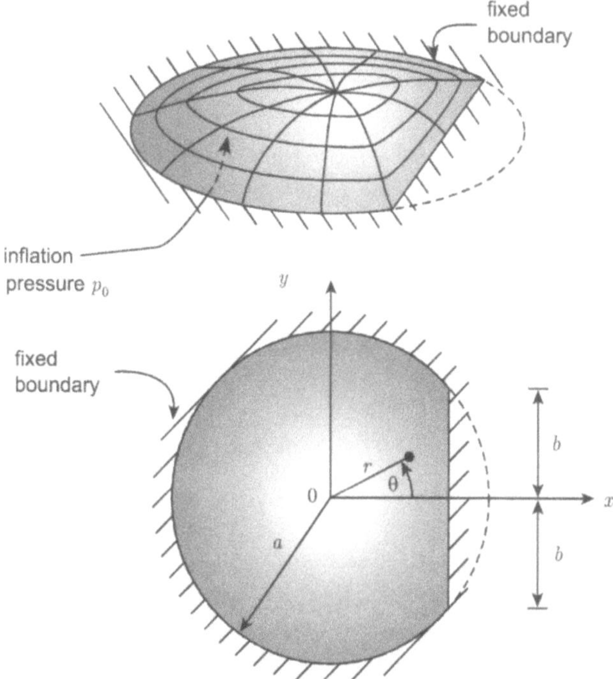

Figure 9.38: Stretched membrane under uniform internal pressure p_o.

9.9 The partial differential equation governing flow of a viscous fluid through a narrow aperture fracture in a geological medium can be derived via the model proposed by H. Hele-Shaw (1854-1941) which is described in Section 8.12. In plane polar coordinates, the partial differential equation takes the form

$$\nabla^2 p(r,\theta) = -\frac{12\eta q(r,\theta)}{(2h)^3}$$

where $p(r,\theta)$ is the fluid pressure distribution in the fracture, $q(r,\theta)$ is the fluid supply to the fracture (in m^3/sec per m^2 of area of fracture), η is the fluid viscosity and $2h$ is the aperture dimension. Figure 9.39 shows an experimental configuration involving a large diameter granite cylinder with a fracture plane. The central portion of the cylinder contains a series of drilled holes for fluid supply. Develop an expression for the distribution of fluid pressure in the fracture if the fluid recharge to the fracture takes the form

$$q(r,\theta) = q_0 \left\{ 1 + \xi\left(\frac{r}{b}\right) \cos\theta \right\} \quad ; \quad 0 < r < b$$

$$q(r,\theta) = 0 \quad ; \quad b < r < a$$

and the fluid pressure at the boundary $r = a$ is maintained at a zero value.

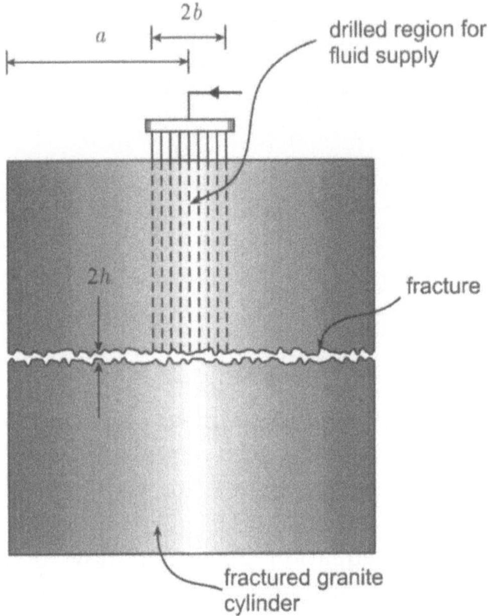

Figure 9.39: Fluid flow in a fracture subjected to recharge.

9.10 Use a double Fourier transform technique to show that the solution to a partial differential equation of the Poisson-type for $\varphi(x,y)$ in a half-space region given by

$$\nabla^2 \varphi(x,y) = -f(x,y) \quad ; \quad -\infty < x < \infty \quad ; \quad 0 \le y < \infty$$

with boundary condition

$$\varphi(x,0) = 0 \quad ; \quad -\infty < x < \infty$$

can be obtained in the form

$$\varphi(x,y) = \frac{1}{8\pi} \int_{-\infty}^{\infty} \int_{-\infty}^{\infty} \log\left\{ \frac{(x-\xi)^2 + (y-\eta)^2}{(x-\xi)^2 + (y+\eta)^2} \right\} f(\xi,\eta)\,d\xi\,d\eta$$

9.11 Use the *"method of images"* to obtain a solution of Poisson's equation for $\varphi(x,y)$, for a quarter-plane region, given by

$$\nabla^2 \varphi(x,y) = -\delta(\mathbf{x} - \mathbf{x}^*) \quad ; \quad 0 \le x < \infty \quad ; \quad 0 \le y < \infty$$

subject to Dirichlet boundary conditions

$$\varphi(x,0) = 0 \quad ; \quad 0 \le x < \infty$$

$$\varphi(0,y) = 0 \quad ; \quad 0 \le y < \infty$$

Show that the method can be extended to obtain solutions to Poisson's equation for the quarter-plane, the boundaries of which are subjected to homogeneous forms of Neumann boundary conditions.

9.12 The Figure 9.40 illustrates the edge boundary configuration at a narrow aperture containing a Newtonian viscous fluid. Axial flow is initiated in the fluid by a constant pressure gradient. The partial differential equation governing the axial velocity field $v_z(x,y)$ is given by a Poisson-type elliptic equation

$$\eta \left(\frac{\partial^2 v_z}{\partial x^2} + \frac{\partial^2 v_z}{\partial y^2} \right) = \frac{\partial p}{\partial z} = -C$$

where η is the fluid viscosity and C is a constant. Derive an expression for the axial velocity field if *non-slip* boundary conditions are applicable at the boundary of the fluid.

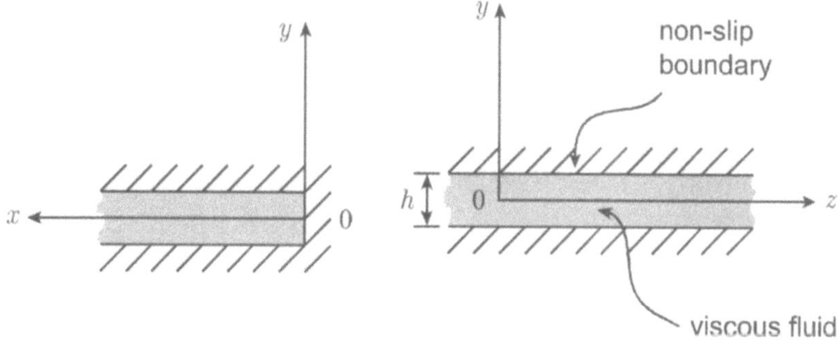

Figure 9.40: Axial flow of a viscous fluid in a semi-infinite aperature.

9.13 Axial flow of a viscous fluid takes place in a pipe with an elliptical cross-section (with major axis $2a$ and minor axis $2b$) under a pressure gradient Λ. Derive an expression for the volumetric flow rate in the pipe if no-slip conditions exist at the boundary of the pipe.

9.14 Use the potential function

$$\psi(r) = C_1 \ell \mathrm{n}\, r + C_2$$

and the Prandtl stress function

$$\phi(r) = \mu \Omega_0 \left[\psi - \frac{r^2}{2} \right]$$

where C_1 and C_2 are arbitrary constants, to develop a solution to the problem of a hollow prismatic cylinder with an outer radius a and an inner radius b.

9.15 Consider a prismatic elastic bar with the cross-section in the form of an equilateral triangle of side length $2a/\sqrt{3}$. The coordinate axes are chosen as indicated in Figure 9.41.

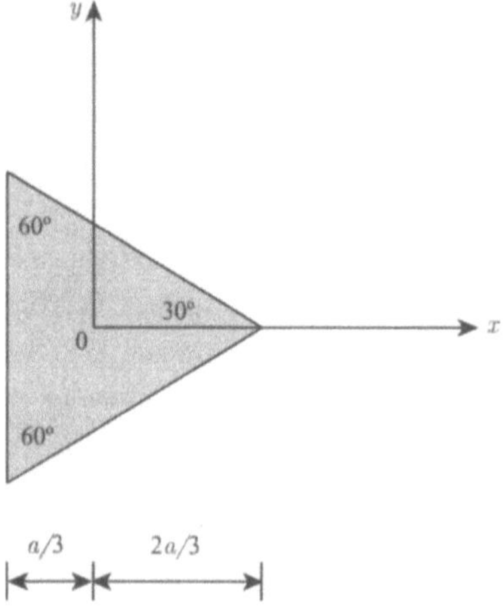

Figure 9.41: Torsion of a prismatic bar with a triangular cross-section.

Consider the harmonic function

$$f_0(x,y) = C_1(x^3 - 3xy^2)$$

where C_1 is a constant to develop a Prandtl stress function of the form

$$\chi(x, y) = \mu\Omega_0 \left[f_0(x, y) - \tfrac{1}{2}(x^2 + y^2) + C_2 \right]$$

where C_2 is a constant, applicable to the bar with the triangular cross-section. Obtain the magnitude and location of the maximum shear stress.

9.16 The Figure 9.42 shows that the cross-section of a prismatic bar in the form of a sector of a circle with subtended angle $2\beta(< \pi)$. Considering a Prandtl stress function of the form

$$\chi(r, \theta) = \mu\Omega_0 \left[\psi(r, \theta) - \frac{r^2}{2} \right]$$

where

$$\psi(r, \theta) = \sum A_m r^m \cos(m\theta)$$

determine a solution for $\chi(r, \theta)$ in series form. Compare the result for the relationship between T_0 and Ω_0 obtained for this problem for the special case of $\beta = 30°$ and $R = 2a/\sqrt{3}$, with the equivalent result for the prismatic elastic bar with a triangular cross-section as shown in Figure 9.41.

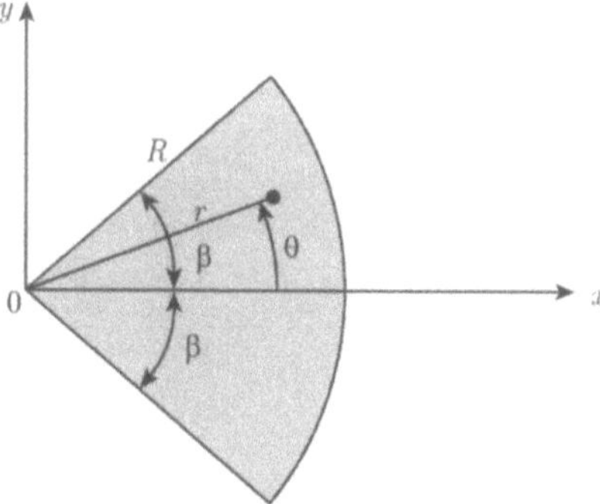

Figure 9.42: Torsion of a prismatic bar with the cross-section in the form of a sector.

9.17 Develop a formal integral expression for the torsional stiffness of a prismatic elastic bar of diameter $2a$ which contains a groove of circular

cross-section of radius b which is placed at the perimeter of the bar
(Figure 9.43). (This problem is attributed to C.H. Weber (1885-1976)).

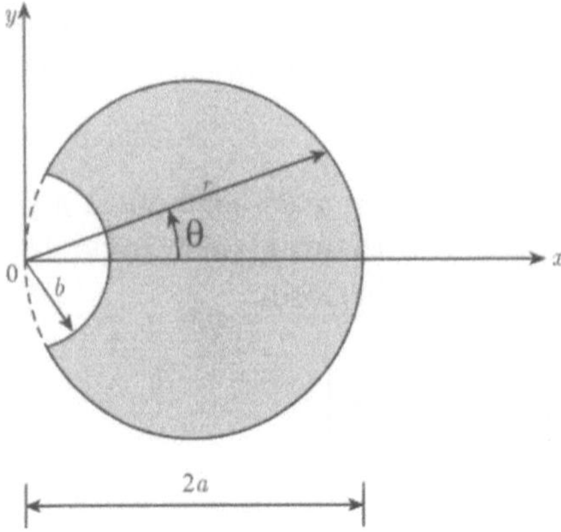

Figure 9.43: Cross-section of a prismatic cylinder with a circular
groove.

9.18 A narrow longitudinal cut is introduced in a hollow cylindrical structural member with external diameter $2a$, internal diameter $2b$ and length
$\ell(a/\ell \ll 1)$. (Figure 9.44) Determine its torsional stiffness and compare
this value with the equivalent result for a complete hollow cylinder with
identical dimensions.

9.19 Consider the problem of torsion of a thin elastic plate with cross-
sectional dimensions $2a \times 2b$ examined in Example 9.11. Maintain the
assumption that $(a/b) \ll 1$ and develop an approximate *corrective solution* for $\chi^C(x, y)$ such that it will satisfy Laplace's equation with
boundary conditions;

$$\chi^C(\pm a, y) = 0$$

and

$$\chi^C(x, \pm b) = -2\mu\Omega_0 \left[1 - \left(\frac{x}{a} \right)^2 \right]$$

yet exhibiting an exponential variation with y. Compare the result for
the torque (T_0) vs. twist (Ω_0) relationship developed through this approximate procedure with the equivalent complete series solution developed in Example 9.10.

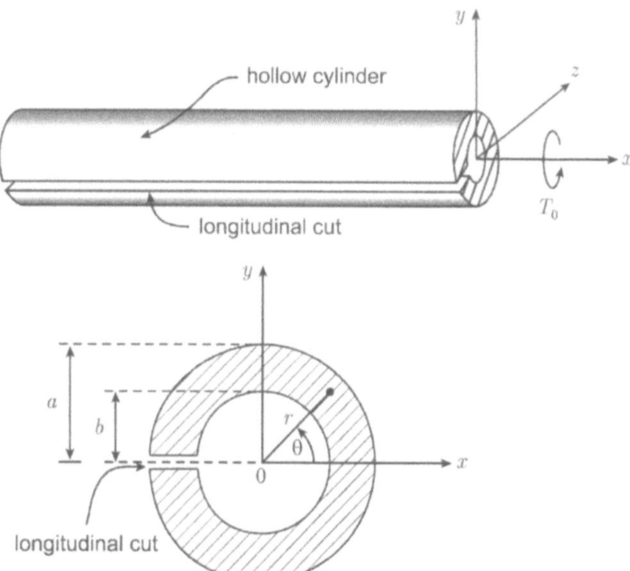

Figure 9.44: Torsion of a hollow cylinder with a narrow longitudinal cut.

9.20 Use a complex variable formulation to develop a solution to the torsion problem described in Problem 9.17. Use a numerical integration technique to evaluate the torsional stiffness of a circular bar with a groove dimension $b = a$. Compare this value with the result developed in Example 9.12.

9.21 Derive from first principles the complex potentials $\omega(z_c)$ which can be used to examine the torsional response of prismatic elastic bars with cross-section having the geometries of (i) an ellipse and (ii) an equilateral triangle.

9.22 The cross-section of a prismatic elastic bar has the shape of a *Cardioid* (Figure 9.45) the equation of which is given by

$$r = a(1 + \cos\theta)$$

(This is the curve described by a point of a circle of radius a as it rolls on the outside of a fixed circle of radius a). Use a conformal transformation technique to derive an expression for the torsional rigidity of the section given that the mapping function is

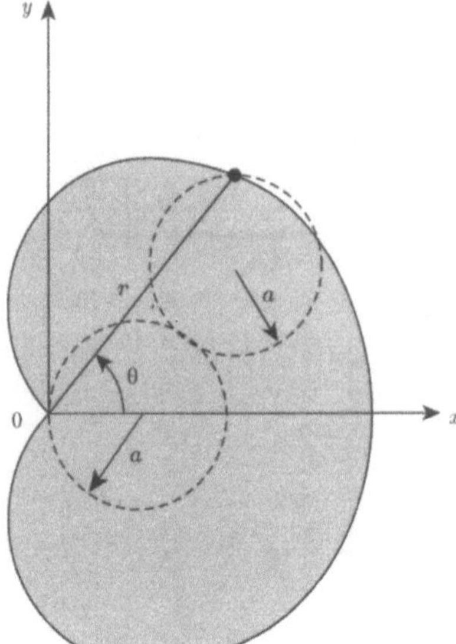

Figure 9.45: Prismatic bar with the cross-sectional shape of a
Cardiod.

$$W(\zeta) = \frac{a}{2}(1 + \zeta)^2$$

Compare the value for the torsional stiffness with the result obtained
in Example 9.12 for the prismatic bar with a semi-circular cross-section
but with an area equal to the area of the *Cardioid.*

9.23 Figure 9.46 shows the cross-section of a prismatic elastic bar which has
a shape bounded by two parabolic curves. The width ξ varies with y
according to

$$\xi = 2a\left(1 - \frac{y^2}{b^2}\right)$$

Develop an approximate estimate for the torsional stiffness of the bar
using a procedure similar to that outlined in Example 9.11. Compare
the result with the exact solution for a prismatic elastic member with
an elliptic cross-section with identical dimensions along the x and y
axes.

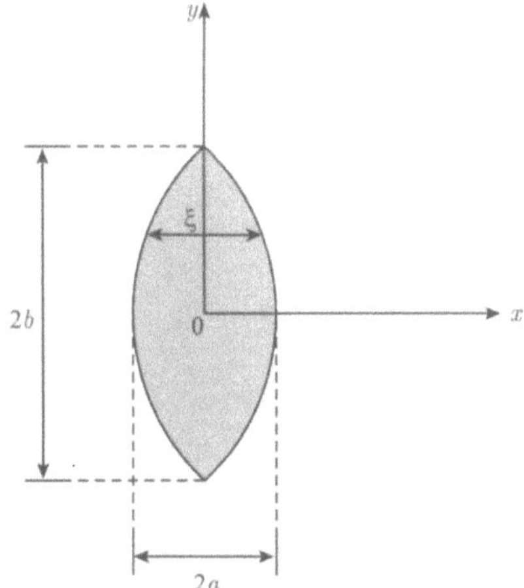

Figure 9.46: Torsion of prismatic bar with parabolic boundaries.

9.24 The electrostatic potential $\mathcal{V}$ is governed by Poisson's equation $\nabla^2\mathcal{V}(\mathbf{x}) = -4\pi\omega$ where ω is the charge density. A point charge of intensity C_0 is placed at the location $(0,0,\ell)$, referred to the rectangular Cartesian coordinate system, which is outside an *earthed* spherical boundary of radius a. Derive an expression for the electrostatic potential $\mathcal{V}(\mathbf{x})$ at locations outside the spherical region. (Hint: Use a spherical polar coordinate formulation of the problem and use solutions of Laplace's equation as a series in Legendre Polynomials).

Bibliography

Advanced Topics in Engineering Mathematics

Bender, C.M. and Orzag, S.A. (1978) *Advanced Mathematical Methods for Scientists and Engineers*, McGraw-Hill, New York.

Boas, M.L. (1966) *Mathematical Methods in the Physical Sciences*, 2nd edition, John Wiley, New York.

Courant, R. and Hilbert, D. (1962) *Methods of Mathematical Physics*,Vols.1 and 2, Interscience, New York.

Dennery, P. and Krzywicki, A. (1967) *Mathematics for Physicists*, Harper and Row, New York.

Dettman, J.W. (1988) *Mathematical Methods in Physics and Engineering*, Dover Publications, New York.

Greenberg, M.D. (1978) *Foundations of Applied Mathematics*, Prentice-Hall, Englewood Cliffs, N.J.

Greenberg, M.D. (1998) *Advanced Engineering Mathematics*, Prentice-Hall, Englewood Cliffs, N.J.

Haug, E. and Choi, K.K. (1993) *Methods of Engineering Mathematics*, Prentice-Hall, Englewood Cliffs, N.J.

Keener, J.P. (1988) *Principles of Applied Mathematics*, Addison Wesley, New York.

Kraut, E.A. (1967) *Fundamentals of Mathematical Physics*, McGraw-Hill, New York.

Kreyszig, E. (1992) *Advanced Engineering Mathematics*, John Wiley, New York.

O'Neil, P.V. (1995) *Advanced Engineering Mathematics*, Brooks/Cole Publishing Company, Boston.

Pipes, L.A. and Harvill, L.R. (1970) *Applied Mathematics for Engineers and Physicists*, McGraw-Hill, New York.

Potter, M.C. (1978) *Mathematical Methods in the Physical Sciences*, Prentice-Hall, Englewood Cliffs, N.J.

Riley, K.F., Hobson, M.P. and Bence, S.J. (1998) *Mathematical Methods for Physics and Engineering*, Cambridge University Press, Cambridge.

Spencer, A.J.M., et al. (1977) *Engineering Mathematics*, Vols. 1 and 2, Van Nostrand Reinhold, New York.

Spiegel, M.R. (1971) *Advanced Mathematics for Engineers and Scientists*, McGraw-Hill, New York.

Strang, G. (1986) *Introduction to Applied Mathematics*, Wellesley-Cambridge Press, Cambridge, Mass.

von Karman, T. and Biot, M.A. (1940) *Mathematical Methods in Engineering*, McGraw-Hill, New York.

Wylie, C. (1958) *Advanced Engineering Mathematics*, McGraw-Hill, New York.

Fourier Series, Integral Transforms, Special Functions and Complex Variables

Andrews, L.C. (1985) *Special Functions for Engineers and Applied Mathematicians*, Macmillan, New York.

Boyce, W.E. and DiPrima, R.C. (1992) *Elementary Differential Equations and Boundary Value Problems*, 5th edition, John Wiley, New York.

Brown, J.W. and Churchill, R.V. (1993) *Fourier Series and Boundary Value Problems*, McGraw-Hill, New York.

Byerly, W.E. (1959) *Fourier Series*, Dover Publications, New York.

Carslaw, H.S. (1980) *An Introduction to the Theory of Fourier Series and Integrals*, Dover Publications, New York.

Churchill, R.V. (1960) *Complex Variables and Applications*, McGraw-Hill, New York.

Davis, H.F. (1963) *Fourier Series and Orthogonal Functions*, Allyn and Bacon, Boston.

Doetsch, G. (1963) *Guide to the Applications of the Laplace Transform*, Van Nostrand, New York.

Edwards, R.E. (1967) *Fourier Series: A Modern Introduction*, Holt, Rinehart and Winston, New York.

Franklin, P. (1949) *An Introduction to Fourier Methods and the Laplace Transformation*, Dover Publications, New York.

Hochstadt, H. (1961) *Special Functions of Mathematical Physics*, Holt Rinehart, New York.

Lebedev, N.N. (1972) *Special Functions and Their Applications* (Translated and Edited by R.A. Silverman), Dover Publications, New York.

Lighthill, M.J. (1958) *An Introduction to Fourier Analysis and Generalized Functions*, Cambridge University Press, Cambridge.

MacRobert, T.M. (1948) *Spherical Harmonics*, Dover Publications, New York.

McLachlan, N.W. (1955) *Bessel Functions for Engineers*, Clarendon Press, Oxford.

Miles, J.W. (1971) *Integral Transforms in Applied Mathematics*, Cambridge University Press, Cambridge.

Rainville, E.D. (1963) *The Laplace Transform: An Introduction*, McMillan, New York.

Seeley, R. (1966) *Introduction to Fourier Series and Integrals*, Benjamin, New York.

Sneddon, I.N. (1951) *Fourier Transforms*, McGraw-Hill, New York.

Sneddon, I.N. (1972) *The Use of Integral Transforms*, McGraw-Hill, New York.

Titchmarsh, E.C. (1967) *Introduction to the Theory of Fourier Integrals*, Clarendon Press, Oxford.

Tolstov, G.P. (1962) *Fourier Series*, Prentice-Hall, Englewood Cliffs, N. J.

Tranter, C.J. (1968) *Bessel Functions with Some Physical Applications*, English Universities Press, London.

Tranter, C.J. (1971) *Integral Transforms in Mathematical Physics*, Chapman and Hall, London.

Watson, E.J. (1981) *Laplace Transforms and Applications*, Van Nostrand Reinhold, New York.

Watson, G.N. (1944) *A Treatise on the Theory of Bessel Functions*, Cambridge University Press, Cambridge.

Whittaker, E.T. and Watson, G.N. (1950) *Modern Analysis*, Cambridge University Press, Cambridge.

Widder, D.V. (1941) *The Laplace Transform*, Princeton University Press, Princeton, N.J.

Ordinary Differential Equations

Agnew, R.P. (1960) *Differential Equations*, McGraw-Hill, New York.

Arnold, V.I. (1982) *Ordinary Differential Equations*, MIT Press, Boston.

Ayres, Jr. F. (1952) *Theory and Problems of Differential Equations*, Schaum Publishing Co., New York.

Bellman, R.E. and Cooke, K.L. (1971) *Modern Elementary Differential Equations*, Addison-Wesley, Reading, Mass.

Birkhoff, G. and Rota, G.-C. (1989) *Introduction to Ordinary Partial Differential Equations*, 4th edition, John Wiley, New York.

Boyce, W.E. and DiPrima, R.C. (1992) *Elementary Differential Equations and Boundary Value Problems*, 5th edition, John Wiley, New York.

Brand, L. (1966) *Differential and Difference Equations*, John Wiley, New York.

Brauer, F. and Nobel, J. (1973) *Ordinary Differential Equations*, 2nd edition, Benjamin, New York.

Braun, M. (1983) *Differential Equations and the Applications: An Introduction to Applied Mathematics*, 3rd Edition, Springer-Verlag, New York.

Burkill, J.C. (1962) *The Theory of Ordinary Differential Equations*, 2nd edition, Interscience Publishers, New York.

Carrier, G.F. and Pearson, C.E. (1968) *Ordinary Differential Equations*, Ginn (Blaisdell), Boston, Mass.

Chorlton, F. (1965) *Ordinary Differential and Difference Equations*, Van Nostrand, London.

Coddington, E.A. (1989) *An Introduction to Ordinary Differential Equations*, Dover Publications, New York.

Coddington, E.A. and Levinson, N. (1984) *Theory of Ordinary Differential Equations*, Krieger Publ., Melbourne, Florida.

Cole, R.H. (1968) *Theory of Ordinary Differential Equations*, Irvington, New York.

Collatz, L. (1986) *Differential Equations: An Introduction with Applications*, John Wiley, New York.

Fattorini, H.O. (1985) *Second Order Linear Differential Equations in Banach Spaces*, North-Holland, New York.

Feineman, G., Garrett, S.J. and Karaus, A.D. (1965) *Applied Differential Equations*, Spartan Books, Washington.

Ford, L.R. (1955) *Differential Equations*, 2nd edition, McGraw-Hill, New York.

Forsyth, A.R. (1921) *Differential Equations*, Macmillan, London.

Golomb, M. and Shanks, M.E. (1965) *Elements of Ordinary Differential Equations*, 2nd edition, McGraw-Hill, New York.

Hagin, F. (1975) *A First Course in Differential Equations*, Prentice-Hall, Englewood Cliffs, N.J.

Hartman, P. (1982) *Ordinary Differential Equations*, 2nd edition, Birkhauser, Boston.

Henrici, P. (1962) *Variable Methods in Ordinary Differential Equations*, John Wiley, New York.

Hill, J.M. (1992) *Differential Equations and Group Methods for Scientists and Engineers*, CRC Press, Boca Raton, Florida.

Hochstadt, H. (1964) *Differential Equations: A Modern Approach*, Holt, New York. (reprinted Dover Publications, New York., 1975).

Hunt, R.W. (1973) *Differential Equations and Related Topics for Science and Engineering*, Brooks-Cole, Monterey, California.

Hurewicz, W. (1958) *Lectures on Ordinary Differential Equations*, M.I.T. Press, Cambridge, Mass.

Ince, E.L. (1956) *Ordinary Differential Equations*, Dover Publications, New York.

Ince, E.L. and Sneddon, I.N. (1987) *The Solution of Ordinary Differential Equations*, 2nd edition, John Wiley, New York.

Jordan, D.W. and Smith, P. (1987) *Nonlinear Ordinary Differential Equations*, 2nd edition, Oxford University Press, Oxford.

Kamke, E. (1948) *Differentialgleichungen Losungsmethoden und Losungen*, Chelsea, New York.

Lambert, J.D. (1973) *Computational Methods in Ordinary Differential Equations*, John Wiley, New York.

Leighton, W. (1966) *Ordinary Differential Equations*, 2nd edition, Wadsworth, Belmont, California.

Lomen, D. and Mark, J. (1986) *Ordinary Differential Equations with Linear Algebra*, Prentice-Hall, Englewood Cliffs, N.J.

Martin, W.T. and Reissner, E. (1961) *Elementary Differential Equations*, Addison-Wesley, Reading, Mass.

Michel, A.N. (1982) *Ordinary Differential Equations*, Academic Press, New York.

Miller, K.S. and Michel, A.N. (1982) *Ordinary Differential Equations*, Academic Press, New York.

Morris, M. and Brown, O.E. (1952) *Differential Equations*, Prentice-Hall, Englewood Cliffs, N.J.

Murphy, G.M. (1960) *Ordinary Differential Equations and Their Solution*, Van Nostrand, Princeton, N.J.

Petrovskii, I.G. (1966) *Ordinary Differential Equations*, Prentice-Hall, Englewood Cliffs, N.J.

Piaggio, H.T.H. (1982) *An Elementary Treatise on Differential Equations and Their Applications*, Bell and Hyman, London.

Pinney, E. (1958) *Ordinary Difference-Differential Equations*, University of California Press, Berkeley.

Rabenstein, A.L. (1972) *Introduction to Ordinary Differential Equations*, Academic Press, New York.

Rainville, E.D. (1964) *Intermediate Differential Equations*, 2nd edition, Macmillan, New York.

Rainville, E.D. and Bedient, P.E. (1989) *Elementary Differential Equations*, Macmillan, New York.

Reid, W. (1980) *Sturmian Theory for Ordinary Differential Equations*, Springer-Verlag, Berlin.

Reid, W.T. (1971) *Ordinary Differential Equations*, John Wiley, New York.

Ross, S. (1964) *Differential Equations*, Blaisdell, New York.

Rubinstein, Z. (1969) *A Course in Ordinary and Partial Differential Equations*, Academic Press, New York.

Spiegel, M.R. (1981) *Applied Differential Equations*, 3rd edition, Prentice-Hall, Englewood Cliffs, N.J

Van Iwaarden, J.L. (1985) *Ordinary Differential Equations with Numerical Techniques*, Harcourt Brace Jovanovitch, Orlando, Florida.

Wilcox, L.R. and Curtis, H.J. (1961) *Elementary Differential Equations*, International Textbook Co., Scranton, Pennsylvania.

Williamson, R.E. (1986) *Introduction to Differential Equations: ODE, PDE and Series*, Prentice-Hall, Englewood Cliffs, N.J.

Partial Differential Equations

Agnew, R.P. (1960) *Differential Equations*, McGraw-Hill, New York.

Andrews, L.C. (1986) *Elementary Partial Differential Equations*, Academic Press, New York.

Bateman, H. (1959) *Partial Differential Equations of Mathematical Physics*, Cambridge University Press, London.

Berg, P.W. and McGregor, J.L. (1966) *Elementary Partial Differential Equations*, Holden-Day, San Francisco.

Bers, L., John, F. and Schetcher, M. (1964) *Partial Differential Equations*, John Wiley, New York.

Bleecker, D. and Csordas, G. (1996) *Basic Partial Differential Equations*, International Press, Cambridge, Massachusetts.

Bluman, G.W. and Kumei, S. (1989) *Symmetries and Differential Equations*, Springer-Verlag, New York.

Boyce, W.E. and DiPrima, R.C. (1992) *Elementary Differential Equations and Boundary Value Problems*, 5th edition, John Wiley, New York.

Braun, M. (1978) *Differential Equations and their Applications*, Springer-Verlag, New York.

Broman, A. (1989) *Introduction to Partial Differential Equations*, Dover Publications, New York.

Carrier, G.F. and Pearson, C.E. (1988) *Partial Differential Equations*, 2nd edition, Academic Press, New York.

Carrol, R.W. (1969) *Abstract Methods in Partial Differential Equations*, Harper and Row, New York.

Chester, C.R. (1971) *Techniques in Partial Differential Equations*, McGraw-Hill, New York.

Colton, D.L. (1976) *Partial Differential Equations in the Complex Domain*, Pitman, San Francisco.

Copson, E.T. (1975) *Partial Differential Equations*, Cambridge University Press, New York.

de Rham, G. (1960) *Varietes Differentiables*, Hermann, Paris.

Dennemeyer, R. (1968) *Introduction to Partial Differential Equations and Boundary Value Problems*, McGraw-Hill, New York.

Dezin, A.A. (1987) *Partial Differential Equations: An Introduction to a General Theory of Linear Boundary Value Problems*, Springer-Verlag, New York

DiBenedetto, E. (1995) *Partial Differential Equations*, Birkhauser, Boston.

Driver, R.D. (1977) *Ordinary and Delay Differential Equations*, Springer-Verlag, Berlin.

Driver, R.D. (1978) *Introduction to Ordinary Differential Equations*, Harper and Row, New York.

DuChateau, P. and Zachmann, D. (1986) *Theory and Problems of Partial Differential Equations*, McGraw-Hill, New York.

DuChateau, P. and Zachmann, D. (1989) *Applied Partial Differential Equations*, Harper and Row, New York.

Duff, G.F.D. (1956) *Partial Differential Equations*, Toronto Univ. Press, Toronto.

El'sgol'ts, L.E. and Norkin, S.B. (1973) *Introduction to the Theory and Application of Differential Equations with Derivating Arguments*, Academic Press, New York.

Epstein, B. (1962) *Partial Differential Equations*, McGraw-Hill, New York.

Evans, L.C. (1993) *Partial Differential Equations*, Berkeley Mathematics Lecture Notes, Vols. 3A 3B, Berkeley, Calif.

Farlow, S.J. (1993) *Partial Differential Equations for Scientists and Engineers*, Dover Publications, New York.

Folland, G.B. (1995) *Introduction to Partial Differential Equations*, 2nd edition, Princeton University Press, N.J.

Friedman, A. (1969) *Partial Differential Equations*, Holt, Rinehart and Winston, New York.

Garabedian, P.R. (1964) *Partial Differential Equations*, John Wiley, New York.

Gilbert, R.P. (1969) *Function Theoretic Methods in Partial Differential Equations*, Academic Press, New York.

Gribben, R.J. (1975) *Elementary Partial Differential Equations*, Van Nostrand Reinhold, New York

Gustafson, K.E. (1987) *Introduction to Partial Differential Equations and Hilbert Space Methods*, 2nd edition, John Wiley, New York.

Haberman, R. (1998) *Elementary Applied Partial Differential Equations*, 3rd edition, Prentice-Hall, Englewood Cliffs, N.J.

Hellwig, G. (1977) *Partial Differential Equations: An Introduction*, B.G. Teubner, Stuttgart.

Hill, J.M. (1992) *Differential Equations and Group Methods for Scientists and Engineers*, CRC Press, Boca Raton, Florida.

Hormander, L. (1964) *Linear Partial Differential Operators*, Springer, Berlin.

Hormander, L. (1985) *The Analysis of Linear Partial Differential Operators*, Vols. 1-4, Springer-Verlag, Berlin.

John, F. (1982) *Partial Differential Equations*, 4th edition, Springer-Verlag, New York.

Kervorkian, J. (1990) *Partial Differential Equations, Analytical Solution Techniques*, Wadsworth, Belmont, California.

Krein, M.G. and Gobberg, L. (1983) *Topics in Differential and Integral Equations and Operator Theory*, Birkhauser Verlag, Boston.

Krzyzanski, M. (1971) *Partial Differential Equations of Second Order* (trans. H. Zorski), Polish Scientific Publishers, Warszawa.

Lamb, Jr. G.L. (1995) *Introductory Applications of Partial Differential Equations*, John Wiley, New York.

Langer, R. (Ed) (1960) *Boundary-Value Problems in Differential Equations*, Univ. of Wisconsin Press, Madison.

Langer, R. (Ed) (1961) *Symp. Partial Differential Equations and Continuum Mechanics*, Univ. of Wisconsin Press, Madison.

Lieberstein, H.M. (1972) *Theory of Partial Differential Equations*, Academic Press, New York.

Lions, J.L. (1957) *Lectures on Elliptic Partial Differential Equations*, Tata, Bombay.

McOwen, R. (1995) *Partial Differential Equations: Methods and Applications*, Prentice-Hall, Englewood Cliffs, N.J.

Miller, F.H. (1941) *Partial Differential Equations*, John Wiley, New York.

Miller, K.S. (1953) *Partial Differential Equations in Engineering Problems*, Prentice-Hall, Englewood Cliffs, N.J.

Moon, P.H. and Spencer, D.E. (1969) *Partial Differential Equations*, D.C. Heath, Lexington, Mass.

Myint-U, T. and Debnath, L. (1987) *Partial Differential Equations for Scientists and Engineers*, 3rd edition, North-Holland, New York.

Noble, B. (1958) *Methods Based on the Wiener-Hopf Technique for the Solution of Partial Differential Equations*, Pergamon Press, New York.

Olver, P.J. (1986) *Applications of Lie Groups to Differential Equations*, Springer-Verlag, New York.

Ovsiannikov, L.V. (1982) *Group Analysis of Differential Equations*, Academic Press, New York.

Paris, R.B. and Wood, A.D. (1986) *Asymptotics of High Order Differential Equations*, John Wiley, New York.

Petrovskii, I.G. (1954) *Lectures on Partial Differential Equations*, Dover Publications, New York.

Pinsky, M.A. (1991) *Partial Differential Equations and Boundary-Value Problems with Applications*, 2nd edition, McGraw-Hill, New York.

Rahman, M. (1991) *Applied Differential Equations for Scientists and Engineers*, Computational Mechanics Publications, Boston.

Renardy, M. and Rogers, R.C. (1993) *An Introduction to Partial Differential Equations*, Springer-Verlag, New York.

Rubinstein, Z. (1969) *A Course in Ordinary and Partial Differential Equations*, Academic Press, New York.

Showalter, R.E. (1977) *Hilbert Space Methods for Partial Differential Equations*, Pitman, London.

Smirnov, M.M. (1966) *Second-Order Partial Differential Equations*, Edited by S. Chomet, Noordhoff, Groningen.

Smith, M.G. (1967) *Introduction to the Theory of Partial Differential Equations*, Van Nostrand, London.

Sneddon, I.N. (1957) *Elements of Partial Differential Equations*, McGraw-Hill, New York.

Sobolev, S.L. (1964) *Partial Differential Equations of Mathematical Physics*, Pergamon Press, Oxford.

Spiegel, M.R. (1958) *Applied Differential Equations*, Prentice-Hall, Englewood Cliffs, N.J.

Stakgold, I. (1968) *Boundary Value Problems of Mathematical Physics*, Vols. I and II, Macmillan, New York.

Stephenson, G. (1996) *Partial Differential Equations for Scientists and Engineers*, Imperial College Press, London.

Strauss, W.A. (1992) *Partial Differential Equations, An Introduction*, John Wiley, New York.

Street, R.L. (1973) *Analysis and Solution of Partial Differential Equations*, Wadsworth, Belmont, California.

Taylor, M. (1996) *Partial Differential Equations - Basic Theory*, Springer-Verlag, Berlin.

Titchmarsh, E.C. (1958) *Eigenfunction Expansions Associated with Second-Order Differential Equations*, Oxford University Press, London.

Treves, F. (1975) *Basic Linear Partial Differential Equations*, Academic Press, New York.

Tricomi, F.G. (1961) *Differential Equations*, Blackie, London.

Trim, D.W. (1990) *Applied Partial Differential Equations*, PWS-KENT Publishing, Boston.

Troutman, J.L. (1994) *Boundary Value Problems of Applied Mathematics*, PWS Publishing, Boston.

Webster, A.G. (1966) *Partial Differential Equations of Mathematical Physics*, 2nd edition, Dover Publications, New York.

Weinberger, H.F. (1965) *A First Course in Partial Differential Equations*, Dover Publications, New York.

Williams, W.E. (1980) *Partial Differential Equations*, Clarendon Press, Oxford.

Williamson, R.E. (1986) *Introduction to Differential Equations: ODE, PDE and Series*, Prentice-Hall, Englewood Cliffs, N.J.

Wloka, J. (1987) *Partial Differential Equations*, Cambridge University Press, Cambridge.

Young, E.C. (1972) *Partial Differential Equations: An Introduction*, Allyn and Bacon, Boston.

Zachmanoglou, E.C. and Thoe, D.W. (1987) *Introduction to Partial Differential Equations with Applications*, Dover Publications, New York.

Zauderer, E. (1998) *Partial Differential Equations of Applied Mathematics*, 2nd edition, John Wiley, New York.

Non-Linear Partial Differential Equations

Bellman, R.E. (1953) *Stability Theory of Differential Equations*, McGraw-Hill, New York.

Cesari, L. (1963) *Asymptotic Behaviour and Stability Problems in Ordinary Differential Equations*, 2nd edition, Springer-Verlag, Berlin.

Davis, H.T. (1961) *Introduction to Nonlinear Differential and Integral Equations*, Dover Publications, New York.

Diaz, J.L. (1985) *Nonlinear Partial Differential Equations and Free Boundaries*, Vol. 1, Pitman, Boston.

Dresner, L. (1983) *Similarity Solutions of Nonlinear Partial Differential Equations*, Pitman, Boston.

Evans, L.C. (1990) *Weak Convergence Methods for Nonlinear Partial Differential Equations*, American Mathematical Society, R.I.

Fuchik, S. and Kufner, A. (1980) *Nonlinear Differential Equations*, Elsevier Scientific Publ., Amsterdam.

Hirsch, W.H. and Smale, S. (1974) *Differential Equations: Dynamical Systems and Linear Algebra*, Academic Press, New York.

Huntley, I. and Johnson, R.M. (1983) *Linear and Nonlinear Differential Equations*, Halsted Press, New York.

Leung, A.W. (1989) *Systems of Nonlinear Partial Differential Equations: Applications to Biology and Engineering*, Kluwer Academic, Boston.

Logan, J.D. (1997) *Introduction to Nonlinear Partial Differential Equations*, John Wiley, New York.

Perko, L. (1991) *Differential Equations and Dynamical Systems*, Springer-Verlag, New York.

Reissig, R., Sansone, G. and Conti, R. (1974) *Non-Linear Differential Equations of Higher Order*, Noordhoff International, Leyden.

Smoller, J.A. (Ed) (1983) *Nonlinear Partial Differential Equations, Contemporary Mathematics*, vol. 17, American Mathematical Society, Providence, R.I.

Struble, R.A. (1962) *Nonlinear Differential Equations*, McGraw-Hill, New York.

Verhulst, F. (1990) *Differential Equations and Dynamical Systems*, Springer-Verlag, New York.

Mathematical Methods and Boundary Value Problems

Ames, W.F., Harrell Jr., E.M. and Herod, J.V. (Eds.) (1993) *Differential Equations with Applications to Mathematical Physics*, Academic Press, New York.

Basmadjian, D. (1999) *The Art of Modelling in Science and Engineering*, Chapman and Hall/CRC Press, Boca Raton, FL.

Bateman, H. (1959) *Partial Differential Equations of Mathematical Physics*, Cambridge University Press, London.

Bick, T.A. (1993) *Elementary Boundary Value Problems*, Marcel Dekker, New York.

Chorlton, F. (1969) *Boundary Value Problems in Physics and Engineering*, Van Nostrand Reinhold, London.

Dennemeyer, R. (1968) *Introduction to Partial Differential Equations and Boundary Value Problems*, McGraw-Hill, New York.

Dixon, C. (1971) *Applied Mathematics of Science and Engineering*, John Wiley and Sons, London.

Duff, G.F.D. and Naylor, D. (1966) *Differential Equations of Applied Mathematics*, John Wiley, New York.

Edelstein-Keshet, L. (1988) *Mathematical Models in Biology*, Random House, New York.

Fowler, A.C. (1997) *Mathematical Models in the Applied Sciences*, Cambridge University Press, Cambridge.

Frank, F. and von Mises, R. (1937) *Differential and Integral Equations of Mathematical Physics*, United Scientific and Technical Press.

Friedman, A. (1990) *Mathematics in Industrial Problems*, Parts1-5, Springer-Verlag, Berlin.

Genin, J. and Maybee, J.S. (1970) *Introduction to Applied Mathematics*, Vol. 1, Holt, Rinehart and Winston, New York.

Guenther, R.B. and Lee, J.W. (1996) *Partial Differential Equations of Mathematical Physics and Integral Equations*, Dover Publications, New York

Hildebrand, F.B. (1965) *Methods of Applied Mathematics*, Prentice-Hall, Englewood Cliffs, N.J.

Irving, J. and Mullineux, N. (1959) *Mathematics in Physics and Engineering*, Academic Press, New York.

Jeffreys, H.J. and Jeffreys, B.S. (1946) *Methods of Mathematical Physics*, Cambridge University Press, Cambridge.

Kovach, L.D. (1984) *Boundary Value Problems*, Addison Wesley, Reading, Mass.

Leis, R. (1983) *Partial Differential Equations of Applied Mathematics*, John Wiley, New York.

Levin, V.L. and Grosberg, Yu.N. (1951) *Differential Equations of Mathematical Physics*, Gostekhizdat, Moscow.

Lin, C.C. and Segel, L.A. (1974) *Mathematics Applied to Deterministic Problems in the Applied Sciences*, Macmillan, New York.

Mackie, A.G. (1965) *Boundary Value Problems*, Oliver and Boyd, London.

Margenau, H. and Murphy, G. (1964) *The Mathematics of Physics and Chemistry*, Vols. I and II, Van Nostrand, New York.

Mathews, J. and Walker, R. (1965) *Mathematical Methods of Physics*, W.A. Benjamin, New York.

Mikhlin, S.G. (1967) *Linear Equations of Mathematical Physics*, Holt, Rinehart and Winston, New York.

Mikhlin, S.G. (1970) *Mathematical Physics, An Advanced Course*, North-Holland Publishing Co., Amsterdam, The Netherlands.

Morse, P.M. and Feshbach, H. (1953) *Methods of Theoretical Physics*, Parts I and II, McGraw-Hill, New York.

Peskin, C.S. (1976) *Partial Differential Equations in Biology*, Courant Institute of Mathematical Sciences, New York.

Powers, D.L. (1979) *Boundary Value Problems*, Academic Press, New York.

Reed, M. and Simon B. (1972) *Methods of Mathematical Physics*, Vols. I and II, Academic Press, New York.

Ritger, P.D. and Rose, N.J. (1968) *Differential Equations and Applications*, McGraw-Hill, New York.

Rubinow, S.I. (1976) *Introduction to Mathematical Biology*, John Wiley, New York.

Rubinstein, I. and Rubinstein, L. (1993) *Partial Differential Equations in Classical Mathematical Physics*, Cambridge University Press, Cambridge.

Sagan, H. (1989) *Boundary Value Problems in Mathematical Physics*, Dover Publications, New York.

Schechter, M. (1977) *Modern Methods in Partial Differential Equations: An Introduction*, 2nd edition, McGraw-Hill, New York.

Segel, L.A. (1987) *Mathematics Applied to Continuum Mechanics*, Dover Publications, New York.

Sobolev, S.L. (1964) *Partial Differential Equations of Mathematical Physics*, Pergamon Press, Oxford.

Sokolnikoff, I.S. and Redheffer, R.M. (1966) *Mathematics of Physics and Modern Engineering*, McGraw-Hill, New York.

Sommerfeld, A. (1964) *Partial Differential Equations in Physics*, Academic Press, New York.

Stakgold, I. (1968) *Boundary Value Problems of Mathematical Physics*, Vols. I and II, Macmillan, New York.

Svobodny, T. (1998) *Mathematical Modelling for Industry and Engineering*, Prentice-Hall, Englewood Cliffs, N.J.

Taylor, A.B. (1986) *Mathematical Models in Applied Mechanics*, Clarendon Press, Oxford.

Tikhonov, A.N. and Samarskii, A.A. (1990) *Equations of Mathematical Physics*, Dover Publications, New York.

Troutman, J.L. (1994) *Boundary Value Problems of Applied Mathematics*, PWS Publishing, Boston.

Walter, W. (1964) *Differential- und Integral-Ungleichungen*, Springer-Verlag, Berlin.

Whitham, G.B. (1976) *Linear and Nonlinear Waves*, John Wiley, New York.

Wong, C.W. (1991) *Introduction to Mathematical Physics: Methods and Concepts*, Oxford University Press, Oxford.

Yosida, K. (1960) *Lectures on Differential and Integral Equations*, Wiley-Interscience, New York.

First-Order Partial Differential Equations: Advective Flows, Traffic Flow, Heat Exchangers

Aris, R. and Amundson, N. (1973) *Mathematical Methods in Chemical Engineering, Vol. II, First Order Partial Differential Equations*, Prentice-Hall, Englewood Cliffs, N.J.

Bird, R.B., Stewart, H.E. and Lightfoot, E.N. (1960) *Heat, Mass and Momentum Transfer*, Prentice-Hall, Englewood Cliffs, N.J.

Botha, J.F. and Pinder, G.F. (1983) *Fundamental Concepts in the Numerical Solution of Differential Equations*, John Wiley, New York.

Caratheodory, C. (1968) *Calculus of Variations and Partial Differential Equations of the First Order*, Vol. I and II, Holden-Day, San Francisco.

Haight, F.A. (1963) *Mathematical Theory of Traffic Flow*, Academic Press, New York.

Lapidus, L. and Pinder, G.F. (1982) *Numerical Solution of Partial Differential Equations in Engineering*, John Wiley, New York.

Moon, P.H. and Spencer, D.E. (1969) *Partial Differential Equations*, D.C. Heath, Lexington, Mass.

Rhee, H., Aris, R. and Amundson, N.R. (1986) *First Order Partial Differential Equations*, Prentice-Hall, Englewood Cliffs, N.J.

Whitham, G.B. (1976) *Linear and Nonlinear Waves*, John Wiley, New York.

Laplace's Equation: General Concepts, Mechanics of Fluid Flow, Flow in Porous Media

Basset, A.B. (1888) *A Treatise on Hydrodynamics*, Vols. 1 and 2, Dover Publications, New York.

Batchelor, G.K. (1967) *Introduction to Fluid Dynamics*, Cambridge University Press, Cambridge.

Bear, J. (1972) *Dynamics of Fluids in Porous Media*, Dover Publications, New York.

Bear, J. (1979) *Hydraulics of Groundwater*, McGraw-Hill, New York.

Bergman, S. and Schiffer, M. (1953) *Kernel Functions and Elliptic Differential Equations in Mathematical Physics*, Academic Press, New York.

Chorlton, F. (1967) *Textbook of Fluid Dynamics*, Van Nostrand, London.

Collins, R.E. (1961) *Flow of Fluids Through Porous Media*, Reinhold, New York.

Dullien, F.A.L. (1992) *Porous Media: Fluid Transport and Pore Structure*, Academic Press, New York.

Fichera, G. (1965) *Linear Elliptic Differential Systems and Eigen-value Problems*, Lecture Notes in Mathematics, 8, Springer-Verlag, Berlin.

Gilbarg, D. and Trudinger, N. (1977) *Elliptic Partial Differential Equations of the Second Order*, Springer-Verlag, Berlin.

Greenkorn, R.A. (1983) *Flow Phenomena in Porous Media*, Marcel-Dekker, New York.

Harr, M.E. (1962) *Groundwater and Seepage*, McGraw-Hill, New York.

Hermance, J.F. (1991) *A Mathematical Primer on Groundwater Flow*, Prentice-Hall, Englewood Cliffs, N.J.

Kellogg, O.D. (1953) *Foundations of Potential Theory*, Dover Publications, New York.

Kirkham, D. and Powers, W.L. (1972) *Advanced Soil Physics*, John Wiley, New York.

Kleinstreuer, C. (1997) *Engineering Fluid Dynamics*, Cambridge University Press, Cambridge.

Lamb, Sir Horace (1993) *Hydrodynamics*, Cambridge University Press, Cambridge.

Lin, C.C. and Segel, L.A. (1974) *Mathematics Applied to Deterministic Problems in the Applied Sciences*, Macmillan, New York.

Lions, J.L. (1957) *Lectures on Elliptic Partial Differential Equations*, Tata, Bombay.

Miranda, C. (1970) *Partial Differential Equations of Elliptic Type*, 2nd edition, Springer-Verlag, New York.

Muskat, M. (1937) *The Flow of Homogeneous Fluids Through Porous Media*, McGraw-Hill, New York.

Nikolaevskii, V.N. (1990) *Mechanics of Porous and Fractured Media*, World Scientific, Singapore.

Ozisik, M.N. (1993) *Heat Conduction*, John Wiley, New York.

Philips, O.M. (1991) *Flow and Reactions in Permeable Rocks*, Cambridge University Press, Cambridge.

Polubarinova-Kochina, P.Ya. (1962) *Theory of Ground Water Movement*, Princeton University Press, Princeton, N.J.

Sahimi, M. (1995) *Flow and Transport in Porous Media and Fractured Rock*, VCH Verlagsgesselschaft, Weinheim.

Scheidegger, A.E. (1974) *The Physics of Flow Through Porous Media*, University of Toronto Press, Toronto.

Strack, O.D.L. (1989) *Groundwater Mechanics*, Prentice-Hall, Englewood Cliffs, N.J.

Verruijt, A. (1970) *Theory of Groundwater Flow*, Gordon and Breach, New York.

Vukovich, M. (1997) *Groundwater Dynamics: Steady Flow*, Water Resources Publications, Littleton, CO.

White, F.M. (1984) *Heat Transfer*, Addison-Wesley, Reading, Mass.

Xiao, S. (Ed) (1992) *Flow and Transport in Porous Media*, World Scientific Publishing, Singapore.

Yih, C.-S. (1969) *Fluid Mechanics*, McGraw-Hill, New York.

Diffusion Equation: General Concepts, Heat Conduction, Pressure Transients, Chemical Diffusion

Banks, R.B. (1994) *Growth and Diffusion Phenomena*, Springer-Verlag, Berlin.

Barenblatt, G.I., Entov, V.M. and Ryzhik, V.M. (1990) *Theory of Fluid Flow Through Natural Rocks*, Kluwer Academic, Dordrecht, The Netherlands.

Carslaw, H.S. and Jaeger, J.C. (1959) *Heat Conduction in Solids*, Clarendon Press, Oxford.

Crank, J. (1975) *The Mathematics of Diffusion*, Clarendon Press, Oxford.

Cussler, E.L. (1984) *Diffusion: Mass Transfer in Fluid Systems*, Cambridge University Press, Cambridge.

Friedman, A. (1964) *Partial Differential Equations of Parabolic Type*, Prentice- Hall, Englewood Cliffs, N.J.

Ghez, R. (1988) *A Primer of Diffusion Problems*, John Wiley, New York.

Hill, J.M. and Dewynne, J. (1987) *Heat Conduction*, Blackwell Scientific Publications, Oxford.

Kirkham, D. and Powers, W.L. (1972) *Advanced Soil Physics*, John Wiley, New York.

Luikov, A.V. (1968) *Analytical Heat Diffusion Theory*, Academic Press, New York.

Mikhailov, M.D. and Ozisik, M.N. (1984) *Unified Analysis of Solutions of Heat and Mass Diffusion*, John Wiley, New York.

Nield, D.A. and Bejan, A. (1999) *Convection in Porous Media*, Springer-Verlag, Berlin.

Nikolaevskii, V.N. (1990) *Mechanics of Porous and Fractured Media*, World Scientific, Singapore.

Okubo, A. (1980) *Diffusion and Ecological Problems: Mathematical Models*, Springer-Verlag, Berlin.

Ozisik, M.N. (1993) *Heat Conduction*, John Wiley, New York.

Philips, O.M. (1991) *Flow and Reactions in Permeable Rocks*, Cambridge University Press, Cambridge.

Sahimi, M. (1995) *Flow and Transport in Porous Media and Fractured Rock*, VCH Verlagsgesselschaft, Weinheim.

Sun, N.-Z. (1996) *Mathematical Modelling in Groundwater Pollution*, Springer-Verlag, Berlin.

Taira, K. (1988) *Diffusion Processes and Partial Differential Equations*, Academic Press, Orlando.

Watson, N.A. (1989) *Parabolic Equations on an Infinite Strip*, Marcel Dekker, New York.

White, F.M. (1984) *Heat Transfer*, Addison-Wesley, Reading, Mass.

Widder, D. (1975) *The Heat Equation*, Academic Press, New York.

Wave Equation: General Concepts, Waves in Strings, Membranes and Elastic Solids

Achenbach, J.D.(1973) *Wave Propagation in Elastic Solids*, North-Holland Publishing Co., Amsterdam, The Netherlands.

Bland, D.R. (1988) *Wave Theory and Applications*, Clarendon Press, Oxford.

Brekhovskikh, L.M and Goncharov, V. (1994) *Mechanics of Continua and Wave Dynamics*, Springer-Verlag, Berlin.

Coulson, C.A. and Jeffrey, A. (1977) *Waves. A Mathematical Approach to the Common Types of Wave Motion*, Longman, London.

Elmore, W.C. and Heald, M.A. (1969) *Physics of Waves*, Dover Publications, New York.

Garabedian, P.R. (1964) *Partial Differential Equations*, John Wiley, New York.

Graff, K.F. (1975) *Wave Motion in Elastic Solids*, Ohio State University Press, OH.

John, F. (1981) *Plane Waves and Spherical Means Applied to Partial Differential Equations*, Springer-Verlag, Berlin.

Lax, P.D. (1963) *Lectures on Hyperbolic Partial Differential Equations*, Stanford University Press, California.

Lighthill, M.J. (1978) *Waves in Fluids*, Cambridge University Press, Cambridge.

Morse, P.M. (1948) *Vibration and Sound*, McGraw-Hill, New York.

Nowacki, W. (1963) *Dynamics of Elastic Systems*, Chapman and Hall, London.

Rayleigh, Lord (1945) *Theory of Sound*, Vols. 1 and 2, Dover Publications, New York.

Sakamoto, R. (1982) *Hyperbolic Boundary Value Problems*, Cambridge University Press, Cambridge.

Whitham, G.B. (1976) *Linear and Nonlinear Waves*, John Wiley, New York.

Witten, M. (1983) *Hyperbolic Partial Differential Equations*, Pergamon Press, New York.

Biharmonic Equation: Mechanics of Elastic Solids, Contact Problems, Theory of Plates, Slow Viscous Flow

Ambartsumyan, S.A. (1991) *Theory of Anisotropic Plates: Strength, Stability and Vibrations*, Hemisphere Publishing Corporation, New York.

Atkin, R.J. and Fox, N. (1980) *An Introduction to the Theory of Elasticity*, Longmans, London.

Basset, A.B. (1888) *A Treatise on Hydrodynamics*, Vols. 1 and 2, Dover Publications, New York.

Batchelor, G.K. (1967) *An Introduction to Fluid Dynamics*, Cambridge University Press, Cambridge.

Bird, R.B., Stewart, H.E. and Lightfoot, E.N. (1960) *Heat, Mass and Momentum Transfer*, Prentice-Hall, Englewood Cliffs, N.J.

Birkhoff, G. (1960) *Hydrodynamics: A Study in Logic, Fact and Similitude*, Princeton University Press, Princeton, N.J.

Boresi, A.P. and Chong, K.P. (1987) *Elasticity in Engineering Mechanics*, Elsevier Science Publ., New York.

Boresi, A.P. and Lynn, P.P. (1974) *Elasticity in Engineering Mechanics*, Prentice-Hall, Englewood Cliffs, N.J.

Chou, P.C. and Pagano, N.J. (1967) *Elasticity, Tensor, Dyadic and Engineering Approaches*, Dover Publications, New York.

Churchill, S.W. (1988) *Viscous Flows*, Butterworths, Boston.

Clements, D.L. (1981) *Boundary Value Problems Governed by Second-Order Elliptic Systems*, Pitman Publ. Ltd., London

Constantinescu, V.N. (1995) *Laminar Viscous Flow*, Springer-Verlag, Berlin.

Davis, R.O. and Selvadurai, A.P.S. (1996) *Elasticity and Geomechanics*, Cambridge University Press, Cambridge,

Drew, D.A. and Passman, S.L. (1999) *Theory of Multicomponent Fluids*, Applied Mathematical Sciences Vol. 135, Springer-Verlag, Berlin.

England, A.H. (1971) *Complex Variable Methods in Elasticity*, Wiley-Interscience. London.

Eringen, A.C. (1967) *Mechanics of Continua*, John Wiley, New York.

Flugge, S. (Ed.) (1959) *Encyclopedia of Physics*, Vol. VIII/1(C.Truesdell, Co-Editor), Springer-Verlag, Berlin.

Freeze, R.A. and Cherry, J.A. (1979) *Groundwater*, Prentice-Hall, Englewood Cliffs, N.J.

Fung, Y.C. (1965) *Foundations of Solid Mechanics*, Prentice-Hall, Englewood Cliffs, N.J.

Gladwell, G.M.L. (1980) *Contact Problems in the Classical Theory of Elasticity*, Sijthoff and Noordhoff, Alphen aan den Rijn, The Netherlands.

Godfrey, D.E.R. (1959) *Theoretical Elasticity and Plasticity for Engineers*, Thames and Hudson, London.

Gould, P.L. (1983) *Introduction to Linear Elasticity*, Springer-Verlag, Berlin.

Green, A.E. and Zerna, W. (1968) *Theoretical Elasticity*, Clarendon Press, Oxford.

Gurtin, M.E.(1972) Linear Theory of Elasticity, in *Handbook of Physics Vol. VIa/2, Mechanics of Solids II*(Ed. C. Truesdell), Springer-Verlag, Berlin

Gurtin, M.E. (1981) *An Introduction to Continuum Mechanics*, Academic Press, New York.

Hahn, H.G. (1984) *Elastizitatstheorie*, B.G.Teubner, Stuttgart.

Happel, J. and Brenner, H. (1973) *Low Reynolds Number Hydrodynamics*, Martinus Nijhoff, The Hague, The Netherlands.

Herrmann, G. (Ed) (1974) *R.D.Mindlin and Applied Mechanics*, Pergamon Press, Oxford.

Hunter, S.C. (1983) *Mechanics of Continuous Media*, Ellis Horwood-Wiley, New York.

Jaeger, L.G. (1964) *Elementary Theory of Elastic Plates*, Pergamon Press, Oxford.

Kalandiya, A.I. (1975) *Mathematical Methods of Two-Dimensional Elasticity*, Mir publishers, Moscow.

Kleinstreuer, C. (1997) *Engineering Fluid Dynamics*, Cambridge University Press, Cambridge.

Karasudhi, P. (1991) *Foundations of Solid Mechanics*, Kluwer Academic Publishers, Dordrecht, The Netherlands.

Knops, R.J. and Payne, L.E. (1971) *Uniqueness Theorems in Linear Elasticity*, Springer-Verlag, Berlin.

Lamb, Sir Horace (1993) *Hydrodynamics*, Cambridge University Press, Cambridge.

Landau, H.G. and Lifshitz, E.M. (1970) *Theory of Elasticity, Course in Theoretical Physics*, Vol. 7, Pergamon Press, Oxford.

Landau, L.D. and Lifshitz, E.M. (1987) *Fluid Mechanics*, 2nd edition, Pergamon Press, Oxford.

Langlois, W.E. (1964) *Slow Viscous Flow*, Macmillan, New York.

Leipholz, H. (1974) *Theory of Elasticity*, Noordhoff, Leyden, The Netherlands.

Lin, C.C. and Segel, L.A. (1974) *Mathematics Applied to Deterministic Problems in the Applied Sciences*, Macmillan, New York.

Little, R.W. (1973) *Elasticity*, Prentice-Hall, Englewood Cliffs, N.J.

Love, A.E.H. (1944) *An Introduction to the Mathematical Theory of Elasticity*, Dover Publicatons, New York.

Lur'e, A.I., (1964) *Three-Dimensional Problems of the Theory of Elasticity*, Wiley-Interscience, New York.

Mal, A.K. and Singh, S.J. (1991) *Deformation of Elastic Solids*, Prentice-Hall, Englewood Cliffs, N.J.

Malvern, L.E. (1969) *Introduction to the Mechanics of a Continuous Medium*, Prentice-Hall, Englewood Cliffs, N.J.

Mansfield, E.H. (1989) *The Theory of Bending and Stretching of Plates*, Cambridge University Press, Cambridge.

Mase, G.E. and Mase, G.T. (1991) *Continuum Mechanics for Engineers*, CRC Press, Boca Raton, Florida.

Milne-Thomson, L.M. (1960) *Plane Elastic Systems*, Springer-Verlag, Berlin.

Muskhelishvili, N.I. (1963) *Some Basic Problems in the Mathematical Theory of Elasticity*, Noordhoff, Groningen, The Netherlands.

Nadeau, G. (1964) *Introduction to Elasticity*, Holt, Rinehart and Winston, New York.

Nield, D.A. and Bejan, A. (1999) *Convection in Porous Media*, Springer-Verlag, Berlin.

Ockendon, H. and Ockendon, J.R. (1995) *Viscous Flow*, Cambridge University Press, New York.

Pai, S.-I. (1959) *Slow Viscous Flow Theory, Vol. 1, Laminar Flow*, Van Nostrand, Princeton, N.J.

Panc, V. (1975) *Theories of Elastic Plates*, Noordhoff, Leyden, The Netherlands.

Panton, R.L. (1984) *Incompressible Flow*, John Wiley, New York.

Parton, V.Z. and Perlin, P.I. (1982) *Integral Equations in Elasticity*, Mir Publishers, Moscow.

Paterson, A.R. (1983) *A First Course in Fluid Dynamics*, Cambridge University Press, Cambridge.

Pedley, T. (1980) *The Fluid Mechanics of Large Blood Vessels*, Cambridge University Press, Cambridge.

Philips, O.M. (1991) *Flow and Reactions in Permeable Rocks*, Cambridge University Press, Cambridge.

Reddy, J.N. (1999) *Theory and Analysis of Elastic Plates*, Taylor and Francis, Philadelphia, Pennsylvania.

Reismann, H. (1988) *Elastic Plates: Theory and Application*, John Wiley, New York.

Reismann, H. and Pawlik, P.S. (1980) *Elasticity, Theory and Applications*, John Wiley, New York.

Scheidegger, A.E. (1974) *The Physics of Flow Through Porous Media*, University of Toronto Press, Toronto.

Sedov, L.I. (1972) *A Course in Continuum Mechanics*, Vols.I-IV, Wolters-Noordhoff, Groningen, The Netherlands.

Segel, L.A. (1987) *Mathematics Applied to Continuum Mechanics*, Dover Publications, New York.

Selvadurai, A.P.S. (1979) *Elastic Analysis of Soil-Foundation Interaction*, Elsevier Scientific Publishing, Amsterdam, The Netherlands.

Sherman, F.S. (1990) *Viscous Flow*, McGraw-Hill, New York.

Sokolnikoff, I.S. (1956) *Mathematical Theory of Elasticity*, McGraw-Hill, New York.

Southwell, R.V. (1969) *An Introduction to the Theory of Elasticity*, Dover Publications, New York.

Spencer, A.J.M. (1980) *Continuum Mechanics*, Longmans, London.

Sun, N.-Z. (1996) *Mathematical Modelling in Groundwater Pollution*, Springer-Verlag, Berlin.

Szilard, R. (1974) *Theory and Analysis of Plates, Classical and Numerical Methods*, Prentice-Hall, Englewood Cliffs, N.J.

Timoshenko, S.P. and Goodier, J.N. (1970) *Theory of Elasticity*, McGraw-Hill, New York.

Timoshenko, S.P. and Woinowsky-Krieger, S. (1959) *Theory of Plates and Shells*, McGraw-Hill, New York.

Volterra, E. and Gaines, J.H. (1971) *Advanced Strength of Materials*, Prentice-Hall, Englewood Cliffs, N.J.

Wang, C.T. (1953) *Applied Elasticity*, McGraw-Hill, New York.

Westergaard, H.M. (1952) *Theory of Elasticity and Plasticity*, Harvard University Press. Cambridge, Mass.

Yih, C.-S. (1969) *Fluid Mechanics*, McGraw-Hill, New York.

Poisson's Equation: Flow in Porous Media, Axial Viscous Flows, Torsion of Prismatic Elastic Solids

Boresi, A.P. and Lynn, P.P. (1974) *Elasticity in Engineering Mechanics*, Prentice-Hall, Englewood Cliffs, N.J.

Chou, P.C. and Pagano, N.J. (1967) *Elasticity, Tensor, Dyadic and Engineering Approaches*, Dover Publications, New York.

Kellogg, O.D. (1953) *Foundations of Potential Theory*, Dover Publications, New York.

Little, R.W. (1973) *Elasticity*, Prentice-Hall, Englewood Cliffs, N.J.

Muskat, M. (1937) *The Flow of Homogeneous Fluids Through Porous Media*, McGraw-Hill, New York.

Nield, D.A. and Bejan, A. (1999) *Convection in Porous Media*, Springer-Verlag, Berlin.

Polya, G. and Szego, G. (1981) *Isoperimetric Inequalities in Mathematical Physics*, Annals of Mathematical Studies, Princeton University Press, N.J.

Protter, M.H. and Weinberger, H.F. (1984) *Maximum Principles in Differential Equations*, Springer-Verlag, Berlin.

Reismann, H. and Pawlik, P.S. (1980) *Elasticity, Theory and Applications*, John Wiley, New York.

Sherman, F.S. (1990) *Viscous Flow*, McGraw-Hill, New York.

Sokolnikoff, I.S. (1956) *Mathematical Theory of Elasticity*, McGraw-Hill, New York.

Southwell, R.V. (1969) *An Introduction to the Theory of Elasticity*, Dover Publications, New York.

Stakgold, I. (1968) *Boundary Value Problems of Mathematical Physics*, Vols. I and II, Macmillan, New York.

Szarski, J. (1965) *Differential Inequalities*, Polish Scientific Publishers, Warsaw.

Timoshenko, S.P. and Goodier, J.N. (1970) *Theory of Elasticity*, McGraw-Hill, New York.

Green's Functions

Barton, G. (1989) *Elements of Green's Functions and Propagation: Potentials, Diffusion and Waves*, Clarendon Press, Oxford.

Greenberg, M.D. (1971) *Application of Green's Functions in Science and Engineering*, Prentice-Hall, Englewood Cliffs, N.J.

Roach, G.F. (1982) *Green's Functions*, Cambridge University Press, Cambridge.

Sneddon, I.N. (1972) *The Use of Integral Transforms*, McGraw-Hill, New York.

Stakgold, I. (1998) *Green's Functions and Boundary Value Problems*, 2nd edition, John Wiley, New York.

Williams, W.E. (1980) *Partial Differential Equations*, Clarendon Press, Oxford.

Zauderer, E. (1998) *Partial Differential Equations of Applied Mathematics*, 2nd edition, John Wiley, New York.

Collections of Problems

Budak, B.M., Samarskii, A.A. and Tikhonov, A.N. (1988) *A Collection of Problems in Mathematical Physics*, Dover Publications, New York.

Lebedev, N.N., Skalskaya, I.P. and Uflyand, Ya.S. (1965) *Problems in Mathematical Physics*, Pergamon Press, Oxford.

Smirnov, M.M. (1967) *Problems on the Equations of Mathematical Physics*, Noordhoff, The Netherlands.

Mathematical Tables and Handbooks

Abramowitz, M. and Stegun, I.A. (Eds.) (1972) *Handbook of Mathematical Functions with Formulas, Graphs and Mathematical Tables*, Dover Publications, New York.

CRC Standard Mathematical Tables, (1991) CRC Press, Boca Raton, Florida.

Dwight, H.B. (1961) *Tables of Integrals and Other Mathematical Data*, Macmillan, New York.

Erdelyi, A. et al. (1954) *Tables of Integral Transforms*, Vols. 1 and 2, McGraw-Hill, New York.

Gradshteyn, I.S. and Ryzhik, I.M. (1980) *Table of Integrals, Series and Products*, Academic Press, New York.

Hughes, W.F. and Gaylord, E.W. (1964) *Basic Equations of Engineering Science*, McGraw-Hill, New York.

Jolley, L.B.W. (1961) *Summation of Series*, Dover Publications, New York.

Korn, G. and Korn, T. (1968) *Mathematical Handbook for Scientists and Engineers*, McGraw-Hill, New York.

Petit Bois, G. (1961) *Tables of Indefinite Integrals*, Dover Publications, New York.

Roberts, G.E. and Kaufman, H. (1966) *Table of Laplace Transforms*, Saunders, Philadelphia, PA.

Spiegel, M.R. (1968) *Mathematical Handbook of Formulas and Tables*, McGraw-Hill, New York.

Wheelon, A.D. (1968) *Tables of Summable Series and Integrals Involving Bessel Functions*, Holden-Day, San Francisco, California.

Zwillinger, D. (1992) *Handbook of Differential Equations*, Academic Press, Boston.

Computer Based Symbolic Mathematics

Abell, M. and Braselton, J. (1994) *Differential Equations with Maple V*, Academic Press, New York.

Abell, M. and Braselton, J. (1994) *Mathematica by Example*, Academic Press, New York.

Articolo, G.A. (1998) *Partial Differential Equations and Boundary Value Problems with Maple V*, Academic Press, San Diego.

Betounes, D. (1998) *Partial Differential Equations for Computational Science: with Maple and Vector Analysis*, Springer-Verlag, New York.

Kreyszig, H.E. (1994) *Maple Computer Manual for Advanced Engineering Mathematics*, John Wiley, New York.

Kythe, P.M., Puri, P., and Schäferkotter, M.R. (1997) *Partial Differential Equations and Mathematica*, CRC Ress, Boca Raton, FL.

Redfern, D. (1994) *Maple Handbook: Maple V Release 3*, Springer-Verlag, Berlin.

Vvedensky, D. (1992) *Partial Differential Equations with Mathematica*, Addison-Wesley, New York.

Wilson, H.B. and Turcotte, L.H. (1994) *Advanced Mathematics and Mechanics Applications using MATLAB*, CRC Press, Boca Raton, Florida.

Numerical Methods for Partial Differential Equations

Ames, W.F. (1977) *Numerical Methods for Partial Differential Equations*, 2nd edition, Academic Press, New York.

Aziz, A.K. (Ed.) (1972) *The Mathematical Foundations of the Finite Element Method with Applications to Partial Differential Equations*, Academic Press, New York.

Bellman, R. and Cooke, K.L (1963) *Differential-Difference Equations*, Academic Press, New York.

Bellman, R.E. and Adomian, G. (1985) *Partial Differential Equations: New Methods for their Treatment and Solution*, Reidel Publishing, Hingham, Mass.

Botha, J.F. and Pinder, G.F. (1983) *Fundamental Concepts in the Numerical Solution of Differential Equations*, John Wiley, New York.

Brand, L. (1966) *Differential and Difference Equations*, John Wiley, New York.

Collatz, L. (1960) *The Numerical Treatment of Partial Differential Equations*, 3rd edition, Springer-Verlag, Berlin.

Colombini, F., Marino, A., Modica, L. and Spagnolo, S. (Eds) (1989) *Partial Differential Equations and the Calculus of Variations*, Vol. II, Birkhauser, Boston.

Crandall, S.H. (1956) *Engineering Analysis - A Survey of Numerical Procedures*, McGraw-Hill, New York.

Forsythe, G.E. (1958) *Numerical Analysis and Partial Differential Equations*, John Wiley, New York.

Forsythe, G.E. and Wasow, W.R. (1967) *Finite-Difference Methods for Partial Differential Equations*, John Wiley, New York.

Gear, C.W. (1971) *Numerical Initial Value Problems in Ordinary Differential Equations*, Prentice-Hall, Englewood Cliffs, N.J.

Gladwell, I. and Wait, R. (Eds) (1979) *A Survey of Numerical Methods for Partial Differential Equations*, Oxford University Press, Oxford.

Johnson, C. (1993) *Partial Differential Equations by the Finite Element Method*, Cambridge University Press, Cambridge.

Kocak, H. (1989) *Differential and Difference Equations through Computer Experiments*, 2nd edition, Springer-Verlag, New York.

Lapidus, L. and Pinder, G.F. (1982) *Numerical Solution of Partial Differential Equations in Engineering*, John Wiley, New York.

Meis, T.H. and Marcowitz, U. (1981) *Numerical Solutions of Partial Differential Equations*, Springer-Verlag, Berlin.

Mitchell, A.R. (1969) *Computational Methods in Partial Differential Equations*, John Wiley, New York.

Mitchell, A.R. and Griffiths, D.F. (1980) *The Finite Difference Method in Partial Differential Equations*, John Wiley, New York.

Mitchell, A.R. and Wait, R. (1977) *The Finite Element Method in Partial Differential Equations*, John Wiley, New York.

Noye, J. (1984) *Computational Techniques for Differential Equations*, North-Holland Mathematical Studies 83, North-Holland, New York.

Salvadori, M.G. and Schwarz, R.J. (1965) *Differential Equations in Engineering*, Prentice-Hall, Englewood Cliffs, N.J.

Sewell, G. (1988) *The Numerical Solution of Ordinary and Partial Differential Equations*, Academic Press, Boston

Shampine, L.F. and Gordon, M.K. (1975) *Computer Solution of Ordinary Differential Equations: The Initial Value Problem*, Freeman, San Francisco.

Smith, G.D. (1985) *Numerical Solution of Partial Differential Equations: Finite Difference Equations*, 3rd edition, Oxford University Press, Oxford.

Vichnevetsky, R. (1981) *Computer Methods for Partial Differential Equations*, Vol. I, Prentice-Hall, Englewood Cliffs, N.J.

von Rosenberg, D.U. (1969) *Methods for the Numerical Solution of Partial Differential Equations*, American Elsevier Publ. Co., New York.

Weinstein, A. (1966) *Numerical Solution of Partial Differential Equations*, Academic Press, New York.

Historical Notes

Abraham, R. and Marsden, J.E. (1978) *Foundations of Mechanics*, Benjamin/Cummings, Reading, Massachusetts.

Albers, D.J. and Alexanderson, G.L. (Eds.) (1985) *Mathematical People. Profiles and Interviews*, Birkhäuser-Verlag, Basel.

Anglin, W.S. (1994) *Mathematics, a Concise History and Philosophy*, Springer-Verlag, Berlin.

Baron, M.E. (1987) *The Origins of the Infinitesimal Calculus*, Dover Publications, New York.

Boyer, C.B. (1968) *A History of Mathematics*, John Wiley, New York.

Bucciarelli, L.L. and Dworsky, N. (1980) *Sophie Germain: An Essay in the History of the Theory of Elasticity*, D.Reidel Publishing, Dordrecht, The Netherlands.

Cajori, F. (1985) *A History of Mathematics*, Chelsea, New York.

Cannell, D.M. (1993) *George Green: Mathematical Physicist, 1793-1841: The Background to His Life and Work*, Athlone Press, Atlantic Highlands, N.J.

Cannon, J.T. and Dostrovsky, S. (1981) *The Evolution of Dynamics: Vibration Theory from 1687 to 1742*, Springer-Verlag, Berlin.

Dugas, R. (1955) *A History of Mechanics*, Central Book Co., New York.

Duhem, P.-M.-M. (1980) *The Evolution of Mechanics*, Sijthoff and Noordhoff, Alphen aan den Rijn, The Netherlands.

Engelsman, S.B. (1984) *Families of Curves and the Origins of Partial Differentiation*, North Holland, Amsterdam, The Netherlands.

Eves, H.W. (1964) *An Introduction to the History of Mathematics*, Holt, Rinehart and Winston, New York.

Gellert, W., Küstner, H., Hellwich, M. and Kästner, H. (1975) *Mathematics at a Glance: A Compendium*, VEB Bibliographisches Institut, Leipzig.

Gellert, W. et al. (1975) *Mathematics at a Glance*, VEB Bibliographisches Institut, Leipzig, Germany.

Gillespie, C.C. (Ed.) (1972) *Dictionary of Scientific Bibliography*, Charles Scribner and Sons, New York.

Girvin, H.F. (1948) *A Historical Appraisal of Mechanics*, International Textbook Co., Scranton, PA.

Heath, Sir Thomas (1981) *A History of Greek Mathematics*, Dover Publications, New York.

Herival, J.W. (1975) *Joseph Fourier, The Man and the Physicist*, Clarendon Press, Oxford.

Kelvin, Lord and Tait, P.G. (1903) *A Treatise on Natural Philosophy*, Cambridge University Press.

Mach, E. (1907) *The Science of Mechanics: A Critical and Historical Account of its Development*, The Open Court Publishing Co., London.

O'Neil, P.V. (1995) *Advanced Engineering Mathematics*, Brooks/Cole, Boston, Mass.

Singer, C. (1959) *A Short History of Scientific Ideas to 1900*, Oxford University Press, Oxford.

Struik, D.J. (1969) *A Source Book in Mathematics 1200-1800*, Harvard University Press, Cambridge, Massachusetts.

Szabo, I. (1987) *Geschichte der Prinzipe der Mechanik*, Birkhauser-Verlag, Basel.

Timoshenko, S.P. (1953) *History of the Strength of Materials*, McGraw-Hill, New York.

Todhunter, I. and Pearson, K. (1886) *A History of the Theory of Elasticity*, Vol. I, Cambridge University Press, Cambridge.

Todhunter, I. and Pearson, K. (1893) *A History of the Theory of Elasticity*, Vol. II, Parts I and II, Cambridge University Press, Cambridge.

Truesdell, C. (1968) *Essays in the History of Mechanics*, Springer-Verlag, Berlin.

Truesdell, C. (1984) *An Idiot's Fugitive Essays on Science. Methods, Criticism, Training, Circumstances*, Springer-Verlag, Berlin.

Young, R.V. (Ed.) (1998) *Notable Mathematicians. From Ancient Times to the Present*, Gale Research, Detroit, Mich.

Index

Abel's Integral Equation, 262, 633
Acceleration in Fluid Motion, 394
Adherence Condition
– Slow Viscous Flows, 411
Airy Stress Function
– Biharmonic Equation, 94
– Biharmonic Equation in Plane Polar
Coordinates, 162
– Boundary Loading of Half-plane, 122
– Complex Variable Formulation, 145
– Complex Variable Representations, 97
– Fourier Transform Solution, 130
– Generalized Solution in Plane Polar
Coordinates, 166, 167
– Periodic Loading of Half-plane, 124
– Periodic Loading of Rectangular Region,
127
– Plane Polar Coordinate Formulation,
151, 161, 164
– Polynomial Representation, 105
– Polynomial Solutions of Biharmonic
Equation, 342
– Strip Loading of Half-plane, 136
– Two-dimensional Problems, 92
Airy Stress Function Approach
– Half-plane Problem, 147
– Plate Containing Circular Hole, 174
– Radial Loading of Fluid Filled Cavity,
170
Analytic Function Representation
– Boundary Conditions for Plate, 379
– Flexural Moments and Shear Force, 376
– Plate Deflection, 376
Anisotropy, Elastic, 62
Annular Thin Plate, Pure Bending, 313
Anticlastic Bending
– of Thin Plate, 336
– Paraboloidal Surface, 338

Averaged Stress-Strain Relations, 86
Axial Loading of Spherical Inclusion in
Elastic Solid, 482
Axial Symmetry
– Elastic Stress-Strain Relations, 196
– Equations of Equilibrium, 195, 196, 211
– Laplace's Operator, 197
Axial Viscous Flow, Poisson's Equation,
563
Axisymmetric Deflections of Thin Plate,
307
Axisymmetric Flexure of Clamped Plate,
487
Axisymmetric Indentation of Elastic
Half-space
– Dual Integral Equation, 259
Axisymmetric Loading of Conical Elastic
Region, 481
Axisymmetric Problems
– Boundary Conditions, 198
– in Elasticity, 182, 197
– in Elasticity, Regions with Cylindrical
Boundaries, 250
– Indentation of Half-space, 198
– Love's Strain Function Approach, 194
– Slow Viscous Flow, 409
Axisymmetric Surface Loading of an
Elastic Layer, 247
Axisymmetric Viscous Flow
– Past Rigid Sphere, 424

Bar with Elliptical Cross-section
– Torque-Twist Relationship, 604
Bar with Semi-Circular Cross-section
– Torque-Twist Relationship, 599
Bending Moments in Plate, 280
Bending of Circular Cantilever, 476

680   Index

Bending of Rectangular Plate, by End
    Couples, 108
Bernoulli Potential, for Viscous Fluid, 452
Bernoulli-Euler Beam Model
– Bending of Beam, 110
Bernoulli-Euler Plate Model
– Cantilever, 118
Betti-Maxwell Reciprocity Relations
– Elastic Continua, 62
Bi-Material Elastic Interface
– Concentrated Load at, 480
Biharmonic Equation, 1
– Airy Stress Function, Plane Polar
    Coordinates, 162
– Complex Variable Formulation, 374
– Flexure of Thin Circular Plate, 305
– for Airy Stress Function, 94
– for Love's Strain Function, 196, 197
– for Plate Deflection, 289
– for Slow Viscous Flow, 393
– for Stokes' Stream Function, 408
– for Thin Plate Flexure, 298
– Formulation of Slow Viscous Flows, 404
– Fourier Transform Solution, 130
– General Solutions of, 339
– Generalized Solution in Plane Polar
    Coordinates, 166, 167
– Generalized Variables Separable
    Solutions, 339
– Integral Transform Solution, 130
– Interior Solution, 199
– Methods of Solution, 104
– Polynomial Solution, 199
– Polynomial Solution for Airy Stress
    Function, 342
– Series Solution, 105
– Slow Viscous Flow, 422
– Slow Viscous Flow Problems, 407
– Solution Based on Integral Transforms,
    366
– Solutions Based on Integral Transforms,
    238
– Stokes' Stream Function, 459
– Three-dimensional Elasticity, 181
– Variables Separable Solution in
    Cartesian Coordinates, 118
– Variables Separable Solution in Complex
    Form, 378
Biharmonic Equation for Galerkin Vector,
    185
Biharmonic Function, General Properties,
    186
Body Force Vector, 40

Borehole in Elastic Solid, Radial
    Pressurization, 486
Borehole, Gravity Stress Field, 466
Boundary Conditions
– Thin Plate with Fixed Edge, 333
– Thin Plate with Free Edge, 333
– Thin Plate with Simply Supported
    Edge, 333
– Deflections of Membranes, 511
– Displacement, 68
– Elastic Support, 68
– Elastic Torsion, Prandtl's Stress
    Function, 573
– Groundwater Flow, 509
– Heat Conduction, 510
– Kirchhoff Shear Force, 299
– Plane Polar Coordinates, 162
– Plate Problems, 293
– Porous Media Flow, 509
– Slow Viscous Flow, 411, 461
– Thin Plates, Complex Variable Form,
    376
– Tractions, 68
Boundary Loading
– Infinite Plate, Integral Transform
    Solution, 367
– Rectangular Cantilever, 113
– Rectangular Plate, 111
– Semi-Infinite Plate by Distributed
    Moments, 491
Boundary Moments
– Loading of Simply Supported Rectangu-
    lar Plate, 360
Boundary Shear Tractions
– on Infinite Elastic Strip, 474
Boundary Tractions on Cantilever Plate,
    473
Boussinesq's Problem
– Displacements in Cylindrical Coordi-
    nates, 231
– Displacements in Spherical Coordinates,
    230
– Lamé Strain Potential for, 228
– Stress Field in Cylindrical Coordinates,
    231
– Stress Field in Spherical Coordinates,
    230
– Surface Loading of Half-plane, 225
– Surface Reinforced Half-space, 243
– Tractions on a Hemispherical Surface,
    229
– Tractions on Spherical Surface, 467
Bulk Modulus, 66

Bulk Viscosity, 398

Cantilever Plate, Boundary Shear Loading, 473
Cantilever, Bernoulli-Euler Plate Model, 118
Cartesian Coordinates
– Transformation to Plane Polar Coordinates, 159
Cauchy's Integral Formula, 606
Cauchy's Stress Principle, 36
Cauchy-Riemann Equations, 96, 99
Cavity in Elastic Medium
– Complex Variable Solution for Radial Loading, 177
Cayley-Hamilton Equation, 24, 49
Centre of Dilatation
– in Elastic Solid, 222
Change in Length, Line Element, 11
Charge Distribution on Disc, 631
Circular Bar
– Torque-Twist Relationship, 584
Circular Cantilever
– Bending of, 476
Circular Cavity in Stressed Elastic Solid
– Complex Variable Approach, 181
– Displacement Fields, 181
– Stress Field, 181
Circular Fluid Domain
– Diffusive Motion, 448
– Unsteady Fluid Motion, 448
Circular Hole
– in Uniformly Stressed Plate, 173
Circular Hole in Infinite Elastic Medium
– Complex Variable Approach, 179
Circular Plate
– Complex Variable Method of Analysis, 325
– Loaded by Non-Uniform Fluid Pressure, 489
– with Diametral Displacement Constraints, 489
Circular Region
– Slow Viscous Flow, 449
Circular Rigid Indentor
– Elastic Force-Displacement Relationship, 266
Circular Solid Plate
– Pure Bending, 203
– Uniform Loading, 206
Circular Thin Plate, 305
Boundary Conditions, 306
Circular Tube, Unsteady Viscous Flow in, 443

Clamped Circular Plate, Radius of Inversion, 326
Clamped Plate
– Axisymmetric Flexure, 487
– Uniqueness of Solution, 390
Clamped Stretched Elastic Sheet, 470
Clamped Thin Circular Plate
– Concentrated Loading, 311
– Green's Function, 324
– Loading by Stiffener, 316
– Uniform Loading, 308
Clamped Thin Plate, Concentrated Loading, 318
Classical Elasticity
– Axisymmetric Problems, 182
– Plane Problems, 81
– Two-dimensional Problems, 81
– Uniqueness Theorem, 72
Classical Theory of Thin Elastic Plates, 297
Compatibility Conditions
– for Existence of Solution for Poisson's Equation, 523
– for Strains, 28
– for Strains in Plate, 279
Compatibility Equations of Saint-Venant, 28
Compatibility of Strains in a Continuum, 454
Compatibility Relationship
– for Strains, 26
– for Stresses, 89
Complex Form
– Solution of Biharmonic Equation, 378
Complex Potentials
– Circular Cavity in Stressed Elastic Solid, 179
Complex Value of Eigenfunctions, 542
Complex Variable Approach
– Circular Hole in Uniaxially Loaded Medium, 179
– Formulation of Biharmonic Equation, 374
– Plane Problem in Elasticity, 176
– Polar Coordinate Formulation, 176
– Uniform Loading of Thin Clamped Plate, 384
– Uniqueness of Solution for Clamped Plate, 390
Complex Variable Form
– Boundary Conditions for Thin Plates, 376
Complex Variable Formulation

– Mapping Function, Elastic Torsion, 605
– Prandtl's Stress Function, 600
– Torque-Twist Relationship, 607
– Torsion of Circular Bar, 607
– Torsion of Elastic Bar with Elliptical Cross-section, 603
– Torsion Problem, 600
Complex Variable Method
– Analytic Function Representation of Deflection, 376
– Applications to Thin Plates, 374
– Applied to Pure Twisting of Plate, 379
– Applied to Thin Circular Plate, 325
– Boundary Loading of Half-plane, 147
– for Plane Elasticity Problems, 145
Complex Variables, Applications in Elasticity, 94
Compression of Elastic Layer, 485
Concentrated Boundary Loading, Elastic Half-plane, 143
Concentrated Force
– Action in Infinite Elastic Solid, 213
– on Surface Reinforced Half-space, 243
– Surface Loading of Half-plane, 225
Concentrated Loading
– Bi-Material Elastic Interface, 480
– Clamped Thin Plate, 311, 318
Conformal Transformation
– Elastic Torsion, Complex Variable Formulation, 605
Conical Elastic Region
– Axisymmetric Loading, 481
Conjugate Displacement Gradient Dyadic, 9
Conjugate Harmonic Function
– in Airy Stress Function Approach, 97
– Stress Dyadic, 44
Conservative Force Field, 616, 618
Consistent Boundary Conditions for Elastic Plate, 296
Constitutive Equations
– for Elastic Solids, 57
– Incompressible Newtonian Viscous Fluid, 399
– Newtonian Viscous Fluid, 397
Constitutive Relationship for Newtonian Viscous Fluid, 398
Constrained Bi-Material Elastic Infinite Space Loaded by Inclusion, 486
Constrained Elastic Solid
– Cylindrical Cavity in, 477
Contact Problem for Elastic Half-space
– Dual Integral Equations, 259

Contact Problems in Elasticity, 258
Contact Stresses Beneath Rigid Indentor, 266
Continuity Equation, 396
Continuum
– Compatibility of Strains, 454
– Displacements, 5
– Strains in, 5
– Stresses in, 36
Continuum Concept, 3
Continuum Mechanics, 3
Corner Forces
– in Thin Plate Boundary Conditions, 303
Couette Flow Around Cylinder, 497
Creeping Flows, Stokes' Paradox, 419, 459
Curl of Strain Dyadic, 26
Cylindrical Boundaries
– Axisymmetric Problems in Elasticity, 250
Cylindrical Cavity
– in Constrained Elastic Solid, 477
– in Infinite Body, Complex Variable Approach, 177
Cylindrical Elastic Bar
– Surface Loading, 254
Cylindrical Polar Coordinates
– Galerkin Vector, 185, 186

Darcy's Law
– Groundwater Flow, 506
– Porous Media Flow, 506
Decomposition Theorem due to Helmholz, 184
Deflection of Beam, by End Couples, 110
Deflection of Stretched Partially Loaded Membrane, 638
Deflections in Thin Plates
– Admissible Solution in Polar Coordinates, 307
– Complete Solution in Polar Coordinates, 307
Deflections of Membranes, Boundary Conditions, 511
Deformation
– of Twisted Prismatic Bar, 568
– Rate of, 395
– Small, 11
– Two-dimensional, 13
Deformation Gradient
– Matrix, 6
– Transpose of, 7
Deformation of Plate Region, 269
Delta Function, Heisenberg's, 257
Density

– at a Point, 4
– within Finite Volume, 4
Derivatives, Time, 71
Deviator Stresses in Viscous Fluid, 397
Deviatoric Strain, 66
Deviatoric Stress, 66
Diffusion Equation
– Similarity Solutions Approach, 500
Diffusion of Velocity, in Viscous Fluid
    Domain, 442
– Laplace Transform Solution, 442
Diffusive Fluid Motion
– Circular Viscous Fluid Domain, 448
Diffusive Motion in Viscous Fluids, 440
– Axial Flow, 446
– Flow in Circular Tube, 444
– Navier-Stokes' Equation, 452
– Rotary Motion, 447
– Vorticity, 451
Diffusive Viscous Flow, Unsteady Motion,
    440
Directions Cosines Matrix, 20
Dirichlet Boundary Value Problem
– Green's Function, 531
– Poisson's Equation, 529
Dirichlet Eigenvalues, Poisson's Equation,
    541
Dirichlet Green's Function
– Poisson's Equation for Half-space, 554
Dirichlet Problem
– for Half-space, Laplace's Equation, 555
– for Poisson's Equation, Quarter Plane
    Region, 641
– for Prandtl's Stress Function, Poisson's
    Equation, 573
– Poisson's Equation, 558
– Poisson's Equation, Eigenfunctions
    Solution, 546
– Prandtl's Stress Function, Multiply
    Connected Region, 578
Disc Inclusion at Bi-Material Elastic
    Infinite Space, 486
Displacement Boundary Conditions, 69
– Axisymmetric Problem, 198
– Plane Polar Coordinates, 162
Displacement Components
– Love's Strain Function, 193, 194
– Multiply Connected Domain, 100
– Simply Connected Domain, 98
Displacement Field
– Circular Cavity in Stressed Elastic Field,
    181
– Galerkin Vector Representation, 184

– Kelvin's Problem, 218
Displacement Field for Boussinesq's
    Problem
– Cylindrical Coordinates, 231
– Spherical Coordinates, 230
Displacement Gradient
– Dyadic, 8
– Matrix, 7, 18
Displacement Vector, 6
– Plane Polar Coordinates, 154
Displacements in a Continuum, 5
Displacements in Infinite Solid
– Influence of Distributed Loads, 219
Displacements in Spherical Coordinates
– Lamé Strain Potential Formulation, 211
– Love's Strain Function Appoach, 212
Distributed Interior Loading of Elastic
    Solid, 219
Distribution of Charge on Disk, 634
Divergence of Strain Dyadic, 26
Double Layer Potential, 529
Drag on Sphere, Slow Viscous Flow, 430
Dual Integral Equations
– Axisymmetric Indentation of Elastic
    Half-space, 259
– Mixed Boundary Value Problem, 633
– Slow Viscous Flow Induced by Moving
    Plate, 438
– Slow Viscous Flow Induced by Rotating
    Plate, 434
Dubois-Reymond Lemma, 44, 76, 292, 392,
    396, 400, 462, 505, 508
Dyadic
– Conjugate, 9
– Displacement Gradients, 8
– of Flexural Strains, 275
– of Stresses, 37, 38
– Rate of Deformation, 395
– Rotation, 10, 277
– Strain, 10, 63
– Stress, 63
– Unit, 63
– Vorticity, 395
Dynamic Equilibrium
– Equations of, 42
– Viscous Fluid Flow, 397
Dynamic Shear Viscosity, 398

Earthed Cavity
– Interaction with a Line Charge, 626
Edge Loading of Clamped Plate, 493
Edge Loading of Infinite Plate
– Action of Distributed Moments, 367
Eigenfunction Expansion Solution

684   Index

– Groundwater Flow, 550
Eigenfunction Solution
– Green's Function for Poisson's Equation, 535
– Poisson's Equation, Rectangular Region, 546
Eigenfunctions
– Complex Valued, 542
– Proof of Orthogonality, 543
Eigenfunctions Expansion Solution, Application to Poisson's Equation, 550
Eigenvalue Problem, Poisson's Equation, 540
Eigenvalues, of Strain Matrix, 24
Einstein's Summation Convention, 25
Elastic Anisotropy, 62
Elastic Body
– Kinetic Energy, 80
– Total Energy, 81
Elastic Constants, 62
– Matrix, 58
– Relaxed Constraints, 78
– Thermodynamic Constraints, 66
Elastic Continua
– Reciprocity Relations, 62
Elastic Cylinder in Viscous Fluid Region Under Shear Flow, 496
Elastic Energy, 58
Elastic Force-Displacement Relationship
– Circular Rigid Indentor on Half-space, 266
Elastic Half-plane
– Boundary Loading, 132, 147
– Concentrated Line Loading, 143
– Green's Function, 143
– Harmonic Boundary Loading, 121
– Periodic Boundary Loading, 124
Elastic Half-space
– Concentrated Force on Reinforced Surface, 243
– Frictionless Axisymmetric Indentation by Cylinder, 259
– Frictionless Indentation, 258
– Indentation Problem, 198
Elastic Half-space with Bonded Boundary
– Localized Loading, 485
Elastic Inclusion Problem, 167
Elastic Incompressible Material, 66
Elastic Layer
– Axisymmetric Surface Loading, 247
– Compression, 485
Elastic Material
– Strain Energy, 60

Elastic Mound
– Gravity Stresses, 468
Elastic Plate
– Consisitent Boundary Conditions, 296
– In-Plane Boundary Tractions, 472
– Strain Energy of, 294
– Torque-Twist Relationship, 594
– Virtual Work, Equation for, 294
Elastic Plug in Rigid Cavity, 167
Elastic Solid
– Centre of Dilatation, 222
– Containing Constrained Cavity, 483
– Containing Fluid Inclusion, 483
– Containing Spherical Cavity, Uniform Tension, 233
– Smoothly Embedded Inclusion, 484
– Spherical Cavity Problem, 233
Elastic Stiffness of Spring Mount, 471
Elastic Strain Energy
– Positive Definiteness, 66
Elastic Stress-Strain Relations
– Axial Symmetry, 196
– for Plate, 282
– Plane Polar Coordinates, 158
Elastic Strip
– Boundary Shear Tractions, 474
– Elastically Supported, 474
Elastic Support Boundary Conditions, 69
Elastic Torsion
– Prismatic Bar, Alternative Derivation, 608
– Bar with Semi-Circular Section, 597
– Circular Bar, 583
– Complex Variable Formulation, 600
– Elliptical Bar, 584
– Equations of Equilibrium, 594
– General Formulation, 568
– Laplace's Equation for Warping Function, 571
– Mapping Function, Complex Variable Formulation, 605
– Neumann Boundary Value Problem, 612
– Poisson's Equation, 565
– Prandtl's Stress Function Representation of Stress Components, 594
– Prandtl's Stress Function, Bar with Semi-Circular Cross-section, 599
– Prismatic Bar with Rectangular Section, 588
– Prismatic Bar, Multiply Connected Section, 577

– Prismatic Bar, Stress-Strain Relations, 570
– Resultant Torque on Multiply Connected Region, 580
– Resultant Torque, Prandtl's Stress Function, 577
– Resultant Torque, Warping Function, 576
– Solution from Equations of Boundary, 581
– Stiffness Bounds, 592
– Stress Function Approach, 573
– Thin Plate, 592
– Tractions on Surface of Prismatic Bar, 571
– Variables Separable Solution, 587
Elastic Torsion of Prismatic Bar
– Strain Components, 570
– Warping Function, 571
Elastically Supported Infinite Elastic Strip, 474
Elasticity
– Airy Stress Function, Problems, 92
– Axisymmetric Problem, 197
– Complex Variable Methods, 94, 325, 374, 600
– Contact Problems, 258
– Isotropic, 62
– Love's Strain Function Approach, 193
– Plane Problems, 81
– Plate Problems, 267
– Three-dimensional Problems, 181
– Torsion Problems, 568
Elasticity Matrix, Symmetry of, 60
Elasticity Problems, Mixed Boundary Conditions, 258
Elliptical Cross-section
– Elastic Torsion, Complex Variable Formulation, 603
Energy Balance Equation
– Viscous Fluids in Motion, 400
Energy Dissipation
– Non-conservative Force Field, 616
Entropy of Viscous Fluid, 401
Equation of Compatibility
– Plane Polar Coordinates, 158
Equation of Continuity, 396
– Groundwater Flow with Recharge, 506
– Porous Media Flow with Recharge, 506
Equations of Boundary
– for Solving Elastic Torsion Problems, 581
Equations of Dynamic Equilibrium, 42

Equations of Equilibrium, 41, 292
– Axial Symmetry for Spherical Coordinates, 211
– Dyadic Notation, 42
– Elastic Plate via Virtual Work, 296
– for Axial Symmetry, 195, 196
– Indicial Notation, 42
– Plane Polar Coordinates, 153, 161
– Plate, 286
– Torsion of Prismatic Bodies, 594
Equations of Motion
– Slow Viscous Flow, 397
– Viscous Fluid, Dyadic Notation, 397
– Viscous Fluid, Indicial Notation, 397
Equilibrium, Equations of, 41
Equivalence of Plane Elastic Stress States, 91
Euler-Type Ordinary Differential Equations, 164
Eulerian Description, Fluid Flow, 395
Existence of Solutions, Poisson's Equation, 521
Exterior Domains, Lamé Strain Potentials, 215, 216

Finite Fourier Cosine Transform, 260
Fixed Boundaries, Infinite Plate, 370
Flat Circular Plate, Slow Viscous Flow Region, 436
Flexural
– Moment at Plate Boundary, 282
– Moments in Thin Circular Plates, 305
– Rigidity of Plate, 283
– Strains, Dyadic, 275
– Stresses, 279
Flexure of
– Circular plate by Bonded Rigid Cylinder, 490
– Plate, Movement-Curvature Relationship, 286
– Plate, Shear Force-Rotation Relationship, 286
– Plate-Column Connection, 491
– Simply Supported Rectangular Plate, 348, 351
– Thick Elastic Plate, 268
– Thin Elastic Plate, 267
Flow Between Parallel Plates
– Fluid Potential, 415
Flow in a Narrow Aperture
– Viscous Thin Film Lubrication, 416
Flow in Fracture, Parallel Plate Model, 415
Flow in Porous Media

– Applications of Poisson's Equation, 504
– Porosity, 504
Flow of Viscous Fluid in a Narrow Aperture, 413
– Averaged Fluid Velocity, 415
Flow Past Sphere, Viscous Fluid, 424
Fluid Filled Cylindrical Cavity
– Loading of Half-plane by, 170
Fluid Filled Cylindrical Cavity in Elastic Solid, 470
Fluid Flow
– Eulerian Description, 395
– Lagrangian Description, 394
Fluid Inclusion in Elastic Solid Subjected to Triaxial Loading, 483
Fluid Potential, Flow Between Parallel Plates, 415
Fluid Pressure
– in Viscous Lubrication Zone, 418
– Loading of Plate, 494
Fluid, Incompressibility Condition, 396
Force Field
– Conservative, 616, 618
– Non-conservative, 616, 619
Force of Attraction, 614
Force-Velocity Relationship
– Circular Plate in Viscous Fluid Region, 439
Fourier Cosine Transform, 260
Fourier Series Solution
– Elastic Torsion of Prismatic Bar, Rectangular Section, 589
Fourier Transform of Green's Function, 533
Fourier Transform Solution
– Half-plane Problem, 132
– Half-plane Subjected to Strip Load, 135
– Infinite Elastic Plane, 138
– Strip Loading of Half-plane, 136
Fourier's Law, Heat Conduction, 507
Frictional Sliding
– Spherical Inclusion in Viscous Flow Domain, 500
Frictionless Indentation of Elastic Half-space, 258
Fundamental Solution for Laplace's Equation, 527
Fundamental Solution for Poisson's Equation
– in Three Dimensions, 553
– in Two Dimensions, 553

Galerkin Vector

– Biharmonic Equation for, 185
– Cylindrical Polar Coordinates, 185, 186
Galerkin Vector Representation of Displacements, 184
General Properties of Biharmonic Functions, 186
General Solutions of Biharmonic Equation, 339
Generalized Hooke's Law, 57
Generalized Plane Stress, 81
Generalized Plane Stress Solution, 85, 87
Generalized Results, Poisson's Equation, 513
Generalized Solution
– Airy Stress Function in Plane Polar Coordinates, 166, 167
Generalized Variables Separable Solutions
– Biharmonic Equation, 339
Generic Point, 6
Gradient of Vector, 8
Gradient Operator
– in Two-Dimensions, 270
Gravitational Law, 614
Gravitational Potential, 614
– Poisson's Equation, 623
Gravity Stresses
– Around Borehole, 466
– in Elastic Mound, 468
Green's Function
– Clamped Thin Circular Plate, 324
– Dirichlet Boundary Value Problem, 531
– Dirichlet Problem, Symmetry of, 551
– Dirichlet, Poisson's Equation, 546
– Eigenfunction Solution for Poisson's Equation, 535
– Elastic Half-plane, 143
– for Thin Plate, 330
– Fourier Transform, 533
– Half-space, Dirichlet Conditions, 554
– Half-space, for Poisson's Equation, 554
– Infinite Elastic Plane, 138
– Reciprocity Principle, Poisson's Equation, 552
– Regularity Condition, 532
– Simply Supported Rectangular Plate, 350
Green's Function for Laplace's Equation, 190
Green's Function for Poisson's Equation
– Dirichlet Boundary Condition, 535
– for Infinite Strip, 532
– Method of Images, 553

– Series Solution for Rectangular Region, 535
– Single Fourier Series Solution, 537
Groundwater Flow
– Applications of Poisson's Equation, 504
– Boundary Conditions, 509
– Darcy's Law, 506
– Eigenfunctions Expansion Solution, 550
– Equation of Continuity, with Recharge, 506
– in Porous Seam, 635

Hagen-Poiseulle Flow, in Circular Tube, 445
Half-plane Problem
– Boundary Loading, 147
– Complex Variable Approach, 147
– Fourier Transform Solution, 132
– State of Stress in, 50
– Subject to Boundary Loading, 132, 135
– Subject to Uniform Strip Loading, 135
– Surface Loading by Concentrated Force, 225
Half-space
– Potential Problem, 629
Hankel Transform Solution of Love's Strain Function, 239
Harmonic Boundary Loading
– Elastic Half-plane, 121
– Semi-Infinite Plane, 121
Harmonic Function
– in Airy Stress Function Approach, 97
– Mean Value Theorem, 517
Heat Conduction
– Boundary Conditions, 510
– Fourier's Law, 507
– Steady State, Poisson's Equation, 508
– with Heat Generation, 507
– with Heat Sources, Poisson's Equation, 507
Heat Gain
– Convected Time Derivative of, 402
Heat Generating
– Concrete, 635
– Irradiated Nuclear Waste, 636
Heat Generation
– Decaying Materials, 507
Heisenberg's Delta Function, 257
Hele-Shaw Model
– Flow Between Parallel Plates, 416
– Flow in Fracture, 640
– Slow Viscous Flows, 413
Helmholz's Decomposition Theorem, 184

Hermitian Inner Product, 541
Hermitian Symmetry, 542
Homogeneity, 62
Hooke's Law, Generalized, 57
Hyperbolic Paraboloidal Surface
– Anticlastic Bending of Thin Plate, 338

Identity Matrix, 7
Image Solution
– Half-space Problem, 629
In-Plane Boundary Tractions on Elastic Plane, 472
Inclusion in Rigid Cavity
– Plane Problem, 167
Incompressibility Condition
– Elastic Solid, 66
– Viscous Fluid, 396
Incompressible Fluid, 396
Indentation Problem, Elastic Half-space, 198
Infinite Elastic Plane
– Fourier Transform Solution, 138
– Green's Function, 138
Infinite Elastic Solid
– Kelvin's Problem, 213
– Spherical Cavity Problem, 209
Infinite Plate
– Clamped Boundaries, 370
– Edge Loading, 493
– of Finite Width, Integral Transform Solution, 367
– of Infinite Width, Simply Supported, 353
Infinite Space
– Interior Loading, 209
Infinite Strip
– Green's Function for Poisson's Equation, 532
Inner Product, 541
– Hermitian, 541
– Symmetry of, 541
Integrability of Strains in Continuum
– Necessary and Sufficient Conditions, 455
Integral Equation
– Abel-Type, 262
– Mixed Boundary Value Problem, 633
Integral Transform Solution
– Biharmonic Equation, 238
– of Poisson's Equation, 531
– of Stokes' Problem for Circular Disc, 438
– Plane Problem in Elasticity, 130
– Potential Problem, 629

– Rotating Flat Plate in Viscous Fluid,
  431
Integral Transform Techniques
– Applications in Flexure of Plates, 366
– for Slow Viscous Flow Problem, 430
Interaction of Earthed Cavity and Line
  Charge, 626
Interior Loading of Infinite Space
– Kelvin's Problem, 209
Internal Energy Dissipation
– Newtonian Viscous Fluid, 399
– Viscous Fluid, 402
Invariants
– of Strain Matrix, 24, 25
– of Stress Matrix, 49
Isotropic Elastic Cylinder
– Radial Loading, 481
Isotropic Elasticity, 62
– Plane Polar Coordinates, 158
Isotropic Stress
– Stress State in Viscous Fluid, 397
Isotropic Viscous Fluids, 398
Isotropy, 62

Kelvin's Problem
– Concentrated Force in Infinite Solid, 213
– Displacement Field, 218
– Interior Loading of Infinite Space, 209
– Neuber-Papkovich Formulation, 218
– Stress Field, 218
Kinetic Energy
– Convected Time Derivative for Slow
  Viscous Flow, 463
– Newtonian Viscous Fluid, 399
– Slow Viscous Flows, 463
Kinetic Energy of Elastic Body, 80
Kirchhoff Boundary Conditions
– for Shear Force, 299
– for Thin Plate, 300
– Rectangular Plate, 333
Kirchhoff's Uniqueness Theorem, 73
Kronecker's Delta Function, 24

Lagrangian Description
– Fluid Flow, 394
Lamé Strain Potential
– Expression for Displacement in Spherical
  Coordinates, 211
– Expression for Stresses, 211
– Exterior Domains, 215, 216
– Solution of Boussinesq's Problem, 228
Lamé's Constants, 62
Lamé's Strain Potential, 182

Laplace Transform Solution of Diffusive
  Motion in Viscous Fluid, 442
Laplace's Equation
– Dirichlet Problem for Half-space, 555
– for Warping Function, 571
– for Warping Function, Polar Coordi-
  nates, 595
– Fundamental Solution, 527
– Green's Function, 190
– Solution for Spherical Symmetry, 526
Laplace's Operator
– Cylindrical Coordinates, 186
– for Axial Symmetry, 197
– Plane Polar Coordinates, 162
– Rectangular Cartesian Coordinates, 93
Legendre's Equation, 200
Legendre's Function
– First Kind, 200
– Recurrence Relation for First Kind, 200
– Recurrence Relation for Second Kind,
  201
– Second Kind, 200
Lemma
– Dubois-Reymond, 44, 76, 292, 392, 396,
  400, 462, 505, 508
Levy's Solution
– Simply Supported Rectangular Plate,
  351
Levy's Theorem
– Plane Elastic Stress State, 91
Lifting a Heavy Plate by Localized Load,
  488
Line Charge
– in Conductive Medium, 625
Line Element
– Change in Length, 11
Line Loading
– Integral Transform Solution, 138
– of Elastic Half-plane, 143
– of Infinite Elastic Plane, 138
– of Infinite Plate, Simply Supported, 353
Line Vortex in Viscous Flow Domain, 454
Linear Elasticity, Constitutive Equations,
  57
Loading
– by Stiffener, Clamped Circular Plate,
  316
– of a Half-plane by Fluid Filled
  Cylindrical Cavity, 169
– of Clamped Semi-Infinite Plate,
  Concentrated Load, 381
– of Plate at Re-Entrant Corner, 475
– of Wedge Shaped Plate, 476

Local Time Derivatives, 71
Localized Loading
- of Two-Material Plate Region, 494
- Thin Plate Simply Supported, 350
Love's Strain Function
- Axisymmetric Problem in Elasticity, 193
- Biharmonic Equation for, 196, 197
- Displacement Components in Terms of, 193, 194
- Expression for Displacement in Spherical Coordinates, 212
- Expression for Stresses, 213
- for Kelvin's Problem, 216
- Hankel Transform Solution, 239
- Legendre Polynomial Solution, 201, 202
- Spherical Polar Coordinate Formulation, 209
- Stress Components in Terms of, 194
Low Reynolds Number Flow, 404
Low Reynolds Number Flows
- Biharmonic Equation, 393
- Past Cylindrical Objects, Non-existence of, 421

Mapping Function
- Complex Variable Formulation, Elastic Torsion, 605
Mass Density, 4
Material
- Homogeneity, 62
- Isotropy, 62
- Time Derivative, 71, 395
Matrix
- Deformation Gradient, 6
- Displacement Gradient, 7, 18
- of Direction Cosines, 20
- of Elastic Constants, 58
- of Stresses, 38
- Rotation, 8, 18
- Strain, 8, 11
- Stress, 37
Matrix Transformation, 20
Maximum Principle
- Subharmonic Function, 519
Mean Value Theorem
- Harmonic Function, 517
- Subharmonic Function, 515
Mechanics of Continua, 3
Mechanics of Elastic Solids, 2
Mechanics of Viscous Fluids, 2
Method of Eigenfunctions
- Solution of Poisson's Equation, 540
Method of Images

- Green's Function for Poisson's Equation, 553
Methods of Solution of Biharmonic Equation, 104
Mixed Boundary Value Problems
- Dual Integral Equation, 633
- in Elasticity, 258
- Potential Problem, 633
- Slow Viscous Flow Induced by Moving Plate, 437, 438
- Slow Viscous Flow Induced by Rotating Plate, 434
Modulus
- Bulk, 66
- Shear, 65
- Young's, 65
Moment-Curvature Relationship
- Cartesian Relationship, 284
- for Plate, Invariant Form, 284
- for Plate, Polar Coordinates, 286
- Rectangular Cartesian Coordinates, 332
Moment-Rotation Relationship
- Plate in a Viscous Fluid Domain, 435
Monotonicity Result
- for Poisson's Equation, 557
Moving Circular Plate, Viscous Fluid Region, 436
Moving Plate on Viscous Fluid, Unsteady Motion, 441
Multiply Connected Domain, Displacement Components, 100
Multiply Connected Sections, Elastic Torsion of, 577

Navier's Solution
- Flexure of Thin Rectangular Plates, 345
- Simply Supported Rectangular Plate, 348
Navier-Stokes Equation
- Diffusive Motion, 452
- Linearized Form, 403
- Newtonian Viscous Fluid, 403
- Non-Linear Form, 403
- Reduction to Slow Viscous Flows, 404
- Viscous Flows, 463
Neuber-Papkovich Formulation
- Kelvin's Problem, 218
Neumann Boundary Value Problem
- Elastic Torsion, 612
Neumann Problem
- Laplace's Equation for Warping Function, 572
Newton's Law of Gravitation, 614
Newtonian Viscous Flow

690 Index

− Biharmonic Equation Approach, 393
Newtonian Viscous Fluid
− Constitutive Equation, Incompressible, 399
− Constitutive Equations, 397
− Constitutive Relationship, 398
− Internal Energy Dissipation, 399
− Kinetic Energy, 399
Non-conservative Force Field, 616, 619
− Energy Dissipation, 616
Non-Existence of Solutions
− Slow Viscous Flows, 419, 459
Norm of Function, 541
Normal Stresses
− Sign Convention, 45

One-Dimmensional State of Strain, 13
Ordinary Differential Equations of the Euler Type, 164
Orthogonality of Eigenfunctions, Proof of, 543
Orthogonality of Function, 541

P-wave Velocity, 78
Paraboloidal Shape
− Pure Bending of Thin Plate, 336
Parallel Plate Model of Flow in Fracture, 415
Partial Differential Equation for Rotation Vector, 289
Partially Loaded Stretched Membrane, 637
Periodic Boundary Tractions on Rectangular Regions, 127
Periodic Loading of Half-plane
− Airy Stress Function Approach, 124
Permeability of a Fracture
− Parallel Plate Model, 416
Physical Interpretation
− Strain Matrix, 16
− Strains, 12
Planar Problems in Slow Viscous Flow, 407
Plane Elastic Stress States, Equivalence, 91
Plane Elasticity Problem, Complex Variable Formulation, 145
Plane Polar Coordinate Formulation
− Airy Stress Function, 151, 161, 164
− in Elasticity, 151
Plane Polar Coordinates
− Boundary Conditions, 162
− Displacement Boundary Condition, 162
− Displacement Vector, 154

− Elastic Stress-Strain Relations, 158
− Equations of Compatibility, 158
− Equations of Equilibrium, 153, 161
− Isotropic Elasticity, 158
− Strain Energy Density, 159
− Strain Matrix, 157
− Stress Matrix, 152
− Traction Boundary Condition, 162
Plane Problem for Half-plane Region, 169
Plane Problem in Elasticity
− Complex Variable Approach, 176
− Fourier Transform Solution, 130
− Polar Coordinate Formulation, 151
− Polar Coordinates, 151
Plane Problems
− in Classical Elasticity, 81
− Polar Coordinate Formulation, 158
Plane Problems in Elasticity
− Complex Variable Approach, 94
Plane Slow Viscous Flows
− Stokes' Paradox, 419, 459
Plane Strain, 81
Plane Stress
− Generalized, 81
− Generalized Solution, 85, 87
− Solution, 84
Plane Viscous Flow
− due to Rotation of Flat Plate, 431
Plate
− Equation of Equilibrium, 286
− Moment-Curvature Relationship, 284
− Strain Energy of, 289
Plate Bending Moments, 280
Plate Boundary
− Flexural and Twisting Moments, 282
Plate Deflection
− Biharmonic Equation for, 289
Plate Flexural Rigidity, 283
Plate Problems, Boundary Conditions, 293
Plate Region, Deformation of, 269
Plate Stress-Strain Relations, 282
Plate Subjected to Fluid Pressure, 494
Plate Twisting Moments, 280
Poisson's Equation
− Applications, 503
− Axial Viscous Flow in Duct, 563
− Compatibility Condition for Existence of Solution, 523
− Dirichlet Boundary Value Problem, 529
− Dirichlet Eigenvalues, 541
− Dirichlet Problem, 558
− Dirichlet Problem, Prandtl's Stress Function, 613

- Eigenfunctions Solution for Rectangular Region, 546
- Eigenvalue Problem, 540
- Elastic Torsion, 568
- Elastic Torsion of Prismatic Bodies, 565
- Elastic Torsion, Rectangular Cross-section, 588
- Elastic Torsion, Variables Separable Solution, 587
- Existence of Solutions, 521
- Flow in Fracture with Recharge, 640
- Flow in Porous Media with Internal Sources, 504
- Fundamental Solution in Three Dimensions, 553
- Fundamental Solution in Two Dimensions, 553
- Generalized Results, 513
- Gravitational Potential, 623
- Green's Function for, 526
- Green's Function for a Half-space, 554
- Green's Function for Infinite Strip, 532
- Green's Function for Rectangular Region, 535
- Green's Function for the Dirichlet Problem, 531
- Groundwater Flow, 506
- Groundwater Flow in Porous Seam, 548, 635
- Heat Conduction with Heat Sources, 507
- Heat Generating Concrete, 635
- Heat Generating Irradiated Nuclear Waste, 636
- Integral Transform Techniques, 531
- Maximum Principle for Subharmonic Function, 524
- Maximum Principle for Superharmonic Functions, 524
- Method of Images, 641
- Method of Images Solution for Green's Function, 553
- Monotonicity Result, 557
- Neumann Boundary Value Problem, 523
- Partially Loaded Stretched Membrane, 637
- Porous Media Flow, 506
- Prandtl's Stress Function, 573, 613
- Prandtl's Stress Function for Multiply Connected Region, 578
- Prandtl's Stress Function, Polar Coordinates, 596
- Reciprocity Principle for Green's Function, 552
- Single Series Solution for Green's Function, 537
- Solution by Method of Eigenfunctions, 540
- Strong Maximum Principle, 558
- Torsion of Bar, Cross-section Bounded by Parabolic Curves, 646
- Torsion of Circular Bar with Groove, 644
- Torsion of Hollow Circular Bar with Longitudinal Cut, 644
- Torsion of Hollow Cylinder, 642
- Torsion of Prismatic Bar with Triangular Section, 643
- Torsion of Prismatic Bar, Cross-section of a Sector, 643
- Torsion of Prismatic Bar, Cross-section of Cardioid, 646
- Transverse Deflection of Membranes, 508
- Uniqueness Theorem, 521
- Viscous Flow in Conduits, 559, 641
- Weak Maximum Principle, 558
Poisson's Ratio, 65
- Thermodynamic Constraints, 68
Polar Coordinate Approach
- Complex Variable Approach, 176
Polya-Inequalities for Torsional Stiffness of Prismatic Bars, 592
Polynomial Representation
- Airy Stress Function, 105
Polynomial Solutions
- Airy Stress Function, 342
- Biharmonic Equation, 199
Porosity, Flow in Porous Media, 504
Porous Media Flow
- Boundary Conditions, 509
- Darcy's Law, 506
- Equation of Continuity, with Recharge, 506
Porous Seam
- Groundwater Flow, Poisson's Equation, 548
Positive Definite Energy Dissipation
- Viscous Fluid, 402
Positive Definiteness
- Elastic Strain Energy, 66
Potential
- Double Layer, 529
- Due to a Disk, 624
- Due to a Line Charge, 625
- Gravitational, 617

– Line Charge-Earthed Cavity Interaction, 628
– Single Layer, 529
– Spatial Distribution of Matter, 620
Potential Problem
– Half-space Region, 629
– Integral Transform Solution, 629
– Mixed Boundary Value Problem, 633
Power and Energy, Theorem of, 75
Prandtl's Solution
– Elastic Torsion of Thin Plate, 592
Prandtl's Stress Function
– Complex Variable Formulation, 600
– Dirichlet Problem for Poisson's Equation, 573
– Elastic Torsion, 573
– Elastic Torsion, General Formulation, 613
– Evaluation of Torque-Twist Relationship, 599
– Poisson's Equation for, 573
– Poisson's Equation for Multiply Connected Region, 578
– Polar Coordinate Representation, 594
– Representation of Torque-Twist Relationship, 614
– Torsion of Bar with Rectangular Section, 589
– Torsion of Bar with Semi-Circular Cross-section, 599
– Torsion of Circular Bar, 583
– Torsion of Elliptical Bar, 584
– Torsion of Hollow Cylinder, 642
Principal Strain Invariants, 25
Principal Strains, 24
Principal Stresses, 49
Principle of Saint-Venant, 567
Principle of Virtual Work, 291
Prismatic Bar with Rectangular Section
– Elastic Torsion, 588
Prismatic Bar, Deformation of, 568
Proof of Real Eigenvalues
– for Poisson's Equation, 542
Pure Bending
– Annular Thin Plate, 313
– Circular Solid Plate, 203
– Rectangular Plane, 108
– Rectangular Plate, 334
– Thin Plate to Paraboloidal Shape, 336
Pure Twisting
– Rectangular Plate, 379
– Thin Rectangular Plate, 336

Radial Loading of Isotropic Elastic Cylinder, 481
Radial Pressurization of Borehole in Elastic Solid, 486
Radially Symmetric Problem, for Elastic Inclusion, 167
Radius of Inversion
– Analysis of Clamped Circular Plate, 326
Rate of Deformation, 395
– Dyadic, 395
Rate of Heat Supply to Viscous Fluid, 400
Real Eigenvalues for Poisson's Equation
– Proof of, 542
Reciprocity Principle
– Green's Function for Poisson's Equation, 552
Reciprocity Relations
– due to Betti and Maxwell, 62
Rectangular Cantilever
– Subject to Boundary Load, 113
Rectangular Duct
– Axial Viscous Flow, 563
Rectangular Plane
– Pure Bending, 108
Rectangular Region
– Green's Function for Poisson's Equation, 535
– Periodic Boundary Loading, 127
Rectangular Thin Plate
– Boundary Loading, 111
– Clamped-Simply Supported, Uniform Loading, 363
– Kirchhoff Boundary Conditions, 333
– Levy's Solution for Simply Supported, 351
– Moment-Curvature Relationship, 332
– Navier's Solution, 345
– Navier's Solution for Simply Supported, 348
– Partial Differential Equation Governing Flexure, 332
– Pure Bending of, 334
– Pure Twisting of, 379
– Simply Supported Action of Boundary Moments, 360
– Simply Supported and Uniformly Loaded, 357
– Simply Supported, Loading by Edge Moments, 364
– Subjected to Pure Twisting, 336
– Thin Plate Boundary Conditions, 302
Recurrence Relationship

– Legendre's Function of the First Kind, 200
– Legendre's Function of the Second Kind, 201
Regularity Conditions
– Green's Function for Poisson's Equation, 532
Relaxed Constraints on Elastic Constants, 78
Residue Theorem, 606
Resultant Torque
– Elastic Torsion, Prandtl's Stress Function, 577
– Elastic Torsion, Warping Function, 576
– Multiply Connected Region, 580
Resultants
– Stress, 279
Rigid Disc Inclusion in Elastic Solid
– Shear Stress Field, 477
Rigid Indentor
– Contact Stresses Beneath, 266
Rotary Motion in Viscous Fluid
– Diffusive Motion, 447
Rotation Dyadic, 10, 277
Rotation Matrix, 18, 183
– Indicial Rotation, 18
– Physical Interpretation, 16
Rotation of Cylinder in Viscous Fluid, 502
Rotation Vector
– in Two-Dimensions, 272
– Partial Differential Equation for, 289
Rotational Motion
– Unsteady Flow of Viscous Fluid, 448

S-wave Velocity, 78
Saint-Venant's Compatibility Equations, 28
Saint-Venant's Principle, 567
Saint-Venant's Theory of Torsion, 565
Semi-Circular Cross-section
– Torsion of Bar with, 597
Semi-Infinite Plane
– Harmonic Boundary Loading, 121
– Periodic Boundary Loading, 124
Semi-Infinite Plate
– Action of Concentrated Load on a Clamped Plate, 381
– Loading by Boundary Moments, 491
Series Solution of Biharmonic Equation, 105
Shear Force in Plate, 281
Shear Force-Rotation Relationship
– Polar Coordinates, 286

Shear Forces
– Thin Circular Plate, 305
Shear Modulus, 65
Shear Stresses
– Sign Convention, 45
Shear Stresses in Elliptical Bar
– Torsion of, 586
Shearing Flow
– Couple Exerted by Viscous Fluid on Cylinder, 423
– of Viscous Fluid Past Cylinder, 421
Sign Convention
– for Normal Stresses, 45
– for Shear Stresses, 45
– for Stresses, 45
Similarity Solution
– of Diffusion Equation, 500
Simply Connected Domain
– Displacement Components, 98
Simply Supported Rectangular Plate
– Loading by Edge Moments, 364
Simply Supported Thin Plate
– Localized Loading, 350
Single Layer Potential, 529
Slider Bearing
– Viscous Fluid Motion in, 496
Slow Viscous Flow, 3
– Adherence Condition, 411
– Application of Integral Transform Techniques, 430
– Axisymmetric Problems, 409
– Between Parallel Planes, 413
– Biharmonic Equation Approach, 393
– Biharmonic Equation for Stokes' Stream Function, 422
– Biharmonic Function Representation, 407
– Boundary Conditions Governing, 411
– Cauchy-Riemann Equation for Plane Flow, 409
– Couple Exerted on Cylinder in, 423
– Double Shear Flow Past Cylinder, 497
– Drag on Sphere in Uniform Flow Field, 430
– Dual Integral Equations for Rotation of Rigid Plate, 434
– Fluid Pressure in Lubrication Zone, 418
– Formulation in Terms of Stokes' Operator, 410
– Induced by Steady Rotation of Flat Plate, 431
– Kinetic Energy, 463
– Laminar Flow in Tube, 445

694   Index

- Low Reynolds Number, 404
- Moment-Rotation Relationship for Plate, 435
- No-Slip Boundary Conditions, 411, 461
- Non-Existence of Solutions, 419, 459
- Past a Viscous Inclusion, 499
- Past Cylindrical Object, Absence of, 421
- Past Flat Circular Plate, 436
- Past Sphere, Frictional Interface, 500
- Past Sphere, Stokes' Stream Function Approach, 426
- Planar Problems, 407
- Reduction to Biharmonic Equation, 404
- Representations of Velocity, 405
- Rotation of Fluid in Circular Domain, 449
- Shearing Flow Past Cylinder, 421
- Shearing Flow Past Elastic Cylinder, 496
- Stokes' Paradox, 419, 459
- Uniform Flow Past Sphere, 424
- Uniqueness Theorem, 462
Small Deformations, 11
Small Strain Analysis, 23
Small Strains, 11, 15, 23
Smoothly Embedded Spherical Inclusion
- in Elastic Solid, 484
Spatial Distribution of Matter
- Potential, 620
Speed
- in Viscous Flow Regions, 452
Spherical Cavity in Elastic Solid, 233
- Constrained Surface, 483
- Stress Concentration due to, 238
- Uniform Tension Field, 233
Spherical Cavity in Infinite Elastic Solid, 209
Spherical Inclusion in Elastic Solid
- Axial Loading, 482
Spherical Inclusion in Viscous Flow Region
- Frictional Sliding, 500
Spherical Symmetry
- Solution for Laplace's Equation, 526
Square Plate, Stress Field in, 106
State of Stress at a Point, 36
State of Stress in Half-plane, 50
Steady Flow of Viscous Fluid in Circular Tube, 445
Steady Rotation of Flat Plate
- in Viscous Flow Region, 431
Stokes' Operator
- Formulation of Axisymmetric Slow Viscous Flow, 410

Stokes' Paradox
- Plane Slow Viscous Flows, 419, 459
Stokes' Problem
- for a Flat Plate, 500
- for a Moving Plate on a Viscous Fluid, 441
Stokes' Result for Viscous Drag on Sphere, 430
Stokes' Stream Function
- Biharmonic Equation, 422
- General Solution in Polar Coordinates, 460
- Integral Representation, 438
- Representation of No-Slip Boundary Conditions, 462
- Slow Viscous Flow Past Sphere, 426
- Slow Viscous Flow Problems, 408
Strain
- Deviatoric, 66
- Dyadic Form, 63
- Indicial Notation, 63
- Stress Relations, 63
- Volumetric, 66
Strain Components
- Elastic Torsion of Prismatic Bar, 570
Strain Dyadic, 10
- Curl of, 26
- Divergence of, 26
- in Two-Dimensions, 270
Strain Energy
- in an Elastic Material, 60
- of Elastic Plate, 294
- of Plate, 289
Strain Energy Density, 58, 60
- Plane Polar Coordinates, 159
Strain Function, Love's, 182
Strain Invariants, 24, 25
Strain Matrix, 8, 11
- Axial Symmetry, 194
- Eigenvalues, 24
- Indicial Notation, 18
- Physical Interpretation, 12
- Plane Polar Coordinates, 157
- Plane Strain Problems, 158
- Small Strains, 157
- Transformation of, 19, 24
- Transformation Rule, 20
Strain Rate, 395
Strain Vectors, 9
Strains
- Compatibility, 28
- Compatibility Relationship, 26
- in a Continuum, 5

– in Plate, Compatibility Conditions, 279
– Principal, 24
– Small, 11, 15
Stream Function Formulation
– Slow Viscous Flow, 408
Stress Components
– Love's Strain Function Representation, 194
Stress Concentration
– at a Hole, Uniformly Stressed Plate, 176
– Cylindrical Cavity, 173
– in Elastic Solid due to Spherical Cavity, 238
Stress Dyadic, 37, 38
– in Two-Dimensions, 270
– Conjugate, 44
Stress Field
– Boussinesq's Problem, Cylindrical Coordinates, 231
– Boussinesq's Problem, Spherical Coordinates, 230
– Circular Cavity in Stressed Elastic Solid, 181
– in Square Plate, 106
– Kelvin's Problem, 218
Stress Function Approach
– Elastic Torsion of Prismatic Bar, 573
Stress Invariants, 49
– Indicial Notation, 49
Stress Matrix, 37, 38
– Plane Polar Coordinates, 152
– Symmetry of, 43
– Transformation of, 47
– Transformation Rule, 49, 160
Stress Principle, Cauchy's, 36
Stress Resultants, 279
Stress State at a Point, 36
Stress State in Plate
– Influence of Circular Hole, 173
Stress State in Viscous Fluid
– Deviatoric Component, 397
– Isotropic Component, 397
Stress Tensor, 38
Stress Vector at a Point, 36
Stress-Strain Relations
– Averaged, 86
– Axial Symmetry in Spherical Coordinates, 211
– Dyadic Form, 62
– Elastic Torsion of Prismatic Bar, 570
– for Elastic Plate, 282
– Plane Polar Coordinates, 158
Stresses

– Compatibility Relationship, 89
– Deviatoric, 66
– Flexural, 279
– in a Continuum, 36
– Principal, 49
– Representation in Terms of Airy Stress Function, 93
– Sign Convention, 45
Stresses in Spherical Coordinates
– Lamé Strain Potential Formulation, 211
– Love's Strain Function Appoach, 213
Stretched Membranes, Poisson's Equation, 508
Strong Maximum Principle, Poisson's Equation, 558
Subharmonic Functions, 513
– Maximum Principle, 519, 524
– Mean Value Theorem, 515
Subharmonic Nature of Shear Stress Field
– Torsional Loading, 587
Substantial Derivative, 395
Summation convention
– Einstein's, 25
Superharmonic Functions
– Maximum Principle, 524
Surface Loading
– of Cylindrical Elastic Bar, 254
– of Viscous Layer, Unsteady Motion, 501
Surface Reinforced Half-space
– Boussinesq's Problem, 243
Symmetric Strain Matrix, 8
Symmetry
– Hermitian, 542
– of Elasticity Matrix, 60
– of Green's Function, Dirichlet Problem, 551
– of Inner Product, 541
– of the Stress Matrix, 43

Theorem for Stiffness Bounds
– Elastic Torsion, Theorem due to Polya, 592
Theorem of Power and Energy, 75
Theory of Elasticity, 2
Theory of Torsion, Saint-Venant, 565
Thermodynamic Bounds, Viscosity Constants, 402
Thermodynamic Constraints
– Newtonian Viscous Fluids, 399
– on Bulk Modulus, 68
– on Elastic Constants, 66
– on Elastic Modulus, 68
– on Poisson's Ratio, 68
– on Shear Modulus, 68

– on Viscosity Parameters, 399
– Viscosity Coefficients, 399
Thick Elastic Plates, Flexure of, 268
Thin Elastic Plates
– Classical Theory of, 297
– Flexure of, 267
Thin Plate
– Action of Concentrated Load on
  Clamped Semi-Infinite Plate, 381
– Admissible Forms of Deflections, 307
– Analtyic Function Representation of
  Flexural Moments, 376
– Analtyic Function Representation of
  Shear Forces, 376
– Analytic Function Representation of
  Deflection, 376
– Annular, Pure Bending, 313
– Anticlastic Bending, 336
– Application of Complex Variable
  Methods, 374
– Application of Integral Transform
  Technique, 366
– Axisymmetric Deflections, 307
– Biharmonic Equation Governing
  Flexure, 298
– Biharmonic Equations in Polar
  Coordinates, 305
– Boundary Condition for Fixed Edge, 333
– Boundary Condition for Free Edge, 333
– Boundary Condition for Simply
  Supported Edge, 333
– Boundary Conditions for Circular Edge,
  306
– Boundary Conditions for Corner Forces,
  303
– Boundary Conditions for Rectangular
  Plates, 302
– Boundary Conditions in Complex
  Variable Form, 376
– Boundary Conditions in Terms of
  Analytic Functions, 379
– Circular, 305
– Circular, Loading by Stiffener, 316
– Clamped-Simply Supported Edges,
  Uniform Loading, 363
– Complete Solution for Plate Deflections,
  307
– Deflection of Clamped Circular Plate by
  Uniform Load, 308
– Deflection of Clamped Circular Plate,
  Concentrated Load, 311
– Deflection of Clamped Plate by
  Concentrated Load, 318

– Elastic Torsion, 592
– Flexural Moments in Circular Plate, 305
– Green's Function, 330
– Green's Function for Clamped Circular,
  324
– Green's Function for Rectangular Plate,
  350
– Infinite Plate with Clamped Boundaries,
  370
– Integral Transform Solution for Infinite
  Plate, 367
– Kirchhoff Boundary Conditions, 300,
  333
– Line Load on Infinite Plate of Infinite
  Width, 353
– Navier's Solution for Rectangular Plate,
  345
– Pure Twisting of, 379
– Radius of Inversion for Circular
  Clamped Plate, 326
– Rectangular, 332
– Shear Forces in Circular Plate, 305
– Simply Supported Rectangular Plate,
  Uniform Loading, 357
– Simply Supported, Action of Boundary
  Moments, 360
– Simply Supported, Loading by Edge
  Moments, 364
– Twisting of Rectangular Plate, 336
– Uniform Loading of Clamped Plate,
  Complex Variable Approach, 384
– Uniqueness of Solution, 386
– Uniqueness of Solution for Clamped
  Plate, 390
Three-dimensional Problems in Elasticity
– Biharmonic Functions, 181
Time Derivatives, 71
– Local, 71
– Material, 71
Torque-Twist Relationship
– Bar with Circular Cross-section, 584
– Bar with Elliptical Cross-section, 586,
  604
– Bar with Rectangular Cross-section, 591
– Bar with Semi-Circular Cross-section,
  599
– Complex Variable Formulation, 607
– Elastic Plate, 594
– General Formulation, Prandtl's Stress
  Function, 614
Torsion
– Traction Boundary Conditions,
  Prandtl's Stress Function, 573

Torsion of Bar with Multiply Connected
    Cross-section, 577
Torsion of Bar with Rectangular Section,
    Fourier Series Solution, 589
Torsion of Elliptical Bar, Shear Stresses in
    Bar, 586
Torsion of Hollow Circular Bar with
    Longitudinal Cut, 644
Torsion of Hollow Cylinder, 642
Torsion of Prismatic Bar
- Circular Cross-section, 583
- Circular Cross-section with Groove, 644
- Circular Cross-section, Complex
    Variable Formulation, 607
- Cross-section Bounded by Parabolic
    Curves, 646
- Cross-section of a Sector, 643
- Cross-section of Cardioid, 646
- Elliptical Cross-section, 584
- Triangular Cross-section, 643
- Warping of Cross-section, 570
Torsion of Prismatic Body
- General Formulation, 568
Torsion of Prismatic Elastic Solids
- Poisson's Equation, 565
Torsional Stresses
- Bar with Elliptical Cross-section, 587
Total Derivative, 395
Total Energy of Elastic Body, 81
Traction Boundary Conditions, 69, 292
- Axisymmetric Problem, 198
- Plane Polar Coordinates, 162
Traction Vector, 37
Tractions
- on a Hemisphere, Boussinesq's Problem,
    229
- on Arbitrary Plane, 39
- on Prismatic Surface, Elastic Torsion of
    Bar, 571
Transformation
- from Cartesian Coordinates, 159
- of Strain Matrix, 19, 20, 24
- of Stress Matrix, 47, 49
Transpose of Deformation Gradient
    Matrix, 7
Transverse Deflection of Membranes
- Poisson's Equation, 508
Transverse Shear in Plate, 281
Twisting Moments
- at Plate Boundary, 282
- in Plate, 280
Twisting of Prismatic Bar, 568
Two-dimensional Deformations, 13

Two-dimensional Forms
- Gradient Operator, 270
- Rotation Vector, 272
- Strain Dyadic, 270
- Stress Dyadic, 270
Two-dimensional Problems
- in Classical Elasticity, 81
- in Slow Viscous Flow, 408
Two-Material Plate Region, Localized
    Loading, 494

Uniaxial Deformation, 13
Uniform Loading
- Circular Solid Plate, 206
- Clamped Thin Plate, 308
- Clamped-Simply Supported Rectangular
    Plate, 363
- Rectangular Simply Supported Plate,
    357
Uniform Strip Loading, Elastic Half-plane,
    135
Uniformly Stressed Plate
- Airy Stress Function Approach, 174
- Circular Hole, 173
- Stress Concentration at Hole, 176
- Stress State due to Circular Hole, 173
Uniqueness of Solution
- Clamped Plate, 390
- Clamped Plate, Complex Variable
    Approach, 390
- Flexure of Thin Plates, 386
Uniqueness Theorem
- Clamped Thin Plate, 390
- Kirchhoff's, 72
- Poisson's Equation, 521
- Slow Viscous Flow, 462
- Thin Plates, 386
Uniqueness Theorem in Linear Elasticity,
    72
Unit Dyadic, 63
Unsteady Flow of Viscous Fluid, 440
- Diffusive Motion, 451
- in Circular Tube, 444, 446
- in Layer, 501
- Induced by Plate Motion, 500
- Induced by Rotating Cylinder, 502
- Rotational Motion, 448

Variables Separable Solution
- Biharmonic Equation in Cartesian
    Coordinates, 118
- Elastic Torsion, 587
Vector
- Displacement, 6

698   Index

– Gradient of, 8
– of Body Force, 40
– of Tractions, 37
Vectors
– Strain, 9
– Stress, 36
Velocity
– P-wave, 78
– S-wave, 78
Velocity Field Representations
– Slow Viscous Flow, 405
Virtual Work
– Equations of Equilibrium via, 296
– Principle of, 291
Virtual Work Equation for Elastic Plate, 294
Viscosity
– Bulk, 398
– Dynamic Shear, 398
Viscosity Coefficients
– Thermodynamic Bounds, 402
– Thermodynamic Constraints, 399
Viscous Flow, 3
– Bernoulli Potential, 452
– Definition of Speed, 452
– Diffusion of Velocity in Semi-Infinite Domain, 442
– Diffusion of Vorticity, 453
– Diffusive Motion, 440
– Diffusive Motion in Circular Region, 444, 451
– Diffusive Motion, Laplace Transform Technique, 442
– Equations of Motion, 397
– in Conduits, Poisson's Equation for, 559
– in Narrow Aperture, 640
– Line Vortex in Viscous Fluid Domain, 454
– Unsteady Flow in Circular Tube, 443, 446
– Unsteady Motion Induced by Moving Plate, 441
– Unsteady Rotational Motion, 447
– Vorticity in Diffusive Motion, 451
Viscous Flow Problems
– Formulation in Terms of Stokes' Stream Function, 408
Viscous Fluid
– Entropy of, 401
– Heat Gain per Unit Volume, 402
– Internal Energy Dissipation, 402
– Navier-Stokes Equation, 403
– Positive Definite Form of Energy's Dissipation, 402
– Rate of Heat Supply, 400
Viscous Fluid Motion in Slider Bearing, 496
Viscous Fluids
– Energy Balance Equation, 400
– Isotropic, 398
Viscous Inclusion, Slow Viscous Flow, 499
Viscous Thin Film Lubrication, Flow in a Narrow Aperture, 416
Volumetric Strain, 66
Vorticity
– Dyadic, 395
– in Diffusive Motion of Viscous Fluid, 451

Warping
– Elastic Torsion of Prismatic Bar, 570
Warping Function
– Laplace's Equation for, 571
– Laplace's Equation in Polar Coordinates, 595
– Neumann Problem, Elastic Torsion, 572
– Torsion of Prismatic Bar, Polar Coordinates, 595
Weak Maximum Principle
– Poisson's Equation, 558
Weight Function, 541

Young's Modulus, 65

Production: Druckhaus Beltz, Hemsbach